Parallel Manipulators
New Developments

Parallel Manipulators
New Developments

Edited by
Jee-Hwan Ryu

I-Tech

Published by I-Tech Education and Publishing

I-Tech Education and Publishing
Vienna
Austria

www.i-techonline.com
Additional copies can be obtained from:
publication@ars-journal.com

First published April 2008
Printed in Croatia

A catalogue record for this book is available from the Austrian Library.
Parallel Manipulators, Edited by Jee-Hwan Ryu
p. cm.
ISBN 978-3-902613-20-2
1. Parallel Manipulators. 2. New Developments. I. Jee-Hwan Ryu

Preface

Parallel manipulators are characterized as having closed-loop kinematic chains. Compared to serial manipulators, which have open-ended structure, parallel manipulators have many advantages in terms of accuracy, rigidity and ability to manipulate heavy loads. Therefore, they have been getting many attentions in astronomy to flight simulators and especially in machine-tool industries.

The aim of this book is to provide an overview of the state-of-art, to present new ideas, original results and practical experiences in parallel manipulators. This book mainly introduces advanced kinematic and dynamic analysis methods and cutting edge control technologies for parallel manipulators. Even though this book only contains several samples of research activities on parallel manipulators, I believe this book can give an idea to the reader about what has been done in the field recently, and what kind of open problems are in this area.

Finally, I would like to thanks all the authors of each chapter for their contribution to make this book possible.

Jee-Hwan Ryu
Korea University of Technology and Education
Republic of Korea
jhryu@kut.ac.kr

Contents

1

On the Robust Dynamics Identification of Parallel Manipulators: Methodology and Experiments

Houssem Abdellatif, Bodo Heimann and Jens Kotlarski
Institute of Robotics, Hannover Center of Mechatronics, Leibniz University of Hannover Germany

1. Introduction

The proposed chapter presents a self-contained approach for the dynamics identification of parallel manipulators. Major feature is the consequent consideration of structural properties of such machines in order to provide an experimentally adequate identification method. Thereby, we aim to achieve accurate model parameterization for control, simulation or analysis purposes. Despite the big progress made on identification of serial manipulators, it is interesting to state the missing of systematic identification methodologies for closed-loop and parallel kinematic manipulators (PKM's). This is due to many factors that are discussed and treated systematically in this chapter.

First, the issue of modelling the dynamics of PKM's in a linear form with respect to the parameters to be identified is addressed. As it is already established in the field of classic serial robotics, such step is necessary to ensure model identifiability and to apply computationally efficient linear estimation (Swevers et al., 1997; Khalil & Dombre, 2002; Abdellatif & Heimann, 2007). The case of parallel manipulators is more complicated, since a multitude of coupled and closed kinematic chains has to be considered (Khalil & Guegan, 2004; Abdellatif et al., 2005a). Beside the rigid-body dynamics, friction plays a central role in modelling, since its accurate compensation yields important improvement of control accuracy. If friction in the passive joints is regarded, the dimension of the parameter vector grows and affects the estimation in a negative way. To cope with such problem, a method for the reduction of the friction parameter number is proposed, which is based on the identifiability analysis for a given manipulator structure and by considering the actual measurement noise. The calculation procedure of a dynamics model in minimal parametrized form is given in section 2.

Another important issue of PKM's is the appropriate design of the identification experiment, in order to obtain reliable estimation results. Two aspects are here crucial: The choice or the definition of the experiment framework at the one hand and its related experiment optimization at the other hand. Regarding the first aspect, the harmonic excitation approach proposed a couple of years ago for serial manipulators is chosen (Swevers et al., 1997). The method provides bounded motion that can be fitted in the usually highly restricted and small workspace of parallel robots. Thus, we propose an appropriate adaptation for PKM's. The experiment optimization is carried out within a statistical frame in order to account for

the cross correlation of measurement noise and the motion dependency of the coupled actuators (Abdellatif et al., 2005b). The Experiment design is discussed in section 3.
The typically non measurable information of the end-effector postures, velocities and accelerations are necessary to calculate the dynamics model and therefore to obtain the regression equation. Since in general only actuator measurements are available, there is a need for an adequate estimation of the executed end-effector motion during the identification experiment. However, the numerical computation of the direct and the differential kinematics yields a spectral distortion and noise amplification in the calculated data. Therefore, an appropriate and simple frequency-domain data processing method is introduced in section 4. An accurate and noise-poor regression model is then provided, which is crucial for bias-free estimation of the model parameters. Additionally, we provide useful relationships to evaluate the resulting parameter uncertainties. Here, uncertainties of single parameters as well as the uncertainties of entire parameter sets are discussed and validated.

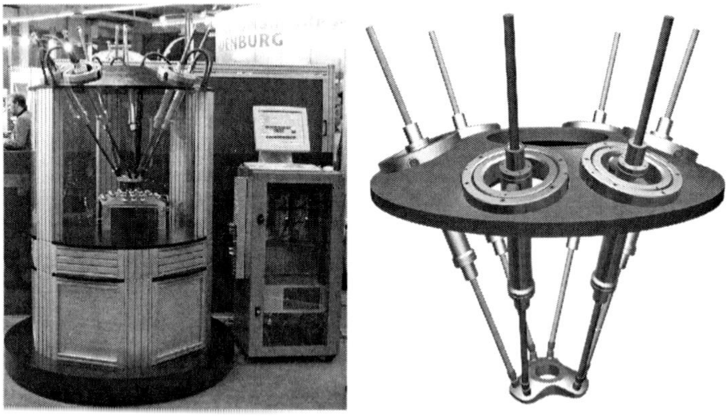

Fig. 1. Case Study: Hexapod PaLiDA; left: Presentation in the Hannover industrial fair, Right: CAD-Model

Finally and in section 5, an important part of the chapter presents the experimental substantiation of the theoretical methods. The effectiveness of our approach is demonstrated on a six degrees-of-freedom (dof's) directly actuated parallel manipulator PaLiDA (see Fig. 1). We address the important issue of exploiting the identification results for model-based control. The impact of accurately identified models on the improvement of control accuracy is illustrated by numerous of experimental investigations.

2. Parameterlinear formulation of the dynamics model

The objective of this section is to derive the inverse dynamics model in a linear form with respect to a set of the parameters to be identified. Such formulation allows for using linear techniques to provide the estimation of model parameters from measurement data. This kind of approach is well established for serial robots (Khalil & Dombre, 2002; Abdellatif & Heimann, 2007). Thereby, the model accounts for the rigid-body as well as for friction dynamics. We consider the case of 6-dof's parallel manipulator, that is constituted of a moving platform (end-effector platform) attached with six serial and non-redundant actuated kinematic chains to the base platform. Fig. 2 shows a general sketch of such robotic manipulators. Let $\mathbf{x},\mathbf{v},\mathbf{a}$ be the 6-dimensional vectors denoting the posture, the velocity, and the acceleration of the end-effector, respectively. The posture vector is composed of the cartesian coordinates of the end-effector platform ${}_{(0)}\mathbf{r}_{\mathrm{E}}^{0} = [x \;\; y \;\; z]^{\mathrm{T}}$ and the tilting angles (ϕ,θ,ψ) according to the Cardan or the Euler formalism. The velocity vector is defined as $\mathbf{v} = \left[{}_{(0)}\mathbf{v}_{\mathrm{E}}^{\mathrm{T}} \;\; {}_{(0)}\boldsymbol{\omega}_{\mathrm{E}}^{\mathrm{T}}\right]^{\mathrm{T}}$ that includes the translational and angular velocities with reference to a cartesian frame. It is known, that $\mathbf{v} \neq \dot{\mathbf{x}}$ holds for systems with two or more rotational dof's (Merlet, 2006). The 6-dimensional vector of actuated joints is denoted by $\mathbf{q}_{\mathrm{a}}$. The passive joint variables are grouped in $\mathbf{q}_{\mathrm{p}}$. The vector $\mathbf{q}$ contains all joint variables.

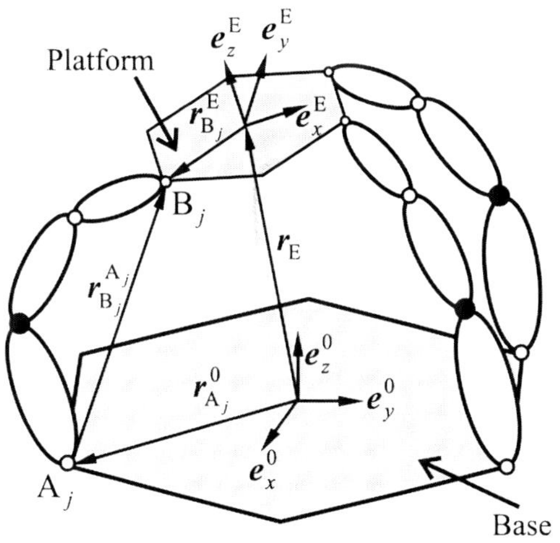

Fig. 2. Scheme of a general parallel manipulator

The major difference between serial and parallel manipulators is the definition of configuration variables or the configuration space. For classic serial manipulators, the actuation variables $\mathbf{q}_{\mathrm{a}}$ are sufficient to determine exactly the system's configuration. This is not the case for PKM's, because the solution of the direct kinematics is ambiguous (Khalil & Dombre, 2002; Merlet, 2006). It is established that the motion of the end effector given by $\mathbf{x},\mathbf{v}$ and $\mathbf{a}$ is used to derive the dynamics of high-mobility parallel robots (Tsai, 2000; Harib & Srinivasan, 2003; Khalil & Guegan, 2004; Abdellatif et al., 2005a). The solution of inverse kinematics is supposed to be already achieved. It means that for a given dynamic motion of the end-effector, all necessary kinematic quantities are available. The latter include: Velocities and accelerations of any body i with respect to a body-fixed frame[1] $\mathbf{v}_i,\mathbf{a}_i$ or those of the center of mass $\mathbf{v}_{\mathrm{S}_i},\mathbf{a}_{\mathrm{S}_i}$; the angular velocities and angular accelerations $\boldsymbol{\omega}_i,\dot{\boldsymbol{\omega}}_i$; the

[1] The body-fixed frames can be defined according to the modified Denavit-Hartenberg (MDH) notation (Khalil & Dombre, 2002).

body Jacobians $\mathbf{J}_{T_i} = \partial \mathbf{v}_i / \partial \mathbf{v}$ and $\mathbf{J}_{R_i} = \partial \boldsymbol{\omega}_i / \partial \mathbf{v}$ and the inverse Jacobian of the manipulator $\mathbf{J}^{-1} = \partial \mathbf{q}_a / \partial \mathbf{v}$ (see (Abdellatif et al., 2005a) and references therein for more details).
The aimed dynamics model consists of the following equation:

$$\mathbf{Q}_a = \mathbf{A}(\mathbf{x},\mathbf{v},\mathbf{a})\mathbf{p} \;\Leftrightarrow\; \mathbf{Q}_a = \mathbf{A}_{a,rb}(\mathbf{x},\mathbf{v},\mathbf{a})\mathbf{p}_{rb} + \mathbf{A}_{a,f}(\mathbf{x},\mathbf{v},\mathbf{a})\mathbf{p}_f\,, \tag{1}$$

with $\mathbf{Q}_a$ being the actuator forces and where the indexes rb and f refer to the rigid-body and friction terms, respectively.

2.1 Parameterlinear formulation of the rigid-body dynamics

Generally, it is recommended to use the Jourdain's principle of virtual power to derive the dynamics in an efficient manner. In analogy to the virtual work, a balance of virtual power can be addressed:

$$\delta \mathrm{v}^{\mathrm{T}} \tau = \delta \dot{\mathrm{q}}_a^{\mathrm{T}} \mathrm{Q}_a \;\Leftrightarrow\; \tau = \left(\frac{\partial \dot{\mathrm{q}}_a}{\partial \mathrm{v}}\right)^{\mathrm{T}} \mathrm{Q}_a\,, \tag{2}$$

where $\boldsymbol{\tau}$ is the vector of the generalized forces, defined with respect to the end-effector generalized velocities $\mathbf{v}$. Equation (2) means that the virtual power resulting in the space of generalized velocities is equal to the actuation power. The power balance can be applied for rigid-body forces:

$$\mathbf{Q}_{a,rb} = \tau\left(\frac{\partial \dot{\mathbf{q}}_a}{\partial \mathbf{v}}\right)^{\mathrm{T}} \boldsymbol{\tau}_{rb} = \mathbf{J}^{\mathrm{T}} \boldsymbol{\tau}_{rb}\,. \tag{3}$$

The generalized rigid-body forces for a manipulator with N bodies are obtained by

$$\boldsymbol{\tau}_{rb} = \sum_{i=1}^{N}\left[\mathbf{J}_{T_i}^{\mathrm{T}}\left(m_i\,{}_{(i)}\dot{\mathbf{v}}_i + {}_{(i)}\dot{\tilde{\boldsymbol{\omega}}}_i \mathbf{s}_i + {}_{(i)}\tilde{\boldsymbol{\omega}}_i\,{}_{(i)}\tilde{\boldsymbol{\omega}}_i \mathbf{s}_i\right) + \mathbf{J}_{R_i}^{\mathrm{T}}\left({}_{(i)}\mathbf{I}_i^{(i)}\,{}_{(i)}\dot{\boldsymbol{\omega}}_i + {}_{(i)}\tilde{\boldsymbol{\omega}}_i\left({}_{(i)}\mathbf{I}_i^{(i)}\,{}_{(i)}\boldsymbol{\omega}_i\right) + \tilde{\mathbf{s}}_i\,{}_{(i)}\dot{\mathbf{v}}_i\right)\right], \tag{4}$$

with the dynamic parameters of each body i, its mass m_i, its statical first moment $\mathbf{s}_i \,\hat{=}\, \begin{bmatrix} s_{i_x} & s_{i_y} & s_{i_z}\end{bmatrix}^{\mathrm{T}} = m_i\,{}_{(i)}\mathbf{r}_{S_i}^i$ (${}_{(i)}\mathbf{r}_{S_i}^i$: Vector from coordinate frame to centre of mass) and its inertia tensor about the corresponding body-fixed coordinate frame ${}_{(i)}\mathbf{I}_i^{(i)}$. New operators $(\;)^*$ and $(\;)^\Diamond$ are defined:

$$\boldsymbol{\omega}_i^* \mathbf{I}_i^\Diamond \,\hat{=}\, {}_{(i)}\mathbf{I}_i^{(i)}\,{}_{(i)}\boldsymbol{\omega}_i\,, \tag{5}$$

with

$$\boldsymbol{\omega}_i^* = \begin{bmatrix} \omega_{i_x} & \omega_{i_y} & \omega_{i_z} & 0 & 0 & 0 \\ 0 & \omega_{i_x} & 0 & \omega_{i_y} & \omega_{i_z} & 0 \\ 0 & 0 & \omega_{i_x} & 0 & \omega_{i_y} & \omega_{i_z} \end{bmatrix} \text{ and } \mathbf{I}_i^\Diamond = \begin{bmatrix} I_{i_{xx}} & I_{i_{xy}} & I_{i_{xz}} & I_{i_{yy}} & I_{i_{yz}} & I_{i_{zz}} \end{bmatrix}^{\mathrm{T}}, \tag{6}$$

which helps the simplification of the generalized rigid-body forces:

$$\boldsymbol{\tau}_{\mathrm{rb}} = \sum_{i=1}^{N} \underbrace{\left[\mathbf{J}_{\mathrm{T}_i}^{\mathrm{T}} \quad \mathbf{J}_{\mathrm{R}_i}^{\mathrm{T}}\right]\boldsymbol{\Omega}_i}_{\mathbf{H}_i} \underbrace{\begin{bmatrix} \mathbf{I}_i^{\Diamond} \\ \mathbf{s}_i \\ m_i \end{bmatrix}}_{\mathbf{p}_i} = \left[\mathbf{H}_1 \quad \mathbf{H}_2 \quad \dots \quad \mathbf{H}_N\right]\left[\mathbf{p}_1^{\mathrm{T}} \quad \mathbf{p}_2^{\mathrm{T}} \quad \dots \quad \mathbf{p}_N^{\mathrm{T}}\right]^{\mathrm{T}}, \tag{7}$$

with

$$\boldsymbol{\Omega}_i = \begin{bmatrix} \mathbf{0} & {}_{(i)}\dot{\tilde{\boldsymbol{\omega}}} + {}_{(i)}\tilde{\boldsymbol{\omega}}_{i\,(i)}\tilde{\boldsymbol{\omega}}_i & {}_{(i)}\dot{\mathbf{v}}_i \\ {}_{(i)}\dot{\boldsymbol{\omega}}_i^{*} + {}_{(i)}\tilde{\boldsymbol{\omega}}_{i\,(i)}\boldsymbol{\omega}_i^{*} & -{}_{(i)}\dot{\mathbf{v}}_i & \mathbf{0} \end{bmatrix}. \tag{8}$$

Equation (7) is already linear with respect to the parameter vector $\mathbf{p}_{\mathrm{rb}}^{*} = \left[\mathbf{p}_1^{\mathrm{T}} \quad \mathbf{p}_2^{\mathrm{T}} \quad \dots \quad \mathbf{p}_N^{\mathrm{T}}\right]^{\mathrm{T}}$. The dimension of the latter has now to be reduced for an efficient calculation and to assure the identifiability of the system. The proposed algorithm in the following is based on former works for serial and parallel manipulators (Grotjahn & Heimann, 2000; Grotjahn et al., 2002). The matrices $\mathbf{H}_i$ in equation (7-8) can be grouped in single serial kinematic chains such that a recursive calculation

$$\mathbf{H}_i = \mathbf{H}_{i-1}\mathbf{L}_i + \mathbf{K}_i \tag{9}$$

can be achieved. The matrices $\mathbf{L}_i$ and $\mathbf{K}_i$ are given in (Grotjahn et al., 2002). The first step considers in eliminating all parameters $\mathbf{p}_{\mathrm{rb},j}^{*}$ that correspond to a zero column $\mathbf{h}_j$ of $\mathbf{H}$, since they do not contribute to the dynamics. The remaining parameters are then regrouped to eliminate all linear dependencies by investigating $\mathbf{H}$. If the contribution of a parameter $\mathbf{p}_{\mathrm{rb},j}^{*}$ depends linearly on the contributions of some other parameters $\mathbf{p}_{\mathrm{rb},1j}^{*},\dots,\mathbf{p}_{\mathrm{rb},kj}^{*}$, the following equation holds:

$$\mathbf{h}_j = \sum_{l=1}^{k} a_{lj}\mathbf{h}_{lj}. \tag{10}$$

Then $\mathbf{p}_{\mathrm{rb},j}^{*}$ can be set to zero and the regrouped parameters $\mathbf{p}_{\mathrm{rb},lj,\mathrm{new}}^{*}$ can be obtained by

$$\mathbf{p}_{\mathrm{rb},lj,\mathrm{new}}^{*} = \mathbf{p}_{\mathrm{rb},lj}^{*} + a_{lj}\mathbf{p}_{\mathrm{rb},j}^{*}. \tag{11}$$

The recursive relationship given in (9) can be used for parameter reduction. If one column or a linear combination of columns of $\mathbf{L}_i$ is constant with respect to the joint variable and the corresponding columns of $\mathbf{K}_i$ are zero columns, the parameters can be regrouped. This leads to the rules which are formulated in (Khalil & Dombre, 2002) and in (Grotjahn & Heimann, 2000).

The rules can be directly applied to the struts or legs of the manipulator, since they are considered as serial kinematic chains. For revolute joints the 9th, the 10th and the sum of the 1st and 4th columns of $\mathbf{L}_i$ and $\mathbf{K}_i$ comply with the mentioned conditions. Thus, the corresponding parameters $I_{i_{yy}}$, s_{i_z} and m_i can be grouped with the parameters of the

antecedent joint $i-1$. For prismatic joints however, the moments of inertia can be added to the carrying antecedent joint, because the orientation between both links remain constant.
The end-effector platform closes the kinematic loops and further parameter reduction is possible. The velocities of the platform joint points $\mathbf{B}_j$ (see Fig. 2) and those of the terminal fixed-body frames of the respective legs are the same, yielding dependencies of the respective energy-functions. The masses of terminal bodies can be grouped to the inertial parameter of the platform according to Steiner's laws.
After applying every possible parameter reduction the generalized rigid-body forces are obtained from (7) with respect to a minimal set of parameters $\boldsymbol{\tau}_{\mathrm{rb}} = \mathbf{A}_{\mathrm{rb}}\mathbf{p}_{\mathrm{rb}}$. In combination with (3) the desired form for the rigid-body part of the actuation forces is obtained as

$$\mathbf{Q}_{\mathrm{a,rb}} = \mathbf{J}^{\mathrm{T}}\mathbf{A}_{\mathrm{rb}}\mathbf{p}_{\mathrm{rb}} = \mathbf{A}_{\mathrm{a,rb}}\mathbf{p}_{\mathrm{a,rb}}. \tag{12}$$

2.2 Parameterlinear formulation of the friction forces

In analogy to the rigid-body dynamics, the Jourdain's principle can be applied for friction forces. By defining an arbitrary steady-state model at joint-level $\mathbf{Q}_{\mathrm{f}} = \mathbf{f}_{\mathrm{r}}(\dot{\mathbf{q}})$, a new power balance can be derived:

$$\mathbf{Q}_{\mathrm{a,f}} = \left(\frac{\partial \dot{\mathbf{q}}}{\partial \dot{\mathbf{q}}_{\mathrm{a}}}\right)^{\mathrm{T}} \mathbf{Q}_{\mathrm{f}} = \mathbf{J}^{\mathrm{T}}\left(\frac{\partial \dot{\mathbf{q}}}{\partial \mathbf{v}}\right)^{\mathrm{T}} \mathbf{Q}_{\mathrm{f}}. \tag{13}$$

Equation (13) means that the friction dissipation power in all joints (passive and active) has to be overcome by an equivalent counteracting actuation power. We notice that the case of classic open-chain robots correspond to the special case, when the joint-Jacobian $\partial\dot{\mathbf{q}}/\partial\dot{\mathbf{q}}_{\mathrm{a}}$ is equal to the identity matrix. In the more general case of parallel mechanisms, friction in passive joints should not be neglected (Abdellatif & Heimann, 2006).
For identification purpose, friction in robotics is commonly modelled as superposition of Coulomb (or dry) friction and viscous damping depending on joint velocities $\dot{q}_i$ (Abdellatif et al., 2007; Swevers et al., 1997):

$$Q_{\mathrm{f}_i} = f_{\mathrm{r}}(\dot{q}_i) = \left[\operatorname{sign}(\dot{q}_i)\;\; \dot{q}_i\right]\left[r_{1_i}\;\; r_{2_i}\right]^{\mathrm{T}}. \tag{14}$$

Regrouping friction forces in all n joints yields to

$$\mathbf{Q}_{\mathrm{f}} = \underbrace{\left[\mathbf{D}_1(\dot{\mathbf{q}})\;\; \mathbf{D}_2(\dot{\mathbf{q}})\right]}_{\mathbf{D}_{\mathrm{f}}}\underbrace{\left[\mathbf{r}_1^{\mathrm{T}}\;\; \mathbf{r}_2^{\mathrm{T}}\right]^{\mathrm{T}}}_{\mathbf{p}_{\mathrm{f}}}, \tag{15}$$

with

$$\mathbf{r}_k^{\mathrm{T}} = \left[r_{k_1}, \ldots, r_{k_n}\right], \tag{16}$$

$$\mathbf{D}_1(\dot{\mathbf{q}}) = \operatorname{diag}\left\{\operatorname{sign}(\dot{q}_1), \ldots, \operatorname{sign}(\dot{q}_n)\right\}, \tag{17}$$

$$\mathbf{D}_2(\dot{\mathbf{q}}) = \operatorname{diag}\left\{\dot{q}_1, \ldots, \dot{q}_n\right\}. \tag{18}$$

Considering (13) and (15) the linear form of the resulting friction forces in the actuation space is obtained:

$$\mathbf{Q}_{\mathrm{a,f}} = \left[\mathbf{J}^{\mathrm{T}} \left(\frac{\partial \dot{\mathbf{q}}}{\partial \mathbf{v}} \right)^{\mathrm{T}} \mathbf{D}_{\mathrm{f}} \right] \mathbf{p}_{\mathrm{f}} = \mathbf{A}_{\mathrm{a,f}}(\mathbf{x},\mathbf{v})\mathbf{p}_{\mathrm{f}} \,. \tag{19}$$

Unlike the rigid-body dynamics, there is no uniform or standard approach for the reduction of the parameter vector dimension. In a former publication, we proposed a method that is highly adequate for identification purposes. Thereby, the expected correlation of the friction parameter estimates is analyzed for a given and statistically known measurement disturbance. Parameters whose effects are beneath the disturbance level are eliminated. Parameters with high correlation are replaced by a common parameter. The interested reader is here referred to (Abdellatif et al., 2005c) and (Abdellatif et al., 2007) for a deep insight.

3. Identification experiment design for parallel manipulators

Almost all identification methods in robotics are based on the parameterlinear form that is given by (1) in combination with (12) and (19) (Swevers et al., 1997; Khalil & Dombre, 2002; Abdellatif & Heimann, 2007). Given experimentally collected and noise corrupted N measurement sets, the estimation problem can be formulated according to (1) as

$$\underbrace{\begin{bmatrix} \mathbf{Q}_{\mathrm{a}_1} \\ \vdots \\ \mathbf{Q}_{\mathrm{a}_N} \end{bmatrix}}_{\Gamma} = \underbrace{\begin{bmatrix} \mathbf{A}(\mathbf{x}_1,\mathbf{v}_1,\mathbf{a}_1) \\ \vdots \\ \mathbf{A}(\mathbf{x}_N,\mathbf{v}_N,\mathbf{a}_N) \end{bmatrix}}_{\Psi} \mathbf{p} + \underbrace{\begin{bmatrix} e_1 \\ \vdots \\ e_N \end{bmatrix}}_{\eta} , \tag{20}$$

with the measurement vector $\mathbf{\Gamma}$, the information or regression matrix $\mathbf{\Psi}$ and the error vector $\boldsymbol{\eta}$ that accounts for disturbances. The most classic and simple solution of the overdetermined equation system (20) can be achieved by the Least-Squares (LS) approach. However, such method assumes that the disurbances of the different actuators are not cross correlated. The assumption does not hold for high-coupled systems like the case of parallel manipulators (Abdellatif et al., 2005b). It is recommended to use the Gauss-Markov (GM) approach that presents a more general case

$$\hat{\mathbf{p}}_{\mathrm{GM}} = \left(\mathbf{\Psi}^{\mathrm{T}} \mathbf{\Sigma}_{\eta\eta}^{-1} \mathbf{\Psi} \right)^{-1} \mathbf{\Psi}^{\mathrm{T}} \mathbf{\Sigma}_{\eta\eta}^{-1} \mathbf{\Gamma} \tag{21}$$

The crosscoupling is regarded by the full covariance matrix $\mathbf{\Sigma}_{\eta\eta} = \mathrm{E}\left(\boldsymbol{\eta}\boldsymbol{\eta}^{T}\right)$ of the measurement disturbances $\boldsymbol{\eta}$. Neglecting this fact by applying the simple LS-method will lead to biased estimates (Abdellatif et al., 2005b).

3.1 Design of the excitation trajectory

An important step in identification is the choice of the measurement data to be collected. A classic choice consists in the so-called excitation trajectory, which ensures that the effects of all considered parameters are contained in the measurement data. A challenging issue with parallel manipulators is their restricted and highly constrained workspace. Such property reduces the possibility of highly dynamic and variable motion that is necessary for the excitation of all parameters to be identified. The appropriate choice should be a trajectory that is naturally bounded to fit into a small workspace. An attractive approach is the harmonic excitation approach originally proposed by Swevers et al. (Swevers et al., 1997) and adapted in the following for the case of parallel manipulators.

For each posture coordinate corresponding to the ith element of $\mathbf{x}$ a respective trajectory with n_{h} harmonics is defined as

$$x_i = x_0^i + \sum_{k=1}^{n_{\mathrm{h}}} \left(\frac{\mu_{\mathrm{k}}^{\mathrm{i}}}{k\omega_f} \sin\left(k\omega_f t\right) - \frac{v_{\mathrm{k}}^{\mathrm{i}}}{k\omega_f} \cos\left(k\omega_f t\right) \right), \tag{22}$$

providing a proper trajectory parameter vector

$$\Xi_i = \left[x_0^i, \mu_1^i, \ldots, \mu_{n_{\mathrm{h}}}^i, v_1^i, \ldots, v_{n_{\mathrm{h}}}^i \right]^{\mathrm{T}} \tag{23}$$

with ω_{f} being the fundamental frequency. The difference to the implementation for serial robots is that the excitation trajectory is now defined with respect to $\mathbf{x}$ (and therefore $\mathbf{v}$ and $\mathbf{a}$) rather than to the actuator coordinates $\mathbf{q}_{\mathrm{a}}$. Such modification is necessary, since the dynamics can determined only in the configuration space defined by $\mathbf{x}$. With the proposed modification, a direct relationship between the dynamics to be excited and the trajectory is available. If the excitation trajectory is defined with respect to the actuated coordinates $\mathbf{q}_{\mathrm{a}}$, the closure constraints of the parallel manipulator and the numerical calculation of the direct kinematics have to be performed while the optimization and design of the trajectory. First ensures a feasible trajectory and second provides the resulting dynamics in form of the regression model. Both operations increase the solution cost and introduce additional numerical errors.

3.2 Optimization of the excitation trajectory

The next step consists in determining the values of all trajectory parameters

$$\Xi = \left[\Xi_1^{\mathrm{T}}, \ldots, \Xi_6^{\mathrm{T}}, \omega_{\mathrm{f}} \right]^{\mathrm{T}} \tag{24}$$

to provide a best possible excitation of the dynamics parameters. Such procedure is called optimal input experiment design. The design is performed by using constrained nonlinear optimization (Swevers et al., 1997; Gevers, 2005). The required constraints are expressed with respect to the actuation variables

$$\begin{aligned} &\mathbf{q}_{\mathrm{a}}^{\min} \le \mathbf{q}_{\mathrm{a}}(t,\Xi) \le \mathbf{q}_{\mathrm{a}}^{\max} \\ &\dot{\mathbf{q}}_{\mathrm{a}}^{\min} \le \dot{\mathbf{q}}_{\mathrm{a}}(t,\Xi) \le \dot{\mathbf{q}}_{\mathrm{a}}^{\max}, \quad \forall \Xi \text{ and } t \in \left[0, T_{\mathrm{f}}\right] \\ &\ddot{\mathbf{q}}_{\mathrm{a}}^{\min} \le \ddot{\mathbf{q}}_{\mathrm{a}}(t,\Xi) \le \ddot{\mathbf{q}}_{\mathrm{a}}^{\max} \end{aligned} \tag{25}$$

to account for actuator limitation and therefore indirectly for workspace constraints and dynamics capabilities of the manipulator. The inverse kinematics has to be performed while the optimization, which does not introduce any significant computational cost due to its simplicity (Khalil & Guegan, 2004; Abdellatif et al., 2005a; Abdellatif & Heimann, 2007). Of course, it is possible to express the constraints *ad-hoc* with respect to $\mathbf{x}$, $\mathbf{v}$ and $\mathbf{a}$. It depends on the considered manipulator, whether such approach is preferable or not, since it results in different constraints than (25), which can accelerate the convergence of the optimization process. The optimization or the experiment design criterion should contribute to the reduction of parameter uncertainty (Gevers, 2005). To account appropriately for disturbances in the information matrix $\mathbf{\Psi}$ it is recommended to opt for the *D*-optimal design

$$\Xi = \arg\left\{\min_{\Xi}\left(-\ln\det\left(\mathbf{\Psi}^{\mathrm{T}}\mathbf{\Sigma}_{\eta\eta}^{-1}\mathbf{\Psi}\right)\right)\right\} \tag{26}$$

that aims increasing the volume of the asymptotic confidence ellipsoid for the parameter estimates, which is equivalent to the determinant of the inverse of the asymptotic parameter covariance matrix $\mathbf{P} = \left(\mathbf{\Psi}^{\mathrm{T}}\mathbf{\Sigma}_{\eta\eta}^{-1}\mathbf{\Psi}\right)^{-1}$ or the Cramér-Rao bound (Gevers, 2005). Due to the complexity of the nonlinear dynamics contained in the regressor $\mathbf{\Psi}$ the optimization is mostly a non-convex one and the obtained results will not correspond to the global minimum. This is however not critical since for experimental identification just a sufficiently good excitation trajectory is needed.

4. Identification procedure: Data processing, implementation and parameter uncertainties

At this stage, the dynamics of the manipulator is available in linear form (section 2). Additionally, the appropriate choice of an excitation experiment is proposed (section 3.1) with a recommended method for its optimal design (section 3.2). Therefore, the experiment can be executed and the data can be collected to achieve an estimation according to (21). Here, the next challenge for parallel manipulators is evident. The measurements are provided in the actuation space in form of actuation forces and actuator positions, whereas the information matrix $\mathbf{\Psi}$ is built up by using $\mathbf{x}$, $\mathbf{v}$ and $\mathbf{a}$ that are not directly measured. Thus, a reconstruction of these variables from the corrupted measurement of $\mathbf{q}_{\mathrm{a}}$ is necessary.

4.1 Data processing

The first step consists in calculating the direct kinematics to provide a first estimate of the posture $\hat{\mathbf{x}}$. The terminal condition of the numerical calculation has to be set less than the resolution of the used sensors (Merlet, 2006). The obtained estimate is of course noisy and has to be filtered. Filtering the measurement in the time-domain (i.e. by using classic low-pass filters) may cause lost of information, since ideal and exact filtering is not possible. More critical is the calculation of $\mathbf{v}$ and $\mathbf{a}$. Numerical differentiation of the posture data is not convenient. Additionally to the measurement noise, possible oscillations of the direct kinematic solution introduces disturbances, such that the resulting data may be not useful at all (Abdellatif et al., 2004).

By taking advantage of the periodic and harmonic nature of the excitation trajectory, exact filtering in the frequency-domain can be achieved. First, it is recommended to calculate the DFT-transform of each component i of the pre-computed posture $\hat{\mathbf{x}}_i \rightarrow \hat{\mathbf{X}}_i(\mathrm{j}\omega)$. Afterwards the spectrum is filtered by a frequency-domain lowpass filter. Ideal filtering can be achieved by means of a rectangular window with a desirable cutoff-frequency f_c. The latter may be chosen (but is not limited to) to correspond the nominal fundamental frequency $f_c \hat{=} n_h \omega_f / 2\pi$. The windowed and filtered spectrum $\mathbf{X}_i(\mathrm{j}\omega)$ is multiplied twice by $\mathrm{j}\omega$

$$\begin{aligned} \dot{\mathbf{X}}_i(\mathrm{j}\omega) &= \mathrm{j}\omega \mathbf{X}_i(\mathrm{j}\omega), \\ \ddot{\mathbf{X}}_i(\mathrm{j}\omega) &= -\omega^2 \mathbf{X}_i(\mathrm{j}\omega). \end{aligned} \tag{27}$$

Transforming back to the time domain yields the filtered signals $\dot{\hat{\mathbf{x}}}_i = \mathrm{DFT}^{-1}\left(\dot{\mathbf{X}}_i(\mathrm{j}\omega)\right)$ and $\ddot{\hat{\mathbf{x}}}_i = \mathrm{DFT}^{-1}\left(\ddot{\mathbf{X}}_i(\mathrm{j}\omega)\right)$. The posture estimate is also updated according to $\hat{\mathbf{x}}_i = \mathrm{DFT}^{-1}\left(\mathbf{X}_i(\mathrm{j}\omega)\right)$. The procedure of data processing in the frequency domain is depicted in Fig. 3. The filtered estimates of the velocities and accelerations of the end-effector are provided by using classic kinematic transformations (Merlet, 2006).

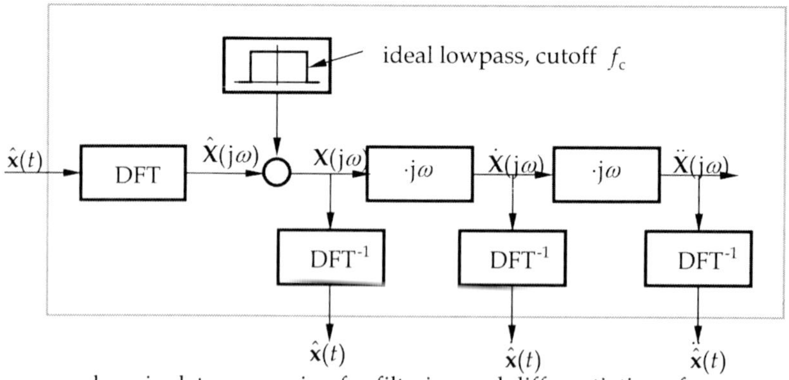

Fig. 3. Frequency-domain data processing for filtering and differentiation of non-measured signals

4.2 Parameter uncertainties

To validate the results of the identification, statements on the uncertainties of the obtained parameters are necessary. For the given linear model structure (20) and by assuming Gaussian disturbance vector $\boldsymbol{\eta}$ (see Abdellatif et al., 2005b), the covariance of the parameter estimate resulting from (21) is

$$\mathbf{P} = \left(\boldsymbol{\Psi}^{\mathrm{T}} \boldsymbol{\Sigma}_{\eta\eta}^{-1} \boldsymbol{\Psi}\right)^{-1}. \tag{28}$$

The confidence area of the estimated parameter set $\hat{\mathbf{p}}_{\mathrm{GM}}$ with respect to the unknown true parameter vector $\mathbf{p}$ can be calculated for a given quantile $\alpha \in [0 \ldots 1]$ as a $100(1-\alpha)\%$ confidence ellipsoid:

$$\mathcal{E}_\alpha = \left\{ \mathbf{p} \in \Re^{n_\mathrm{p}}, \; (\mathbf{p} - \hat{\mathbf{p}}_{\mathrm{GM}})^\mathrm{T} \mathbf{P}^{-1} (\mathbf{p} - \hat{\mathbf{p}}_{\mathrm{GM}}) \le \chi^2_\alpha(n_\mathrm{p}) \right\}, \tag{29}$$

where $\chi^2_\alpha(n_\mathrm{p})$ denotes the value of the χ^2 distribution with n_p degrees of freedom at the quantile α and n_p is the dimension of the parameter vector (Gevers, 2005). Consequently, the estimate of the single parameter $\hat{p}_k$ is normally distributed $N(p_k, P_{kk})$ with variance P^2_{kk}, where p_k is the true parameter value and P_{kk} is the kth diagonal element of $\mathbf{P}$. A 95% confidence interval can be determined as

$$C_k^{95\%} = \left[\hat{p}_k - 2\rho_k, \; \hat{p}_k + 2\rho_k\right] = \left[\hat{p}_k - 2\sqrt{P_{kk}}, \; \hat{p}_k + 2\sqrt{P_{kk}}\right]. \tag{30}$$

Equations (29) and (30) are useful to evaluate the confidence of the estimate results for the complete parameter set or for the single parameters, respectively.

5. Experimental results for model-based control

This section is dedicated to the experimental results achieved on the hexapod PaLiDA.

5.1 Description and modelling of the hexapod

The parallel robot PaLiDA (see Fig. 1) was developed by the Institute of Production Engineering and Machine Tools at the University of Hannover as a Stewart-Gough platform. It is designed with electromagnetic linear direct drives used as extensible struts for use in fast handling and light cutting machining like deburring. The actuation principle has several advantages compared to conventional ball screw drives: Fewer mechanical components, no backlash, low inertia with a minimized number of wear parts. Furthermore, higher control bandwidth and extremely high accelerations can be achieved. A commercial electromagnetic linear motor originally designed for fast lifting motions is improved for use in the struts. Each strut of the hexapod is composed of three bodies as depicted in Fig. 4. Thus, the system is modelled with 19 bodies: The movable platform (index E), 6 identical movable cardan rings (index 1), 6 identical stators (index 2) and 6 identical sliders (index 3).

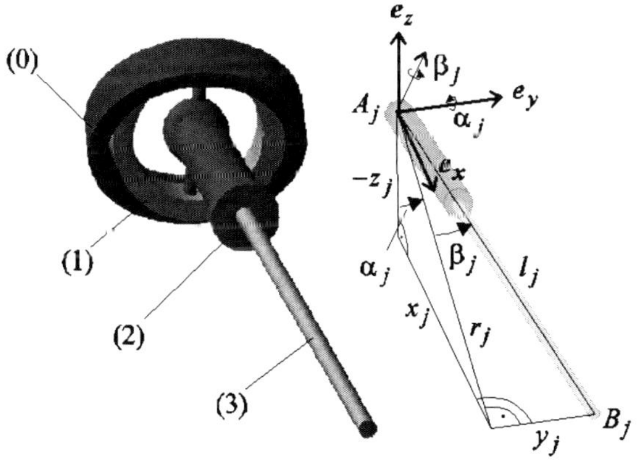

i	ϑ_i	d_i	a_i	α_i
1	$\alpha_i - \frac{\pi}{2}$	0	0	$\frac{\pi}{2}$
2	$\beta_i - \frac{\pi}{2}$	0	0	$-\frac{\pi}{2}$
3	0	l_i	0	$-\frac{\pi}{2}$

Fig. 4. MDH-frames and parameters of the struts

The dynamics model in parameterlinear form results by applying the rules discussed in section 2. The rigid-body part contains 10 base parameters (see Table 1). According to the friction modelling approach (14) the actuated joints $\mathbf{q}_a$ correspond to 6 different dry friction and also 6 different viscous damping coefficients. Friction in the passive joints is modelled only as dry friction with a common parameter for all α_j and another one for all β_j-joints. The friction model contains therefore 14 different parameters. Its structure was optimized according to the statistical analysis mentioned in section 2.2 and presented in (Abdellatif et al., 2005c).

rigid-body	friction
$p_1 = I_{zz_1} + I_{yy_2} + I_{zz_3}$ $[\mathrm{kgm}^2]$	$p_{11} = r_\alpha$ $[\mathrm{N}]$
$p_2 = I_{xx_2} + I_{xx_3} - I_{yy_2} - I_{zz_3}$ $[\mathrm{kgm}^2]$	$p_{12} = r_\beta$ $[\mathrm{N}]$
$p_3 = I_{zz_2} + I_{yy_3}$ $[\mathrm{kgm}^2]$	$p_{13} = r_{1_1}$ $[\mathrm{N}]$
$p_4 = s_{y_2}$ $[\mathrm{kgm}]$	$p_{14} = r_{1_2}$ $[\mathrm{N}]$
$p_5 = s_{y_3}$ $[\mathrm{kgm}]$	$p_{15} = r_{1_3}$ $[\mathrm{N}]$
$p_6 = I_{xx_E} + m_3 \sum_{j=1}^{6} \left(r_{B_{y_j}}^2 + r_{B_{z_j}}^2 \right)$ $[\mathrm{kgm}^2]$	$p_{16} = r_{1_4}$ $[\mathrm{N}]$
$p_7 = I_{yy_E} + m_3 \sum_{j=1}^{6} \left(r_{B_{x_j}}^2 + r_{B_{z_j}}^2 \right)$ $[\mathrm{kgm}^2]$	$p_{17} = r_{1_5}$ $[\mathrm{N}]$
$p_8 = I_{zz_E} + m_3 \sum_{j=1}^{6} \left(r_{B_{x_j}}^2 + r_{B_{y_j}}^2 \right)$ $[\mathrm{kgm}^2]$	$p_{18} = r_{1_6}$ $[\mathrm{N}]$
$p_9 = s_{z_E} + m_3 \sum_{j=1}^{6} r_{B_{z_j}}$ $[\mathrm{kgm}]$	$p_{19} = r_{2_1}$ $[\mathrm{Nsm}^{-1}]$
$p_{10} = m_E + 6 m_3$ $[\mathrm{kg}]$	$p_{20} = r_{2_2}$ $[\mathrm{Nsm}^{-1}]$
	$p_{21} = r_{2_3}$ $[\mathrm{Nsm}^{-1}]$
	$p_{22} = r_{2_4}$ $[\mathrm{Nsm}^{-1}]$
	$p_{23} = r_{2_5}$ $[\mathrm{Nsm}^{-1}]$
	$p_{24} = r_{2_6}$ $[\mathrm{Nsm}^{-1}]$

Table 1. Rigid-body and friction model parameters for the parallel robot PaLiDA

5.2 Experiment design and data processing

The experiment design has been carried out according to the method given in section 3. An example of a resulting excitation trajectory with the order $n_h = 5$ is depicted by Fig. 5. The obtained measurements of the actuator lengths are transformed numerically by the direct kinematics. The resulting estimation of posture elements are then filtered and differentiated in the frequency domain as proposed in section 4.1. Fig. 6 illustrates exemplarily such procedure for the reconstruction of the second translational degree of freedom y corresponding to the excitation trajectory, shown in Fig. 5.

The left side of Fig. 5 depicts the frequency-discrete spectral amplitudes of the signals along with the used selection window that corresponds to an ideal lowpass filter. The respective signals in the time-domain are given on the right side of the picture. The effectiveness of the proposed filter is obvious, since the calculated signals exhibit almost no noise or disturbance corruption. Such property is a central requirement for a robust and reliable identification of parallel manipulators, because the necessary but non-measurable information has to be extracted from corrupted and limited measurements of the actuator displacements.

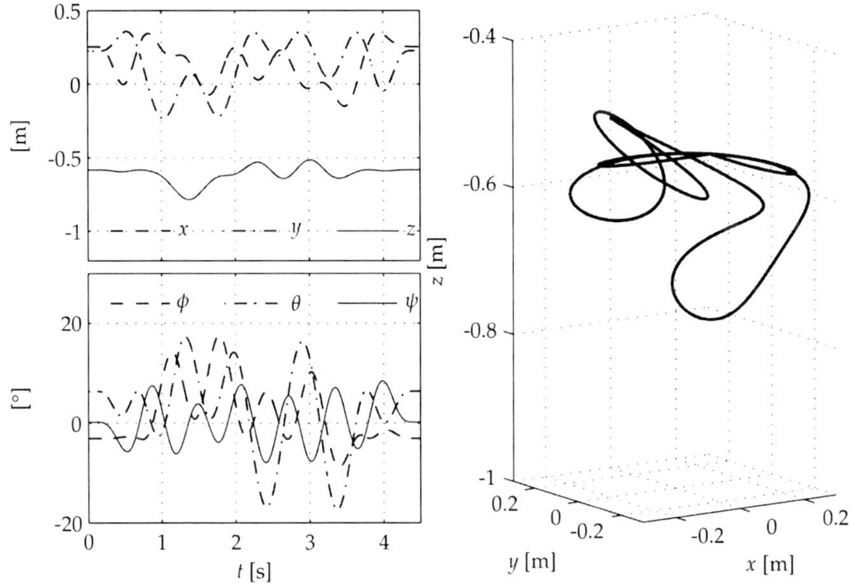

Fig. 5. Example of a periodic excitation trajectory $n_h = 5$; top left: Translational coordinates, bottom left: Rotational coordinates, right: 3-D presentation

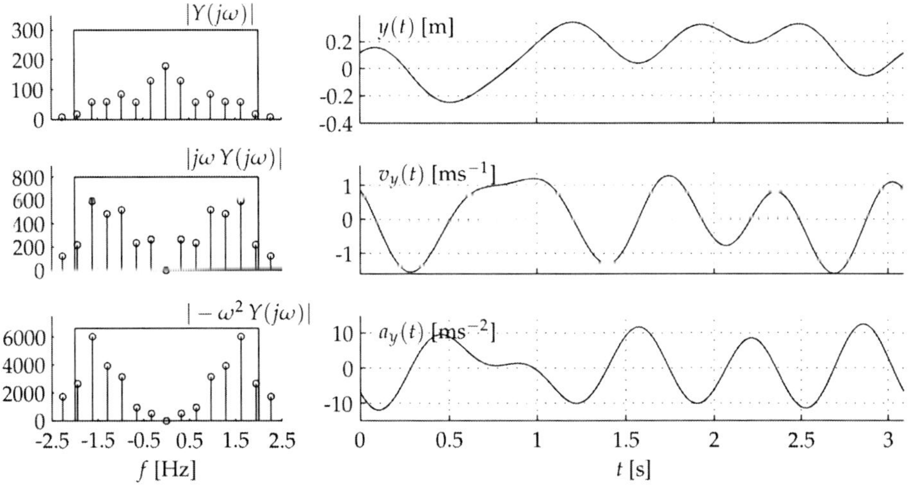

Fig. 6. Reconstruction of the end-effector displacement, velocity, and acceleration, with respect to the inertial y -axis by using frequency-domain filtering

In the following three models are compared, that all result from the identification using the same trajectory but after implementing three different data-processing techniques. The first one results directly from rough data without any filtering. For the second, the measurements of the actuator displacements were filtered in the time domain. The third model has been identified according to the proposed frequency domain method. The validation of the models on a circular bench-mark trajectory, that was not used for identification, is depicted in Fig. 7. The frequency-domain processing yields the best prediction quality corresponding to the smallest error variance σ^2. Time-domain filtering is not accurate enough to extract all information at the relevant frequencies.

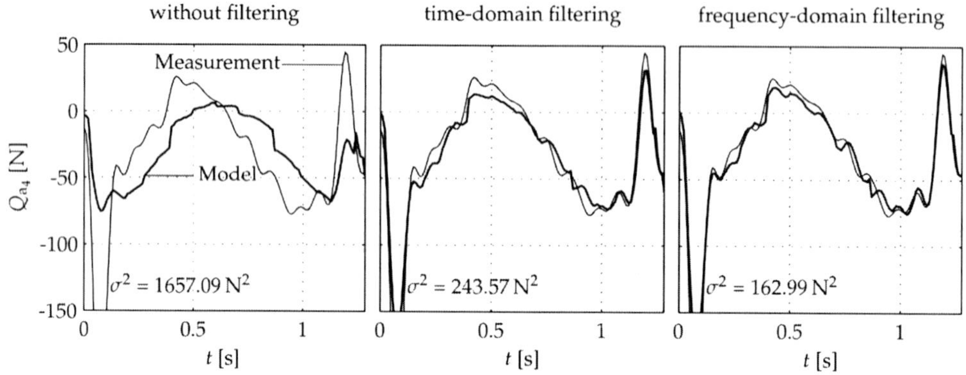

Fig. 7. Prediction accuracy of three different models for an arbitrarily chosen actuator; left: By using rough data, middle: By using time-domain filtering, right: By using frequency-domain filtering

5.3 Estimation results and parameter uncertainties

The filtered data resulting from the investigated trajectory (Fig. 5) are used to compute the regressor matrix $\boldsymbol{\Psi}$. The corresponding actuation forces can be obtained from the measurement of the motor currents. The case of PaLiDA reveals high noisy and cross-correlated force measurements (Abdellatif et al., 2005b). Therefore, the Gauss-Markov estimate has been proposed earlier (see (21)) that yields the parameter set given in Table 2. It is important to notice, that the provided a priori values do not present the true parameters, since they were calculated by using uncertain CAD-Data. The quality of the results is in general very high, despite that the parameters with small values exhibit higher uncertainties. This is however a known and general problem of experimental estimation in practice. We refer to former publications for detailed discussions on the different aspects of the estimation results (Abdellatif et al., 2005b; Abdellatif et al., 2005c).

The validation of the parameter estimation robustness can be provided, e.g. after repeating the identification experiment 100 times. The resulted parameter sets are compared to the 95% confidence intervals (see eq. (30)). Such investigation is depicted for some exemplarily chosen parameters in Fig. 8. The history of the weighted parameter estimate $\hat{p}_i/\overline{p}_i$ are illustrated over the measurement trials, where $\overline{p}_k$ is the mean value of all estimates. The corresponding weighted upper and lower bounds C_{M} and C_{m} of the confidence intervals are additionally shown. The robustness of the identification is proven, since the estimates remain mostly within the confidence intervals. Some exceptions are observed though, such the very small first rigid-body parameter and the first few measurement trials. The latter is

p_k	$\hat{p}_k$	$\rho_k = \sqrt{P_{kk}}$	$C_k^{95\%}$	a priori
p_1 $[kgm^2]$	-0.0447	0.0039	[-0.0526 -0.0369]	0.0074
p_2 $[kgm^2]$	1.0892	0.0070	[1.0753 1.1032]	0.9439
p_3 $[kgm^2]$	1.0077	0.0045	[0.9988 1.0166]	0.9458
p_4 $[kgm]$	0.5995	0.0036	[0.5922 0.6068]	0.6201
p_5 $[kgm]$	-1.2885	0.0056	[-1.2998 -1.2772]	1.2295
p_6 $[kgm^2]$	0.3078	0.0061	[0.3049 0.3106]	0.2878
p_7 $[kgm^2]$	0.3021	0.0014	[0.2996 0.3045]	0.2878
p_8 $[kgm^2]$	0.1176	0.0012	[0.1152 0.1201]	0.1217
p_9 $[kgm]$	1.8896	0.0012	[1.8774 1.9017]	1.9012
p_{10} $[kg]$	16.3081	0.0460	[16.2161 16.4002]	16.1920
r_α $[Nm]$	0.5756	0.0158	[0.5440 0.6072]	-
r_β $[Nm]$	0.9195	0.0179	[0.8837 0.9552]	-
r_{1_1} $[N]$	11.9772	0.2485	[11.4803 12.4742]	-
r_{1_2} $[N]$	4.8071	0.1861	[4.4350 5.1793]	-
r_{1_3} $[N]$	20.1528	0.3226	[19.5075 20.7980]	-
r_{1_4} $[N]$	5.1518	0.1817	[4.7884 5.5151]	-
r_{1_5} $[N]$	1.5857	0.2618	[1.0620 2.1094]	-
r_{1_6} $[N]$	5.0057	0.3519	[4.3018 5.7096]	-
r_{2_1} $[Nsm^{-1}]$	16.8771	0.5268	[15.8235 17.9307]	-
r_{2_2} $[Nsm^{-1}]$	16.7406	0.3712	[15.9981 17.4830]	-
r_{2_3} $[Nsm^{-1}]$	6.3408	0.5720	[5.1968 7.4848]	-
r_{2_4} $[Nsm^{-1}]$	23.1662	0.3799	[22.4065 23.9259]	-
r_{2_5} $[Nsm^{-1}]$	26.4675	0.4461	[25.5754 27.3596]	-
r_{2_6} $[Nsm^{-1}]$	22.8053	0.5539	[21.6974 23.9131]	-

Table 2. Estimated dynamics parameters of the hexapod parameters $\hat{p}$ with corresponding standard deviations, confidence intervals and a priori values for the rigid-body model parameters

due to the variation of friction at the beginning of the measurement process until a nearly stationary state is reached. Additionally to the single parameters, the confidence of the entire parameter set can be validated. The outer bound of the 95% confidence ellipsoid $\mathcal{E}_{5\%}$

is given by $\chi^2_{5\%}(\dim(\mathbf{p})=24)=36.42$. Its comparison with distribution $\chi^2(\hat{\mathbf{p}}_i)$ of the vector estimates $\hat{\mathbf{p}}_i$ over the measurement trials is given by Fig. 9. Excepting the first trial, the set of all parameters lays clearly within the confidence ellipsoid, which demonstrates the effectiveness and robustness of the estimation.

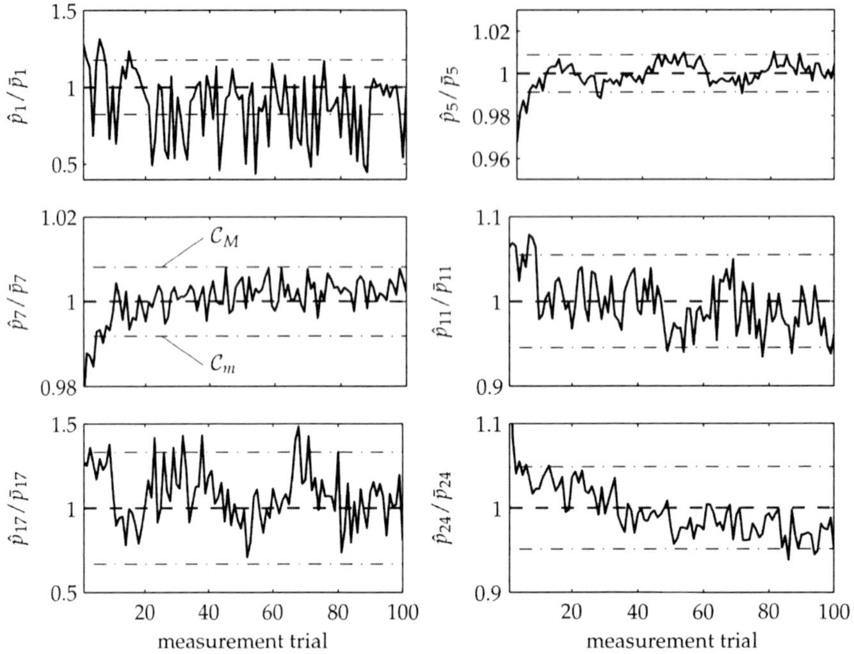

Fig. 8. History of parameter estimates over different measurement trials: For the sake of uniform illustration, the parameters are given as weighted terms with respect to their respective mean values.

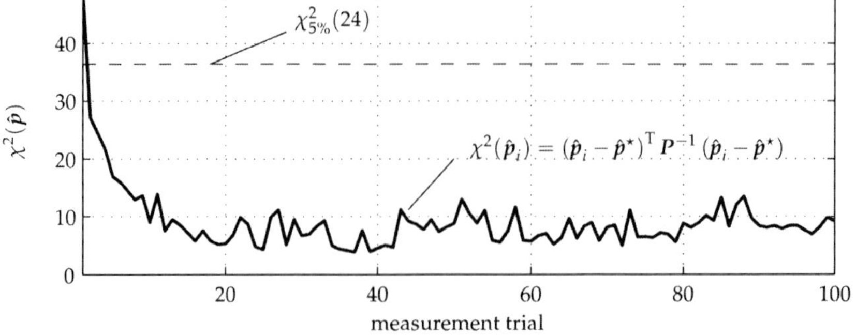

Fig. 9. Comparison of the χ^2 distribution of the estimated parameter sets with the radius $\chi^2_{5\%}(24)$ of the 95% confidence ellipsoid

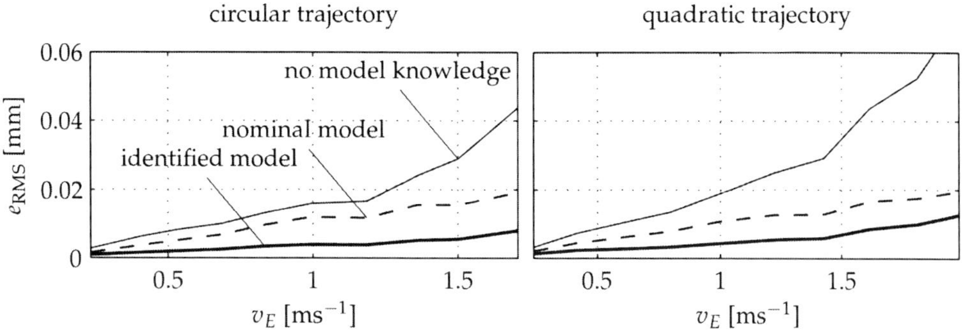

Fig. 10. Control errors for both test trajectories at increasing end-effector velocity and by implementing different control strategies

5.4 Identification and model-based control

The impact of identification on the control and tracking accuracy of the hexapod PaLiDA is studied in the following. Hereby three control strategies are investigated. The first variation passes on any model knowledge, i.e. by implementing only linear controller for the single actuators. The second uses the inverse dynamics model to compensate for the nonlinear dynamics by considering only nominal parameter values. The third variation uses the identified model for the feedforward compensating control. All approaches are substantiated experimentally on two different trajectories: The first trajectory is a circular one and allows reaching high actuation forces, whereas the second is quadratic and is characterized with high actuator velocities.

Both trajectories were executed at different velocities v_E of the end-effector and for the three mentioned control variations. The evolution of the rooted mean squares errors e_{RMS} of all actuator deviations is depicted in Fig. 10 with respect to v_E.

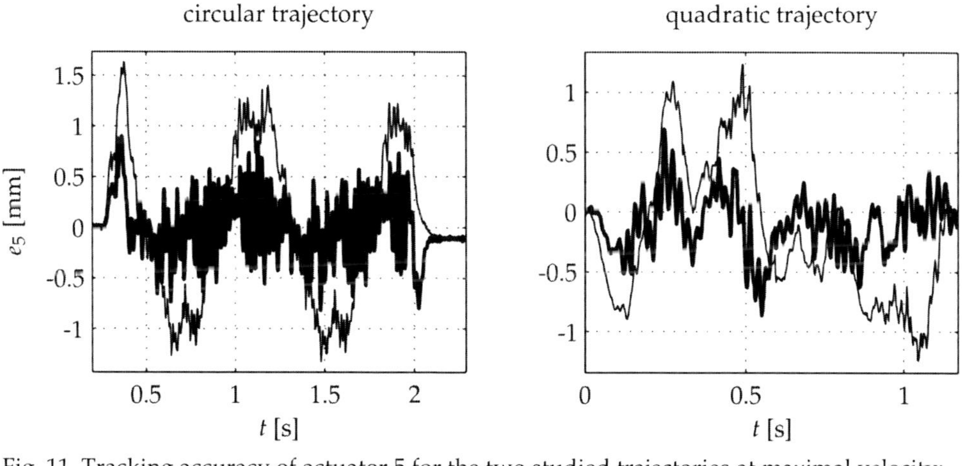

Fig. 11. Tracking accuracy of actuator 5 for the two studied trajectories at maximal velocity; comparison between the compensation of nominal model (thin line) and identified model (thick line)

As expected, the use of standard linear control (variation 1) exhibits a significant decreasing accuracy with increasing speeds, since the impact of nonlinear and coupled dynamics increases with higher velocities and accelerations. Using model-knowledge (variation 2 and 3) improves always the tracking performance. Furthermore, the compensation of identified model (variation 3) outperforms clearly variation 2 that just uses the nominal parameter values. The latter statement can be proven at the level of actuator tracking accuracy like depicted in Fig. 11. For the same arbitrarily chosen actuator, the tracking accuracy is higher if the identified model is implemented. The same results are noticeable for the cartesian tracking accuracy Δx, like depicted in Fig. 12. It may be concluded that only accurately identified model allows keeping good tracking performance over a wide range of the robot dynamics.

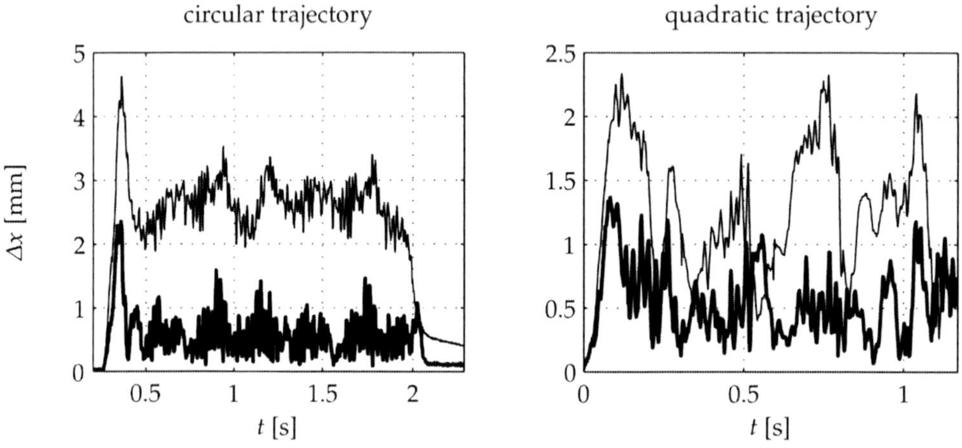

Fig. 12. Calculated cartesian tracking accuracy Δx for the two studied trajectories at maximal velocity; comparison between the compensation of nominal model (thin line) and identified model (thick line)

6. Conclusions

The present chapter discussed most significant aspects to achieve accurate and robust dynamics identification for parallel manipulators with 6 dof's. Hereby, the adequate consideration of structural properties of such systems has been stressed out. First, an efficient methodology to determine the inverse dynamics in a parameterlinear form has been presented, which enables the use of linear estimation techniques. Periodic excitation has been proved to be a powerful method for parallel robots, since it allows for appropriate consideration of hard workspace constraints. Due to measurement noise and cross coupling between the actuators, the achievement of the identification in a statistical framework is recommended. This includes the consideration of disturbance covariances in the experiment design, the use of Gauss-Markov estimation approach as well as the frequency-domain filtering to extract non measurable information from rough data. The robustness of the results has been substantiated on a direct driven hexapod. The obtained estimates have presented high confidence in terms of single parameters, as well as in terms of the whole parameter set. Additionally, the benefits of accurate identification on the enhancement of control performance have been clearly and experimentally demonstrated.

7. References

Swevers, J.; Gansemann, C. & Tükel, D. & Schutter, J. d. & Brussel, H. v. (1997). Optimal robot excitation and identification, *IEEE Transactions on Robotics and Automation*, 13, 5, pp. 730-740

Khalil, W. & Dombre, E. (2002). *Modelling, Identification and Control of Robots*, Hermes, London

Abdellatif, H. & Heimann, B. (2007). *Industrial Robotics: Theory, Modelling and Control*, Pro-Literatur Verlag, pp. 523-556.

Khalil, W. & Guegan, S. D. (2004). Inverse and direct dynamics modelling of Gough-Stewart robots, *IEEE Transactions on Robotics*, 20, 4, pp. 754-762.

Abdellatif, H.; Grotjahn, M. & Heimann B. (2005a). High efficient dynamics calculation approach for computed-force control of robots with parallel structures, *Proceedings of the 44th IEEE Conf. on Decision and Control and the 2005 European Control Conference*, pp. 2024-2029 , Seville, 2005.

Abdellatif H.; Heimann, B. & Grotjahn M. (2005b). Statistical approach for bias-free identification of a parallel manipulator affected with high measurement noise, *Proceedings of the 44th IEEE Conf. on Decision and Control*, pp. 3357-3362, Seville, 2005.

Merlet, J.-P. (2006). *Parallel Robots, 2nd Edition*, Springer, Netherlands.

Tsai, L.-W. (2000). Solving the inverse dynamics of a Stewart-Gough manipulator by the principle of virtual work, *ASME Journal of Mechanical Design*, 122, 5, pp. 3-9.

Harib, K. & Srinivasan, K. (2003). Kinematic and dynamics analysis of Stewart platform-based machine tool structures, *Robotica*, 21, 5, pp. 541-554.

Grotjahn, M. & Heimann, B. (2000). Determination of dynamic parameters of robots by base sensor measurements, *Proceedings of the 6th IFAC Symposium on Robot Control*, pp. 277-282, Vienna, 2000.

Grotjahn, M.; Kuehn, J. & Heimann, B. & Grendel, H. (2002). Dynamic equations of parallel robots in minimal dimensional parameter-linear form, *Proceedings of the 14th CISM-IFToMM Symp. on the Theory and Practice of Robots and Manipulators*, pp. 67-76, Udine, 2002.

Abdellatif, H. & Heimann, B. (2006). On compensation of passive joint friction in robotic manipulators: Modeling, detection and identification, *Proceedings of the 2006 IEEE International Conf. on Control Applications*, pp. 2510 – 2515, Munich, 2006.

Abdelllatif, H.; Grotjahn, M. & Heimann, B. (2007). Independent identification of friction characteristics for parallel manipulators, *ASME Journal of Dynamic Systems, Measurment and Control*, 129, 3, pp. 294-302.

Abdellatif, H.; Heimann, B. & Hornung, O. & Grotjahn, M. (2005c). Identification and appropriate parametrization of parallel robot dynamic models by using estimation statistical properties, *Proceedings of the 2005 IEEE/RSJ International Conf. on Intelligent Robots and Systems*, pp. 444-449, Edmonton, 2005.

Gevers, M. (2005). Identification for control: From the early achievements to the revival of experiment design, *European Journal of Control*, 11, 4-5, pp. 335-352.

Abdellatif, H.; Benimeli, F. & Heimann, B. & Grotjahn, M. (2004). Direct identification of dynamic parameters for parallel manipulators, *Proceedings of the International Conf. on Mechatronics and Robotics 2004*, pp. 999-1005, Aachen, 2004.

2

Application of Neural Networks to Modeling and Control of Parallel Manipulators

Ahmet Akbas
Marmara University
Turkey

1. Introduction

There are mainly two types of the manipulators: serial manipulators and parallel manipulators. The serial manipulators are open-ended structures consisting of several links connected in series. Such a manipulator can be operated effectively in the whole volume of its working space. However, as the actuator in the base has to carry and move the whole manipulator with its links and actuators, it is very difficult to realize very fast and highly accurate motions by using such manipulators. As a consequence, there arise the problems of bad stiffness and reduced accuracy.

Unlike serial manipulators their counterparts, parallel manipulators, are composed of multiple closed-loop chains driving the end-effector collectively in a parallel structure. They can take a large variety of form. However, most common form of the parallel manipulators is known as platform manipulators having architecture similar to that of flight simulators in which two special links can be distinguished, namely, the base and moving platform. They have better positioning accuracy, higher stiffness and higher load capacity, since the overall load on the system is distributed among the actuators.

The most important advantage of parallel manipulators is certainly the possibility of keeping all their actuators fixed to base. Consequently, the moving mass can be much higher and this type of manipulators can perform fast movements. However, contrary to this situation, their working spaces are considerably small, limiting the full exploitation of these predominant features (Angeles, 2007).

Furthermore, for the fast and accurate movements of parallel manipulators it is required a perfect control of the actuators. To minimize the tracking errors, dynamical forces need to be compensated by the controller. In order to perform a precise compensation, the parameters of the manipulator's dynamic model must be known precisely.

However, the closed mechanical chains make the dynamics of parallel manipulators highly complex and the dynamic models of them highly non-linear. So that, while some of the parameters, such as masses, can be determined, the others, particularly the firiction coefficients, can't be determined exactly. Because of that, many of the control methods are not efficient satisfactorly. In addition, it is more difficult to investigate the stability of the control methods for such type manipulators (Fang et al., 2000).

Under these conditions of uncertainty, a way to identify the dynamic model parameters of parallel manipulators is to use a non-linear adaptive control algorithm. Such an algorithm

can be performed in a real-time control application so that varying parameters can continuously be updated during the control process (Honegger et al., 2000).
Another way to identify the dynamic system parameters may be using the artificial intelligence (AI) techniques. This approach combines the techniques from the fields of AI with those of control engineering. In this context, both the dynamic system models and their controller models can be created using artificial neural networks (ANN).
This chapter is mainly concerned with the possible applications of ANNs that are contained within the AI techniques to modeling and control of parallel manipulators. In this context, a practical implementation, using the dynamic model of a conventional platform type parallel manipulator, namely Stewart manipulator, is completed in MATLAB simulation environment (www.mathworks.com).

2. ANN based modeling and control

Intelligent control systems (ICS) combine the techniques from the fields of AI with those of control engineering to design autonomous systems. Such systems can sense, reason, plan, learn and act in an intelligent manner, so that, they should be able to achieve sustained desired behavior under conditions of uncertainty in plant models, unpredictable environmental changes, incomplete, inconsistent or unreliable sensor information and actuator malfunction.
An ICS comprises of perception, cognition and actuation subsystems. The perception subsystem collects information from the plant and the environment, and processes it into a form suitable for the cognition subsystem. The cognition subsystem is concerned with the decision making process under conditions of uncertainty. The actuation subsystem drives the plant to some desired states.
The key activities of cognition systems include reasoning, using knowledge-based systems and fuzzy logic; strategic planning, using optimum policy evaluation, adaptive search, genetic algorithms and path planning; learning, using supervised or unsupervised learning in ANNs, or adaptive learning (Burns, 2001).
In this chapter it is mainly concerned with the application of ANNs that are contained within the cognition subsystems to modeling and control of parallel manipulators.

2.1 ANN overwiev

ANN is a network of single neurons jointed together by synaptic connections. Such that they are organized as neuronal layers. Each neuron in a particular layer is connected to neurons in the subsequent layer with a weighted synaptic connection. They attempt to emulate their biological counterparts.

2.1.1 Perceptrons

McCulloch and Pitts was started first study on ANN in 1943. They proposed a simple model of neuron. In 1949 Hebb described a technique which became known as Hebbian learning. In 1961 Rosenblatt devised a single layer of neurons, called a perceptron that was used for optical pattern recognition (Burns, 2001)
Perceptrons are early ANN models, consisting of a single layer and simple threshold functions. The architecture of a perceptron consisting of multiple neurons with *Nx1* inputs and *Mx1* outputs is shown in Fig. 1. As seen in this figure, the output vector of the

perceptron is calculated by summing the weighted inputs coming from its input links, so that

$$\mathbf{u} = \mathbf{W}\,\mathbf{p} + \mathbf{b} \tag{1}$$

$$\mathbf{q} = \mathbf{f}(\mathbf{u}) \tag{2}$$

where **p** is $Nx1$ input vector ($p_1, p_2, \ldots p_N$), **W** is MxN weighting coefficients matrix ($w_{11}, w_{12}, \ldots w_{1N}; \ldots; w_{j1}, w_{j2}, \ldots, w_{jN}; \ldots; w_{M1}, w_{M1}, \ldots, w_{MN}$), **b** is $Mx1$ bias factor vector, **u** is $Nx1$ vector including the sum of the weighted inputs ($u_1, u_2, \ldots u_M$) and bias vector, **q** is $Mx1$ output vector ($q_1, q_2, \ldots q_M$),, and **f(.)** is the activation function.

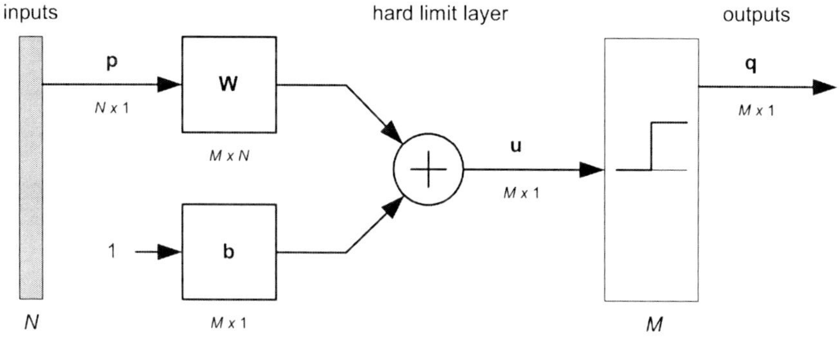

Fig. 1. The architecture of a perceptron

In early perceptron models, the activation function was selected as hard-limiter (unit step) given as follows:

$$q_i = \begin{matrix} 0 & , f(u_i) < 0 \\ 1 & , f(u_i) \geq 0 \end{matrix} \tag{3}$$

where $i = 1,2,\ldots,M$ denotes the number of neuron in the layer, u_i weighted sum of its particular neuron, and q_i its output. However, in any ANN the activation function $f(u_i)$ can take many forms, such as, linear (ramp), hyperbolic tangent and sigmoid forms. The equation for sigmoid function is:

$$f(u_i) = 1 \,/\, (1 + e^{-u_i}) \tag{4}$$

The sigmoid activation function given in Equation (4) is popular for ANN applications since it is differantiable and monolithic, both of which are a requirement for training algorithms like as the backpropagation algorithm.

Perceptrons must include a training rule for adjusting the weighting coefficients. In the training process, it compares the actual network outputs to the desired network outputs for each epoch to determine the actual weighting coefficients:

$$\mathbf{e} = \mathbf{q}^d - \mathbf{q} \tag{5}$$

$$\mathbf{W}^{new} = \mathbf{W}^{old} + \mathbf{e}\,\mathbf{p}^T \tag{6}$$

$$\mathbf{b}^{new} = \mathbf{b}^{old} + \mathbf{e} \tag{7}$$

where **e** is $Mx1$ error vector, $\mathbf{q}^d$ is $Mx1$ target (desired) vector, the upscripts T , *old* and *new* denotes the transpose, the actual and previous (old) representation of the vector or matrix, respectively (Hagan et al., 1996).

2.1.2 Network architectures

There are mainly two types of ANN architectures: feedforward and recurrent (feedback) architectures. In the feedforward architecture, all neurons in a particular layer are fully connected to all neurons in the subsequent layer. This generally called a fully connected multilayer network. Recurrent networks are based on the work of Hopfield and contain feedback paths. A recurrent network having two inputs and three outputs is shown in Fig. 2. In Fig. 2, the inputs occur at time (kT) and the outputs are predicted at time $(k+1)T$, where k is discrete time index and T is sampling time, respectively.

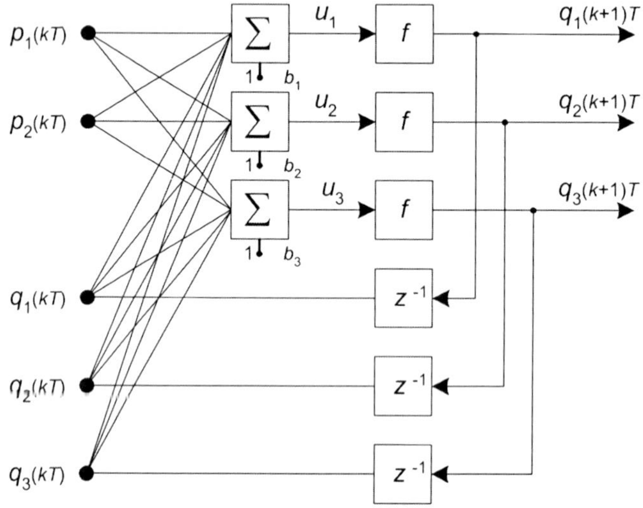

Fig. 2. Recurrent neural network architecture

Then the network can be represented in matrix form as:

$$\mathbf{q}(k+1)T = \mathbf{f}\,(\mathbf{W}_1\,\mathbf{p}(kT) + \mathbf{W}_2\,\mathbf{q}(kT) + \mathbf{b}) \tag{8}$$

where **b** is bias vector, **f(.)** is activation function, $\mathbf{W}_1$ and $\mathbf{W}_2$ are weight matrix for inputs and feedback paths, respectively.

2.1.3 Learning

Learning in the context of ANNs is the process of adjusting the weights and biases in such a manner that for given inputs, the correct responses, or outputs are achieved. Learning algorithms include supervised learning and unsupervised learning.

In the supervised learning the network is presented with training data that represents the range of input possibilities, together with associated desired outputs. The weights are adjusted until the error between the actual and desired outputs meets some given minimum value.

Unsupervised learning is an open-loop adaption because the technique does not use feedback information to update the network's parameters. Applications for unsupervised learning include speech recognition and image compression.
Important unsupervised learning include the Kohonen self-organizing map (KSOM), which is a competitive network, and the Grossberg adaptive resonance theory (ART), which can be for on-line learning.
There are multitudes of different types of ANN models for control applications. The first one of them was by Widrow and Smith (1964). They developed an Adaptive LINear Element (ADLINE) that was taught to stabilize and control an inverted pendulum. Kohonen (1988) and Anderson (1972) investigated similar areas, looking into associative and interactive memory, and also competitive learning (Burns, 2001).
Some of the more popular of ANN models include the multi-layer perceptron (MLP) trained by supervised algorithms in which backpropagation algorihm is used.

2.1.4 Backpropagation

The backpropagation algorithm was investigated by Werbos (1974) and futher developed by Rumelhart (1986) and others, leading to the concept of the MLP. It is a training method for multilayer feedforward networks. Such a network including N inputs, three layers of perceptrons, each has $L1$, $L2$, and M neurons, respectively, with bias adjustment is shown in Fig. 3.

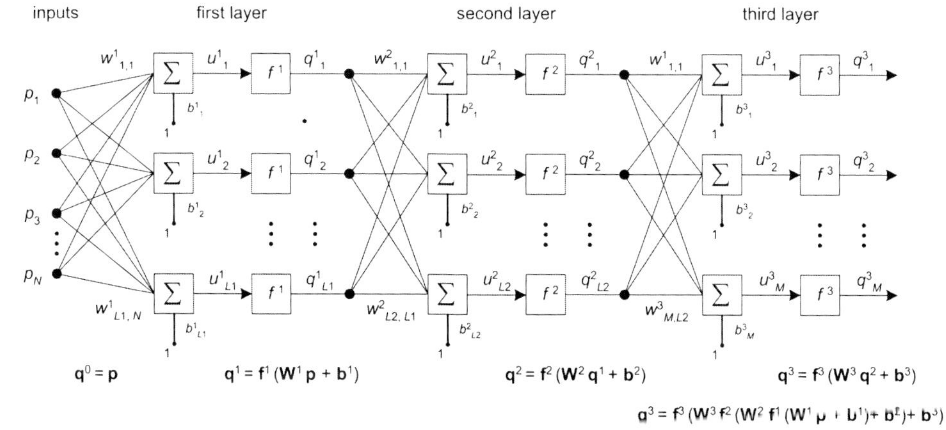

Fig. 3. Three-layer feedforward network

First step in backpropogation is propagating the inputs towards the forward layers through the network. For L layer feedforward network, training process is stated from the output layer:

$$\mathbf{q}^0 = \mathbf{p}$$

$$\mathbf{q}^{l+1} = \mathbf{f}^{l+1}(\mathbf{W}^{l+1}\mathbf{q}^l + \mathbf{b}^{l+1}), \quad l = 0, 1, 2, \ldots, L-1 \tag{9}$$

$$\mathbf{q} = \mathbf{q}^L$$

where l is particular layer number; $\mathbf{f}^l$ and $\mathbf{W}^l$ represent the activation function and weighting coefficients matrix related to the layer l, respectively.
Second step is propagating the sensivities (**s**) from the last layer to the first layer through the network: $\mathbf{s}^L, \mathbf{s}^{L-1}, \mathbf{s}^{L-2}, \ldots, \mathbf{s}^l \ldots, \mathbf{s}^2, \mathbf{s}^1$. The error calculated for output neurons is propagated to the backward through the weighting factors of the network. It can be expressed in matrix form as follows:

$$\mathbf{s}^L = -2\dot{\mathbf{F}}^L(\mathbf{u}^L)\,(\mathbf{q}^d - \mathbf{q}) \quad , \quad \mathbf{s}^l = \dot{\mathbf{F}}^l(\mathbf{u}^l)\,(\mathbf{W}^{l+1})^T \mathbf{s}^{l+1} \;, \quad \text{for } l = L\text{-}1, \ldots, 2, 1 \tag{10}$$

where $\dot{\mathbf{F}}^l(\mathbf{u}^l)$ is Jacobian matrix which is described as follows

$$\dot{\mathbf{F}}^l(\mathbf{u}^l) = \begin{bmatrix} \frac{\partial f^l(u_1^l)}{\partial u_1^l} & 0 & \cdots & 0 \\ 0 & \frac{\partial f^l(u_2^l)}{\partial u_2^l} & \cdots & 0 \\ \vdots & \vdots & & \vdots \\ 0 & 0 & \cdots & \frac{\partial f^l(u_N^l)}{\partial u_N^l} \end{bmatrix} \tag{11}$$

Here N denotes the number of neurons in the layer l. The last step in backpropagation is updating the weighting coefficients. The state of the network always changes in such a way that the output follows the error curve of the network towards down:

$$\mathbf{W}^l\,(k+1) = \mathbf{W}^l\,(k) - \alpha\, \mathbf{s}^l\,(\mathbf{q}^{l-1})^T \tag{12}$$

$$\mathbf{b}^l\,(k+1) = \mathbf{b}^l\,(k) - \alpha\, \mathbf{s}^l \tag{13}$$

where α represents the training rate, k represents the epoch number (k=1,2,...,K). By the algorithmic approach known as gradient descent algorithm using approximate steepest descent rule, the error is decreased repeatedly (Hagan, 1996).

2.2 Applications to parallel manipulators

ANNs can be used for modeling various non-linear system dynamics by learning because of their non-linear system modelling capability. They offer highly parallel, adaptive models that can be trained by using system input-output data.
ANNs have the potential advantages for modeling and control of dynamic systems, such that, they learn from experience rather than by programming, they have the ability to generalize from given training data to unseen data, they are fast, and they can be implemented in real-time.
Possible applications using ANN to modeling and control of parallel manipulators may include:

- Modeling the manipulator dynamics,
- Inverse model of the manipulator,
- Controller emulation by modeling an existing controller,
- Various intelligent control applications using ANN models of the manipulator and/or its controller. Such as, ANN based internal model control (Burns, 2001).

2.2.1 Modeling the manipulator dynamics

Providing input/output data is available, an ANN may be used to model the dynamics of an unknown parallel manipulator, providing that the training data covers whole envelope of the manipulator operation (Fig. 4).

However, it is difficult to imagine a useful non-repetitive task that involves making random motions spanning the entire control space of the manipulator system. This results an intelligent manipulator concept, which is trained to carry out certain class of operations rather than all virtually possible applications. Because of that, to design an ANN model of the chosen parallel manipulator training process may be implemented on some areas of the working volume, depending on the structure of chosen manipulator (Akbas, 2005). For this aim, the manipulator(s) may be controlled by implementation of conventional control algorithms for different trajectories.

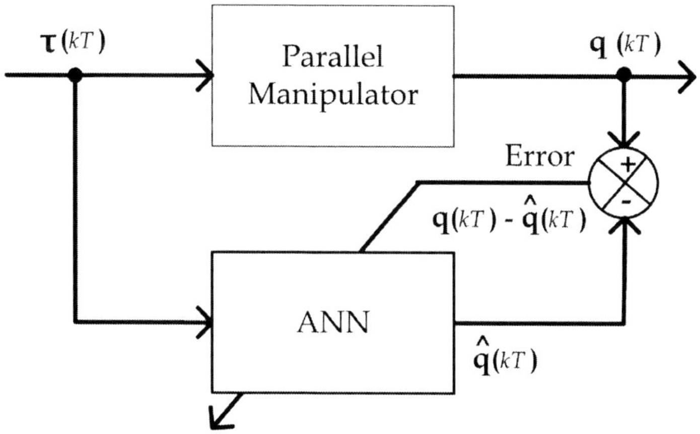

Fig. 4. Modelling the forward dynamics of a parallel manipulator

If the ANN in Fig. 4 is trained using backpropagation, the algorithm will minimize the following performance index:

$$PI = \sum_{n=1}^{N} \left(\left(q(kT) - \hat{q}(kT) \right)^{t} \left(q(kT) - \hat{q}(kT) \right) \right) \tag{14}$$

where $\mathbf{q}$ and $\hat{\mathbf{q}}$ denote the output vector of the manipulator and ANN model, respectively.

2.2.2 Inverse model of the manipulator

The inverse model of a manipulator provides a control vector $\tau(kT)$, for a given output vector $\mathbf{q}(kT)$ as shown in Fig. 5. So, for a given parallel manipulator model, the inverse model could be trained with the parameters reflecting the forward dynamic characteristics of the manipulator, with time.

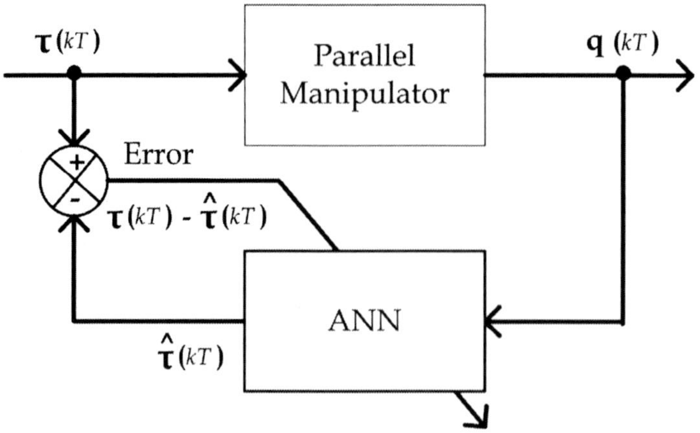

Fig. 5. Modelling the inverse dynamics of parallel manipulator

As indicated above, the training process may be implemented using input-output data obtained by manipulating certain class of operations on some areas of the working volume depending on the structure of chosen manipulator.

2.2.3 Controller emulation

A simple application in control is the use of ANNs to emulate the operation of existing controllers (Fig. 6).

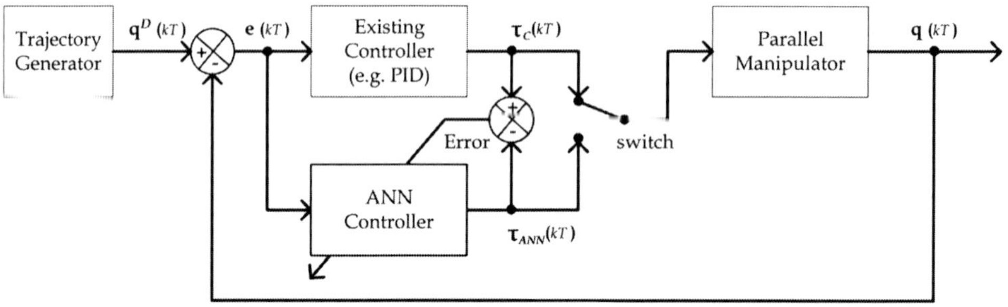

Fig. 6. Training the ANN controller and its implementation to the control system

It may be require several tuned PID controllers to operate over the constrained range of control actions. In this context, some manipulators may be required more than one emulated controllers that can be used in parallel form to improve the reliability of the control system by error minimization approach.

2.2.4 IMC implementation

ANN control can be implemented in various intelligent control applications using ANN models of the manipulator and/or its controller. In this context the internal model control

(IMC) can be implemented using ANN model of parallel manipulataor and its inverse model (Fig. 7).

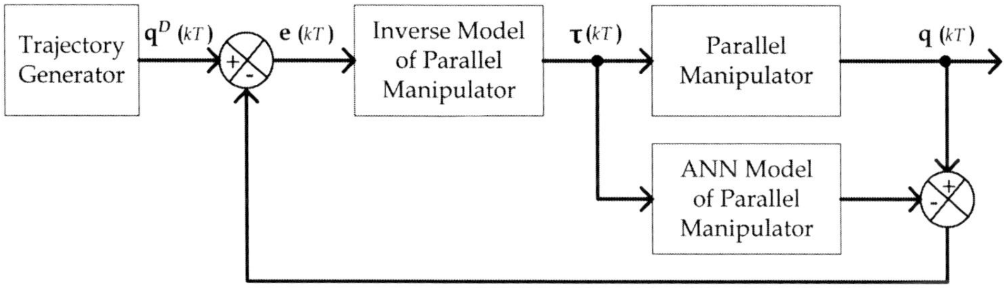

Fig. 7. IMC application using ANN models of parallel manipulator

In this implementation an ANN model model replaces the manipulator model, and an inverse ANN model of the manipulator replaces the controller as shown in Fig. 7.

2.2.5 Adaptive ANN control

All closed-loop control systems operate by measuring the error between desired inputs and actual outputs. This does not, in itself, generate control action errors that may be backpropagated to train an ANN controller. However, if an ANN of the manipulator exists, backpropagation through this network of the system error will provide the necessary control action errors to train the ANN controller as shown in Fig.8.

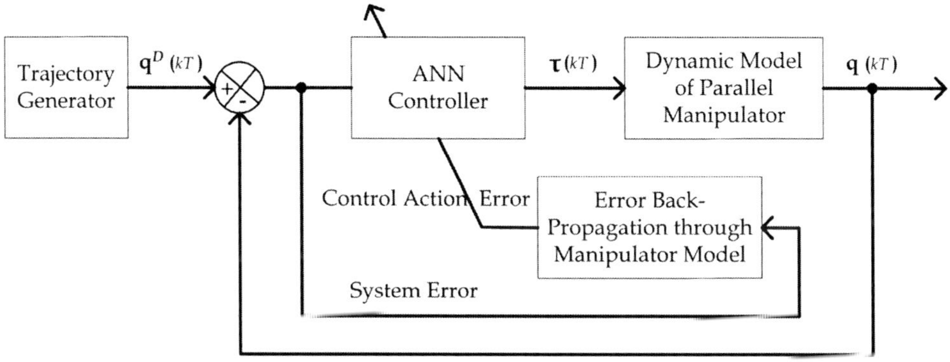

Fig. 8. Control action generated by adaptive ANN controller

3. The structure of Stewart manipulator

Six degrees of freedom (6-dof) simple and practical platform type parallel manipulator, namely Stewart manipulator, is sketched in Fig. 9. These type manipulators were first introduced by Gough (1956-1957) for testing tires. Stewart (1965) suggested their use as flight simulators (Angeles, 2007).

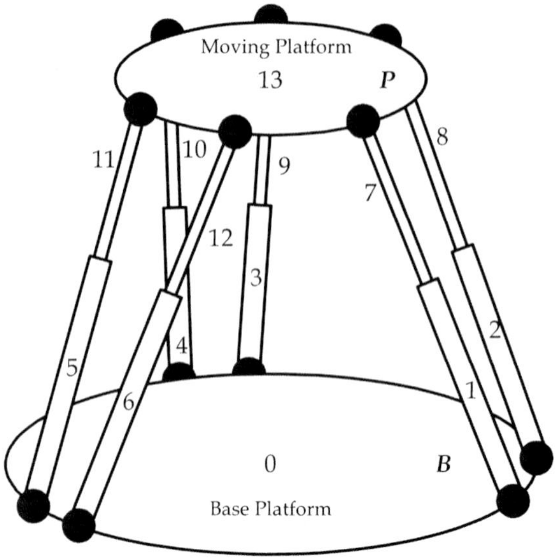

Fig. 9. A sketch of the 6-dof Stewart manipulator

In Fig. 9, the upper rigid body forming the moving platform, ***P***, is connected to the lower rigid body forming the fixed base platform, ***B***, by means of six legs. Each leg in that figure has been represented with a spherical joint at each end. Each leg has upper and lower rigid bodies connected with a prismatic joint, which is, in fact, the only active joint of the leg. So, the manipulator has thirteen rigid bodies all together, as denoted by 1,2.....13 in Fig. 9.

3.1 Kinematics

Motion of the moving platform is generated by actuating the prismatic joints which vary the lengths of the legs, $q^L{}_i$, i = 1....6. So, trajectory of the center point of moving platform is adjusted by using these variables.

For modeling the Stewart manipulator, a base reference frame F_B ($O_Bx_By_Bz_B$) is defined as shown in Fig. 10. A second frame F_P ($O_Px_Py_Pz_P$) is attached to the center of the moving platform, O_P , and the points linking the legs to the moving platform are noted as Q_i , i = 1....6, and each leg is attached to the base platform at the point B_i, i = 1....6.

The pose of the center point, O_P , of moving platform is represented by the vector

$$\mathbf{x} = [x_B \; y_B \; z_B \; \alpha \; \beta \; \gamma]^T \tag{15}$$

where x_B, y_B, z_B are the cartesian positions of the point O_P relative to the frame F_B and a, β, γ are the rotation angles, namely Euler angles, representing the orientation of frame F_P relative to the frame F_B by three successive rotations about the x_P, y_P and z_P axes, given by the matrices $R_x(a)$, $R_y(\beta)$, $R_z(\gamma)$ respectively (Spong & Vidyasagar, 1989). Thus, the rotation matrix between the F_B and F_P frames is given as follows:

$$R_B^P = R_x(a)\, R_y(\beta)\, R_z(\gamma) \tag{16}$$

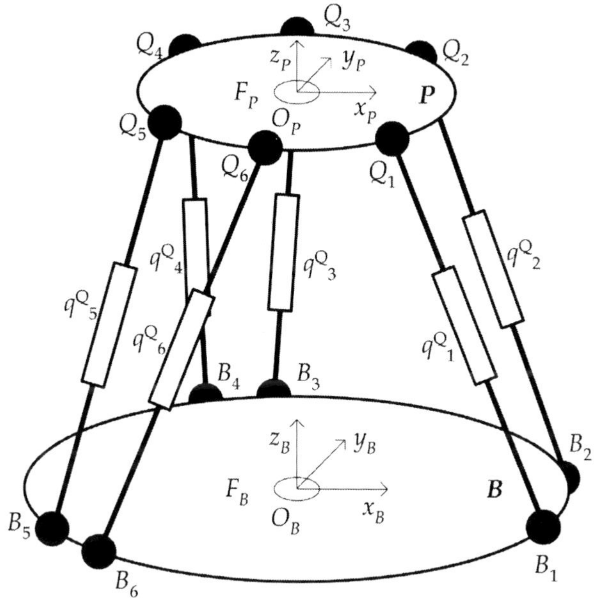

Fig. 10. Assignments for kinematic analysis of the Stewart manipulator

Then we can analyze the inverse kinematics of Stewart manipulator by the representation of any one of its legs. For a given pose of the center point of moving platform, O_P, the defining vectors are shown in Fig. 11.

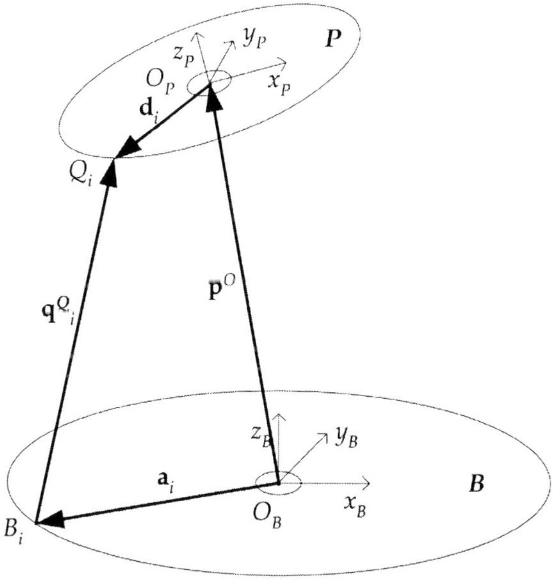

Fig. 11. Defining the vectors for a given pose of the moving platform

By using the rotation matrix given by equation (16), the position vector of the upper joint position, Q_i, connecting the moving platform to the leg i, $\mathbf{q}^Q{}_i$ can be transformed to the frame F_B as follows:

$$\mathbf{q}_i^Q = \mathbf{p}^O + R_B^P\,\mathbf{d}_i \qquad i = 1....6 \tag{17}$$

where $\mathbf{p}^O$ represents the position vector of the center point of moving platform, O_P, relative to the frame F_B, $\mathbf{d}_i$ is the position vector of the point Q_i, $i = 1....6$, relative to the frame F_P. Then the vector $\mathbf{q}^A{}_i$ representing the leg legths between the joint points B_i and Q_i can be transformed to the frame F_B as follows:

$$\overrightarrow{B_iQ_i} = \mathbf{q}_i^A = -\mathbf{a}_i + \mathbf{q}_i^Q \qquad i = 1....6 \tag{18}$$

where $\mathbf{a}_i$ represents the position vector of the point B_i, $i = 1....6$, relative to the frame F_B. The leg lengths $q^A{}_i$, $i = 1....6$, is then obtained by Euclidean norm of the leg vector given above. So, using equation (17) and (18) we can write (Zanganeh et al., 1997)

$$(q_i^A)^2 = (\mathbf{a}_i + \mathbf{p}^O + R_B^P\,\mathbf{d}_i)^T(\mathbf{a}_i + \mathbf{p}^O + R_B^P\,\mathbf{d}_i)\ ,\ i = 1....6 \tag{19}$$

The leg lengths related to a given pose of moving platform can be obtained for a trajectory defined by the pose vector, $\mathbf{x}$, given in equation (15). Considering a circular motion depicted as in Fig. 12, the trajectory of moving platform with zero rotation angles ($[\alpha\ \beta\ \gamma] = [0\ 0\ 0]$) is given as follows:

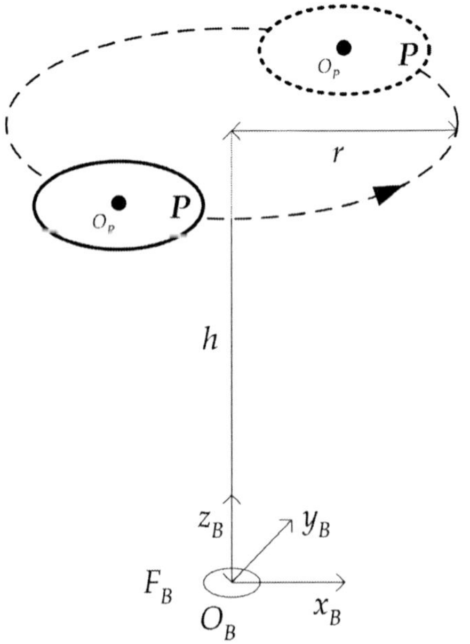

Fig. 12. A circular motion trajectory of the moving platform

$$\mathbf{x} = [(\mathbf{p}^O)^T\ 0\ 0\ 0]^T = \mathbf{A}(t)\ \mathbf{x}_0 \tag{20}$$

where $\mathbf{p}^O = [x_B\ \ y_B\ \ z_B\ ]^T$ denotes the 3x1 position vector of the center point of moving platform, $\mathbf{A}(t)$ is a 6x6 matrix and $\mathbf{x}_0$ is a 6x1 coeeficient vector given as below

$$A(t)=\begin{bmatrix} \cos[\theta(t)] & -\sin[\theta(t)] & 0 & \\ \sin[\theta(t)] & \cos[\theta(t)] & 0 & 0 \\ 0 & 0 & 1 & \\ & 0 & & 0 \end{bmatrix} \tag{21}$$

$$\mathbf{x}_0 = [0\ r\ h\ 0\ 0\ 0]^T \tag{22}$$

where **O** denotes the 3x3 zero matrix, h is the hight of the center point of moving platform with respect to base frame, and r is the radius of the circle.
The Jacobian matrix that gives the relation between the prismatic joint velocities and the velocity of the center point of moving platform, O_P, can be derived using the partial differentiation of the inverse geometric model of the manipulator given in equation (19).

3.2 Dynamics

As descripted in Fig. 9, Stewart manipulator has thirteen rigid bodies. The Newton-Euler equations of the manipulator can be derived in a more compact form as described below (Fang et al., 2000; Khan et al., 2005):
Let the 6x6 matrix $\mathbf{M}_i$, denoting the mass and moment of inertia properties of the rigid body i be

$$M_i = \begin{bmatrix} I_i & 0 \\ 0 & m_i \times 1 \end{bmatrix} \quad , \ i = 1....13 \tag{23}$$

where **O** and **1** denote the 3x3 zero and identity matrices; $\mathbf{I}_i$ is inertia matrix defined with respect to the mass center, C_i, of the body i; m_i is the mass of the body i. Let $\mathbf{c}_i$ and $\dot{\mathbf{c}}_i$ denote the position and velocity vectors of C_i, and $\boldsymbol{\omega}_i$ denote the angular velocity vector of C_i. Then the wrench vector $\mathbf{t}_i$ is defined in terms of the angular and linear velocities as follows:

$$t_i = \begin{bmatrix} \omega_i \\ \cdot \\ c_i \end{bmatrix} \quad , \ i = 1....13 \tag{24}$$

Let the 6x6 matrix $\boldsymbol{\Omega}_i$, denoting the angular velocity of the rigid body i be

$$\Omega_i = \begin{bmatrix} \omega_i & 0 \\ 0 & 0 \end{bmatrix} \quad , \ i = 1....13 \tag{25}$$

where, **O** denotes the 3x3 zero matrix. The generalized matrices given in equation (23) and (25) are block symmetrical, as follows:

$$M = diag(M_1, M_2, ..., M_{13}), \Omega = diag(\Omega_1, \Omega_2, ..., \Omega_{13}) \tag{26}$$

Then, the generalized wrench matrix **t** can be expressed as follows

$$t=[t_1^T t_2^T ...... t_{13}^T]^T \tag{27}$$

For the system having constraint on velocity, the constraint of velocity can be expressed by following equation:

$$Dt=0 \tag{28}$$

Let **T** be the natural orthogonal complement (NOC) of the coefficient matrix **D** related to the constraint equation (28) of velocity. Hence, employing the joint coordinates $\mathbf{q} \in R^6$ as generalized coordinate vector, we can get the dynamic model of system, which don't contain the constraint forces.

$$\mathbf{M}(\mathbf{q})\ddot{\mathbf{q}} + \mathbf{C}(\mathbf{q},\dot{\mathbf{q}})\dot{\mathbf{q}} + \mathbf{G}(\mathbf{q}) = \tau \quad (29)$$

where **M(q)** is a symmetrical and positive definition matrix as given below;

$$\mathbf{M}(\mathbf{q}) = \mathbf{T}^T\mathbf{M}\,\mathbf{T} \in R^{6x6} \quad (30)$$

C is the coefficient matrix of the vectors of Coriolis and centripetal force as given below;

$$\mathbf{C} = \mathbf{C}(\mathbf{q},\dot{\mathbf{q}}) = \mathbf{T}^T\mathbf{M}\,\dot{\mathbf{T}} + \mathbf{T}^T\Omega\mathbf{M}\mathbf{T} \quad (31)$$

q is the generalized coordinate vector, $\tau \in R^6$ is the generalized force (driving force) vector, respectively. **G(q)** is the gravity vector as given below;

$$\mathbf{G}(\mathbf{q}) = \tau^g(\mathbf{q}) = \mathbf{T}^T\mathbf{W}^g \quad (32)$$

where $\mathbf{W}^g$ are wrenches vector due to gravity:

$$\mathbf{W}^g = [\mathbf{W}_1{}^{g^T}, \mathbf{W}_2{}^{g^T}, \ldots \mathbf{W}_{13}{}^{g^T}]^T = [0, m_1\mathbf{g}^T, \ldots 0, m_{13}\mathbf{g}^T]^T \quad (33)$$

where **0** is 3x1 zero vector, **g** is the vector of acceleration of gravity.

4. Controller emulation by using Elman networks

In this stage, it is aimed to implement an application of ANN to emulate the operation of an existing PID controller in a Stewart manipulator control system. This system is given as a control system example for MATLAB applications (www.mathworks.com). The block diagram of the control system is given in Fig. 13.

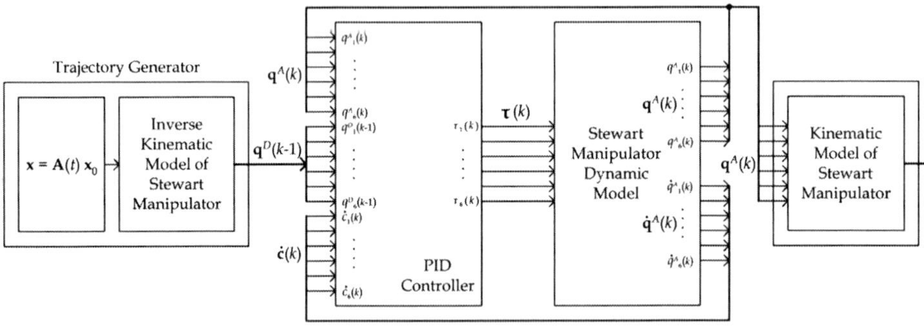

Fig. 13. Srewart manipulator control system using PID controller

As shown in this figure, trajectory generator calculates the leg lengths, which are desired leg lengths formed as a 6x1 $\mathbf{q}^D$ vector feeding the PID controller input, by using the inverse kinematic model of Stewart manipulator. PID controller produces a 6x1 control vector, τ, consisting of the leg forces applied to the prismatic joint actuators of the manipulator. In response, the dynamic model of the manipulator produces two 6x1 output vectors, $\mathbf{q}^A$ and $\dot{\mathbf{c}}=\cdot\mathbf{q}^A$, which include actual leg lengths and actual linear leg velocities, respectively. These are fed back to the controller. So, the controller has 18 inputs and 6 outputs totally. PID controller compares the actual and desired leg lengths to generate the error vector feeding its proportional and integral inputs. In the same time, the velocity feedback vector feeds the derivative input of the controller.

Designing an ANN emulation of controller generalized for the whole area of working space is more difficult task. It is also difficult to imagine a useful non-repetitive task that involves making random motions spanning the entire control space of the manipulator system. This results an intelligent manipulator concept, which is trained to carry out certain class of operations rather than all virtually possible applications (Akbas, 2005).

On the other hand, since the parallel manipulators have more complex dynamic structures, training process may be required much more data then other type plants. So, it can be taught to design more than one ANN controller trained by different input-output data sets, and use them in a parallelly formed controller structure instead of unique ANN controller structure. This can improve the reliability of the controller. Because of that, three ANN controllers are trained and they are used in parallel form in this case study.

4.1 Training

Due to its profound impact on the learning capability and networking performance, Elman network having recurrent structure is selected for training. Three of them, each have 18 inputs and 6 outputs, are trained by using PID controller input-output data. For this aim, input-output data are prepared during the implementation of the PID controller to the Stewart manipulator.

During the data log phase, manipulator is operated in a constrained area of its working space. For this aim, the manipulator is controlled by implementation of different trajectories selected uniformly in a planar sub-space, created as given example in equations (21) and (22) also as given in Fig. 12. Load variations are taken into consideration to generate the training data.

Three sets of input-output data each have 5000 vectors are generated by MATLAB simulations for each of Elman networks. MATLAB ANN toolbox is used for off-line training of Elman networks. Conventional backpropagation algorithm, which uses a threshold with a sigmoidal activation function and gradient descent error-learning, is used. Learning and momentum rates are selected optimally by MATLAB program. The numbers of neurons in the hidden layers are selected experimentally during the training. These are used as 40, 30 and 50, respectively for each network.

4.2 Implementation

After the off-line training, three of Elman networks are prepared as embedded Simulink blocks with obtained synaptic weights. To improve the reliability of the controller by error minimization approach, they are used in a parallel structure and embedded to the control system block diagram (Fig. 14). In this figure, parallely-implemented Elman ANN controller is represented in a block form. Its detailed representation is given in Fig. 15.

In this implementation, the force values generated by three Elman networks are applied to the inputs of the corresponding manipulator's dynamic model. Error vector is computed for each of the ANN by using the difference between the actual leg lengths generated by manipulator's dynamic model and the desired leg lengths. The results are evaluated to select the network generating the best result. Then it is assigned as the ANN controller for actual time step, and its output is assigned as the force output of the parallely-implemented Elman ANN controller output driving the manipulator's dynamic model (instead of a real manipulator, in this case).

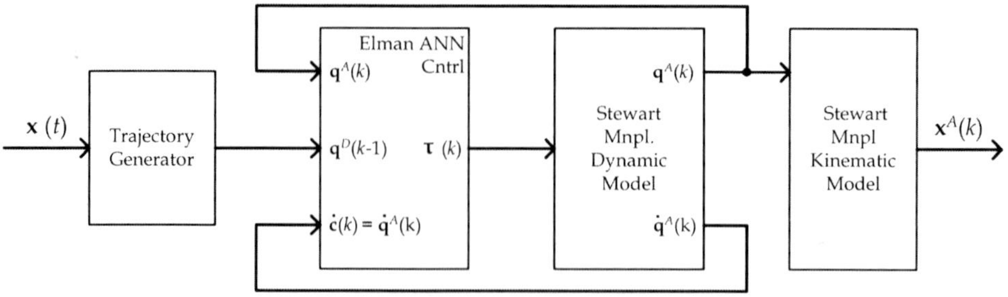

Fig. 14. ANN controller implementation to the manipulator control system

4.3 Simulation results

To compare the performance of the created ANN controller, the Srewart manipulator control system is operated both by the PID controller, and the parallelly-implemented Elman ANN controller for T=4 s. simulations. For these operations, a trajectory like as given with equations (21) and (22) is created with the parameter assignments: h = 2 m, r = 0.02 m. Also $\theta(t)$ parameter is used as follows:

$$\theta(t) = \frac{2\pi}{\mathrm{T}} t \quad , \quad 0 \le t \le \mathrm{T} \tag{34}$$

During the simulations, the sampling period is chosen, as 0.001 s. So, totally 4000 steps are included in each simulation.

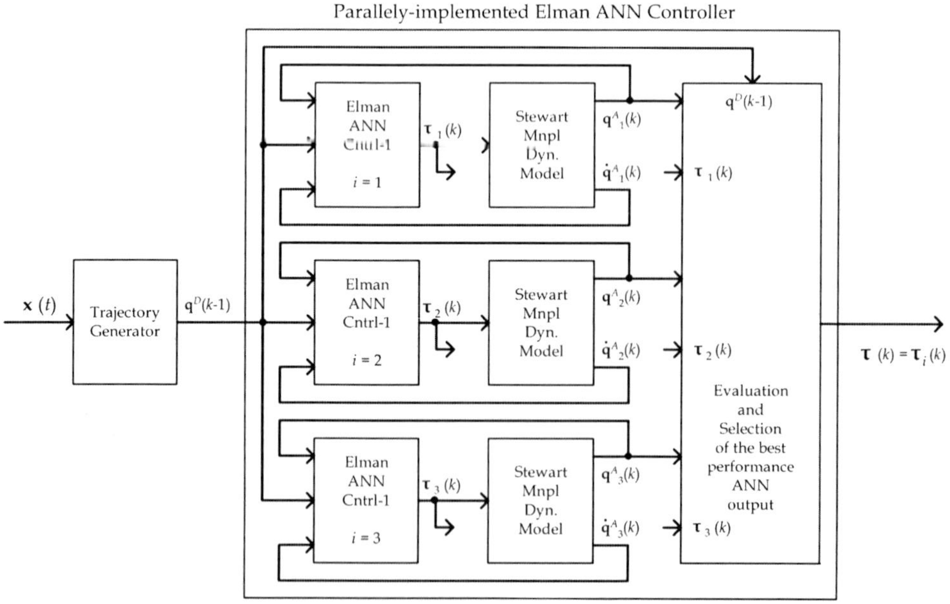

Fig. 15. The structure of parallelly-implemented Elman ANN controller

An example of the variations of the force outputs generated by both controllers is shown in Fig. 16, for the first leg of the manipulator. Fig. 16a and Fig. 16b show the force output of the PID controller and parallely-implemented Elman ANN controller, respectively. In these simulations, it has been observed that, the error between the two controller outputs is a little more at the starting phase of the simulations then the remaining times.
However, it can be said that, ANN controller emulates the PID controller successfully as a whole for the given trajectory.

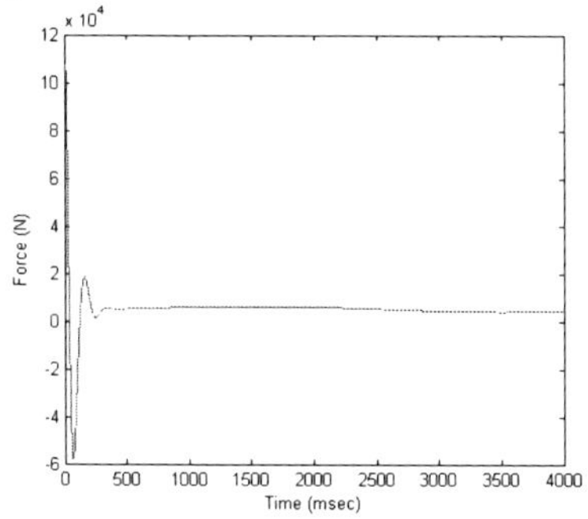

(a)

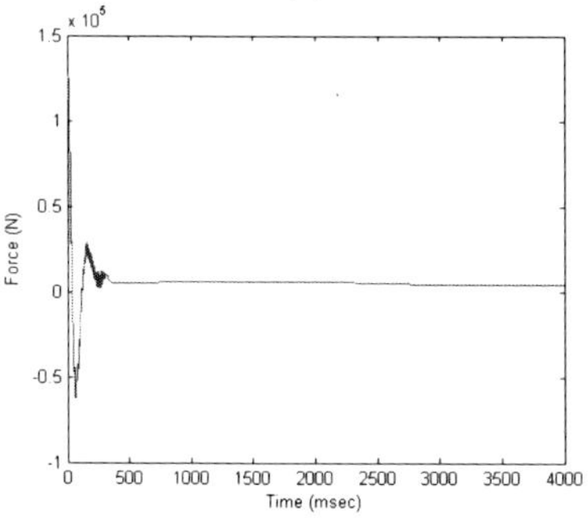

(b)

Fig.16. Force outputs of the controllers applied to the first leg of the Stewart manipulator (a)-PID controller output, (b)-ANN controller output

Similar adaptations are obtained for the control system output. For the given trajectory, position errors obtained by averaging the sum of the square errors relative to the desired position of the center point of moving platform both for the PID controller and ANN controller is given in Table 1. As seen in this table obtained position error values due to the x_B, y_B and z_B variations have too small changes.

Axis	Average Position Errors (m)	
	PID controller	ANN controller
x_B	0.0000799	0.0000807
y_B	0.0002594	0.0002618
z_B	0.0000087	0.0000086

Table 1. The sum of the squares of the position errors obtained by PID and ANN

During simulations, variations of the x_B, y_B and z_B positions of the center point of moving platform are given in Fig. 17, so that, Fig. 17a and Fig. 17b show the variation of actual x_B, y_B and z_B positions obtained simulations using PID controller and parallely-implemented Elman ANN controller, respectively. As seen, the tracing error between the two control modes is a little more at the starting phase only. This is due to instantaneous big difference between the desired y_B position and its starting value. However, tracing the desired positions by PID controller is well emulated by parallely-implemented Elman ANN controller, as a whole.

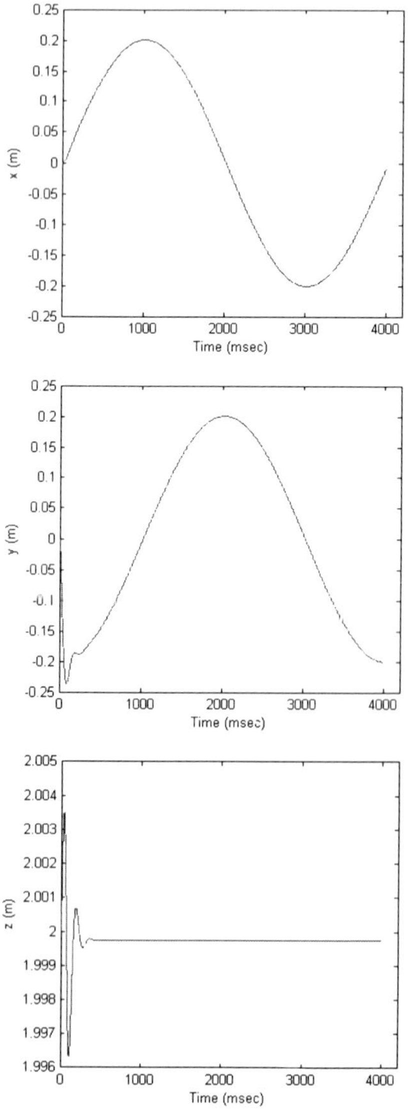

(a)

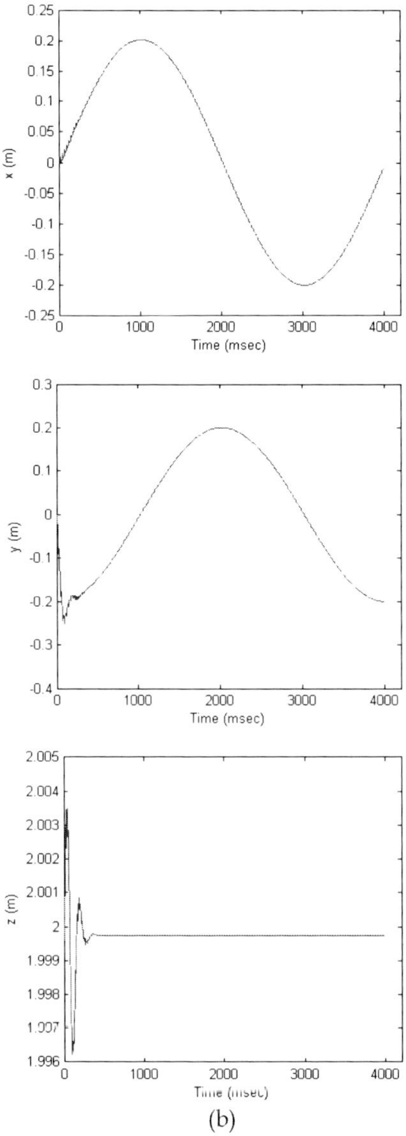

(b)

Fig.16. Variation of actual position of the center point of moving platform, in simulations (a)-Obtained by the PID controller, (b)- Obtained by the ANN controller

5. Conclusion

This chapter is mainly concerned with the application of ANNs to modeling and control of parallel manipulators. A practical implementation is completed to emulate the operation of

an existing PID controller in a Stewart manipulator control system. It can be said that, excepted results has been achieved for this case study.
Since the parallel manipulators have more complex dynamic structures, depending on the chosen type of applications training process it may be required much more data then in this case. So, designing an ANN for applications including the whole area of working space is more difficult task. It is also difficult to imagine a useful non-repetitive task that involves making random motions spanning the entire control space of the manipulator system.
However, for a succesfull study, it may have an important role selecting the type and structure of ANN by experience, depending on the requirements of the chosen application.

6. References

Akbas, A. (2005). Intelligent predictive control of a 6-dof robotic manipulator with reliability based performance improvement, *Proceedings of the 6th International Conference on Intelligent Data Engineering and Automated Learning*, LNCS Vol. 3578, pp. 272-279, ISBN: 3-540-26972-X, Brisbane, Australia, July 2005, Springer

Angeles, J. (2007). *Fundamentals of Robotic Mechanical Systems, Theory, Methods, and Algorithms*, Springer Science+Business Media, LLC, ISBN: 0 387 29412 0, NY, USA

Burns, R.S. (2001). *Advanced Control Engineering*, Butterworth-Heinemann, ISBN: 0 7506 5100 8, Oxford

Fang, H.; Zhou, B.; Xu, H. & Feng, Z. (2000). Stability analysis of trajectory tracing control of 6-dof parallel manipulator, *Proceedings of the 3d World Congress on Intelligent Control and Automation, IEEE*, Vol. 2, pp. 1235-1239, ISBN: 0-7803-5995-X, Hefei, China, June 28-July 2, 2000

Hagan, M.T.; Demuth, H.B. & Beale, M. (1996). *Neural Network Design*, PWS Publishing Company- Thompson Learning, ISBN: 7 111 10841 8, USA

Honegger, M.; Brega, R. & Schweitzer, G. (2000). Application of a nonlinear adaptive controller to a 6 dof parallel manipulator, *Proceedings of the 2000 IEEE International Conference on Robotics and Automation, ICRA 2000*, pp. 1930-1935, ISBN: 0-7803-5889-9, San Francisco, CA, USA, April 24-28, 2000

Khan, W.A.; Krovi, V.N.; Saha, S.K. & Angeles, J. (2005). Modular and recursive kinematics and dynamics for parallel manipulators. *Multibody System Dynamics*, Vol. 14, No.3-4, Nov. 2005, pp. 419-455, ISSN: 1384-5640, Springer

Spong, M.W. & Vidyasagar, M. (1989). Robot Dynamics and Control. pp. 39-50, John Wiley & Sons

Zanganeh, K.E.; Sinatra, R. & Angeles, J. (1997). Kinematics and dynamics of a six-degree-of-freedom parallel manipulator with revolute legs. *Robotica*, Vol. 15, No.04, Jule 1997, pp. 385-394, ISSN: 0263-5747, Cambridge University Press

3

Asymptotic Motions of Three-Parametric Robot Manipulators with Parallel Rotational Axes

Ján Bakša
Technical University in Zvolen
Slovak Republic

1. Introduction

In this paper we deal with the properties of 3-parametric robot manipulators (in short robots) with parallel rotational axes. We describe motions of the robot effector by using the theory of Lie groups and Lie algebras which is applied to the Lie group $E(3)$ of all orientation preserving congruences of the Euclidean space E_3. By the concept of an n-parametric robot we will understand the map $\Upsilon_{A_n}: R^n \rightarrow E(3)$, see (Karger, 1988), where the robot Υ_{A_n} is viewed as an immersed submanifold Υ_{A_n} of the Lie group $E(3)$. We classify 3-parametric robots into four classes. The classification criterion is the *spherical rank* of the robot, which is the number of independent directions of revolute joints axes. Robots of the spherical rank 1 are robots whose axes of revolute joints are mutually parallel and different. The main aim of the paper is to introduce *asymptotic robot motions*. The notion of asymptotic motions is connected with the theory of connections. On a pseudo-Riemannian manifold ($E(3)$ has pseudo-Riemannian structure), there is a canonical connection called the Levi-Civita connection. As a connection on the tangent bundle, it provides a well defined method for differentiating all kinds of tensors. The Levi-Civita connection is a torsion-free connection on the tangent bundle and it can be used to describe many intrinsic geometric objects. For instance, a geodesic path, a parallel transport for vector fields, a curvature and so on.

On the Lie group $E(3)$ there is the Levi-Civita connection ∇ induced by the Klein form KL. If the restriction $KL|_{\Upsilon_{A_n}}$ is regular then there is the Levi-Civita connection $\widetilde{\nabla}$ on Υ_{A_n} such that $\nabla_{\dot\gamma}\dot\gamma = \widetilde{\nabla}_{\dot\gamma}\dot\gamma + V$, where V lies in KL-orthogonal complement to the tangent bundle $T\Upsilon_{A_n}$. If $V = \overline{0}$ then motions on Υ_{A_n} is *asymptotic*, see (Karger 1993). We will introduce asymptotic robot motions without explicit use of the Levi-Civita connection. A robot motion is asymptotic, if the Coriolis acceleration is tangential to $T_e\Upsilon_{A_n}$. Obviously, robot motions with zero Coriolis accelerations are asymptotic. The simple examples of the asymptotic motions are motions when only one joint work. The properties of the acceleration operator are important for the dynamic of the robot especially in singular positions where they can affect the behaviour of the robot expressively. We will introduce the notion of the Coriolis and Klein subspaces and show that they are closely associated with asymptotic motions. In this paper we describe all asymptotic motions by systems of differential equations for all 3-parametric robot manipulators with parallel rotational axes. Future research: to describe all asymptotic motions for all 3- ,4- ,5-parametric robot manipulators with revolute and prismatic joints only, practical purposes of the asymptotic motions.

2. Basic notions of robot manipulators

Common commercial industrial robots are serial robot-manipulators consisting of a sequence of links connected by joints, see Fig. 1. Each joint has one degree of freedom, either prismatic or revolute. For a robot with n joints, numbered from 1 to n, there are $n+1$ links, numbered from 0 to n. The link 0 or n will be called respectively the base or the effector of the robot. The base will be fixed (non movable). Joint i connects links i and $i+1$. We view a link as a rigid body defining the relationship between two neighbouring joints. In the concept of the Denavit-Hartenberg conventions (Denavit & Hartenberg, 1955) the base coordinate system S_0 is firmly connected with the base. The base axis z_0 is the axis o_1 of 1st joint. The effector begins in n th joint and is firmly connected to the coordinate system S_n .

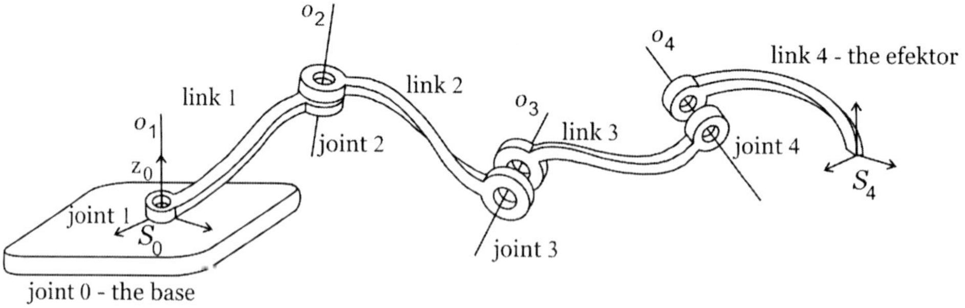

Figure 1. n -parametric robot, $n=4$

A congruence in the Euclidean space E_3 is determined by the base coordinate system S_0 and by the effector coordinate system S_n in each position of the robot (i.e., at time t). Therefore a motion of the effector determines a curve on the Lie group $E(3)$. We assume a fixed choice of the base orthonormal coordinate system $S_0=\{O;\bar{i}_0,\bar{j}_0,\bar{k}_0\}$ with respect to which we will relate all elements.

Let us recall basic facts about the Lie group $E(3)$ and its Lie algebra $e(3)$. Elements of the Lie group $E(3)$ will be considered in the matrix form 4×4, which will be written in the form $\begin{pmatrix} A & P \\ 0 & 1 \end{pmatrix}$, where A is an orthogonal matrix of the form 3×3, $\det A = 1$ and P is a column matrix of the form 3×1 (a translation vector).

Let V_3 be the vector space associated with the Euclidean space E_3 and let $\gamma(t)=H(t)$ be a curve on $E(3)$ which is going through the unit element I of the group $E(3)$; i.e., $H(t_0)=I$, where I is the unit matrix. Then the motion of the effector point L determined by the curve $\gamma(t)$ can be expressed by

$$\begin{pmatrix} x(t) \\ y(t) \\ z(t) \\ 1 \end{pmatrix} = \begin{pmatrix} A(t) & P(t) \\ 0 & 1 \end{pmatrix} \begin{pmatrix} x \\ y \\ z \\ 1 \end{pmatrix},$$

where $(x,y,z,1)^T$ are the homogeneous coordinates of the point L at t_0 and $(x(t),y(t),z(t),1)^T$ are the homogeneous coordinates of the point L at any t. The coordinates of the point L are related to the base coordinate system S_0. $A(t)$ is an orthogonal matrix; i.e., $A(t)A^T(t) = I$, where $A^T(t)$ is the transposed matrix to the matrix $A(t)$. The inverse matrix to the matrix $H(t)$ is $H^{-1}(t) = \begin{pmatrix} A^T(t) & -A^T(t)P(t) \\ 0 & 1 \end{pmatrix}$. We suppose $A(t_0) = I$. The derivative of the equation $A(t)A^T(t) = I$ at $t = t_0$ is $\dot{A}(t_0) = -\dot{A}^T(t_0)$; i.e., $\dot{A}(t_0)$ is a skew-symmetric matrix. All skew-symmetric matrices have the form $\dot{A}(t_0) = \begin{pmatrix} 0 & -\omega_3 & \omega_2 \\ \omega_3 & 0 & -\omega_1 \\ -\omega_2 & \omega_1 & 0 \end{pmatrix}$ and we can associate them with vectors $\overline{\omega} := (\omega_1, \omega_2, \omega_3) \in V_3$. If we denote $\dot{P}^T(t_0) := \overline{b} = (\beta_1, \beta_2, \beta_3)$, then the tangent vector

$$\dot{\gamma}(t_0) = \dot{H}(t_0) = \begin{pmatrix} \dot{A}(t_0) & \dot{P}(t_0) \\ 0 & 0 \end{pmatrix} = \begin{pmatrix} 0 & -\omega_3 & \omega_2 & \beta_1 \\ \omega_3 & 0 & -\omega_1 & \beta_2 \\ -\omega_2 & \omega_1 & 0 & \beta_3 \\ 0 & 0 & 0 & 0 \end{pmatrix} \tag{1}$$

of the curve $\gamma(t)$ at $t = t_0$ can be associated with the element $(\overline{\omega}, \overline{b}) \equiv X \in V_3 \times V_3$ and we call it the twist. Hence the Lie algebra $e(3)$ can be represented in the matrix form (1) or by twists in $V_3 \times V_3$, where addition and the Lie bracket are defined as follows:

$$k_1(\overline{\omega}_1, \overline{b}_1) + k_2(\overline{\omega}_2, \overline{b}_2) = (k_1\overline{\omega}_1 + k_2\overline{\omega}_2, k_1\overline{b}_1 + k_2\overline{b}_2),$$

$$[(\overline{\omega}_1, \overline{b}_1), (\overline{\omega}_2, \overline{b}_2)] = (\overline{\omega}_1 \times \omega_2, \overline{\omega}_1 \times \overline{b}_2 - \overline{\omega}_2 \times \overline{b}_1),$$

where $(\overline{\omega}_i, \overline{b}_i) \in V_3 \times V_3$, $k_i \in R$, $i = 1,2$ and $\times$ denotes the vector product in V_3. The line p determined by the point C, $\overline{OC} = (1/\overline{\omega}^2)\overline{\omega} \times \overline{b}$ and by the direction $\overline{\omega}$ will be called the axis of the twist $X = (\overline{\omega}, \overline{b})$, $\overline{\omega} \neq \overline{0}$. If $\overline{\omega} = \overline{0}$, then the axis of the element $X = (\overline{0}, \overline{b})$ is the line at infinity of the plane in the projective space P_3 (P_3 is E_3 together with the points at infinity) which is perpendicular to the vector $\overline{b}$.

In the algebra $V_3 \times V_3$ we have the *Klein form* given by

$$KL(X_1, X_2) \overset{\text{def}}{:=} \overline{\omega}_1 \cdot \overline{b}_2 + \overline{\omega}_2 \cdot \overline{b}_1$$

where $X_1 = (\overline{\omega}_1, \overline{b}_1), X_2 = (\overline{\omega}_2, \overline{b}_2)$ are twists from $V_3 \times V_3$ and the dot $\cdot$ denotes the scalar product in V_3. If $KL(X_1, X_2) = 0$, then the twists X_1, X_2 will be called KL *-orthogonal*. The Klein form is a symmetric regular bilinear form.

A subspace $A \subset V_3 \times V_3$ is called KL -orthogonal to a subspace $B \subset V_3 \times V_3$, if $KL(X,Y) = 0$ for every $X \in A$ and every $Y \in B$. There is a unique subspace $A^K \subset V_3 \times V_3$ which is KL -orthogonal to the subspace $A \subset V_3 \times V_3$; i.e., if any arbitrary vector subspace B is KL -orthogonal to A, then $B \subset A^K$.

Definition 1. Let $A \subset V_3 \times V_3$. The subspace $K \overset{\text{def}}{:=} A \cap A^K$ will be called the *Klein subspace* of the space A. If $K = A$, then A is isotropic.

Let us recall that the matrix form of the exponential map from the Lie algebra $e(3)$ to the Lie group $E(3)$, $\exp : e(3) \to E(3)$, is given by $\exp(S) = \sum_{n=0}^{\infty} \frac{1}{n!} S^n$, where $S \in e(3)$ is the matrix of the form (1) and S^n is n th power of the matrix S. The matrix $\exp(S)$ is a regular matrix, $(\exp(S))^{-1} = \exp(-S)$, for further properties see (Helgason, 1962). For the motion determined by the curve $\gamma(t) = \exp(t(\overline{\omega}, \overline{b}))$, where $(\overline{\omega}, \overline{b}) \in e(3)$ and exp is exponential map, we have:

(1) If $\overline{\omega} = \overline{0}$ then the curve $\gamma(t) = \exp(t(\overline{0}, \overline{b}))$ determines a translation with velocity $\overline{b}$.

(2) If $\overline{\omega} \neq \overline{0}$ then the curve $\gamma(t) = \exp(t(\overline{\omega}, \overline{b}))$ determines a uniform screw motion in E_3 with the axis p of the twist $(\overline{\omega}, \overline{b})$, the angular velocity $\overline{\omega}$ and with the translation $h\overline{\omega}$, where $h = (\overline{\omega} \cdot \overline{b}) / \omega^2$, see Fig. 2.

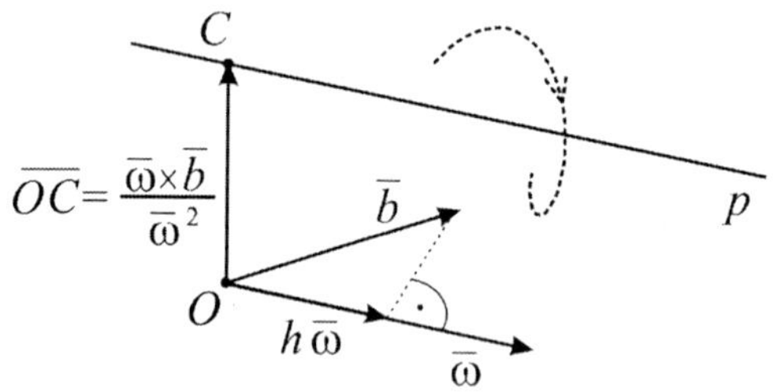

Figure 2. Screw motion determined by $\exp(t(\overline{\omega}, \overline{b}))$

If $h = 0$ (i.e., $\overline{\omega} \cdot \overline{b} = 0$) then it is a rotational motion.

From the mathematical point of view, we can define a robot by the exponential map which is applied to the elements of the Lie algebra $e(3)$, see (Karger, 1988), as follows:

Definition 2. Let $X_i \in e(3), i = 1,2,\ldots,n$. Then a robot with n degrees of freedom is a map $\Upsilon_{X_1,\ldots,X_n} : R^n \to E(3)$ given by

$$\Upsilon_{X_1,\dots,X_n}(u_1,u_2,\dots,u_n) = \exp u_1 X_1 \exp u_2 X_2 \dots \exp u_n X_n.$$

Let us deal with the velocity and the acceleration of an effector point L. Let $\Upsilon_{X_1,\dots,X_n}$ be any n-parametric robot given by twists $X_1, X_2, \dots, X_n$, respectively. Let the motion of the effector be given by a curve $\gamma(t) = \exp u_1(t)X_1 \exp u_2(t)X_2 \dots \exp u_n(t)X_n \equiv H(t)$ and let $L(t_0)$ be the homogeneous coordinates of the effector point L at t_0. Then the homogeneous coordinates $L(t)$ of the point L at any t are given by $L(t) = H(t)L(t_0)$. So its velocity is given by $\dot{L}(t) = \dot{H}(t)H^{-1}(t)L(t)$. The element $\dot{H}(t)$ determines the tangent vector at $H(t)$ and $\dot{H}(t)H^{-1}(t)$ is a right translation by $H^{-1}(t)$. Then $Y(t) \overset{\text{def}}{:=} \dot{H}(t)H^{-1}(t)$ belongs to the Lie algebra $e(3)$. The velocity of the motion $L(t) = H(t)L(t_0)$ determined by $H(t)$ at t_0 and the velocity of the motion $L(s) = \exp(sY(t_0))L(t_0)$ determined by $\exp(sY(t_0))$ at $s = 0$ are the same. The twist $Y(t)$ is called the *velocity operator* or shortly the *velocity twist*.

Remark 1. For simplicity we will use u instead $u(t)$.

As $H = \exp u_1 X_1 \dots \exp u_n X_n$ we get

$$Y = \dot{H}H^{-1} = \dot{u}_1 Y_1 + \dot{u}_2 Y_2 + \dots + \dot{u}_n Y_n \tag{2}$$

see (Karger, 1989), where $Y_1 = X_1$, $Y_i = g_{i-1} X_i g_{i-1}^{-1}$, $g_{i-1} = \exp u_1 X_1 \dots \exp u_{i-1} X_{i-1}$ and g_{i-1}^{-1} is the inverse element of g_{i-1}, $i = 2,\dots,n$. Elements Y_i belong to the Lie algebra $e(3)$. The space $A_n(u) \overset{\text{def}}{:=} span(Y_1, Y_2, \dots, Y_n)$ will be called the *space of velocity twists*, where $(u) = (u_1, u_2, \dots, u_n)$. If $\dim A_n(u) = n$ then we call the point (u) *regular* or we say that the robot is in a *regular* position. If $\dim A_n(u) < n$ then we call the point (u) *singular* or we say that the robot is in a *singular* position. If every point of the curve $\gamma(t) = H(t) \subset \Upsilon_{X_1,\dots,X_n}$ is singular then this motion of the robot determined by the curve $\gamma(t) = H(t)$ will by called *singular*. A robot $\Upsilon_{X_1,\dots,X_n}$ is of *rank* m if m is the maximal dimension of the velocity twists spaces; i.e., $m = \max_{(u)}\{\dim A_n(u)\}$.

Remark 2. In what follows we confine ourselves to $n = m$. Without loss of generality we will assume that $A_n := span(X_1, X_2, \dots, X_n)$ and $\dim A_n = n = m$ is the rank of the robot. Then there is a neighborhood $\Omega_0 \subset R^n$ of the point $O = (0,0,\dots,0) \in R^n$ that $\Upsilon_{X_1,\dots,X_n}\big|_{\Omega_0} =: \Upsilon_{A_n}$ is an immersed submanifold of the Lie group $E(3)$ and $(u_1, u_2, \dots, u_n)$ is a local coordinate system of Υ_{A_n}.

Let us consider the acceleration of any effector point L. The velocity of the point L at a time t is determined by $\dot{L}(t) = \dot{H}(t)H^{-1}(t)L(t) = Y(t)L(t)$. Let us differentiate the last equation. We get the relation for the acceleration of the effector point L at the time t: $\ddot{L}(t) = \dot{Y}(t)L(t) + Y(t)\dot{L}(t) = (\dot{Y}(t) + Y(t)Y(t))L(t)$. The derivative of the equation (2) is $\dot{Y}(t) = \sum_{i=1}^{n} \dot{u}_i \dot{Y}_i + \sum_{i=1}^{n} \ddot{u}_i Y_i$, where $\dot{Y}_i = \sum_{k=1}^{i-1} \frac{\partial Y_i}{\partial u_k} \dot{u}_k$. All elements Y_i are in the matrix form, therefore we can use the Lie bracket in the matrix form: $[A,B]=AB-BA$. Then we get $\dot{Y}_1=\bar{0}$, $\dot{Y}_i=[Y_1,Y_i]\dot{u}_1+[Y_2,Y_i]\dot{u}_2+\cdots+[Y_{i-1},Y_i]\dot{u}_{i-1}$, $i=2,\dots n$. The acceleration of the point L at any t is of the form

$$\ddot{L}(t)=\left(\sum_{k<i}\dot{u}_k\dot{u}_i[Y_k,Y_i]+\sum_{i=1}^{n}\ddot{u}_iY_i+YY\right)L(t),\quad k,i=1,\dots n.$$

The expression $\sum_{i=1}^{n}\ddot{u}_iY_i$ represents the acceleration caused by joint accelerations $\ddot{u}_i$, the expression YY represents centrifugal or centripetal components of acceleration and the expression $\dot{Y}_c\overset{\text{def}}{:=}\sum_{k<i}\dot{u}_k\dot{u}_i[Y_k,Y_i]$ stands for the so-called Coriolis acceleration.

Definition 3. The subspace $CA\overset{\text{def}}{:=}span([Y_1,Y_2],[Y_1,Y_3],\dots,[Y_{n-1},Y_n])$ of the space $e(3)$ will be called the *Coriolis subspace*.

Definition 4.

(1) If $CA\subset A_n(u)$ then the point (u) of the robot $\Upsilon_{X_1,\dots X_n}$ will be called *flat.*

(2) If every point of the robot is flat then the robot will be called *flat*.

(3) If $\dot{Y}_c(u(t_0))\in A_n(u(t_0))$ for a point $u(t_0)$ then the motion of the robot determined by the curve $\gamma(t)=H(t)$ will by called *asymptotic at the point* $u(t_0)$.

(4) If $\dot{Y}_c(t)\in A_n(u)$ for every t then this motion of the robot determined by the curve $\gamma(t)=H(t)$ will by called *asymptotic*.

Examples.

I. Robot motions with the zero Coriolis acceleration are asymptotic. Then:

a) If only prismatic joints work then the robot motion is asymptotic.

b) If only one joint works then the robot motion is asymptotic.

II. If $A_n(u)$ is a subalgebra; i.e., iff $CA\subset A_n(u)$ then (u) is flat and thus every motion is asymptotic at (u).

If $A_n(u)$ is a subalgebra or only translational joints work or just one joint works then these motions will be called *trivial* asymptotic motions. In the next part we will deal only with nontrivial asymptotic motions.

The base coordinate system S_0 is connected with the 1 st joint. Its axis z_0 is the axis o_1 of the first joint of the robot. The axis of the twist $Y_i=(\bar{\omega}_i,\bar{b}_i)$ is the axis o_i of the i th joint. If the i th joint is revolute or helical then its axis is the axis of Y_i. If it is prismatic then the direction of its axis is $\bar{b}_i$, $Y_i=(\bar{0},\bar{b}_i)$.

Remark 3. If $X=(\bar{0},\bar{b})$ is translational then $\bar{b}$ is orthogonal to the plane (O,o) where O is the origin of the coordinate system S_0 and o is the axis of X.

We will deal with robots which have no helical joints. The capital letter R will indicate a revolute joint, T a translational (prismatic) joint. Then, for example, RRT denotes a 3-parametric robot the first and second joints of which are revolute and the third is prismatic.

Definition 5. A robot Υ_{A_n} is the robot of *spherical rank r* if r is the maximal number of linearly independent directions of revolute joint axes.

It is obvious that 3-parametric robots have spherical rank 0, 1, 2 or 3. In the next part we will deal with 3-parametric robots of spherical rank 1.

Remark 4. It is interesting deal with the problem whether 3-parametric robots Υ_{A_3} of spherical rank 1 lie within any subgroups of the Lie group $E(3)$ the dimension of which is less than 6. Common knowledge is that there is only one connected 4-dimensional Lie subgroup H_4 of $E(3)$ (up to conjugacy), see (Karger 1990), (Selig, 1996). The subgroup H_4 is the group generated by a one parametric rotation around straight line ($SO(2)$) and all translations (R^3), (i.e., $H_4 = SO(2)\times R^3$). Hence 3-parametric robot manipulators of spherical rank 1 belong to 4-dimensional Lie subgroups. This fact does not affect our own work except that we know that all Coriolis space elements of the 3-parametric robots of spherical rank 1 belong to the Lie algebra h_4 of the Lie group H_4.

3. Three parametric robots of spherical rank 1

Now $\dim A_3 = 3$ and the revolute joint axes are parallel at any position u. Therefore Y_i is of the form $Y_i = (k_i\overline{\omega},\overline{m}_i)$ where at least one of the k_i is not zero. We assume $\|\overline{\omega}\| = 1$ and $k_i \in \{0,1\}$. We can always choose twists $B_1 = (\overline{\omega},\overline{b}_1)$, $B_2 = (\overline{0},\overline{b}_2)$, $B_3 = (\overline{0},\overline{b}_3)$, $\overline{\omega}\cdot\overline{b}_1 = 0$ in the space $A_3(u) = span(Y_1,Y_2,Y_3)$ such that $A_3(u) = span(B_1,B_2,B_3)$.

Remark 5. The robot Υ_{A_3} is in a singular position iff $\dim A_3(u) < 3$ and this occurs $\overline{b}_2\times\overline{b}_3 = \overline{0}$.

Let us determine conditions for an arbitrary twist $B = t_1B_1 + t_2B_2 + t_3B_3$ of the space $A_3(u)$, $t_1,t_2,t_3 \in R$ to be a rotational or translational twist. A twist B is rotational or translational iff $KL(B,B) = 0$ and this is equivalent to $t_1(t_2(\overline{\omega}\cdot\overline{b}_2) + t_3(\overline{\omega}\cdot\overline{b}_3)) = 0$.

A twist B is *translational* iff $t_1 = 0$; i.e., iff $B \in span(B_2,B_3) \overset{\text{def}}{=:} \tau$. Therefore $\dim\tau \le 2$. In singular positions, $\dim\tau = 1$.

A twist B is *rotational* if and only if $t_2(\overline{\omega}\cdot\overline{b}_2) + t_3(\overline{\omega}\cdot\overline{b}_3) = 0,\quad t_1 \ne 0$. If $\overline{\omega}\cdot\overline{b}_2 = 0$, $\overline{\omega}\cdot\overline{b}_3 = 0$ then there are no screw elements in $A_3(u)$; i.e., $A_3(u) - \tau$ is the space of all rotational twists. If at least one number of $\overline{\omega}\cdot\overline{b}_2$, $\overline{\omega}\cdot\overline{b}_3$ is not equal to 0 then there is a two-dimensional space of rotational twists. For example, if $\overline{\omega}\cdot\overline{b}_2 \ne 0$ then $B = (t_1\overline{\omega}, t_1\overline{b}_1 + (\overline{b}_3 - (\overline{\omega}\cdot\overline{b}_3/\overline{\omega}\cdot\overline{b}_2)\overline{b}_2)t_3)$. The axes of rotational twists generate a bundle of parallel lines with the direction $\overline{\omega}$.

The matrix of the Klein form has the form

$$KL|_{A_3(u)} = \begin{pmatrix} 0 & \overline{\omega}\cdot\overline{b}_2 & \overline{\omega}\cdot\overline{b}_3 \\ \overline{\omega}\cdot\overline{b}_2 & 0 & 0 \\ \overline{\omega}\cdot\overline{b}_3 & 0 & 0 \end{pmatrix}$$

in the basis B_1,B_2,B_3. The rank of the Klein form is 0 or 2; i.e., $KL|_{A_3(u)}$ is singular.

The rank is 0 if and only if $\overline{\omega}\cdot\overline{b}_2=0$, $\overline{\omega}\cdot\overline{b}_3=0$; i.e., if the vector $\overline{\omega}$ is perpendicular to $\tau_2 \overset{\text{def}}{:=} span(\overline{b}_2,\overline{b}_3)$; i.e., if there are no screw elements in $A_3(u)$. In this case the Klein space is $K = A_3(u)$, i.e., $A_3(u)$ is isotropic and the robot is planar.
The rank is 2 iff the direction of the revolute joints axes is not perpendicular to τ_2; i.e., at least one number $\overline{\omega}\cdot\overline{b}_2$, $\overline{\omega}\cdot\overline{b}_3$ is not equal to 0. Let us determine the Klein subspace K. The twist $B=t_1B_1+t_2B_2+t_3B_3, t_1,t_2,t_3\in R$ is KL-orthogonal to $A_3(u)$ if and only if $KL(B,A_3(u))=0$ and this is equivalent to $t_1=0$, $t_2=k(\overline{\omega}\cdot\overline{b}_3)$, $t_3=-k(\overline{\omega}\cdot\overline{b}_2)$, $k\in R$. Therefore the Klein subspace is determined by the element $\hat{Y}=(\overline{0},(\overline{\omega}\cdot\overline{b}_3)\overline{b}_2-(\overline{\omega}\cdot\overline{b}_2)\overline{b}_3)$ (i.e., $K=span(\hat{Y})\subset\tau$), its direction is perpendicular to $\overline{\omega}$ and it belongs to τ_2.
Let us summarize previous considerations.
Proposition 1. Let Υ_{A_3} be a robot of spherical rank 1. Then
a) if the direction $\overline{\omega}$ of revolute joints is perpendicular to the space τ_2 of translational elements directions then the rank of $KL|_{A_3(u)}$ is 0, the Klein space K is $K=A_3(u)$; i.e., $A_3(u)$ is isotropic and the robot is planar;
b) if $\overline{\omega}$ is not perpendicular to the space τ_2 then the rank of $KL|_{A_3(u)}$ is 2 and the Klein space K is $K=span\left((\overline{0},(\overline{\omega}\cdot\overline{b}_3)\overline{b}_2-(\overline{\omega}\cdot\overline{b}_2)\overline{b}_3)\right)$, $K\subset\tau$ and its direction is perpendicular to $\overline{\omega}$.
The Coriolis subspace is $CA=span([B_1,B_2],[B_1,B_3],[B_2,B_3])$, where $[B_1,B_2]=(\overline{0},\overline{\omega}\times\overline{b}_2)$, $[B_1,B_3]=(\overline{0},\overline{\omega}\times\overline{b}_3)$, $[B_2,B_3]=(\overline{0},\overline{0})$. It means that elements of CA are translational and their directions are perpendicular to $\overline{\omega}$ and $\dim CA\le 2$. The following cases are possible:
(1) $\dim CA\le 1$ if and only if $(\overline{\omega}\times\overline{b}_2)\times(\overline{\omega}\times\overline{b}_3)=\left((\overline{\omega}\times\overline{b}_2)\cdot\overline{b}_3\right)\overline{\omega}=\overline{0}$ and this is equivalent to $\omega\cdot(\overline{b}_2\times\overline{b}_3)=0$. We have the following cases:

a) $\overline{b}_2\times\overline{b}_3=\overline{0}$; i.e., the robot is in a singular position, $\dim\tau=1$. Let $\overline{b}_3=k\overline{b}_2$, $k\in R$. Then $A_3(u)\cap CA\ne\overline{0}$ if and only if $\overline{\omega}\times\overline{b}_2=c\overline{b}_2$ and $\overline{\omega}\times\overline{b}_2\ne\overline{0}$, $c\in R$. It is impossible. Therefore $A_3(u)\cap CA=\overline{0}$.

b) $\overline{b}_2\times\overline{b}_3\ne\overline{0}$ and $\overline{\omega}\in span(\overline{b}_2,\overline{b}_3)$; i.e., the robot is in a regular position and the vectors $\overline{\omega},\overline{b}_2,\overline{b}_3$ are linearly dependent. We can write $\overline{\omega}=c_2\overline{b}_2+c_3\overline{b}_3$, $c_2,c_3\in R$. Then $CA=span((\overline{0},\overline{b}_2\times\overline{b}_3))$. The vector $\overline{b}_2\times\overline{b}_3$ does not belong to τ_2, therefore $CA\cap A_3(u)=\overline{0}$ and $\dim CA=1$.

(2) $\dim CA=2$ if and only if the position of the robot is regular (i.e., $\overline{b}_2\times\overline{b}_3\ne\overline{0}$) and the direction $\overline{\omega}$ of the revolute joints axes is not complanar with the space τ_2. It means that the vectors $\overline{\omega}\times\overline{b}_2$, $\overline{\omega}\times\overline{b}_3$ are linearly independent. The twists $[B_1,B_2]$, $[B_1,B_3]$ determine the basis of the Coriolis space CA. This basis will be called the *canonical basis* of the space CA. In this case the space CA is the space of all translational elements, whose directions are perpendicular to the direction of the revolute joints $\overline{\omega}$. We have following cases:

a) $CA = \tau$. In this case, the vector $\overline{\omega}$ is perpendicular to the space τ_2; i.e., $\overline{\omega} \cdot \overline{b}_2 = 0$, $\overline{\omega} \cdot \overline{b}_3 = 0$ and the rank of the Klein form is 0. If $CA = \tau$ then $CA \subset A_3(u)$; i.e., $A_3(u)$ is a Lie subalgebra. A reverse assertion is also valid. If $A_3(u)$ is a subalgebra then $CA \subset A_3(u)$, $\dim CA = 2$ and the elements of CA are translational, therefore $CA = \tau$.

b) $CA \neq \tau$. In this case at least one of the numbers $\overline{\omega} \cdot \overline{b}_2$, $\overline{\omega} \cdot \overline{b}_3$ is not equal to 0 and the rank of the Klein form is 2. Now

$$K = span((\overline{\omega} \cdot \overline{b}_3)\overline{B}_2 - (\overline{\omega} \cdot \overline{b}_2)\overline{B}_3) = CA \cap A_3(u).$$

Let us summarize the above reflections.

Proposition 2. If a robot of spherical rank 1 is in a singular position or the direction of the revolute joint axes is complanar with the space τ_2 in a regular position then $\dim CA = 1$ and $CA \cap A_3(u) = \overline{0}$. There are asymptotic motions with zero Coriolis acceleration only in these positions.

If a robot of spherical rank 1 is in a regular position and $\overline{\omega} \notin \tau_2$ then $\dim CA = 2$ and there are two cases.

a) If $CA = \tau_2 \subset A_3(u)$; i.e., if $\overline{\omega}$ is perpendicular to τ_2; i.e., if $A_3(u)$ is a subalgebra of $e(3)$ then all motions are asymptotic in this position. The point (u) is flat.

b) If $CA \neq \tau_2$ then $CA \cap A_3(u) = span(\hat{Y}) = K$ and there are asymptotic motions with nonzero Coriolis acceleration in this regular position.

Revolute joints axes of spherical rank 1 robots are in all positions parallel, therefore perpendicularity of revolute joints axes and prismatic joints is preserved. Therefore if $A_3(u)$ is a subalgebra in one regular position then it is a subalgebra in all regular positions. Then $A_3(u)$ is a subalgebra in a regular position iff $A_3 = span(X_1, X_2, X_3)$ is a subalgebra.

3.1 Robots with 2 prismatic and 1 revolute joints

There are the following possibilities with respect to the configuration.

a$_1$) For RTT we have $Y_1 = (\overline{\omega}, \overline{0})$, $Y_2 = (\overline{0}, \overline{m}_2)$, $Y_3 = (\overline{0}, \overline{m}_3)$. Now $B_i = Y_i$, $[Y_i, Y_j] = [B_i, B_j]$, $i, j = 1,2,3$ and $\tau_2 = span(\overline{m}_2, \overline{m}_3)$.

a$_2$) For TRT we have $Y_1 = (\overline{0}, \overline{m}_1)$, $Y_2 = (\overline{\omega}, \overline{m}_2)$, $Y_3 = (\overline{0}, \overline{m}_3)$ and $B_1 = Y_2$, $B_2 = Y_1$, $Y_3 = B_3$. Therefore $[Y_1, Y_2] = -[B_1, B_2]$, $[Y_1, Y_3] = [B_2, B_3]$, $[Y_2, Y_3] = [B_1, B_3]$ and $\tau_2 = span(\overline{m}_1, \overline{m}_3)$.

a$_3$) For TTR we have $Y_1 = (\overline{0}, \overline{m}_1)$, $Y_2 = (\overline{0}, \overline{m}_2)$, $Y_3 = (\omega, \overline{m}_3)$ and $B_1 = Y_3$, $B_2 = Y_2$, $B_3 = Y_1$. Now $[Y_1, Y_2] = -[B_2, B_3]$, $[Y_1, Y_3] = -[B_1, B_3]$, $[Y_2, Y_3] = -[B_1, B_3]$ and $\tau_2 = span(\overline{m}_1, \overline{m}_2)$.

A singular position exists only in the case TRT provided there is $u_2(t_0) = \tilde{u}_2$ such that $o_1(t_0) \| o_3(t_0)$, i.e., $\overline{b}_2(t_0) \times \overline{b}_3(t_0) = \overline{0}$. This is possible iff $\angle(o_1, o_2) = \angle(o_2, o_3)$. $A_3(u)$ is a subalgebra iff $CA = \tau_2$ in a regular position is valid and this is possible iff $\overline{\omega} \cdot \overline{b}_2 = 0$, $\overline{\omega} \cdot \overline{b}_3 = 0$ in a regular position, so we have the following statements.

Proposition 3. All positions of robots RTT, TTR are regular.

There are singular positions in the case of TRT iff $\angle(o_1,o_2)=\angle(o_2,o_3)$.

$A_3(u)$ is an algebra iff the axis of the revolute joint is perpendicular to the axes of the prismatic joints in a regular position.

Remark 6. Robots RTT, TTR are homogeneous spaces. In the case TRT, this robot is a homogeneous space if A_3 is a subalgebra (the planar robot).

Let us investigate asymptotic motions of robots RTT, TRT, TTR. In the case when $A_3(u)$ is a subalgebra then all motions through the point (u) are asymptotic. We have the following cases:

(1) In a singular position; i.e., only for TRT, when $\overline{m}_3=c\overline{m}_1, c\in R$, the subspace CA is defined by $[Y_1,Y_2]=(\overline{0},-\overline{\omega}\times\overline{m}_1)$, $[Y_1,Y_3]=(\overline{0},\overline{0})$, $[Y_2,Y_3]=(\overline{0},c\overline{\omega}\times\overline{m}_1)$ and the Coriolis acceleration is $\dot{Y}_c=\sum_{i<j}\dot{u}_i\dot{u}_j[Y_i,Y_j]=(-\dot{u}_1\dot{u}_2+c\dot{u}_2\dot{u}_3)(\overline{0},\overline{\omega}\times\overline{m}_1)$, where $\overline{\omega}\times\overline{m}_1\neq\overline{0}$.

A motion is asymptotic at a singular point $u(t_0)$ if and only if $\dot{u}_2(t_0)(-\dot{u}_1(t_0)+c\dot{u}_3(t_0))=0$; i.e., $\dot{u}_2(t_0)=0$ so that the revolute joint is not working at t_0 or the joint velocities of the prismatic joints satisfy the relationship $\dot{u}_1(t_0):\dot{u}_3(t_0)=c$.

If every position of the robot motion is singular (i.e., $u_2(t)=\hat{u}_2=\text{const}$, $\dot{u}_2=0$) then this motion is the trivial asymptotic motion (only prismatic joints work).

Proposition 4. A motion of the robot TRT is nontrivial asymptotic in a singular position $u(t_0)$ iff all joints work and the joint velocities of the prismatic joints satisfy the relationship $\dot{u}_1(t_0):\dot{u}_3(t_0)=c$. The singular motion of the robot TRT is trivial asymptotic.

(2) Let us investigate asymptotic motions of the robot in a regular position when the subspace CA is one-dimensional. We know that $\overline{\omega}=c_2\overline{b}_2+c_3\overline{b}_3$ and $CA\cap A_3(u)=\overline{0}$. A motion is asymptotic when the Coriolis acceleration $\dot{Y}_c=\sum\dot{u}_i\dot{u}_j[Y_i,Y_j]=\overline{0}$ and this occurs

a_1) if $\dot{u}_1\dot{u}_2(\overline{0},-c_3\overline{m}_2\times\overline{m}_3)+\dot{u}_1\dot{u}_3(\overline{0},c_2\overline{m}_2\times\overline{m}_3)=0$; i.e., $\dot{u}_1(-\dot{u}_2c_3+\dot{u}_3c_2)=0$ in the case of RTT,

a_2) if $\dot{u}_1\dot{u}_2(\overline{0},c_3\overline{m}_1\times\overline{m}_3)+\dot{u}_2\dot{u}_3(\overline{0},c_2\overline{m}_1\times\overline{m}_3)=0$; i.e., $\dot{u}_2(\dot{u}_1c_3+\dot{u}_3c_2)=0$ in the case of TRT,

a_3) if $\dot{u}_1\dot{u}_3(\overline{0},-c_3\overline{m}_1\times\overline{m}_2)+\dot{u}_2\dot{u}_3(\overline{0},c_2\overline{m}_1\times\overline{m}_2)=0$; i.e., $\dot{u}_3(\dot{u}_1c_3-\dot{u}_2c_2)=0$ in the case of TTR.

In the cases of RTT, TTR, if the equation $\overline{\omega}=c_2\overline{b}_2+c_3\overline{b}_3$ is valid in one position then it is valid for all positions.

In the case TRT, the equation $\overline{\omega}=c_2\overline{b}_2+c_3\overline{b}_3$, $c_2\cdot c_3\neq 0$ is valid only if 3 rd axis turns around the axis o_2 to the position complanar with axes o_1,o_2 (i.e., the directions of the joint axes are linear dependent). If $c_2\cdot c_3=0$; i.e., $o_3\|o_2$ or $o_1\|o_2$ then the equation $\overline{\omega}=c_2\overline{b}_2+c_3\overline{b}_3$ is valid for all positions of the axes.

Let us recall that we are interested only in nontrivial asymptotic motions. Then the Coriolis acceleration is zero in the case RTT if $-\dot{u}_2c_3+\dot{u}_3c_2=0$, in the case TRT if $\dot{u}_1c_3+\dot{u}_3c_2=0$ and in the case TTR if $-\dot{u}_1c_3+\dot{u}_2c_2=0$. We have the following cases:

a) Let $c_2\cdot c_3\neq 0$ at (u). Then a motion through the point (u) is nontrivial asymptotic iff all joints work and the joint velocities of the prismatic joints satisfy the relationship $c_2:c_3$ in the cases RTT, TTR and $-c_2:c_3$ in the case TRT.

b) Let $c_2 \cdot c_3 = 0$ at (u). Then a motion through the point (u) is nontrivial asymptotic iff the revolute joint and only the prismatic joint whose axis is parallel to the axis of the revolute joint, work.

Proposition 5. Let Υ_{A_3} be a robot of spherical rank 1 with two prismatic joints and let the directions of the joint axes be linear dependent at $u(t_0)$; i.e., $\overline{\omega} = c_2\overline{b}_2 + c_3\overline{b}_3$. Then:

a) The zero Coriolis acceleration is a necessary condition for the motion to be asymptotic at $u(t_0)$.

b) In the case that no two axes of joints are parallel at (u): a motion through the point (u) is nontrivial asymptotic iff all joints work and the joint velocities of the prismatic joints satisfy the relationship $c_2 : c_3$ in the cases RTT, TTR and $-c_2 : c_3$ in the case TRT.

c) In the case that the axis of the revolute joint is parallel to one axis of a prismatic joint: a motion is nontrivial asymptotic iff the revolute joint and only the prismatic joint whose axis is parallel to the axis of the revolute joint, work.

(3) Let us investigate asymptotic robot motions in a regular position, when $\dim CA = 2$ and $A_3(u)$ is not a subalgebra. Then $CA \cap A_3(u) = K = span(((\overline{\omega}\cdot\overline{b}_3)B_2 - (\overline{\omega}\cdot\overline{b}_2)B_3))$; i.e., the equation: $(\overline{\omega}\cdot\overline{b}_3)\overline{b}_2 - (\overline{\omega}\cdot\overline{b}_2)\overline{b}_3 = k_2(\overline{\omega}\times\overline{b}_2) + k_3(\overline{\omega}\times\overline{b}_3)$, $k_2, k_3 \in R$, is valid.

In this case the motion is asymptotic at the point (u) if and only if

a$_1$) for RTT $\dot{u}_1\dot{u}_2(\overline{0},\overline{\omega}\times\overline{m}_2) + \dot{u}_1\dot{u}_3(\overline{0},\overline{\omega}\times\overline{m}_3) = \lambda(\overline{0},k_2(\overline{\omega}\times\overline{m}_2) + k_3(\overline{\omega}\times\overline{m}_3))$, $\lambda \in R$; i.e., $\dot{u}_1\dot{u}_2 = k_2\lambda$, $\dot{u}_1\dot{u}_3 = k_3\lambda$,

a$_2$) for TRT $\dot{u}_1\dot{u}_2(\overline{0},\overline{\omega}\times\overline{m}_1) + \dot{u}_2\dot{u}_3(\overline{0},\overline{\omega}\times\overline{m}_3) = \lambda(\overline{0},k_2(\overline{\omega}\times\overline{m}_1) + k_3(\overline{\omega}\times\overline{m}_3))$, $\lambda \in R$; i.e., $\dot{u}_1\dot{u}_2 = k_2\lambda$, $\dot{u}_2\dot{u}_3 = k_3\lambda$,

a$_3$) for TTR $\dot{u}_1\dot{u}_3(\overline{0},\overline{\omega}\times\overline{m}_1) + \dot{u}_2\dot{u}_3(\overline{0},\overline{\omega}\times\overline{m}_2) = \lambda(\overline{0},k_2(\overline{\omega}\times\overline{m}_1) + k_3(\overline{\omega}\times\overline{m}_2))$, $\lambda \in R$; i.e., $\dot{u}_1\dot{u}_3 = k_2\lambda$, $\dot{u}_2\dot{u}_3 = k_3\lambda$.

We summarize the previous results.

Proposition 6. Let Υ_{A_3} be a robot of spherical rank 1 with two prismatic joints and let the directions of the joint axes be independent at t_0; i.e., $\overline{\omega} \neq c_2\overline{b}_2 + c_3\overline{b}_3$. Then:

A motion is nontrivial asymptotic at t_0 iff joint velocities at t_0 satisfy $\dot{u}_1\dot{u}_2 = k_2\lambda$, $\dot{u}_1\dot{u}_3 = k_3\lambda$ for RTT, $\dot{u}_1\dot{u}_2 = k_2\lambda$, $\dot{u}_2\dot{u}_3 = k_3\lambda$ for TRT and for TTR $\dot{u}_1\dot{u}_3 = k_2\lambda$, $\dot{u}_2\dot{u}_3 = k_3\lambda$, where $\lambda \in R$ and k_2, k_3 are the coefficients of the linear combination of $\hat{\Upsilon} = (\overline{0},(\omega\cdot\overline{b}_3)\overline{b}_2 - (\overline{\omega}\cdot b_2)b_3)$ in the canonical basis of the Coriolis space. If these relations are true for any admissible t then the motion is asymptotic.

In this case there are nontrivial asymptotic motions with the nonzero Coriolis acceleration.

3.2 Robots with 1 prismatic and 2 revolute joints

Let ξ be the plane determined by the axes of the revolute joints. There are three possibilities with respect to the configuration.

b$_1$) RRT: then $Y_1 = (\overline{\omega},\overline{0})$, $Y_2 = (\overline{\omega},\overline{m}_2)$, $Y_3 = (\overline{0},\overline{m}_3)$, where $\overline{m}_2 \neq \overline{0}$ and $\overline{\omega}\cdot\overline{m}_2 = 0$. Now $B_1 = Y_1 = (\overline{\omega},\overline{0})$, $B_2 = Y_2 - Y_1 = (\overline{0},\overline{b}_2 = \overline{m}_2)$, $B_3 = Y_3 = (\overline{0},\overline{b}_3 = \overline{m}_3)$. We know, see Remark 3 that the vector $\overline{m}_2$ is perpendicular to the plane ξ. We have $[Y_1,Y_2]=[B_1,B_2]$, $[Y_1,Y_3]=[B_1,B_3]$, $[Y_2,Y_3]=[B_1,B_3]$ and $\tau_2 = span(\overline{m}_2,\overline{m}_3)$.

b$_2$) RTR: then $Y_1=(\overline{\omega},\overline{0})$, $Y_2=(\overline{0},\overline{m}_2)$, $Y_3=(\overline{\omega},\overline{m}_3)$, where $\overline{m}_3 \neq \overline{0}$ and $\overline{\omega}\cdot\overline{m}_3=0$. Now $B_1=Y_1=(\overline{\omega},\overline{0})$, $B_2=Y_2=(\overline{0},\overline{m}_2)$, $B_3=Y_3-Y_1=(\overline{0},\overline{b}_3=\overline{m}_3)$. The vector $\overline{m}_3$ is perpendicular to the plane ξ. We have $[Y_1,Y_2]=[B_1,B_2]$, $[Y_1,Y_3]=[B_1,B_3]$, $[Y_2,Y_3]=-[B_1,B_2]$ and $\tau_2 = span(\overline{m}_2,\overline{m}_3)$.

b$_3$) TRR: then $Y_1=(\overline{0},\overline{m}_1)$, $Y_2=(\overline{\omega},\overline{m}_2)$, $Y_3=(\overline{\omega},\overline{m}_3)$, $\overline{m}_2 \neq \overline{m}_3$, $\overline{\omega}\cdot\overline{m}_2=0$, $\overline{\omega}\cdot\overline{m}_3=0$. Now $B_1=Y_2$, $B_2=Y_1$, $B_3=Y_3-Y_2$. It is easy to show that the vector $\overline{m}_3-\overline{m}_2$ is perpendicular to the plane ξ. We have $[Y_1,Y_2]=-[B_1,B_2]$, $[Y_1,Y_3]=-[B_1,B_2]$, $[Y_2,Y_3]=[B_1,B_3]$ and $\tau_2 = span(\overline{m}_1,\overline{m}_3-\overline{m}_2)$.

So we have

Proposition 7. Let ξ be the plane determined by the axes of the revolute joints. The space τ_2 of the directions of the translational velocity elements is generated by the direction of the prismatic joint and the normal vector of the plane ξ. If the axis of the prismatic joint is perpendicular to the plane ξ then the robot is in the singular position. The robot has a singular position iff A_3 is a subalgebra.

The subspace $A_3(u)$ is a subalgebra iff the axes of the revolute joints are perpendicular to the axis of the prismatic joint in a regular position.

In the next part we will investigate asymptotic robot motions of RRT, RTR, TRR. If $A_3(u)$ is a subalgebra then all motions through the point (u) are asymptotic. Let $\overline{n}_\xi$ be the normal vector of the plane ξ. By our previous considerations we have the following cases:

(1) Let $u(t_0)$ be a singular position (A_3 is a subalgebra). Then $\tau_2 = span(\overline{n}_\xi)$ and $A_3(u(t_0))$ is not a subalgebra. We have at t_0: for RRT $\overline{m}_3 = c\overline{m}_2, 0 \neq c \in R$, $\dot{Y}_c = (\dot{u}_1\dot{u}_2 + c(\dot{u}_1\dot{u}_3 + \dot{u}_2\dot{u}_3))(\overline{0},\overline{\omega}\times\overline{m}_2)$, $\overline{\omega}\cdot\overline{m}_2=0$, for RTR $\overline{m}_3 = c\overline{m}_2, 0 \neq c \in R$, $\dot{Y}_c = (\dot{u}_1\dot{u}_2 + c\dot{u}_1\dot{u}_3 - \dot{u}_2\dot{u}_3)(\overline{0},\overline{\omega}\times\overline{m}_2)$, $\overline{\omega}\cdot\overline{m}_2=0$ and for TRR $\overline{m}_3-\overline{m}_2 = c\overline{m}_1, 0 \neq c \in R$, $\dot{Y}_c = (\dot{u}_1\dot{u}_2 + \dot{u}_1\dot{u}_3 - c\dot{u}_2\dot{u}_3)(\overline{0},\overline{\omega}\times\overline{m}_1)$, $\overline{\omega}\cdot\overline{m}_1=0$. We know that a motion is asymptotic at a singular position $u(t_0)$ only if the Coriolis acceleration is zero. A singular motion ($u_2(t)=u_2(t_0)=const, \dot{u}_2(t)=0$) can be only trivial asymptotic when only one joint works. Thus we get

Proposition 8. Let Υ_{A_3} be a robot of spherical rank 1 with two revolute joints. Then a motion is nontrivial asymptotic at the singular position $u(t_0)$ iff at t_0 all joints work and for RRT, RTR, TRR we have $(\dot{u}_1\dot{u}_2 + c(\dot{u}_1\dot{u}_3 + \dot{u}_2\dot{u}_3))=0$, $(\dot{u}_1\dot{u}_2 + c\dot{u}_1\dot{u}_3 - \dot{u}_2\dot{u}_3)=0$, $(\dot{u}_1\dot{u}_2 + \dot{u}_1\dot{u}_3 - c\dot{u}_2\dot{u}_3)=0$ at t_0 respectively. The singular motion is trivial asymptotic.

(2) Let us assume that $u(t_0)$ is a regular position, $\overline{\omega}\in\tau_2$ and $A_3(u)$ is not a subalgebra. Then $\overline{\omega} = c_1\overline{m} + c_2\overline{n}_\xi$, $c_1,c_2 \in R, c_1 \neq 0$, where $\overline{m}$ is the direction of the axis of the prismatic

joint and $\overline{n}_\xi$ is the normal vector of the plane ξ. The axis of the prismatic joint is parallel to the axes of the revolute joints iff $c_2 = 0$. This position does not vary to time. If the axis of the prismatic joint is not parallel to the axes of the revolute joints then always $u_2 = \tilde{u}_2$, when $\overline{\omega} = c_1\overline{m} + c_2\overline{n}_\xi$.

b$_1$) For RRT: if $\overline{\omega} = \overline{m}_3$ then $\dot{Y}_c = \dot{u}_1\dot{u}_2(\overline{0}, \overline{\omega} \times \overline{m}_2)$, $\overline{\omega} \cdot \overline{m}_2 = 0$ for every (u). If $\overline{\omega} \neq \overline{m}_3$ then there is the position ($u_2 = \tilde{u}_2$) so that the axis o_3 turning around the axis o_2 gets into the position complanar with the space $span(\overline{\omega}, \overline{m}_2)$; i.e., $\overline{m}_3 = c_1\overline{\omega} + c_2\overline{m}_2$. Then $\dot{Y}_c = (\dot{u}_1\dot{u}_2 + c_2\dot{u}_1\dot{u}_3 + c_2\dot{u}_2\dot{u}_3)(\overline{0}, \overline{\omega} \times \overline{m}_2)$.

b$_2$) For RTR: if $\overline{\omega} = \overline{m}_2$ then $\dot{Y}_c = \dot{u}_1\dot{u}_3(\overline{0}, \overline{\omega} \times \overline{m}_3)$, $\overline{\omega} \cdot \overline{m}_3 = 0$ for every (u). If $\overline{\omega} \neq \overline{m}_2$ then there is the position ($u_2 = \tilde{u}_2$) so that the normal $\overline{m}_3$ of the plane ξ is complanar with the space $span(\overline{\omega}, \overline{m}_2)$; i.e., $\overline{m}_3 = c_1\overline{\omega} + c_2\overline{m}_2$. Then $\dot{Y}_c = (\dot{u}_1\dot{u}_2 + c_2\dot{u}_1\dot{u}_3 - \dot{u}_2\dot{u}_3)(\overline{0}, \overline{\omega} \times \overline{m}_2)$.

b$_3$) For TRR: if $\overline{m}_1 = \overline{\omega}$ then $\dot{Y}_c = \dot{u}_2\dot{u}_3(\overline{0}, \overline{\omega} \times (\overline{m}_3 - \overline{m}_2))$, for every (u). If $\overline{m}_1 \neq \overline{\omega}$ then there is the position ($u_2 = \tilde{u}_2$) so that the normal $\overline{m}_3 - \overline{m}_2$ of the plane ξ is complanar with the space $span(\overline{\omega}, \overline{m}_1)$; i.e., $\overline{m}_3 - \overline{m}_2 = c_1\overline{\omega} + c_2\overline{m}_1$. Then $\dot{Y}_c = (\dot{u}_1\dot{u}_2 + \dot{u}_1\dot{u}_3 - c_2\dot{u}_2\dot{u}_3)(\overline{0}, \overline{\omega} \times \overline{m}_1)$. We know, see Proposition 2, that in the case when $\overline{\omega} \in \tau_2$ the motion is asymptotic iff $\dot{Y}_c = 0$. We get

Proposition 9. Let Υ_{A_3} be a robot of spherical rank 1 with two revolute joints and let the axis of the prismatic joint is complanar with the space $span(\overline{\omega}, \overline{n}_\xi)$ at t_0 i.e $\overline{m} = c_1\overline{\omega} + c_2\overline{n}_\xi$. Then we have:

a) The zero Coriolis acceleration is a necessary condition for a motion to be asymptotic at t_0.

b) A motion of the robot Υ_{A_3} is nontrivial asymptotic at the point $u(t_0)$ iff in the cases of RRT, RTR, TRR the equalities $(\dot{u}_1\dot{u}_2 + c_2\dot{u}_1\dot{u}_3 + c_2\dot{u}_2\dot{u}_3) = 0$, $(\dot{u}_1\dot{u}_2 + c_2\dot{u}_1\dot{u}_3 - \dot{u}_2\dot{u}_3) = 0$, $(\dot{u}_1\dot{u}_2 + \dot{u}_1\dot{u}_3 - c_2\dot{u}_2\dot{u}_3) = 0$ are valid at t_0, respectively.

c) A motion of the robot Υ_{A_3}, whose all axes are parallel to each other $(c_2 = 0)$, is nontrivial asymptotic iff the prismatic joint and only one revolute joint work.

(3) Let $\dim CA = 2$ and $A_3(u)$ be not a subalgebra. Then $CA \cap A_3(u) = K$ is the Klein subspace, $K = span(\hat{Y})$, $\hat{Y} \in \iota$ and the direction of $\hat{Y}$ is perpendicular to $\overline{\omega}$. A motion is asymptotic at the point (u), iff $\dot{u}_1\dot{u}_2[Y_1, Y_2] + \dot{u}_1\dot{u}_3[Y_1, Y_3] + \dot{u}_2\dot{u}_3[Y_2, Y_3] = \lambda\hat{Y}$, $\lambda \in R$. We get

b$_1$) for RRT: $[Y_2, Y_3] = [Y_1, Y_3]$ and $[Y_1, Y_2], [Y_1, Y_3]$ are the basis elements of the space CA and $\hat{Y} = k_2[Y_1, Y_2] + k_3[Y_1, Y_3]$, $k_2, k_3 \in R$. Then the motion is asymptotic iff $\dot{u}_1\dot{u}_2[Y_1, Y_2] + (\dot{u}_1\dot{u}_3 + \dot{u}_2\dot{u}_3)[Y_1, Y_3] = \lambda(k_2[Y_1, Y_2] + k_3[Y_1, Y_3])$ and this occurs if and only if $\dot{u}_1\dot{u}_2 = \lambda k_2$, $(\dot{u}_1 + \dot{u}_2)\dot{u}_3 = \lambda k_3$.

b$_2$) for RTR: $[Y_2,Y_3]=-[Y_1,Y_2]$ and $[Y_1,Y_2],[Y_1,Y_3]$ are the basis elements of the space CA and $\hat{Y}=k_2[Y_1,Y_2]+k_3[Y_1,Y_3]$, $k_2,k_3\in R$. Then the motion is asymptotic iff $(\dot{u}_1\dot{u}_2-\dot{u}_2\dot{u}_3)[Y_1,Y_2]+\dot{u}_1\dot{u}_3[Y_1,Y_3]=\lambda(k_2[Y_1,Y_2]+k_3[Y_1,Y_3])$ and this occurs if and only if $\dot{u}_2(\dot{u}_1-\dot{u}_3)=\lambda k_2$, $\dot{u}_1\dot{u}_3=\lambda k_3$.

b$_3$) for TRR: $[Y_1,Y_3]=[Y_1,Y_2]$ and $[Y_1,Y_2],[Y_2,Y_3]$ are the basis elements of the space CA and $\hat{Y}=k_2[Y_1,Y_3]+k_3[Y_2,Y_3]$, $k_2,k_3\in R$. Then the motion is asymptotic iff $(\dot{u}_1\dot{u}_2+\dot{u}_1\dot{u}_3)[Y_1,Y_2]+\dot{u}_2\dot{u}_3[Y_2,Y_3]=\lambda(k_2[Y_1,Y_3]+k_3[Y_2,Y_3])$ and this occurs if and only if $\dot{u}_1(\dot{u}_2+\dot{u}_3)=\lambda k_2$, $\dot{u}_2\dot{u}_3=\lambda k_3$.

So we have

Proposition 10. Let Υ_{A_3} be a robot of spherical rank 1 with two revolute joints and let the axis of the prismatic joint be not complanar with the space $span(\overline{\omega},\overline{n}_\xi)$ at t_0 i.e $\overline{m}\neq c_1\overline{\omega}+c_2\overline{n}_\xi$. Then a motion is asymptotic at t_0 iff the joint velocities at t_0 satisfy $\dot{u}_1\dot{u}_2=\lambda k_2$, $(\dot{u}_1+\dot{u}_2)\dot{u}_3=\lambda k_3$ for RRT, $\dot{u}_2(\dot{u}_1-\dot{u}_3)=\lambda k_2$, $\dot{u}_1\dot{u}_3=\lambda k_3$ for RTR and for TRR: $\dot{u}_1(\dot{u}_2+\dot{u}_3)=\lambda k_2$, $\dot{u}_2\dot{u}_3=\lambda k_3$, where $\lambda\in R$ and k_2,k_3 are the coefficients of the linear combination of $\hat{Y}=(\overline{0},(\overline{\omega}\cdot\overline{b}_3)\overline{b}_2-(\overline{\omega}\cdot\overline{b}_2)\overline{b}_3)$ in the canonical basis of the Coriolis space CA. If these relations are true for any admissible t then the motion is asymptotic.
In this case there are nontrivial asymptotic motions with nonzero Coriolis acceleration.

3.3 Robots with 3 revolute joints

These robots have the axes of the joints parallel and different from each other (the robots are planar). The elements Y_i satisfy $Y_1=(\overline{\omega},\overline{0})$, $Y_2=(\overline{\omega},\overline{m}_2)$, $Y_3=(\overline{\omega},\overline{m}_3)$, $\overline{\omega}\cdot\overline{m}_2=0$, $\omega\cdot\overline{m}_3=0$, $\overline{m}_3\neq\overline{m}_2\neq\overline{0}$. Let us denote planes $\xi_2=(o_1,o_2)$ and $\xi_3=(o_1,o_3)$. Then $\overline{m}_2$ is the normal vector to the plane ξ_2 and $\overline{m}_3$ is the normal vector to the plane ξ_3. For the elements B_i we have $B_1=Y_1$, $B_2=Y_2-Y_1=(\overline{0},\overline{b}_2=\overline{m}_2)$, $B_2=Y_3-Y_1=(\overline{0},\overline{b}_3=\overline{m}_3)$. Because $\tau_2=span(\overline{m}_2,\overline{m}_3)$, $\overline{\omega}\cdot\tau_2=0$ and $[Y_1,Y_2]=(\overline{0},\overline{\omega}\times\overline{m}_2)$, $[Y_1,Y_3]=(\overline{0},\overline{\omega}\times\overline{m}_3)$, $[Y_1,Y_2]=(\overline{0},\overline{\omega}\times\overline{m}_3-\overline{\omega}\times\overline{m}_2)$ we conclude that $A_3(u)$ is a subalgebra in a regular position. If the plane ξ_3 turning around the axis o_2 coincides with the plane ξ_2 then the robot is in a singular position at t_0; i.e., $\overline{m}_3=c\overline{m}_2, c\in R$. In regular positions we have $\dim CA=2$ and all motions are asymptotic while $\dim CA=1$ in singular positions and the Coriolis acceleration satisfies: $\dot{Y}_C=\dot{u}_1(\dot{u}_2+c\dot{u}_3)(\overline{0},\overline{\omega}\times\overline{m}_2)$.

Proposition 11. Let RRR be a robot the revolute joint axes of which are parallel. Then its position $u(t_0)$ is singular if all axes of the joints lie in a plane. $A_3(u)$ is a subalgebra in the regular position and $K=A_3(u)$. $A_3(u)$ is not a subalgebra in the singular position. A motion through the singular position $u_2(t_0)=\hat{u}_2$ is asymptotic at $u_2(t_0)=\hat{u}_2$

a) if 1 st joint does not work or

b) the ratio of the joint velocities of the 2 nd and 3 rd joints at t_0 is $-c$.

A singular motion ($u_2(t) = \hat{u}_2$) can be only trivial asymptotic.

Let us present a survey of all nontrivial asymptotic motion of the robots of spherical rank one.

1. The robots with one revolute joint (RTT, TRT, TTR).

   a) Let the directions of the joint axes be dependent (i.e., $\overline{\omega} = c_2\overline{b}_2 + c_3\overline{b}_3$) and $c_2c_3 \neq 0$ in the cases RTT, TTR. Then a robot motion is nontrivial asymptotic iff all joints work and the ratio of the joint velocities of the prismatic joints is $c_2 : c_3$.

   b) Let the axis of the revolute joint be paralel to one axis of the prismatic joint (i.e., $c_2c_3 = 0$). Then a robot motion is nontrivial asymptotic iff the revolute joint and only the prismatic joint whose axis is parallel to the revolute joint axis work.

   c) Let the directions of the joint axes be independent (i.e., $\overline{\omega} \neq c_2\overline{b}_2 + c_3\overline{b}_3$). Then a robot motion is nontrivial asymptotic iff the joint velocities satisfy for any admissible t: $\dot{u}_1\dot{u}_2 = k_2\lambda$, $\dot{u}_1\dot{u}_3 = k_3\lambda$ for RTT, $\dot{u}_1\dot{u}_2 = k_2\lambda$, $\dot{u}_2\dot{u}_3 = k_3\lambda$ for TRT, $\dot{u}_1\dot{u}_3 = k_2\lambda$, $\dot{u}_2\dot{u}_3 = k_3\lambda$ for TTR, where k_2, k_3 are the coefficients of the linear combination of the Klein direction in the canonical basis of the Coriolis space.

2. The robots with two revolute joints (RRT, RTR, TRR).

   a) Let the joint axes be parallel. Then a robot motion is nontrivial asymptotic iff one revolute joint does not work.

   b) Let the axis of the prismatic joint be not complanar with the space $span(\overline{\omega}, \overline{n}_\xi)$. Then a robot motion is nontrivial asymptotic iff for the joint velocities and any admissible t we have: $\dot{u}_1\dot{u}_2 = \lambda k_2$, $(\dot{u}_1 + \dot{u}_2)\dot{u}_3 = \lambda k_3$ for RRT, $\dot{u}_2(\dot{u}_1 - \dot{u}_3) = \lambda k_2$, $\dot{u}_1\dot{u}_3 = \lambda k_3$ for RTR and $\dot{u}_1(\dot{u}_2 + \dot{u}_3) = \lambda k_2$, $\dot{u}_2\dot{u}_3 = \lambda k_3$ for TRR, where k_2, k_3 are coefficients of the linear combination of $\hat{Y} = (\overline{0}, (\overline{\omega} \cdot \overline{b}_3)\overline{b}_2 - (\overline{\omega} \cdot \overline{b}_2)\overline{b}_3)$ in the canonical basis of the Coriolis space.

3. The robots with three revolute joints (RRR).

In this case, there are only trivial asymptotic motions.

4. References

Denavit, J.; Hartenberg, R. S. (1955) A kinematics notation for lower-pair mechanisms based on matrices, *Journal of Applied Mechanics*, Vol. 22, June 1955

Helgason, S. (1962). *Differential geometry and symmetric spaces*, American Mathematical Society, ISBN 0821827359, New York, Russian translation

Karger, A. (1988). Geometry of the motion of robot manipulators. *Manuscripta mathematica*. Vol. 62, No. 1, March 1988, 1 130, ISSN 0025-2611

Karger, A. (1989). Curvature properties of 6-parametric robot manipulators. *Manuscripta mathematica,* Vol. 65, No. 3, September 1989, 257-384, ISSN 0025-2611

Karger, A. (1990). Classification of Three-Parametric Spatial Motions with transitive Group of Automorphisms and Three-Parametric Robot Manipulators, *Acta Applicandae Mathematicae,* Vol. 18, No. 1, January 1990, 1-97, ISSN 0167-8019

Karger, A. (1993). Robot-manipulators as submanifold, *Mathematica Pannonica,* Vol. 4, No. 2, 1993, pp. 235-247 , ISSN 0865-2090

Samuel, A. E.; McAree, P. R.; Hunt, K. H. (1991). Unifying Screw Geometry and Matrix Transformations. *The International Journal of Robotics Research*, Vol. 10, No. 5, October 1991, 439-585, ISSN 0278-3649

Selig, J. M. (1996). *Geometrical Methods in Robotics,* Springer-Verlag, ISBN 0387947280, New York

4

Topology and Geometry of Serial and Parallel Manipulators

Xiaoyu Wang and Luc Baron
Polytechnique of Montreal
Canada

1. Introduction

The evolution of requirements for mechanical products toward higher performances, coupled with never ending demands for shorter product design cycle, has intensified the need for exploring new architectures and better design methodologies in order to search the optimal solutions in a larger design space including those with greater complexity which are usually not addressed by available design methods. In the mechanism design of serial and parallel manipulators, this is reflected by the need for integrating topological and geometric synthesis to evaluate as many potential designs as possible in an effective way.

In the context of kinematics, a mechanism is a kinematic chain with one of its links identified as the base and another as the end-effector (EE). A manipulator is a mechanism with all or some of its joints actuated. Driven by the actuated joints, the EE and all links undergo constrained motions with respect to the base (Tsai, 2001). A serial manipulator (SM) is a mechanism of open kinematic chain while a parallel manipulator (PM) is a mechanism whose EE is connected to its base by at least two independent kinematic chains (Merlet, 1997). The early works in the manipulator research mostly dealt with a particular design; each design was described in a particular way. With the number of designs increasing, the consistency, preciseness and conciseness of manipulator kinematic description become more and more problematic. To describe how a manipulator is kinematically constructed, no normalized term and definition have been proposed. The words architecture (Hunt, 1982a), structure (Hunt, 1982b), topology (Powell, 1982), and type (Freudenstein & Maki, 1965; Yang & Lee, 1984) all found their way into the literature, describing kinematic chains without reference to dimensions. However, some kinematic properties of spatial manipulators are sensitive to certain kinematic details. The problem is that with the conventional description, e.g. the topology (the term topology is preferred here to other terms), manipulators of the same topology might be too different to even be classified in the same category. The implementation of the kinematic synthesis shows that the traditional way of defining a manipulator's kinematics greatly limits both the qualitative and quantitative designs of spatial mechanisms and new method should be proposed to solve the problem. From one hand, the dimension-independent aspect of topology does not pose a considerable problem to planar manipulators, but makes it no longer appropriate to describe spatial manipulators especially spatial PMs, because such properties as the degree of freedom (DOF) of a manipulator and the degree of mobility (DOM) of its EE as well as the mobility nature are highly dependent on some geometric elements. On the other hand, when performing geometric synthesis, some dimensional and geometric constraints should be imposed in order for the design space to have a good correspondence with the set of manipulators which can satisfy the basic design requirements (the DOF, DOM and the mobility nature), otherwise, a large proportion of the design space may have nothing to do with the design problem in hand. As for the kinematic representation of PMs, one can hardly find a method which is adequate for a wide range of manipulators and commonly accepted and used in the literature. However, in the classification (Balkan et al., 2001; Su et al., 2002), comparison studies (Gosselin et al., 1995; Tsai & Joshi, 2001) (equivalence, isomorphism, similarity, difference, etc.) and manipulator kinematic synthesis, an effective kinematic representation is essential. The first part of this work will be focused on the topology issue.

Manipulators of the same topology are then distinguished by their kinematic details. Parameter (Denavit & Hartenberg, 1954), dimension (Chen & Roth, 1969; Chedmail, 1998), and geometry (Park & Bobrow, 1995) are among the terms used to this end and the ways of defining a particular manipulator are even more diversified. When performing kinematic synthesis, which parameters should be put under what constraints are usually dictated by the convenience of the mathematic formulation and the synthesis algorithm implementation instead of by a good delimitation of the searching space. Another problematic is the numeric representation of the topology and the geometry which is suitable for the implementation of global optimization methods, e.g. genetic algorithms and the simulated annealing. This will be the focus of the second part of this work.

2. Preliminary

Some basic concepts and definitions about kinematic chains are necessary to review as a starting point of our discussion on topology and geometry. A kinematic chain is a set of rigid bodies, also called links, coupled by kinematic pairs. A kinematic pair is, then, the coupling of two rigid bodies so as to constrain their relative motion. We distinguish upper pairs and lower pairs. An upper kinematic pair constrains two rigid bodies such that they keep a line or point contact; a lower kinematic pair constrains two rigid bodies such that a surface contact is maintained (Angeles, 2003). A joint is a particular mechanical implementation of a kinematic pair (IFToMM, 2003). As shown in Fig. 1, there are six types of joints corresponding to the lower kinematic pairs - spherical (S), cylindrical (C), planar (E), helical (H), revolute (R) and prismatic (P) (Angeles, 1982). Since all these joints can be obtained by combining the revolute and prismatic ones, it is possible to deal only with revolute and prismatic joints in kinematic modelling. Moreover, all these joints can be represented by elementary geometric elements, i.e., point and line. To characterize links, the notions of simple link, binary link, ternary link, quaternary link and n-link were introduced to indicate how many other links a link is connected to. Similarly, *binary joint*, *ternary joint* and *n-joint* indicate how many links are connected to a joint. A similar notion is the connectivity of a link or a joint (Baron, 1997). These basic concepts constitute a basis for kinematic analysis and kinematic synthesis.

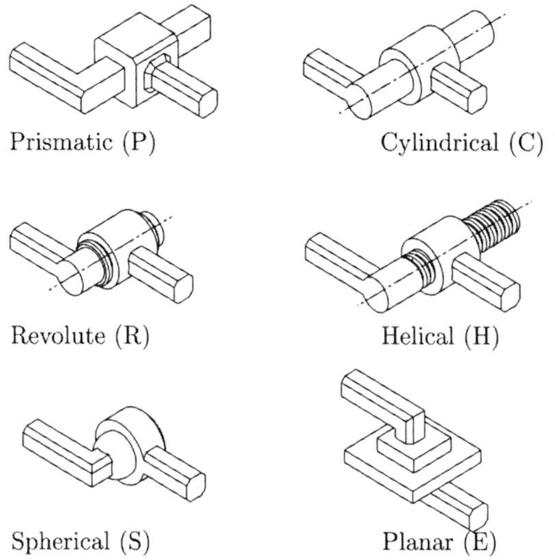

Figure 1: Lower Kinematic Pairs

3. Topology

For kinematic studies, the kinematic description of a mechanism consists of two parts, one is qualitative and the other quantitative. The qualitative part indicates which link is connected to which other links by what types of joints. This basic information is referred to as structure, architecture, topology, or type, respectively, by different authors. When dealing with complex spatial mechanisms, the qualitative description alone is of little interest, because the kinematic properties of the corresponding mechanisms can vary too much to characterize a mechanism. This can be demonstrated by the single-loop 4-bar mechanisms shown in Fig. 2. Without reference to dimensions, all mechanisms shown in Fig. 2 are of the same kinematic structure but have very distinctive kinematic properties and therefore are used for different applications— mechanism a) generates planar motion, mechanism b) generates spherical motion, mechanism c) is a Bennett mechanism (Bennett, 1903), while mechanism d) permits no relative motion at any joints. Fig. 3 shows an example of parallel mechanisms having the same kinematic structure—mechanism a) has 3 DOFs whose EE has no mobility, mechanism b) has 3 DOFs whose EE has 3 DOMs in translation, mechanism c) permits no relative motion at any joints.

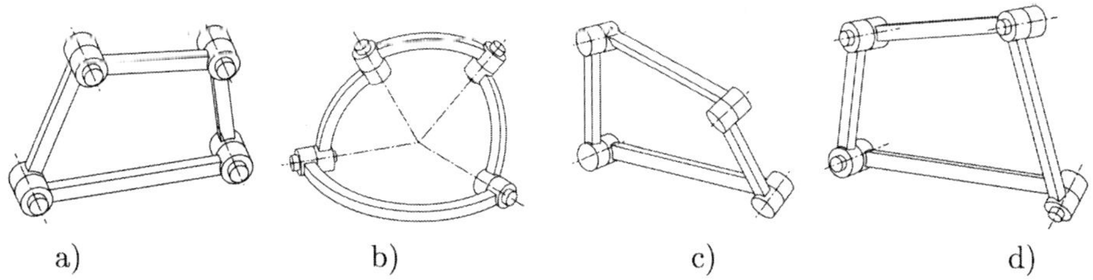

Figure 2: 4-bar mechanisms of different geometries

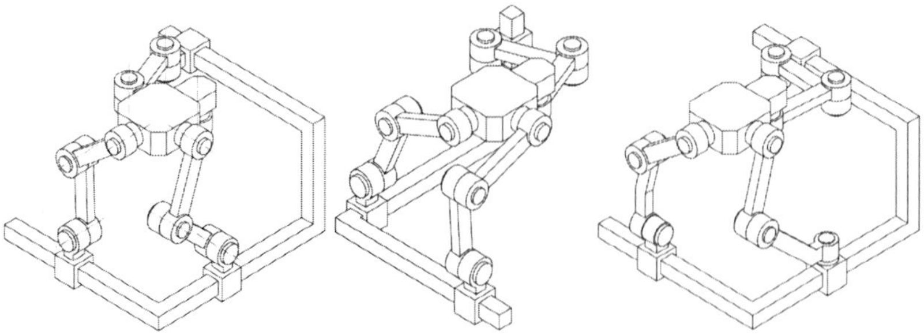

Figure 3: 3-PRRR parallel mechanisms

A particular mechanism is thus described, in addition to the basic information, by a set of parameters which define the relative position and orientation of each joint with respect to its neighbors. For complex closed-loop mechanisms, an often ignored problem is that certain parameters must take particular values or be under certain constraints in order for the mechanism to be functional and have the intended kinematic properties. In absence of these special conditions, the mechanisms may not even be assembled. More attention should be payed to these particular conditions which play a qualitative role in determining some important kinematic properties of the mechanism. For kinematic synthesis, not only do the eligible mechanisms have particular kinematic structures, but also they feature some particular relative positions and orientations between certain joints. If this particularity is not taken into account when formulating the synthesis model, a great number of mechanisms generated with the model will not have the required kinematic properties and have to be discarded. This is why the topology and geometry issue should be revisited, the special joint dispositions be investigated and an adapted definition be proposed.

Since the 1960s, a very large number of manipulator designs have been proposed in the literature or disclosed in patent files. The kinematic properties of these designs were studied mostly on a case by case basis; characteristics of their kinematic structure were often not investigated explicitly; the constraints on the relative joint locations which are essential for a manipulator to meet the kinematic requirements were rarely treated in a topology perspective.

Constraints are introduced mainly to meet the functional requirements, to simplify the kinematic model, to optimize the kinematic performances, or from manufacturing considerations. These constraints can be revealed by investigating the underlying design ideas.

For a serial manipulator to generate planar motion, all its revolute joints need to be parallel and all its prismatic joints should be perpendicular to the revolute joints. For a serial manipulator to generate spherical motion, the axes of all its revolute joints must be concurrent (McCarthy, 1990). For a parallel manipulator with three identical legs to produce only translational motion, the revolute joints of the same leg must be arranged in one or two directions (Wang, 2003).

A typical example of simplifying the kinematic model is the decoupling of the position and orientation of the EE of a 6-joint serial manipulator. This is realized by having three consecutive revolute joint axes concurrent. A comprehensive study was presented in (Ozgoren, 2002) on the inverse kinematic solutions of 6-joint serial manipulators. The study reveals how the inverse kinematic problem is simplified by making joint axes parallel, perpendicular or intersect.

Based on the analysis of the existing kinematic design, the definition of the manipulator topology and geometry is proposed as the following:

- the *kinematic composition* of a manipulator is the essential information about the number of its links, which link is connected to which other links by what types of joints and which joints are actuated;
- the *characteristic constraints* are the minimum conditions for a manipulator of given kinematic composition to have the required kinematic properties, e.g. the DOF, the DOM;
- the *topology* of a manipulator is its kinematic composition plus the characteristic constraints;
- The *geometry* of a manipulator is a set of constraints on the relative locations of its joints which are unique to each of the manipulators of the same topology.

Hence, topology also has a geometric aspect such as parallelism, perpendicularity, coplanar, and even numeric values and functions on the relative joint locations which used to be considered as geometry. By definition, geometry no longer includes relative joint locations which are common to all manipulators of the same topology because the later are the characteristic constraints and belong to the topology category. A manipulator can thus be much better characterized by its topology.

Taking the basic ideas of graph representation (Crossley, 1962; Crossley, 1965) and layout graph representation (Pierrot, 1991), we propose that the kinematic composition be represented by a diagram having the graph structure so as to be eventually adapted for automatic synthesis. The joint type is designated as an upper case letter, *i.e.*, **R** for revolute, **P** for prismatic, **H** for helical, **C** for cylindrical, **S** for spherical and **E** for planar. Actuated joints are identified by a line under the corresponding joint. The letters denoting joint types are placed at the vertices of the diagram, while the links are represented by edges. Fig. 4 and Fig. 5 are two examples of representation of kinematic composition. Each joint has two joint elements, to which element a link is connected is indicated by the presence or absence of the arrow. Any link connected to the same joint element is actually rigidly attached and no relative motion is possible. The most left column represents the base carrying three actuated revolute joints while the most right column the EE. The EE is connected to the base by three identical kinematic chains composed of three revolute joints respectively. It is noteworthy that the two different manipulators have exactly the same kinematic composition. The diagram must bear additional information in order to appropriately represent the topology.

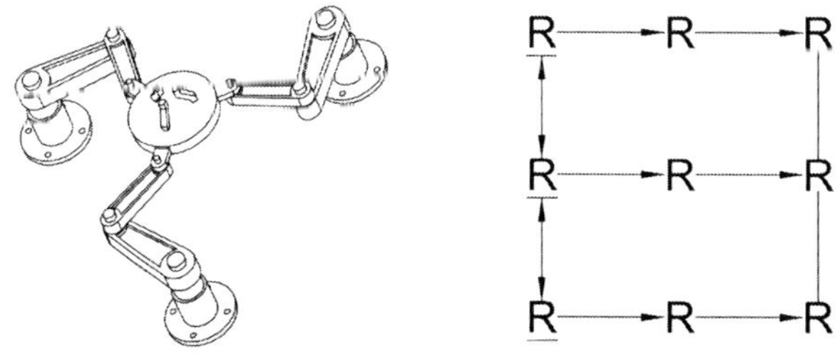

a) Physical manipulator b) Diagram

Figure 4: Kinematic Composition of a Planar 3-RRR parallel manipulator

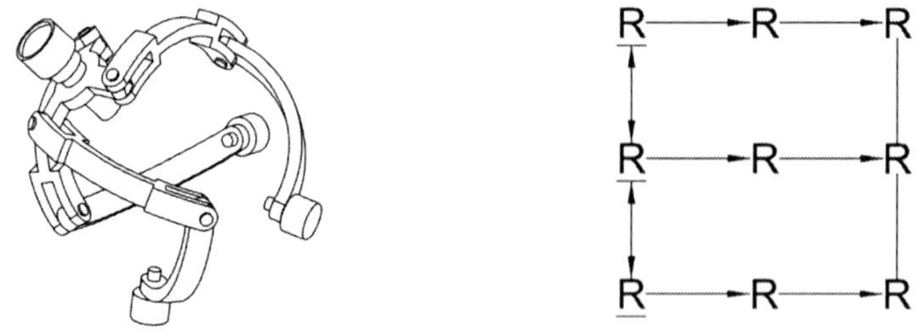

a) Physical manipulator b) Diagram

Figure 5: Kinematic Composition of a Spherical 3-$\underline{R}$RR parallel manipulator

When dealing with manipulators composed of only lower kinematic pairs, the characteristic constraints are the relative locations between lines. Constraints on relative joint axis locations can be summarized as the following six and only six possible situations shown in Fig.6. Superimposing the characteristic constraint symbols on the kinematic composition diagrams shown in Fig. 4 and 5, we get the diagrams shown in Fig. 7 and 8.

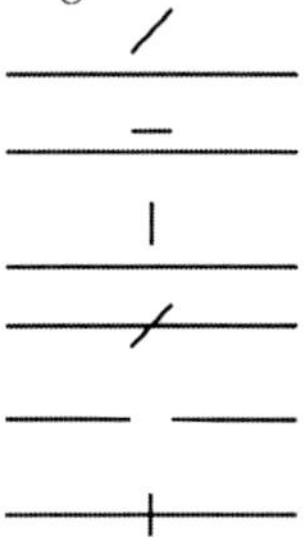

Joint axes are not parallel and do not intersect;

Joint axes are parallel and do not intersect;

Joint axes are perpendicular and do not intersect;

Joint axes intersect at an arbitrary angle;

Joint axes intersect at zero angle or aligned;

Joint axes intersect at right angle.

Figure 6: Graphic symbols for characteristic constraints

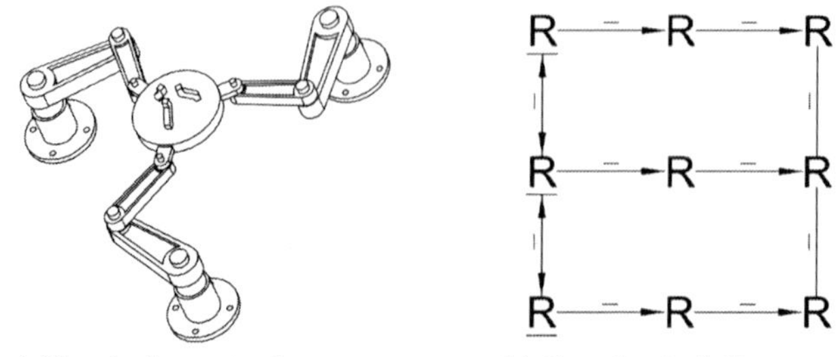

a) Physical manipulator b) Topological diagram

Figure 7: Diagram of a planar parallel manipulator with characteristic constraints

When implementing the automatic topology generation of a SM composed of only revolute and prismatic joints, the topology is represented by 6 integers, *i.e.*

- n: number of joints.
- x_0: kinematic composition. Its bits 0 to n − 1 represent respectively the joint type of joints 1 to n with 1 for revolute and 0 for prismatic.

- x_1: bits 0 to n − 2 indicate respectively whether the axes of joints 2 to n − 1 intersect the immediate preceding joint axis.
- x_2: each two consecutive bits characterize the orientation of the corresponding joint relative to the immediate preceding joint with 00 for parallel, 01 for perpendicular, and 10 for the general case.
- x_3: supplementary constraint identifying joints whose axes are concurrent. All joint axes whose corresponding bits are set to 1 are concurrent.
- x_4: supplementary constraint identifying joints whose axes are parallel. All joint axes whose corresponding bits are set to 1 are parallel.

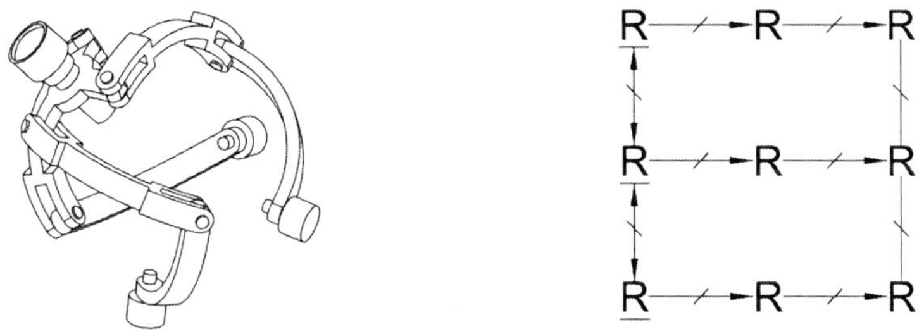

a) Physical manipulator b) Diagram

Figure 8: Diagram of a spherical parallel manipulator with characteristic constraints

With this numerical representation, topological constraint can be imposed on a general kinematic model to carry out geometric synthesis to ensure that the search is performed in designs with the intended kinematic properties. The binary form makes the representation very compact. No serial kinematic chain should have more than 3 prismatic joints, so all values for x_0 of 6 joint kinematic chains take only 42B (byte) storage. Those for x_1 take 31B while those for x_2 243B. Without supplementary constraints which are applied between non adjacent joints, the maximum number of topologies is 316386 (some topologies, those with two consecutive parallel prismatic joints for example, will not be considered for topological synthesis purpose). All topologies without supplementary constraint can be stored in a list, making the walk through quite straightforward. Applying supplementary constraints while walking through the list provides a systematic way for automatic topology generation.

4. Geometry

In the kinematic synthesis of SMs, the most successively employed geometric representation is the Denavit-Hartenberg notation (Denavit & Hartenberg, 1954). For PMs, the Denavit-Hartenberg notation is more or less adapted to suit the particularity of the manipulator being studied, especially for reducing the number of parameters and simplifying the formulation and solution of the kinematic model (Baron et al., 2002). One major problem of the later in implementing computer aided geometric synthesis is the computation of the initial configuration. Once a new set of parameters are generated, the assembly of each design take too much computation and sometimes the computation don't converge at all. This may be du to the complexity of the kinematic model or that the set of parameters correspond to no manipulator in the real domain. It also arrives that only within a subspace of the entire workspace, a particular design possesses the desired kinematic properties, making the computation useless outside the subspace. A PM (Fig. 9) presented in (Zlatanov et al., 2002) is a good example of this kind. Depending on the initial configuration, the manipulation can be a translational one or spherical one. Another problem encountered when performing computer aided synthesis is that the entire set of equations is underdetermined, while a subset of the set is overdetermined. It seems that the set of parameters correspond to no functional manipulator. But manipulators having such mathematic equations do exist. The PM shown in Fig. 10 has 8 DOF for the system on the whole and its EE has 3 DOM. The two *PRRR* legs form an overdetermined system, but the system on the whole is underdetermined.

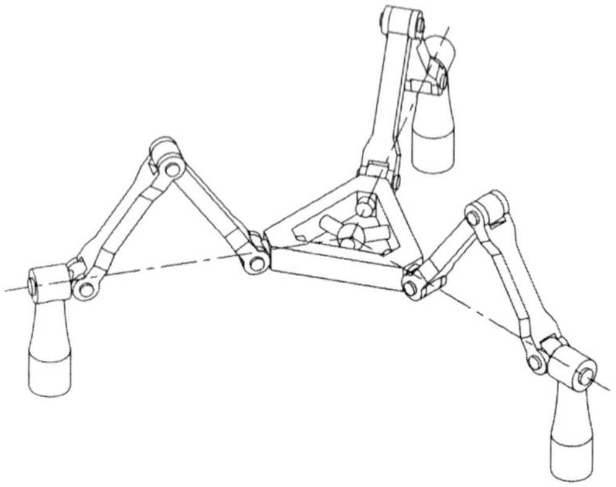

Figure 9: 3-RRRRR [28]

To improve the efficiency of the computation algorithms, an initial configuration seems to be an effective solution. So, for PMs, we proposed that the geometry definition be always accompanied by an initial configuration to start with and the evaluation computation is carried out mainly in certain neighborhood of the initial configuration.

The most challenging part of the kinematic synthesis is the integration of the topological synthesis and geometric synthesis. From the best of knowledge of the authors, the most systematic study in this regard is that presented in (Ramstein, 1999). In (Ramstein, 1999), the synthesis problem is formulated as an global optimization problem with genetic algorithms as solution tools. The joint type is represented by boolean numbers with 1 for prismatic and 0 for revolute. The synthesis results are far from what were expected. The problem is that the population does not migrate as much as expected from one topology region to another, making the synthesis concentrate on a very few topologies.

Since the joint type is represented by discrete numbers, a joint can only be either prismatic or revolute, nothing in between, which greatly limites the diversity and the migration of the solution population. With the simulated annealing techniques, similar situations have been observed by the authors.

Inspired by this observation, the basic concept of fuzzy logic and the fact that a prismatic joint is actually a revolute joint at infinity, we introduce the concept: **joint nature** which is a non negative real number to characterize the level of the "revoluteness" of a joint. This allows us to deal with the prismatic joints and the revolute ones in the same way and permit a joint to evolve between revolute and prismatic. Although a joint in between is meaningless in real application, this increases the migration channels for the solution populations and probability of finding the global optima. Before proposing the joint nature definition, it should be inspected how a revolute joint mathematically evolves toward prismatic joint.

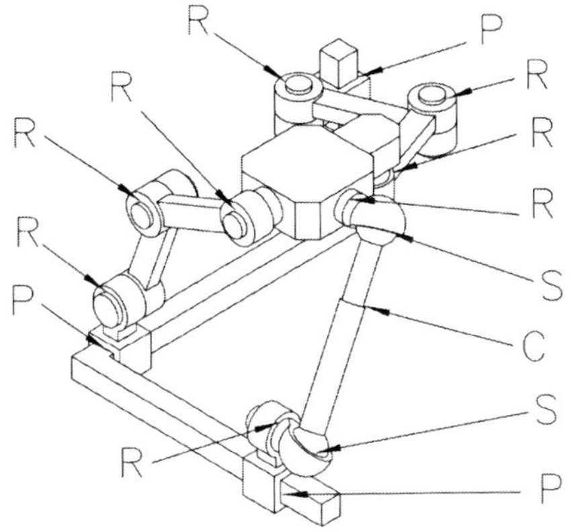

Figure 10: An overconstrained mechanism with redundant joints

Nomenclature

- b : subscript to identify the base;
- e : subscript to identify the end-effector;
- F_i : reference frame attached to *link i;*
- G_i : 3 × 3 orientation matrix of F_i with respect to F_{i-1} at the initial configuration;
- G_{hi} : 4 × 4 homogeneous orientation matrix of F_i with respect to F_{i-1} at the
- initial configuration;
- ${}^d\rho_c$: 3 × 1 position vector of the origin of F_c in F_d;
- ρ_i : 3 × 1 position vector of the origin of F_i in F_{i-1};
- p_i : 3 × 1 position vector of the origin of F_i in F_b
- A_i : 3 × 3 orientation matrix of F_i with respect to F_{i-1};
- ${}^dQ_c$: 3 × 3 orientation matrix of F_c with respect to F_d;
- Q_c : 3 × 3 orientation matrix of F_c with respect to F_b;
- $R_z(\theta)$: 3 × 3 rotation matrix about z axis with θ being the rotation angle:

$$R_z(\theta)=\begin{bmatrix}\cos(\theta) & -\sin(\theta) & 0\\ \sin(\theta) & \cos(\theta) & 0\\ 0 & 0 & 1\end{bmatrix};$$

- $\mathbf{R}_{hz}(\theta)$: 4 × 4 homogeneous rotation matrix about z axis with θ being the rotation angle;
- $\mathbf{B}_x$ (r) : 4 × 4 homogeneous translation matrix along x axis with r being the translation distance;
- $\mathbf{C}_i$: 4 × 4 homogeneous transformation matrix of F_i in F_{i-1};
- $\mathbf{H}_i$: 4 × 4 homogeneous transformation matrix of F_i in F_b;
- ${}^d\mathbf{H}_c$: 4 × 4 homogeneous transformation matrix of F_c in F_d;
- e_i : the k_{th} canonical vector which is defined as

$$e_k = \left[\underbrace{0 \dots 0}_{k-1} 1 \underbrace{0 \dots 0}_{n-k}\right]^T$$

whose dimension is implicit and depends on the context;

- ${}^{d}\mathbf{T}_c$: tangent operator of F_c in F_d expressed in F_b;
- ${}^{f,d}\mathbf{T}_c$: tangent operator of F_c in F_d expressed in F_f ;
- ${}^{d}t_c$: tangent vector of F_c in F_d expressed in F_b;
- ${}^{f,d}t_c$: tangent vector of F_c in F_d expressed in F_f ;
- t_c : tangent vector of F_c in F_b expressed in F_b.

Suppose two links coupled by a revolute joint and a reference frame is attached to each of them; at an initial configuration, the origins of the two reference frames F_{i-1} and F_i coincide; the joint axis is parallel to the z-axis of F_{i-1} and intersects the negative side of the x-axis of F_{i-1} at right angle (Fig. 11).
The relative orientation and position are given as

$$A_i = \mathbf{R}_z(\theta_i)G_i \tag{1}$$

$$\rho_i = -r_i e_1 + r_i \mathbf{R}_z(\theta_i) e_1 \tag{2}$$

$$\rho_i = \begin{bmatrix} r_i \cos(\theta_i) - r_i \\ r \ \sin(\theta_i) \\ 0 \end{bmatrix} = \begin{bmatrix} -2r_i \sin^2(\theta_i/2) \\ r_i \sin(\theta_i) \\ 0 \end{bmatrix} \tag{3}$$

Instead of taking θ_i as joint variable, we define

$$q_i = r_i \theta_i \tag{4}$$

to measure the relative pose of the two links and q_i is referred to as normalized joint variable. In addition, we define

$$w_i = \frac{1}{r_i} \tag{5}$$

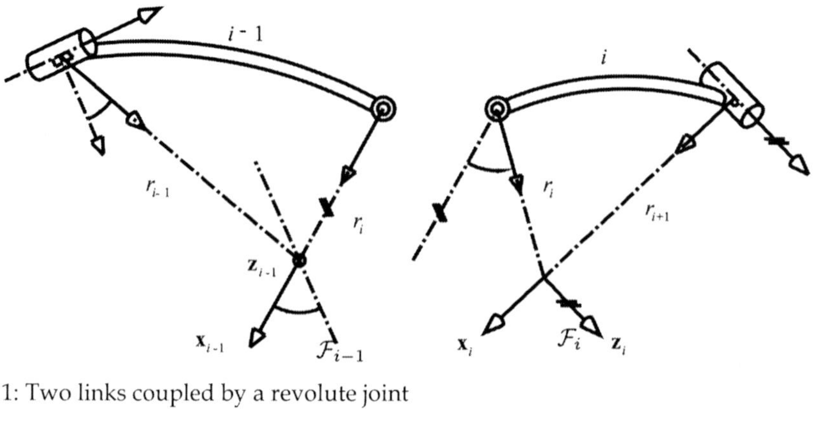

Figure 11: Two links coupled by a revolute joint

Then from equations (3), (4), and (5), we have

$$\rho_i = \begin{bmatrix} -2\sin^2(w_i q_i/2)/w_i \\ \sin(w_i q_i)/w_i \\ 0 \end{bmatrix} \quad (6)$$

It is evident that

$$\lim_{w_i \to 0} \rho_i = \begin{bmatrix} 0 \\ q_i \\ 0 \end{bmatrix} \quad (7)$$

$$\lim_{w_i \to 0} A_i = \lim_{w_i \to 0} [R_z(w_i q_i) G_i] = G_i$$

Equation (7) is just the relative pose of the two links when they are coupled by a prismatic joint. With the above formulation, revolute joints and prismatic ones can be treated in a unified way and the normalization of the joint variable is the key to achieve this.

Definition: *the nature of a joint in a kinematic chain is represented by a pair (k,w) where k is a natural number identifying its orientation from other joints, while w is a non negative number characterizing its membership to revolute joint.*

In fact, w characterizes the distance of a revolute joint with respect to the origin of the global reference and represent a prismatic joint when it is equal to 0.

The topology of a fully parallel mechanism of n-DOF is represented by n matrices with each matrix representing a subchain from the base to the end-effector:

$$\begin{bmatrix} k_{j,1} & k_{j,2} & \cdots & k_{j,m_j-1} & k_{j,m_j-1} \\ w_{j,1} & w_{j,2} & \cdots & w_{j,m_j-1} & w_{j,m_j-1} \end{bmatrix}, \text{j} = 1, 2, \ldots, n \quad (8)$$

where m_j is the total number of joints of jth subchain.

This numerical representation is aimed at simultaneous synthesis of both topology and geometry.

For geometric representation, instead of describing separately the geometry of each link, we describe an initial configuration. This is done by giving the coordinates of all joint axes with respect to the global reference frame.

Definition: *the location of a joint axis at an initial configuration is represented by a triple (* $\hat{n}$, $\hat{m}$ *,w) where* $\hat{n}$ *is a unit vector defining the orientation of the joint axis,* $\hat{m}$ *is a unit vector indicating the direction of the moment of* $\hat{n}$ *with respect to the origin of the global reference frame, w is the nature of the joint.*

It is here that the topology information is integrated into the geometric definition.

The *Plücker* coordinates of the joint axis is simply

$$l = \begin{bmatrix} w\hat{n} \\ \hat{m} \end{bmatrix} \quad (9)$$

With this representation, it should be avoided to position the joint such that its axis is too close to the origin of the global reference frame, because this will lead to parameter

singularity, that is w will approach infinity. This does not limit the representation method, because it is the relative location of the joints that defines the geometry, changing the reference frame does not change the geometry.
The topology and geometry of a fully parallel mechanism of n-DOF is represented by n matrices with each matrix representing a subchain from the base to the EE:

$$\begin{bmatrix} \hat{n}_{j,1} & \hat{n}_{j,2} & \cdots & \hat{n}_{j,m_j-1} & \hat{n}_{j,m_j} \\ \hat{m}_{j,1} & \hat{m}_{j,2} & \cdots & \hat{m}_{j,m_j-1} & \hat{m}_{j,m_j} \\ w_{j,1} & w_{j,2} & \cdots & w_{j,m_j-1} & w_{j,m_j} \end{bmatrix}, j = 1, 2, \ldots, n \tag{10}$$

where m_j is the total number of joints of jth subchain.
Those are the design parameters, they are continuous and suffer from no parameter singularity problem.

5. Kinematic modelling of general PMs

The reference frames for all links are defined at the initial configuration and this is done by following the rules given below:

1. Locate the reference frame for the EE such that no joint axis passes through its origin (Fig. 12);

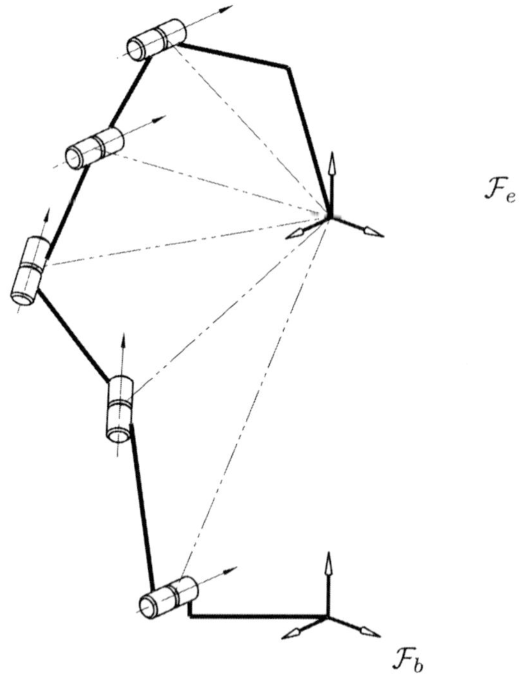

Figure 12: Frame assignment for the EE

2. Change the reference frame of the topological and geometric parameters to the EE frame: recall that ${}^b\rho_e$ and ${}^b\mathbf{Q}_e$ denote respectively the position and the orientation of the EE frame in the base frame. For every joint (the subscript is dropped off for simplicity), if ${}^bw = 0$ then

$$
\begin{aligned}
{}^e\hat{n} &= {}^e\mathbf{Q}_b\,{}^b\hat{n} \\
{}^e\hat{\boldsymbol{m}} &= {}^e\mathbf{Q}_b\,{}^b\hat{\boldsymbol{m}} \\
{}^e w &= 0
\end{aligned} \tag{11}
$$

otherwise, let P be a point on the axis, ${}^br$ and ${}^er$ denote its positions in the base frame and in the EE frame respectively, we then have

$$
\begin{aligned}
{}^e\hat{\boldsymbol{n}} &= {}^e\mathbf{Q}_b\,{}^b\hat{\boldsymbol{n}} \\
{}^e\boldsymbol{r} &= {}^e\mathbf{Q}_b({}^b\boldsymbol{r} - {}^b\boldsymbol{\rho}_e) \\
{}^e\boldsymbol{m} &= {}^e\boldsymbol{r} \times {}^e\hat{\boldsymbol{n}} = {}^e\mathbf{Q}_b({}^b\boldsymbol{r} \times {}^b\hat{\boldsymbol{n}} - {}^b\boldsymbol{\rho}_e \times {}^b\hat{\boldsymbol{n}})
\end{aligned} \tag{12}
$$

Let $[{}^b\rho_e \times]$ denote the cross product matrix associated with ${}^b\rho_e$, since

$$
{}^b\boldsymbol{r} \times {}^b\hat{\boldsymbol{n}} = {}^b\boldsymbol{m} = {}^b\hat{\boldsymbol{m}}/{}^bw \tag{13}
$$

by substituting equation (13) into (12), we have

$$
{}^e\boldsymbol{m} = -{}^e\mathbf{Q}_b\left[{}^b\boldsymbol{\rho}_e\times\right]{}^b\hat{\boldsymbol{n}} + {}^e\mathbf{Q}_b\,{}^b\hat{\boldsymbol{m}}/{}^bw \tag{14}
$$

then, the *Plücker* coordinates of the axis in the EE frame can be computed as

$$
\begin{bmatrix} {}^e\hat{\boldsymbol{n}} \\ {}^e\boldsymbol{m} \end{bmatrix} = \begin{bmatrix} {}^e\mathbf{Q}_b & \mathbf{O} \\ -{}^e\mathbf{Q}_b\left[{}^b\boldsymbol{\rho}_e\times\right] & {}^e\mathbf{Q}_b \end{bmatrix} \begin{bmatrix} {}^b\hat{\boldsymbol{n}} \\ {}^b\hat{\boldsymbol{m}}/{}^bw \end{bmatrix} \tag{15}
$$

Finally, ${}^ew = 1/\left\|{}^em\right\|_2$ and ${}^e\hat{m} = {}^em/{}^ew$.

3. Links of subchain j from the base to the EE are identified by *link*(j, 0) to *link*(j,m_j), the base being *link*(j, 0) and the EE being *link*(j,m_j); joint coupling *link*(j, i-1) and *link*(j, i) is identified by *joint*(j, i); frame $F_{j,i}$ is attached to *link*(j, i)(*Fig.* 13); the base and the EE have multiple rigidly attached frames with each of them corresponding to an individual subchain;
4. The reference frame for *link*(j, i) is defined such that

$$
{}^e\mathbf{Q}_{j,i} = \left[\ {}^e\hat{\boldsymbol{m}}_{j,i+1} \times {}^e\hat{\boldsymbol{n}}_{j,i+1},\ \ {}^e\hat{\boldsymbol{m}}_{j,i+1},\ \ {}^e\hat{\boldsymbol{n}}_{j,i+1}\ \right] \tag{16}
$$

$$
{}^e\rho_{j,i} = 0 \tag{17}
$$

the z-axis of $F_{j,i}$ being parallel to the axis of *joint*(j, i + 1) and the x-axis intersecting the the axis of *joint*(j, i + 1) and pointing from the intersecting point to the origin of the EE frame (Fig. 14). The y-axis is determined as usual by the right-hand rule.

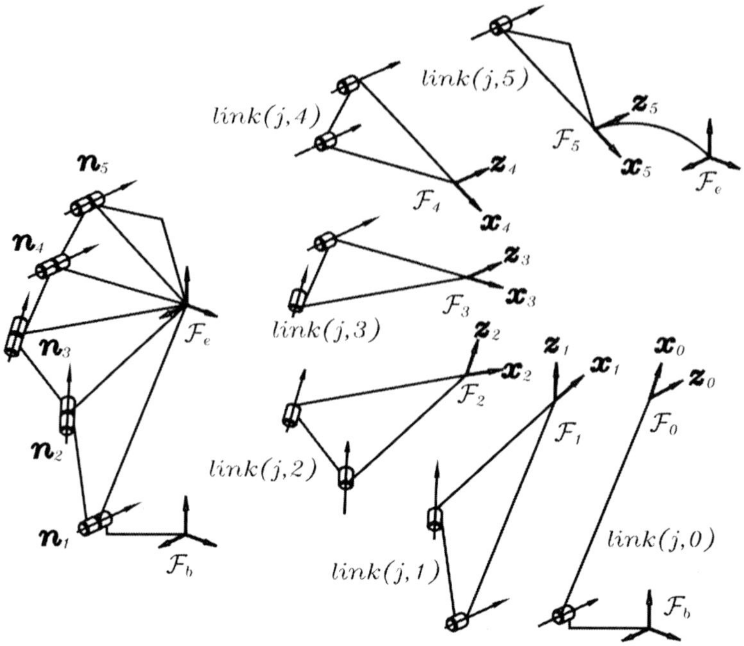

Figure 13: Link reference frames

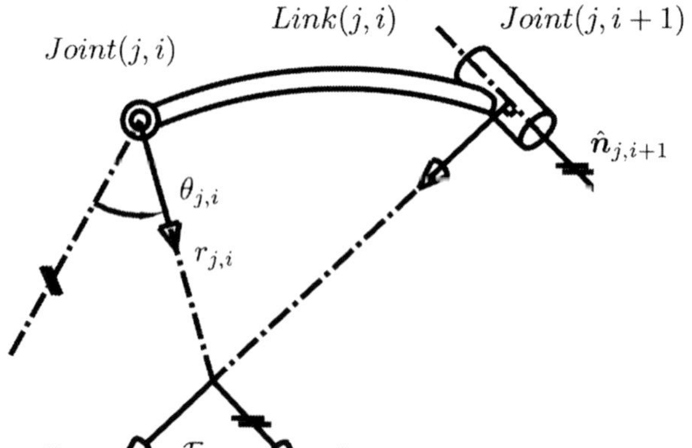

Figure 14: Reference frame definition for *link(i, j)*

5. The normalized joint variable of *joint(j, i)* is denoted by $q_{j,i}$, the rotation angle with respect to the initial configuration is denoted by $\theta_{j,i}$ and

$$\theta_{j,i} = w_{j,i} q_{j,i} \tag{18}$$

6. Compute the link geometry matrices from ${}^{b}Q_{e}$, ${}^{e}Q_{j,0}$, $\cdots$, and ${}^{e}Q_{j,mj}$:
for $G_{j,1}$ to $G_{j,mj-1}$

$$G_{i,j} = {}^{j,i-1}Q_e\,{}^{e}Q_{j,i} \tag{19}$$

$G_{j,0}$, $G_{j,mj}$, and $G_{j,e}$ are treated differently, *i.e.*

$$G_{j,0} = {}^{b}Q_e\,{}^{e}Q_{j,0} \tag{20}$$

$$G_{j,mj} = 1 \tag{21}$$

$$G_{j,e} = {}^{j,mj}Q_e \tag{22}$$

The sequence of links in each subchain has a corresponding sequence of homogeneous transformations that defines the pose of each link relative to its neighbor in the chain. The pose of the EE is therefore constrained by the product of these transformations through every subchain. With the above frame assignment, the pose of *link*(*j*, *i*) with respect to *link*(*j*, *i* − 1) is given as

$$\mathbf{C}_{j,i} = \mathbf{B}_{\mathbf{x}}(-\frac{1}{w_{j,i}})\mathbf{R}_{\mathbf{hz}}(w_{j,i}q_{j,i})\mathbf{B}_{\mathbf{x}}(\frac{1}{w_{j,i}})\mathbf{G}_{\mathbf{h}j,i} \tag{23}$$

The corresponding 3 × 3 orientation matrix is given as

$$\mathbf{A}_{j,i} = \mathbf{R}_{\mathbf{z}}(w_{j,i}q_{j,i})\mathbf{G}_{j,i} \tag{24}$$

The corresponding position is given as

$$\boldsymbol{\rho}_{j,i} = -\frac{\boldsymbol{e}_1}{w_{j,i}} + \mathbf{R}_{\mathbf{z}}(w_{j,i}q_{j,i})\frac{\boldsymbol{e}_1}{w_{j,i}} \tag{25}$$

This leads to

$$\boldsymbol{\rho}_{j,i} = \begin{bmatrix} \frac{1}{w_{j,i}}\cos(w_{j,i}q_{j,i}) - \frac{1}{w_{j,i}} \\ \frac{1}{w_{j,i}}\sin(w_{j,i}q_{j,i}) \\ 0 \end{bmatrix} = \begin{bmatrix} -\frac{2}{w_{j,i}}\sin^2\left(\frac{w_{j,i}q_{j,i}}{2}\right) \\ \frac{1}{w_{j,i}}\sin(w_{j,i}q_{j,i}) \\ 0 \end{bmatrix} \tag{26}$$

When $w_{j,i}$ approaches 0, we have

$$\lim_{w_{j,i}\to 0}\mathbf{A}_{j,i} = \mathbf{G}_{j,i} \tag{27}$$

$$\lim_{w_{j,i}\to 0}\boldsymbol{\rho}_{j,i} = \begin{bmatrix} 0 \\ q_{j,i} \\ 0 \end{bmatrix} \tag{28}$$

This corresponds to the situation of a prismatic joint.
The pose of the EE under the structure constraint of subchain *j* is

$$\mathbf{H}_e = \mathbf{H}_{j,0}\left(\prod_{i=1}^{m_j}\mathbf{C}_{j,i}\right)\mathbf{C}_{j,e},\ i = 1,\ 2,\ \cdots,\ m_j \tag{29}$$

In terms of orientation and position, equation (29) can be written as

$$\mathbf{Q}_e = \mathbf{Q}_{j,\,0}\left(\prod_{i=1}^{m_j}\mathbf{A}_{j,\,i}\right)\mathbf{A}_{j,\,e},\ i = 1,\ 2,\ \cdots,\ m_j \tag{30}$$

$$\mathbf{Q}_{j,\,i} = \mathbf{Q}_{j,\,0}\prod_{k=1}^{i}\mathbf{A}_{j,\,k},\ k = 1,\ 2,\ \cdots,\ i \tag{31}$$

$$\boldsymbol{p}_{j,\,i} = \boldsymbol{p}_{j,\,0} + \sum_{k=1}^{i}\left(\mathbf{Q}_{j,\,k-1}\boldsymbol{\rho}_{j,\,k}\right),\ k = 1,\ 2,\ \cdots,\ i \tag{32}$$

$$\boldsymbol{p}_{j,\,e} = \boldsymbol{p}_{j,\,0} + \sum_{i=1}^{m_j}\left(\mathbf{Q}_{j,\,i-1}\boldsymbol{\rho}_{j,\,i}\right) + \mathbf{Q}_{j,\,m_j}\boldsymbol{\rho}_{j,\,e},\ i = 1,\ 2,\ \cdots,\ m_j \tag{33}$$

Equations (31) and (32) are used to compute the orientation and position of links other than the base and the EE.

For a PM of n degree of freedom, the n subchains are closed by rigidly attaching together their fist link frames and last link frames respectively. The structure equations are obtained by equating the transformation products defined by equation (29) of all subchains, *i.e.*, $\forall\, j, k = 1, 2, \cdots, n$ and $j \neq k$

$$\mathbf{H}_{j,\,0}\left(\prod_{i=1}^{m_j}\mathbf{C}_{j,\,i}\right)\mathbf{C}_{j,\,e} = \mathbf{H}_{k,\,0}\left(\prod_{i=1}^{m_k}\mathbf{C}_{k,\,i}\right)\mathbf{C}_{k,\,e} \tag{34}$$

It is obvious that this kinematic formulation is not aimed at simplifying the forward or inverse kinematic solutions, but for the simultaneous topological and geometric synthesis with numeric method, genetic algorithms in particular. The initial population will be generated using the numeric topological representation proposed in Section 3 and the reproduction performed while respecting the characteristic constraints. The implementation of the synthesis for translational PMs is being carried out in our laboratory.

6. Conclusion

By introducing characteristic constraints, kinematic chains of serial and parallel manipulators can be better characterized. This is essential for both topology synthesis and geometry synthesis. On the one hand, topology synthesis of spatial manipulator is no longer dimension-independent; most of the topology syntheses are actually the search for some special geometric constraints which play a key role in determining the fundamental kinematic properties. On the other hand, it is necessary to identify the characteristic constraints when performing geometry synthesis in order for the design space to correspond appropriately to the manipulators having the intended kinematic properties. The graph structure of the proposed topological representation makes it possible to implement computer algorithms in order to perform systematic enumeration, comparison and classification of serial and parallel manipulators. The geometric representation is well adapted for computer aided simultaneous topological and geometric synthesis by introducing the concepts of initial configuration and the joint nature, making it possible to represent revolute joints and prismatic joints in a unified way. Then a singularity-free parametrization of both topology and geometry was proposed. After that, joint variables were normalized, which enables the joint type to be seamlessly incorporated into kinematic model, it is no longer necessary to reformulate the kinematic model when a revolute joint is replaced by a prismatic one or vice versa. The effectiveness of the propose kinematic modelling remains to be evaluated.

7. Acknowledgment

The authors acknowledge the financial support of NSERC (National Science and Engineering Research Council of Canada) under grants OGPIN-203618 and RGPIN- 138478.

8. References

L.-W. Tsai, *Mechanism design: enumeration of kinematic structures according to function*. Mechanical engineering series: CRC mechanical engineering series, CRC Press, 2001.

J.-P. Merlet, *Les robots paralleles*. Paris: Hermes, c1997.

K. H. Hunt, "Geometry of robotics devices," *Mechanical Engineering Transactions*, vol. 7, no. 4, pp. 213–220, 1982. Record Number: 2700 1982.

K. H. Hunt, "Structural kinematics of in parallel actuated robot arms," 1982. Record Number: 2710 Proceedings Title: Design and Production Engineering Technical Conference Place of Meeting: Washington.

I. Powell, "The kinematic analysis and simulation of the parallel topology manipulator," *Marconi Rev. (UK)*, vol. 45, no. 226, pp. 121 - 38, 1982.

F. Freudenstein and E. Maki, "On a theory for the type synthesis of mechanism," in *Proceedings of the 11th International Congress of Applied Mechanics*, (Springer, Berlin), pp. 420–428, 1965.

D. Yang and T. Lee, "Feasibility study of a platform type of robotic manipulator from a kinematic viewpoint," *Journal of Mechanisms, Transmissions and Automation in Design*, vol. 106, pp. 191–198, 1984.

T. Balkan, M. Kemal Ozgoren, M. Sahir Arikan, and H. Murat Baykurt, "A kinematic structure-based classification and compact kinematic equations for sixdof industrial robotic manipulators," *Mechanism and Machine Theory*, vol. 36, no. 7, pp. 817 -832, 2001.

H. Su, C. Collins, and J. McCarthy, "Classification of rrss linkages," *Mechanism and Machine Theory*, vol. 37, no. 11, pp. 1413 - 1433, 2002.

C. M. Gosselin, R. Ricard, and M. A. Nahon, "Comparison of architectures of parallel mechanisms for workspace and kinematic properties," *American Society of Mechanical Engineers, Design Engineering Division (Publication) DE*, vol. 82, no. 1, pp. 951 - 958, 1995.

L.-W. Tsai and S. Joshi, "Comparison study of architectures of four 3 degree offreedom translational parallel manipulators," *Proceedings - IEEE International Conference on Robotics and Automation*, vol. 2, pp. 1283 - 1288, 2001.

J. Denavit and R. S. Hartenberg, "Kinematic notation for lower-pair mechanisms based on matrices," in *American Society of Mechanical Engineers (ASME)*, 1954.

P. Chen and B. Roth, "A unified theory for finitely and infinitesimally seperated position problems of kinematic synthesis," *ASME Journal of Engineering for Industry, Series B*, vol. 91, pp. 203–208, 1969. Record Number: 2800.

P. Chedmail, "Optimization of multi-dof mechanisms," in *Computational Methods in Mechanisms System* (J. Angeles and E. Zakhariev, eds.), pp. 97–130, Springer Verlag, 1998.

F. C. Park and J. E. Bobrow, "Geometric optimization algorithms for robot kinematic design," *Journal of Robotic Systems*, vol. 12, no. 6, pp. 453 - 463, 1995.

J. Angeles, *Fundamentals of robotic mechanical systems : theory, methods, and algorithms*. Mechanical engineering series: Mechanical engineering series (Springer), New York: Springer, 2nd ed. ed., c2003.

IFToMM, "Iftomm terminology," *Mechanism and Machine Theory*, vol. 38, pp. 913–912, 2003.

J. Angeles, *Spatial Kinematic Chains. Analysis, Synthesis, Optimization*. Berlin: Springer-Verlag, 1982.

L. Baron, *Contributions to the estimation of rigid-body motion under sensor redundancy*. PhD thesis, McGill University, c1997.

G. Bennett, "A new mechanism," *Engineering*, 1903.

J. M. McCarthy, *An Introduction to Theoretical Kinematics*. Cambridge, Massachusetts, London, England: The MIT Press, 1990.

X. Wang, L. Baron, and G. Cloutier, "Design manifold of translational parallel manipulators," in *Proceedings of 2003 CCToMM Symposium on Mechanisms, Machines, and Mechatronics* (l'Agence spatiale canadienne, ed.), (Montreal, Quebec, Canada), pp. 231-239, 2003.

M. Ozgoren, "Topological analysis of 6-joint serial manipulators and their inverse kinematic solutions," *Mechanism and Machine Theory*, vol. 37, no. 5, pp. 511 - 547, 2002.

F. Crossley, "Contribution to gruebler's theory in number synthesis of plane mechanisms," *American Society of Mechanical Engineers - Papers*, pp. 5 –, 1962.

F. Crossley, "Permutations of kinematic chains of eight members or less from graph - theoretic viewpoint," *Developments in Theoretical and Applied Mechanics*, vol. 2, pp. 467 - 486, 1965.

F. Pierrot, *Robots pleinement paralleles legers: Conception, Modelisation et Commande*. PhD thesis, Universite Montpellier II, Montpellier, France, 1991.

L. Baron, X. Wang, and G. Cloutier, "The isotropic conditions of parallel manipulators of delta topology," in *Advances in Robot Kinematics, Theory and Applications* (J. Lenarcic and F. Thomas, eds.), pp. 357–367, Kluwer Academic Publishers, 2002.

D. Zlatanov, I. Bonev, and C. Gosselin, "Constraint singularities as configuration space singularities." ParalleMIC - the Parallel Mechanisms Information Center, http://www.parallemic.org/Reviews/Review008.html, 2002.

E. Ramstein, *Contribution a la formation generale d'un probleme de synthese de mcanismes et resolution*. PhD thesis, Universite de Nantes, Ecole doctorale science pour l'ingenieur de Nantes, France, 1999.

5

Conserving Integrators for Parallel Manipulators

Stefan Uhlar and Peter Betsch
Chair of Computational Mechanics, Department of Mechanical Engineering, University of Siegen
Germany

1. Introduction

The present work deals with the development of time stepping schemes for the dynamics of parallel manipulators. In particular, we aim at energy and momentum conserving algorithms for a robust time integration of the differential algebraic equations (DAEs) which govern the motion of closed-loop multibody systems. It is shown that a rotationless formulation of multibody dynamics is especially well-suited for the design of energy-momentum schemes. Joint coordinates and associated forces can still be used by applying a specific augmentation technique which retains the advantageous algorithmic conservation properties. It is further shown that the motion of a manipulator can be partially controlled by appending additional servo constraints to the DAEs.

Starting with the pioneering works by Simo and co-workers [SW91, STW92, ST92], energy-momentum conserving schemes and energy-decaying variants thereof have been developed primarily in the context of nonlinear finite element methods. In this connection, representative works are due to Brank et al. [BBTD98], Bauchau & Bottasso [BB99], Crisfield & Jelenić [CJ00], Ibrahimbegović et al. [IMTC00], Romero & Armero [RA02], Betsch & Steinmann [BS01a], Puso [Pus02], Laursen & Love [LL02] and Armero [Arm06], see also the references cited in these works.

Problems of nonlinear elastodynamics and nonlinear structural dynamics can be characterized as stiff systems possessing high frequency contents. In the conservative case, the corresponding semi-discrete systems can be classified as finite-dimensional Hamiltonian systems with symmetry. The time integration of the associated nonlinear ODEs by means of energy-momentum schemes has several advantages. In addition to their appealing algorithmic conservation properties energy-momentum schemes are known to possess enhanced numerical stability properties (see Gonzalez & Simo [GS96]). Due to these advantageous properties energy-momentum schemes have even been successfully applied to penalty formulations of multibody dynamics, see Goicolea & Garcia Orden [GGO00]. Indeed, the enforcement of holonomic constraints by means of penalty methods again yields stiff systems possessing high frequency contents. The associated equations of motion are characterized by ODEs containing strong constraining forces. In the limit of infinitely large penalty parameters these ODEs replicate Lagrange's equations of motion of the first kind (see Rubin & Ungar [RU57]), which can be identified as index-3 differential-algebraic equations (DAEs). This observation strongly supports the expectation that energy-momentum methods are also beneficial to the discretization of index-3 DAEs (see G´eradin & Cardona [GC01, Chapter 12] and Leyendecker et al. [LBS04]).

The specific formulation of the equations of motion strongly affects the subsequent time discretization. In the context of multibody systems the main distinguishing feature of alternative formulations is the choice of coordinates for the description of the orientation of the individual rigid bodies. For this purpose some kind of rotational variables (e.g. joint-angles, Euler angles or other 3-parameter representations of finite rotations) are often employed. In general, the equations of motion in terms of rotational variables are quite cumbersome. In the case of systems with tree structure one is typically confronted with highly-nonlinear ODEs. Further challenges arise in the case of closed-loop systems due to the presence of algebraic loop-closure constraints leading to index-3 DAEs. As a consequence of their inherent complexity, the design of energy-momentum conserving schemes is hardly conceivable for formulations of general multibody systems involving rotations.

In the present work the use of rotational variables is completely circumvented in the formulation of the equations of motion. Our formulation turns out to be especially well-suited for the energy-momentum conserving integration of both open-loop and closed-loop multibody systems. In our approach the orientation of each rigid body is characterized by the elements of the rotation matrix (or the direction cosine matrix). This leads to a set of redundant coordinates which are subject to holonomic constraints. In this connection two types of constraints may be distinguished (see also Betsch & Steinmann [BS02b]): (i) Internal constraints which are intimately connected to the assumption of rigidity and, (ii) external constraints due to the interconnection of the bodies constituting the multibody system. Item (ii) implies that loop-closure constraints can be taken into account without any additional difficulty. The resulting DAEs exhibit a comparatively simple structure which makes possible the design of energy-momentum conserving schemes. Another advantage of the present rotationless formulation of multibody systems lies in the fact that planar motions as well as spatial motions can be treated without any conceptual differences. That is, the extension from the planar case to the full three-dimensional case can be accomplished in a straightforward way, which is in severe contrast to formulations employing rotations, due to their non-commutative nature in the three-dimensional setting. It is worth mentioning that the present rotationless approach resembles to some degree the natural coordinates formulation advocated by Garcia de Jalon et al. [JUA86].

As pointed out above the rotationless formulation of multibody systems benefits the design of energy-momentum schemes. On the other hand, the advantages for the discretization come at the expense of a comparatively large number of unknowns. In addition to that, joint-angles and associated torques are often required in practical applications, for example, if a joint is actuated. The size of the algebraic system to be solved can be systematically reduced by applying the discrete null space method developed in [Bet05a]. Indeed, the present treatment of planar multibody dynamics fits into the framework proposed in [BL06,LBS]. The main new contributions presented herein are (i) a coordinate augmentation technique which facilitates to incorporate rotational degrees of freedom along with associated torques and, (ii) the incorporation of control constraints in order to perform a controled movement of fully and underactuated multibody systems.

An outline of the rest of the paper is as follows: In Section 2 the formulation of constrained mechanical systems is outlined and the energy-momentum conserving discretization is introduced. Section 3 contains the advocated description of rigid bodies in terms of redundant coordinates. Section 4 deals with two basic kinematic pairs, i.e. the revolute and prismatic pair as building blocks of multibody systems. In addition to that, the newlyproposed coordinate augmentation technique for the incorporation of joint coordinates and associated torques or forces is presented. The application of the above mentioned features will be carried out with the example of a planar parallel manipulator of RPR type (Section 5). Conclusions are drawn in Section 6.

2. Dynamics of constrained mechanical systems

In the present work we focus on discrete mechanical systems subject to constraints which are holonomic and scleronomic. Due to the specific formulation of rigid bodies (see Section 3) the equations of motion for multibody systems can be written in the form

$$\boxed{\begin{aligned} \dot{\boldsymbol{q}} - \boldsymbol{v} &= \boldsymbol{0} \\ \boldsymbol{M}\dot{\boldsymbol{v}} - \boldsymbol{F} + \boldsymbol{G}^T\boldsymbol{\lambda} &= \boldsymbol{0} \\ \boldsymbol{\Phi}(\boldsymbol{q}) &= \boldsymbol{0} \end{aligned}} \tag{1}$$

where $\mathbf{q}(t) \in \mathbb{R}^n$ specifies the configuration of the mechanical system at time t, and $\mathbf{v}(t) \in \mathbb{R}^n$ is the velocity vector. Together $(\mathbf{q}, \mathbf{v})$ form the vector of state space coordinates (see, for example, Rosenberg [Ros77]). A superposed dot denotes differentiation with respect to time and $\mathbf{M} \in \mathbb{R}^{n\times n}$ is a *constant* and symmetric mass matrix, so that the kinetic energy can be written as

$$T(\boldsymbol{v}) = \frac{1}{2}\boldsymbol{v} \cdot \boldsymbol{M}\boldsymbol{v} \tag{2}$$

Moreover, $\mathbf{F} \in \mathbb{R}^n$ is a load vector which in the present work is decomposed according to

$$\boldsymbol{F} = \boldsymbol{Q} - \nabla V(\boldsymbol{q}) \tag{3}$$

Here, $V(\mathbf{q}) \in \mathbb{R}$ is a potential energy function and $\mathbf{Q} \in \mathbb{R}^n$ accounts for loads which can not be derived from a potential. Moreover, $\phi(\mathbf{q}) \in \mathbb{R}^m$ is a vector of geometric constraint functions, $\mathbf{G} = D\,\phi(\mathbf{q}) \in \mathbb{R}^{m\times n}$ is the constraint Jacobian and $\boldsymbol{\lambda} \in \mathbb{R}^m$ is a vector of multipliers which specify the relative magnitude of the constraint forces. In the above description it is tacitly assumed that the m constraints are independent.

Due to the presence of holonomic (or geometric) constraints $(1)_3$, the configuration space of the system is given by

$$\mathrm{Q} = \{\boldsymbol{q}(t) \in \mathbb{R}^n | \, \boldsymbol{\Phi}(\boldsymbol{q}) = \boldsymbol{0}\} \tag{4}$$

The equations of motion (1) form a set of index-3 differential-algebraic equations (DAEs) (see, for example, Kunkel & Mehrmann [KM06]). They can be directly derived from the classical Lagrange's equations.

2.1 Energy-momentum discretization

'Experience indicates that the best results can generally be obtained using a direct discretization of the equations of motion.' Leimkuhler & Reich [LR04, Sec. 7.2.1]

2.1.1 The basic energy-momentum scheme

For the direct discretization of the DAEs (1), we employ the methodology developed by Gonzalez [Gon99]. Consider a representative time interval $[t_n, t_{n+1}]$ with time step $\Delta t = t_{n+1} - t_n$, and given state space coordinates $\mathbf{q}_n \in Q$, $v_n \in \mathbb{R}^n$ at t_n. The discretized version of (1) is given by

$$\begin{aligned} \boldsymbol{q}_{n+1} - \boldsymbol{q}_n &= \frac{\Delta t}{2}(\boldsymbol{v}_n + \boldsymbol{v}_{n+1}) \\ \boldsymbol{M}(\boldsymbol{v}_{n+1} - \boldsymbol{v}_n) &= \Delta t \mathbf{F}(\boldsymbol{q}_n, \boldsymbol{q}_{n+1}) - \Delta t \mathbf{G}(\boldsymbol{q}_n, \boldsymbol{q}_{n+1})^T \bar{\boldsymbol{\lambda}} \\ \boldsymbol{\Phi}(\boldsymbol{q}_{n+1}) &= \mathbf{0} \end{aligned} \tag{5}$$

with

$$\mathbf{F}(\boldsymbol{q}_n, \boldsymbol{q}_{n+1}) = \mathbf{Q}(\boldsymbol{q}_n, \boldsymbol{q}_{n+1}) - \bar{\nabla} V(\boldsymbol{q}_n, \boldsymbol{q}_{n+1}) \tag{6}$$

In the sequel, the algorithm (5) will be called the basic energy-momentum (***BEM***) scheme. The advantageous algorithmic conservation properties (see Remark 2.1 below) of the BEM scheme are linked to the notion of a discrete gradient (or derivative) of a function $f: \mathbb{R}^n \rightarrow \mathbb{R}$. In the present work $\overline{\nabla} f(\mathbf{q}_n, \mathbf{q}_{n+1})$ denotes the discrete gradient of f. It is worth mentioning that if f is at most quadratic then the discrete gradient coincides with the standard gradient evaluated in the mid-point configuration $\mathbf{q}_{n+1/2} = (\mathbf{q}_n + \mathbf{q}_{n+1})/2$, that is, in this case $\overline{\nabla} f(\mathbf{q}_n, \mathbf{q}_{n+1}) = \nabla f(\mathbf{q}_{n+1/2})$. In $(5)_2$ the discrete gradient is applied to the potential energy function V as well as to the constraint functions ϕ_i. In particular, the discrete constraint Jacobian is given by

$$\mathbf{G}(\boldsymbol{q}_n, \boldsymbol{q}_{n+1})^T = \left[\bar{\nabla}\Phi_1(\boldsymbol{q}_n, \boldsymbol{q}_{n+1}), \ldots, \bar{\nabla}\Phi_m(\boldsymbol{q}_n, \boldsymbol{q}_{n+1})\right] \tag{7}$$

Concerning (6), for the present purposes it suffices to set $\mathbf{Q}(\mathbf{q}_n, \mathbf{q}_{n+1}) = \mathbf{Q}(\mathbf{q}_{n+1/2})$. The BEM scheme can be used to determine $\mathbf{q}_{n+1} \in \mathbf{Q}$, $v_{n+1} \in \mathbb{R}^n$ and $\bar{\lambda} \in \mathbb{R}^m$. To this end, one may substitute for v_{n+1} from $(5)_1$ into $(5)_2$ and then solve the remaining system of nonlinear algebraic equations for the $n + m$ unknowns $(\mathbf{q}_{n+1}, \bar{\lambda})$. We refer to [Bet05a] for further details of the implementation.

Remark 2.1 The algorithm (5) inherits fundamental mechanical properties from the underlying continuous formulation such as (i) conservation of energy, and (ii) conservation of momentum maps that are at most quadratic in (q, v). While algorithmic conservation of linear momentum is a trivial matter, algorithmic conservation of angular momentum and total energy is made possible by the specific formulation of rigid bodies and multibody systems proposed in the present work.

3. The planar rigid body

In the present work we make use of six redundant coordinates for the description of the placement of the planar rigid body. In particular, the vector of redundant coordinates is given by

$$\boldsymbol{q} = \begin{bmatrix} \boldsymbol{\varphi} \\ \boldsymbol{d}_1 \\ \boldsymbol{d}_2 \end{bmatrix} \tag{8}$$

where $\boldsymbol{\varphi} \in \mathbb{R}^2$ is the position vector of the center of mass and $\mathbf{d}_\alpha \in \mathbb{R}^2, \alpha \in \{1, 2\}$, are two directors which specify the orientation of the rigid body (Fig. 1). In the sequel, all of the coordinates in (8) are referred to a right-handed orthonormal basis $\{\mathbf{e}_1, \mathbf{e}_2\}$, which plays the role of an inertial frame. The directors are assumed to constitute a right-handed body frame which coincides with the principal axis of the rigid body. Since the directors are fixed in the body and moving with it, they have to stay orthonormal for all times $t \in \mathbb{R}^+$. This gives rise to three independent geometric (or holonomic) constraints $\phi^i_{\text{int}}(\mathbf{q}) = 0$, which may be termed internal constraints since they are intimately connected with the assumption of rigidity. The functions $\phi^i_{\text{int}} : \mathbb{R}^6 \rightarrow \mathbb{R}$ may be arranged in the vector of internal constraint functions

$$\boldsymbol{\Phi}_{int}(\boldsymbol{q}) = \begin{bmatrix} \frac{1}{2}[\boldsymbol{d}_1^T \boldsymbol{d}_1 - 1] \\ \frac{1}{2}[\boldsymbol{d}_2^T \boldsymbol{d}_2 - 1] \\ \boldsymbol{d}_1^T \boldsymbol{d}_2 \end{bmatrix} \tag{9}$$

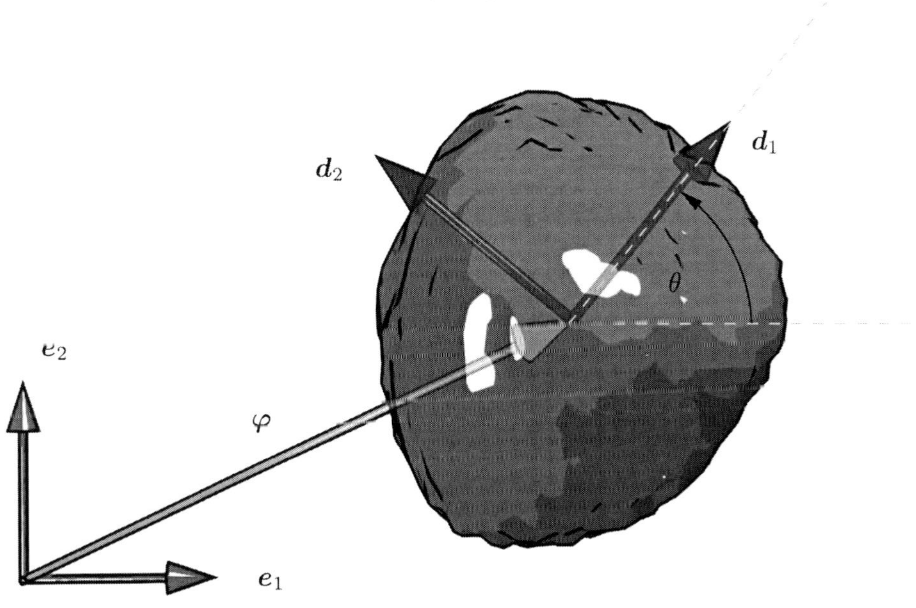

Figure 1: The planar rigid body.

With regard to the internal constraints the configuration space of the free rigid body may now be written in the form

$$Q_{\text{free}} = \{\boldsymbol{q}(t) \in \mathbb{R}^6 \mid \boldsymbol{\Phi}_{int}(\boldsymbol{q}) = \boldsymbol{0}\,,\ (\boldsymbol{d}_1 \times \boldsymbol{d}_2) \cdot \boldsymbol{e}_3 = +1\} \tag{10}$$

Note that the director frame $\{\mathbf{d}_1, \mathbf{d}_2\}$ can be connected with a rotation matrix $\mathbf{R} \in SO(2)$, through the relationship $\mathbf{d}_\alpha = \mathbf{R}e_\alpha$. In this connection,

$$SO(2) = \{\boldsymbol{R} \in \mathbb{R}^{2\times 2} \mid \boldsymbol{R}^T\boldsymbol{R} = \boldsymbol{I}_2\,,\ \det \boldsymbol{R} = +1\} \tag{11}$$

is the special orthogonal group of $\mathbb{R}^2$. Accordingly, $\mathbf{R}_{\alpha\beta} = e_\alpha \cdot d_\beta$, such that the directors coincide with the columns of the rotation matrix. Alternatively, the configuration space of the free rigid body may be written as

$$Q_{\text{free}} = \mathbb{R}^2 \times SO(2) \subset \mathbb{R}^6$$

The motion of the free rigid body can now be described by means of the DAEs (1). To this end, we have to provide the mass matrix $\mathbf{M} \in \mathbb{R}^{6\times 6}$, which is given by

$$\boldsymbol{M} = \begin{bmatrix} M\boldsymbol{I}_2 & \boldsymbol{0} & \boldsymbol{0} \\ \boldsymbol{0} & E_1\boldsymbol{I}_2 & \boldsymbol{0} \\ \boldsymbol{0} & \boldsymbol{0} & E_2\boldsymbol{I}_2 \end{bmatrix} \tag{12}$$

Here, M is the total mass of the rigid body and E_1, E_2 are the principal values of the Euler tensor relative to the center of mass. With respect to a reference configuration β with material points $X = (X_1, X_2) \in \beta$ these quantities are given by

$$\begin{aligned} M &= \int_{\mathcal{B}} \varrho(X)\, d^2X \\ E_\alpha &= \int_{\mathcal{B}} (X_\alpha)^2 \varrho(X)\, d^2X \end{aligned} \tag{13}$$

where $\rho(X)$ is the local mass density. Note that E_1, E_2 can be related to the classical polar momentum of inertia about the center of mass, J, via the relationship

$$J = E_1 + E_2 \tag{14}$$

Furthermore, in view of the constraint functions (9), the constraint Jacobian pertaining to the free rigid body is given by $\mathbf{G}_{int} = D\,\phi_{int}(q)$. Thus

$$\boldsymbol{G}_{int}(\boldsymbol{q}) = \begin{bmatrix} \boldsymbol{0}^T & \boldsymbol{d}_1^T & \boldsymbol{0}^T \\ \boldsymbol{0}^T & \boldsymbol{0}^T & \boldsymbol{d}_2^T \\ \boldsymbol{0}^T & \boldsymbol{d}_2^T & \boldsymbol{d}_1^T \end{bmatrix} \tag{15}$$

To summarize, the motion of the planar free rigid body is governed by the DAEs (1), with $n = 6$ and $m = 3$. This rigid body formulation is the cornerstone of the present approach to the energy-momentum integration of arbitrary multibody systems. Additional details about the present rigid body formulation may be found in [BS01b,BL06].

4. Kinematic pairs

This section deals with basic kinematic pairs which are fundamental for building complex multibody systems. Here we will present the revolute and the prismatic pair which represent the basic pairs necessary to model common planar parallel manipulators. Within this chapter we will also introduce a specific coordinate augmentation technique for both pairs in order to incorporate joint variables into the present rigid body formulation.

4.1 The planar revolute pair

Each rigid body of the multibody system depicted in Fig. 2 is modelled as constrained mechanical system as described in Section 3. Accordingly, body A is characterized by 6 redundant coordinates

$$\boldsymbol{q}^A = \begin{bmatrix} \boldsymbol{\varphi}^A \\ \boldsymbol{d}_1^A \\ \boldsymbol{d}_2^A \end{bmatrix} \tag{16}$$

along with internal constraints $\boldsymbol{\phi}_{\text{int}}^{A}(\mathbf{q}^A) \in \mathbb{R}^3$ of the form (9), associated constraint Jacobian $\mathbf{G}_{\text{int}}^{A}(\mathbf{q}^A) \in \mathbb{R}^{3\times 6}$ of the form (15), and mass matrix $\mathbf{M}^A \in \mathbb{R}^{6\times 6}$ of the form (12).

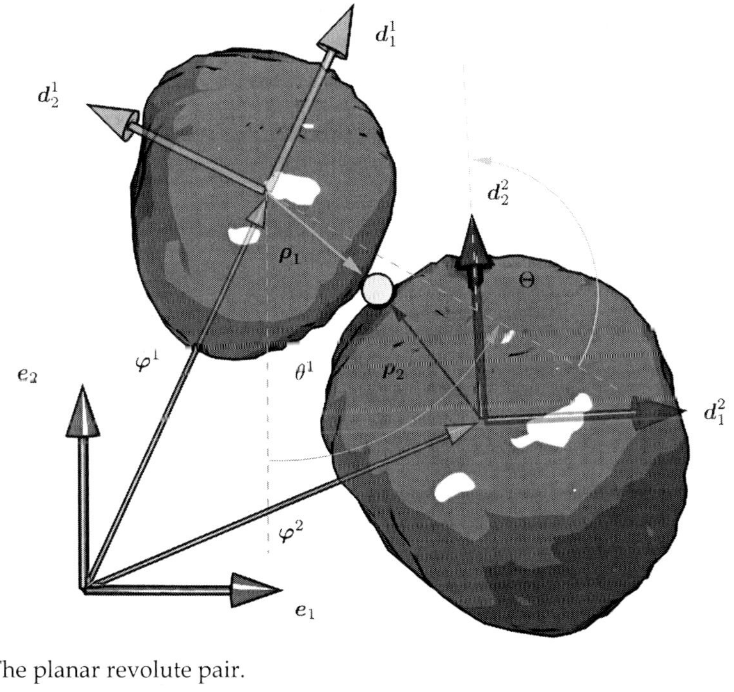

Figure 2: The planar revolute pair.

The description of the whole multibody system relies on the assembly of the individual bodies. The assembly procedure consists of the following steps. (i) The contributions of each individual body are collected in appropriate system vectors/matrices. For example, in the case of the present 2-body system (Fig. 2) we get the vector of redundant coordinates

$$\boldsymbol{q} = \begin{bmatrix} \boldsymbol{q}^1 \\ \boldsymbol{q}^2 \end{bmatrix} \tag{17}$$

along with the mass matrix

$$\boldsymbol{M} = \begin{bmatrix} \boldsymbol{M}^1 & \boldsymbol{0}_{6\times 6} \\ \boldsymbol{0}_{6\times 6} & \boldsymbol{M}^2 \end{bmatrix} \tag{18}$$

which, in view of (12), is diagonal and constant. Moreover, the constraints of rigidity are collected in the vector

$$\boldsymbol{\Phi}_{int} = \begin{bmatrix} \boldsymbol{\Phi}^1_{int} \\ \boldsymbol{\Phi}^2_{int} \end{bmatrix} \tag{19}$$

with corresponding constraint Jacobian

$$\boldsymbol{G}_{int} = \begin{bmatrix} \boldsymbol{G}^1_{int} & \boldsymbol{0}_{3\times 6} \\ \boldsymbol{0}_{3\times 6} & \boldsymbol{G}^2_{int} \end{bmatrix} \tag{20}$$

(ii) The interconnection between the rigid bodies in a multibody system is accounted for by external constraints.
For the revolute pair we get two additional constraint functions of the form

$$\boldsymbol{\Phi}_{ext}(\boldsymbol{q}) = \boldsymbol{\varphi}^2 - \boldsymbol{\varphi}^1 + \boldsymbol{\varrho}^2 - \boldsymbol{\varrho}^1 \tag{21}$$

where the vector

$$\boldsymbol{\varrho}^A = \sum_{\alpha=1}^{2} \varrho^A_\alpha \boldsymbol{d}^A_\alpha \tag{22}$$

specifies the position of the joint on body A. The constraints (21) give rise to the Jacobian

$$\boldsymbol{G}_{ext}(\boldsymbol{q}) = D\boldsymbol{\Phi}_{ext}(\boldsymbol{q}) = \begin{bmatrix} -\boldsymbol{I} & -\varrho^1_1\boldsymbol{I} & -\varrho^1_2\boldsymbol{I} & \boldsymbol{I} & \varrho^2_1\boldsymbol{I} & \varrho^2_2\boldsymbol{I} \end{bmatrix} \tag{23}$$

Accordingly, the present 2-body system is characterized by a total of m = 8 independent constraints

$$\boldsymbol{\Phi}(\boldsymbol{q}) = \begin{bmatrix} \boldsymbol{\Phi}_{int}(\boldsymbol{q}) \\ \boldsymbol{\Phi}_{ext}(\boldsymbol{q}) \end{bmatrix} \tag{24}$$

with corresponding 8×12 constraint Jacobian

$$\boldsymbol{G}(\boldsymbol{q}) = \begin{bmatrix} \boldsymbol{G}_{int}(\boldsymbol{q}) \\ \boldsymbol{G}_{ext}(\boldsymbol{q}) \end{bmatrix} \tag{25}$$

To summarize, the present description of the revolute pair makes use of $n = 12$ redundant coordinates subject to $m = 8$ constraints. This complies with the fact that the system at hand has $n - m = 4$ degrees of freedom. Obviously, the configuration space of the revolute pair, Q_{revolute}, can be written in the form (4).

4.1.1 Discrete constraint Jacobian

Since the constraint functions in (24) are at most quadratic, the associated discrete derivative coincides with the mid-point evaluation of the continuous constraint Jacobian (25), i.e.

$$\mathbf{G}(\boldsymbol{q}_n, \boldsymbol{q}_{n+1}) = \boldsymbol{G}(\boldsymbol{q}_{n+\frac{1}{2}}) \tag{26}$$

4.1.2 Coordinate augmentation

In many practical applications rotational variables along with associated torques are required for the description of a multibody system. Although the present approach circumvents the use of rotational variables throughout the discretization procedure, rotations can be easily incorporated into the present method. To this end, we next propose a coordinate augmentation technique. The idea is to incorporate a joint torque into the revolute pair (Fig. 2). Therefore we extend the original configuration vector

$$\boldsymbol{q} = \begin{bmatrix} \boldsymbol{q}^1 \\ \boldsymbol{q}^2 \\ \Theta \end{bmatrix} \tag{27}$$

The new coordinate Θ is connected with the original ones by introducing an additional constraint function of the form

$$\Phi_{aug}^R(\boldsymbol{q}) = \boldsymbol{d}_2^2 \cdot \boldsymbol{d}_1^1 + \sin\Theta + \boldsymbol{d}_2^2 \cdot \boldsymbol{d}_2^1 - \cos\Theta \tag{28}$$

In anticipation of the subsequent treatment of the discretization we write (28) in partitioned form

$$\Phi_{aug}^R(\boldsymbol{q}) = \Phi_{aug}^1(\boldsymbol{q}_{ori}) + \Phi_{aug}^2(\Theta) \tag{29}$$

with the original coordinates

$$q_{ori} = \begin{bmatrix} q^1 \\ q^2 \end{bmatrix} \tag{30}$$

and

$$\begin{aligned} \Phi^1_{aug}(q_{ori}) &= d^2_2 \cdot d^1_1 + d^2_2 \cdot d^1_2 \\ \Phi^2_{aug}(\Theta) &= \sin\Theta - \cos\Theta \end{aligned} \tag{31}$$

Additionally, we get the Jacobian

$$G_{aug}(q) = D\Phi_{aug}(q) = \begin{bmatrix} 0^T & d_2^{2^T} & d_2^{2^T} & 0^T & 0^T & (d^1_1 + d^1_2)^T & (\sin\Theta + \cos\Theta) \end{bmatrix} \tag{30}$$

With regard to (29), we decompose (32) according to

$$G_{aug}(q) = \begin{bmatrix} G^1_{aug}(q_{ori}) & G^2_{aug}(\Theta) \end{bmatrix} \tag{33}$$

with

$$\begin{aligned} G^1_{aug}(q_{ori}) &= \begin{bmatrix} 0^T & d_2^{2^T} & d_2^{2^T} & 0^T & 0^T & (d^1_1 + d^1_2)^T \end{bmatrix} \\ G^2_{aug}(\Theta) &= \sin\Theta + \cos\Theta \end{aligned} \tag{34}$$

To summarize, we now have n = 13 coordinates subject to m = 9 geometric constraints. In order to completely specify the DAEs (1) for the augmented system at hand one simply has to extend the relevant matrices of the revolute pair in Section 4.1. Accordingly, the mass matrix of the augmented system is given by

$$M = \begin{bmatrix} M^1 & 0_{6\times6} & 0_{6\times1} \\ 0_{6\times6} & M^2 & 0_{6\times1} \\ 0_{1\times6} & 0_{1\times6} & 0 \end{bmatrix} \tag{35}$$

In view of (28), the augmentation gives rise to an extended vector of constraint functions of the form

$$\Phi(q) = \begin{bmatrix} \Phi_{ori}(q_{ori}) \\ \Phi_{aug}(q) \end{bmatrix} \tag{36}$$

where ϕ_{ori} stands for the original constraints given by (24). The augmented constraint Jacobian assumes the form

$$\boldsymbol{G}(\boldsymbol{q}) = \begin{bmatrix} \boldsymbol{G}_{ori}(\boldsymbol{q}_{ori}) & \mathbf{0}_{8\times 1} \\ \boldsymbol{G}^1_{aug}(\boldsymbol{q}_{ori}) & G^2_{aug}(\Theta) \end{bmatrix} \tag{37}$$

where $\mathbf{G}_{\text{ori}}$ represents the original constraint Jacobian given by (25).

4.1.3 Discrete constraint Jacobian

The discrete version of (37) can be written as

$$\mathrm{G}(\boldsymbol{q}_n, \boldsymbol{q}_{n+1}) = \begin{bmatrix} \boldsymbol{G}_{ori}\big((\boldsymbol{q}_{ori})_{n+\frac{1}{2}}\big) & \mathbf{0}_{8\times 1} \\ \boldsymbol{G}^1_{aug}\big((\boldsymbol{q}_{ori})_{n+\frac{1}{2}}\big) & \mathrm{G}^2_{aug}(\Theta_n, \Theta_{n+1}) \end{bmatrix} \tag{38}$$

Since the constraint functions $\phi_{\text{ori}}(q_{\text{ori}})$ and $\phi^1_{aug}(q_{\text{ori}})$ (cf. (24) and $(31)_1$, respectively) are at most quadratic, the associated discrete gradient coincides with the mid-point evaluation of the respective continuous constraint Jacobians. This is in contrast to the constraint function $\phi^2_{aug}(\Theta)$, see $(31)_2$. In this case we choose

$$\mathrm{G}^2_{aug}(\Theta_n, \Theta_{n+1}) = \frac{\Phi^2_{aug}(\Theta_{n+1}) - \Phi^2_{aug}(\Theta_n)}{\Theta_{n+1} - \Theta_n} \tag{39}$$

If

$$\Theta_{n+1} = \Theta_n, \text{ then } \mathrm{G}^2_{aug}(\Theta_n, \Theta_{n+1}) = \left(\Phi^2_{aug}\right)'(\Theta_n).$$

Remark 4.1 *Formula (39) can be interpreted as G-equivariant discrete derivative of the corresponding constraint function in the sense of Gonzalez [Gon96]. In this connection G represents the group acting by translations and rotations, respectively. In the present case (39) coincides with Greenspan's formula [Gre84].*

4.1.4 Numerical example

To demonstrate the numerical performance of the present formulation we investigate the free flight of our institute logo NM (Numerical Mechanics[1]). Both letters are modelled as rigid bodies which are connected by a revolute joint. (Fig 3).
The inertial parameters for the numerical example are summarized in Table 1. The location of the joint relative to each body is specified by (22) with
The inertial parameters for the numerical example are summarized in Table 1. The location of the joint relative to each body is specified by (22) with

$$[\varrho^1_\alpha] = \begin{bmatrix} 0 \\ -0.4 \end{bmatrix} \quad \text{and} \quad [\varrho^2_\alpha] = \begin{bmatrix} 0 \\ 0.4 \end{bmatrix} \tag{40}$$

---

[1] http://www.uni-siegen.de/fb11/nm

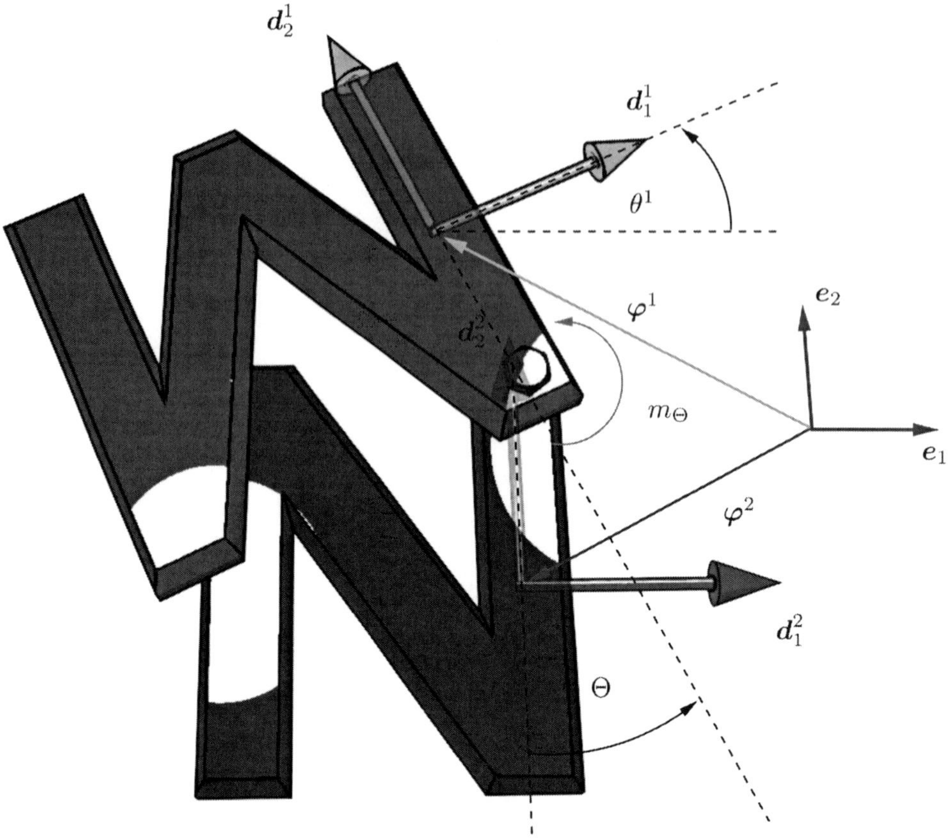

Figure 3: The NM-logo as 2-body system. Arbitrary configuration of both connected letters.

The initial configuration of the system is given by the following generalized coordinates (see Fig. 3)

$$\boldsymbol{u}_0 = \begin{bmatrix} \varphi_0^1 \\ \theta_0^1 \\ \Theta_0 \end{bmatrix} = \begin{bmatrix} \mathbf{0} \\ 0 \\ \pi \end{bmatrix} \tag{41}$$

Initial generalized velocities can be written as

$$\boldsymbol{\nu}_0 = \begin{bmatrix} \boldsymbol{e}_1 \cdot (\boldsymbol{v}_\varphi^1)_0 \\ \boldsymbol{e}_2 \cdot (\boldsymbol{v}_\varphi^1)_0 \\ \omega_0^1 \\ \dot{\Theta}_0 \end{bmatrix} \tag{42}$$

In the present example the system is initially at rest, i.e. $v_0 = 0$. Since it is a free flight, we neglect the gravitational forces, having no potential energy in the system. To initialize the motion, external loads $\mathbf{Q} \in R^{13}$ are acting on the system. Specifically,

$$\boldsymbol{Q} = \begin{bmatrix} \mathbf{0}_{12\times 1} \\ m_\Theta(t) \end{bmatrix} \tag{43}$$

This means that we only apply an external joint torque, which is directly acting on the newly introduced rotational component Θ. The torque itself is applied in the form of a hat function over time (cf. Fig. 4), where $t_1 = 0.25$, $t_2 = 0.5$, $\overline{m} = 5$. Accordingly, for $t > t_2$ no external forces act on the system anymore. The system can thus be classified as an autonomous Hamiltonian system with symmetry. Consequently, the Hamiltonian (or the total energy) represents a conserved quantity for $t > t_2$. The angular momentum remains equal for all times, since it is an internal joint torque acting on the system. The present energy-momentum scheme does indeed satisfy these conservation properties for any time step Δt, see Fig. 5. The simulated motion is illustrated with some snapshots at discrete times in Fig. 6. Moreover, the evolution of the angle $\Theta(t)$, calculated with different time steps $\Delta t \in \{0.1, 0.05, 0.01\}$, is depicted in Fig. 7.

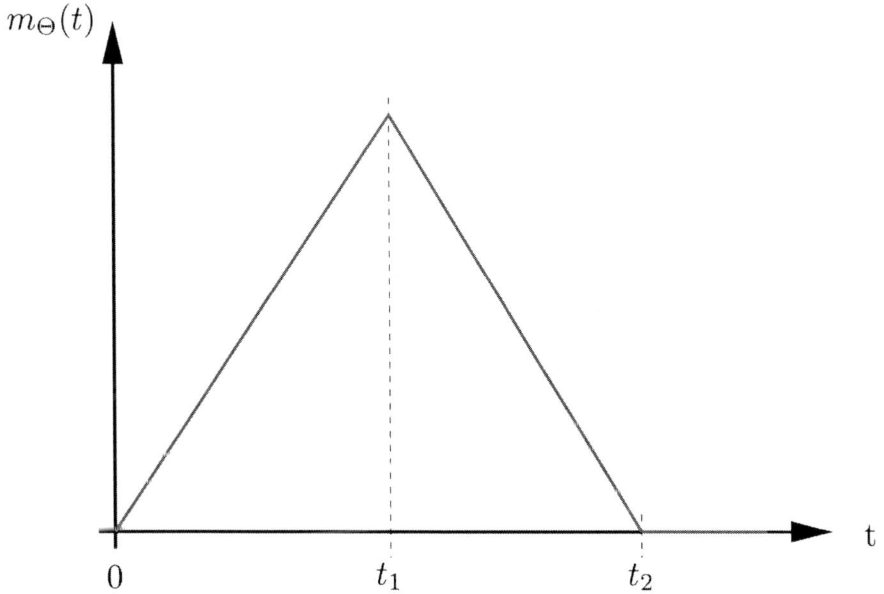

$$m_\Theta(t) = \begin{cases} \frac{\bar{m}}{t_1} t & \text{for} \quad 0 \le t \le t_1 \\ \frac{\bar{m}}{t_1 - t_2}(t - t_2) & \text{for} \quad t_1 \le t \le t_2 \end{cases}$$

Figure 4: Magnitude of the torque during the initial load period.

body	M	E_1	E_2
1	1.1	0.004	0.0917
2	2	0.0073	0.1667

Table 1: Inertial parameters for the 2-body system.

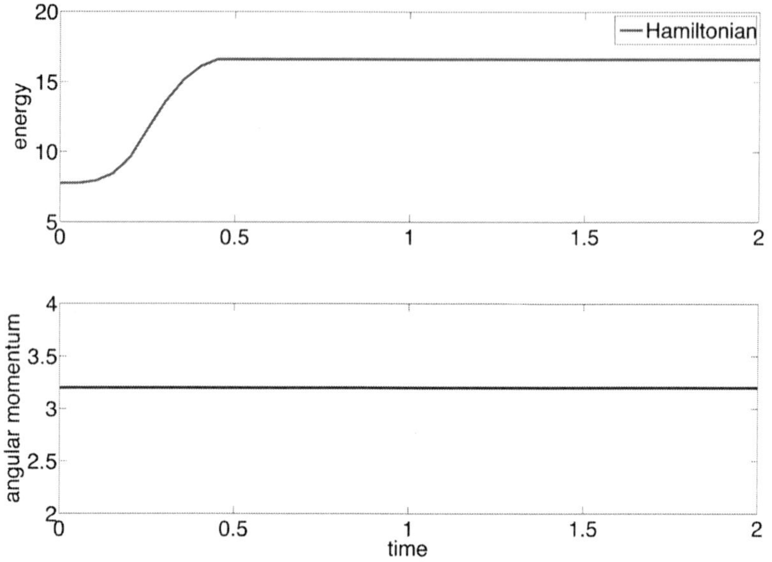

Figure 5: Algorithmic conservation of energy and angular momentum, $\Delta t = 0.05$.

Figure 6: Snapshots of the free flying NM-logo. The two curves correspond to the trajectories of the mass centers of the individual bodies constituting the present multibody system ($t \in \{0, 1, 2\}$s).

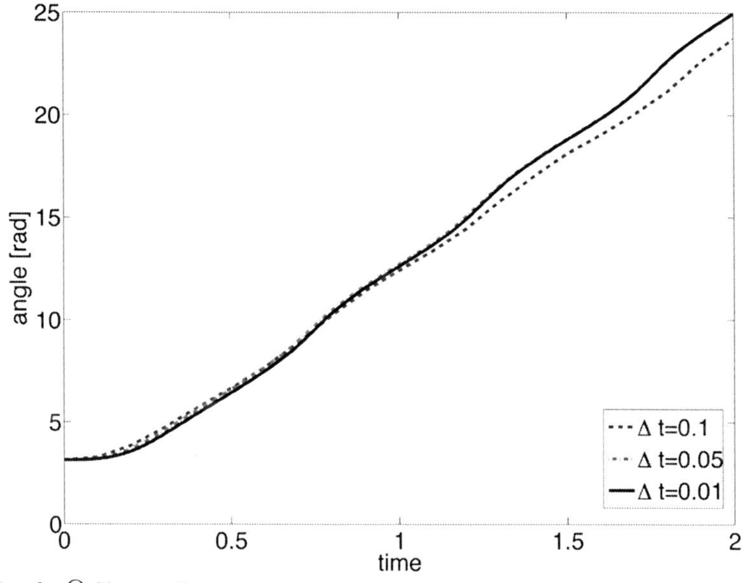

Figure 7: Angle $\Theta\ (t)$ over time.

4.2 The planar prismatic pair

Analogous to the previously presented revolute pair, we now focus on the prismatic pair. The procedure is similar to the prismatic pair, we will present the necessary constraints and their Jacobians. A coordinate augmentation for the prismatic pair will measure the distance between both rigid bodies. The example will deal with a planar linear motion guide.

The prismatic pair (Fig. 8) will again be considered as a constrained mechanical systems. Since the number of bodies and their internal description corresponds to the revolute pair, the configuration vector (17), the mass matrix (18) and the internal constraints as well as their Jacobians (19), (20) have the same structure as already presented for the revolute pair. The interconnection between both bodies characterizes the prismatic joint and can be written as:

$$\boldsymbol{\Phi}_{ext}(\boldsymbol{q}) = \begin{bmatrix} (\boldsymbol{m}^1) \cdot (\boldsymbol{p}^2 - \boldsymbol{p}^1) \\ \boldsymbol{d}_1^1 \cdot \boldsymbol{d}_2^2 - \eta \end{bmatrix} \tag{44}$$

with the vectors

$$\boldsymbol{m} = \sum_{\alpha=1}^{2} m_\alpha \boldsymbol{d}_\alpha^1 \quad \text{and} \quad \boldsymbol{p}^i = \boldsymbol{\varphi}^i + \boldsymbol{\rho}^i \tag{45}$$

The vector $\boldsymbol{\rho}^i$ has already been defined in eq. (22). The value of η in (44) needs to be prescribed initially. The corresponding constraint Jacobian yields:

$$\boldsymbol{G}_{ext}(\boldsymbol{q}) = \begin{bmatrix} -(\boldsymbol{m})^T & \boldsymbol{G}_1^1 & \boldsymbol{G}_2^1 & (\boldsymbol{m})^T & \rho_1^2(\boldsymbol{m})^T & \rho_2^2(\boldsymbol{m})^T \\ \mathbf{0}^T & (\boldsymbol{d}_2^2)^T & \mathbf{0}^T & \mathbf{0}^T & \mathbf{0}^T & (\boldsymbol{d}_1^1)^T \end{bmatrix} \tag{46}$$

with

$$\boldsymbol{G}_i^1 = m_i(\boldsymbol{p}^2 - \boldsymbol{p}^1)^T - \rho_i^1(\boldsymbol{m})^T \quad \text{for} \quad i = 1, 2 \tag{47}$$

This leads again to m = 8 independent constraints, the global constraint Jacobian has the form of eq. (25). The number of unknowns is the same as for the revolute pair, since we only have one relative coordinate between both bodies (u).

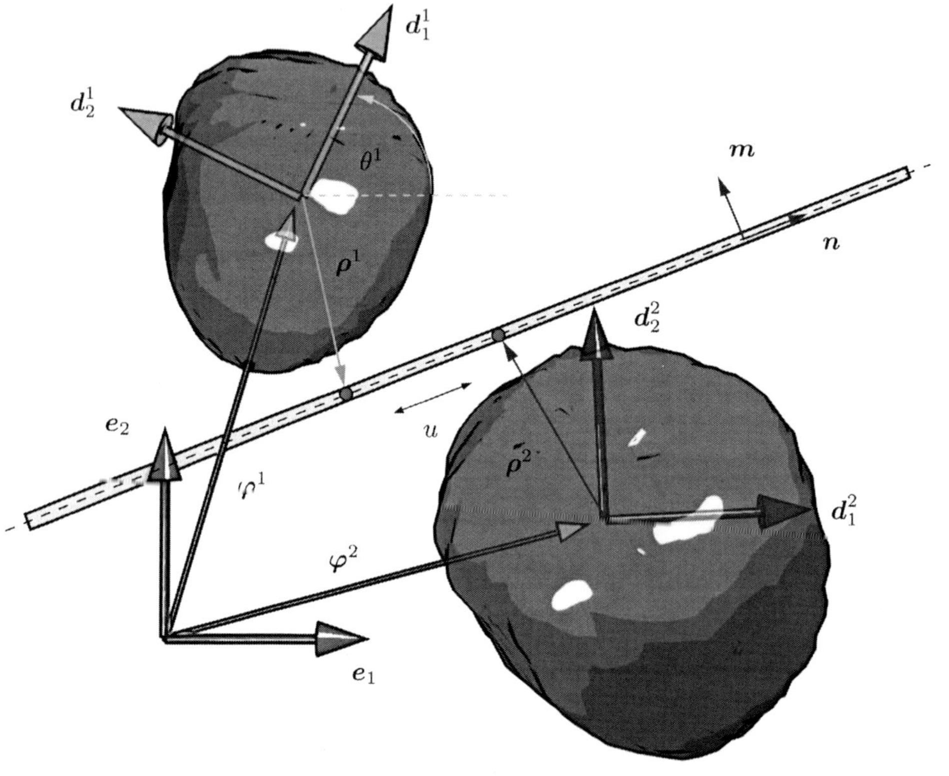

Figure 8: The planar prismatic pair.

4.2.1 Discrete constraint Jacobian

A closer investigation of (44) reveals that the constraint functions are quadratic, which means that the discrete derivative coincides with the mid-point evaluation of the constraint Jacobian (46). Therefore the discrete version of the constraint Jacobian is given by:

$$\mathbf{G}(\boldsymbol{q}_n, \boldsymbol{q}_{n+1}) = \boldsymbol{G}(\boldsymbol{q}_{n+\frac{1}{2}})$$

4.2.2 Coordinate augmentation

As already outlined for the revolute pair, for practical issues it is vital to incorporate augmented values into our rotationless formulation for multibody systems. Similar to the introduction of a relative angle for the revolute pair, we now account for the translational displacement between both rigid bodies. This time we will augment the system by the variable u which represents a generalized coordinate measuring the distance between the center of masses of both bodies.

Accordingly we start with the extension of our configuration vector by the new coordinate:

$$\boldsymbol{q} = \begin{bmatrix} \boldsymbol{q}^1 \\ \boldsymbol{q}^2 \\ u \end{bmatrix} = \begin{bmatrix} \boldsymbol{q}_{ori} \\ u \end{bmatrix} \tag{49}$$

The incorporation of a new redundant coordinate needs also a corresponding constraint. In this case we can write:

$$\Phi^P_{aug}(\boldsymbol{q}) = (\boldsymbol{p}^2 - \boldsymbol{p}^1) \cdot \boldsymbol{n} - u \tag{50}$$

As outlined before, **n** represents the axis of sliding and can also be described as

$$\boldsymbol{n} = \sum_{\alpha=1}^{2} n_\alpha \boldsymbol{d}^1_\alpha \tag{51}$$

Again we decompose the constraint vector in two parts. One depending on the original coordinates and a second one depending on the newly introduced coordinate u

$$\Phi^P_{aug}(\boldsymbol{q}) = \Phi^1_{aug}(\boldsymbol{q}_{ori}) + \Phi^2_{aug}(u) \tag{52}$$

The same will be done with its corresponding constraint Jacobian:

$$\boldsymbol{G}_{aug}(\boldsymbol{q}) = \left[\boldsymbol{G}^1_{aug}(\boldsymbol{q}_{ori}) \quad G^2_{aug}(u)\right] \tag{53}$$

For both parts we obtain:

$$\begin{aligned} \boldsymbol{G}^1_{aug}(\boldsymbol{q}_{ori}) &= \left[-\boldsymbol{n}^T \quad n_1(\boldsymbol{p}^2 - \boldsymbol{p}^1)^T - \rho^1_1 \boldsymbol{n}^T \quad n_2(\boldsymbol{p}^2 - \boldsymbol{p}^1)^T - \rho^1_2 \boldsymbol{n}^T \quad \boldsymbol{n}^T \quad \rho^2_1 \boldsymbol{n}^T \quad \rho^2_2 \boldsymbol{n}^T\right] \\ G^2_{aug}(u) &= -1 \end{aligned} \tag{54}$$

As already presented in section (4.1.2), extending the configuration vector means also to expand the mass matrix (35) and the global constraint Jacobian (37). These steps are equivalent to the revolute pair.

4.2.3 Discrete constraint Jacobian

The discrete version of (37) for the prismatic pair can be written as

$$\mathbf{G}(\boldsymbol{q}_n, \boldsymbol{q}_{n+1}) = \begin{bmatrix} \boldsymbol{G}_{ori}\left((\boldsymbol{q}_{ori})_{n+\frac{1}{2}}\right) & \mathbf{0}_{8\times 1} \\ \boldsymbol{G}^1_{aug}\left((\boldsymbol{q}_{ori})_{n+\frac{1}{2}}\right) & \mathrm{G}^2_{aug}(u_{n+\frac{1}{2}}) \end{bmatrix} \tag{55}$$

Since the augmented constraint is at most quadratic, a simple mid-point evaluation is sufficient.

body	M	E_1	E_2	Length	Width
1	1.1	0.0229	5.8667	8	0.5
2	2	0.1667	0.1667	1	1

Table 2: Inertial parameters for the prismatic 2-body system.

4.2.4 Numerical example

In order to demonstrate the performance of the prismatic pair, we consider a linear motion guide (Fig. 9). It consists of two rigid bodies connected via a prismatic joint. The pair moves freely with given initial velocities in space.

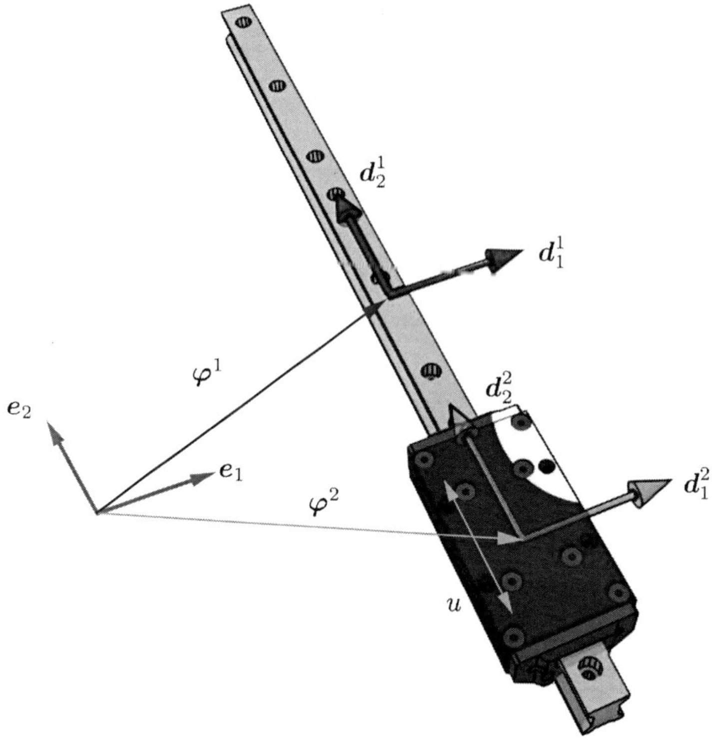

Figure 9: The linear motion guide as a 2 body system.

The inertial parameters for the numerical example are summarized in Table 2. The initial configuration of the system is given by (cf. Section 4.2 and Fig. 8):

$$\boldsymbol{u}_0 = \begin{bmatrix} \varphi_0^1 \\ \theta_0^1 \\ u_0 \end{bmatrix} = \begin{bmatrix} 0 \\ 0 \\ -3 \end{bmatrix} \tag{56}$$

Initial velocities can again be set in a generalized form:

$$\boldsymbol{\nu}_0 = \begin{bmatrix} \boldsymbol{e}_1 \cdot (\boldsymbol{v}_\varphi^1)_0 \\ \boldsymbol{e}_2 \cdot (\boldsymbol{v}_\varphi^1)_0 \\ \omega_0^1 \\ \dot{u}_0 \end{bmatrix} = \begin{bmatrix} 4 \\ 1 \\ \pi/2 \\ -6.5 \end{bmatrix} \tag{57}$$

Since there are no loads applied on the system, the total energy (Hamiltonian) and the angular momentum shall be conserved quantities. Once again the present energy-momentum scheme does indeed satisfy these conservation properties for any time step Δt, see Fig. 10. Some specific positions of the motion are displayed in Fig. 11. The evolution of the augmented coordinate u for different time steps $\Delta t \in \{0.1, 0.05, 0.01\}$, is depicted in Fig. 12.

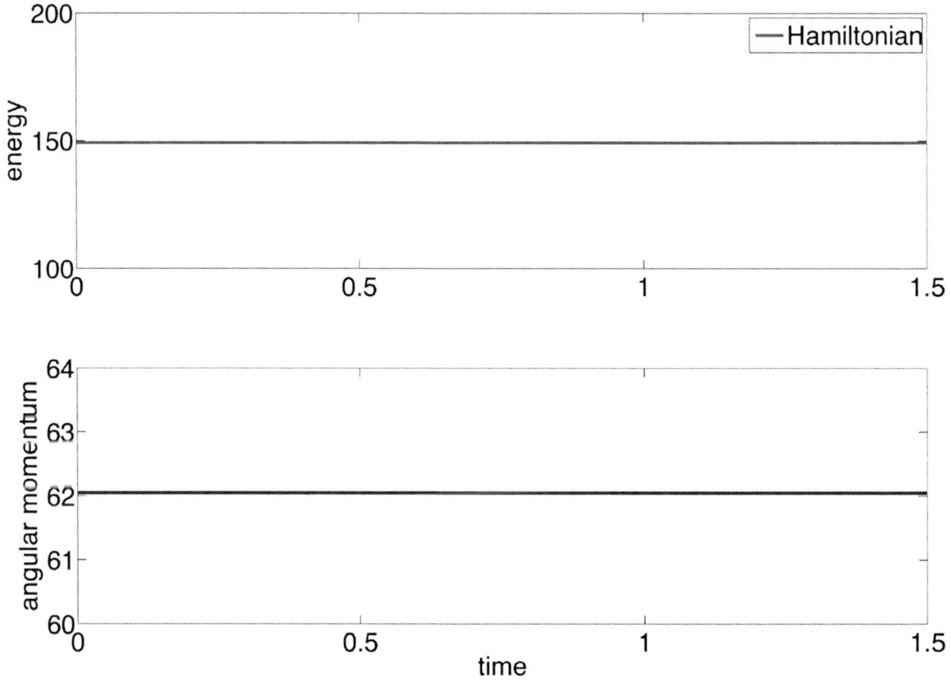

Figure 10: Algorithmic conservation of energy and angular momentum, $\Delta t = 0.1$.

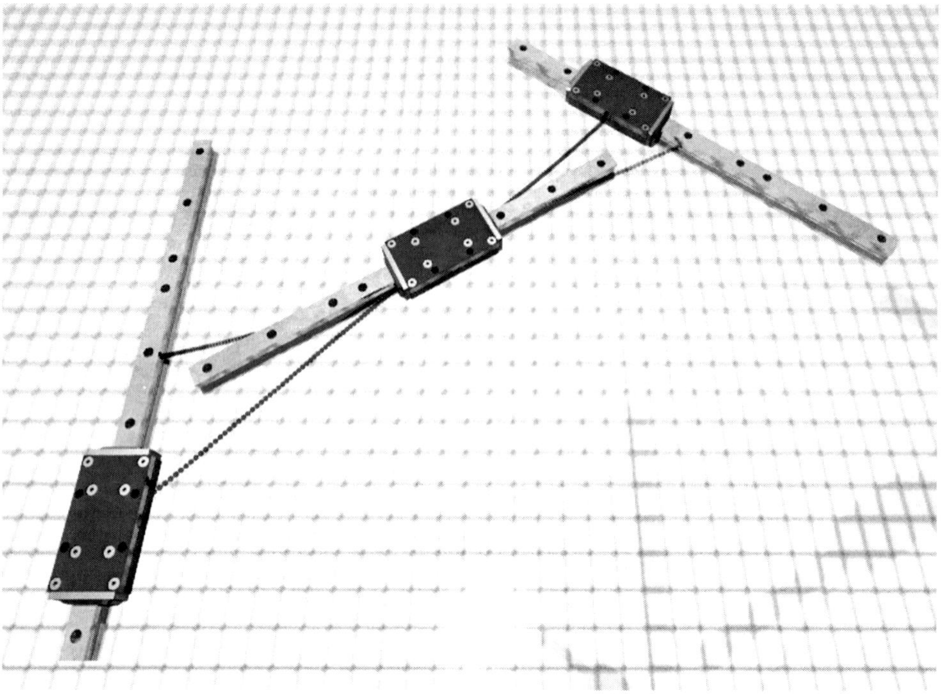

Figure 11: Snapshots of the free flight of the prismatic pair. Trajectories mark the movement of the center of masses ($t \in \{0, 0.8, 1.5\}$s).

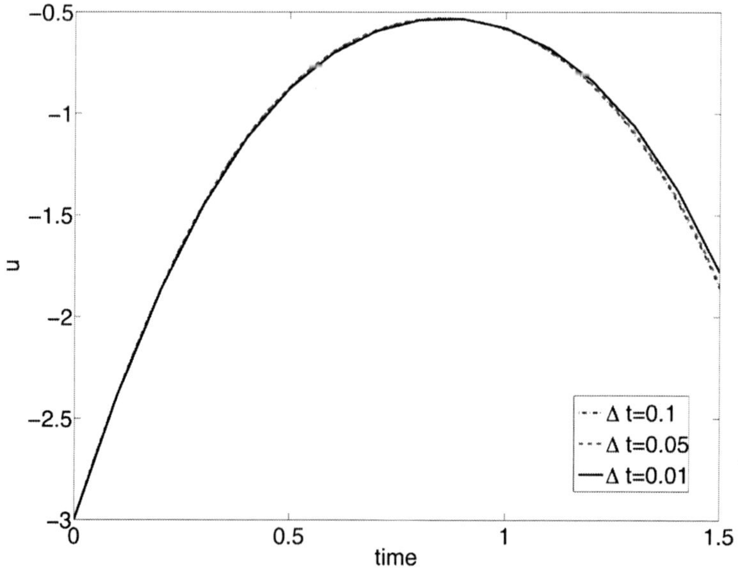

Figure 12: Translational displacement u over time.

5. Planar parallel manipulator

In this section we will combine all previous features in the example of a planar parallel manipulator. Since we have presented the revolute and prismatic pair, we will build a model of a RPR-manipulator, where the letters mark the kind of joints the mechanism consists of (**R**evolute-**P**rismatic-**R**evolute). The Figure below shows the configuration of the RPR-manipulator:

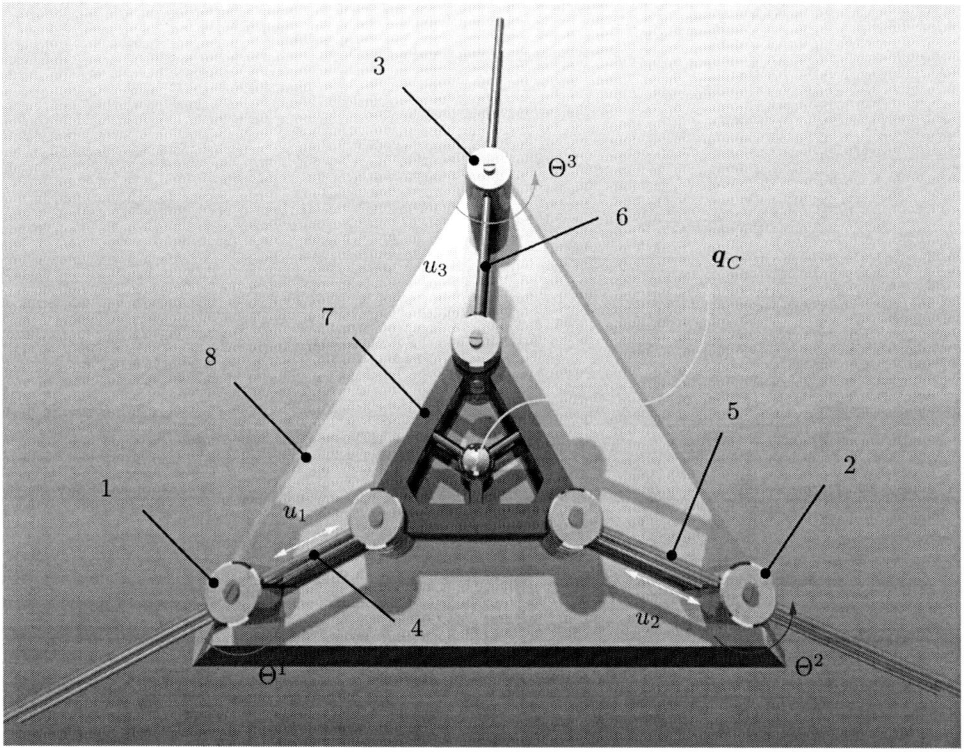

Figure 13: Schematics of the RPR-manipulator.

The goal in this example is to perform a controlled motion (vector $\mathbf{q}_C$ in upper Figure) of the inner triangle (body 7). Therefore we need to augment our original BEM-scheme (1) by control constraints and their corresponding constraint Jacobian. The enhanced continuous DAE structure yields to:

$$\boxed{\begin{array}{rcl} \dot{\boldsymbol{q}} - \boldsymbol{v} & = & \mathbf{0} \\ \boldsymbol{M}\dot{\boldsymbol{v}} - \boldsymbol{F} + \boldsymbol{G}^T\boldsymbol{\lambda} + \boldsymbol{B}^T\bar{\boldsymbol{m}} & = & \mathbf{0} \\ \boldsymbol{\Phi}(\boldsymbol{q}) & = & \mathbf{0} \\ \boldsymbol{\Phi}_C(\boldsymbol{q}) & = & \mathbf{0} \end{array}} \tag{58}$$

Here $\phi_C(\mathbf{q})$ accounts for the newly introduced control constraints. Their corresponding Jacobian is **B**, while its product with $\bar{m}$ represents the necessary control forces.

A direct discretization of the equations above leads to an enhanced BEM-scheme for the presented underactuated system:

$$\begin{aligned} \boldsymbol{q}_{n+1} - \boldsymbol{q}_n &= \frac{\Delta t}{2}\left(\boldsymbol{v}_n + \boldsymbol{v}_{n+1}\right) \\ \boldsymbol{M}\left(\boldsymbol{v}_{n+1} - \boldsymbol{v}_n\right) &= \Delta t\mathbf{F}(\boldsymbol{q}_n, \boldsymbol{q}_{n+1}) - \Delta t\mathbf{G}(\boldsymbol{q}_n, \boldsymbol{q}_{n+1})^T\bar{\boldsymbol{\lambda}} - \Delta t\mathbf{B}(\boldsymbol{q}_n, \boldsymbol{q}_{n+1})^T\bar{\boldsymbol{m}} \\ \boldsymbol{\Phi}(\boldsymbol{q}_{n+1}) &= \mathbf{0} \\ \boldsymbol{\Phi}_C(\boldsymbol{q}_{n+1}) &= \mathbf{0} \end{aligned} \tag{59}$$

5.1 Rotationless formulation for the RPR manipulator

Here we will present the rotationless formulation for the RPR manipulator. The incorporation of rotational redundant coordinates plays a crucial role for the desired control problem. Additionally, as already presented in the sections before, we will also introduce translational redundant coordinates which measure the movement of the prismatic pairs. The mechanism presented herein consists of 8 rigid bodies. Bodies 1, 2 and 3 are connect via revolute joints to the free floating platform (body 8). The connection between body 1, 2, 3 and 4, 5, 6 is established by prismatic pairs. Finally 4, 5 and 6 are connected to the small triangle (body 7) via revolute joints. This structure consists of two closed loops, which means to formulate corresponding loop-closure constraints. The system at hand can then be characterized by the following configuration vector:

$$\boldsymbol{q}_{ori} = \begin{bmatrix} \boldsymbol{q}^1 \\ \boldsymbol{q}^2 \\ \boldsymbol{q}^3 \\ \boldsymbol{q}^4 \\ \boldsymbol{q}^5 \\ \boldsymbol{q}^6 \\ \boldsymbol{q}^7 \\ \boldsymbol{q}^8 \end{bmatrix}_{48\times 1} \quad \text{where} \quad \boldsymbol{q}_I = \begin{bmatrix} \boldsymbol{\varphi}^I \\ \boldsymbol{d}_1^I \\ \boldsymbol{d}_2^I \end{bmatrix} \quad \text{with} \quad I = 1, 2, ..., 8 \tag{60}$$

The upper vector has a size of 48, having eight rigid bodies means to invoke another m_{int} = 18 internal constraints and having nine joints at hand leads to m_{ext} = 24 external constraints. The difference n - m_{int} - m_{ext} = 6 means that the system at hand has a total of 6 DOF, since the platform (body 8) moves completely free and the inner triangle has another three DOF.

The necessary constraints for building the individual joints can be directly derived from chapter 4.1 and 4.2. This leads automatically to the closure of both loops. Here we neglect a detailed description of each individual joint and their constraint Jacobians, and only refer to the two previous chapters.

5.2 Coordinate augmentation

We now focus on the augmentation technique which is vital for the present application. As already outlined for both pairs (4.1 and 4.2), we incorporate rotational DOF (relative angles in-between body 8 and body 1, 2, 3) as well as translational DOF (distance between center of mass of body 1, 2, 3 and 4, 5, 6).

5.2.1 Rotational DOF

As indicated in Fig. 13, the first three joints of the parallel manipulator (with corresponding joint-rates $\dot{\Theta}^1$, $\dot{\Theta}^2$ and $\dot{\Theta}^3$) are actuated. To incorporate into the underlying rotationless formulation the possibility of imposing joint-torques ($\overline{m}_1$, $\overline{m}_2$, $\overline{m}_3$), we apply the coordinate augmentation technique proposed in Section 4.1.2. Indeed, the application of the coordinate augmentation technique to the present closed-loop system follows from a straight-forward extension of the treatment of the revolute pair in Section 4.1.

Similar to (27), we augment the originally used redundant coordinates $q_{ori} \in \mathbb{R}^{48}$ with the joint-angles

$$\Theta = \begin{bmatrix} \Theta^1 \\ \Theta^2 \\ \Theta^3 \end{bmatrix} \tag{61}$$

such that the augmented configuration vector reads

$$q = \begin{bmatrix} q_{ori} \\ \Theta \end{bmatrix} \tag{62}$$

Accordingly, we now have $n = 51$ redundant coordinates. The three additional coordinates (61) are linked to the original ones through the introduction of three additional constraint functions. Similar to (36), the extended vector of constraint functions reads

$$\Phi(q) = \begin{bmatrix} \Phi_{ori}(q) \\ \Phi_{aug}(q) \end{bmatrix} \tag{63}$$

where, similar to (29), the additional constraints are specified by

$$\Phi_{aug}(q) = \Phi_{aug}^{I}(q_{ori}) + \Phi_{aug}^{II}(\Theta) \tag{64}$$

where

$$\Phi_{aug}^{I}(q_{ori}) = \begin{bmatrix} (\Phi_{aug}^{I})_1 \\ (\Phi_{aug}^{I})_2 \\ (\Phi_{aug}^{I})_3 \end{bmatrix} \quad \text{with} \quad (\Phi_{aug}^{I})_j = d_2^j \cdot (d_1^\beta + d_2^8) \tag{65}$$

and

$$\Phi_{aug}^{II}(\Theta) = \begin{bmatrix} (\Phi_{aug}^{II})(\Theta^1) \\ (\Phi_{aug}^{II})(\Theta^2) \\ (\Phi_{aug}^{II})(\Theta^3) \end{bmatrix} \quad \text{with} \quad (\Phi_{aug}^{II})(\Theta^j) = \sin \Theta^j - \cos \Theta^j \tag{66}$$

We thus have a total of $m = 45$ constraints. Consequently, the BEM scheme relies on $n + m = 96$ unknowns. Similar to (37), the augmented constraint Jacobian is given by

$$G(q) = \begin{bmatrix} G_{ori}(q_{ori}) & 0_{48\times 3} \\ G^{I}_{aug}(q_{ori}) & G^{II}_{aug}(\Theta) \end{bmatrix} \tag{67}$$

The 3×48 matrix $\mathbf{G}^{I}_{aug}(\mathbf{q}_{ori})$ has the same structure as $(34)_1$, and $\mathbf{G}^{II}_{aug}(\Theta)$ is given by

$$G^{II}_{aug} = \begin{bmatrix} \sin\Theta^1 + \cos\Theta^1 & 0 & 0 \\ 0 & \sin\Theta^2 + \cos\Theta^2 & 0 \\ 0 & 0 & \sin\Theta^3 + \cos\Theta^3 \end{bmatrix} \tag{68}$$

Similar to (55) the discrete counterpart of (67) can be written in the form

$$G(q_n, q_{n+1}) = \begin{bmatrix} G_{ori}((q_{ori})_{n+\frac{1}{2}}) & 0_{48\times 3} \\ G^{I}_{aug}((q_{ori})_{n+\frac{1}{2}}) & G^{II}_{aug}(\Theta_n, \Theta_{n+1}) \end{bmatrix} \tag{69}$$

Here, the discrete version of (68) assumes the form

$$G^{II}_{aug}(\Theta_n, \Theta_{n+1}) = \begin{bmatrix} G^{II}_{aug}(\Theta^1_n, \Theta^1_{n+1}) & 0 & 0 \\ 0 & G^{II}_{aug}(\Theta^2_n, \Theta^2_{n+1}) & 0 \\ 0 & 0 & G^{II}_{aug}(\Theta^3_n, \Theta^3_{n+1}) \end{bmatrix} \tag{70}$$

with

$$G^{II}_{aug}(\alpha, \beta) = \begin{cases} \dfrac{(\Phi^{II}_{aug})(\beta) - (\Phi^{II}_{aug})(\alpha)}{\beta - \alpha} & \text{if} \quad \alpha \neq \beta \\ (\Phi^{II}_{aug})'(\alpha) & \text{if} \quad \alpha = \beta \end{cases} \tag{71}$$

5.2.2 Translational DOF

As already outlined for the prismatic pair in section 4.2.2, we apply the coordinate augmentation technique to incorporate translational DOF in the prismatic connection for the RPR manipulator. This means that additionally to the angle augmentation, we again augment the configuration vector by another three redundant coordinates:

$$u = \begin{bmatrix} u^1 \\ u^2 \\ u^3 \end{bmatrix} \tag{72}$$

taking into account the augmented part from section 5.2.1 such that the new augmented configuration vector reads

$$\boldsymbol{q} = \begin{bmatrix} \boldsymbol{q}_{ori} \\ \boldsymbol{\Theta} \\ \boldsymbol{u} \end{bmatrix} \tag{73}$$

Thus the number of redundant coordinates raises to $n = 54$. Once again, the new redundant coordinates require additional constraint functions. Similar to (64), the constraint functions are specified by

$$\boldsymbol{\Phi}_{aug}(\boldsymbol{q}) = \boldsymbol{\Phi}^{I}_{aug}(\boldsymbol{q}_{ori}) + \boldsymbol{\Phi}^{II}_{aug}(\boldsymbol{u}) \tag{74}$$

where

$$\boldsymbol{\Phi}^{I}_{aug}(\boldsymbol{q}_{ori}) = \begin{bmatrix} \left(\Phi^{I}_{aug}\right)_1 \\ \left(\Phi^{I}_{aug}\right)_2 \\ \left(\Phi^{I}_{aug}\right)_3 \end{bmatrix} \quad \text{with} \quad \left(\Phi^{I}_{aug}\right)_j = \left(\varphi^k - \varphi^j\right) \cdot \boldsymbol{d}^j_2 \quad \text{with} \quad k = 4, 5, 6 \tag{75}$$

and

$$\boldsymbol{\Phi}^{II}_{aug}(\boldsymbol{u}) = \begin{bmatrix} \left(\Phi^{II}_{aug}\right)\left(u^1\right) \\ \left(\Phi^{II}_{aug}\right)\left(u^2\right) \\ \left(\Phi^{II}_{aug}\right)\left(u^3\right) \end{bmatrix} \qquad \text{with} \qquad \left(\Phi^{II}_{aug}\right)\left(u^j\right) = -u_j \tag{76}$$

The corresponding augmented constraint Jacobian in a decomposed fashion (67) is given by

$$\boldsymbol{G}^{II}_{aug}(\boldsymbol{u}) = \begin{bmatrix} -1 & 0 & 0 \\ 0 & -1 & 0 \\ 0 & 0 & -1 \end{bmatrix} = -\boldsymbol{I}_3 \tag{77}$$

For the sake of simplicity $\mathbf{G}^{I}_{aug}(q)$ will not be treated detailed, because its structure has already been presented in 4.2, (46).
The discrete counterpart of the equation above equals the expression itself.

5.3 Numerical example

As mentioned before our intention is to let body number 7 move upon a prescribed trajectory and calculate the necessary driving torques (input values) acting in the revolute joints. The desired trajectory shall follow a figure-8 pattern as similarly proposed in [MR06]:

$$\boldsymbol{\Phi}_{traj}(\boldsymbol{q}) = \begin{bmatrix} q^7_x(t_0) + \frac{1}{12}\sin(\omega(t)) \\ q^7_y(t_0) + \frac{1}{16}\sin(2\omega(t)) \end{bmatrix} \tag{78}$$

while $\omega(t)$ describes the angular velocity which for this example is defined as a 9th order polynomial. The polynomial was proposed in [BK04] and is well suited for control problems due to its continuous and steady character. In this example it is defined as followed:

$$\theta(t) = \begin{cases} s_1(t) & \text{for} \quad 0 \leq t \leq t_1 \\ s_2(t) & \text{for} \quad t_1 \leq t \leq t_2 \\ s_3(t) & \text{for} \quad t_2 \leq t \leq t_3 \end{cases} \tag{79}$$

where

$$\begin{aligned} s_1(t) &= \left[\tfrac{126}{6} t_1 \left(\tfrac{t}{t_1}\right)^6 - \tfrac{420}{7} t_1 \left(\tfrac{t}{t_1}\right)^7 + \tfrac{540}{8} t_1 \left(\tfrac{t}{t_1}\right)^8 - \tfrac{315}{9} t_1 \left(\tfrac{t}{t_1}\right)^9 + \tfrac{70}{10} t_1 \left(\tfrac{t}{t_1}\right)^{10}\right] \cdot \omega_0 \\ s_2(t) &= s_1(t_1) + \omega_0 (t - t_1) \\ s_3(t) &= s_2(t) - (t_3 - t_2) \cdot \left[\tfrac{126}{6} \left(\tfrac{t-t_2}{t_3-t_2}\right)^6 - \tfrac{420}{7} \left(\tfrac{t-t_2}{t_3-t_2}\right)^7 + \tfrac{540}{8} \left(\tfrac{t-t_2}{t_3-t_2}\right)^8 \right. \\ &\quad \left. - \tfrac{315}{9} \left(\tfrac{t-t_2}{t_3-t_2}\right)^9 + \tfrac{70}{10} \left(\tfrac{t-t_2}{t_3-t_2}\right)^{10}\right] \cdot \omega_0 \end{aligned} \tag{80}$$

Specifically we choose here

$$t_1 = 1s, \quad t_2 = 2s, \quad t_3 = 3s \quad \text{and} \quad \omega_0 = \pi \tag{81}$$

Since during this motion the inner triangle (body 7) shall not rotate we also have to implement another constraint suppressing the rotation

$$\mathbf{\Phi}^3(\boldsymbol{q}) = \boldsymbol{e}_2 \cdot \boldsymbol{d}_1^7 \tag{82}$$

The whole control constraint for the desired motion can then be written as:

$$\mathbf{\Phi}_C(\boldsymbol{q}) = \begin{bmatrix} \mathbf{\Phi}_{traj}(\boldsymbol{q}) \\ \mathbf{\Phi}^3(\boldsymbol{q}_0) \end{bmatrix} \tag{83}$$

The corresponding constraint Jacobian for the new control constraints yields:

$$\boldsymbol{B} = \begin{bmatrix} \mathbf{0}_{3\times 48} & \boldsymbol{I}_{3\times 3} & \mathbf{0}_{3\times 3} \end{bmatrix} \tag{84}$$

Since no external forces act on the system, its center of mass does not have to move. Moreover, since no external torques act on the system, the total angular momentum shall be a conserved quantity. The necessary driving torques to perform the desired motion are computed directly.

body	M	E_1	E_2	length	width
1	3	0.0125	0.0275	0.25	0.05
2	3	0.0125	0.0275	0.25	0.05
3	3	0.0125	0.0275	0.25	0.05
4	4	0.04	0.08	0.35	0.05
5	4	0.04	0.08	0.35	0.05
6	4	0.04	0.08	0.35	0.05

Table 3: Inertial and geometric properties pertaining to the six legs of the manipulator.

body	M	E_1	E_2	L
7	3	0.0408	0.0408	0.4
8	8	0.1	0.1	1.0

Table 4: Inertial and geometric properties pertaining to the two platforms of the manipulator.

Inertial and geometric properties of the rigid bodies constituting the parallel manipulator are summarized in Tables 3 and 4. In this connection, the two platforms (bodies 7 and 8) coincide with isosceles triangles of side-length L (Table 4).
The initial configuration of the closed-loop system can completely be specified by its generalized coordinates, accordingly

$$\boldsymbol{u}_0 = \begin{bmatrix} \boldsymbol{\varphi}^8(0) \cdot \boldsymbol{e}_1 \\ \boldsymbol{\varphi}^8(0) \cdot \boldsymbol{e}_2 \\ \theta^8(0) \\ \Theta^1(0) \\ \Theta^2(0) \\ \Theta^3(0) \\ \Theta^4(0) \\ \Theta^5(0) \\ \Theta^6(0) \\ \Theta^7(0) \end{bmatrix} = \begin{bmatrix} 0 \\ 0.2887 \\ 0 \\ \frac{\pi}{6} \\ \frac{2}{3}\pi \\ -\frac{\pi}{2} \\ \frac{\pi}{6} \\ \frac{2}{3}\pi \\ -\frac{\pi}{2} \\ 0 \end{bmatrix} \tag{85}$$

where the value of the initial posture of the small triangle (body 7) has been rounded for simplicity of exposition. As expected, the present energy-momentum schemes does indeed satisfy the above-mentioned conservation properties for any time step Δt, see Fig. 14. The simulated motion of the manipulator is illustrated in Fig. 16 by showing snapshots of the multibody system at subsequent points of time. The conservation of the total angular momentum also indicates that the position of the center of mass does not move for all times. The red glowing path in Fig. 16 corresponds to the trajectory of the center of mass of the small platform (body 7), representing the prescribed trajectory. Moreover, the evolution of the joint-angles $\Theta^1(t)$, $\Theta^2(t)$ and $\Theta^3(t)$, the translational displacements of the prismatic pairs $u_1(t)$, $u_2(t)$ and $u_3(t)$ calculated with a time step of $\Delta t = 0.02$, are depicted in Fig. 15 and Fig. 17. The necessary driving torques to perform the prescribed motion are displayed in Fig. 18.

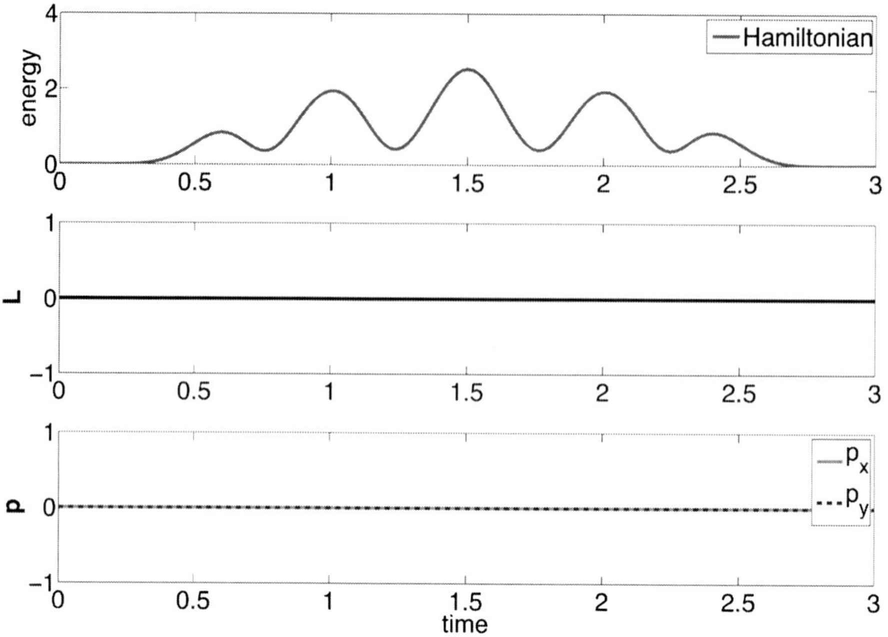

Figure 14: Total energy, conservation of angular and linear momentum ($\Delta t = 0.02$).

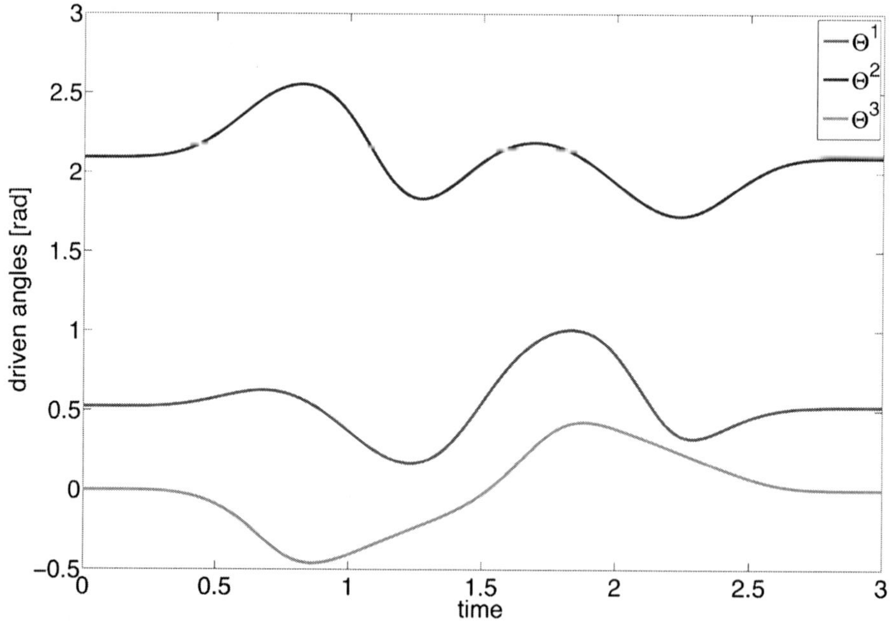

Figure 15: Joint-angles over time ($\Delta t = 0.02$).

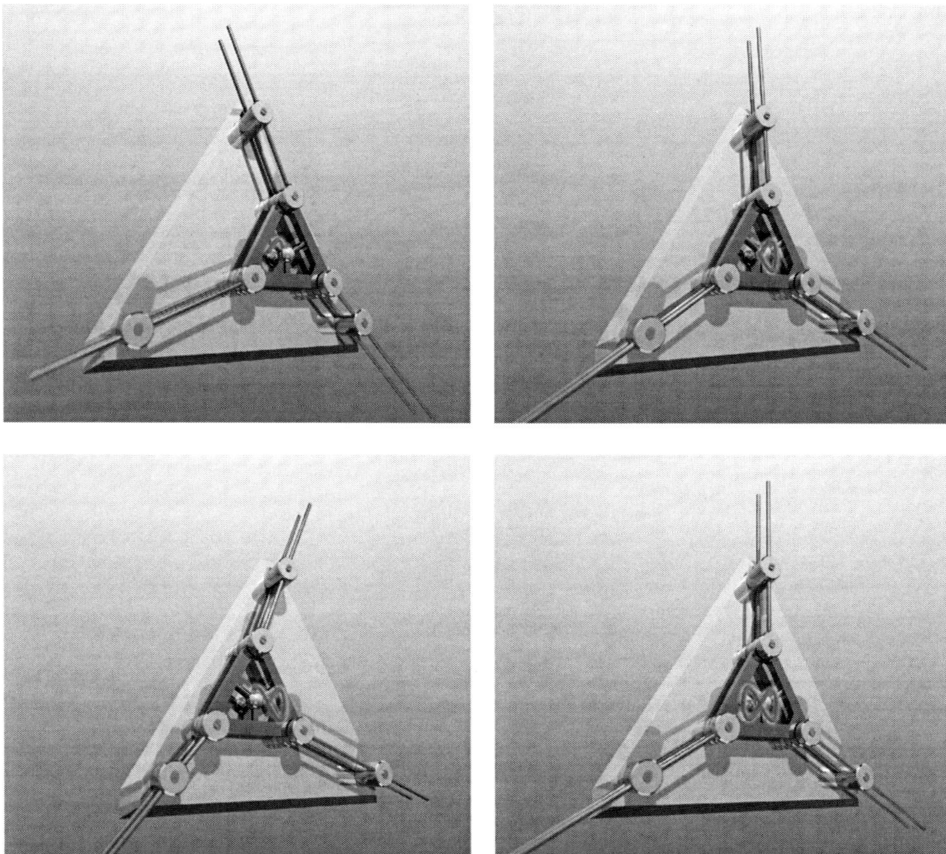

Figure 16: Snapshots of the motion of the free floating parallel manipulator for t $\in$ {1, 1.5, 2, 3}s.

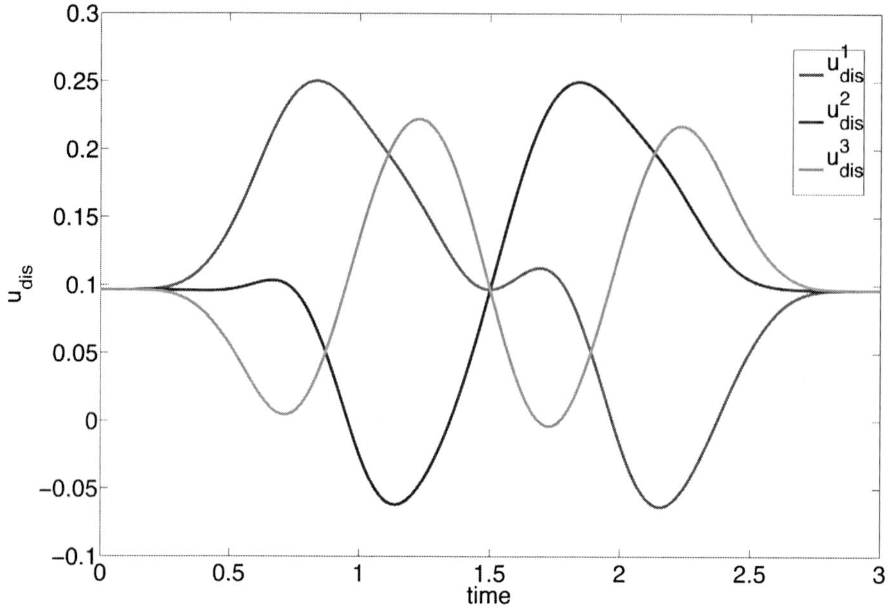

Figure 17: Augmented translational displacement over time (Δ t = 0.02).

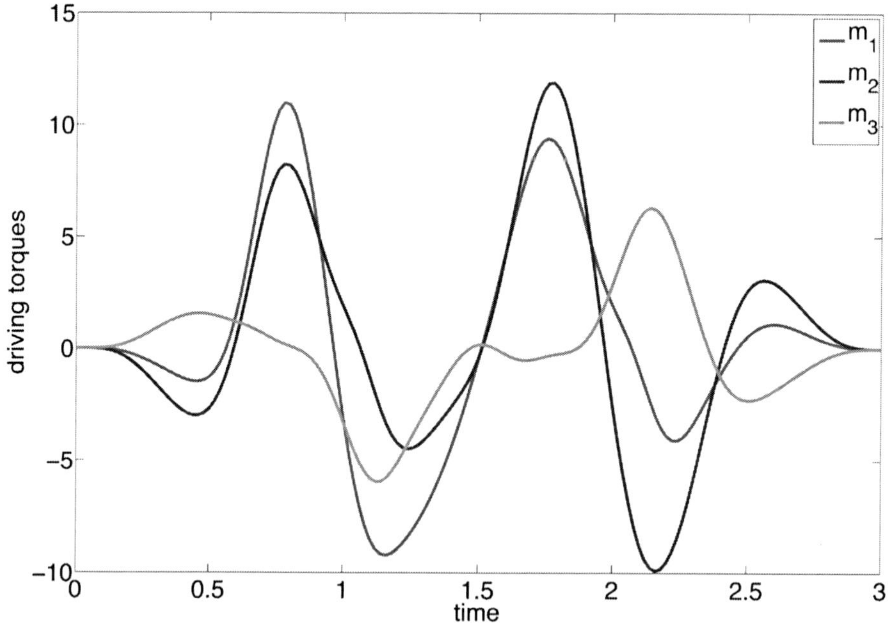

Figure 18: Driven joint-torques over time (Δ t = 0.02).

6. Conclusions

We have shown that the proposed rotationless formulation of multibody dynamics is well-suited for the energymomentum conserving integration of both open-loop and closed-loop multibody systems. Although the use of rotations has been completely circumvented throughout the whole discretization, joint-forces can still be applied to a specific multibody system by resorting to the proposed coordinate augmentation technique.
The present developments have been restricted to the planar case. However, it is important to note, that the extension to the three-dimensional setting can be performed without any conceptual differences. Similarly, alternative types of joints belonging to the class of lower kinematic pairs such as cylindric joints can be easily incorporated into the present approach. Both aforementioned issues have been addressed in [BL06].
The numerical examples presented herein have been specifically designed to check the algorithmic conservation properties. Within computational accuracy, the present approach facilitates the algorithmic conservation of energy as well as linear and angular momentum. Energy-momentum preserving schemes meet the specific demands on the stable numerical integration of the underlying index-3 DAEs. While the BEM scheme employed herein (cf. Section 2.1.1) is second-order accurate in the state space coordinates, higher-order energy-momentum schemes may be designed as set forth in [BS02a,GBS05]. The ostensible disadvantage of using redundant coordinates can be remedied by applying the size reduction techniques proposed in [BU07,BL06]. Specifically, it is shown in [BU07] that these techniques can be systematically applied to closed loop systems. Accordingly, they can be directly used in the example of the parallel manipulator dealt with in Section 5.
We have also presented the incorporation of servo / control constraints into our BEM scheme. This makes possible to perform a direct discretization for fully or underactuated systems and computing directly the necessary input values in order to control a system, without solving the standard inverse dynamics problem. Similar work has also been published in [BUQ].
It is further worth mentioning that semi-discrete formulations of flexible bodies such as nonlinear continua, beams and shells perfectly fit into the present framework provided by the DAEs (1). Accordingly, the present approach can be directly extended to flexible multibody dynamics (see [Bet06,Bet05b,LBS,SB]).

7. References

Armero, F.: Energy-dissipative momentum-conserving time-stepping algorithms for finite strain multiplicative plasticity. In: *Comput. Methods Appl. Mech. Engrg.* 195 (2006), S. 4862–4889

Bauchau, O.A. ; Bottasso, C.L.: On the design of energy preserving and decaying schemes for flexible, nonlinear multi-body systems. In: *Comput. Methods Appl. Mech. Engrg.* 169 (1999), S. 61–79

Brank, B. ; Briseghella, L. ; Tonello, N. ; Damjanic, F.B.: On Non-Linear Dynamics of Shells: Implementation of Energy-Momentum Conserving Algorithm for a Finite Rotation Shell Model. In: *Int. J. Numer. Methods Eng.* 42 (1998), S. 409–442

Betsch, P.: The discrete null space method for the energy consistent integration of constrained mechanical systems. Part I: Holonomic constraints. In: *Comput. Methods Appl. Mech. Engrg.* 194 (2005), Nr. 50-52, S. 5159–5190

Betsch, P.: On the discretization of geometrically exact shells for flexible multibody dynamics. In: *Proceedings of the ECCOMAS Thematic Conference on Multibody Dynamics (on CD).* Madrid, Spain, 21-24 June 2005, S. 1–13

Betsch, P.: Energy-consistent numerical integration of mechanical systems with mixed holonomic and nonholonomic constraints. In: *Comput. Methods Appl. Mech. Engrg.* 195 (2006), S. 7020–7035

Blajer, W. ; Ko!lodziejczyk, K.: A Geometric Approach to Solving Problems of Control Constraints: Theory and a DAE Framework. In: *Multibody System Dynamics* 11 (2004), Nr. 4, S. 343–364

Betsch, P. ; Leyendecker, S.: The discrete null space method for the energy consistent integration of constrained mechanical systems. Part II: Multibody dynamics. In: *Int. J. Numer. Methods Eng.* 67 (2006), Nr. 4, S. 499–552

Betsch, P. ; Steinmann, P.: Conservation Properties of a Time FE Method. Part II: Time-Stepping Schemes for Nonlinear Elastodynamics. In: *Int. J. Numer. Methods Eng.* 50 (2001), S. 1931–1955

Betsch, P. ; Steinmann, P.: Constrained Integration of Rigid Body Dynamics. In: *Comput.Methods Appl. Mech. Engrg.* 191 (2001), S. 467–488

Betsch, P. ; Steinmann, P.: Conservation Properties of a Time FE Method. Part III: Mechanical systems with holonomic constraints. In: *Int. J. Numer. Methods Eng.* 53 (2002), S. 2271–2304

Betsch, P. ; Steinmann, P.: A DAE approach to flexible multibody dynamics. In: *Multibody System Dynamics* 8 (2002), S. 367–391

Betsch, P. ; Uhlar, S.: Energy-momentum conserving integration of multibody dynamics. In: *Multibody System Dynamics* 17 (2007), Nr. 4, S. 243–289

Betsch, P. ; Uhlar, S. ; Quasem, M.: *On the incorporation of servo constraints into a rotationless formulation of flexible multibody dynamics.* In Proceedings of the ECCOMAS Thematic Conference on Multibody Dynamics, 25-28 June 2007, Milano, Italy,

Crisfield, M.A. ; Jeleni´c, G.: Energy/Momentum Conserving Time Integration Procedures with Finite Elements and Large Rotations. In: Ambr´osio, J. (Hrsg.) ; Kleiber, M. (Hrsg.): *NATOARW on Comp. Aspects of Nonlin. Struct. Sys. with Large Rigid Body Motion.* Pultusk, Poland, July 2-7, 2000, S. 181–200

Groß, M. ; Betsch, P. ; Steinmann, P.: Conservation properties of a time FE method. Part IV: Higher order energy and momentum conserving schemes. In: *Int. J. Numer. Methods Eng.* 63 (2005), S. 1849–1897

G´eradin, M. ; Cardona, A.: *Flexible multibody dynamics: A finite element approach.* John Wiley & Sons, 2001

Goicolea, J.M. ; Garcia Orden, J.C.: Dynamic analysis of rigid and deformable multibody systems with penalty methods and energy-momentum schemes. In: *Comput. Methods Appl. Mech. Engrg.* 188 (2000), S. 789–804

Gonzalez, O.: Time Integration and Discrete Hamiltonian Systems. In: *J. Nonlinear Sci.* 6 (1996), S. 449–467

Gonzalez, O.: Mechanical Systems Subject to Holonomic Constraints: Differential-Algebraic Formulations and Conservative Integration. In: *Physica D* 132 (1999), S. 165–174

Greenspan, D.: Conservative Numerical Methods for $\ddot{x} = f(x)$. In: *Journal of Computational Physics* 56 (1984), S. 28–41

Gonzalez, O. ; Simo, J.C.: On the Stability of Symplectic and Energy-Momentum Algorithms for non-linear Hamiltonian Systems with Symmetry. In: *Comput. Methods Appl. Mech. Engrg.* 134 (1996), S. 197–222

Ibrahimbegovi´c, A. ; Mamouri, S. ; Taylor, R.L. ; Chen, A.J.: Finite Element Method in Dynamics of Flexible Multibody Systems: Modeling of Holonomic Constraints and Energy Conserving Integration Schemes. In: *Multibody System Dynamics* 4 (2000), Nr. 2-3, S. 195–223

Jalon, J. Garcia d. ; Unda, J. ; Avello, A.: Natural coordinates for the computer analysis of multibody systems. In: *Comput. Methods Appl. Mech. Engrg.* 56 (1986), S. 309–327

Kunkel, P. ; Mehrmann, V.: *Differential-Algebraic Equations.* European Mathematical Society,2006

Leyendecker, S. ; Betsch, P. ; Steinmann, P.: *The discrete null space method for the energy consistent integration of constrained mechanical systems. Part III: Flexible multibody dynamics.* Accepted for publication in Multibody System Dynamics,

Leyendecker, S. ; Betsch, P. ; Steinmann, P.: Energy-conserving integration of constrained Hamiltonian systems - a comparison of approaches. In: *Computational Mechanics* 33 (2004), Nr. 3, S. 174–185

Laursen, T.A. ; Love, G.R.: Improved implicit integrators for transient impact problems-geometric admissibility withing the conserving framework. In: *Int. J. Numer. Methods Eng.* 53 (2002), S. 245–274

Leimkuhler, B. ; Reich, S.: *Simulating Hamiltonian Dynamics.* Cambridge University Press, 2004

McPhee, J.J. ; Redmond, S.M.: Modelling multibody systems with indirect coordinates. In: *Comput. Methods Appl. Mech. Engrg.* 195 (2006), S. 6942–6957

Puso, M.A.: An energy and momentum conserving method for rigid-flexible body dynamics. In: *Int. J. Numer. Methods Eng.* 53 (2002), S. 1393–1414

Romero, I. ; Armero, F.: An objective finite element approximation of the kinematics of geometrically exact rods and its use in the formulation of an energy-momentum conserving scheme in dynamics. In: *Int. J. Numer. Methods Eng.* 54 (2002), S. 1683–1716

Rosenberg, R.M.: *Analytical dynamics of discrete systems.* Plenum Press, 1977

Rubin, H. ; Ungar, P.: Motion under a strong constraining force. In: *Commun. Pure Appl. Math.* 10 (1957), Nr. 1, S. 65–87

S¨anger, N. ; Betsch, P.: *A uniform rotationless formulation of flexible multibody dynamics: Conserving integration of rigid bodies, nonlinear beams and shells.* In Proceedings of the ECCOMAS Thematic Conference on Multibody Dynamics, 25 28 June 2007, Milano, Italy,

Simo, J.C. ; Tarnow, N.: The Discrete Energy-Momentum Method. Conserving Algorithms for Nonlinear Elastodynamics. In: *Z. angew. Math. Phys. (ZAMP)* 43 (1992), S. 757–792

Simo, J.C. ; Tarnow, N. ; Wong, K.K.: Exact Energy-Momentum Conserving Algorithms and Symplectic Schemes for Nonlinear Dynamics. In: *Comput. Methods Appl. Mech. Engrg.* 100 (1992), S. 63–116

Simo, J.C. ; Wong, K.K.: Unconditionally Stable Algorithms for Rigid Body Dynamics that Exactly Preserve Energy and Momentum. In: *Int. J. Numer. Methods Eng.* 31 (1991), S. 19–52

6

Wire Robots Part I Kinematics, Analysis & Design

Tobias Bruckmann, Lars Mikelsons, Thorsten Brandt, Manfred Hiller and Dieter Schramm
University Duisburg-Essen (Chair for Mechatronics)
Germany

1. Introduction

One drawback of classical parallel robots is their limited workspace, mainly due to the limitation of the stroke of linear actuators. Parallel wire robots (also known as Tendon-based Steward platforms or cable robots) face this problem through substitution of the actuators by wires (or tendons, cables, . . .). Tendon-based Steward platforms have been proposed in (Landsberger & Sheridan, 1985). Although these robots share the basic concepts of classical parallel robots, there are some major differences:

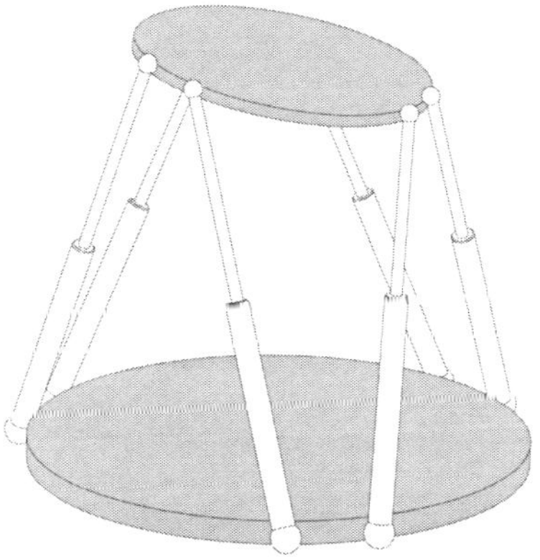

Fig. 1(a) Conventional parallel manipulator

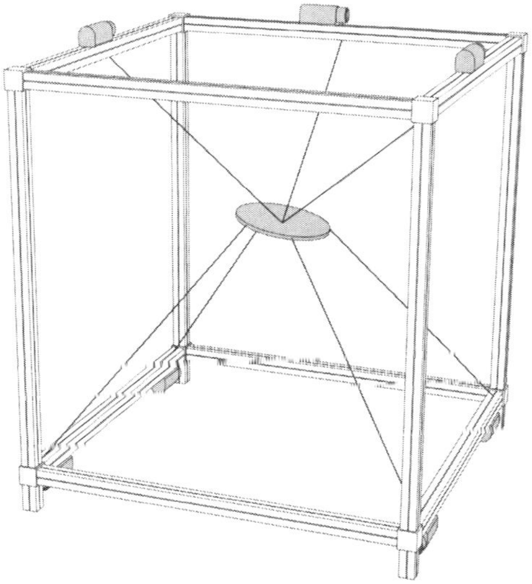

Fig. 1(b) Parallel Wire Robot

- The flexibility of wires allows large changes in the length of the kinematic chain, for example by coiling the tendons onto a drum. This allows to overcome the purely geometric workspace limitation factor of classical robots.

- Wires can be coiled by very fast drums while the moving mass of the robot is extremely low, which allows the robot to reach very high end effector speeds and accelerations.
- Wires are modeled as unilateral constraints, i.e. wires can only transmit pulling forces.
- The number of wires m can be increased to modify the workspace, to carry higher loads or to increase safety due to redundancy. Thus, having an end effector (in the following called platform) with n degrees-of-freedom (d.o.f.), more than n parallel links are used to connect the platform to the base frame.

This contribution is organized as follows: In section 2 the classification of wire robots, based on several approaches is presented. Furthermore, the kinematic calculations for wire robots are described which is followed by the description of the force equilibrium in section 3. Based on the force equilibrium, methods for workspace analysis and robot design are proposed in section 4 and 5, respectively. This contribution is extended in Part 2 (Bruckmann et al., 2008a) by the description of dynamics, control methods and application examples. Within this and the next chapter, the following abbreviations are used:

$_B\boldsymbol{r}$	vector $\boldsymbol{r}$ denoted in coordinate system $\mathcal{B}$
r_i	i-th component of vector $\boldsymbol{r}$
$\boldsymbol{A}$	matrix $\boldsymbol{A}$
${}^B\boldsymbol{R}_P$	transformation matrix from coordinate system $\mathcal{P}$ to $\mathcal{B}$
$\boldsymbol{A}^T$	shorthand for the transpose of $\boldsymbol{A}$
$\boldsymbol{A}^{-T}$	shorthand for $(\boldsymbol{A}^{-1})^T$
$\dot{x}$	derivation of x with resprect to time, $\dot{x} = \frac{dx}{dt}$

2. Kinematics

2.1 Classification

For wire robots, different classifications based on the difference between the number of wires m and the number d.o.f. n have been proposed. Further on, this difference is called the redundancy $r = m - n$. According to (Ming & Higuchi, 1994) wire robots can be categorized based on the redundancy as follows:

- CRPM (Completely Restrained Parallel Manipulator): The pose of the robot is completely determined by the unilateral kinematic constraints defined by the tensed wires. For a CRPM at least $m = n + 1$ wires are needed.
- IRPM (Incompletely Restrained Parallel Manipulator): In addition to the unilateral constraints induced by the tensed wires at least one dynamical equation is required to describe the pose of the end effector.

In (Verhoeven, 2004) the category of CRPMs is further divided into two categories. The class of the CRPMs is restricted to robots with $m = n+1$ wires. Wire robots with $m > n + 1$ are called RRPMs (Redundantly Restrained Parallel Manipulator). Note that within this definition CRPM and RRPM robots can convert into IRPM robots if they are used at poses where external wrenches (inertia and generalized forces and torques applied onto the platform) are necessary to find completely positive wire forces. Therefore in (Verhoeven, 2004) another classification is proposed based on the number of controlled d.o.f. which is listed below.

- 1T: linear motion of a point
- 2T: planar motion of a point
- 1R2T: planar motion of a body
- 3T: spatial motion of a point
- 2R3T: spatial motion of a beam
- 3R3T: spatial motion of a body

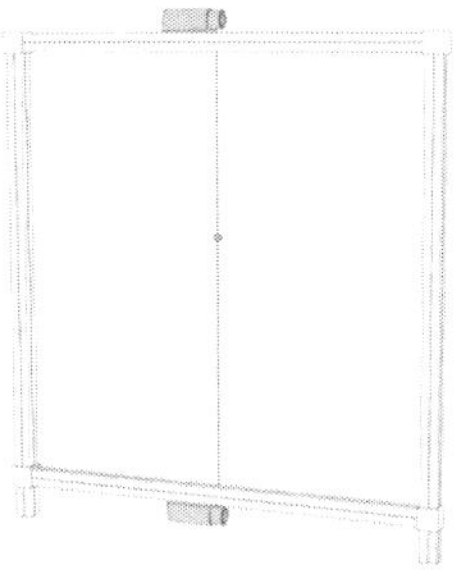

Fig. 2(a) class 1T

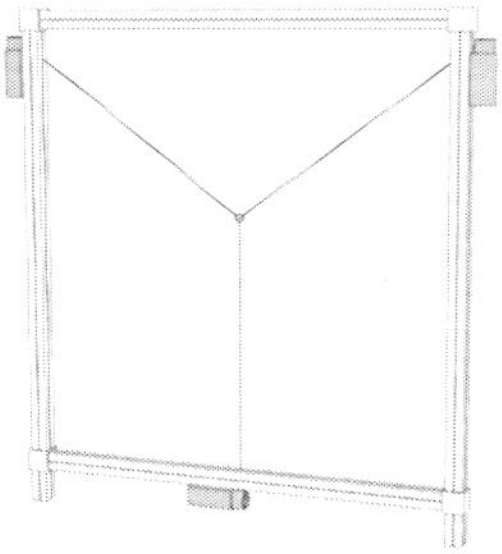

Fig. 2(b) class 2T

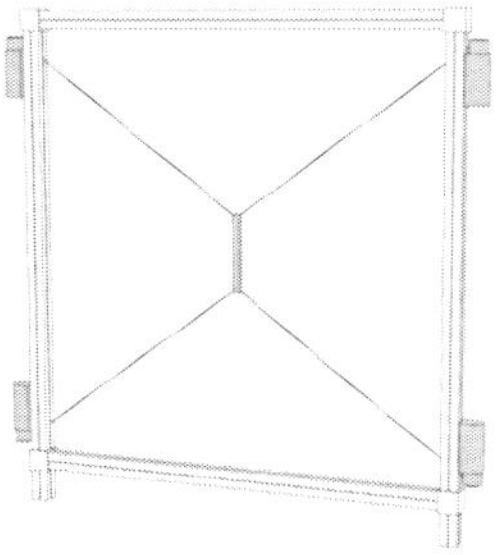

Fig. 2(c) class 1R2T

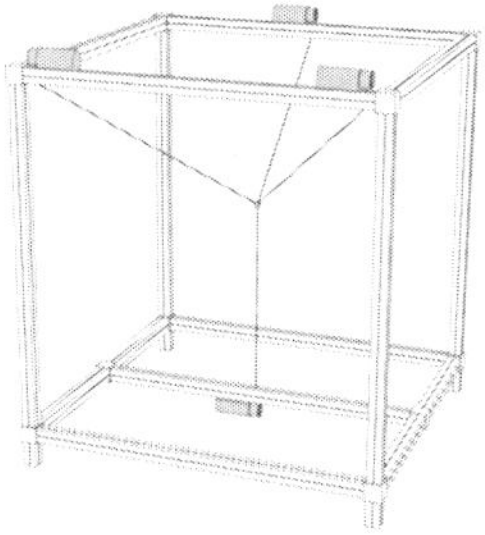

Fig. 2(d) class 3T

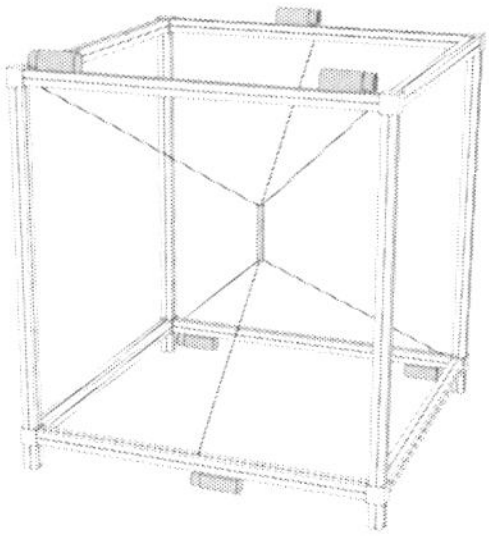

Fig. 2(e) class 2R3T

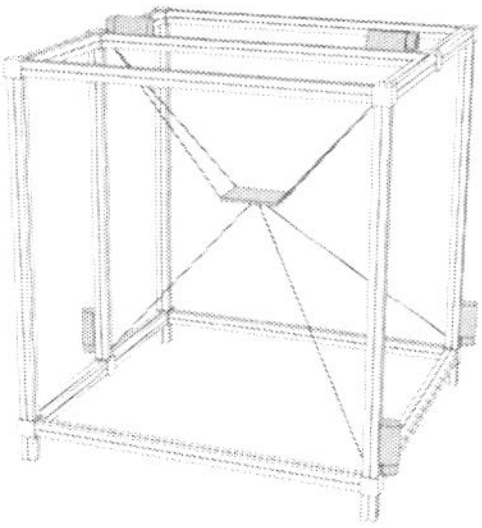

Fig. 2(f) class 3R3T

Here T stands for translational and R for rotational d.o.f.. It is notable that this definition is complete and covers all wire robots. The classification of (Fang, 2005) is similar to Verhoeven's approach. Here, three classes are defined as:

- IKRM (Incompletely Kinematic Restrained Manipulators), where $m < n$
- CKRM (Completely Kinematic Restrained Manipulators), where $m = n$
- RAMP (Redundantly Actuated Manipulators), where $m > n + 1$

This chapter as well as the next one focuses on CRPM and RRPM robots. For IRPM see e.g. (Maier (2004)).

2.2 Inverse kinematics

Inverse kinematics refers to the problem of calculating the joint variables for a given end-effector pose. For the class of robots under consideration those are the lengths of the wires, comparable to the strokes of linear actuators. Therefore, the kinematical description of a wire robot resembles the kinematic structure of a Stewart-Gough platform, presuming the wires are always tensed and can thus be treated as line segments representing bilateral constraints. Modeling a wire robot as a platform, which is connected to m points on the base by m bilateral constraints, it is reasonable to denote the platform pose $\boldsymbol{x} = [\ {}^B\boldsymbol{r}^T\ \varphi\ \vartheta\ \psi\ ]$ and the base points ${}^B\boldsymbol{b}_i$, $i = 1 \leq i \leq m$, referenced in the inertial frame $\mathcal{B}$. Besides that, the platform connection points $\boldsymbol{p}^i$ are referenced in the platform-fixed coordinate frame $\mathcal{P}$.The orientation of the platform in the base frame is represented by the rotation matrix ${}^B\boldsymbol{R}_P$. Note that throughout this chapter roll-pitch-yaw angles are used. Assuming the wires are led by point-shaped guidances (e.g. small ceramic eyes) from the winches to the platform, the base vectors ${}^B\boldsymbol{b}_i$ are constant. Now the vector chain pictured in fig. 3 delivers

$$ {}^B\boldsymbol{l}_i = {}^B\boldsymbol{b}_i - \underbrace{{}^B\boldsymbol{r} - {}^B\boldsymbol{R}_P\,{}^P\boldsymbol{p}_i}_{{}^B\boldsymbol{p}_i}, \quad 1 \leq i \leq m \tag{1} $$

immediately. Hence, the length of the i_{th} wire can be calculated by

$$ l_i = \left\| {}^B\boldsymbol{b}_i - {}^B\boldsymbol{p}_i \right\|_2, \quad 1 \leq i \leq m \tag{2} $$

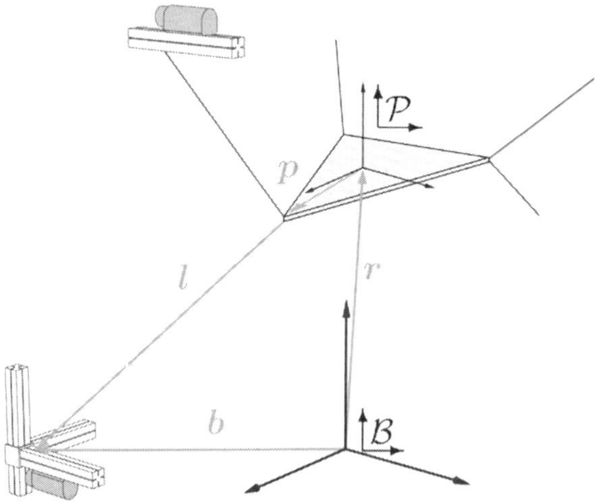

Fig. 3: Kinematics of a wire robot

Based on the relatively simple inverse kinematics, a position control in joint space can be designed for a wire robot which already may deliver satisfying results. Note, this simple calculation only holds for the described simple guidance. While it may be sufficient for simple prototypes, it suffers from a very high wear and abrasion. Thus it is not feasible for practical applications. An alternative concept is the roller-based guidance which is e.g. widely used in theatre and stage technology, see fig. 4. As a drawback, the kinematical description becomes more difficult due to the pose dependent exit points points ${}^B\boldsymbol{s}_i$ of the wires. The roller with radius ρ is mounted onto a pivot arm. To calculate the exit points ${}^B\boldsymbol{s}_i$, two angles have to be known: the pivoting angle θ_i and the wrap angle α_i(see fig. 4). The pivoting angle can be calculated using a projection onto the plane D whose normal vector is the rotation axis (without loss of generality the z-axis of the inertial frame) of the pivoting angle as:

$$\tan \Theta_i = \frac{{}^B p_{i,y} - {}^B b_{i,y}}{{}^B p_{i,x} - {}^B b_{i,x}}, \quad 1 \le i \le m. \tag{3}$$

Here ${}^B\boldsymbol{b}_i$ denotes the vector to the point, at which the wire enters the roller. With this knowledge the vector ${}^B\boldsymbol{m}_i$ to the midpoint of the i−th roller can be constructed

$${}^B\boldsymbol{m}_i = {}^B\boldsymbol{b}_i + \boldsymbol{R}_{z_B,\Theta_i} \cdot \rho \cdot {}^B\boldsymbol{e}_x \tag{4}$$

Where R_{z^B,Θ_i} is a rotation matrix for angle Θ_i around the z-axis of the inertial frame. Note that without loss of generality the projection of ${}^B\boldsymbol{b}_i - {}^B\boldsymbol{m}_i$ onto the $x - z$-plane of $\mathcal{B}$ is parallel to the x-axis in the reference orientation of the roller. Then the wrap angle α_i is according to fig. 4 given by

$$\alpha_i = \pi - (\alpha_{i,1} + \alpha_{i,2}), \tag{5}$$

where

$$\sin \alpha_{i,1} = \frac{{}^B p_{i,z} - {}^B m_{i,z}}{\|{}^B\boldsymbol{p}_i - {}^B\boldsymbol{m}_i\|_2}, \quad \cos \alpha_{i,2} = \frac{\rho}{\|{}^B\boldsymbol{p}_i - {}^B\boldsymbol{m}_i\|_2}, \quad 1 \le i \le m. \tag{6}$$

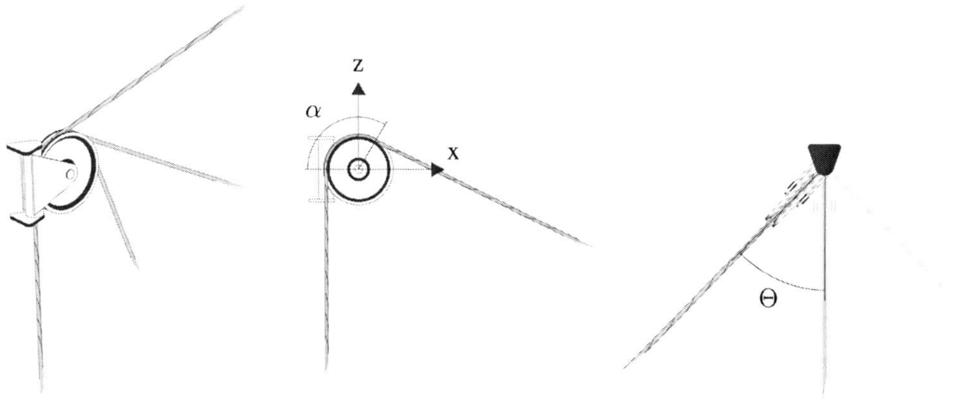

Fig. 4: Roller-based guidance

In a projection onto the plane D, $\alpha_{i,1}$ describes the angle between the x-y-plane of the inertial frame and the vector $\boldsymbol{q}$ from ${}^B\boldsymbol{m}_i$ to the platform connection point ${}^B\boldsymbol{p}_i$. The angle $\alpha_{i,2}$ is the angle between the vector from ${}^B\boldsymbol{m}_i$ to the exit point and vector $\boldsymbol{q}$. Furthermore the exit point ${}^B\boldsymbol{s}_i$ of the i-th wire can be found as

$${}^B\boldsymbol{s}_i = \begin{bmatrix} {}^B m_{i,x} + \rho \cos \alpha_1 \cdot \cos \Theta_1 \\ {}^B m_{i,y} + \rho \cos \alpha_1 \cdot \sin \Theta_1 \\ {}^B m_{i,z} + \rho \cdot \sin \alpha_1 \end{bmatrix}, \quad 1 \le i \le m. \tag{7}$$

Therefore the wire length can be calculated by

$$l_i = \rho \cdot (\pi - \alpha_i) + \left\| {}^B\boldsymbol{s}_i - {}^B\boldsymbol{p}_i \right\|_2 . \tag{8}$$

Analog to the Stewart-Gough platform, the forward kinematics is much more complicated, in particular for the case of roller guidances.

2.3 Forward kinematics

In opposite to the inverse kinematics, where the equations are decoupled and therefore straight forward to solve, the forward kinematics problem is more involved. In general the forward kinematics are not analytically solveable. However, in some cases a geometrical approach allows a closed solution. To be more precise, a setup with three base points connected to one platform connection points leads to the task of finding the intersection points of three spheres where the radii of the spheres represent the measured lengths of the wires and the centers of the spheres are the base points b_i. Hence, the spheres represent possible positions of the endpoints of the wires. Note, that a point-shaped wire guidance is presumed. More details can be found in (Williams et al., 2004). Nevertheless, in general no analytical solution is at hand. Thus, numerical approaches have to be employed to find the solution, which is disadvantageous in terms of computation time, especially when the computation has to be done in real-time. The forward kinematics problem is generally described by m nonlinear equations in n unknown variables.

$$\rho \cdot (\pi - \alpha_i) + \left\| {}^B\boldsymbol{s}_i - {}^B\boldsymbol{p}_i \right\|_2 - l_i = 0 \tag{9}$$

If point-shaped wire guidances are used, ρ becomes zero. In case of $m = n$, (Fang, 2005) proposes to apply a Newton-Raphson solver while for CRPMs and RRPMs, one has to consider an overdetermined system. A standard approach to this class of problems is the use of a least square method which minimizes the influence of measurement errors. However, the Newton-Raphson approach can also be used for the case of $m \geq n + 1$ as shown in the following, denoting the vector of wire lengths $\boldsymbol{l} = [\, l_1 \ldots l_m]^T$ (Fang, 2005):

$$\dot{\boldsymbol{l}}(t) = \boldsymbol{J}_{inv}(\boldsymbol{l}(t)) \cdot \dot{\boldsymbol{x}}(t) \quad \boldsymbol{J}_{inv}(\boldsymbol{l}(t)) = \frac{\partial \boldsymbol{l}}{\partial \boldsymbol{x}} \in \mathbb{R}^{m \times n}. \tag{10}$$

Since in kinematics positive wire tensions are assumed, the wires are modeled as bilateral constraints, already six constraints fix the platform, i.e. r rows of the inverse Jacobian $\boldsymbol{J}_{inv}$ can be removed, resulting in $\tilde{\boldsymbol{J}}_{inv}$. Assuming $\boldsymbol{J}_{inv}$ having full rank, in case of a CRPM, any arbitrary choice of a row leads to full ranked $\tilde{\boldsymbol{J}}_{inv}$. In case of a RRPM, this does not hold in general. Thus, one has to test for a feasible choice of r rows which allows to calculate the reduced Jacobian of the forward kinematics $\tilde{\boldsymbol{J}}_{forw} = \tilde{\boldsymbol{J}}_{inv}^{-1}$. Without loss of generality, let n wire lengths $l_1, \ldots, l_n$ be chosen. Thus,

$$\dot{\boldsymbol{x}} = \tilde{\boldsymbol{J}}_{forw}(\boldsymbol{x}(t)) \underbrace{\begin{bmatrix} l_1 & l_2 & \ldots & l_n \end{bmatrix}^T}_{\boldsymbol{l}_{red}} \tag{11}$$

holds. The position at the time t_1 can be calculated by forward integration in time

$$\boldsymbol{x}(t) = \boldsymbol{x}(t_0) + \int_{t_0}^{t} \dot{\boldsymbol{x}}(\tau)\, d\tau \;\Leftrightarrow\; \boldsymbol{x}(t) = \boldsymbol{x}(t_0) + \int_{t_0}^{t} \tilde{\boldsymbol{J}}_{forw}(\boldsymbol{x}(\tau))\dot{\boldsymbol{l}}_{red}(\tau)\, d\tau. \tag{12}$$

Taylor expansion of the second term around t_0 delivers

$$\int_{t_0}^{t} \tilde{\boldsymbol{J}}_{forw}(\boldsymbol{x}(\tau))\dot{\boldsymbol{l}}_{red}(\tau)\, d\tau = \sum_{k=0}^{\infty} \frac{1}{k!}\Big(\int_{t_0}^{t} \tilde{\boldsymbol{J}}_{forw}(\boldsymbol{x}(\tau))\dot{\boldsymbol{l}}_{red}(\tau)\, d\tau\Big)^{(k)}_{|t=t_0} \cdot (t - t_0) \tag{13}$$

Neglecting terms of second order and higher leads to

$$\int_{t_0}^{t} \tilde{\boldsymbol{J}}_{forw}(\boldsymbol{x}(\tau))\dot{\boldsymbol{l}}_{red}(\tau)\, d\tau \approx \tilde{\boldsymbol{J}}_{forw}(\boldsymbol{x}(t_0))\dot{\boldsymbol{l}}_{red}(t_0) \cdot (t - t_0). \tag{14}$$

Approximating the differential quotient by the difference quotient gives

$$\tilde{\boldsymbol{J}}_{forw}(\boldsymbol{x}(t_0))\dot{\boldsymbol{l}}_{red}(t_0) \cdot (t - t_0) \approx \tilde{\boldsymbol{J}}_{forw}(\boldsymbol{x}(t_0))\Delta \boldsymbol{l}_{red}, \tag{15}$$

where

$$\Delta \boldsymbol{l}_{red} = \boldsymbol{l}_{red}(t) - \boldsymbol{l}_{red} \cdot (t_0) \tag{16}$$

Using these simplified expressions, the platform pose x can be approximated by $\boldsymbol{x}_{app}$:

$$\boldsymbol{x}_{app}(t) = \boldsymbol{x}(t_0) + \tilde{\boldsymbol{J}}_{forw}(\boldsymbol{x}(t_0))\Delta \boldsymbol{l}_{red} \tag{17}$$

For $\boldsymbol{x}_{app}$ (t), the inverse kinematics and the pose estimation error Δ $\boldsymbol{x}$ (t) can be calculated, delivering the wire lengths l_{app} for the approximated pose. Now the difference Δ $l(t)$ between the measured and approximated wire lengths can be calculated, giving a measure for the pose error:

$$\Delta \boldsymbol{l}(t) = \boldsymbol{l}_{app}(t) - \boldsymbol{l}(t) \quad \wedge \quad \Delta \boldsymbol{x}(t) - \boldsymbol{x}_{app}(t) \quad \boldsymbol{x}(t) \Leftrightarrow \boldsymbol{x}(t) = \boldsymbol{x}_{app}(t) - \Delta \boldsymbol{x}(t) \tag{18}$$

Once again using the approximations

$$\Delta \boldsymbol{x} = \tilde{\boldsymbol{J}}_{forw}(\boldsymbol{x}_{app}(t))\Delta \boldsymbol{l}_{red} \quad \text{for} \quad \dot{\boldsymbol{x}} = \tilde{\boldsymbol{J}}_{forw}(\boldsymbol{x}(t))\dot{\boldsymbol{l}}_{red} \tag{19}$$

it follows

$$\boldsymbol{x}(t) \approx \boldsymbol{x}_{app}(t) - \tilde{\boldsymbol{J}}_{forw}(\boldsymbol{x}_{app}(t))\Delta \boldsymbol{l}_{red} = \boldsymbol{x}_{app}(t) - \tilde{\boldsymbol{J}}_{forw}(\boldsymbol{x}_{app}(t))(\boldsymbol{l}_{app}(t) - \boldsymbol{l}(t)) \tag{20}$$

where $l_{app}(t)$ is calculated by the inverse kinematics for $\boldsymbol{x}_{app}(t)$. Noteworthy, this approach works only for small pose displacements. When displacements become larger, an iteration can improve the precision of the calculated pose by using $\boldsymbol{x}(t)$ as the estimate $\boldsymbol{x}_{app}(t)$ for the next step (Merlet, 2000). In (Williams et al., 2004), the authors show an iterative algorithm for a roller-based wire guidance neglecting the pivoting angle.

3. Force equilibirum

The end effector of wire robots is guided along desired trajectories by tensed wires. This design is superior to classical parallel kinematic designs in terms of workspace size - due to the practically unlimited actuator stroke creating potentially large workspaces - and mechanical simplicity. On the other hand and caused by the unilateral constraints of the wires, the workspace of wire robots is primarily limited by the forces which may be exerted by the wires. The unilateral constraints necessitate positive forces. Practically, long wires will sag at low tensions which makes kinematical computations more complicated and may lead to vibration problems. Hence, the minimum allowed forces in the wires should never fall below a predefined positive value. Against, high forces lead to increased wear and elastic deformations. Therefore the working load of wires is bounded between predefined values $\boldsymbol{f}_{\min} \in \mathbb{R}^m$ and $\boldsymbol{f}_{\max} \in \mathbb{R}^m$ and wire forces must remain between these limits. Thus, a description of the force distribution in the wires for given end effector poses and wrenches is needed. Here a convenient description of the force distribution will be presented, while in (Bruckmann et al., 2008a) three different methods for the force calculation are shown. The force and torque equilibrium at the end effector gives according to figure 5

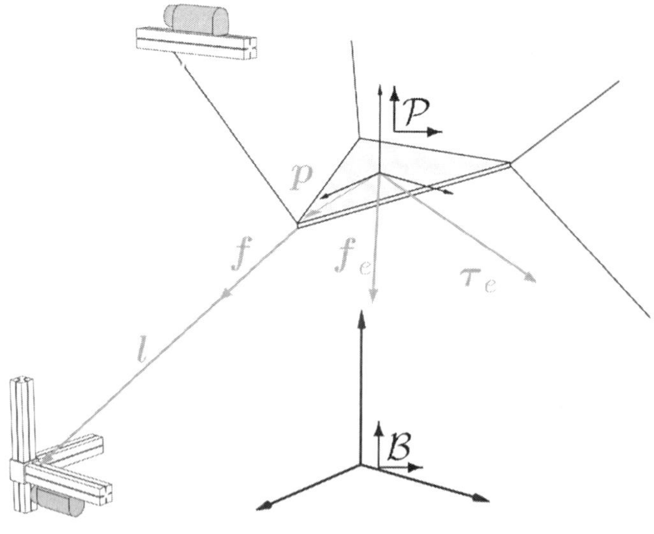

Fig. 5: Forces for a wire robot

$$\sum_{i=1}^{m} \boldsymbol{f}_i + \boldsymbol{f}_p = 0, \text{ and } \sum_{i=1}^{m} \boldsymbol{p}_i \times \boldsymbol{f}_i + \boldsymbol{\tau}_p = 0 \tag{21}$$

The force vectors $\boldsymbol{f}_i$ can be written as

$$\boldsymbol{f}_i = f_i \cdot \frac{\boldsymbol{l}_i}{\|\boldsymbol{l}_i\|_2} = f_i \cdot \boldsymbol{\nu}_i, \quad (1 \leq i \leq m) \tag{22}$$

since the forces act along the wires. Hence, the force and torque equilibrium can be written in matrix form

$$\begin{bmatrix} \boldsymbol{\nu}_1 & \cdots & \boldsymbol{\nu}_m \\ \boldsymbol{p}_1 \times \boldsymbol{\nu}_1 & \cdots & \boldsymbol{p}_m \times \boldsymbol{\nu}_m \end{bmatrix} \begin{bmatrix} f_1 \\ \vdots \\ f_m \end{bmatrix} + \begin{bmatrix} \boldsymbol{f}_p \\ \boldsymbol{\tau}_p \end{bmatrix} = 0 \tag{23}$$

with

$$\boldsymbol{f}_{\max} \geq \boldsymbol{f} \geq \boldsymbol{f}_{\min} > 0 \tag{24}$$

or in a more compact form as

$$\boldsymbol{A}^{\mathrm{T}} \boldsymbol{f} + \boldsymbol{w} = \boldsymbol{0} \tag{25}$$

$$\boldsymbol{f}_{\max} \geq \boldsymbol{f} \geq \boldsymbol{f}_{\min} > 0. \tag{26}$$

In the following the matrix A^{T} is called structure matrix. It is noteworthy that the structure matrix can also be derived as the transpose of the Jacobian of the inverse kinematics, but generally, it is easier to construct it based on the force approach (Verhoeven, 2004).

4. Workspace analysis

In practical applications knowledge of the workspace of the robot under consideration is essential. In contrast to conventional parallel manipulators using rigid links, the workspace of a wire robot is not mainly limited by the actuator strokes, since the length of the wires is not the main limiting factor, just restricted by the drum capacity. In fact, the workspace of a wire robot is limited anyway by the wire force limits $f_{\min}$ and $f_{\max}$. A pose $\boldsymbol{r}$ is said to be part of the workspace if a wire force distribution $\boldsymbol{f}$ exists, such that $f_{\min} \leq \boldsymbol{f} \leq f_{\max}$ holds. Additionally further criteria, like stiffness or wire collisions, can be taken into account. Different methods to calculate the workspace of a wire robot are available. Here discrete methods as well as a continuous method using interval analysis are discussed. Further methods exist as for example presented in (Bosscher & Ebert-Uphoff, 2004), where the workspace boundaries are computed.

4.1 Discrete analysis

In order to perform a discrete workspace analysis at first an assumed superset of the workspace is discretized. Mostly an equidistant discretization is desired. This leads to a set of points, which is then tested with respect to the chosen workspace requirements. This is a widely used approach, but nevertheless, some considerations should be taken into account:

- The calculation of the workspace conditions for the grid points generally requires the verification of a valid wire force distribution. Since it is sufficient to identify any valid distribution, fast calculation methods as presented in section (Bruckmann et al., 2008a) can be employed.
- For some parallel kinematic mechanisms, typically symmetrical configurations are singular, leading to uncontrollable d.o.f. of the end effector. Thus, it is recommended to explicitly test at symmetrical poses of the end effector.
- Generally, it is desired to rule out gaps in the workspace. Using a discrete approach, this is intrinsically impossible, but for practical usage, one may try to increase the grid resolution. Clearly this leads to a dramatical increase of the number of points to be checked and thus to extremely long computation times. To come up against this, parallelisation of the calculation by partitioning the workspace and allocation to different processing units is helpful and especially for this problem very efficient due to the independency of the workspace parts. Nevertheless, up from a specific resolution, continuous methods as presented in the next section should be considered.

4.2 Continuous analysis

In this section a method to compute the workspace of a wire robot, formulating this task as a constraint satisfaction problem (CSP), is shown. The CSP can be solved using interval analysis. However, other solving algorithms are also conceivable. The presented formulation can also be used for design just by interchanging the roles of the variables (Bruckmann et al., 2007), (Bruckmann et al., 2008b). This fact simplifies the generally complicated and complex task of robot design. For details see section 5. In (Gouttefarde et al., 2007) also interval analysis is used to determine the workspace of a wire robot. A criteria for the solvability of the interval formulation of eqn. 24 is given. In particular, the interval formulation is reduced to 2^n $n \times m$ systems of linear inequalities in the form of eqn. 24. The solvability of those 2^n systems of linear inequalities guarantees the existance of at least one valid wire force distribution. Based on this criteria a bisection algorithm is presented. This approach is beneficial in terms of the number of variables on which bisections are performed since no verification or existance variables are required. Here, however the CSP approach is presented due to its straight forward transferability to robot design.

4.2.1 Constraint satisfaction problems (CSP)

A constraint satisfaction problem (CSP) is the problem of determining all $c \in \mathcal{X}_c$ such that

$$\boldsymbol{\Phi}(\boldsymbol{c},\boldsymbol{v}) > 0, \forall \boldsymbol{v} \in \mathcal{X}_v, \tag{27}$$

where Φ is a system of real functions defined on a real domain representing the constraints. It will be shown later that for a description of the workspace, this problem can to be extended to

$$\boldsymbol{\Phi}(\boldsymbol{c}, \boldsymbol{v}, \boldsymbol{e}) > 0, \forall \boldsymbol{v} \in \mathcal{X}_v, \exists \boldsymbol{e} \in \mathcal{X}_e. \tag{28}$$

Within this definition

- c is the vector of the calculation variables,
- v is the vector of the verification and,
- e is the vector of the existance variables.

The solution set for calculaton variables of a CSP is called $\mathcal{X}_S$ i.e.

$$\boldsymbol{\Phi}(\boldsymbol{c}, \boldsymbol{v}, \boldsymbol{e}) > 0, \forall \boldsymbol{c} \in \mathcal{X}_S \subset \mathcal{X}_c, \forall \boldsymbol{v} \in \mathcal{X}_v, \exists \boldsymbol{e} \in \mathcal{X}_e, \tag{29}$$

where $\mathcal{X}_c$ is the so-called search domain, i.e. the range of the calculation variables wherein for solutions is searched.

4.2.2 Workspace analysis as CSP

Examining eqn. 25, the structure matrix A^T needs to be inverted to calculate the wire forces f from a given platform pose and given external forces w. Since A^T has a non-squared shape, this is usually done using the Moore-Penrose pseudo inverse. Thus, the calculated forces will be a least squares solution. In fact, not a least squares result but a force distribution within predefined tensions is demanded. To overcome this problem, the structure matrix is divided into a squared $n \times n$ matrix A^T_{pri} and a second matrix A^T_{sec} with $r = m - n$ columns. Now, the resulting force distribution can be calculated as

$$\boldsymbol{f}_{pri} = -{\boldsymbol{A}_{pri}^T}^{-1}(\boldsymbol{w} + \boldsymbol{A}_{sec}^T \boldsymbol{f}_{sec}). \tag{30}$$

In this equation, f_{sec} is unknown. Every point and wrench satisfying

$$\boldsymbol{f}_{\min} \leq \boldsymbol{f}_{sec} \leq \boldsymbol{f}_{\max} \tag{31}$$

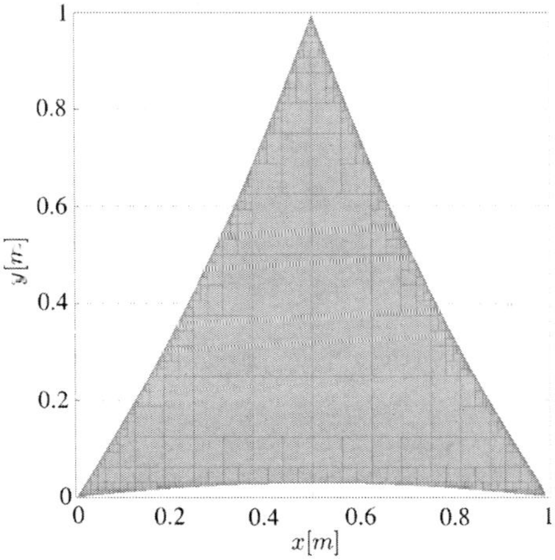

Fig. 6: Force equilibrium workspace of plain manipulator, 2 translational d.o.f., $w^T = (0, 0)N$, $f_{\min} = 10N$, $f_{\max} = 90N$

and leading to primary wire forces

$$\boldsymbol{f}_{\min} \leq \boldsymbol{f}_{pri} \leq \boldsymbol{f}_{\max} \tag{32}$$

belongs to the workspace. Hence eqns. 31 and 32 represent a CSP of the form of eqn. 28 with $\boldsymbol{f}_{sec}$ as existence an variable. To calculate a workspace for a specific robot, the following variable set for the CSP is used:

- The platform coordinates are the *calculation variables.*
- The wire forces $\boldsymbol{f}_{sec}$ are the *existence variables.*
- Optionally, the exerted external wrench w and desired platform orientations can be set as *verification variables.* The workspace for a fix orientation of the platform is called *constant orientation workspace* according to (Merlet, 2000). On the other hand, sometimes free orientation of the platform within given ranges must be possible within the whole workspace. The resulting workspace is called the *total orientation workspace.*

In fig. 6, the workspace of a simple plain manipulator is shown, based on the force equilibrium condition. In fig. 7, the workspace under a possible external load range is shown. Fig. 8(b) shows an example of the workspace of a spatial CRPM robot prototype while fig. 9(b) is the same protoype in a RRPM configuration with 8 wires. Additionally, the RRPM workspace was calculated with a verification range of ±3° for φ and θ, i.e. $\varphi = \theta = [-3, 3]$ °.

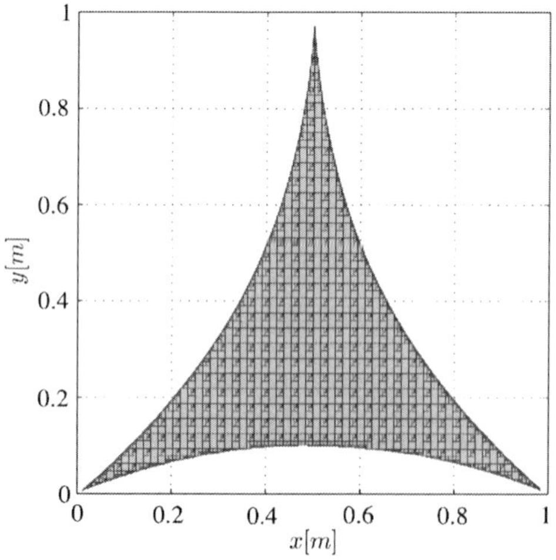

Fig. 7: Force equilibrium workspace of plain manipulator, 2 translational d.o.f., w^{T} = *([−20, 20]N, [−20, 20]N),* $\boldsymbol{f}_{\min}$ *= 10N,* $\boldsymbol{f}_{\max}$ *= 90N*

4.2.3 Interval analysis

Interval Analysis is a powerful tool to solve CSPs. Therefore a short introduction is given in the following section. For two real numbers a, b an interval $\boldsymbol{I} = [a, b]$ is defined as follows

$$[a, b] := \{r \mid a \leq r \leq b\}\,, \tag{33}$$

where

$$a \leq b \tag{34}$$

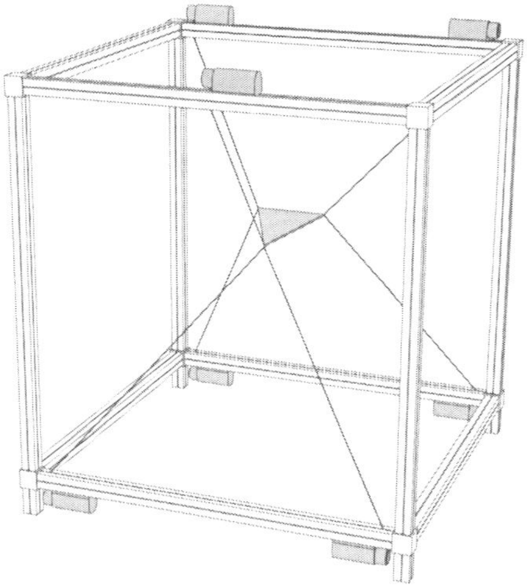

Fig. 8(a) SEGESTA prototype with 7 wires prototype

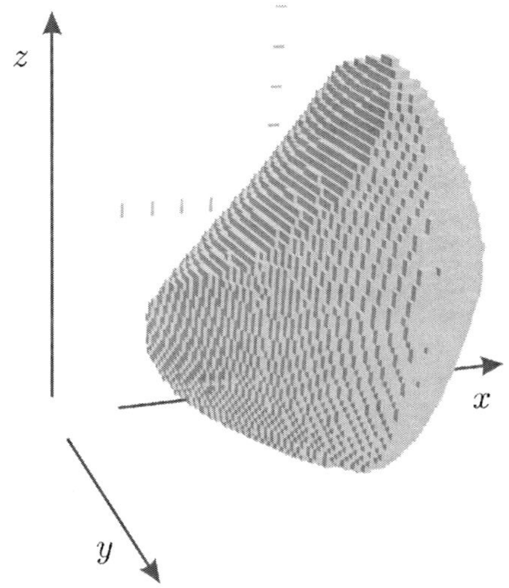

Fig. 8(b)Workspace of the SEGESTA with 7 wires

Then b is called the supremum and a the infimum of $\boldsymbol{I}$. A n-tupel of intervals is called box or interval vector. It is possible to define every operation $\circ$ on $\mathbb{R}$ on the set of intervals $\boldsymbol{I} = \{[a, b] \mid a, b \in \mathbb{R}, a \leq b\}$, such that the following holds:
Let $\boldsymbol{I}_0, \boldsymbol{I}_1 \in \boldsymbol{I}$ be two intervals. Then

$$\forall u \in I_0,\ \forall v \in I_1\ \exists z \in I_0 \circ I_1, \tag{35}$$

where

$$z = u \circ v. \tag{36}$$

Hence

$$\max_{u \in I_0, v \subset I_1} u \circ v \leq \mathrm{Sup}(I_0 \cap I_1), \tag{37}$$

where < occurs if one variable appears more than once. This phenomenon is called overestimation and causes additional numerical effort to get sharp boundaries. For sure the same holds for min and Inf. Thus for input intervals $\boldsymbol{I}_0, \ldots, \boldsymbol{I}_n$ interval analysis delivers evaluations for the domain $\boldsymbol{I}_0 \times \boldsymbol{I}_1 \times \ldots \times \boldsymbol{I}_n$. This evaluation is guaranteed to include all possible solutions, e.g.

$$[1, 3] + [1, 3] \cdot [-2, 1] = [-5, 6] \tag{38}$$

while

$$[1, 3] \cdot (1 + [-2, 1]) = [-3, 6] \, . \quad (39)$$

As shown in detail in (Pott, 2007), a CSP can be solved using interval analysis which guarantees reliable solutions (Hansen, 1992),(Merlet, 2004b),(Merlet, 2001). Solving the CSP with interval analysis delivers a list of boxes $\mathcal{L}_S$ representing an inner approximation of $\mathcal{X}_S$. According to eqn. 29, the solutions in $\mathcal{L}_S$ hold for total $\mathcal{X}_v$ and a subset of $\mathcal{X}_e$. Additionally, available implementations for interval analysis computations are robust against rounding effects. The following CSP solving algorithms have been proposed in (Pott, 2007) and (Bruckmann et al., 2008b). To use it for the special problem of analyzing wire robots, they have been extended. Details are described in the next sections.

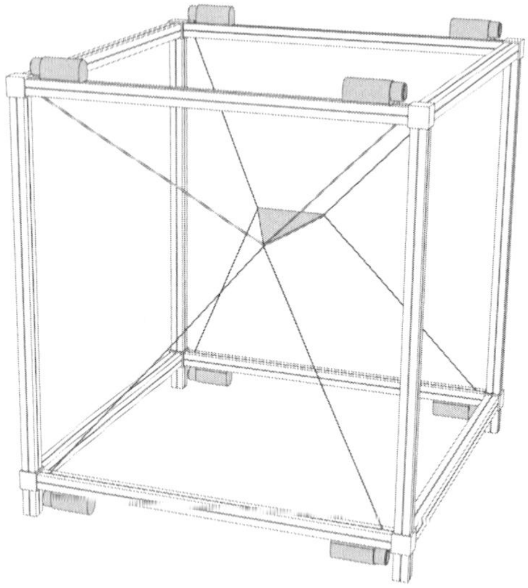

Fig. 9(a) SEGESTA prototype with 8 wires

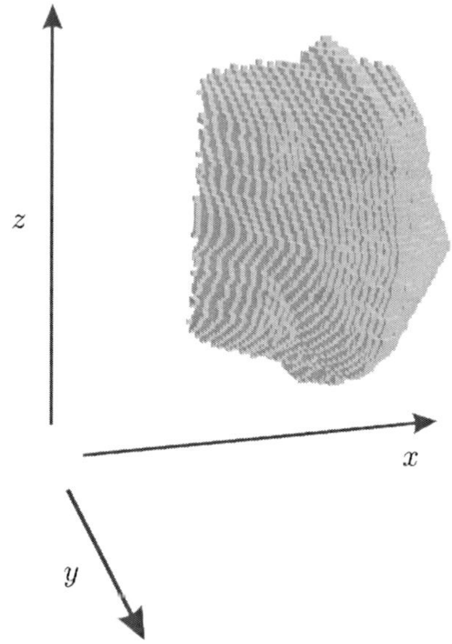

Fig. 9(b)Workspace of the SEGESTA prototype with 8 wires

Algorithm Verify

Verify is called with a box $\hat{\boldsymbol{c}}$ and checks whether

$$\begin{aligned} &\Phi(\hat{c}, v, e) > 0 \\ &\forall\, v \in \mathcal{X}_v \\ &\exists\, e \in \mathcal{X}_e \end{aligned} \quad (40)$$

is valid for the given box $\hat{\boldsymbol{c}}$. Here the domain $\mathcal{X}v$ is represented by the list of boxes $\mathcal{L}_T^v$. Thus, the result can be *valid, invalid, undefined* or *finite*. If at least one box is invalid, the whole search domain does not fulfill the required properties and is therefore invalid. Algorithm *Verify*

1. Define a search domain in the list $\mathcal{L}_T^v$. In the simplest case, $\mathcal{L}_T^v$ contains one search box.

2. If $\mathcal{L}_T^v$ is empty, the algorithm is finished with *valid*.
3. Take the next box $\hat{v}$ from the list $\mathcal{L}_T^v$.
4. If the diameter of the box $\hat{v}$ is smaller than a predefined value $\in_v$ return with *finite*.
5. If existence variables are present, call *Existence* with $\hat{c}$ and $\hat{v}$. If the result is *valid*, goto (2). If the box is invalid, return with *invalid*. If the box is *finite*, goto (10).
6. Evaluate $\hat{h} = \Phi(\hat{c}, \hat{v})$.
7. If Inf $\hat{h} > 0$, the infimum of $\hat{h}$ is greater than 0 in all its components. Thus, the box is valid. Goto (2).
8. If Sup $\hat{h} < 0$, the supremum of $\hat{h}$ is smaller than 0 in at least one component. Thus, the box is invalid. Return with *invalid*.
9. If Inf $\hat{h} < 0 <$ Sup $\hat{h}$, $\hat{h}$ is rated as *undefined*.
10. Divide the box on a verification variable and add the parts to $\mathcal{L}_T^v$. Goto (2).

Algorithm Existence

Existence is a modification of *Verify*. It is called with the boxes $\hat{c}$, $\hat{v}$ and checks whether

$$\begin{aligned} &\Phi(\hat{c}, \hat{v}, e) > 0 \\ &\exists\; e \in X_e \end{aligned} \tag{41}$$

is valid. Here the domain $\mathcal{X}_e$ is represented by the list of boxes $\mathcal{L}_T^e$ The result can be *valid*, *invalid* or *finite*. If at least one box is valid, the whole search domain fulfills the required properties and is therefore valid. Algorithm *Existence*

1. Define a search domain in the list $\mathcal{L}_T^e$. In the simplest case, $\mathcal{L}_T^e$ contains one search box.
2. If $\mathcal{L}_T^e$ is empty, the algorithm is finished with *invalid*.
3. Take the next box $\hat{e}$ from the list $\mathcal{L}_T^e$.
4. If the diameter of the box $\hat{e}$ is smaller than a predefined value $\in_e$, return with *finite*.
5. *Evaluate* $\hat{h} = \Phi(\hat{c}, \hat{v}, \hat{e})$.
6. If Inf $\hat{h} > 0$, the infimum of $\hat{h}$ greater than 0 in all its components. Thus, the box is valid. Return with *valid*.
7. If Sup $\hat{h} < 0$, the supremum of $\hat{h}$ smaller than 0 in at least one component. Goto (2).
8. If Inf $\hat{h} < 0 <$ Sup $\hat{h}$, $\hat{h}$ is rated as *undefined*. Divide the box on an existence variable and add the parts to $\mathcal{L}_T^e$. Goto (2).

Algorithm Calculate

Calculate is called with a search domain for c represented by a list of boxes $\mathcal{L}_T^c$. It uses *Existence* or *Verify* to identify valid boxes within the search domain. Thus, the result is a list $\mathcal{L}_S$ of valid boxes (and optionally the lists $\mathcal{L}_I$ for invalid boxes and $\mathcal{L}_F$ for finite boxes, respectively). Algorithm *Calculate*

1. Define a search domain in the list $\mathcal{L}_T^c$. In the simplest case, $\mathcal{L}_T^c$ contains one search box.

2. Create the lists
   (a) $\mathcal{L}_S$ for solution boxes,
   (b) $\mathcal{L}_I$ for invalid boxes,
   (c) $\mathcal{L}_F$ for finite boxes.
3. If $\mathcal{L}_T^c$ is empty, the algorithm is finished.
4. Take the next box $\hat{c}$ from the list $\mathcal{L}_T^c$.
5. If the diameter of the box $\hat{c}$ is smaller than a predefined value ϵ_c the box is treated as *finite* and thus moved to the list $\mathcal{L}_F$. Goto (3).
6. If verification variables are present, call *Verify* with $\hat{c}$. Otherwise call *Existence* with $\hat{c}$ and an empty box for $\hat{v}$.
7. If the result of *Verify* is *valid*, move the box to the solution list $\mathcal{L}_S$. Goto (3).
8. If the result of *Verify* is *invalid*, move the box to the invalid list $\mathcal{L}_I$. Goto (3).
9. If the result of *Verify* is *finite*, move the box to the finite list $\mathcal{L}_F$. Goto (3).

Calling Sequence

Let $\mathcal{X}_c$, $\mathcal{X}_v$, $\mathcal{X}_e \neq 0$ be given and represented as lists of boxes $\mathcal{L}_T^c, \mathcal{L}_T^v, \mathcal{L}_T^e$. In order to determine $\mathcal{L}_S$, *Calculate* is called with the search domain $\mathcal{L}_T^c$. Within *Calculate, Verify* is called. Since existence variables are present, *Existence* is called in order to validate the current calculation box (Otherwise in *Verify* the CSP would be directly evaluated). In the *Existence* algorithm the CSP is evaluated and the result is rated. In case that the result is *undefined,* the current box is divided on an existence variable. In case that the *Existence* algorithm returns with *finite,* the calling algorithm divides on its own variables and calls *Existence* again. If the result is *valid* or *invalid,* the result is directly returned to the calling algorithm. If *valid* is returned, the result is valid for all values within $\hat{c}$ and $\hat{v}$. The same calling sequence and return behaviour is used in *Calculate* calling *Verify*. For an effective CSP solver the return scheme should be more advanced in the way that not one variable is bisected until the box under consideration is finite, but a more sophisticated bisection distribution is used. It is noteworthy that the calculation time increases considerably with the number of variables and decreasing ϵ_i, $i \in \{c, v, e\}$.

Preliminary Checks

Since solving the force equilibrium is a computationally expensive task, favorable prechecks are demanded to reduce computation time. An effective check is to examine the interval evaluation of $\tau_{check} := A^T f_{check} + w$ for f_{check} being the box with infimum f_{min} and supremum f_{max}. If

$$\exists i \in 1,\ldots, m \; 0 \notin \tau_{check,i}\,, \tag{42}$$

one can conclude that the poses under consideration do not belong to the workspace under the given load w due to the non-existance of valid wire force distributions. The resulting preliminary workspace is an outer estimate and excludes poses which are not treated furthermore. Another possibility to reduce the computation time is to take symmetries into account. If symmetry axes as well as a symmectrical load range are present it is sufficient to compute only one part of the workspace and to complete the workspace by proper mirroring.

4.3 Further criteria
4.3.1 Stiffness

Besides the force equilibrium, additional workspace conditions can be applied. Due to the high elasticity of the wires (using plastic material, e.g. polyethylene), the stiffness may be low in parts of the workspace. Thus, for practical applications, especially if a predefined precision is required, it may be necessary to guarantee a given stiffness for the whole workspace. Otherwise, the compensation of elasticity effects by control may be required. Generally, this should be avoided as far as possible by an appropriate design. As shown in (Verhoeven, 2004), the so-called passive stiffness can be described as the reaction of a mechanical system onto a small pertubation, described by a linear equation:

$$\delta \boldsymbol{w} = \boldsymbol{K}(\boldsymbol{x})\delta \boldsymbol{x} \tag{43}$$

where

$$\boldsymbol{K}(\boldsymbol{x}) = k' \boldsymbol{A}^T(\boldsymbol{x})\boldsymbol{L}^{-1}(\boldsymbol{x})\boldsymbol{A}(\boldsymbol{x}). \tag{44}$$

Here, $\boldsymbol{L}$ is the diagonal matrix of the wire lengths and k' is the proportionality factor (force per relative elongation), treating the wires as linear springs. For the calculation, the inverse problem

$$\delta \boldsymbol{x} = \boldsymbol{K}(\boldsymbol{x})^{-1}\delta \boldsymbol{w} \tag{45}$$

is solved and evaluated where only domains having a position pertubation within the predefined limits $\delta \boldsymbol{x}_{min}$ and $\delta \boldsymbol{x}_{max}$ under predefined loads between $\delta \boldsymbol{w}_{min}$ and $\delta \boldsymbol{w}_{max}$ are considered as workspace. This equation can again be treated as a CSP. However, stiffness can also be checked performing a discrete workspace analysis. The stiffness workspace for a simple plain manipulator with 2 translational d.o.f. is shown in fig. 10(a). The parameters k' = 1000**N**, f_{min} = 10**N** and f_{max} = 90**N** were set. For a given load of $\delta \boldsymbol{w}$ = ([−20, 20]**N**, [−20, 20]**N**) the platform was allowed to sag elastically in the ranges $\delta \boldsymbol{x}$ = ([−0.015, 0.015]**N**, [−0.015, 0.015]**N**).

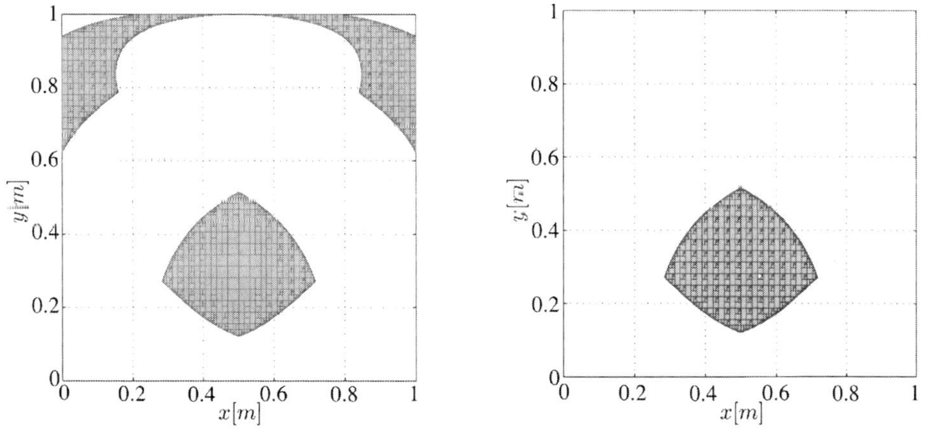

Fig. 10(a) Stiffness workspace of plain manipulator

Fig. 10(b) Combined force equilibrium and stiffness workspace of plain manipulator

4.3.2 Singularities

A pose of a wire robot is said to be singular if and only if

$$rank\ \boldsymbol{A}^T < n. \tag{46}$$

Therefore all wire robots with pure translational d.o.f. are singularity free except those, which are always singular (Verhoeven, 2004). For a wire robot with rotational and translational d.o.f. the workspace certainly has be to checked for singularities. Since within the workspace analysis (discrete or continuous) typically a system of linear equations is solved, the singularity criteria eqn.46 can be checked implicitly. Mechanically, at singular poses certain d.o.f. become uncontrollable (overmobility). Often this happens in symmetrical configurations.

4.3.3 Wire collision

In analogy to the problem of link collisions for conventional parallel manipulators, wire collisions have to be avoided. Due to their normally small diameter one possibility is to consider the wires as lines. In (Merlet, 2004a) an algorithm is proposed to determine the regions in which collisions between wires as well as the collisions between wires and the end-effector occur. Practically, wires have certain diameter and thus, a predefined minimum distance (at least the wire diameter) should be always ensured. Therefore, the well-known problem of determining the smallest distance between two lines arises. Since the lines are known after solving the inverse kinematics this is a very basic task but may be computational expensive. Clearly, the distance condition has to be formulated as a inequality. Hence, this criteria can be easily included in the CSP formulation.

5. Robot design

While workspace analysis examines the properties of already parametrized manipulators which allows to determine the applicable use cases, robot design describes the opposite task of finding the optimal robot for a given task. Generally, the task is abstracted e.g. as a desired workspace or a desired path or trajectory. To identify the optimal robot, usually different designs have to be compared with respect to the desired properties which makes the design process generally a computationally expensive task. Finally, one or more designs turn out as most favourable. In parallel to the analysis methods, again both discrete as well as continuous methods are available and show differences in the analysis quality and the calculation effort. For the continuous approach the CSP formulation can be used again which is amongst others advantageous in terms of implementation effort. The interchanging of the roles of the variables turns the workspace analysis just into a design task. According to (Merlet, 2005), the design (or synthesis) task can be divided into two separated subtasks:

- structure synthesis: This step includes the determination of the topology of the mechanical structure. In particular, the number and type of d.o.f. of the joints and their interconnection is identified.
- dimensional synthesis: Here position and orientation of the joints as well as the length of the links is specified.

For the special case of a wire robot, the structure synthesis covers different aspects: While the link topology itself is fixed, one has to choose the number of wires wisely.

Additionally, the concurrence of at least two (in the planar case) or three (in the spatial case) platform connection points may be prudential:

- Forward kinematic calculations become much easier (see section 2.3).
- The number of design parameters is reduced, which is beneficial in terms of computation time.
- The occurence of wire collisions is reduced since wires can intersect in at most one point.
- The workspace is comparably large (Fang, 2005).

After completion of the structure synthesis a dimensional synthesis can be performed. For a wire robot this is nothing but the identification of feasible base points. This section is addressed to dimensional synthesis mainly.

5.1 Discrete synthesis

Discrete methods are widely used for wire robot design. In (Fattah & Agrawal, 2005) and (Pusey et al., 2004) both the parameter set and an assumed superset of the workspace are discretized. Then for every point on the resulting parameter grid the discretized workspace is computed and its volume is determined by counting the points on the grid fulfilling all workspace conditions. The approaches share the same concept:

1. Build up an equidistant Grid of the design variables and loop through all parameter sets.
2. For every parameter set, specify a superset of the workspace and discretize it by an equidistant grid.
3. Loop through all grid points of step 2. For every point, determine if a valid wire force distribution according to eqn. 25 and 26 exists.
4. Count all points belonging to the workspace and store the number for every parameter set.
5. Obtain the maximum volume workspace, i.e., the maximum of all workspace volumes that are counted in step 4, and the associated optimized design variables.

Instead of the volume of the workspace a different optimization criterion can be employed. To increase the practical usability and the robustness of the design, a dexterity criterion is proposed, which uses the condition number of the structure matrix A^T. These approaches have two drawbacks. Since the design variables are discretized, every combination of parameters is checked. Hence, this method is computationally intensive. Furthermore, no desired workspace can be guaranteed by the obtained design. Hay and Snyman use a special optimizer instead of a grid of the design variables (Hay & Snyman, 2004), (Hay & Snyman, 2005). Again, in this approach a desired workspace is not guaranteed by the obtained optimal design.

5.2 Continuous synthesis (Design-To-Workspace)

Examining eqn.28, eqn.31 and eqn.32, the roles of the variables can arbitrary be assigned. An imaginable choice is

- The winch poses and platform fixation points are the *calculation variables*. Thus, the calculation delivers robot designs solving the CSP.
- The platform coordinates are *verification variables*. Hence, the workspaces of all resulting robot designs will cover the set given in $\mathcal{X}_v$ for the platform coordinates for sure.

- Optionally, the exerted external wrench w and desired platform orientations can be set as *verification variables* to extend the applicability of the emerged designs for certain process wrenches and tasks.
- The wire forces f_{sec} are the *existence variables*.

The suggested choice of variables leads to a CSP, whose solutions are robot designs. Furthermore, each obtained robot can reach every point given in $\mathcal{X}_v$ for the platform coordinates with every orientation and wrench given in $\mathcal{X}_v$. Generally, the design task is deemed to be more complicated than the analysis. Here, the methods and formulations are inherited and just adapted to the design problem. Nevertheless, robot design is a computationally intensive task. The use of parallel computations is strongly advised. Solving the CSP is advantageous due to the following reasons:

- The workspaces of the resulting designs are guaranteed to have no holes or singularities.
- The design process can be extended by a global optimization step.
- The interval CSP solver can be effectively parallelized.

5.3 Continuous optimization

Optimization is always performed with respect to a cost function. In industrial application usually the term optimal is used with respect to economic aspects, i.e. costs. In the case of wire robots, the most cost-driving factor are the wire winch units. However, optimizing the number of winches is part of the structure synthesis. Thus, here another cost function has to be chosen. This choice is generally arbitrary. Nevertheless, a reasonable choice is the volume expansion. On one hand, reducing the expansion of the robot saves space within a production facility which reduces costs, on the other hand, the required wire lengths are minimized. In literature, usually the optimization is performed with respect to the size (or volume) of the workspace or the integral of workspace indices over the workspace. This gives finally the robot with optimal (e.g. largest) workspace with respect to some criterion, but it says nothing about its shape and its usability for applications. Thus, here another approach is used (Pott, 2007): Not a maximum size of the workspace is demanded, but the guaranteed enclosure of a predefined domain is desired. The optimization is performed using interval analysis. Let a list $\mathcal{L}$ of n boxes of robot designs, e.g. a solution of the according CSP be given. The following algorithm performs the required steps for a minimization (maximization is performed analogously):

1. Set $i = 0$ and $F_{opt} = [\infty,\infty]$.
2. Set $i = i + 1$. If $i > n$ the algorithm finishes.
3. Take the i-th element l_i of $\mathcal{L}$ and compute its cost function $F(l_i)$.
4. If Sup$(F(l_i))$ < Sup(F_{opt}), set $F_{opt} = F(l_i)$.
   - If Sup$(F(l_i))$ < Inf(F_{opt}) delete all elements of the solution list and initialize it with l_i. Goto 2.
   - Store l_i in the solution list. Goto 2.
5. If Inf$(F(l_i))$ < Sup(F_{opt}) store l_i in the solution list.
6. Discard l_i and goto 2

For performance reasons the optimization can be included in the CSP Solver. This will reduce computation time drastically since non-optimal designs are discarded at an early stage. An example for the optimization of an 1R2T robot is shown in fig. 11(b). For the upper winches, y-positions are free, for the lower ones, the x-positions are the free optimization parameters.

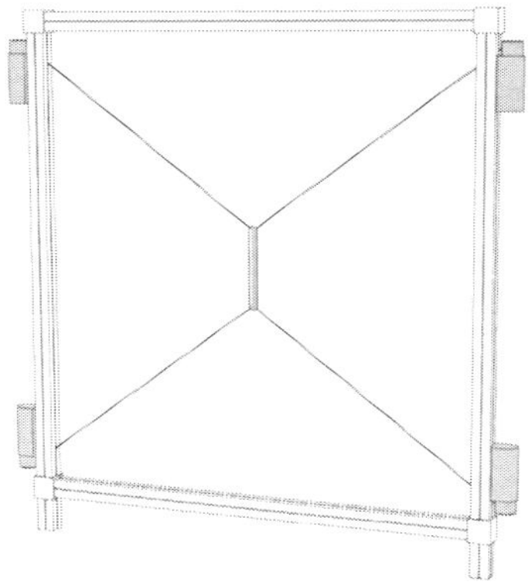

Fig. 11(a) 1R2T example

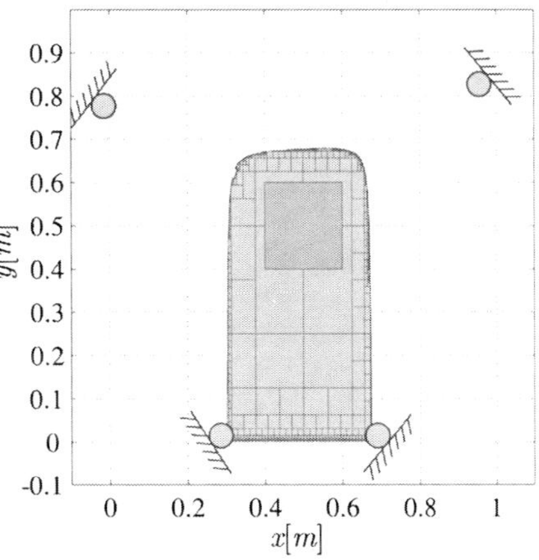

Fig. 11(b) 1R2T robot optimized for shown desired quadratic workspace.

5.4 Design-To-Task

The Design-to-Workspace method results in manipulators, guaranteed to have a desired workspace. Thus, the manipulator is able to perform every task within this workspace. Nevertheless, from the economic point of view, there is a need for manipulators which perform a specific task in minimum time, with minimum energy consumption or with lowest possible power. A typical industrial application is e.g. the pick-and-place task, moving a load from one point to another. Usually, this task is performed within series production, i.e. it is repeated many times. In such an application the optimal manipulator for sure finishes the job in minimal time with respect to the technical constraints (here, the term optimal is used with respect to minimal time without loss of generality). Thus, the set-up of a specialized (i.e. taskoptimized) manipulator can be profitable. When using classical industrial robots, the freedom to modify the mechanical setup of the robot is very limited. Thus, only the trajectories can be modified and optimized with respect to the task. Due to the modular design of a wire robot, the task-specific optimization can be seperated into two tasks:

- Optimization of the robot: within all suitable designs, the robot which performs the task in shortest time is chosen.
- Optimization of the trajectory: within all possible trajectories, the trajectory which connects the points in shortest time is chosen. The concepts needed for this step are partly explained in (Bianco & Piazzi, 2001b),(Bianco & Piazzi, 2001a) and (Merlet, 1994).

By treating this task as a CSP, both claims can be optimized at the same time. In particular, the final result contains the robot which is able to perform the task quickest *and* the corresponding trajectory description. To perform an optimization of the wire robot and the trajectory simultaneously, the latter is planned first. Afterwards it is checked whether the complete trajectory belongs to the workspace. The robot designer may provide a predefined trajectory or leave this up to the optimizer. The parameters of the trajectory are therefore either fixed or *calculation variables*. Hence, the CSP looks the same as in eqn.31 and eqn.32 except the previous trajectory generation. For integrated optimization, the variables are assigned as follows. Note, that also a separate optimization of robot and trajectory is possible:

- Robot optimization
    - The robot base is described by the positions of the winches. To optimize the robot, the winches can be moved. Therefore, $\boldsymbol{b}_i$ are *calculation variables*
    - The end effector is described by the positions of the platform anchor points $\boldsymbol{p}_i$. To optimize the robot, these points can be moved on the platform. Therefore, $\boldsymbol{p}_i$ are *calculation variables*
- Trajectory optimization
    - The path is described by a polynomial of fourth order without loss of generality. Besides the start and end poses, also the velocities are predefined. This leaves one free parameter, e.g. the start acceleration for translational d.o.f. or the orientation at half travel time for rotational d.o.f.. These can be set as *calculation variables.*
    - To describe the trajectory, additionally the travel time T has to be defined. To calculate the minimum time, T is a *calculation variable.*
    - For the whole trajectory, a path parameter t is assigned. Usually, it is normalized between zero and one. Since the whole trajectory shall betraced for validity, t is a *verification variable*

Optionally, the exerted external wrenches $\boldsymbol{w}$ can be set as *verification variables.* Note, that within the trajectory verification the dynamics of the robot are taken into account by adding the inertia loads resulting from the calculated accelerations to the platform loads $\boldsymbol{w}$. The example in fig. 12(b) shows the result of an optimization for a point-to-point (PTP) movement. A $n = 3$ d.o.f. wire robot with $m = 4$ wires is considered (see 12(a)). It consists of a bar-shaped platform of $0.1m$ length, connected by four winches to the base frame. Free optimization parameters were the y-position of the upper right winch, the travel time and the intermediate acceleration of the rotation angle at $T = 0.5$ s.

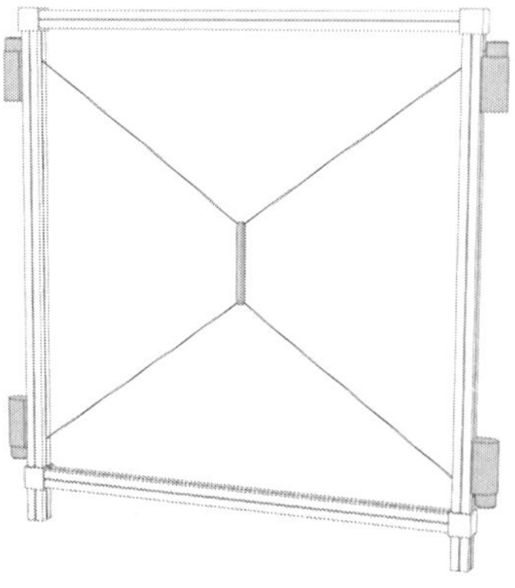

Fig. 12(a) 1R2T example

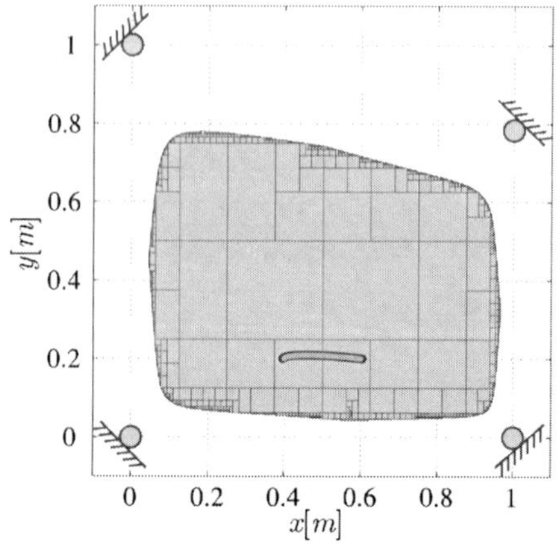

Fig. 12(b) 1R2T robot optimized for shown desired PTP trajectory

6. Conclusion

In this chapter, the analysis and design of wire robots was discussed. The required basics like kinematics and the force equilibrium - which is the one of the main workspace criteria - were introduced as well as serveral classification approaches. The analysis of wire robots was described as a CSP task which can be solved by interval analysis. Besides reliable results, the same CSP can be used for robot design by a variable exchange, which is generally a challenging problem. In addition to this continuous approach, also the more straightforward discrete methods are shortly introduced. The next chapter is dedicated to the application and control of wire robots. Therefore, the dynamical description as well as different methods to calculate a force distribution for a given pose and platform wrench are presented. Based on this, some control concepts are described. The use of wire robots for several fields of application is demonstrated by a number of examples.

7. Acknowledgements

This work is supported by the German Research Council (Deutsche Forschungsgemeinschaft) under HI370/24-1, HI370/19-3 and SCHR1176/1-2. The authors would like to thank Martin Langhammer for contributing the figure design.

8. References

Bianco, C. G. L. and Piazzi, A. (2001a). A hybrid algorithm for infinitely constrained optimization. *International Journal of Systems Science*, 32(1):91–102.

Bianco, C. G. L. and Piazzi, A. (2001b). A semi-infinite optimization approach to optimal spline trajectory planning of mechanical manipulators. In Goberna, M. A. and Lopez, M. A., editors, *Semi-Infinite Programming: Recent Advances*, chapter 13, pages 271–297. Kluwer Academic Publisher.

Bosscher, P. and Ebert-Uphoff, I. (2004). Wrench-based analysis of cable-driven robots. *Proceedings of the 2004 IEEE International Conference on Robotics & Automation*, pages 4950–4955.

Bruckmann, T., Mikelsons, L., Brandt, T., Hiller, M., and Schramm, D. (2008a). Wire robots part II - dynamics, control & application. In Lazinica, A., editor, *Parallel Manipulators*, ARS Robotic Books. I-Tech Education and Publishing, Vienna, Austria. ISBN 978-3-902613-20-2.

Bruckmann, T., Mikelsons, L., and Hiller, M. (January 9-11, 2008b). A design-to-task approach for wire robots. In Kecskeméthy, A., editor, *Conference on Interdisciplinary Applications of Kinematics 2008*, Lima, Peru.

Bruckmann, T., Mikelsons, L., Schramm, D., and Hiller, M. (2007). Continuous workspace analysis for parallel cable-driven stewart-gough platforms. *to appear in Proceedings in Applied Mathematics and Mechanics.*

Fang, S. (2005). *Design, Modeling and Motion Control of Tendon-based Parallel Manipulators.* Ph. D. dissertation, Gerhard-Mercator-University, Duisburg, Germany. Fortschritt-Berichte VDI, Reihe 8, Nr. 1076, Düsseldorf.

Fattah, A. and Agrawal, S. K. (2005). On the design of cable-suspended planar parallel robots. *ASME Transactions, Journal of Mechanical Design*, 127(5):1021– 1028.

Gouttefarde, M., Merlet, J.-P., and Daney, D. (2007). Wrench-feasible workspace of parallel cable-driven mechanisms. *2007 IEEE International Conference on Robotics and Automation, ICRA 2007, 10-14 April 2007, Roma, Italy*, pages 1492–1497.

Hansen, E. *(1992). Global Optimization using Interval Analysis.* Marcal Dekker, Inc.

Hay, A. and Snyman, J. (2004). Analysis and optimization of a planar tendon-driven parallel manipulator. In Lenarcic, J. and Galetti, C., editors, *Advances in Robot Kinematics*, pages 303–312, Sestri Levante.

Hay, A. and Snyman, J. (2005). Optimization of a planar tendon-driven parallel manipulator for a maximal dextrous workspace. In *Engineering Optimization*, volume 37 of 20, pages 217–236.

Landsberger, S. and Sheridan, T. (1985). A new design for parallel link manipulator. In *International Conference on Cybernetics and Society*, pages 812–814,, Tucson, Arizona.

Maier, T. (2004). *Bahnsteuerung eines seilgeführten Handhabungssystems - Modellbildung, Simulation und Experiment.* PhD thesis, Universität Rostock, Brandenburg. Fortschritt-Berichte VDI, Reihe 8, Nr. 1047, Düsseldorf.

Merlet, J.-P. (1994). Trajectory verification in the workspace for parallel manipulators. The *International Journal of Robotics Research*, 13(4):326–333.

Merlet, J.-P. (2000). *Parallel Robots.* Kluwer Academic Publishers, Norwell, MA, USA.

Merlet, J.-P. (2001). A generic trajectory verifier for the motion planning of parallel robots. *Journal of Mechanical Design*, 123:510–515.

Merlet, J.-P. (2004a). Analysis of the influence of wires interference on the workspace of wire robots. *On Advances in Robot Kinematics*, pages 211–218.

Merlet, J.-P. (2004b). Solving the forward kinematics of a gough-type parallel manipulator with interval analysis. *Int. J. of Robotics Research*, 23(3):221–236.

Merlet, J.-P. (2005). Optimal design of robots. In *Robotics: Science and Systems*, Boston.

Ming, A. and Higuchi, T. (1994). Study on multiple degree of freedom positioning mechanisms using wires, part 1 - concept, design and control. *International Journal of the Japan Society for Precision Engineering*, 28:131–138.

Pott, A. (2007). *Analyse und Synthese von Parallelkinematik-Werkzeugmaschinen.* Ph. D. dissertation, Gerhard-Mercator-University, Duisburg, Germany. Fortschritt-Berichte VDI, Reihe 20, Nr. 409, Düsseldorf.

Pusey, J., Fattah, A., Agrawal, S. K., and Messina, E. (2004). Design and workspace analysis of a 6-6 cable-suspended parallel robot. *Mechanism and Machine Theory.* Vol. 39, No.7, pp.761-778.

Verhoeven, R. (2004). Analysis *of the Workspace of Tendon-based Stewart Platforms.* PhD thesis, University of Duisburg-Essen.

Williams, R. L., Albus, J. S., and Bostelman, R. V. (2004). 3d cable-based cartesian metrology system. *Journal of Robotic Systems*, 21(5):237–257.

7

Wire Robots Part II Dynamics, Control & Application

Tobias Bruckmann, Lars Mikelsons, Thorsten Brandt,
Manfred Hiller and Dieter Schramm
University Duisburg-Essen (Chair for Mechatronics)
Germany

1. Introduction

In (Bruckmann et al., 2008) the kinematics, analysis and design of wire robots were presented. This chapter focuses on control and applications of wire robots. Wire robots are a very recent area of research. Nevertheless, they are well studied and already in application (see section 5). Due to their possible lightweight structure, wire robots can operate at very high velocities. Hence, as can be seen by experiment, only positioning control using the inverse kinematics is not sufficient. In particular, slackness in the wires can be observed at highly dynamic motions. To overcome this problem, force control can be employed. In section 4 different control schemes are proposed. The required dynamical model is obtained in section 2, while for the calculation of feasible wire force distributions are proposed in section 3. Since wire robots are kinematically redundant the latter is not straightforward, but requires advanced approaches. The same holds for the control schemes, since a CRPM as well as a RRPM is a non-linear, coupled, redundant system (Ming & Higuchi, 1994).

2. Dynamics

According to figure 1 a wire robot can be considered as a multibody system with m unilateral constraints. In contrast to the generally complicated forward kinematics (Bruckmann et al., 2008) the dynamical equations of motion are comparably easy to formulate with respect to the base frame $\mathcal{B}$. The wrench $\boldsymbol{w}_{\mathbf{wire}}$ of the wires acting on the platform can be written as (see Fig. 2)

$$\boldsymbol{w}_{wire} = \begin{bmatrix} \boldsymbol{f}_{wire} & \boldsymbol{\tau}_{wire} \end{bmatrix}^T = \begin{bmatrix} \sum\limits_{i=1}^{m} \boldsymbol{f}_i & \sum\limits_{i=1}^{m} \boldsymbol{p}_i \times \boldsymbol{f}_i \end{bmatrix}^T . \tag{1}$$

Since the forces act along the wires

$$\boldsymbol{f}_i = f_i \cdot \frac{\boldsymbol{l}_i}{\|\boldsymbol{l}_i\|_2} = f_i \cdot \boldsymbol{\nu}_i, \quad (1 \leq i \leq m) \tag{2}$$

holds. It follows

$$\boldsymbol{w}_{wire} = \begin{bmatrix} \boldsymbol{\nu}_1 & \cdots & \boldsymbol{\nu}_m \\ \boldsymbol{p}_1 \times \boldsymbol{\nu}_1 & \cdots & \boldsymbol{p}_m \times \boldsymbol{\nu}_m \end{bmatrix} \begin{bmatrix} f_1 \\ \vdots \\ f_m \end{bmatrix} \boldsymbol{f} = \boldsymbol{A}^T \boldsymbol{f} \tag{3}$$

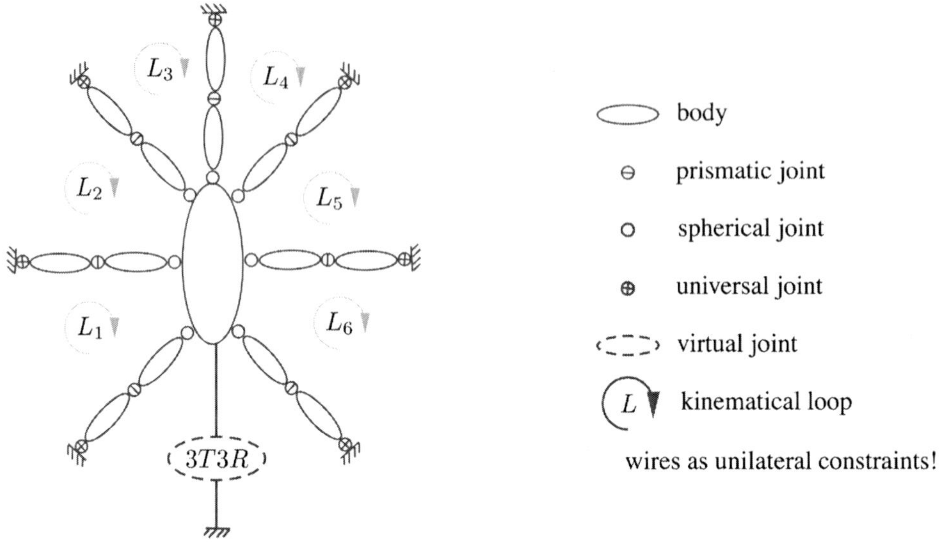

Fig. 1: Topological structure of a CRPM with n = 6.

The Newton-Euler equations lead to

$$m_p \ddot{\boldsymbol{r}} = \boldsymbol{f}_E + \boldsymbol{f}_{wire} \tag{4}$$

$$\boldsymbol{I}\ddot{\boldsymbol{\Omega}} + \dot{\boldsymbol{\Omega}} \times (\boldsymbol{I}\dot{\boldsymbol{\Omega}}) = \boldsymbol{\tau}_{wire} + \boldsymbol{\tau}_E, \tag{5}$$

with

m_p: the mass of platform,

$I \in \mathbb{R}^{3\times3}$: inertia tensor defined with respect to the inertial system $\mathcal{B}$ which is an expression of rotation angles,

$\Omega = [\ \varphi\ \vartheta\ \psi\ ]^T$: orientation of the platform in $\mathcal{B}$,

f_E: vector of external forces,

τ_E: vector of external torques.

The equations eqn. 4 can be rewritten by

$$\underbrace{\underbrace{\begin{bmatrix} m_p\boldsymbol{E} & \boldsymbol{0} \\ \boldsymbol{0} & \boldsymbol{I} \end{bmatrix}}_{\boldsymbol{M}_p} \underbrace{\begin{bmatrix} \ddot{\boldsymbol{r}} \\ \ddot{\boldsymbol{\Omega}} \end{bmatrix}}_{\ddot{\boldsymbol{x}}} + \underbrace{\begin{bmatrix} \boldsymbol{0} \\ \dot{\boldsymbol{\Omega}} \times (\boldsymbol{I}\dot{\boldsymbol{\Omega}}) \end{bmatrix}}_{\boldsymbol{g}_C} - \underbrace{\begin{bmatrix} \boldsymbol{f}_E \\ \boldsymbol{\tau}_E \end{bmatrix}}_{\boldsymbol{g}_E}}_{-\boldsymbol{w}} = \boldsymbol{A}^T \boldsymbol{f}. \tag{6}$$

with
M_p : mass matrix of platform,
E : identity matrix,
$g_C \in \mathbb{R}^{n\times 1}$: Cartesian space vector of Coriolis and centrifugal forces and torques,
$g_E \in \mathbb{R}^{n\times 1}$: vector of the generalized applied forces and torques, not including the resultants of wire tensions.

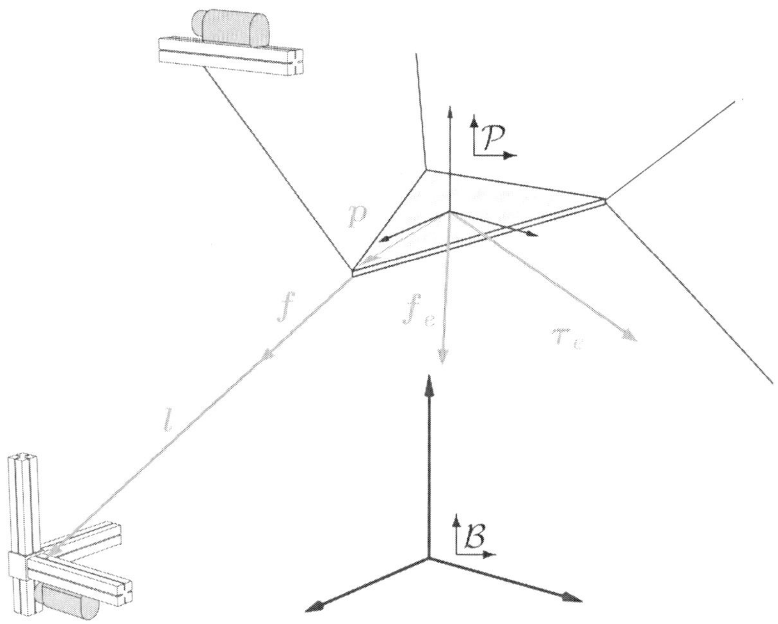

Fig. 2: Forces for a wire robot

Taking wire force limits f_{min} and f_{max} (see (Bruckmann et al., 2008)) into account it follows

$$A^T f + w = 0 \text{ with} \tag{7}$$

$$f_{min} \leq f \leq f_{max} \tag{8}$$

3. Wire force calculation

In section 2 a description of the force equilibrium was presented. Here methods for the calculation of a feasible force distribution f, i.e. a force distribution f which satisfies eqn. 7 and the constraints in eqn. 8, are presented. Obviously eqn. 7 represents an underdetermined system of linear equations. Its solution space is r-dimensional. Hence isolating the force distribution $\boldsymbol{f}$ leads to

$$\boldsymbol{f} = -\boldsymbol{A}^{+\mathrm{T}}\boldsymbol{w} + \boldsymbol{H}\,\boldsymbol{\lambda}, \tag{9}$$

where $\boldsymbol{A}^{+\mathrm{T}}$ denotes the Moore-Penrose Pseudo-Inverse of A^{T} . Thus the task of finding a feasible wire force distribution has been transformed to the task of finding $\lambda \in \mathbb{R}^r$ such that $\boldsymbol{f} > 0$ holds. Note that $\boldsymbol{H}$ is the nullspace or kernel of A^{T} defined as

$$\boldsymbol{H} := \begin{bmatrix} \boldsymbol{h}_1 & \dots & \boldsymbol{h}_r \end{bmatrix}, \tag{10}$$

where

$$\boldsymbol{A}^{\mathrm{T}}\boldsymbol{h}_i = \boldsymbol{0}, \quad 1 \leq i \leq r. \tag{11}$$

In other words, a linear combination of the columns of $\boldsymbol{H}$ describes force distributions creating an inner tension in the system without applying wrenches w_{wire} onto the end effector. In case of an homogenious problem, i.e. $w = 0$, it describes the possible solutions of eqn. 7 for $\boldsymbol{f}$. Now the problem of satisfying the constraints of eqn. 8 arises, i.e. the force limits also have to be considered. Thus plugging eqn. 9 into eqn. 8 leads to

$$\boldsymbol{f}_{\mathrm{min}} + \boldsymbol{A}^{+\mathrm{T}}\boldsymbol{w} \leq \boldsymbol{H}\boldsymbol{\lambda} \leq \boldsymbol{f}_{\mathrm{max}} + \boldsymbol{A}^{+\mathrm{T}}\boldsymbol{w}. \tag{12}$$

Therefore the task of identifying a feasible force distribution is equivalent to the problem of identifying $\boldsymbol{\lambda} \in \mathbb{R}^r$ such that eqn. 12 holds. In other words, the boundaries of the wire forces form a m-dimensional hypercube $\mathcal{C} \subset \mathbb{R}^m$. All force distributions satisfying eqn. 9 obviously form a r-dimensional subspace $\mathcal{S} \subset \mathbb{R}^m$ spanned by the kernel of the structure matrix (see fig. 3). Hence, if the intersection $\mathcal{F}$ of the hypercube $\mathcal{C}$ and the subspace $\mathcal{S}$ is non-empty, feasible solutions $\boldsymbol{f}$ exist, i.e. $\mathcal{F} = \mathcal{C} \cap \mathcal{S} \neq 0$, where $\mathcal{F}$ is a r-dimensional manifold in the $\mathbb{R}^m$. A more detailed introduction is given in (Oh & Agrawal, 2005) and (Mikelsons et al., 2008). Noteworthy, the r-dimensional solution space generally allows to compute force distributions with different characteristics: While for fast motion, smallest possible forces are demanded, for applications requiring a high stiffness, high forces are advantageous (Kawamura et al., 2000), (Fang, 2005).

3.1 Linear optimization

Looking at the geometric interpretation of finding feasible force distributions, the most intuitive way is to search for a convenient characterization of the manifold $\mathcal{F}$. Since $\mathcal{F}$ is completely determined by its vertices, the computation of those seems to be a promising way. In this work, two approaches following this idea are shown: In section 3.3, a method using the kernel as a transformation is presented. This leads to $\binom{m}{r}$ r-dimensional linear systems of equations. Alternatively, the approach presented in this section presumes no knowledge of the kernel but solves $\binom{m}{r}$ n dimensional linear systems of equations. Hence, the method to be applied has to be chosen depending on m and n.

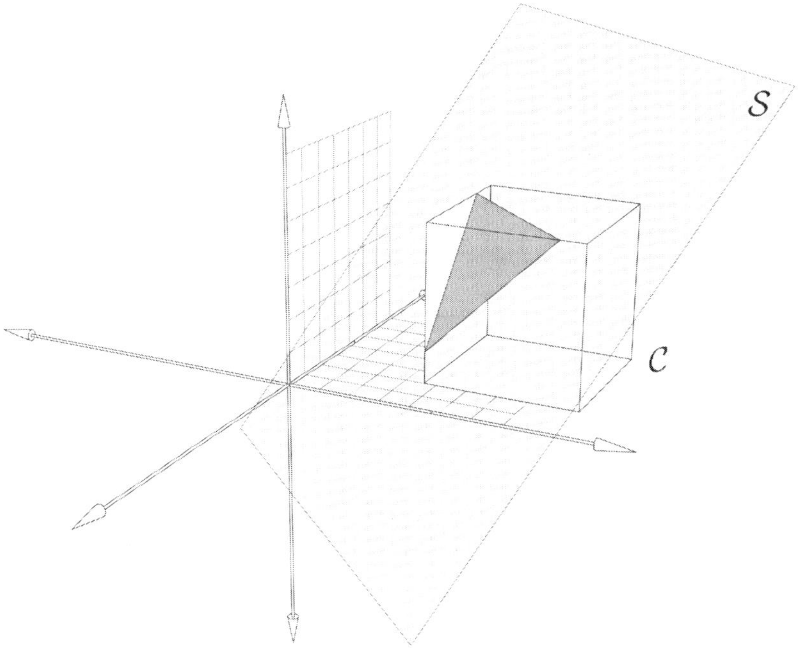

Fig. 3: The subset $\mathcal{S}$ intersecting the hypercube $\mathcal{C}$ in the case of $n = 1$ and $m = 3$.

Examining eqn. 7, one needs to set r forces in the wire force distribution to get a quadratic system. Obviously the desired points are located on the faces of the cube $\mathcal{C}$. It can be shown that a point belongs to the workspace if and only if a valid wire force distribution f that satisfies [1]

$$\exists A \subset \{1,\dots,m\},\ |A| = r, \text{ such that } \|f_i\| = f_{\max} \vee \|f_i\| = f_{\min} \ \forall i \in A \tag{13}$$

exists[2]. Therefore, r wire forces can be set to their minimum or maximum value, respectively. It is unknown in advance which wire forces have to be preset to get a feasible distribution. Thus, in the worst case all combinations of r wires have to be tested, leaving $m \times m$ systems of linear equations to be solved for every combination. For sure every vertex represents a valid wire force distribution. Choosing the vertex, which minimizes the 1-norm

---

[1] For a set A, $|A|$ denotes the cardinal number of A

[2] Using the kernel as a transformation from the $\mathbb{R}^r$ into the $\mathbb{R}^m$ (see section 3.3), the feasible force distribution form a polyhedron bounded by the force limits. r force limits determine a vertex. This finishes the proof.

could be an appropriate procedure. The resulting procedure can formally be expressed as a Linear Optimization Problem

$$minimize \quad [1 \quad \dots \quad 1]\, \boldsymbol{f} \quad subject\ to \quad \boldsymbol{f}_{\min} \le \boldsymbol{f} \le \boldsymbol{f}_{\min} \quad \wedge \quad \boldsymbol{A}^T \boldsymbol{f} + \boldsymbol{w} = \boldsymbol{0}.$$

In (Oh & Agrawal, 2005) a Linear Programming approach is presented to solve the problem in the $\mathbb{R}^r$. Note that for control purposes, the Linear Optimization approach may deliver inadequate results since along a trajectory through the workspace, the result may be discontinuous.

3.2 Nonlinear optimization

Due to the formulation of the cost function, the Linear Programming method may deliver discontinuous solutions along a continuous trajectory. This leads to jumps in the time history of the wire forces, causing stability problems and additional mechanical wear. In (Verhoeven, 2004) it is proven that cost functions using a p-norm $(1 < p < \infty)$, lead to guaranteed continuous wire forces along a continuous trajectory. The resulting formulation of the optimization problem is as follows:

$$minimize \quad \|\boldsymbol{f}\|_p = \sqrt[p]{\sum_{\mu=1}^{m} f_\mu^p} \quad subject\ to \quad \boldsymbol{f}_{\min} \le \boldsymbol{f} \le \boldsymbol{f}_{\min} \quad \wedge \quad \boldsymbol{A}^{\mathrm{T}} \boldsymbol{f} + \boldsymbol{w} = \boldsymbol{0}.$$

In (Verhoeven, 2004), also an effective algorithm is presented which solves the problem employing the knowledge of the solution structure, based on an iterative approximation of the optimal solution. However, this algorithm has the drawback to fail in specific configurations, i.e. solutions might be not found although they exist. To obtain the lowest possible force distribution (according to a p-norm), the unbounded polyhedron $\mathcal{P}_{low}$ is introduced, which is limited by the lower wire force limits:

$$\mathcal{P}_{low} := \{\boldsymbol{f} \in \mathbb{R}^m : \boldsymbol{A}^T \boldsymbol{f} + \boldsymbol{w} = \boldsymbol{0} \wedge \boldsymbol{f} \ge \boldsymbol{f}_{\min}\} \tag{14}$$

Furthermore, the wire force distribution f_{low} is introduced, which has minimal p-norm:

$$\|\boldsymbol{f}_{low}\|_p = \min_{\boldsymbol{f} \in \mathcal{P}_{low}} \|\boldsymbol{f}\|_p \tag{15}$$

It should be mentioned that for $1 < p < \infty$ f_{low} is unique, which is essential for the continuity of f_{low}. The algorithm works as follows

1. Compute an initial guess $\tilde{f}_{low}$ for f_{low}.
2. If $\tilde{f}_{low}$ is not contained in $\mathcal{P}_{low}$, move $\tilde{f}_{low}$ towards $\mathcal{P}_{low}$ until it is placed on the polyhedron.
3. Minimize the p-norm of $\tilde{f}_{low}$.

The initial guess is obtained by the orthogonal projection $\tilde{f}_{low}$ of $f_{\min}$ onto the manifold of feasible force distributions F. Note that $\tilde{f}_{low}$ is not always contained in $\mathcal{P}_{low}$. The second step of the algorithm is performed by moving along the negative gradient of the distance between the polyhedron $\mathcal{P}_{low}$ and $\tilde{f}_{low}$. The distance is measured in the squared 2-norm. Finally, the minimization of $\tilde{f}_{low}$ is done using a gradient based method again. Analogously, a vector f_{high} representing the highest possible solution in the chosen p-norm can be obtained. Hence, choosing a wire force distribution on the line between f_{low} and f_{high} allows either fast motions due to low wire forces or high stiffness due to high wire forces. This approach is very effective in terms of computation time since the initial guess is often already a feasible solution, but suffers from the fact that a solution is not always found.

3.3 Barycentric force calculation

The shown approaches require the usage of an optimizer to deliver continuous results as shown in ((Verhoeven, 2004),(Nahon & Angeles, 1991), (Bruckmann et al., 2006), (Voglewede & Ebert-Uphoff, 2004) and (Bosscher & Ebert-Uphoff, 2004)). Standard optimizer implementations as LAPACK or the NAG® library require iterative computations, which may not be used within a realtime control system due to their normally non-predictable worst-case runtime. In this section, a non-iterative algorithm is shown, which provides continuous force distributions furthermost from the force limits. The algorithm provides a force distribution, which lies in the center of gravity (CoG or barycenter) of the intersection manifold $\mathcal{F}$.

The structure matrix A^T has the dimension $n \times m$. Hence, within the workspace, the kernel can be computed as $H = (h_1 \ldots h_r) \in \mathbb{R}^{m \times r}$. Here, the kernel is used to define a map from the $\mathbb{R}^r$ to $\mathcal{S} \subset \mathbb{R}^m$, i.e. for all $\lambda \in \Lambda$, eqn. 12 must hold, where Λ is the (convex) polyhedron-shaped preimage of the manifold $\mathcal{F}$ under the mapping $\gamma : \mathbb{R}^r \rightarrow \mathbb{R}^m$, $\lambda \rightarrow -A^{+T}w + H\lambda$. In other words, since γ maps the $\mathbb{R}^r$ onto the solution subspace $\mathcal{S}$, it maps the polyhedron $\Lambda \subset \mathbb{R}^r$ onto the solution manifold $\mathcal{F}$. Since there is no explicit expression for Λ, a convenient representation is sought. As mentioned above Λ is a polyhedron. Thus, its vertices determine Λ completely. Componentwise evaluation of both sides of eqn. 12 gives $2m$ hyperplanes in $\mathbb{R}^r$. The vertices of Λ are intersection points of r hyperplanes. Hence, all those intersection points are calculated and examined with respect to their compatibility with all inequalities. Obviously a vertex of the polyhedron Λ has to satisfy all inequalities of eqn. 12. In order to compute the center of gravity of the obtained polyhedron, Λ is triangulated, i.e. splitted into r-simplexes. In the case of $r = 2$ this just means dividing into triangles. Advanced techniques as shown in (Cignoni et al., 1998) are required in the case of higher dimensions. Triangulation delivers a list of n_s simplexes P_k with each having $r + 1$ vertices v_{k_j} with $k = 1 \ldots n_s$ and $j = 1 \ldots r + 1$. The volumes V_k of the simplexes can be determined by integration (Hammer et al., 1956). Furthermore their CoG λ_{s_k} are computed by the equation

$$\lambda_{s_k,i} = \frac{\sum\limits_{\nu=1}^{r+1} v_{k_\nu,i}}{r+1}, \quad 1 \leq i \leq r, \quad 1 \leq k \leq n_s \tag{16}$$

which is used to calculate the CoG λ_s of the polyhedron via

$$\lambda_s^i = \frac{\sum_{\mu=1}^{n_s} (\lambda_{s_\mu,i} \cdot V_\mu)}{\sum_{\mu=1}^{n_s} V_\mu}. \tag{17}$$

Finally, the solution is transformed back using the mapping γ

$$\boldsymbol{f}_s = -\boldsymbol{A}^{+\mathrm{T}}\boldsymbol{w} + \boldsymbol{H}\,\boldsymbol{\lambda}_s \tag{18}$$

where $\boldsymbol{f}_s$ is the center of gravity of the manifold $\mathcal{F}$.

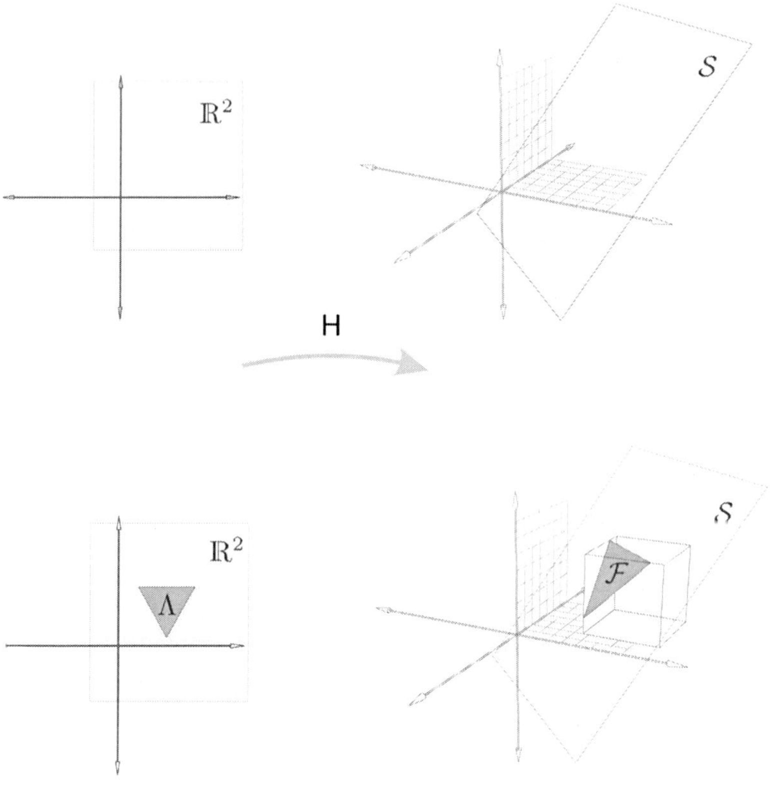

Fig. 4: Visualisation of the map $\boldsymbol{H}$ in the case of $m = 3$ and $n = 1$

3.3.1 Proof-of-Concept

In this section we prove that the CoG of the manifold $\mathcal{F}$ can be computed by calculating the CoG of the convex polyhedron. Without loss of generality $w = 0$ is assumed. The CoG of the manifold $\mathcal{F}$ can be computed componentwise as

$$f_{s,i} = \frac{\int_{\mathcal{F}} x_i d\mathcal{F}}{V(\mathcal{F})}, \quad 1 \le i \le m. \tag{19}$$

The theorem for integration on manifolds states

$$f_{s,i} = \frac{\int_{\Lambda} x_i \circ \boldsymbol{H}^\star \sqrt{det((\boldsymbol{DH})^{\star T}(\boldsymbol{DH})^\star)} \;\; d\boldsymbol{\lambda}}{\int_{\Lambda} 1 \circ \boldsymbol{H}^\star \sqrt{det((\boldsymbol{DH})^{\star T}(\boldsymbol{DH})^\star)} \;\; d\boldsymbol{\lambda}} \tag{20}$$

where $H^* : \Lambda \to \mathcal{F}, \lambda \to H\lambda$ is a map from Λ to $\mathcal{F}$ and $(DH)^*$ is the Jacobian of H^* which is equal to H itself since it is linear. Furthermore, $\sqrt{det(H^T H)}$ is independent from λ and can hence be canceled in the next step. Additionally splitting Λ into the simplexes gives:

$$f_{s,i} = \frac{\sum_{\nu=1}^{n_s} \int_{P_\nu} x_i \circ \boldsymbol{H}^\star \;\; d\boldsymbol{\lambda}}{\sum_{\nu=1}^{n_s} \int_{P_\nu} 1 \;\; d\boldsymbol{\lambda}} = \frac{\sum_{\nu=1}^{n_s} \int_{P_\nu} \sum_{\mu=1}^{r} \boldsymbol{h}_{\mu,i} \lambda^\mu \;\; d\boldsymbol{\lambda}}{\sum_{\nu=1}^{n_s} V_\nu} \tag{21}$$

Since H is independent from λ, it can be moved out of the integral:

$$f_{s,i} = \frac{\begin{bmatrix} h_{1,i} & \dots & h_{r,i} \end{bmatrix}}{\sum_{\nu=1}^{n_s} V_\nu} \cdot \begin{bmatrix} \sum_{\nu=1}^{n_s} \int_{P_\nu} \lambda_1 \;\; d\boldsymbol{\lambda} \\ \vdots \\ \sum_{\nu=1}^{n_s} \int_{P_\nu} \lambda_r \;\; d\boldsymbol{\lambda} \end{bmatrix} \tag{22}$$

Using eqn. 19 and eqn. 17 this can be rewritten as

$$f_{s,i} = \begin{bmatrix} h_{1,i} & \dots & h_{r,i} \end{bmatrix} \begin{bmatrix} \lambda_{s,1} \\ \vdots \\ \lambda_{s,r} \end{bmatrix} = \begin{bmatrix} h_{1,i} & \dots & h_{r,i} \end{bmatrix} \boldsymbol{\lambda}_s \tag{23}$$

Therefore $f_s = H\lambda_s$ holds where λ_s denotes the CoG of Λ in $\mathbb{R}^r$.

3.3.2 Continuity of solution

In this section the continuity of of the solution of the developed algorithm in the p-norm $\|\cdot\|_p$ $(p \neq 1, \infty)$ is proven, i.e. the function $\Gamma : \mathbb{R}^{m \cdot n} \to \mathbb{R}^n$, which maps a matrix $A \in \mathbb{R}^{m \times n}$ (considered as a vector in $\mathbb{R}^{m \cdot n}$) onto the center of gravity as described before, is continuous on the set of points of the workspace.

Proof

Again without loss of generality $w = 0$ is assumed. First Γ is splitted into two mappings $Ker : \mathbb{R}^{m n} \to \mathbb{R}^{n \cdot r}$ and $Grav^C : \mathbb{R}^{n \cdot r} \to \mathbb{R}^n$. The latter maps a vector $\boldsymbol{p}$ from $\mathbb{R}^{n \cdot r}$ onto the center of gravity of the manifold $\mathcal{F}$ spanned by the r n-dimensional downwards listed vectors in $\boldsymbol{p}$. $Ker : \mathbb{R}^{m \cdot n} \to \mathbb{R}^{n \cdot r}$ maps a matrix $\boldsymbol{A}$ on its kernel $\boldsymbol{H}$ represented as a vector $\boldsymbol{p}$ in $\mathbb{R}^{n \cdot r}$. In calculations the kernel is still denoted with $\boldsymbol{H}$ for simplicity. Continuity of Ker and $Grav^C$ implies continuity of Γ, since $\Gamma = Grav^C \circ Ker$.

First the continuity of $Grav^C$ will be proven. Therefore $\Lambda \neq 0$ is assumed (i.e. the intersection of hypercube $\mathcal{C}$ and subspace $\mathcal{S}$ is non-empty and thus also the CoG exists), since continuity inside of $\mathcal{C}$ is to be proven. The CoG λ_s is considered:

$$\lambda_{s,i} = \frac{\int\limits_{\Lambda} \lambda_i d\boldsymbol{\lambda}}{V(\Lambda)} \qquad 1 \leq i \leq r \tag{24}$$

Let $\tilde{\lambda}_s$ be the CoG of $\tilde{\Lambda}$, where $\tilde{\Lambda}$ is the preimage of $\tilde{F}$, which is obtained from $\tilde{\boldsymbol{H}} = \boldsymbol{H} + \mathbf{E}$. The matrices $\boldsymbol{H} = [h_1 \ . \ . \ . \ h_r]^T \in \mathbb{R}^{n \times r}$ and $\boldsymbol{E} = [e_1 \ . \ . \ . \ e_r]^T \in \mathbb{R}^{n \times r}$ are considered as vectors in $\mathbb{R}^{n \cdot r}$. Then the p-norm of $\boldsymbol{H}$ is $\left\| [h_1 \quad h_2 \quad \dots \quad h_{n \cdot r}]^T \right\|_p$. It follows

$$\lim_{\|E\|_p \to 0} \left| \tilde{\lambda}_{s,i} - \lambda_{s,i} \right| = \lim_{\|E\|_p \to 0} \left| \frac{\int\limits_{\tilde{\Lambda}} \lambda_i d\boldsymbol{\lambda}}{V(\tilde{\Lambda})} - \frac{\int\limits_{\Lambda} \lambda_i d\boldsymbol{\lambda}}{V(\Lambda)} \right| \tag{25}$$

$$= \lim_{\|E\|_p \to 0} \left| \frac{V(\Lambda) \int\limits_{\tilde{\Lambda} \setminus \Lambda} \lambda_i d\boldsymbol{\lambda} + V(\tilde{\Lambda}) \int\limits_{\Lambda \setminus \tilde{\Lambda}} \lambda_i d\boldsymbol{\lambda}}{V(\tilde{\Lambda}) V(\Lambda)} \right|, \qquad 1 \leq i \leq r. \tag{26}$$

Since the vertices of the polyhedron $\tilde{\lambda}$ are obtained from the inequality

$$\boldsymbol{f}_{\min} \leq \tilde{\boldsymbol{H}} \boldsymbol{\lambda} \leq \boldsymbol{f}_{\max} \tag{27}$$

$$\Leftrightarrow \boldsymbol{f}_{\min} \leq \boldsymbol{H} \boldsymbol{\lambda} + \boldsymbol{E} \boldsymbol{\lambda} \leq \boldsymbol{f}_{\max} \tag{28}$$

and the vertices of the polyhedron Λ are obtained from (12), it is obvious that

$$\lim_{\|E\|_p \to 0} V(\tilde{\Lambda} \setminus \Lambda) = 0 \tag{29}$$

$$\lim_{\|E\|_p \to 0} V(\Lambda \setminus \tilde{\Lambda}) = 0. \tag{30}$$

Hence

$$\lim_{\|E\|_p \to 0} \left| \tilde{\lambda}_{s_i} - \lambda_{s_i} \right| = 0, \quad 1 \le i \le r \tag{31}$$

holds, because $\tilde{\Lambda}$ and Λ are bounded. This yields together with eqn. (18)

$$\lim_{\|E\|_p \to 0} \left| \boldsymbol{f}_{s,i} - \tilde{\boldsymbol{f}}_{s,i} \right| = \lim_{\|E\|_p \to 0} \left| \boldsymbol{h}_{i,r} \boldsymbol{\lambda}_s - \tilde{\boldsymbol{h}}_{i,r} \tilde{\boldsymbol{\lambda}}_s \right| \tag{32}$$

$$= \lim_{\|E\|_p \to 0} \left| \boldsymbol{h}_{i,r} \boldsymbol{\lambda}_s - \boldsymbol{h}_{i,r} + \boldsymbol{e}_{i,r} \tilde{\boldsymbol{\lambda}}_s \right| = \lim_{\|E\|_p \to 0} \left| \boldsymbol{h}_{i,r} (\boldsymbol{\lambda}_s - \tilde{\boldsymbol{\lambda}}_s) - \boldsymbol{e}_{i,r} \tilde{\boldsymbol{\lambda}}_s \right| = 0, \quad 1 \le i \le r. \tag{33}$$

This implies the continuity of $Grav^C$.
The continuity of *Ker* follows from the fact that the solution of a full ranked linear system of equations depends continuously on the coefficient matrix.

4. Control

Wire robots allow for very high velocities and accelerations when handling lightweight goods. In this case, wire robots benefit from their lightweight structure and low moved masses. Contrariwise, wire-based mechanisms like cranes, winches or lifting blocks are used widely to move extremely heavy loads. Thus, the wide range of application demands for a robust and responsive control. To move the platform along a trajectory precisely, position control is mandatory. On the other hand, the usage of wires claims for a careful observation and control of the applied tensions to guarantee a safe and accurate operation. Pure force control suffers from the drawbacks of model based control, e.g. model mismatch and parameter uncertainties. Thus force control is not sufficient and a combined force and position control is advised. Beside this, the relatively high elasticity of the wires may demand for a compensation by control. (Fang, 2005) shows more details of the shown concepts.

4.1 Elastic wire compensation

Compared to a conventional parallel kinematic machine (e.g. Stewart platform), a wire robot has generally a higher elasticity in the kinematic chains connecting the base and the platform. This is both due to the stiffness of the wire material as well as due to the wire construction (e.g. laid/twisted, braided or plaited)(Feyrer, 2000). Approximating the dynamical characteristics of the wires by a linear spring-damper model and considering the unilateral constraint, the wire model can be described as

$$\boldsymbol{f}_i = \begin{cases} c_i \Delta l_i + d_i \Delta \dot{l}_i & f_i > 0 \\ 0 & f_i \le 0 \end{cases} \tag{34}$$

with $1 < i < m$, $\boldsymbol{c}_i$ and $\boldsymbol{d}_i$ denoting the stiffness and damping coefficients, respectively and $\Delta\, l_i$ denoting the length change due to elasticity. Assuming the untensed wire length is $l_{i,0}$, $\Delta\, l_i$

can be computed as $\Delta l_i = l_i - l_{i,0}$. The stiffness coefficient c_i depends on the actual wire length. Using the wire cross section A and Young's modulus E, c_i can be calculated as

$$c_i = \frac{E \cdot A}{q_{i,0}} = \frac{E \cdot A}{q_i(1 - \epsilon_i)} \tag{35}$$

with

$$\epsilon_i = \frac{\Delta q_i}{q_i} \tag{36}$$

Note that this is only a linear approach. Taking into account long and heavy wires, a specific wire composition and applied tensions close to the admittible work load, advanced non-linear models have to be utilized. Especially the damping coefficient d_i may be hard to estimate (Wehking et al., 1999) and thus, experiments have to be carried out (Vogel & Götzelmann, 2002).

4.2 Motion control in joint space

The idea of motion control in joint space is to use a feedback position control and a feedforward force controller. The feedforward control employs an inverse dynamics model to calculate the winch torques necessary for the accelerations belonging to the desired trajectory. Since the used dynamic model usually will not cover all mechanical influences (e.g. friction), the remaining position errors can be compensated by the position control which employs the inverse kinematics. Noteworthy, the inverse dynamics is calculated for the desired platform position. Optionally, one may think of tracking control to guide the platform along the desired trajectory for the price of additional calculations. Referring to eqn. 6, the inverse system dynamics (i.e. the wire force distribution) can be computed by methods shown in section 3 (where the loads w include the inertia and gravity loads). Assuming the winch drives are adressable by desired torques (which is normally the case for DC/EC motors, preferably with digital current control), the motor dynamics can be modeled as

$$\boldsymbol{M}_M \ddot{\boldsymbol{\Theta}} + \boldsymbol{T}_f(\dot{\boldsymbol{\Theta}}) + \eta \boldsymbol{D} \boldsymbol{f} = \boldsymbol{u}, \tag{37}$$

where $M_M \in \mathbb{R}^{m \times m}$ is the inertial matrix of the drive units, η is the radius of the drums and $D \in \mathbb{R}^{m \times m}$ depends on the structure of the motors. Combining the feedforward force control and the feedback position control leads to the following controller output:

$$\boldsymbol{u} = \underbrace{\boldsymbol{M}_M \ddot{\boldsymbol{\Theta}} + \boldsymbol{T}_f(\dot{\boldsymbol{\Theta}}) + \boldsymbol{r} \boldsymbol{D} \boldsymbol{f}}_{\text{feedforward force control}} + \underbrace{\boldsymbol{K}_p(\boldsymbol{\Theta}_d - \boldsymbol{\Theta}) + \boldsymbol{K}_d(\dot{\boldsymbol{\Theta}}_d - \dot{\boldsymbol{\Theta}})}_{\text{feedback position control}} \tag{38}$$

denoting the feedback gain matrices $K_p \in \mathbb{R}^{m \times m}$ and $K_d \in \mathbb{R}^{m \times m}$ and the actual and desired motor angles Θ and Θ_d, respectively. Due to the decoupled position controllers, these may be designed as decentralized, simple and high control rate devices. To compensate for elastic tendons, the following correction may be applied:

$$\Theta_{d,i} = \hat{\Theta}_{d,i} + \Delta\Theta_i = \frac{l_{d,i} + \frac{f_i}{c_i}}{\eta} \tag{39}$$

where $\hat{\Theta}_{d,i}$ corresponds to the uncompensated drum angle ($1 \leq i \leq m$).

4.3 Motion control in operational space

Observing the sections above, independent linear PD controllers are applied. Practical experiences show that this is possible even though the system dynamics are described by a nonlinear, coupled system of equations due to the parallel topology of the robot, represented by the pose dependent structure matrix. Nevertheless, it is difficult to determine stable or even optimal controller parameters since the usual tools of the linear control theory may only be applied for locally linearized configurations of the robot. For predefined trajectories, this may be possible (e.g. by defining a cost function accumulating the control errors in simulation and applying a nonlinear optimizer to obain values for K_p and K_d), but is is desirable to have a globally linear system to avoid this only locally valid approach. From literature (Schwarz, 1991) (Woernle, 1995), exact linearization approaches are known which eliminate the nonlinear system characteristics by feedback. Using this as an inner loop, an outer linear controller may now be applied to the resulting linear system. Eqns. 37 and 6 deliver

$$\boldsymbol{M}_p\ddot{\boldsymbol{x}} + \boldsymbol{g}_C - \boldsymbol{g}_E + (\boldsymbol{M}_M\ddot{\boldsymbol{\Theta}} + \boldsymbol{T}_f(\dot{\boldsymbol{\Theta}}))\frac{\boldsymbol{A}^T\boldsymbol{D}^{-1}}{\eta} = \frac{\boldsymbol{A}^T\boldsymbol{D}^{-1}}{\eta}\boldsymbol{u}. \tag{40}$$

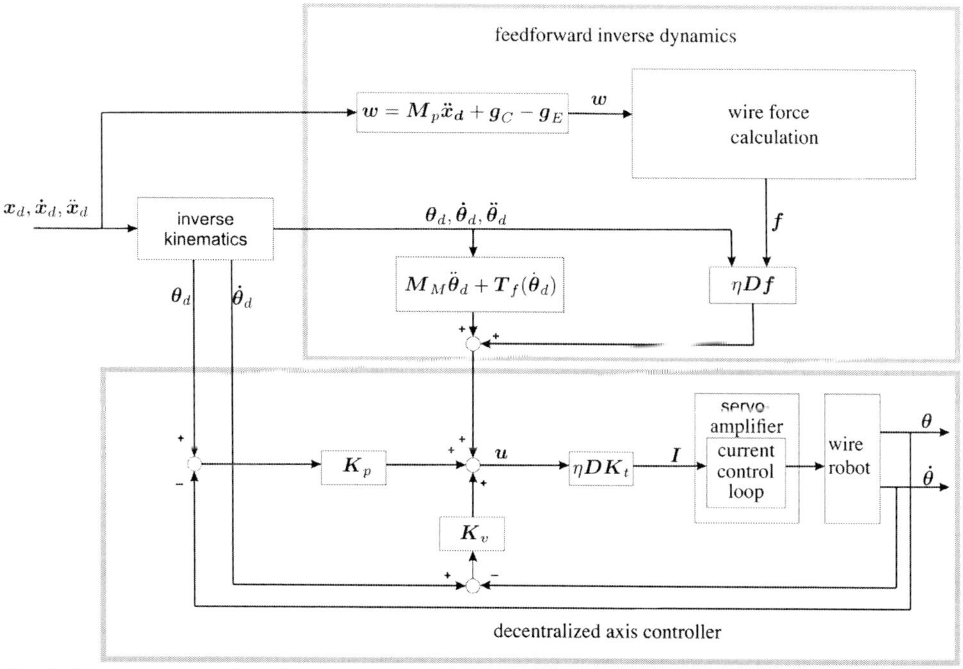

Fig. 5: Block scheme of motion control in joint space (Fang, 2005)

Since the final control law is formulated in the operational space, this equation is transformed into cartesian coordinates using the inverse kinematics relations

$$\dot{\boldsymbol{\theta}} = \frac{\boldsymbol{A}^T \dot{\boldsymbol{x}}}{\eta} \tag{41}$$

$$\ddot{\boldsymbol{\theta}} = \frac{\boldsymbol{A}^T \ddot{\boldsymbol{x}}}{\eta} + \frac{\dot{\boldsymbol{A}}^T \dot{\boldsymbol{x}}}{\eta}. \tag{42}$$

In cartesian coordinates the dynamical equations are then given by

$$\underbrace{\boldsymbol{M}_p + \frac{\boldsymbol{A}^T \boldsymbol{D}^{-1} \boldsymbol{M}_m \boldsymbol{A}}{\eta^2}}_{\boldsymbol{M}_{eq}} \ddot{\boldsymbol{x}} + \underbrace{\boldsymbol{g}_C + \boldsymbol{g}_E + \frac{\boldsymbol{A}^T \boldsymbol{D}^{-1} \boldsymbol{M}_m \dot{\boldsymbol{A}}}{\eta^2} \dot{\boldsymbol{x}} + \frac{\boldsymbol{A}^T \boldsymbol{D}^{-1} \boldsymbol{M}_m \boldsymbol{T}_f(\dot{\boldsymbol{x}})}{\eta}}_{\boldsymbol{N}} = \underbrace{\frac{\boldsymbol{A}^T \boldsymbol{D}^{-1}}{\eta} \boldsymbol{u}}_{\boldsymbol{F}_\nu}. \tag{43}$$

Instead of using the motor torques $\boldsymbol{u}$ as the system input, the resulting forces and torques acting onto the platform $\boldsymbol{F}_\nu$ are chosen to represent the actuator torques. Now a global linearization is desired. Setting $\boldsymbol{F}_\nu = \boldsymbol{M}_{eq} \nu + \boldsymbol{N}$ delivers

$$\ddot{\boldsymbol{x}} = \boldsymbol{\nu}, \tag{44}$$

and is therefore a proper choice. This linear system is now controlled by a PD controller for the position. Thus, the new system input is extended by

$$\boldsymbol{\nu} = \ddot{\boldsymbol{x}}_d + \boldsymbol{K}_p(\boldsymbol{x}_d - \boldsymbol{x}) + \boldsymbol{K}_d(\dot{\boldsymbol{x}}_d - \dot{\boldsymbol{x}}) \tag{45}$$

Substituting eqn. 45 into eqn. 43, $\boldsymbol{F}_\nu$ can be found as

$$\boldsymbol{F}_\nu = \boldsymbol{M}_{eq}(\ddot{\boldsymbol{x}}_d + \boldsymbol{K}_p(\boldsymbol{x}_d - \boldsymbol{x}) + \boldsymbol{K}_d(\dot{\boldsymbol{x}}_d - \dot{\boldsymbol{x}})) + \boldsymbol{N} \tag{46}$$

which describes the required wrench onto the platform $\boldsymbol{w}$ which allows to calculate the desired wire forces by the methods shown in section 3. Optionally, the desired forces can be controlled by an outer feedback loop to enhance the control precision.

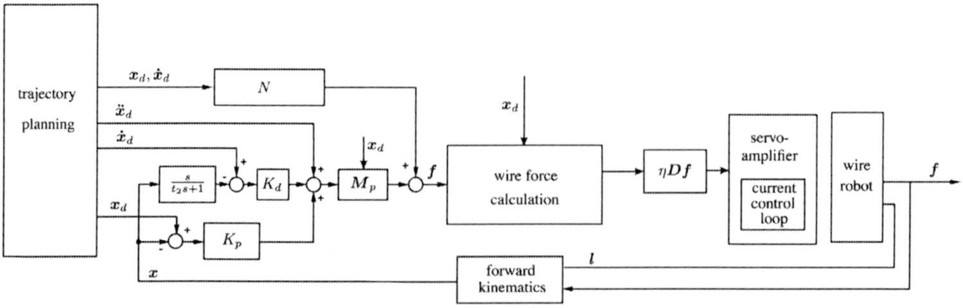

Fig. 6: Block scheme of motion control in operational space

5. Applications

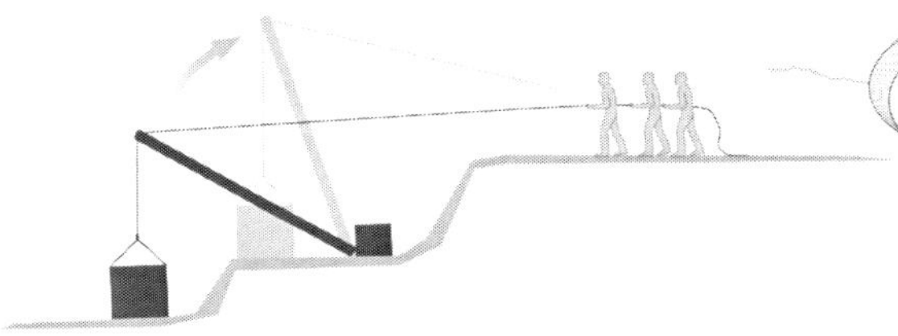

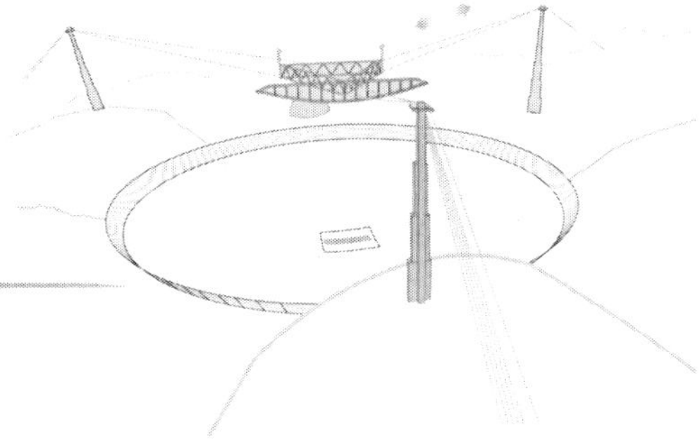

Fig. 7(a) Early wire manipulation Fig. 7(b) Arecibo telescope

As already mentioned before, wire-based manipulation and construction is used since millenia, mostly taking advantage from the principle of the lifting block. In ancient civilisations like the Egypt of the Pharaos, probably wires and winches were applied to build the pyramids - wether using ramps or lifting mechanisms (see fig. 7(a)). Crane technology was only possible due to the usage of wires and especially the old Romans deleloped this technology to a remarkable state - they already lifted loads around 7 tons with cranes driven by 4 workers. With industrialisation, the transport and manipulation of heavy goods became very important, and hence, cranes using steel cables completed the transport chain for cargo handling. In the last few years, the automatisation of crane technology was subject to extensive research, e.g. in the project RoboCrane ® by the National Institute of Standards and Technology (NIST) (Bostelman et al., 2000). At the University of Rostock, the prototype CABLEV (Cable Levitation) (Maier, 2004),(Heyden, 2006) was build up, see fig. 8. It uses a gantry crane and three wires to guide the load along a trajectory. Thew load is stabilized by a tracking control for IRPM systems which eliminates

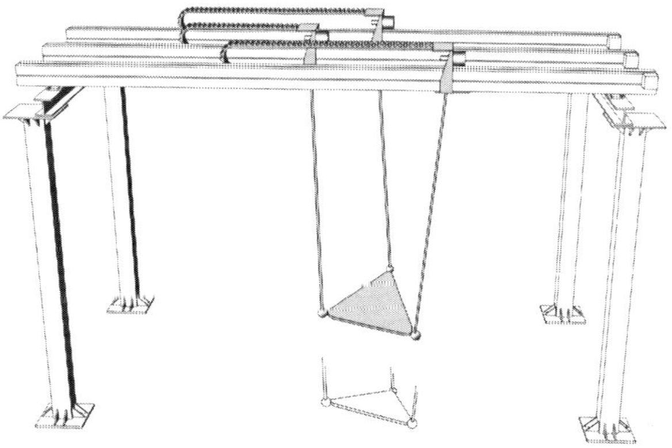

Fig. 8: CABLEV protoype

oscillations. In Japan, the Tadokoro Laboratory of the Tohoku University in Japan proposes the application of wires for rescue robots (Takemura et al., 2005) (Maeda et al., 1999). A problem solved very smart by usage of wires is the positioning of a large telescope. Several projects, e.g. the world's largest telescope at Arecibo (fig. 7(b)), deal with the usage of wires to place the receiver module. The Arecibo project (900t receiver, approximately 300m satellite dish diameter) uses three wires guided by three mast heads while other projects use an inverse configuration, lifting the receiver by balloons (see (Su et al., 2001), (Taghirad & Nahon, 2007a), (Taghirad & Nahon, 2007b)). Another popular application of wire robots is the usage as a manipulator for aerodynamical models in wind tunnels as proposed in (Lafourcade et al., 2002), (Zheng, 2006) and (Yaqing et al., 2007). Here, the experiments take advantage from the very thin wires since undisturbed air flow is mandatory. On the other hand, the wire robot can perform high dynamical motion as for example the FALCON (Fast Load Conveyance) robot (Kawamura et al., 1995). In the past few years at the Chair for Mechatronics at the University of Duisburg-Essen the testbed for wire robots SEGESTA (Seilgetriebene Stewart-Plattformen in Theorie und Anwendung) (Hiller et al., 2005b) has been developed. It is currently operated with seven (see fig. 9) wires in an CRPM configuration or eight wires for a RRPM setup. Focus of research is the development of fast and reliable methods for workspace calculation (Verhoeven & Hiller, 2000) and robot design. Another focus is the development of robust and realtime-capable control concepts (Mikelsons et al., 2008). Since the teststand is available, the theoretical results can be tested and verified (Hiller et al., 2005a). The system performs accelerations up to *10g* and velocities around *10m/s*.

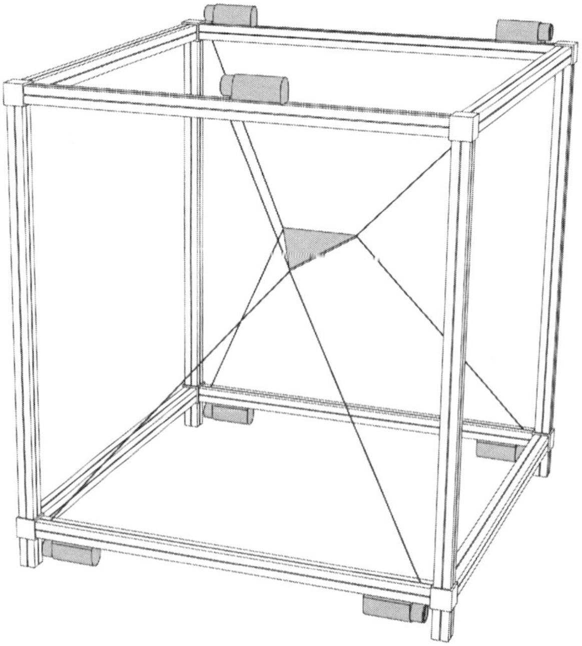

Fig. 9: SEGESTA protoype

Another very recent application area has been created by Visual Act AB®. As pictured in fig. 10. a snowboard simulator was built up. The snowboarder is connected to four wires leading to three translational d.o.f.. Hence, the snowboarder can be guided along a trajectory in a setting consisting of ramps to grind on while he is moving freely in the air. (Visualact AB, 2006). A completely different field is the application of wire robots for rehabilitation which was demonstrated by the system String Man by the Fraunhofer-Institut für Produktionsanlagen und Konstruktionstechnik (IPK) in Berlin, Germany (Surdilovic et al., 2007). Another prototype for rehabilitation is described in (Frey et al., 2006). The application of wire robots as a tracking device was proposed in (Ottaviano & Ceccarelli, 2006), (Thomas et al., 2003) and (Ottaviano et al., 2005). Here, the wire robot is not actively supporting a load but attached to an object which is tracked by the robot.

Fig. 10: Snowboard Simulator

6. Acknowledgements

This work is supported by the German Research Council (Deutsche Forschungsgemeinschaft) under HI370/24-1, HI370/19-3 and SCHR1176/1-2. The authors would like to thank Martin Langhammer for contributing the figure design.

7. References

Bosscher, P. and Ebert-Uphoff, I. (2004). Wrench-based analysis of cable-driven robots. *Proceedings of the 2004 IEEE International Conference on Robotics & Automation*, pages 4950–4955.

Bostelman, R., Jacoff, A., and Proctor, F. (2000). Cable-based reconfigurable machines for large scale manufacturing. In *Japan/USA Flexible Automation Conference Proceedings*, University of Michigan, Ann Arbor, MI.

Bruckmann, T., Mikelsons, L., Brandt, T., Hiller, M., and Schramm, D. (2008). Wire robots part I - kinematics, analysis & design. In Lazinica, A., editor, *Parallel Manipulators*, ARS Robotic Books. I-Tech Education and Publishing, Vienna, Austria. ISBN 978-3-902613-20-2.

Bruckmann, T., Pott, A., and Hiller, M. (2006). Calculating force distributions for redundantly actuated tendon-based Stewart platforms. In Lenarcic, J. and Roth, B., editors, *Advances in Robot Kinematics - Mechanisms and Motion*, pages 403– 413,

Ljubljana, Slowenien. Advances in Robotics and Kinematics 2006, Springer Verlag, Dordrecht, The Netherlands.

Cignoni, P., Montani, C., and Scopigno, R. (1998). Dewall: A fast divide and conquer delaunay triangulation algorithm in ed. *Computer-Aided Design*, 30(5):333–341.

Fang, S. (2005). *Design, Modeling and Motion Control of Tendon-based Parallel Manipulators*. Ph. D. dissertation, Gerhard-Mercator-University, Duisburg, Germany. Fortschritt-Berichte VDI, Reihe 8, Nr. 1076, Düsseldorf.

Feyrer, K. (2000). *Drahtseile*. Springer Verlag Berlin.

Frey, M., Colombo, G., Vaglio, M., Bucher, R., Jörg, M., and Riener, R. (2006). A novel mechatronic body weight support system. *IEEE Transactions on Neural Systems and Rehabilitation Engineering*, 14(3):311–321.

Hammer, P. C., Marlowe, O. P., and Stroud, A. H. (1956). Numerical integration over simplexes and cones. *Math. Tables Aids Comp.*, 10(55):130–137.

Heyden, T. (2006). *Bahnregelung eines seilgeführten Handhabungssystems mit kinematisch unbestimmter Lastführung*. PhD thesis, Universität Rostock. ISBN: 3-18- 510008-5, Fortschritt-Berichte VDI, Reihe 8, Nr. 1100, Düsseldorf.

Hiller, M., Fang, S., Hass, C., and Bruckmann, T. (2005a). Analysis, realization and application of the tendon-based parallel robot segesta. In Last, P., Budde, C., and Wahl, F., editors, *Robotic Systems for Handling and Assembly*, volume 2 of *International Colloquium of the Collaborative Research Center SFB 562*, pages 185–202, Braunschweig, Germany. Aachen, Shaker Verlag.

Hiller, M., Fang, S., Mielczarek, S., Verhoeven, R., and Franitza, D. (2005b). Design, analysis and realization of tendon-based parallel manipulators. *Mechanism and Machine Theory*, 40.

Kawamura, S., Choe, W., Tanaka, S., and Pandian, S. R. (1995). Development of an ultrahigh speed robot falcon using wire drive system. *IEEE International Conference on Robotics and Automation*, pages 215–220.

Kawamura, S., Kino, H., and Won, C. (2000). High-speed manipulation by using parallel wire-driven robots. *Robotica*, 18(1):13–21.

Lafourcade, P., Llibre, M., and Reboulet, C. (October 3-4, 2002). Design of a parallel wire-driven manipulator for wind tunnels. In Gosselin, C. M. and Ebert-Uphoff, I., editors, *Workshop on Fundamental Issues and Future Research Directions for Parallel Mechanisms and Manipulators*.

Maeda, K., Tadokoro, S., Takamori, T., Hattori, M., Hiller, M., and Verhoeven, R. (1999). On design of a redundant wire-driven parallel robot warp manipulator. *Proceedings of IEEE International Conference on Robotics and Automation*, pages 895–900.

Maier, T. (2004). *Bahnsteuerung eines seilgeführten Handhabungssystems - Modellbildung, Simulation und Experiment*. PhD thesis, Universität Rostock, Brandenburg. Fortschritt-Berichte VDI, Reihe 8, Nr. 1047, Düsseldorf.

Mikelsons, L., Bruckmann, T., Hiller, M., and Schramm, D. (2008). A real-time capable force calculation algorithm for redundant tendon-based parallel manipulators. *appears in Proceedings on IEEE International Conference on Robotics and Automation*.

Ming, A. and Higuchi, T. (1994). Study on multiple degree of freedom positioning mechanisms using wires, part 1 - concept, design and control. *International Journal of the Japan Society for Precision Engineering*, 28:131–138.

Nahon, M. and Angeles, J. (1991). Real-time force optimization in parallel kinematics chains under inequality constraints. In *IEEE International Conference on Robotics and Automation*, pages 2198–2203, Sacramento.

Oh, S. R. and Agrawal, S. K. (2005). Cable suspended planar robots with redundant cables: Controllers with positive tensions. In *IEEE Transactions on Robotics*.

Ottaviano, E. and Ceccarelli, M. (2006). Numerical and experimental characterization of singularities of a six-wire parallel architecture. *Robotica*, 25(3):315–324.

Ottaviano, E., Ceccarelli, M., Paone, A., and Carbone, G. (April 18-22 2005). A low-cost easy operation 4-cable driven parallel manipulator. In *Proceedings of the 2005 IEEE International Conference on Robotics and Automation*, pages 4008–4013, Barcelona, Spain.

Schwarz, H. (1991). *Nichtlineare Regelungssysteme*. Oldenbourg, München. ISBN- 13: 978-3486218336.

Su, Y. X., Duan, B. Y., Nan, R. D., and Peng, B. (2001). Development of a large parallel-cable manipulator for the feed-supporting system of a next-generation large radio telescope. In *Journal of Robotic Systems*, volume 18, pages 633–643.

Surdilovic, D., Zhang, J., and Bernhardt, R. (13-15 June 2007). String-man: Wirerobot technology for safe, flexible and human-friendly gait rehabilitation. In *Proccedings of IEEE 10th International Conference on Rehabilitation Robotics, 2007*, pages 446–453, Noordwijk, Netherlands. ISBN: 978-1-4244-1320-1.

Taghirad, H. and Nahon, M. (2007a). Forward kinematics of a macro-micro parallel manipulator. In *Proceedings of the 2007 IEEE/ASME International Conference on Advanced Intelligent Mechatronics (AIM2007)*, Zurich, Switzerland.

Taghirad, H. and Nahon, M. (2007b). Jacobian analysis of a macro-micro parallel manipulator. In *Proceedings of the 2007 IEEE/ASME International Conference on Advanced Intelligent Mechatronics (AIM2007)*, Zurich, Switzerland.

Takemura, F., Enomoto, M., Tanaka, T., Denou, K., Kobayashi, Y., and Tadokoro, S. (2005). Development of the balloon-cable driven robot for information collection from sky and proposal of the search strategy at a major disaster. In *Proceedings on IEEE/ASME International Conference on Advanced Intelligent Mechatronics*, pages 658–663, Monterey.

Thomas, F., Ottaviano, E., Ros, L., and Ceccarelli, M. (September 14-19, 2003). Coordinate-free formulation of a 3-2-1 wire-based tracking device using cayleymenger determinants. In *Proceedings of the 2003 IEEE International Conference on Robotics and Automation*, pages 355–361, Taipei, Taiwan.

Verhoeven, R. (2004). *Analysis of the Workspace of Tendon-based Stewart Platforms*. PhD thesis, University of Duisburg-Essen.

Verhoeven, R. and Hiller, M. (2000). Estimating the controllable workspace of tendonbased Stewart platforms. In *Proceedings of the ARK'00: 7th. International Symposium on Advances in Robot Kinematics*, pages 277–284, Portoroz, Slovenia.

Visualact AB (2006). Visual act 3d. http://www.visualact.net/.

Vogel, W. and Götzelmann, B. (2002). Kraft in Faserseilen bei ausgewählten stossartigen Beanspruchungen. *EUROSEIL*, 121(3):44/45.

Voglewede, P. and Ebert-Uphoff, I. (2004). On the connections between cable-driven robots, parallel manipulators and grasping. In *IEEE International Conference on Robotics and Automation*, volume 5, pages 4521- 4526, New Orleans. IEEE.

von Zitzewitz, J., Duschau Wicke, A., Wellner, M., Lünenburger, L., and Riener, R. (2006). Path control: A new approach in patient-cooperative gait training with the rehabilitation robot lokomat. *Gemeinsame Jahrestagung der Deutschen, Österreichischen und Schweizerischen Gesellschaften für Biomedizinische Technik*. Zürich, Schweiz.

Wehking, K.-H., Vogel, W., and Schulz, R. (1999). Dämpfungsverhalten von Drahtseilen. *F+H Fördern und Heben*, 49(1-2):60–61.

Wellner, M., Guidali, M., Zitzewitz, J., and Riener, R. (June 12-15, 2007). Using a robotic gait orthosis as haptic display - a perception-based optimization approach. *Proceedings of the 2007 IEEE 10th International Conference on Rehabilitation Robotics*, pages 81–88. Noordwijk, The Netherlands.

Woernle, C. (1995). *Regelung von Mehrkörpersystemen durch externe Linearisierung*. Number 517 in Fortschrittberichte VDI, Reihe 8. VDI Verlag, Düsseldorf.

Yaqing, Z., Qi, L., and Xiongwei, L. (June18-21, 2007). Initial test of a wiredriven parallel suspension system for low speed wind tunnels. In *Proceedings on 12thIFToMM World Congress*, Besançon, France.

Zheng, Y.-Q. (2006). Feedback linearization control of a wire-driven parallel support system in wind tunnels. *Sixth International Conference on Intelligent Systems Design and Applications*, 3:9–13.

8

Parallel Robot Scheduling with Genetic Algorithms

Tarık Cakar[1], Harun Resit Yazgan[1] and Rasit Koker[2]
[1]Sakarya University Industrial Engineering Department
[2]Sakarya University Computer Engineering Department
Sakarya Turkey

1. Introduction

There are some main goals in parallel robot scheduling. Those are total completion time, maximum earliness, and maximum tardiness. According to the theoretical viewpoint, parallel robot scheduling is a generalization of the single robot scheduling and a special study of the flow shop. From the practical viewpoint, solution techniques are useful in the real-world problems. Parallel robot scheduling has to deal with balancing the load in practice. Scheduling parallel robot may be considered as a double-step. First, which jobs are allocated to which Robot. Second, allocated jobs sequence. Also, preemption plays a more important role in parallel robot scheduling. Robots may be identical or not. Jobs have a precedence constraint. For all problem structures may be applied different solution techniques for instance algorithms, search algorithms or artificial intelligence techniques. In this chapter we interest in different solution techniques for parallel robot scheduling.

In this chapter, first, a genetic algorithm is used to schedule jobs that have precedence constraints minimizing the total earliness and tardiness cost and maximum flow time on n-number of job and m-number of identical parallel robots. The second one is without precedence constraint. There are many algorithms and heuristics related to the scheduling problem of parallel machines and robots. In this study, a genetic algorithm has been used to find the job schedule, which minimizes maximum flow time. We know that this problem is in the class of NP-hard combinatorial problem.

(Kanjo & Ase, 2003) studied about scheduling in a multi robot welding system. (Sun & Zhu, 2002) applied a genetic algorithm for scheduling dual resources with robots. (Zacharia & Asparagatos, 2005) proposed a method on GAs for optimal robot task scheduling. In this study, the job with n-number of precedence constraints is assigned minimizing mean tardiness on m-number of parallel robot using genetic algorithms.

(Koulamas,1997) developed a heuristic noted hybrid simulated annealing (HAS) based on simulated annealing. (Chen et al.,1997) has developed highes priority job first (HPJF) method, which is based on extension of the WI method extended with various priority rules such as minimum processing time first (priority = 1/processing time), maximum processing time first (priority=processing time), minimum deadline first (priority=1/due date) and maximum deadline first (priority = Due date). (Alidaee & Rosa, 1997) proposed a heuristic which is based on extending the modified due date (MDD) method belonging (Baker &

Bertrand, 1982). Their method is quite effective for parallel machine problem according to their reports. (Azizoglu & Kirca, 1998) proposed a branch and bound (BAB) approach to solve the same problem mentioned in this paper. Another example can be given by considering identical due dates and processing times, (Elmaghraby & Park, 1974), developed an algorithm based on a branch and bound to minimize a function of penalties belonging to tardiness. (Barnes & Brennan,1977) evaluated and improved their method again.

In addition to these previous studies, there are a few more studies, which deal with parallel machine scheduling problem. But these studies are interested in alternatives. A few examples are given in the following for the minimization of the total weighted tardiness: (Emmons & Pinedo, 1990), (Arkin & Roundy, 1991); for uniform or unspecified parallel machines scheduling, the example studies are: (Emmons, 1987) or (Guinet, 1995). (Karp, 1972) has shown that even the total tardiness minimization in two identical machine scheduling problem was NP-hard. A branch and bound algorithm to minimize maximum lateness considering due dates, family setup times and release dates have been presented by (Shutten & Leussink, 1996). A genetic algorithm was used to find a scheduling policy for identical parallel machine with setup times in (Tamimi & Rajan, 1997). (Armento Yamashita , 2000) applied tabu search into parallel machine scheduling. A scheduling problem for unrelated parallel machine with sequence dependent setup times was studied by (Kim et al. , 2002) using simulated annealing. SA was used to determine a scheduling policy to minimize total tardiness. (Min & Cheng, 1995) proposed an algorithm for identical parallel machine problem. Their algorithm is based on using GA and SA to minimize makespan. According to their studies, it is seen that GA proposed is efficient and fit for larger scale identical machine scheduling problem to minimize the makespan.

(Kashara and Narita, 1985) developed a heuristic algorithm and optimization algorithm for parallel processing of robot arm control computation on a multiprocessor system. (Chen et al., 1988) developed a state-space search algorithm coupled with a heuristic for robot inverse dynamics computation on a multiprocessor system. An assignment rule noted traffic priority index (TPI) was built in 1991 by (Ho & Chang, 1991). In this method, SPT and EDD rules are combined using by using a new measurement named as traffic congestion ratio (TCR). Then, for the cases with one or identical machine they built heuristics. Their heuristics consist of building a first solution by scheduling jobs in increasing order of their priority index. Then they improved this solution using permutation technique of WI method, which was developed previously by (Wilkerson & Irwin, 1971).

2. Definition of the problems

In this study, the job with n-number of precedence constraints is scheduled minimizing total earliness and tardiness cost and maximum flow time on m-number of parallel robots. There are process time and due date for each job. There is not any ready time that belongs to jobs. A robot can do just one job at the same time. The processing is non-preemptive. The target function, which will be minimized, is given below in Eq. (1).

$$Total\_earlines\_tardines\_\cos t = w_e \sum_{j=1}^{n} e_i + w_T \sum_{i=1}^{n} T_i \quad (1)$$

Here, T_j = max {0, C_j - d_j} is the tardiness of job j. e_j = max {0, d_j - C_j} is the earliness of job j. C_j being the completion time and d_j being due date for job j. R(i,j), represents processing or unprocessing of j job on i robot. w_e is unit earliness cost, w_T is unit tardiness cost. If j job is being processed on i robot, R(i,j)=1, otherwise (if not being processed) R(i,j)=0. Fmax is maxsimum flow time. Pj is processing time.

$$Fmax = \max \left(F_i = \sum_{i=1}^{m}\sum_{j=1}^{n} R(i,j)p_j\right) \quad (2)$$

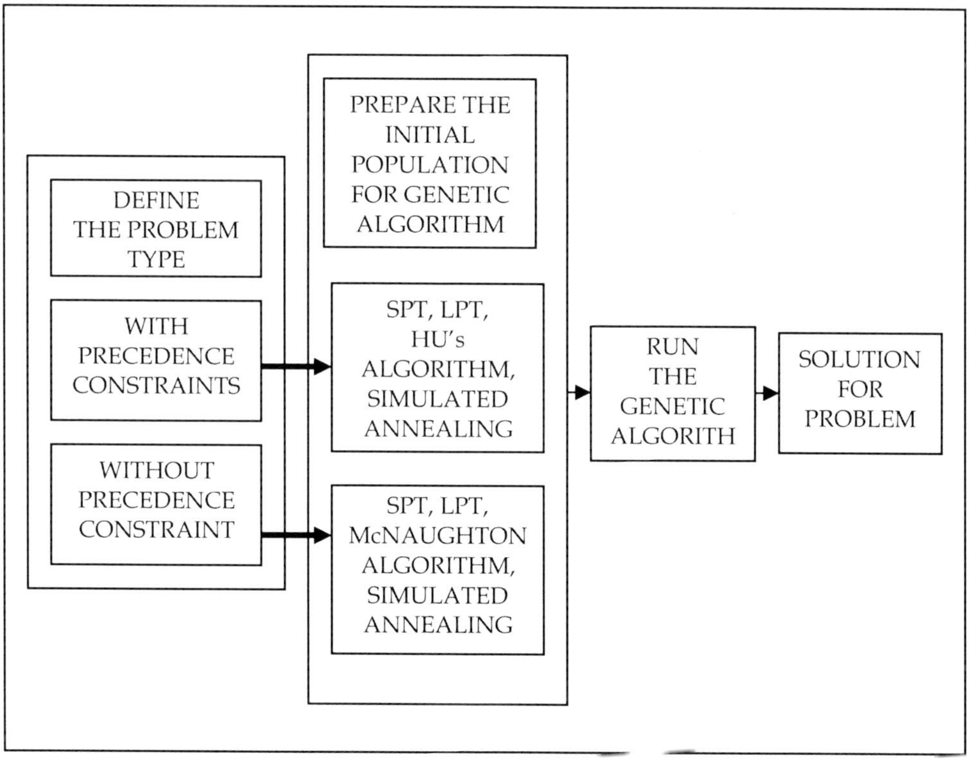

Figure 1. Proposed solution system for the parallel machine scheduling problem.

3. Genetic algorithm

The advantages of the genetic algorithms have been mentioned in the previous section. In this section, the modeling and the application of the GA are explained. From the view point of the working principle, genetic algorithms firstly needs the coding of the problem with the condition that it should be fitting with the GA. After coding process, GA operators are applied on chromosomes. It is not guaranteed that the obtained new offsprings are good solutions by the working of crossover and mutation operators. Feasible solutions are evaluated, and others are left out of evaluation. The feasible ones of the obtained offsprings are taken and new populations are formed by reproduction process using these offsprings. Crossover, mutation and reproduction processes go on until an optimal solution is found. The modeling of the defined problem using genetic algorithm has been presented below with its details.

3.1 Coding for problem statement

The scheduling of the jobs on each robot forms the chromosomes. Here, the chromosomes give the number of robots too. The gene code are c_1, c_2, c_3,.., c_j,... , c_n, where $c_j \in [1,m]$. c_j is positive integer number. Here, each parallel robot represents a chromosome; and gene in chromosome, represents ordered jobs on a robot. The assigned of jobs on robots when forming initial population is done randomly, and while this ordering is done, precedence constraints are taken under care. For instance, let us suppose that there are 8 jobs and 2 robots, and their precedence constraints are given in Figure 3. Sample list representation of the schedule of the jobs on M1 and M2 robots has been given in figure 3. The sample schedule gives also a sample gene code.

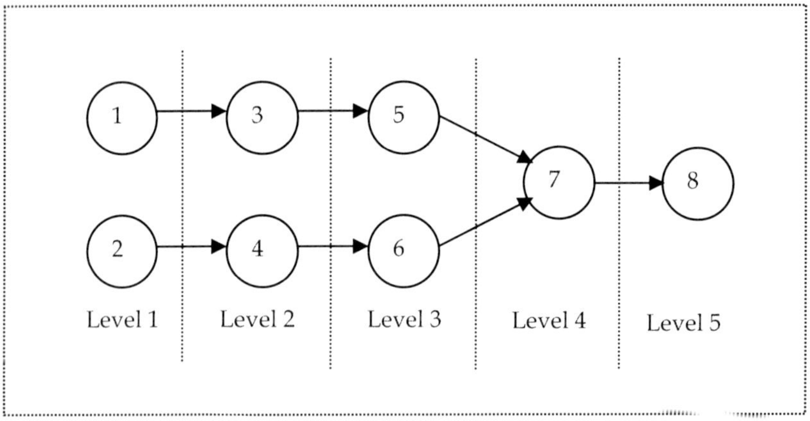

Figure 2. The jobs with precedence constraints

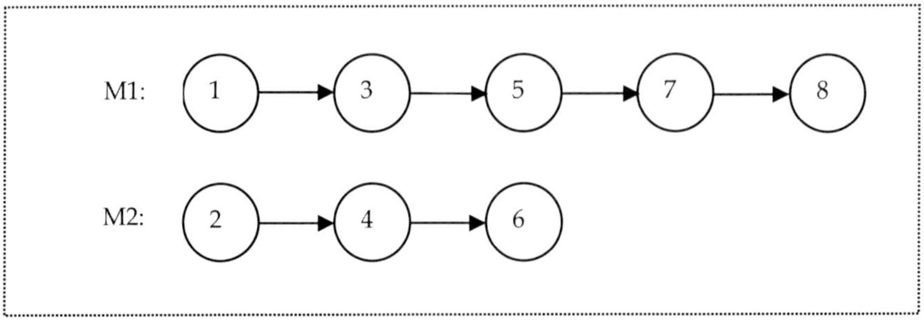

Figure 3. List representation of the schedule

Here, the scheduling of the jobs on robots also shows chromosomes code. M job can be scheduled on N robots in different combinations. But, because of the fact that some of the obtained schedules will be precedence constraints in problem definition, they will not be possible solution. For example, the solution given in figure 4 is not a feasible solution for the precedence constraints in Figure 4. Because the precedence constraints have not been taken under care.

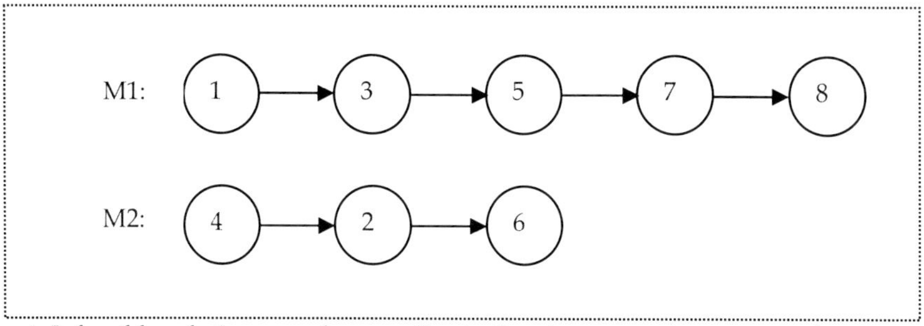

Figure 4. Infeasible solution sample according to the given precedence constraints

3.2 Preparing initial population

Initial population is not produced randomly, fully. In initial population, the solutions for problem with precedence constraints, which are obtained from SPT and EDD heuristics, Simulated Annealing, Hu's algorithm (Baker, 1974) exist and the other. In initial population, the solutions for problem without precedence constraints, which are obtained from SPT and EDD heuristics, Simulated Annealing, McNaughton's algorithm (Baker, 1974) exists. The chromosomes out of these are generated randomly. The jobs are randomly let (determined or given) on robots. However, because of the precedence constraints, in other words, there are some situations like that some jobs may be done before others; some of obtained solutions will not be feasible. These solutions, which are not feasible, will be thrown and the new solutions will be tried to be obtained, randomly.

3.3 Applying crossover operator for the problem

The crossover process is crossing obliquely from cut points of randomly determined two chromosomes. At the end of this operation, two new chromosomes are obtained. In this problem when chromosomes are crossed with, cross is taken care to the chromosomes in the same robots. For instance, number 1 robot in the first chromosomes and number 1 robot in the second chromosomes are crossed. Then, the second robot in the first chromosomes and the second robot in the second chromosomes are crossed. Let us explain this with an example;

By taking care of the given precedence constraints given in Figure 2, let us crossover the given two chromosomes in figure5.

CHROMOSOME #1	M1 1 3 5 7 8	,	CHROMOSOME #2	M1 2 4 6 7 8
	M2 2 4 6			M2 1 3 5

Figure 5. Two different chromosomes for crossover process

As it is seen above, the jobs in the first chromosomes on the first robot have been scheduled as 1-3-5-7-8 and in the second robot they have been scheduled as 2-4-6. The schedule in the second chromosomes on the first robot is as 2-4-6-7-8, and on the second robot as 1-3-5. When crossover process is applied to these chromosomes, the first robot in the first chromosome and the first robot in the second chromosome and the second robot in the first chromosome and the second robot in the second chromosome gene will be crossed from randomly determined points. The result of the crossover operation has been given in figure 6.

CHROMOSOME #1		CHROMOSOME #2		OFFSPRING #1		OFFSPRING #2
1 3 | 5 7 8	X	2 4 | 6 7 8	⇒	1 3 6 7 8	,	2 4 5 7 8
2 4 | 6		1 3 | 5		2 4 5		1 3 6

Figure 6. Crossover process and obtained offsprings

Here, the sign " | " refers to randomly selected crossover point. On the other hand, the sign "X" represents the crossover operation. At the end of crossover operation, two new chromosomes are obtained. The selected crossover point is the same on the parts representing M1 and M2 parallel robots of chromosomes in the example given in figure 5, and it is after than second gene. But, for instance, the point after than second gene for M1 part may be crossover point, likewise the point after than the first gene may be crossover point for M2 part. Here, there is the possibility of obtaining unfeasible solutions when there are precedence constraints between jobs.

3.4 Applying mutation operator for the problem

In the mutation operation, a gene is randomly selected from inside of the chromosomes in the population according to the given mutation rate. This gene will represent a job. This job will be swapped with any other job, which has the same precedence constraint on another robot or on the same robot with it. If there is more than one job, which is on the same level with it, one of them will be selected randomly. At the end of the mutation operation, a new chromosome will be obtained. For example, let us apply mutation operation to the chromosome given in figure 7;

SELECTED CHROMOSOME AND GEN	MUTATION	OFFSPRING
1 3 **5** 7 8	The job, which is on the same level with number 5 job, will be replaced with number 6 job so two jobs will be swapped.	1 3 6 7 8
2 4 6		2 4 **5**

Figure 7. Mutation process and obtained offspring

3.5 Reproduction

A copy of each gene is made by the reproduction operator in the population and it is added to the list of candidate genes. Fundamentally, this warrants that each chromosome in the current population remains a candidate to be selected for the next population. In this problem, the aim is to find the solution that minimizes the given fitness function. As it is known the fitness function is a tardiness value function. Here, the obtained chromosomes are scheduled from low tardiness value to high tardiness value in every population. GA may have better chances to survive chromosomes with quite higher fitness. The living good chromosomes stay in the population. This process will be kept going until an optimal solution is found in each population.

4. Simulated annealing

In this study, two operators have been used in the application of SA. The first operator is that a randomly selected job has been swapped with another job, which is on the same level, and then, a new offspring has been obtained. The second operator is that a randomly selected job has been again swapped with another job and then, a new solution alternative has been obtained. If these obtained solution alternatives are valid, they are taken into consideration. Used first operator does the same operation with the mutation operation in GA. The working mechanism of these used operators has been revealed in figure 8 and 9.
SA begins with an initial solution (A), and initial temperature (B), and an iteration number (C). The duty of temperature (T) is controlling the possibility of the acceptance of a disturbing solution, and an iteration number (C) is used in the decision of the number of repetitions until a solution has a stable state under the temperature. The T may have the following implicit meaning of flexibility index. At high temperature situation, namely, early in the search, there is some flexibility to move to a worse solution situations, on the other hand, at lower temperature, in other words later in the search, less of this flexibility exists. A new neighborhood solution (N) is generated based on these B, C through a heuristic perturbation on the existing solutions. If the change of an objective function is improved, the neighborhood solution (N) becomes a good solution. Even though it is not improved, the neighborhood solution will be a new solution with a convenient probability which is based on $e^{-\Delta/T}$. This situation leaves the possibility of finding a global optimal solution out of a local optimum. The algorithm will be stopped when there is no change after C iterations. Otherwise, the algorithm will be continuing with a new temperature value (T).

4.1. Simulated annealing algorithm

```
Begin;
        INITIALIZE (A,B,C);
Repeat
        For I=1 to C do
            N= PERTURB (A); {generate new neighborhood solution}
            D= C(N)-C(A)
            If((C(N)<-C(A) or (exp(-D/T)>RANDOM(0,1))
               Then A=N; {Accept the movement}
            Endif
      Endfor;
        UPDATE (T, C);
Until (Stop-Criterion)
End
```

In order to apply SA to practical problems, there are several factors to be decided initially. Firstly, the definition of a procedure to generate neighborhood solutions from a current solution is necessary. To generate these solutions efficiently, some parameters should be decided appropriately. Some examples to these parameters can be given as an initial

temperature, the number of repetitions, conditions for completion and the ratio of temperature change. The combination of these parameters should be adjusted according to the problem to obtain a good solution.
SA has some weak points such as long running time and difficulty in selecting cooling parameter when the problem size becomes larger. A geometric ratio was used in SA as $T_{k+1} = \alpha T_k$, where T_k and T_{k+1} are the temperature values for k and k+1 steps, respectively. Geometric ratio is used more commonly in practice. In this study, the initial temperature was taken 10000 and 0.95 was used for cooling ratio (α).

OLD SOLUTION	SA OPERATOR-1	NEW SOLUTION
1 3 **5** 7 8	Only number 6 work is on the same level with number 5 work that is selected randomly; so two works will be exchanged.	1 3 6 7 8
2 4 6		2 4 **5**

Figure 8. The first new solution generation operator used in SA

OLD SOLUTION	SA OPERATOR-2	NEW SOLUTION
1 3 5 7 8	Randomly selected number 1 work will be swapped with again randomly selected number 4 work	4 3 6 7 8
2 4 6		2 **1** 5

Figure 9. The second new solution generation operator used in SA

5. Comparison of GA and SA

GA and SA are not much different algorithms; theoretically, both of them are quite relative algorithms. However, their formulations are done using very different terminology. In a problem solution with SA, the costs, neighbors and moves of the solutions are talked (discussed), however, in a problem solution with GA, one discusses about chromosomes, their crossover, fitness and mutation. Another difference; a chromosome is considered as a genotype, which only indicates a solution. This is a traditional feature of GA and there is not any reason about that why a resembling approach could not be used in SA in the same way.
Fundamentally, for the situation of that the population size is only one, SA can be considered as GA. Because there is only chromosome, and there is not any crossover, but only mutation. Indeed, this the most important difference between GA and SA. SA generates a new solution by modifying only one solution with a local move; however, GA generates solutions by using the different solutions in a combination. It is not exactly known that if this actually makes the algorithm better or worse, however, it is clear that it depends on the problem and the representation. The principles of these two algorithms are based on the same basic supposition that convenient solutions are mode probably found "near" already known convenient solutions than by randomly selecting from the whole solution space. If this were not the case with a particular problem or representation, they

would not perform better than random sampling. The difference in the action of the GA is treating combinations of two existing solutions as being "near", supposing that such combinations (children) significantly share the properties of their parents, so that a child of two suitable solutions is more probably good solution than a random one. It should us significantly emphasized that this is just valid for a particular problem or representation; otherwise GA will not have an advantage over SA.

6. Example problem-I

Seven jobs and two parallel machines problem is given as an example below. The process and due dates belongs the works in table 1 and additionally, the precedence constraints in figure 9 were given. The solution, which minimizes maximum flow time, was obtained by considering these data. The problem was solved by using three different methods, which are SPT heuristic, SA and GA. The data and the results were given below.

Job i	Processing time	Due date
1	3	9
2	2	8
3	4	3
4	6	7
5	7	4
6	5	5
7	8	6

Table 1. Processing time and Due date of every job

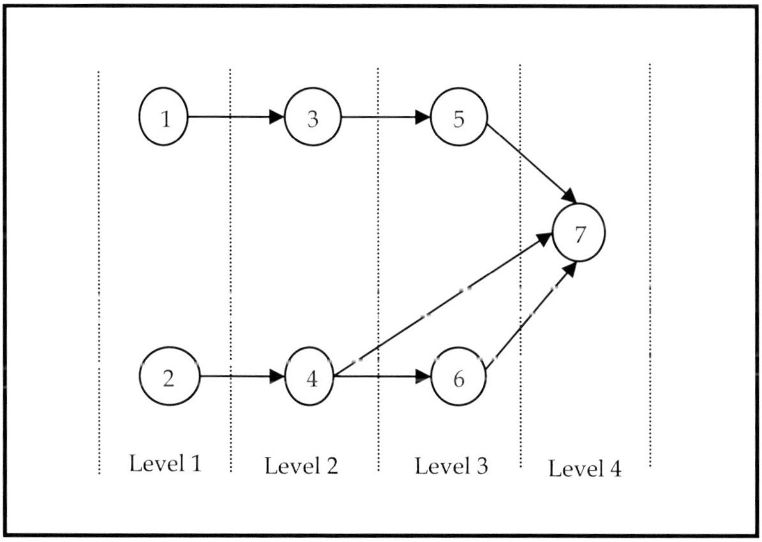

Figure 10. Precedence constraints of every job for example problem

The result of the implementation of GA and SA to the problem stated above has been given in Table 2. Furthermore, in figure 10, the view of the obtained solution from GA on Gannt

Chart has been given. The complementing (finishing) time of each job has been shown on Gannt chart. For example, the finishing time of the number 5 job and number 7 job are 14 and 22, respectively. 7x2 refers to 7 jobs and 2 machines.

Heuristic	Schedule	Maximum Flow Time
SPT	M1: 2-4-6 M2: 1-3-5-7	22
EDD	M1: 2-1-3-5-7 M2: 4-6	24
GA	M1: 1-3-5 M2: 2-4-6-7	22
SA	M1: 1-3-5 M2: 2-4-6-7	22

Table 2. The result of calculation for 7 X 2 problem size

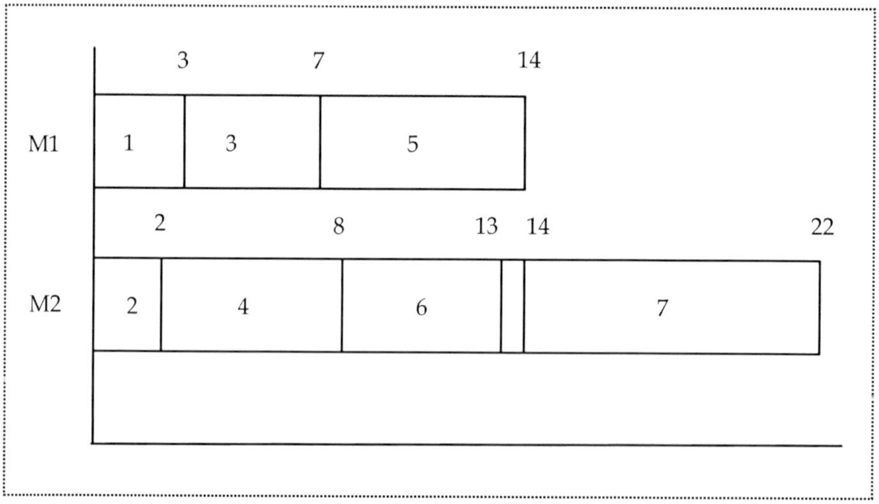

Figure 11. A schedule for two machines displayed as Gannt Chart

7. Example problem-II

As another example, a parallel machine problem with 12 jobs and 2 parallel machines was taken under consideration below. The process times, delivery times and precedence constraints of jobs were given. The solution, which minimizes the total earliness and tardiness cost, was obtained by considering these data. The problem was solved by using SPT, EDD, SA and GA. The data and the results were given below. In Table 3, the jobs with process and due dates belonging to them were given. The precedence constraints of the jobs were given in Figure 11. In Table 4, the solutions obtained from GA, SA, SPT and EDD were given. Tardiness cost and earliness cost have been taken as 1 and 0.5, respectively.

Job i	Processing time	Due date

1	2	1
2	4	3
3	5	2
4	3	8
5	8	7
6	7	4
7	10	12
8	12	14
9	9	11
10	3	8
11	5	9
12	9	15

Table 3. Processing time and Due date of every job for example problem-II

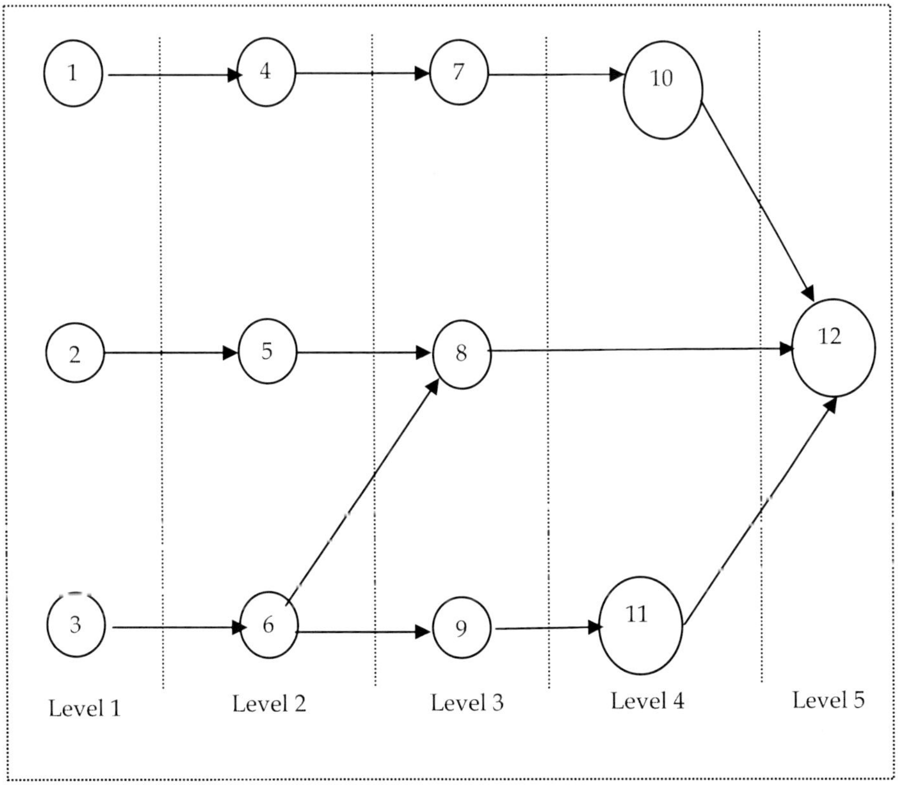

Figure 12. Precedence constraints of every job for example problem-II

Heuristic	Schedule	Total Earliness and Tardiness Cost
SPT	M1: 1-4-5-7-10-8 M2: 2-3-6-9-11-12	148,5
EDD	M1: 1-2-5-9-11-8-12 M2: 3-6-4-7-10	153
GA	M1: 1-4-7-10-9-11 M2: 3-2-6-5-8-12	144,5
SA	M1: 1-4-5-7-10-8-12 M2: 3-2-6-9-11	169,5

Table 4. The result of calculation for 12 X 2 problem size

8. Computational experimentation for scheduling with precedence constraints

The number of jobs used in the problems in this study were given in Table 5. In this table, i denotes the jobs and p_i is an integer processing time and w_i is an integer weight, which were generated from two uniform distributions. The function of [1, 10] and [1, 100] are to create low or high variations, respectively. TF, which is the relative range of due dates, RDD and Average tardiness factor, were selected from the set [0.1, 0.3, 0.5, 0.7, 0.9]. Here, d_i is an integer due date from the uniform distribution [P (1-TF-RDD/2), [P(1-TF+RDD/2)] and it was generated for each job i. In these expressions, P denotes total processing time. As summarized in Table 5, 1700 examples set were considered, totally. The problems were considered in 17 different sizes and for each size 100 different samples were examined. The parameters of the GA were given below. These parameters are firstly tried with different

Population size : 20,	Crossover rate :%100,
Max generation : 100,	Mutation rate :0.05.

Factors	**Settings**
Number of jobs	[10],[20],[30],[40],[50],[60],[70],[80],[90],[100] [120],[150],[170],[200],[220],[250],[300]
Processing time variability	[1-10] [1-100]
Weight variability	[1-10] [1-100]
Relative range of due dates	0.1, 0.3, 0.5, 0.7, 0.9
Average tardiness factor	0.1, 0.3, 0.5, 0.7, 0.9

Table 5. Experimental design

values and according the results of these experimental studies these parameters were determined as the best ones. In different studies, these parameters are determined like the ones obtained in this study. The obtained optimal solutions for different population sizes were given below in Figure 12 for the problem defined with 100x8 sizes. In Figure 13, the cost values for initial population, generation 50 and generation 100 were presented. These figures give clearly information about the selected parameters of GA. As seen in Figure 12, to obtain optimal solution, the different population size values were applied. When the population size is selected as 20, the obtained optimal solution is found better than ones examined with other population sizes.

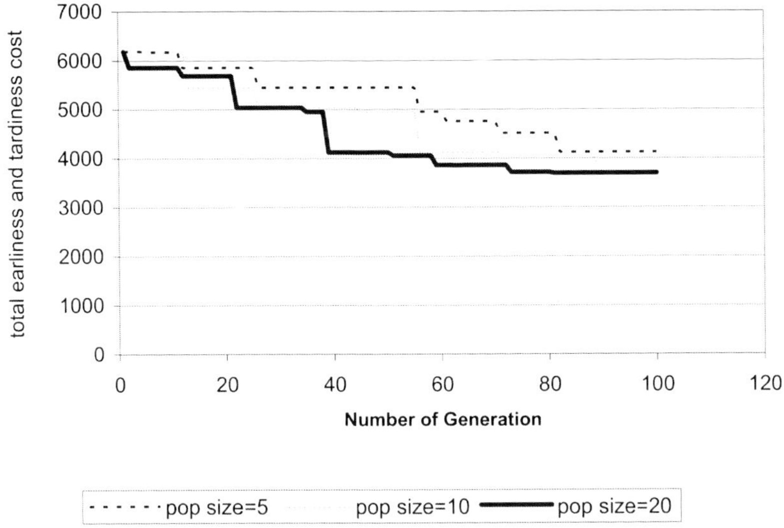

Figure 13. The obtained near optimal solutions according to the different population sizes

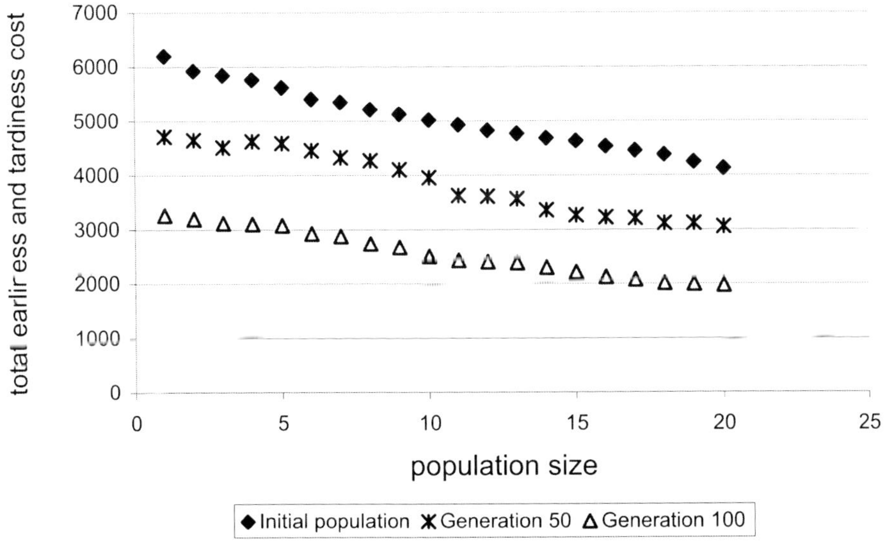

Figure 14. The obtained cost values for initial population, generation 50-100

The results have shown that GA has given better results than SA in large-size problems. SA has some weak points such as long running time and difficulty in selecting cooling parameter when the problem size becomes larger. A geometric ratio was used in SA as $T_{k+1} = \alpha T_k$, where T_k, and T_{k+1} are the temperature values for k and k+1 steps, respectively. Geometric ratio is used more commonly in practice. In this study, the initial temperature was taken 10000 and 0.95 was used for cooling ratio (α). In Table 6 and Table 8, the obtained solutions for different problem sizes were given

Problem size	Number of example	Average value of GA for total earliness and tardiness cost	Average Value of SA for total earliness and tardiness cost	CPU time for GA for an example (s)	CPU time for SA for an example (s)	t statistics
60 X 7	100	756.4	921.2	28.07	34.15	11.30
70 X 7	100	890.0	1056.7	32.71	44.19	12.47
80 X 8	100	1018.1	1263.6	39.03	48.22	15.21
90 X 8	100	1293.0	1512.8	43.35	54.17	16.23
100 X 8	100	1650.8	2004.2	62.28	73.05	17.46
120 X 8	100	1926.2	2137.9	78.05	91.33	19.33
150 X 8	100	2184.4	2410.5	92.17	102.09	21.96
170 X 8	100	2432.7	2985.0	100.02	114.43	22.07
200 X 8	100	3257.3	3863.3	118.34	136.57	24.97
220 X 8	100	3469.2	4112.4	127.28	151.48	25.35
250 X 8	100	3966.4	4698.9	139.11	178.12	29.46
300 X 8	100	5469.6	7282.7	152.22	196.47	31.45

Table 6. The results of the problems in different sizes for total earliness and tardiness cost

In Table 7 and Table 9, the 100 samples given for each problem size were evaluated and how many of the obtained results by using GA are better or equal to SA.. For each problem size, 100 different samples were used. GA and SA were applied to these samples. The average value of the obtained optimal solutions was revealed in the table. According to the average value, it is clearly seen that GA has given the better result. From the viewpoint of evaluating CPU time, the obtained result with GA is again better. All algorithms were coded in C++ and implemented on a Pentium IV 2.4 GHz computer.

Problem size	Number of examples	Number of examples for that GA is better than SA	Number of examples for that GA is equal to SA
60 X 8	100	91	9
70 X 7	100	95	5
80 X 8	100	98	2
90 X 8	100	100	0
100 X 8	100	100	0
120 X 8	100	100	0
150 X 8	100	100	0
170 X 8	100	100	0
200 X 8	100	100	0
220 X 8	100	100	0
250 X 8	100	100	0
300 X 8	100	100	0

Table 7. Comparison of the results of the examples according to the optimal values for total earliness and tardiness cost

Problem size	Number of example	Average value of GA for maximum flow time	Average Value of SA for maximum flow time	CPU time for GA for an example (s)	CPU time for SA for an example (s)	t statistics
60 X 7	100	72	78	18.03	22.21	13.45
70 X 7	100	85	93	26.07	30.18	15.68
80 X 8	100	96	108	32.09	39.43	17.13
90 X 8	100	119	134	39.01	51.19	19.86
100 X 8	100	132	142	45.15	63.05	18.94
120 X 8	100	148	161	54.45	71.45	19.73
150 X 8	100	176	183	62.22	76.39	22.12
170 X 8	100	189	202	70.56	83.55	24.28
200 X 8	100	217	230	81.30	90.57	24.88
220 X 8	100	239	255	92.12	103.49	27.35
250 X 8	100	264	292	102.37	114.42	29.49
300 X 8	100	286	305	129.21	142.47	33.57

Table 8. The results of the problems in different sizes for maximum flow time

Problem size	Number of examples	Number of examples for that GA is better than SA	Number of examples for that GA is equal to SA
60 X 8	100	94	6
70 X 7	100	95	5
80 X 8	100	100	0
90 X 8	100	100	0
100 X 8	100	100	0
120 X 8	100	100	0
150 X 8	100	100	0
170 X 8	100	100	0
200 X 8	100	100	0
220 X 8	100	100	0
250 X 8	100	100	0
300 X 8	100	100	0

Table 9. Comparison of the results of the examples according to the optimal values for maximum flow time

9. Conclusions

The genetic algorithms (GA) have the great advantage and success in the solution of NP problems. There are various important applications on this way. In this study, the job with n-number of precedence constraints is assigned minimizing total earliness and tardiness and maximum flow time on m-number of parallel machine. Genetic algorithms and simulated annealing methods were used to find the solutions, which minimizes the total earliness and tardiness costs. In GA, the solution alternatives, which were obtained by using genetic operators, were investigated to understand that if they are feasible or not and the feasible ones according to precedence constraints were considered. The way, trying to make infeasible solutions feasible, was not selected. Likewise, obtained infeasible solutions were not evaluated. Again any study about making these infeasible solutions feasible was not done. According to the results obtained by using GA and SA methods, it was evidently observed that GA algorithm is more successful. Especially for larger problem sizes, it is seen that GA gives results better than SA.

10. References

Kasahara H., Narita S., Parallel processing of robot arm control computation on a multi microprocessor system, *IEEE J. of Robotics Automation* vol. RA-1, no.2, pp. 104-113, June 1985.

Kanjo, -Y., Ase, -H., Robust scheduling in a multi robot welding system, Transaction of the Institute of systems, *Control and Information Engineers.* 16(8), pp.369-376, 2003.

Jun, S.Z., Ying, Z.J., A genetic algorithm based approach to intelligent optimization for scheduling dual resources with robots, *Robot*, 24(4), pp.342-357, 2002.

Zacharia, P.T., Asparagathos, N.A., Optimal robot task scheduling based on genetic algorithms, *Robotics and Computer Integrated Manufacturing*, 21(1), pp. 67-79, 2005.

E.M. Arkin, R.O. Roundy, Weighted-Tardiness scheduling on parallel machines with proportional weights, *Operational Research*, 39, pp. 64-81. 1991.

H. Emmons, Scheduling to a common due date on parallel uniform processors, *Naval Research Logistics*, 24, pp. 803-810, 1987.

A. Guinet, Scheduling independent jobs on uniform parallel machines to minimize tardiness criteria, *Journal of Intelligent Manufacturing*, 6, pp. 95-103, 1995.

R.M. Karp, Reductibility among combinatorial problems: complexity of computer computations, New york, Plenum press, pp. 85-103, 1972.

J.M.J. Schutten R.A.M. Leussink, Parallel Machine scheduling with release dates, and familiy setup times, International Journal of Production Economy, 46-47, pp. 119-125,1996.

S.A. Tamimi, V.N. Rajan, Reduction of Total weighted tardiness on uniform machines with sequence dependent setups, *Industrial Engineering Research – Conference Proceedings*, pp. 181-185, 1997.

V.A. Armentano, D.S. Yamashita, Tabu Search for scheduling on identical parallel machines to minimize mean tardiness, *Journal of Intelligent Manufacturing*, 11, pp. 453-460, 2000.

Kim K.H., Kim D.W., Unreleated parallel machine scheduling with setup times using simulated annealing, *Robotics and computer Integrated Manufacturing*, Volume 18, Issues 3-4, Pages 223-231,2002.

Min L., Cheng W., A genetic algorithm for minimizing the makespan in case of scheduling identical parallel machines, *Artificial Intelligence in Enginering*, 13, pp. 399-403, 1995

Chen C.L., Lee C.S.G., & Hou, E.S.H., Efficient scheduling Algorithm algorithms for robot inverse dynamics computation on a multiprocessor system, *IEEE Transaction on Systems, Man, and Cybernetics*, Vol..18, pp. 729-743, Dec. 1988.

J.C.Ho, Y.-L.Chang, Heuristics for minimizing mean tardiness for m parallel machines, *Naval Research Logistics*, 38, pp. 367-381, 1991.

L.J. Wilkerson, J.D. Irwin, An improved algorithm for scheduling independent jobs, *AIIE Transactions*, 3, pp. 239-245, 1971.

C. Koulamas, Decomposition and hybris Simulated annealing heuristics for the parallel machine total tardiness problem, *Naval Research Logistics*, 44, pp. 109-125, 1997.

K. Chen, J.S. Wong & J.C. Ho, A heuristic algorithm to minimize tardiness for parallel machines *Proceedings of ISMM\International Conference*. ISSM-ACTA Press, Anaheim CA, USA pp. 118-121,1997.

B. Alidaee, D. Rosa, Scheduling parallel Machines to weighted and un-weight tardiness, *Computers Operational Research*, 24 (8), pp. 775-788, 1997.

K.R.Baker, J.W. Bertrand, A Dynamic priority rule for sequencing against due-date, *Journal of Operational Management*, 3, pp. 37-42, 1982.

M. Azizoğlu, O.Kirca, Tardiness minimization on parallel machines, *International Journal of Production Economics*, 55, pp. 163-168, 1998.

S.E. Elmaghraby, S.H. Park, Scheduling jobs on a number of identical machines, *AIIE Transactions*, 6, pp. 1-13, 1974.

J.W. Barnes, J.J. Brennan, An Improved algorithm for independent jobs to reduce mean finishing time , *AIIE Transactions*, 17, pp. 382-387, 1977.

H. Emmons, M. Pinedo, Scheduling stochastic jobs, with due dates on parallel machines, *Europan Journal of Operational Research*, 47, pp. 49-55, 1990.

T.Cakar , R.Koker & H.I.Demir, Parallel Robot Scheduling to Minimize Mean Tardiness with Precedence Constraints Using a Genetic Algorithm, *Advance in Engineering Software,* Vol 39, No 1, pp. 47-54, January 2008.

K.R.Baker, *Introduction to Sequencing and Scheduling,* John Wiley & Sons, New York,1974

9

Design and Prototyping of a Spherical Parallel Machine Based on 3-CPU Kinematics

Massimo Callegari
*Dipartimento di Meccanica, Università Politecnica delle Marche
Via Brecce Bianche, Ancona,
Italy*

1. Introduction

Parallel kinematics machines, PKMs, are known to be characterised by many advantages like a lightweight construction and a high stiffness but also present some drawbacks, like the limited workspace, the great number of joints of the mechanical structure and the complex kinematics, especially for 6-dof machines. Therefore Callegari *et al.* (2007) proposed to decompose full-mobility operations into elemental sub-tasks, to be performed by separate minor mobility machines, like done already in conventional machining operations. They envisaged the architecture of a mechatronic system where two parallel robots cooperate in order to perform a complex assembly task: the kinematics of both machines is based upon the 3-CPU topology but the joints are differently assembled so as to obtain a translating parallel machines (TPM) with one mechanism and a spherical parallel machine (SPM) with the other.

In one case, joints' axes are set in space so that the mobile platform can freely translate (without rotating) inside its 3D workspace: this is easily obtained by arranging the universal joint of each limb so that the axis of the outer revolute joint is parallel to the base cylindrical joint; such three directions are mutually orthogonal to maximise the workspace and grant optimal manipulability. With a different setting of the joints, three degrees of freedom of pure rotation are obtained at the terminal of the spherical wrist: in this case the axes of the cylindrical joints and those of the outer revolute pairs in the universal joints all intersect at a common point, which is the centre of the spherical motion.

This solution, at the cost of a more sophisticated controller, would lead to the design of simpler machines that could be used also stand-alone for 3-dof tasks and would increase the modularity and reconfigurability of the robotised industrial process. The two robots have been developed till the prototypal stage by means of a virtual prototyping environment and a sketch of the whole system is shown in Fig. 1: while the translating machine has been presented already elsewhere (Callegari & Palpacelli, 2008), the present article describes the design process of the orienting device and the outcoming prototype.

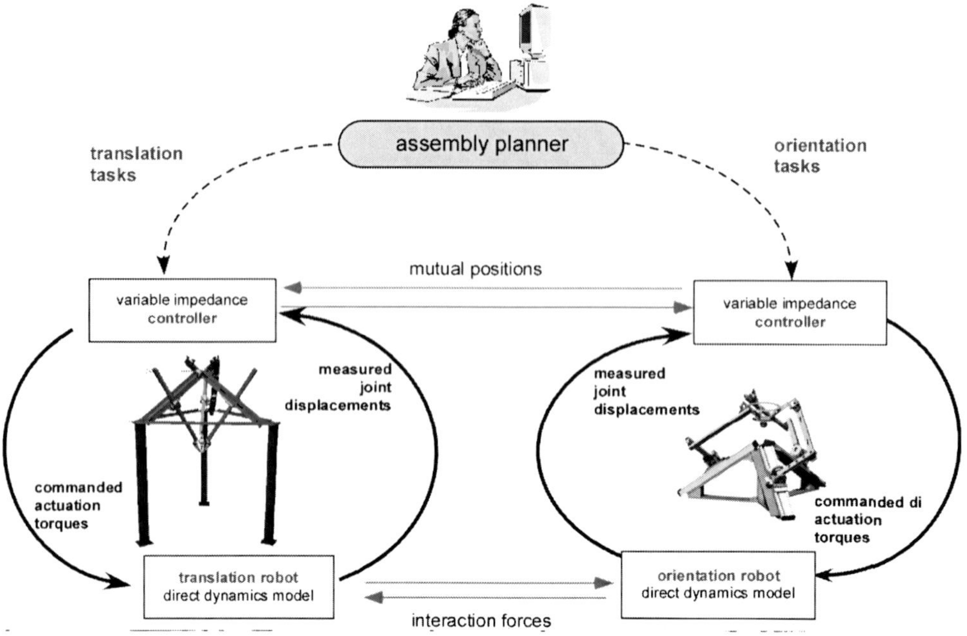

Fig. 1. Architecture of the assembly system based on two cooperating parallel robots

2. Kinematic synthesis

The design of parallel kinematics machines able to perform motions of pure rotation, also called Spherical Parallel Machines, SPM's, is a quite recent research topics: besides the pioneering researches by Asada and Granito (1985), the most important mechanism of this type is the agile eye by Gosselin and Angeles (1989), upon which many prototype machines have been designed since then. Few other studies on the subject are available during the 90's, among which the work of Lee and Chang (1992), Innocenti and Parenti-Castelli (1993) and Alizade *et al.* (1994). In the new millennium, however, a growing interest on spherical parallel wrists produced many interesting results, as new kinematic architectures or powerful design tools. The use of synthesis methods based on or screw theory, for instance, has been exploited by Kong and Gosselin (2004a and 2004b) that provide comprehensive listings of both overconstrained and non-overconstrained SPM's; Hervé and Karouia, on the other hand, use the theory of Lie group of displacements to generate novel architectures, as the four main families in (Karouia & Hervè, 2002) or the 3-RCC, 3-CCR, 3-CRC kinematics specifically treated in (Karouia & Hervè, 2005); Fang and Tsai (2004) use the theory of reciprocal screws to present a systematic methodology for the structural synthesis of a class of 3-DOF rotational parallel manipulators. More interesting architectures, as the 3-URC, the 3-RUU or the 3-RRS, have been studied by Di Gregorio (2001a, 2001b and 2004) and also by other researchers.
Following the approach outlined in (Karouia & Hervè, 2000), Callegari *et al.* (2004) proposed a new wrist architecture, based on the 3-CPU structure; it is noted also that the 3-CRU variant is characterised by a much more complex kinematics but can be useful in view of a

possible prototyping at a mini- or micro- scale, as shown by Callegari *et al.* (2008). The main synthesis steps of the 3-CPU parallel wrist are outlined in the following paragraphs.

First of all, it is noted that only non-overconstrained mechanisms have been searched in order to avoid the strict dimensional and geometric tolerances needed by overconstrained machines during manufacturing and assembly phases. Moreover, the use of passive spherical pairs directly joining the platform to the base has been avoided as well and for economic reasons only modular solutions characterised by three identical legs have been considered. It must be said that these advantages are usually paid with a more complex structure and the possible presence of singular configurations (translation singularities) in which the spherical constraint between platform and base fails.

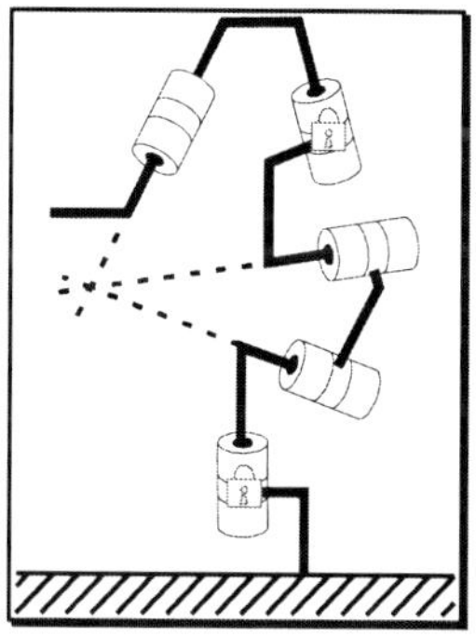

Fig. 2. Limb of connectivity 5 able to generate a spherical motion of the platform

Aiming at this kind of spherical machines, a simple mobility analysis shows that a parallel mechanism able to generate 3-dof motions must be composed by three limbs of connectivity 5. Without losing generality, it is supposed that each single limb consists of 4 links and 5 revolute (R) or prismatic (P) joints that connect the links among them and the limb itself to the fixed frame and to the mobile platform. If each limb's kinematic chain has 3 revolute pairs whose axes intersect at a common point, that is the centre O of the SPM, therefore the moving platform can rotate around the fixed point O: in this way, each limb generates a 5-dimensional manifold that must contain the 3-dimensional group of spherical motions around the point O. If the other two lower pairs are locked, the kinematic chain of the overconstrained Gosselin and Angeles wrist (1989) is obtained, see Fig. 2.

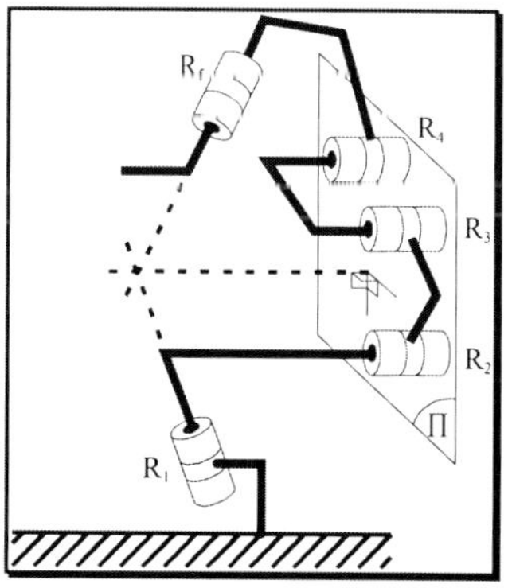

Fig. 3. Limb with subgroup RRR able to generate the subgroup of planar displacements

By analysing the described configuration, it is seen that the spherical motion can be obtained also by using 5 revolute pairs R_1-R_5 where the axes of the joints R_1, R_3 and R_5 still intersect at a common point while the axes of pairs R_2 and R_4 are parallel to the direction of R_3. In such a way, the 3 joints R_2, R_3 and R_4 will generate the 3-dimensional subgroup of planar displacements $G(\Pi)$, i.e. the set of translations lying in Π and rotations around axes perpendicular to Π. The same subgroup $G(\Pi)$ is generated also in case the axis of revolute joint R_3 is still perpendicular to plane Π but does not cross the rotation centre O, as shown in Fig. 3, therefore also with this limb kinematics a spherical wrist can be obtained.
On the other hand, by following the same line of reasoning, the same subgroup of planar displacements $G(\Pi)$ can be generated by substituting one or two revolute joints among the R_2, R_3, R_4 set with prismatic pairs whose axes lie in the plane Π, thus obtaining limbs whose central joints are characterised by one of the sequences PRR, RPR, PPR, PRP, RRP, RPP. Of course, two adjacent joints in limbs kinematics can be merged to yield simpler architectures with fewer links: for instance two revolute joints with orthogonal axes can be superimposed to give a universal (U) joint, while the set of one revolute joint and one prismatic pair with the same axes are equivalent to a cylindrical (C) joint, as shown in Fig. 4.

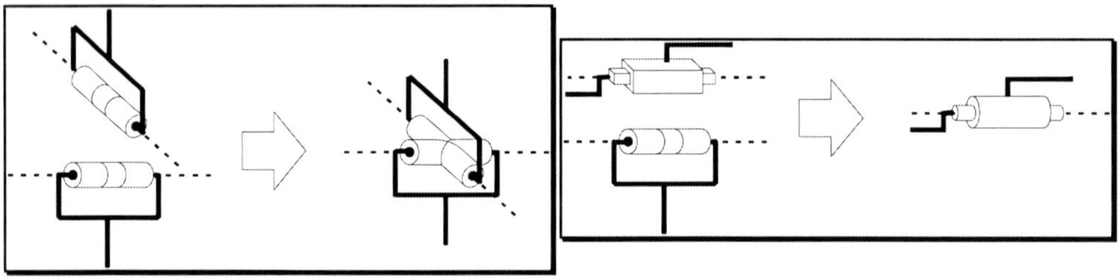

(a) (b)

Fig. 4. Merge of two adjacent joints able to yield universal (a) or cylindrical (b) pairs

The kinematic chains described above prevent the ith limb's end from translating in the direction normal to the plane Π_i, $i=1,2,3$; therefore, if three such chains are used for the limbs and the three normals to the planes Π_i, are linearly independent, all the possible translations in space are locked and the mobile platform, attached to the three limbs, can only rotate around a fixed point.
In this way, seven alternative design concepts have been considered, which are: 3-URU, 3-CRU, 3-URC, 3-UPU, 3-CPU, 3-UPC, 3-CRC. Figures 5-9 show the mentioned synthesis steps leading to the specific limb topology (a) and sketch a first guess arrangement of the introduced joints (b). In particular, the second picture in each one of these figures, labelled (b), shows the simplest possible setting of the limbs, that all lie within vertical planes: unfortunately in this case the 3 normals to limbs' planes are all parallel to the horizontal plane and therefore result linearly dependent, allowing the platform to translate along the vertical direction, see Fig. 10a. Among all the possible setting of these normal axes in space that grant them to be linearly independent, it has been chosen to tilt the limbs' planes so that they are mutually orthogonal in the initial configuration (or "home" position of the wrist), Fig. 10b, thus greatly simplifying the kinematics relations that will be worked out later on; moreover, even if this arrangement changes during operation of the machine, this configuration is the most far from the singular setting previously outlined, therefore granting a better kinematic manipulability of the wrist. The sketch of the outcoming mechanisms are drawn in Fig. 11-13.

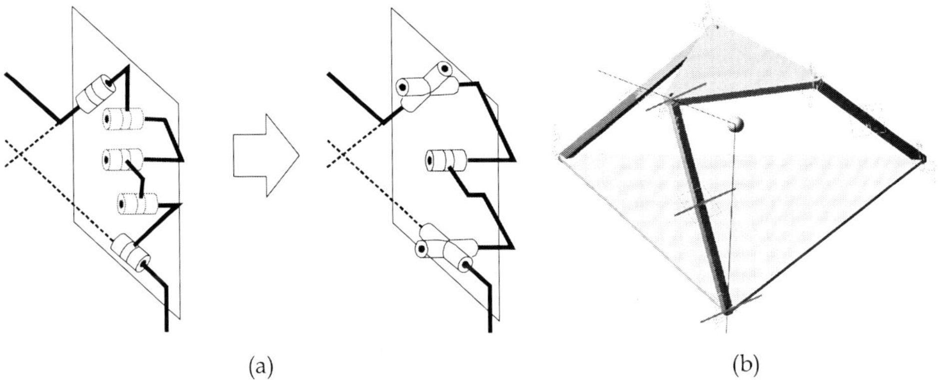

Fig. 5. Synthesis of URU limbs (a) and sketch of the 3-URU mechanism (b)

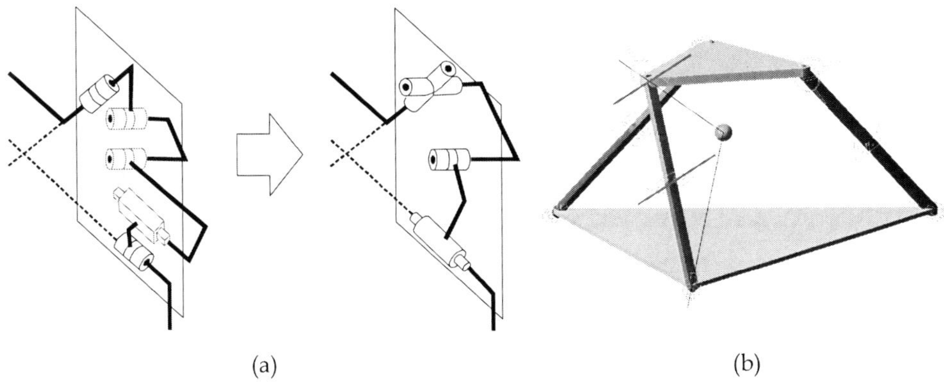

Fig. 6. Synthesis of CRU and URC limbs (a) and sketch of the 3-CRU mechanism (b)

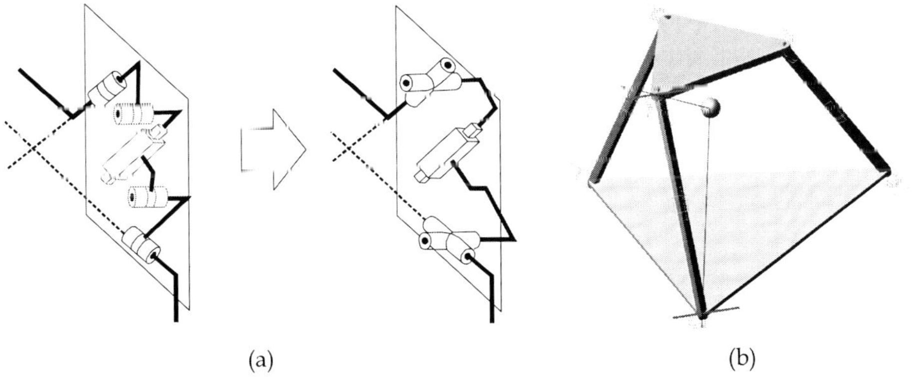

Fig. 7. Synthesis of UPU limbs (a) and sketch of the 3- UPU mechanism (b)

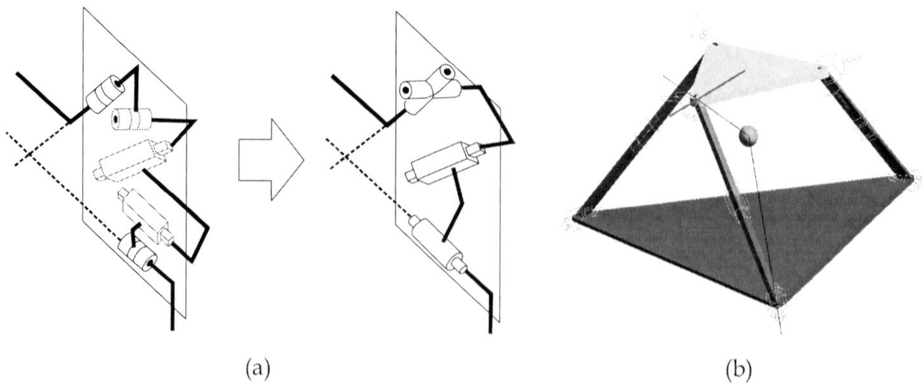

Fig. 8. Synthesis of CPU and UPC limbs (a) and sketch of the 3- CPU mechanism (b)

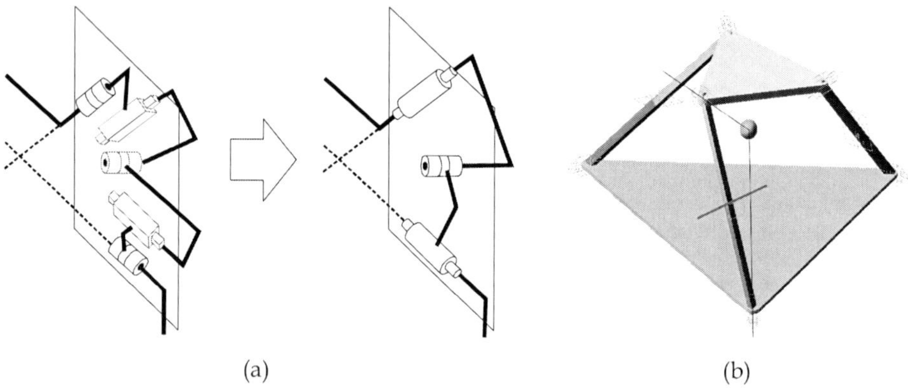

Fig. 9. Synthesis of CRC limbs (a) and sketch of the 3- CRC mechanism (b)

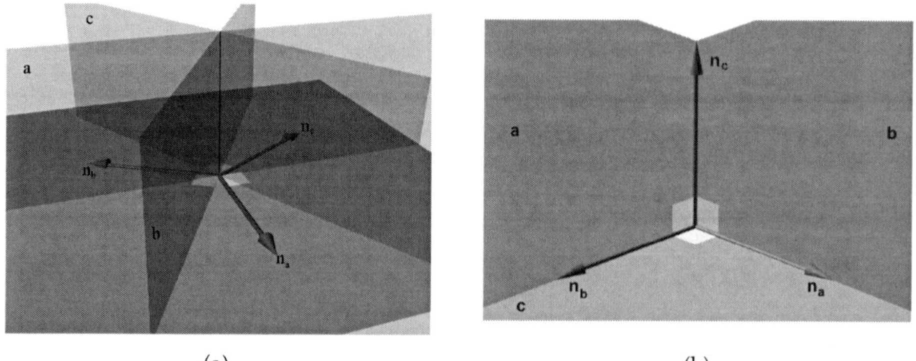

Fig. 10. Setting of the 3 axes normal to limbs' planes: coplanar (a) and orthogonal (b)

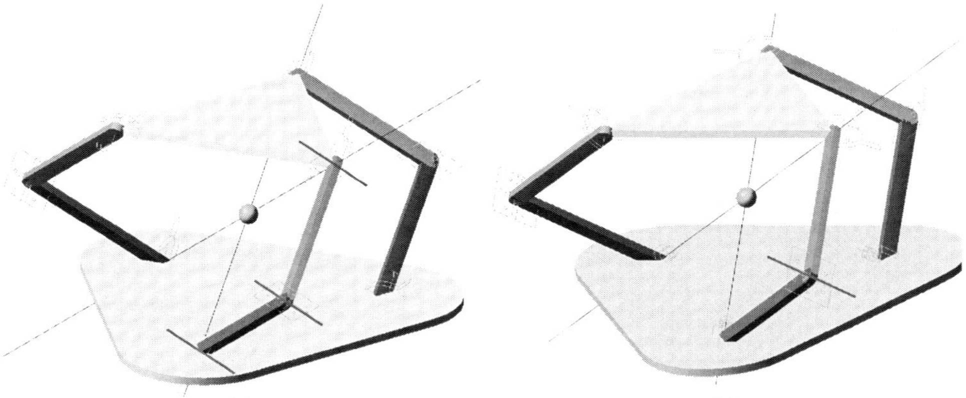

(a) (b)

Fig. 11. Concept of a 3-URU (a) and 3-CRU (b) spherical parallel machine (home pose)

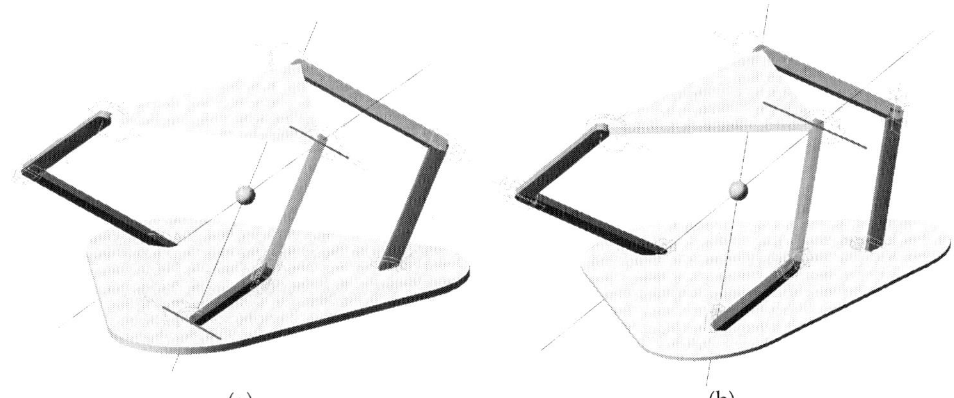

(a) (b)

Fig. 12. Concept of a 3-UPU (a) and 3-CPU (b) spherical parallel machine (home pose)

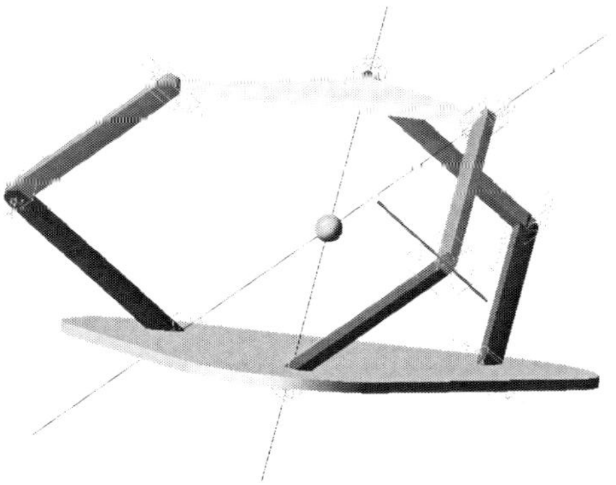

Fig. 13. Concept of a 3-CRC spherical parallel machine (home pose)

The kinematics of such machines has been investigated and in view of the design of a physical prototype the 3-CPU concept has been retained, see Fig. 14: this has been mainly due to the relative simplicity of the kinematics relations that will be worked out in next section, to the compactness of the concept, that allows an easy actuation and finally to the novelty of the kinematics, that has been proposed by Olivieri first (2003) and then studied by Callegari *et al.* (2004). Before studying the kinematics of the 3-CPU SPM it is marginally noted that the same limb's topology, with a different joints arrangement, is able to provide motions of pure translation (Callegari *et al.*, 2005); moreover, the 3-CRU mechanism is extensively studied in (Callegari *et. al.*, 2008) in view of the realisation of a SPM for miniaturized assembly tasks.

3. Kinematic analysis

3.1 Description of geometry and frames setting

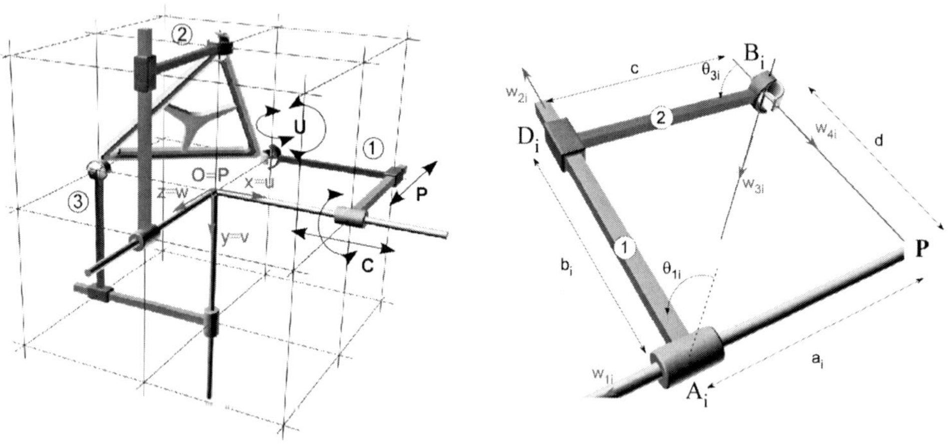

(a) (b)

Fig. 14. Placement of reference frames (home pose) (a) and geometry of a single limb (b)

Making reference to Fig. 14, the axes of cylindrical joints A_i, $i=1,2,3$ intersect at point O (centre of the motion) and are aligned to the axes **x**, **y**, **z** respectively of a (*fixed*) Cartesian frame located in O. The first member of each link *(1)* is perpendicular to A_i and has a variable length b_i due to the presence of the prismatic joint D_i: the second link *(2)* of the leg is set parallel the said cylindrical pair. The universal joint B_i is composed by two revolute pairs with orthogonal axes: one is perpendicular to leg's plane while the other intersects at a common point P with the corresponding joints of the other limbs; such directions, for the legs $i=1,2,3$ orderly, are aligned to the axes **u**, **v**, **w** respectively of a (*mobile*) Cartesian frame, located in P and attached to the rotating platform. For a successful functioning of the mechanism, such *manufacturing conditions* must be accompanied by a proper *mounting condition*: assembly should be operated in such a way that the two frames $O(\mathbf{x,y,z})$ and $P(\mathbf{u,v,w})$ come to coincide. Finally, it is assumed an initial configuration such that the linear displacements a_i of the cylindrical joints are equal to the constant length c (that is the same for all the legs): in this case also the linear displacements b_i of the prismatic joints are equal to the constant length d. It is also evident that, for practical design considerations, SPM's based on the 3-CPU concept are efficiently actuated by driving the linear displacements of the cylindrical pairs coupling the limbs with the frame: therefore in the following kinematic analysis it will be made reference to this case (i.e. joint variables a_i, $i=1,2,3$ will be considered the actuation parameters).

3.2 Analysis of mobility

From the discussion of previous section, it is now evident that in case the recalled manufacturing and assembly conditions are satisfied, the mobile platform is characterised by motions of pure rotation; the mentioned conditions can be geometrically expressed by:

i. $\hat{\mathbf{w}}_{1i}$ and $\hat{\mathbf{w}}_{4i}$ incident in P;

ii. $\hat{\mathbf{w}}_{3i}$ perpendicular to the plane $<\hat{\mathbf{w}}_{1i},\hat{\mathbf{w}}_{4i}>$, i.e. $\hat{\mathbf{w}}_{3i}\cdot\hat{\mathbf{w}}_{4i}=0$ and $\hat{\mathbf{w}}_{3i}\cdot\hat{\mathbf{w}}_{1i}=0$;

iii. $\hat{\mathbf{w}}_{2i}$ lying on the plane $<\hat{\mathbf{w}}_{1i},\hat{\mathbf{w}}_{4i}>$, i.e. $\hat{\mathbf{w}}_{3i}\cdot\hat{\mathbf{w}}_{2i}=0$; due to condition (ii) must also hold: $\hat{\mathbf{w}}_{3i}=\hat{\mathbf{w}}_{1i}\times\hat{\mathbf{w}}_{2i}$;

iv. $\hat{\mathbf{w}}_{2i}$ not parallel to $\hat{\mathbf{w}}_{1i}$ and therefore: $\hat{\mathbf{w}}_{1i}\times\hat{\mathbf{w}}_{2i}\neq\hat{\mathbf{0}}$ (for simplicity, the condition $\hat{\mathbf{w}}_{1i}\cdot\hat{\mathbf{w}}_{2i}=0$ has been posed).

Making reference to Fig. 14b, if the point P is considered belonging to the ith leg, its velocity can be written in three different ways as follows:

$$\dot{\mathbf{P}}=\dot{\mathbf{P}}_{2i}+\dot{\mathbf{P}}_{ri} \quad \text{for } i=1,2,3 \tag{1}$$

where $\dot{\mathbf{P}}_{2i}$ is the velocity of point P if considered fixed to link 2:

$$\dot{\mathbf{P}}_{2i}=\dot{\mathbf{B}}_i+\boldsymbol{\omega}_{2i}\times(P-B_i)=\dot{\mathbf{B}}_i+\boldsymbol{\omega}_{2i}\times d\,\hat{\mathbf{w}}_{4i} \tag{2}$$

and $\dot{\mathbf{P}}_{ri}$ is the velocity of point P relative to a frame fixed to link 2 and with origin in B_i:

$$\dot{\mathbf{P}}_{ri}=\dot{\theta}_{3i}\hat{\mathbf{w}}_{3i}\times(P-B_i)=\dot{\theta}_{3i}\hat{\mathbf{w}}_{3i}\times d\hat{\mathbf{w}}_{4i} \tag{3}$$

In (2), $\boldsymbol{\omega}_{2i}$ is the angular velocity of link 2:

$$\boldsymbol{\omega}_{2i}=\dot{\theta}_{1i}\hat{\mathbf{w}}_{1i} \tag{4}$$

In the same way, with obvious meaning of the symbols, the vector $\dot{\mathbf{B}}_i$ can be expressed as:

$$\dot{\mathbf{B}}_i=\dot{\mathbf{B}}_{1i}+\dot{\mathbf{B}}_{ri} \quad \text{for i=1,2,3} \tag{5}$$

where:

$$\dot{\mathbf{B}}_{1i}=\dot{a}_i\hat{\mathbf{w}}_{1i}+\boldsymbol{\omega}_{1i}\times(B_i-A_i)=\dot{a}_i\hat{\mathbf{w}}_{1i}+\dot{\theta}_{1i}\hat{\mathbf{w}}_{1i}\times(a_i\hat{\mathbf{w}}_{1i}-d\hat{\mathbf{w}}_{4i})=\dot{a}_i\hat{\mathbf{w}}_{1i}-\dot{\theta}_{1i}\hat{\mathbf{w}}_{1i}\times d\hat{\mathbf{w}}_{4i} \tag{6}$$

$$\dot{\mathbf{B}}_{ri}=\dot{b}_i\hat{\mathbf{w}}_{2i} \tag{7}$$

If (2)-(7) are substituted back in (1), it is found:

$$\dot{\mathbf{P}}=\dot{b}_i\hat{\mathbf{w}}_{2i}+\dot{a}_i\hat{\mathbf{w}}_{1i}+\dot{\theta}_{3i}\hat{\mathbf{w}}_{3i}\times d\hat{\mathbf{w}}_{4i} \quad \text{for i=1,2,3} \tag{8}$$

By dot-multiplying (8) by $\hat{\mathbf{w}}_{3i}$ and by taking into account the conditions (i)-(iv), it is finally obtained:

$$\hat{\mathbf{w}}_{3i} \cdot \dot{\mathbf{P}} = 0 \tag{9}$$

that can be differentiated to yield:

$$\hat{\mathbf{w}}_{3i} \cdot \ddot{\mathbf{P}} + \dot{\hat{\mathbf{w}}}_{3i} \cdot \dot{\mathbf{P}} = 0 \tag{10}$$

Equations (9-10), written for the 3 legs, build up a system of 6 linear algebraic equations in 6 unknowns, the scalar components of $\dot{\mathbf{P}}$ and $\ddot{\mathbf{P}}$. Such a system can be written in matrix form as follows:

$$\mathbf{M}\begin{bmatrix} \dot{\mathbf{P}} \\ \ddot{\mathbf{P}} \end{bmatrix} = \mathbf{0} \tag{11}$$

where the 6x6 matrix **M** can be partitioned as:

$$\mathbf{M} = \begin{bmatrix} \mathbf{H} & \mathbf{O} \\ \dot{\mathbf{H}} & \mathbf{H} \end{bmatrix} \tag{12}$$

with:

$$\mathbf{H} = \begin{bmatrix} \hat{\mathbf{w}}_{31}^T \\ \hat{\mathbf{w}}_{32}^T \\ \hat{\mathbf{w}}_{33}^T \end{bmatrix} = \begin{bmatrix} w_{31i} & w_{31j} & w_{31k} \\ w_{32i} & w_{32j} & w_{32k} \\ w_{33i} & w_{33j} & w_{33k} \end{bmatrix} \tag{13}$$

and **O** being the 3x3 null matrix.
If the matrix M is not singular, the system (11) only admits the trivial null solution:

$$\dot{\mathbf{P}} = \ddot{\mathbf{P}} = \mathbf{0} \tag{14-15}$$

which means that the point *P* does not move in space, i.e. the moving platform only rotates around *P*. The singular configurations, on the other hand, can be identified by posing:

$$\det(\mathbf{M}) = [\det(\mathbf{H})]^2 = 0 \tag{16}$$

that leads to:

$$\det(\mathbf{H}) = \hat{\mathbf{w}}_{31} \cdot \hat{\mathbf{w}}_{32} \times \hat{\mathbf{w}}_{33} = 0 \tag{17}$$

Equation (17) is satisfied only when the three unit vectors $\hat{\mathbf{w}}_{31}$, $\hat{\mathbf{w}}_{32}$, $\hat{\mathbf{w}}_{33}$ are linearly dependent; therefore the platform incurs in a *translation singularity* if and only if:

- the planes containing the three legs are simultaneously perpendicular to the base plane;
- such planes are coincident with the base plane (configuration not reachable);
- at least two out of the three aforementioned planes admit parallel normal unit vectors.

This justifies the choice previously operated of having the legs laid on mutual orthogonal planes: in fact this configuration is the most far from singularities.

3.3 Orientation kinematics

Orientation kinematics is based on the definition of the relative rotation between fixed frame $O(\mathbf{x,y,z})$ and the mobile frame $P(\mathbf{u,v,w})$, where is always $P \equiv O$, see Fig. 14; to this aim the following set of Cardan angles is used:

$$ {}^{O}_{P}\mathbf{R}(\alpha,\beta,\gamma)=\mathbf{R}_x(\alpha)\cdot\mathbf{R}_y(\beta)\cdot\mathbf{R}_z(\gamma)=\begin{bmatrix} c\beta c\gamma & -c\beta s\gamma & s\beta \\ s\alpha s\beta c\gamma + c\alpha s\gamma & -s\alpha s\beta s\gamma + c\alpha c\gamma & -s\alpha c\beta \\ -c\alpha s\beta c\gamma + s\alpha s\gamma & c\alpha s\beta s\gamma + s\alpha c\gamma & c\alpha c\beta \end{bmatrix} \quad (18) $$

Moreover, a local frame $O_i(\mathbf{x}_i, \mathbf{y}_i, \mathbf{z}_i)$, $i=1,2,3$ is defined for each leg, as shown in Fig. 15: the $\mathbf{x}_i$ axis is aligned with cylindrical joint's axis and the $\mathbf{y}_i$ axis is chosen parallel to limb's first link, when it is laid in the initial configuration.

One loop-closure equation can be written for each leg as follows:

$$ (A_i - P) + (D_i - A_i) + (B_i - D_i) + (P - B_i) = 0 \quad \text{for i=1,2,3} \quad (19) $$

Equation (19) can be easily expressed in the local frame $O_i(\mathbf{x}_i, \mathbf{y}_i, \mathbf{z}_i)$, $i=1,2,3$:

$$ \begin{bmatrix} a_i \\ 0 \\ 0 \end{bmatrix} + \begin{bmatrix} 0 \\ -b_i \cdot c\theta_{1i} \\ -b_i \cdot s\theta_{1i} \end{bmatrix} + \begin{bmatrix} -c \\ 0 \\ 0 \end{bmatrix} + {}^{i}(P - B_i) = 0 \quad \text{for i=1,2,3} \quad (20) $$

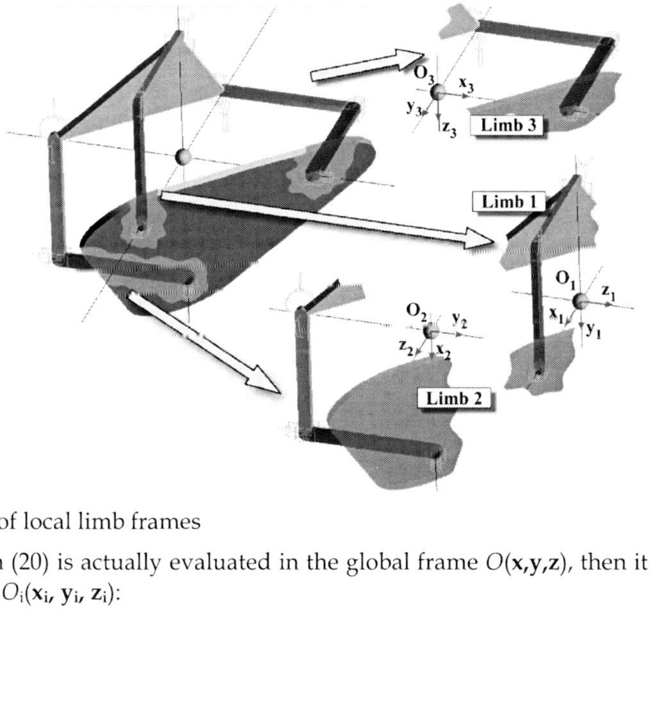

Fig. 15. Setting of local limb frames

The last term in (20) is actually evaluated in the global frame $O(\mathbf{x,y,z})$, then it is transported to limb's frame $O_i(\mathbf{x}_i, \mathbf{y}_i, \mathbf{z}_i)$:

$$ {}^{i}(P-B_i)={}^{i}_{P}\mathbf{R}\cdot{}^{P}(P-B_i)={}^{i}_{O}\mathbf{R}\cdot{}^{O}_{P}\mathbf{R}\cdot{}^{P}(P-B_i) \quad \text{for i=1,2,3} \tag{21} $$

where the introduced terms assume the following values:

$$ {}^{1}_{O}\mathbf{R}=\begin{bmatrix}1&0&0\\0&1&0\\0&0&1\end{bmatrix} \qquad {}^{2}_{O}\mathbf{R}=\begin{bmatrix}0&1&0\\0&0&1\\1&0&0\end{bmatrix} \qquad {}^{1}_{O}\mathbf{R}=\begin{bmatrix}0&0&1\\1&0&0\\0&1&0\end{bmatrix} \tag{22-24} $$

$$ {}^{P}(P-B_1)=d\cdot[0\ \ 1\ \ 0]^T \qquad {}^{P}(P-B_2)=d\cdot[0\ \ 0\ \ 1]^T \qquad {}^{P}(P-B_3)=d\cdot[1\ \ 0\ \ 0]^T \tag{25-27} $$

In inverse kinematics the values of α, β, γ Cardan angles (or equivalently the elements r_{ij} of the rotation matrix ${}^{O}_{P}\mathbf{R}$) are know and the joint variables a_i must be found; loop closure equations (21) for i=1,2,3 represent three decoupled systems of non linear algebraic equations in the unknowns a_i, θ_{1i} and b_i, that can be solved to find the single solution:

$$ \begin{cases} a_1=c-d\cdot r_{12}\\ \theta_{11}=\text{atan}\,2(r_{32},r_{22})\\ b_1=\dfrac{d\cdot r_{22}}{c\theta_{11}} \end{cases} \qquad \begin{cases} a_2=c-d\cdot r_{23}\\ \theta_{12}=\text{atan}\,2(r_{13},r_{33})\\ b_2=\dfrac{d\cdot r_{33}}{c\theta_{12}} \end{cases} \qquad \begin{cases} a_3=c-d\cdot r_{31}\\ \theta_{13}=\text{atan}\,2(r_{21},r_{11})\\ b_3=\dfrac{d\cdot r_{11}}{c\theta_{13}} \end{cases} \tag{28-30} $$

The direct kinematic problem, on the other hand, assumes the knowledge of joint variables a_i, i=1,2,3 and aims at finding the corresponding attitudes of the platform in the space. The analysis is performed by means of simple trigonometric manipulations: by substituting in (28-30) the expression of r_{ij} given in (18), it is obtained:

$$ \begin{cases} c\beta s\gamma=\dfrac{c-a_1}{d}=k_1\\ s\alpha c\beta=\dfrac{c-a_2}{d}=k_2\\ c\alpha s\beta c\gamma-s\alpha s\gamma=\dfrac{c-a_3}{d}=k_3 \end{cases} \tag{31} $$

where the k_i, i=1,2,3 are known values. The 3 equations in (31) can be solved to find up to 4 admissible values for $s\gamma$:

$$ \left(2\frac{k_3k_2}{k_1}-k_2^2+1+\frac{k_2^2}{k_1^2}\right)s^4\gamma+\left(k_3^2-k_2^2-k_1^2-1\right)s^2\gamma+k_1^2=0 \tag{32} $$

For each angle γ that solves (32), 2 different values can be found for angles β and α:

$$ c\beta=\frac{k_1}{s\gamma} \qquad s\alpha=\frac{k_2}{k_1}s\gamma \tag{33-34} $$

therefore system (31) admits up to 16 different solutions: direct kinematics of the mechanism, however, is characterised by a maximum number of 8 different configurations, since angle β can be restricted in the range $[-\pi/2, \pi/2]$ without any loss of information.

3.4 Differential kinematics

By direct differentiation of the first 3 equations in (28-30), the expression of the *analytic Jacobian* $\mathbf{J}_A$ is directly derived:

$$\begin{bmatrix} \dot{a}_1 \\ \dot{a}_2 \\ \dot{a}_3 \end{bmatrix} = d \cdot \begin{bmatrix} 0 & -s\beta s\gamma & c\beta c\gamma \\ c\alpha c\beta & -s\alpha s\beta & 0 \\ -s\alpha s\beta c\gamma - c\alpha s\gamma & c\alpha c\beta c\gamma & -c\alpha s\beta s\gamma - s\alpha c\gamma \end{bmatrix} \cdot \begin{bmatrix} \dot{\alpha} \\ \dot{\beta} \\ \dot{\gamma} \end{bmatrix} = \mathbf{J}_A \begin{bmatrix} \dot{\alpha} \\ \dot{\beta} \\ \dot{\gamma} \end{bmatrix} \tag{35}$$

The *geometric Jacobian* $\mathbf{J}_G$ can be worked out by expressing the relation between the derivatives of Cardan angles and the components of angular velocity $\boldsymbol{\omega}$:

$$\begin{bmatrix} \omega_x \\ \omega_y \\ \omega_z \end{bmatrix} = \begin{bmatrix} 1 & 0 & s\beta \\ 0 & c\alpha & -s\alpha c\beta \\ 0 & s\alpha & c\alpha c\beta \end{bmatrix} \cdot \begin{bmatrix} \dot{\alpha} \\ \dot{\beta} \\ \dot{\gamma} \end{bmatrix} \tag{36}$$

$$\begin{bmatrix} \dot{a}_1 \\ \dot{a}_2 \\ \dot{a}_3 \end{bmatrix} = d \cdot \begin{bmatrix} 0 & -c\alpha s\beta s\gamma - s\alpha c\gamma & c\alpha c\gamma - s\alpha s\beta s\gamma \\ c\alpha c\beta & 0 & -s\beta \\ -s\alpha s\beta c\gamma - c\alpha s\gamma & c\beta c\gamma & 0 \end{bmatrix} \begin{bmatrix} \omega_x \\ \omega_y \\ \omega_z \end{bmatrix} = \mathbf{J}_G \begin{bmatrix} \omega_x \\ \omega_y \\ \omega_z \end{bmatrix} \tag{37}$$

It is noted that the geometric Jacobian $\mathbf{J}_G$ is not a function of geometric parameters, therefore machine's manipulability cannot be optimised by a proper selection of functional dimensions.

3.5 Analysis of singular poses

Limbs' structure does not allow for inverse kinematics singularities, while **direct kinematics singularities** can be found by letting the determinant of $\mathbf{J}_G$ vanish:

$$\det(\mathbf{J}_G) = d^3 \left[s\beta^2 - (c\alpha c\gamma - s\alpha s\beta s\gamma)^2 \right] \tag{38}$$

The zeros of (38) all lie on closed surfaces in the 3-dimensional space α, β, γ their intersections with the coordinate planes are straight lines (see also Fig. 16), as given by:

$$\begin{cases} \alpha = 0 \rightarrow \beta \pm \gamma = \pm \pi/2 \\ \beta = 0 \rightarrow \alpha = \pm \pi/2, \ \gamma = \pm \pi/2 \\ \gamma = 0 \rightarrow \alpha \pm \beta = \pm \pi/2 \end{cases} \tag{39}$$

The analysis of singular configurations has been performed also by means of numerical simulations. Figure 17 shows the value of the determinant of the geometric Jacobian matrix, normalised within the range *[-1, +1]* after division by the constant d^3: the black regions are characterised by determinant values in the range *[-0,05, +0,05]*. All the singularity maps are plot against the β and γ angles, α being a parameter of the representation; the configuration of the mechanism for $\beta=\gamma=0$ is represented aside. Figure 18a plots the singularity surface in the α,β,γ space but it is a hardly readable graph. In Fig. 18b, on other hand, the workspace volumes whose determinant assumes values in the range *[-0,05, +0,05]* have been taken out of the representation, while the colour map still represents the local determinant value: it is now more appreciable the extent of singularity-free regions inside the workspace, Fig. 18c, where the planning of a motion could be performed: e.g. for the mechanism under design a sphere with a radius of about 50° can be internally inscribed.

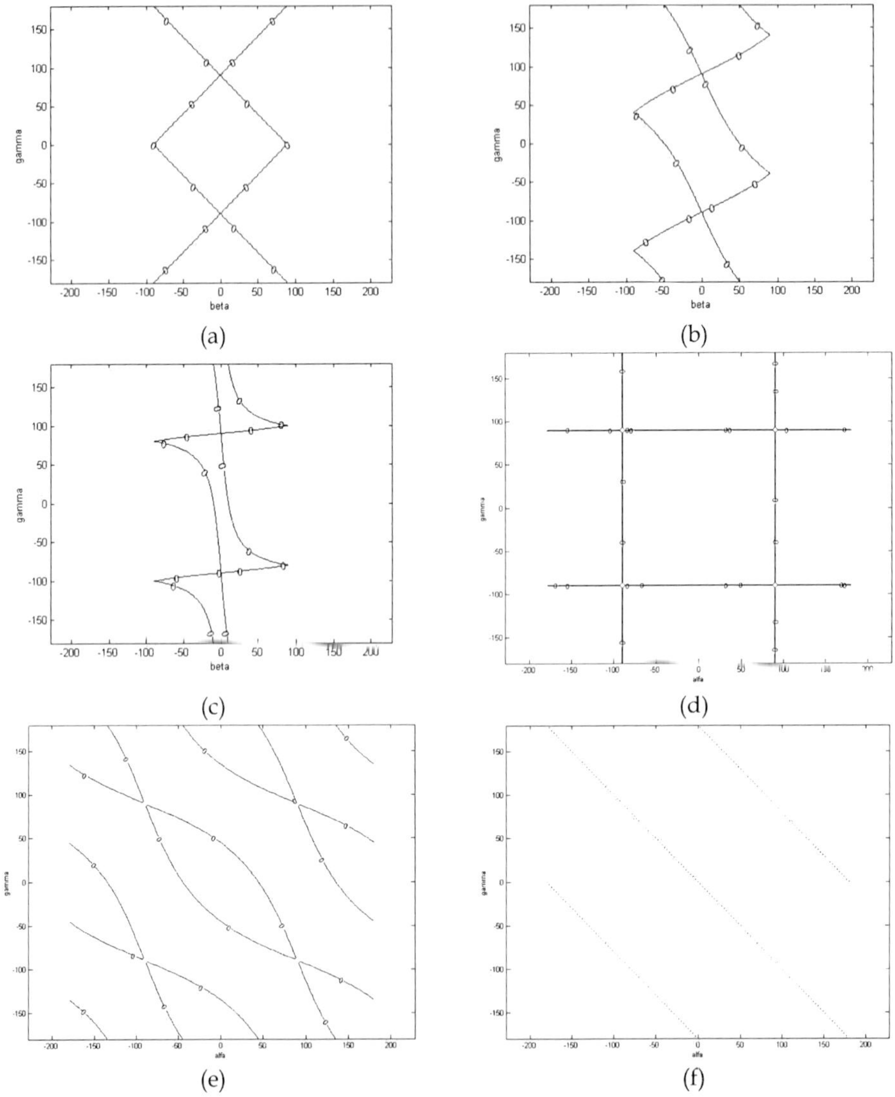

Fig. 16. Projection of direct kinematics singularity surface on several coordinate planes: $\alpha=0°$ (a), $\alpha=40°$ (b), $\alpha=80°$ (c), $\beta=0°$ (d), $\beta=45°$ (e), $\beta=89°$ (f)

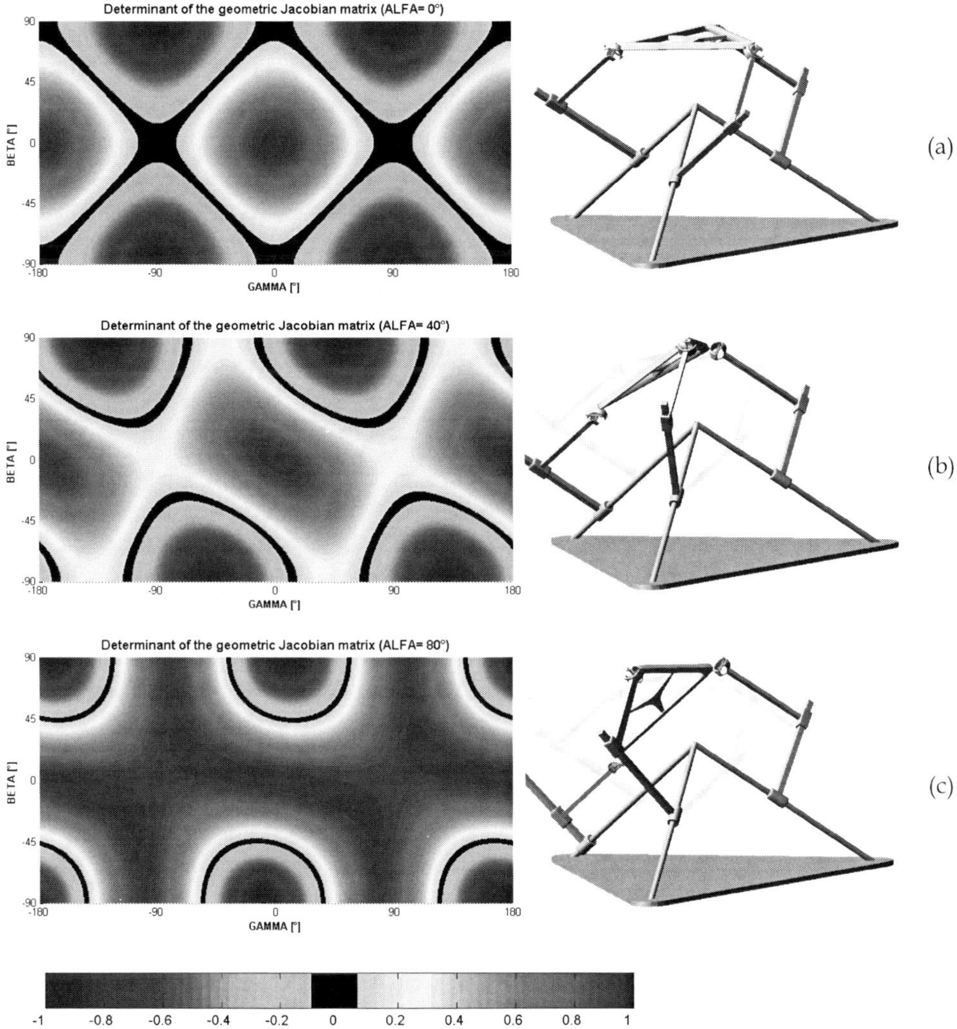

Fig. 17. Determinant of the geometric Jacobian matrix on the planes α=0° (a), α=40° (b), α=80° (c) and representation of manipulator configurations.

The sphere representation of singularity-free regions given in Fig. 18c is suggestive but it is expressed in a space (the α,β,γ Cardan angles) whose geometrical meaning is rather obscure. For many industrial tasks, on the other hand, it may be useful to use the spherical parallel machine for orienting a device or a part within a possibly large 2-dimensional space, identified by the axis of finite rotation, while the need for a further twist around the axis itself may not be urgent or at least only limited rotations may be required. In this case, the geometric Jacobian may be readily represented by a colour map on the surface of a unit sphere. Figure 19, for instance, uses lighter colours to render higher determinant values while black regions represent almost singular configurations; in this figure the orientation of the platform can be easily read through its elevation and azimuth, with the twist around the central axis is taken as a parameter of the representation: it is noted that in this case, at the expense of reduced twist rotations, greater pointing motions can be accomplished in the other 2 space directions.

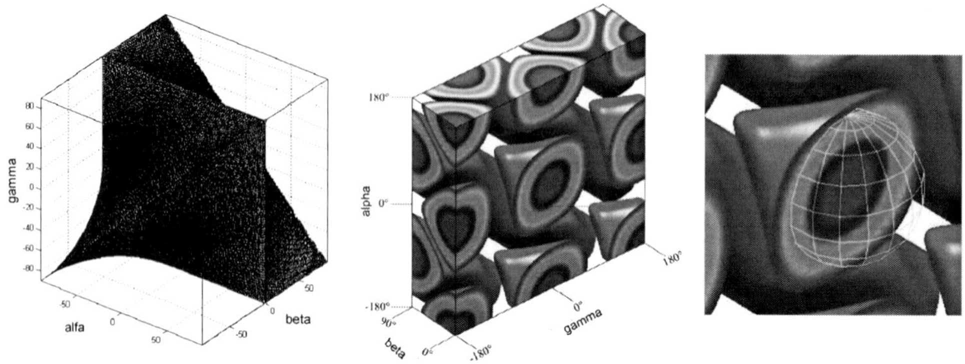

(a) (b) (c)

Fig. 18. Singularity surface in the α,β,γ space (a); colour map representing local determinant values (b) and close-up view of a connected singularity-free region.

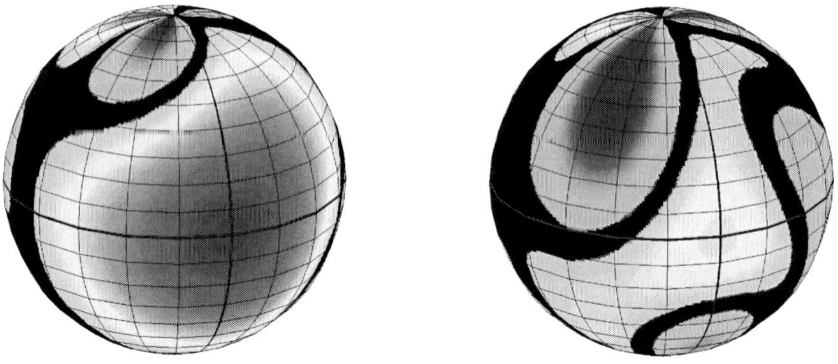

(a) (b)

Fig. 20. Singularity-free regions inside workspace for twist angle equal to 20° (a) and 60° (b)

Turning to **translation singularities**, the singular configurations found in (17) can be easily expressed as a function of articular coordinates θ_{1i}:

$$s\theta_{11}s\theta_{12}s\theta_{13} = c\theta_{11}c\theta_{12}c\theta_{13} \tag{40}$$

and taking into consideration inverse kinematics (28-30) it is obtained:

$$r_{32}r_{13}r_{21} = r_{22}r_{33}r_{11} \tag{41}$$

Equation (41) is a useful expression of translation singularities in task space, where the elements of the rotation matrix are used; by using the definition of the rotation matrix in (18) and after some trigonometric manipulation, an alternative expression can be obtained in function of Cardan angles α, β, γ:

$$s\beta^2 - (c\alpha c\gamma - s\alpha s\beta s\gamma)^2 = 0 \tag{42}$$

It is noted that (42) vanishes in the same configurations of (38), therefore translation singularities coincide with direct kinematics singularities, i.e. no additional singular surfaces are present inside workspace.

3.5 Analysis of static loads

The static analysis is useful in the first phases of machine design for the selection of machine's motors and for a first design of the links, with the related connecting bearings.
The base relation is provided as usual by the well known duality between kinematics and statics, which allows a straightforward assessment of the actuation efforts $\boldsymbol{\tau}$ needed to balance a moment $\mathbf{n}_{\mathrm{pl}}$ applied at the mobile platform:

$$\begin{bmatrix} \tau_1 \\ \tau_2 \\ \tau_3 \end{bmatrix} = \mathbf{J}_G^{-T} \begin{bmatrix} n_{plx} \\ n_{ply} \\ n_{plz} \end{bmatrix} \tag{43}$$

It must be noted that the application of a force $\mathbf{f}_{\mathrm{pl}}$ at the centre of the spherical motion does not require balancing forces by the actuators but it is entirely born by frame bearings: the internal reactions at the bearings caused by the application of the mentioned external wrench have been evaluated as well and used during structural design.

4. Dynamics

4.1 Inverse dynamics model

In this section an inverse dynamics model of the 3-CPU mechanism is worked out by using the virtual work principle: it is assumed that frictional forces at the joints are negligible, therefore the work produced by the constraint forces at the joints is zero and only active forces (including the gravitational effects) must be accounted in the developments.
In the derivation of the model, the notation is based on Fig. 14b and the second subscript i (i=1,2,3) indicates the ith limb while the first subscript j (j=1,2) refers to the first or second link respectively. Namely, m_{ji} and I_{ji} are the mass and (central) inertia tensor of the jth member of the ith limb; ω_{ji} is its angular velocity and v_{ji} is the linear velocity of its centre of mass; m_{pl}, I_{pl}, ω_{pl}, v_{pl} are the same quantities referred to the mobile platform.
The total wrench of active and inertial effects acting on the centre of mass of jth member of the ith limb is written as:

$$\mathbf{F}_{\mathrm{ji}} \doteq \begin{bmatrix} \mathbf{f}_{\mathrm{ji}} \\ \mathbf{n}_{\mathrm{ji}} \end{bmatrix} = \begin{bmatrix} m_{\mathrm{ji}}\mathbf{g} - m_{\mathrm{ji}}\dot{\mathbf{v}}_{\mathrm{ji}} \\ -\mathbf{I}_{\mathrm{ji}}\dot{\boldsymbol{\omega}}_{\mathrm{ji}} - \boldsymbol{\omega}_{\mathrm{ji}} \times (\mathbf{I}_{\mathrm{ji}}\boldsymbol{\omega}_{\mathrm{ji}}) \end{bmatrix} \tag{44}$$

In the same manner, the total wrench acting on the centre of mass of the mobile platform is:

$$\mathbf{F}_{\text{pl}} \doteq \begin{bmatrix} \mathbf{f}_{\text{pl}} \\ \mathbf{n}_{\text{pl}} \end{bmatrix} = \begin{bmatrix} m_{\text{pl}}(\mathbf{g} - \dot{\mathbf{v}}_{\text{pl}}) + \mathbf{f}_{\text{e}} \\ -\mathbf{I}_{\text{pl}}\dot{\boldsymbol{\omega}}_{\text{pl}} - \boldsymbol{\omega}_{\text{pl}} \times (\mathbf{I}_{\text{pl}}\boldsymbol{\omega}_{\text{pl}}) + \mathbf{n}_{\text{e}} \end{bmatrix} \tag{45}$$

where $\mathbf{f}_e$ and $\mathbf{n}_e$ are the external force and moment applied to its centre of mass; it is accidentally noted that the centre of mass of the platform does not coincide with the fixed point O. If $\boldsymbol{\tau}$ is the vector of the actuation forces and $\mathbf{q}$ are the corresponding displacements, the principle of virtual work can be written for the present case:

$$(\delta\mathbf{q})^T \cdot \boldsymbol{\tau} + (\delta\mathbf{x}_{\text{pl}})^T \cdot \mathbf{F}_{\text{pl}} + \sum_{i=1}^{3}\left(\sum_{j=1}^{2}\left((\delta\mathbf{x}_{\text{ji}})^T \mathbf{F}_{\text{ji}}\right)\right) = 0 \tag{46}$$

where the vector $\mathbf{x}_{ji}$ gathers the position of the centre of mass of jth member of the ith limb and the orientation of the same link and $\mathbf{x}_{pl}$ expresses the position of the centre of mass of the mobile platform and its orientation. It is noted that all the infinitesimal rotations appearing in (46) must be expressed as functions of the angular velocity of the respective link, e.g. for the platform:

$$\delta\mathbf{x}_{\text{pl}} = \begin{bmatrix} v_{plx} & v_{ply} & v_{plz} & \omega_{plx} & \omega_{ply} & \omega_{plz} \end{bmatrix}^T \cdot \delta t \tag{47}$$

Since all the virtual displacements in (47) must be compatible with the constraints, they are not independent but can rather be expressed as functions of an independent set of Lagrangian coordinates; if the Cardan angles $\boldsymbol{\varphi}=[\alpha, \beta, \gamma]^T$ of the mobile platform are chosen for this purpose, the following relations hold between the introduced virtual displacements:

$$\delta\mathbf{q} = \mathbf{J} \cdot \delta\boldsymbol{\varphi} \qquad \delta\mathbf{x}_{\text{ji}} = \mathbf{J}_{\text{ji}} \cdot \delta\boldsymbol{\varphi} \qquad \delta\mathbf{x}_{\text{pl}} = \mathbf{J}_{\text{pl}} \cdot \delta\boldsymbol{\varphi} \tag{48-50}$$

where $\mathbf{J}$, $\mathbf{J}_{ji}$ and $\mathbf{J}_{pl}$ are proper Jacobian matrices that can be found through the usual velocity analysis of the mechanism. Equation (46) can be written again as:

$$\delta\boldsymbol{\varphi}^T \cdot \left[\mathbf{J}^T \cdot \boldsymbol{\tau} + \mathbf{J}_{\text{pl}}^T \cdot \mathbf{F}_{\text{pl}} + \sum_{i=1}^{3}\left(\sum_{j=1}^{2} \mathbf{J}_{\text{ji}}^T \cdot \mathbf{F}_{\text{ji}}\right)\right] = 0 \tag{51}$$

Since (51) is valid for any virtual displacement $\delta\boldsymbol{\varphi}$ of the platform, in non-singular configurations it is:

$$\boldsymbol{\tau} = -\mathbf{J}^{-T} \cdot \left(\mathbf{J}_{\text{pl}}^T \cdot \mathbf{F}_{\text{pl}} + \sum_{i=1}^{3}\left(\sum_{j=1}^{2} \mathbf{J}_{\text{ji}}^T \cdot \mathbf{F}_{\text{ji}}\right)\right) \tag{52}$$

Equation (52) completely describes manipulator's dynamics; all the elements in it have been worked out and the resulting model has been proofed by comparison with commercial packages' output, see (Callegari & Marzetti, 2006).

4.2 Dynamic analysis in the task space

The dynamic expression (52) is usefully re-worked in order to explicit the dependency on a proper set of Lagrangian coordinates and its derivatives. In the case of parallel kinematics machines, the dynamic model results quite naturally written in the task space, due to the (usually) difficult expression of DKP; therefore in the present case, after some cumbersome manipulation, it is obtained:

$$\boldsymbol{\tau}_{\varphi} - \mathbf{J}_{\mathrm{pl}}^{T}\mathbf{h} = \mathbf{M}_{\phi}(\boldsymbol{\varphi})\ddot{\boldsymbol{\varphi}} + \mathbf{C}_{\phi}(\boldsymbol{\varphi}, \dot{\boldsymbol{\varphi}})\dot{\boldsymbol{\varphi}} + \mathbf{G}_{\phi}(\boldsymbol{\varphi}) \tag{53}$$

with: $\boldsymbol{\tau}_{\varphi} = \mathbf{J}^{T} \cdot \boldsymbol{\tau}$, moments acting at the end-effector and corresponding to actual forces $\boldsymbol{\tau}$ at actuated joints; $\mathbf{M}_{\varphi}(\boldsymbol{\varphi})$, Cartesian mass matrix of the manipulator; $\mathbf{C}_{\varphi}(\boldsymbol{\varphi}, \dot{\boldsymbol{\varphi}})$, vector of centrifugal and Coriolis terms; $\mathbf{G}_{\varphi}(\boldsymbol{\varphi})$, vector of gravity moments; $\mathbf{h}$, vector of external forces and moments acting at the centre of mass of the mobile platform.

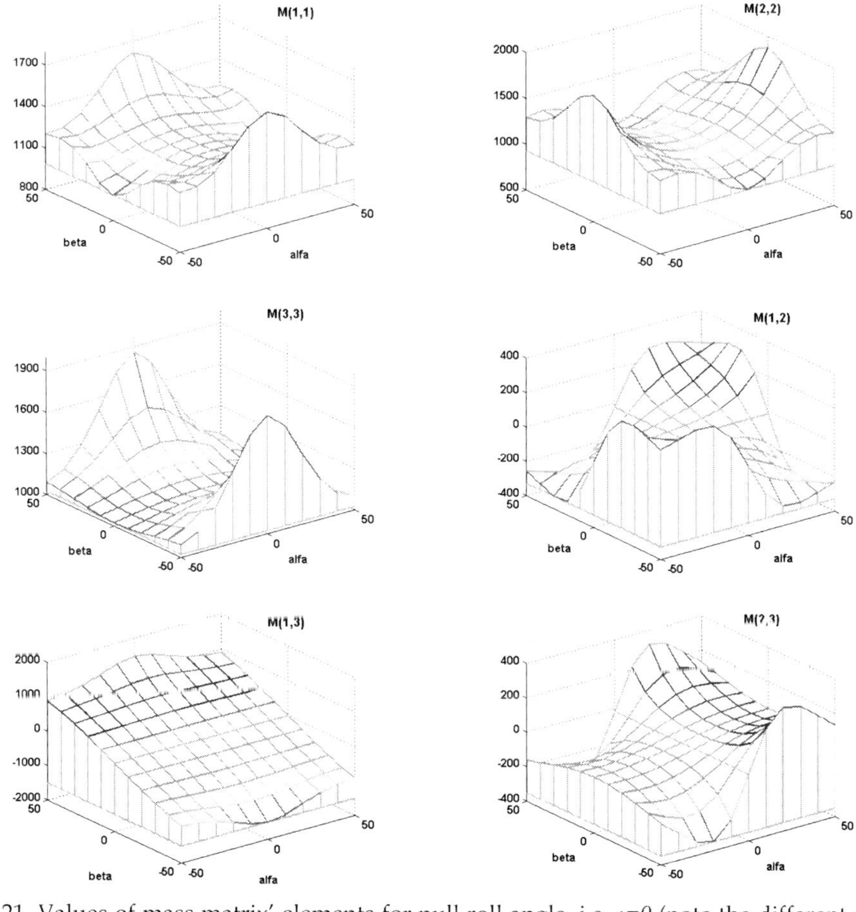

Fig. 21. Values of mass matrix' elements for null roll angle, i.e. $\gamma=0$ (note the different scales of the plots)

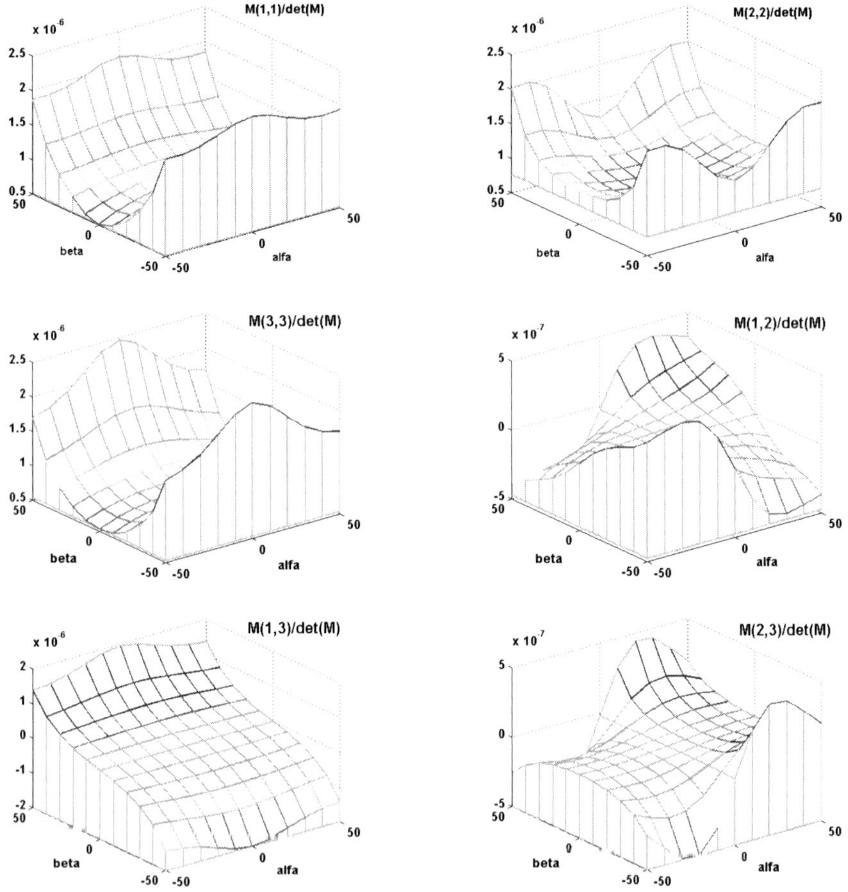

Fig. 22. Plots of mass matrix' elements, normalised by determinant value, for null roll angle, i.e. $\gamma=0$ (note that all the scales of the graphs are multiplied by 10^{-6} but M(1,2) and M(2,3) which are multiplied by 10^{-7})

In view of the realisation of possible control schemes based on the inversion of manipulator's dynamics, it is useful to study the variability of mass matrix throughout the workspace. In fact, a major simplification of the model would be yielded by neglecting the 6 non-diagonal terms of the mass matrix, whether actually allowed by their comparative magnitude; otherwise, all the elements in $\mathbf{M}_\phi$ and $\mathbf{C}_\phi$ could be considered constant. First simulation results show that in this case both simplifying assumptions could be taken into consideration, even if the validity of the reduced models weakens when the operating trajectories get closer to singularity surfaces, as expected.

Figure 21, for instance, shows the values of mass matrix' elements in different workspace configurations characterised by null roll angle, i.e. $\gamma=0$: for robot's parameters it has been made reference to the virtual prototype, whose mass properties, presented in the following Tab. 2, are very similar to physical prototype. In Fig. 22 the same plots have been normalised by dividing the matrix element by the (local) value of matrix determinant, to allow a relative comparison among elements that have very different magnitudes. It can be seen that near the isotropic point ($\alpha=\beta=\gamma=0$) the diagonal elements are dominant and matrix variability is limited, while off-diagonal elements show a stronger influence when getting closer to workspace boundaries; moreover, element M(3,1) is generally an order of magnitude greater than M(2,1) and M(2,3). Such behaviour gets even more evident if one moves away from the plane $\gamma=0$. The plots have been traced for pitch and yaw angles varying between $-50°$ and $+50°$ because the sphere of $50°$ radius in the Cardan angles space is completely free of singularities, as shown already in Fig. 18.

Other kinds of tests have been performed, aiming at identifying the relative contribute of various dynamic terms: for instance it seems that, even for high dynamics manoeuvres, the contribute of gravity is never negligible, while Coriolis and centrifugal forces account for *10%-16%* maximum; on the other hand, the mass and inertia of the mobile platform affect very slightly the overall dynamic behaviour of the machine, possibly allowing for a major simplification of system's model.

4.3 Dynamic manipulability

The dynamic manipulability ellipsoids introduced by Yoshikawa (1985, 2000) are a useful means to study the dynamic properties of a mechanism: they express graphically the capability of a given device to yield accelerations in all the directions stemming from one attitude of its workspace. As a matter of fact, many other measures of manipulability have been proposed by different researchers since that pioneering work but very few applications dealt with orienting devices.

Let us consider all the actuation forces $\boldsymbol{\tau}$ with unit norm:

$$\boldsymbol{\tau}^T \cdot \boldsymbol{\tau} = 1 \tag{54}$$

By manipulating (53) in order to work out $\boldsymbol{\tau}$, it is obtained:

$$\boldsymbol{\tau} = \mathbf{J}^{-T}\mathbf{M}_{\varphi}\left(\ddot{\boldsymbol{\varphi}} + \ddot{\boldsymbol{\varphi}}_{\text{bias}}\right) \tag{55}$$

having defined:

$$\ddot{\boldsymbol{\varphi}}_{\text{bias}} = \mathbf{M}_{\varphi}^{-1}\left(\mathbf{C}_{\varphi}\dot{\boldsymbol{\varphi}} + \mathbf{G}_{\varphi} + \mathbf{J}_{\text{pl}}^{T}\mathbf{h}\right) \tag{56}$$

A meaningful formulation of dynamic manipulability must be expressed as a direct function of the angular acceleration $\dot{\boldsymbol{\omega}}$, therefore the mapping between the rate of change of the Cardan angles $\dot{\boldsymbol{\varphi}}$ and the angular velocity $\boldsymbol{\omega}$ must be made explicit:

$$\begin{bmatrix} \omega_x \\ \omega_y \\ \omega_z \end{bmatrix} = \begin{bmatrix} 1 & 0 & s\beta \\ 0 & c\alpha & -s\alpha c\beta \\ 0 & s\alpha & s\alpha c\beta \end{bmatrix} \begin{bmatrix} \dot{\alpha} \\ \dot{\beta} \\ \dot{\gamma} \end{bmatrix} \rightarrow \boldsymbol{\omega} = \mathbf{E}(\boldsymbol{\varphi})\dot{\boldsymbol{\varphi}} \tag{57}$$

$$\dot{\boldsymbol{\omega}} = \mathbf{E}(\boldsymbol{\varphi})\ddot{\boldsymbol{\varphi}} + \dot{\boldsymbol{\omega}}_1(\boldsymbol{\varphi}, \dot{\boldsymbol{\varphi}}) \tag{58}$$

If $\ddot{\boldsymbol{\varphi}}$ is taken out of (58) and substituted in (55) it is then obtained:

$$\boldsymbol{\tau} = \mathbf{J}^{-T}\mathbf{M}_{\varphi}\mathbf{E}^{-1}\left(\dot{\boldsymbol{\omega}} + \dot{\boldsymbol{\omega}}_{\text{bias}}\right) \tag{59}$$

having defined:

$$\dot{\boldsymbol{\omega}}_{\text{bias}} = -\dot{\boldsymbol{\omega}}_1 + \mathbf{E}\mathbf{M}_{\varphi}^{-1}\mathbf{J}^{T}\mathbf{M}_{\varphi}^{-1}\left(\mathbf{C}_{\varphi}\dot{\boldsymbol{\varphi}} + \mathbf{G}_{\varphi} + \mathbf{J}_{\text{pl}}^{T}\mathbf{h}\right) \tag{60}$$

The constraint expressed by (54) can be finally written in the following quadratic form:

$$\dot{\boldsymbol{\Omega}}^{T} \cdot \boldsymbol{\Gamma}(\boldsymbol{\varphi}) \cdot \dot{\boldsymbol{\Omega}} = 1 \tag{61}$$

with obvious meaning of the introduced terms:

$$\dot{\boldsymbol{\Omega}} = \dot{\boldsymbol{\omega}} + \dot{\boldsymbol{\omega}}_{\text{bias}} = \dot{\boldsymbol{\omega}} - \dot{\boldsymbol{\omega}}_1 + \mathbf{E}\mathbf{M}_{\varphi}^{-1}\mathbf{J}^{T}\mathbf{M}_{\varphi}^{-1}\left(\mathbf{C}_{\varphi}\dot{\boldsymbol{\varphi}} + \mathbf{G}_{\varphi} + \mathbf{J}_{\text{pl}}^{T}\mathbf{h}\right) \tag{62}$$

$$\boldsymbol{\Gamma}(\boldsymbol{\varphi}) = \mathbf{E}^{-T}\mathbf{M}_{\varphi}^{T}\mathbf{J}^{-1}\mathbf{J}^{-T}\mathbf{M}_{\varphi}\mathbf{E}^{-1} \tag{63}$$

The inspection of (62-64) shows that gravity merely induces a translation of the dynamic manipulability ellipsoid while in general velocity has a complex, non-negligible effect on manipulability. Making reference to the remarkable case of a fixed platform ($\dot{\boldsymbol{\varphi}} = \mathbf{0}$) with no external or gravity action applied ($\mathbf{h}$=$\mathbf{G}_{\varphi}$=0), (61) provides:

$$\dot{\boldsymbol{\omega}}^{T} \cdot \boldsymbol{\Gamma}(\boldsymbol{\varphi}) \cdot \dot{\boldsymbol{\omega}} = 1 \tag{64}$$

The quadratic form (64) represents an ellipsoid in the Cartesian space of the angular accelerations: its eigenvalues express the square root of the maximum and minimum accelerations that can be developed with unit actuator forces while the eigenvectors represent the associated directions in the orientation space. Figure 23 represents graphically some dynamic manipulability ellipsoids of the robot in the poses sketched aside.

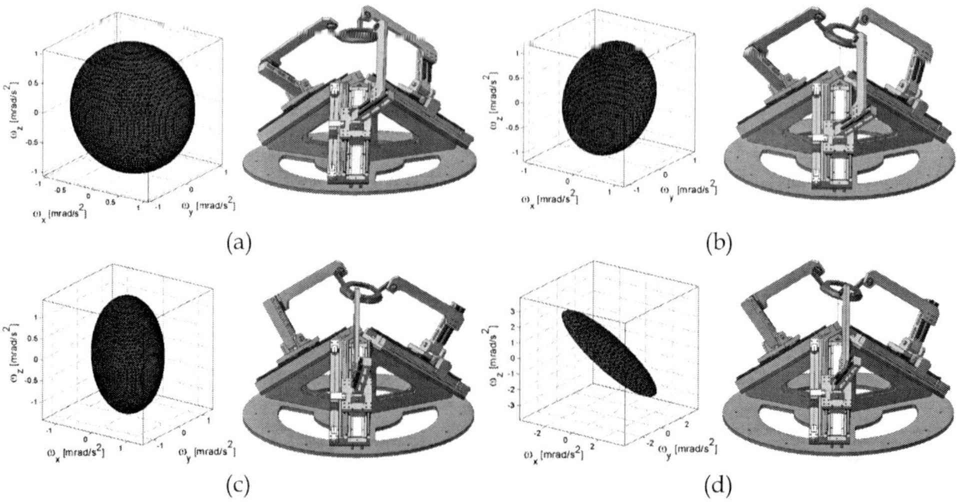

Fig. 23. Dynamic manipulability ellipsoids at different poses (α,β,γ): (0°,0°,0°) (a), (20°,20°,-5°) (b), (40°,40°,10°) (c), (54°,53°,10°) (d)

5. Prototype design

The design of a first prototype has been developed, aiming at obtaining high dynamics performances; as reference figures, the following requirements have been posed:

- orientation range (elevation and azimuth): *150°*
- maximum angular velocity: *500 °/s*
- maximum angular acceleration: *5 000 °/s²*
- spatial resolution: *0.01 °*
- overall dimensions of the machine: maximum volume of *1 m³*.

The particular form of the Jacobian matrix (35) does not allow for a mechanical design based on the optimisation of kinematic properties, since $\mathbf{J}_G$ is not function of robot's geometry, therefore heuristic considerations have been made in a first phase, in order to limit wrist's overall dimensions. By looking at Fig. 24 and taking into consideration (31), it is noted that the value of length *c* does not affect actuators' stroke but only their initial position. The value of length *d*, instead, is directly proportional to the motors' run needed to attain an assigned configuration in space and by decreasing its value a more compact design is obtained: on the other hand, a lower limit is provided by the need to accommodate the universal joints on the mobile platform and to grant a limit positioning accuracy in the task space. By means of computer simulation, all the geometrical parameters represented in Fig. 24 have been made to change, in order to take into account the above considerations and to assess the resulting geometry; in the end, it has been decided to refine the mechanical design by taking into account the concept of dynamic optimisation, enabled by the availability of the inverse dynamics model.

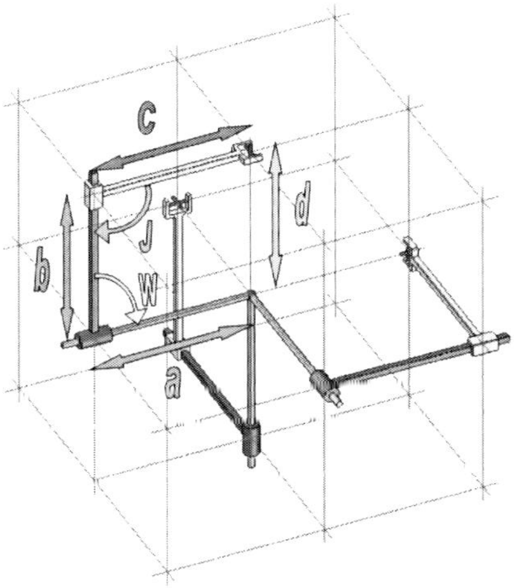

Fig. 24. Main geometrical parameters

Two dynamic figures have been used to drive the design of the machine. The *measure of the dynamic manipulability*, *w*, defined as:

$$w = \sqrt{\det(\mathbf{\Gamma}(\boldsymbol{\varphi}))} \tag{65}$$

results proportional to the volume of the manipulation ellipsoid and therefore yields an overall information on the global manipulation capabilities, but fails to capture the closeness to singular configurations or even the anisotropy of local dynamics. On the other hand, the *index of dynamic manipulability*, *i*, can be defined as:

$$i = \sqrt{\lambda_{\min} / \lambda_{\max}} \tag{66}$$

with λ_{min}, λ_{max} minimum and maximum eigenvalues of the matrix $\Gamma(\varphi)$: the index (66) is independent from the volume of the ellipsoid and vanishes close to singular configurations.

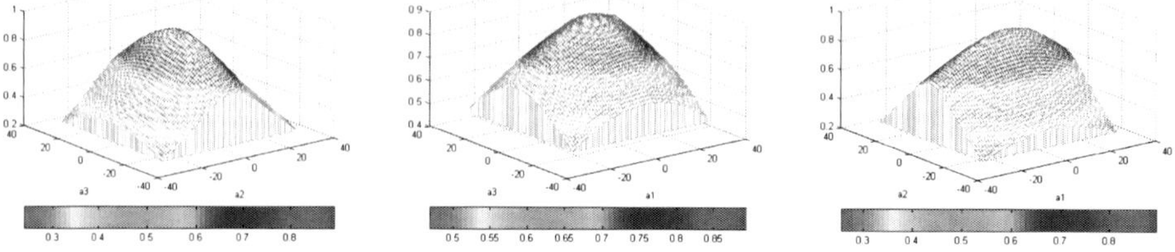

(a) (b) (c)

Fig. 25. Plots of the index of dynamic manipulability as a function of actuators strokes on the three coordinate planes $a_1=0$ (a), $a_2=0$ (b), $a_3=0$ (c)

Figure 25 shows sample plots of the index of dynamic manipulability as a function of actuators strokes a_i on the three coordinate planes for the final design. With specific Matlab routines, a dynamic optimisation of the design has been performed, trying to maximise the global dynamic manipulability of the wrist while still guaranteeing a minimum threshold of the local features. For instance, in the configurations shown in Fig. 23a-23d the indexes assume the values: *0.7755, 0.1374, 0.3571, 0.0341* respectively, while it has been obtained a mean value of $i_{ave}=0.502$ over the central ±*30°* span of the workspace. Table 1 summarises the final geometrical values used for the design, with *h* being the total length of the lower part of the three limbs. It must be said that, as a general rule, in this case the optimisation routines tend to concentrate all the masses in the centre of the spherical motion, that is only too natural.

Figure 26 on the left shows a sketch of the design of final prototype meeting the posed requirements; on the right side, a picture of the machine is presented. The limbs are made of *avional* (an aluminium-copper alloy) in order to join good mechanical properties with a lightweight construction. The mobile platform is made of bronze, therefore allowing the precise machining in a single placement of the 3 journal bearings that have to meet orthogonally in a single point: in this way it has been a high stiffness together with precise geometrical alignments. It must be noted that such revolute joints are idle, since no rotation occurs at all if all the manufacturing and mounting conditions are correctly satisfied. In order to allow the precise mounting of the robot in the initial (home) configuration, the special fixture shown in Fig. 27 has been realised.

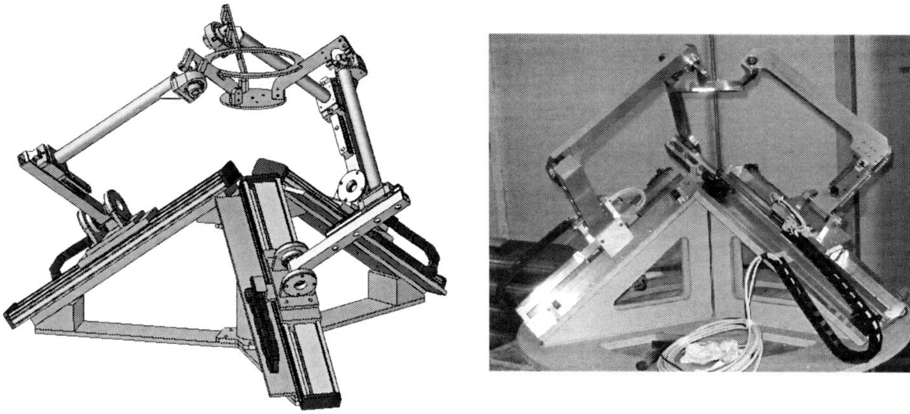

(a) (b)

Fig. 26. CAD model of spherical wrist (a) and picture of first laboratory prototype (b)

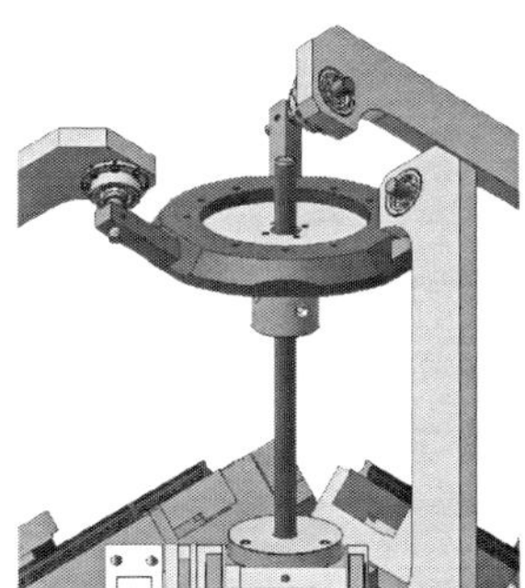

Fig. 27. Sketch of the fixture for axes alignment during machine assembly

The actuation is based on 3 induction linear motors Phase WVS 20.6.3, able to provide a maximum thrust of 184 N at the speed of 6 m/s, with a maximum acceleration of 14.3 g and is controlled by Nation Instrument hardware (Flexmotion/PXI architecture). The first tests of motion are currently under development, while wrist's controller is under design.

d [mm]	c [mm]	h [mm]	a_{imin} [mm]	a_{imax} [mm]	b_{imin} [mm]	b_{imax} [mm]
210	490	280	319	661	130	210

Tab. 1. Main geometrical data of spherical wrist design

link	**m** (mass, kg)	**I_{11}** (x-x moment of inertia, kg m2)	**I_{22}** (y-y moment of inertia, kg m2)	**I_{33}** (z-z moment of inertia, kg m2)
upper limb	2.50	0.016	0.016	0.0013
lower limb	7.50	0.070	0.070	0.0014
platform	5.35	0.030	0.030	0.060

Tab. 2. Mass properties of spherical wrist design

6. Conclusions

The article has described an innovative spherical parallel wrist developed at the Polytechnic University of Marche in Ancona, revisiting all the main design steps, from kinematic synthesis up to physical prototyping.

Machine kinematics has been worked out in closed form and all the singularity surfaces have been analysed: it has been pointed out that the mechanism does not possess inverse kinematics singularities, while direct kinematics singularities and translation singularities lie on the same closed surface. The inner space, where motion paths can be safely planned, has been identified and unfortunately it cannot be enlarged by kinematics optimisation because machine's Jacobian does not depend on geometrical parameters.

For this reason, it was decided to drive machine design by dynamic optimisation concepts and an inverse dynamics model has been developed: the study of machine's dynamic manipulability, by means of different algebraic tools, led to the final design of the wrist, that has been also verified with structural analysis packages. The availability of the dynamic model, on the other hand, will be useful for the development of model based control systems, able to exploit the high potentials of direct drive actuation: a first dynamic analysis, moreover, shows that simplified models could be used, since the non-diagonal terms of mass matrix are much smaller than diagonal terms and platform's inertia could be neglected, at least when manipulator is far from singular configurations.

All design steps have been performed in a virtual prototyping environment, that allowed to take into consideration simultaneously the constraints of the mechanics and the problems of the controller, allowing to assess the performances of the closed-loop system. The physical prototyping of the machine, however, allowed to validate the good properties envisaged during the design phase but also to experience the disadvantages of the concept itself: they are mainly due to the scarce accessibility of the centre of the spherical motion, which is common to most parallel wrists, and to the difficult assembly, which requires a precise alignment of joints axes: this problem has been partially overcome by the manufacturing of specific fixtures that are characterised by very high accuracy and are used while assembling the machine.

The machine has been moved so far only through motors drives and a conventional PID position controller is actually being developed: more advanced control systems, able to exploit the high dynamics of the design and the power of direct actuation, will be studied soon.

7. References

Alizade, R.I.; Tagiyev, N.R. & Duffy, J. (1994) A forward and reverse displacement analysis of an in-parallel spherical manipulator, *Mechanism and Machine Theory*, Vol. 29, No. 1, pp.125-137, ISSN 0094-114X

Asada, H. & Granito, C. (1985) Kinematic and static characterization of wrist joints and their optimal design, *Proc. IEEE Conf. Robotics and Automation*, pp.244-250, St. Louis, USA, March 25-28.

Callegari, M. & Marzetti, P. (2006) Inverse Dynamics Model of a Parallel Orienting Device, *Proc. 8th Intl. IFAC Symposium on Robot Control: SYROCO 2006*, Bologna, Italy, Sept. 6-8, 2006.

Callegari, M. & Palpacelli, M.-C. (2008) Prototype design of a translating parallel robot, *Meccanica* (available on-line at: DOI 10.1007/s11012-008-9116-8), ISSN: 0025-6455.

Callegari, M.; Gabrielli, A. & Ruggiu, M. (2008) Kineto-Elasto-Static Synthesis of a 3-CRU Spherical Wrist for Miniaturized Assembly Tasks, *Meccanica,* ISSN: 0025-6455.

Callegari, M.; Gabrielli, A.; Palpacelli, M.-C. & Principi, M. (2007) Design of Advanced Robotic Systems for Assembly Automation, *Intl J. of Mechanics and Control,* Vol. 8, No. 1 (Dec. 2007). pp.3-8, ISSN 1590-8844

Callegari, M.; Marzetti, P. & Olivieri B. (2004) Kinematics of a Parallel Mechanism for the Generation of Spherical Motions, In: *On Advances in Robot Kinematics,* J. Lenarcic and C. Galletti (Eds), pp.449-458, Kluwer, ISBN 1-4020-2248-4, Dordrecht.

Callegari, M.; Palpacelli, M.C. & Scarponi, M. (2005) Kinematics of the 3-CPU parallel manipulator assembled for motions of pure translation, *Proc. Intl. Conf. Robotics and Automation,* pp 4031-4036, Barcelona, Spain, April 18-22.

Di Gregorio, R. (2001a) Kinematics of a new spherical parallel manipulator with three equal legs: the 3-URC wrist, *J. Robotic Systems,* Vol. 18, No. 5 (Apr. 2001), pp.213-219, ISSN 0741-2223

Di Gregorio, R. (2001b) A new parallel wrist using only revolute pairs: the 3-RUU wrist, *Robotica,* Vol.19, No.3 (Apr. 2001), pp. 305-309, ISSN 0263-5747

Di Gregorio, R. (2004) The 3-RRS Wrist: A New, Simple and Non-Overconstrained Spherical Parallel Manipulator, *J. of Mechanical Design,* Vol. 126, No. 5, (Sept. 2004) pp.850-855, ISSN 1050-0472.

Fang, Y. & Tsai, L.-W. (2004) Structure synthesis of a class of 3-DOF rotational parallel manipulators, *IEEE Trans. on Robotics and Automation,* Vol.20, No.1, (Feb. 2004) pp.117-121, ISSN 0882-4967

Gosselin, C. & Angeles, J. (1989) The optimum kinematic design of a spherical three-degree-of-freedom parallel manipulator, *J. Mechanisms, Transmissions and Automation in Design,* Vol. 111, No 2. pp.202-207, ISSN 0738-0666

Innocenti, C. & Parenti-Castelli, V. (1993) Echelon form solution of direct kinematics for the general fully-parallel spherical wrist, *Mechanism and Machine Theory,* Vol. 28, No. 4, pp.553-561, ISSN 0094-114X

Karouia, M. & Hervè, J.M. (2000) A three-dof tripod for generating spherical rotation, In: *Advances in Robot Kinematics,* J.L. Lenarcic & M.M. Stanisic (Eds.), pp.395-402, Kluwer, ISBN 0-7923-6426-0, Dordrecht.

Karouia, M. & Hervè, J.M. (2002) A Family of Novel Orientational 3-DOF Parallel Robots, *Proc. 14th RoManSy,* pp 359-368, Udine, Italy, July 1-4.

Karouia, M. & Hervé, J.M.. (2006) Non-overconstrained 3-dof spherical parallel manipulators of type: 3-RCC, 3-CCR, 3-CRC, *Robotica,* Vol. 24, No. 1, January 2006, pp.85-94, ISSN 0263-5747

Kong, X. & Gosselin, C.M. (2004) Type synthesis of 3-DOF spherical parallel manipulators based on screw theory, *J. of Mechanical Design,* Vol.126, No.1, (Jan. 2004), pp.101-108, ISSN 1050-0472

Kong, X. & Gosselin, C.M. (2004) Type synthesis of three-degree-of-freedom spherical parallel manipulators, *Intl J. of Robotics Research,* Vol.23, No.3, (March 2004) pp.237-245, ISSN 0278-3649

Lee, J.J. & Chang, S.-L. (1992) On the kinematics of the UPS wrist for real time control, *Proc. 22nd ASME Biennal Mechanisms Conference: Robotics, Spatial Mechanisms and Mechanical Systems*, pp.305-312, Scottsdale, USA, Sept. 13-16.

Olivieri, B. (2003) *Study of a novel parallel kinematics spherical robot for cooperative applications*, Tesi di laurea (in Italian), Università Politecnica delle Marche, Ancona, Italy.

Yoshikawa, T. (1985). Dynamic Manipulability of Robot Manipulators. *J. Robotic Systems*, Vol. 2, pp.113-124, ISSN 0741-2223.

Yoshikawa, T. (2000). Erratum to "Dynamic Manipulability of Robot Manipulators". *J. Robotic Systems*, Vol. 17, No. 8, (Aug. 2000), pp.449, ISSN 0741-2223.

10

Quantitative Dexterous Workspace Comparison of Serial and Parallel Planar Mechanisms

Geoff T. Pond and Juan A. Carretero
University of New Brunswick
Canada

1. Introduction

The dexterity analysis of complex degree of freedom (DOF) mechanisms has thus far been problematic. A well accepted method of measuring the dexterity of spherical or translational manipulators has been the Jacobian matrix condition number as in (Gosselin & Angeles, 1989) and (Badescu & Mavroidis, 2004). Unfortunately, the inconsistent units between elements within the Jacobian of a complex-DOF parallel manipulator do not allow such a measure to be generally made as discussed in (Tsai, 1999) and (Angeles, 2003). In the following section, the mathematical meaning of singular values and the condition number of a matrix are reviewed. Their application to studying robotic dexterity follows next. Later in this chapter, these principles are applied to the study and comparison of the dexterous workspace of both serial and parallel manipulators.

1.1 Mathematical background

The condition number of a matrix is defined as the ratio of the maximum and minimum singular values of the matrix. A brief explanation of the significance of the matrix's singular values is important and is therefore provided here. Strang (Strang, 2003) shows that any matrix or transform, *e.g.*, **J**, may be broken into three components through singular value decomposition:

$$\mathbf{J} = \mathbf{U}\Sigma\mathbf{V}^T \tag{1}$$

where **V** contains the eigenvectors of $\mathbf{J}^T\mathbf{J}$, **U** contains the eigenvectors of $\mathbf{J}\mathbf{J}^T$ ($\mathbf{u}_1$ and $\mathbf{u}_2$ for the two dimensional case shown) and Σ is a diagonal matrix containing the singular values of **J**. Both the matrices **V** and **U** are composed of unit vectors which are mutually perpendicular within each matrix. Figure 1 is adapted from Strang (2003), and graphically depicts the transform described in equation (1) for the two dimensional case.

In terms of dexterity, the most interesting of the three component matrices of **J** is Σ consisting of the singular values of **J** each denoted by σ_i. Consider the conventional relation $\dot{\mathbf{q}} = \mathbf{J}\dot{\mathbf{x}}$, where in more general terms, $\dot{\mathbf{x}}$ corresponds to some unit system output depicted in the furthest left side of Figure 1, $\dot{\mathbf{q}}$, the system input depicted in the furthest right of Figure 1, and **J**, the system transform between them. Generally, the maximum and minimum singular values of **J** indicate a range within which the magnitude of vector $\dot{q}$ must lie, for any unit output in $\dot{\mathbf{x}}$, *i.e.*, $|\dot{\mathbf{x}}| = 1$. The condition number κ is then the ratio of the largest and smallest singular values:

$$\kappa = \frac{\sigma_{max}}{\sigma_{min}} \tag{2}$$

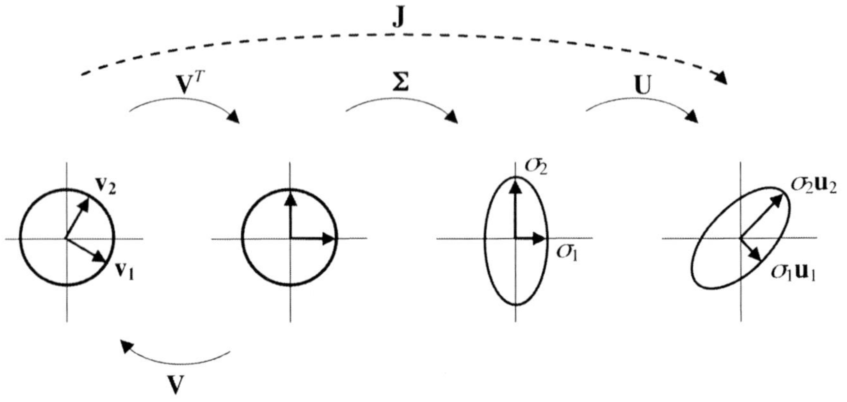

Figure 1: The three steps in any matrix transformation: rotation, scaling, rotation (or reflection).

Now, let the system output $\dot{\mathbf{x}}$ correspond to the velocity vector of a manipulator's end effector and $\dot{\mathbf{q}}$, the vector of actuator velocities. 'Ideal dexterity' occurs at isotropic conditions, that is at the lowest possible Jacobian condition number, *i.e.*, 1 (Angeles, 2003). At such positions, a unit velocity in any feasible direction for the manipulator requires the same total effort in the actuators, *i.e.*, the resolution of end effector pose is the same in each DOF. On the other hand, a condition number of ∞ corresponds to a rank deficiency within the Jacobian matrix. At such configurations, some level of control over the system is lost.

1.2 Application to robotics

In robotics, the Jacobian, and hence its singular values and condition number, are dependant on the architecture of the manipulator as well as the position and orientation, together referred to as pose, of the manipulator's end effector. As a result, the manipulator's level of dexterity changes as it travels through its reachable workspace. A manipulator's dexterous workspace is often defined as poses resulting in a Jacobian matrix condition number below a specified threshold. The higher level of dexterity required, or as conventionally defined, the lower the condition number, the smaller the dextrous workspace will be. This is due to an increasing Jacobian matrix condition number as the reachable workspace boundary is approached. Manipulator singularities exist when the Jacobian condition number becomes infinite, that is, either a) an instantaneously infinite actuator input velocity results in no change in the end effector pose, or b) the end effector pose may be altered without having changed the actuator inputs.

However, using the Jacobian condition number alone may provide misleading results, particularly when comparing multiple manipulators, as this chapter will later do. Consider two 2-DOF manipulators, of the same architecture but of different scale, and in the same pose. The first having Jacobian matrix singular values of 1 and 2, the second being 100 times larger having singular values of 100 and 200. Both result in the same condition number as they both require twice the effort in the second direction as they do to move in the first. That is, the magnitude of the vector $\dot{\mathbf{q}}$ required to perform the motion in the second direction is twice as large in magnitude as the magnitude required to perform the motion in the first direction. In the case of the first system, the end effector pose is far more sensitive to the system inputs (recall that the sensitivity is indicated by the singular values, the condition number only indicates the ratio of this sensitivity for the fastest and slowest directions in the task space). For this reason, the entries of the Jacobian matrix must all share the same units, *e.g.*, distances may be measured in *m* but not by a mix of *m*, *cm*, *mm*, *etc.*

Larger singular values correspond to a better resolution over the pose of the end effector, hence better position control over the mechanism end effector pose is achieved. However, having small singular values also has a benefit. Having smaller singular values suggests that the same system outputs are achieved at lower system inputs when compared to a system with large singular values. This corresponds to higher end effector velocities for the same actuator input magnitude. Therefore, there is a trade-off between high end effector velocities (a Jacobian having small singular values), and fine resolution over the end effector pose which provides better stiffness and accuracy (a Jacobian having large singular values).

In terms of dexterity, higher end effector velocities are generally of greater concern. In terms of either accuracy or stiffness / compliance, a finer resolution over the end effector pose is of greater importance. Therefore, examination of the Jacobian matrix condition number alone, does not fully describe the capabilities of a manipulator in the studied pose.

1.3 Issues with using the Jacobian matrix condition number

It is well known that the use of the condition number of a manipulator's Jacobian matrix to measure dexterity may only be made when all the entries that constitute such a Jacobian matrix share the same units (Tsai, 1999; Angeles, 2003; Doty et al., 1995). This limits the use of the Jacobian condition number to manipulators that have only one type of actuator (*i.e.*, either revolute or prismatic, but not a combination of both). Furthermore, use of the Jacobian condition number is restricted to manipulators having only degrees of freedom (DOF) in either Cartesian or rotational directions only, but not combinations of both. The only mechanisms that fall into this category are 3-DOF (or less) rotational and 3-DOF (or less) translational manipulators. Otherwise, if the manipulator has a mix of revolute and prismatic actuators, or has complex degrees of freedom, their associated Jacobian matrix is dimensionally inconsistent.

As stated earlier, the Jacobian condition number has been a popular measure of dexterity in many works for either of these types of rotational or translational mechanisms (Gosselin & Angeles, 1989; Tsai & Joshi, 2000; Badescu & Mavroidis, 2004). For manipulators outside of this category, the condition number of conventional Jacobian matrices developed by methods such as screw theory or by partial derivatives, is not suitable for dexterity measurement due to their inherent mixture of units between the different columns of **J** (Tsai, 1999; Angeles, 2003; Doty et al., 1995). This leaves no method for the general algebraic formulation of dimensionally homogeneous Jacobian matrices. Therefore, no method is left for reliably measuring or quantifying the dexterity of a vast majority of mechanisms introduced in the literature that have mobility in both translational and rotational DOF, *i.e.*, complex DOF mechanisms (*e.g.*, Stewart, 1965; Lee & Shah, 1988; Siciliano, 1999; Carretero et al., 2000).

Gosselin (1992) introduced a method for formulating a dimensionally-homogeneous Jacobian matrix for both planar and some spatial mechanisms. Planar mechanisms have two translational and one rotational DOF. For the planar case, this Jacobian matrix relates the actuator velocities to the x and y components of the velocities of two points on the end effector platform. Kim and Ryu (2003) furthered this work by developing a general method using the x, y and z velocity components of three points (as opposed to two in (Gosselin, 1992)) on the end effector platform (A_1, A_2 and A_3) to formulate a Jacobian matrix which maps m actuator velocities (where m denotes the number of actuators) to the nine Cartesian velocity components of the three points A_i (*i.e.*, three for each point A_i). Assuming all actuators are of the same type, this $m\times9$ Jacobian is dimensionally-homogeneous, regardless of the conventionally defined independent end effector variables (*i.e.*, translational and/or angular velocities). However, of the total nine x, y and z velocity components (three for each point), at most only n are independent for a mechanism whose task space is n-DOF, where $n \leq 6$. This suggests that $(9 - n)$ terms of the end effector velocity vector may be defined as dependent variables. As this velocity vector and therefore the associated Jacobian includes dependent motions, it is not evident what physical significance the singular values of such a Jacobian matrix might have (Kim & Ryu, 2003). Therefore, using the ratio of maximum and minimum singular values (*i.e.*, the condition number) of the Jacobian matrix seems ill-advised.

In (Pond & Carretero, 2006), the authors present a methodology for obtaining a constrained and dimensionally homogeneous Jacobian based on an extension of the work in (Kim & Ryu, 2003). The singular values of such Jacobians may be used in dexterity analyses as their physical interpretation is typically clear. In the following section, the development of this type of Jacobian matrix is presented for the 3-RRR planar parallel manipulator.

2. The 3-<u>R</u>RR planar parallel manipulator

The symmetrical 3-<u>R</u>RR manipulator depicted in Figure 2 has been the subject of many studies. For example, inverse kinematics including velocity and acceleration, as well as singularity analysis, are provided by (Gosselin, 1988). It is a relatively simple, planar parallel manipulator, as described in the following section.

2.1 Mechanism architecture

As seen in Figure 2, the symmetrical 3-<u>R</u>RR manipulator consists of three identical limbs. Each limb is connected to the base at point G_i by an actuated revolute joint. This is followed by a *proximal* link of length $|\mathbf{b}_i|$ which connects to the *distal* link of length $|\mathbf{c}_i|$ through a passive revolute joint at B_i. Finally, a second passive revolute joint connects each limb to the end effector platform at point A_i. For the symmetric case, points G_i and A_i may each be used to form the corners of equilateral triangles.

For the planar 3-<u>R</u>RR manipulator, all joint axes are parallel and normal to the xy-plane. It can be easily demonstrated using the Grübler-Kutzbach mobility criterion that the mobility of the 3-<u>R</u>RR equals 3.

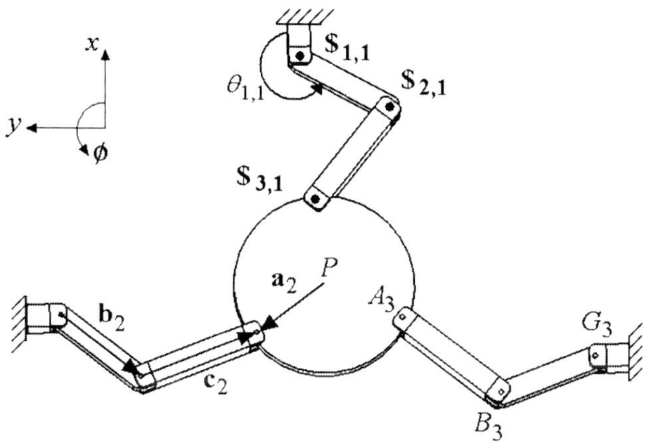

Figure 2: Basic architecture of the 3-RRR parallel manipulator.

The degrees of freedom at the end effector are translations in the x and y directions and a rotation ϕ, around an axis normal to the xy-plane. Note that the base frame's origin is placed coincident with the centre of a circle intersecting each of the three points G_i located at the base of each branch. The x-axis of the base frame is oriented such that point G_1 lies on that axis.

As the inverse displacement solution of this manipulator are previously published, no further discussion on the subject will be provided here. The Jacobian formulation provided for this manipulator in (Gosselin, 1988) and (Arsenault & Boudreau, 2004) is developed by differentiating the various inverse displacement equations, with respect to time. In (Tsai, 1999), the Jacobian matrix was obtained through the method of cross-products. In what follows, the conventional inverse and direct Jacobian matrices will instead, be obtained through screw theory.

2.2 Jacobian analysis using screw theory

The Jacobian developed here will relate the Cartesian velocities of the end effector in $\dot{x}$, $\dot{y}$ and $\dot{\phi}$ (or ω_z in conventional screw coordinate notation) to the actuator velocities.

Three screws $\$_{1,i}$, $\$_{2,i}$ and $\$_{3,i}$, with directions normal to the xy-plane, represent the three joints of each limb i for i = 1, 2, 3 (depicted in Figure 2 for i = 1):

$$\$_{1,i} = \begin{bmatrix} \mathbf{s}_{1,i} \\ (\mathbf{a}_i - \mathbf{c}_i - \mathbf{b}_i) \times \mathbf{s}_{1,i} \end{bmatrix} \tag{3}$$

$$\$_{2,i} = \begin{bmatrix} \mathbf{s}_{2,i} \\ (\mathbf{a}_i - \mathbf{c}_i) \times \mathbf{s}_{2,i} \end{bmatrix} \tag{4}$$

$$\$_{3,i} = \begin{bmatrix} \mathbf{s}_{3,i} \\ \mathbf{a}_i \times \mathbf{s}_{3,i} \end{bmatrix} \tag{5}$$

Each screw is represented with respect to a frame whose origin is coincident with that of the moving frame, *i.e.*, at point P, but whose axes are parallel to those of the fixed frame. The direction of all screws ($\mathbf{s}_{j,i}$) is the same for all of them as all are aligned with the z-axis. Therefore, the screw corresponding to the platform's motion is:

$$\$_p = \dot{\theta}_{1,i}\$_{1,i} + \dot{\theta}_{2,i}\$_{2,i} + \dot{\theta}_{3,i}\$_{3,i} \tag{6}$$

where angle $\theta_{j,i}$ corresponds to the rotation around the j-th revolute joint (j = 1, 2, 3) of the i-th limb (i = 1, 2, 3).

A screw must now be identified that is reciprocal to all screws representing the passive joints of limb i, *i.e.*, the revolute joints at points A_i and B_i. Such a screw may be zero pitch and oriented anywhere on the plane containing vectors $\mathbf{c}_i$ and $\mathbf{s}_{2,i}$ (or $\mathbf{s}_{3,i}$ corresponding to screws $\$_{2,i}$ and $\$_{3,i}$ in Figure 2). Such a reciprocal screw is:

$$\$_{r,i} = \begin{bmatrix} \mathbf{s}_{r,i} \\ \mathbf{a}_i \times \mathbf{s}_{r,i} \end{bmatrix} = \begin{bmatrix} \hat{\mathbf{c}}_i \\ \mathbf{a}_i \times \hat{\mathbf{c}}_i \end{bmatrix} = \frac{1}{|\mathbf{c}_i|}\begin{bmatrix} \mathbf{c}_i \\ \mathbf{a}_i \times \mathbf{c}_i \end{bmatrix} \tag{7}$$

$$= \frac{1}{|\mathbf{c}_i|}\begin{bmatrix} c_{i_x} & c_{i_y} & 0 & 0 & 0 & a_{i_x}c_{i_y} - a_{i_y}c_{i_x} \end{bmatrix}^T \tag{8}$$

where $\hat{\mathbf{c}}_i$ is a unit vector in the direction of $\mathbf{c}_i$. Taking the orthogonal product (here denoted by $\otimes$) of $\$_{r,i}$ with both sides of equation (6), yields:

$$\$_{r,i} \otimes \$_p = \$_{r,i} \otimes \$_{1,i}\dot{\theta}_{1,i} \tag{9}$$

where $\$_p = [\omega_x\ \omega_y\ \omega_z\ \dot{x}\ \ \dot{y}\ \ \dot{z}\ ]^T$. Since an orthogonal product involving screw $\$_{r,i}$ is on both sides of equation (9), the coefficient $1/|\mathbf{c}_i|$ shown in equation (8) may be dropped. To simplify notation, recognising that $\omega_x = \omega_y = \dot{z} = 0$ (since motion only occurs on xy-plane), $\$_{r,i}$ and $\$_p$ may be reduced to three dimensional vectors, *i.e.*, $\$_{r,i} = [\ c_{ix}\ c_{iy}\ (a_{ix}c_{iy} - a_{iy}\ c_{ix})\ ]^T$ and $\$_p = [\omega_z\ \dot{x}\ \ \dot{y}\ ]^T$.

Examining the right side of equation (9), and reducing $\$_{1,i}$ in equation (3), the orthogonal product $\$_{r,i} \otimes \$_{1,i}$ may be expressed as:

$$\$_{r,i} \otimes \$_{1,i} = \begin{bmatrix} (a_{i_x}c_{i_y} - a_{i_y}c_{i_x}) & c_{i_x} & c_{i_y} \end{bmatrix} \begin{bmatrix} 1 \\ (a_i - c_i - b_i)_y \\ -(a_i - c_i - b_i)_x \end{bmatrix} = b_{i_x}c_{i_y} - b_{i_y}c_{i_x} \tag{10}$$

Therefore, writing equation (9) three times corresponding to each of the mechanism's limbs yields the following direct ($\mathbf{J}_x$) and inverse ($\mathbf{J}_q$) Jacobians expressed as:

$$\mathbf{J}_x = \begin{bmatrix} a_{1_x}c_{1_y} - a_{1_y}c_{1_x} & c_{i_x} & c_{i_y} \\ a_{2_x}c_{2_y} - a_{2_y}c_{2_x} & c_{i_x} & c_{i_y} \\ a_{3_x}c_{3_y} - a_{3_y}c_{3_x} & c_{i_x} & c_{i_y} \end{bmatrix}_{3\times3} \tag{11}$$

$$\mathbf{J}_q = \begin{bmatrix} b_{1_x}c_{1_y} - b_{1_y}c_{1_x} & 0 & 0 \\ 0 & b_{2_x}c_{2_y} - b_{2_y}c_{2_x} & 0 \\ 0 & 0 & b_{3_x}c_{3_y} - b_{3_y}c_{3_x} \end{bmatrix}_{3\times3} \tag{12}$$

The results of $\mathbf{J}_x$ and $\mathbf{J}_q$ correspond exactly with those obtained by (Tsai, 1999) through the cross product method and by (Arsenault & Boudreau, 2004) through calculus. The resulting overall Jacobian matrix $\mathbf{J} = \mathbf{J}_q^{-1}\mathbf{J}_x$ is a square 3 × 3 matrix. The relation between end effector and actuator velocities is $\dot{\mathbf{q}} = \mathbf{J}\dot{\mathbf{x}}$ where $\dot{\mathbf{q}} = [\dot{\theta}_{1,1}\ \ \dot{\theta}_{1,2}\ \ \dot{\theta}_{1,2}]^{\mathrm{T}}$ and $\dot{\mathbf{x}} = [\omega_z\ \dot{x}\ \dot{y}]^{\mathrm{T}}$.
In the following section, the Jacobian matrix **J** will be used as a verification tool to evaluate whether the Jacobian matrices formulated the more novel introduced in (Pond & Carretero, 2006) methods are correct.

2.3 Constrained dimensionally-homogeneous Jacobian matrix formulation

As mentioned, the Jacobian matrix **J** developed in the previous section is dimensionally inconsistent. In (Tsai, 1999) and (Angeles, 2003), the authors have outlined the importance in having a dimensionally-homogeneous Jacobian matrix in dexterity analyses.
In (Kim & Ryu, 2003), the following velocity relation was developed:

$$\dot{\mathbf{q}} = (\mathbf{J}'_q)^{-1}\mathbf{J}'_x\dot{\mathbf{x}}' \tag{13}$$

Where, letting $\vec{\mathbf{k}} = [0\ \ 0\ \ 1]^T$, $\dot{\mathbf{q}} = [\dot{\theta}_{1,1}\ \ \dot{\theta}_{1,2}\ \ \dot{\theta}_{1,2}]^T$ and $\dot{\mathbf{x}}' = [\dot{A}_{1x}\ \dot{A}_{1y}\ \dot{A}_{2x}\ \dot{A}_{2y}\ \dot{A}_{3x}\ \dot{A}_{3y}]^T$:

$$\mathbf{J}'_q = \begin{bmatrix} \mathbf{c}_1^T\left(\vec{\mathbf{k}}\times\mathbf{b}_1\right) & 0 & 0 \\ 0 & \mathbf{c}_2^T\left(\vec{\mathbf{k}}\times\mathbf{b}_2\right) & 0 \\ 0 & 0 & \mathbf{c}_3^T\left(\vec{\mathbf{k}}\times\mathbf{b}_3\right) \end{bmatrix}_{3\times3} \tag{14}$$

$$\mathbf{J}'_x = \begin{bmatrix} k_{1,1}\mathbf{c}_i^T & k_{1,2}\mathbf{c}_i^T & k_{1,3}\mathbf{c}_i^T \\ k_{2,1}\mathbf{c}_i^T & k_{2,2}\mathbf{c}_i^T & k_{2,3}\mathbf{c}_i^T \\ k_{3,1}\mathbf{c}_i^T & k_{3,2}\mathbf{c}_i^T & k_{3,3}\mathbf{c}_i^T \end{bmatrix}_{3\times6} \tag{15}$$

Parameters $k_{i,j}$ (for i = 1, 2, 3 and j = 1, 2, 3) are dimensionless parameters defining the parametric equation of a plane containing the three points on the end effector platform and constrained by $k_{i,1} + k_{i,2} + k_{i,3} = 1$. It can be shown (Pond, 2006) that when using the Jacobian formulation as presented in this section, $k_{i,j} = 1$ when $i = j$ and $k_{i,j} = 0$ otherwise.

The multiplication of $(\mathbf{J}'_q)^{-1}\mathbf{J}'_x$ using the dimensionally homogeneous Jacobian matrices above produces the overall Jacobian matrix $\mathbf{J}'$ which is equivalent to:

$$\mathbf{J}' = \begin{bmatrix} \frac{\partial\theta_{1,1}}{\partial A1_x} & \frac{\partial\theta_{1,1}}{\partial A1_y} & \frac{\partial\theta_{1,1}}{\partial A2_x} & \frac{\partial\theta_{1,1}}{\partial A2_y} & \frac{\partial\theta_{1,1}}{\partial A3_x} & \frac{\partial\theta_{1,1}}{\partial A3_y} \\ \frac{\partial\theta_{1,2}}{\partial A1_x} & \frac{\partial\theta_{1,2}}{\partial A1_y} & \frac{\partial\theta_{1,2}}{\partial A2_x} & \frac{\partial\theta_{1,2}}{\partial A2_y} & \frac{\partial\theta_{1,2}}{\partial A3_x} & \frac{\partial\theta_{1,2}}{\partial A3_y} \\ \frac{\partial\theta_{1,3}}{\partial A1_x} & \frac{\partial\theta_{1,3}}{\partial A1_y} & \frac{\partial\theta_{1,3}}{\partial A2_x} & \frac{\partial\theta_{1,3}}{\partial A2_y} & \frac{\partial\theta_{1,3}}{\partial A3_x} & \frac{\partial\theta_{1,3}}{\partial A3_y} \end{bmatrix} \tag{16}$$

It is important to only map a set of independent end effector velocities to the actuator velocities. The mapping being done in equation (13) maps six end effector velocities of which only three are independent (for the 3-DOF mechanism) to the three actuator velocities. Similar to what is presented in (Pond & Carretero, 2006), a constraining matrix mapping the independent end effector velocities to the full set of both independent and dependent end effector velocities may be obtained.

If a constraining matrix $\mathbf{P}$ that maps the Cartesian velocities $\dot{A}_{1x}, \dot{A}_{2x}, \dot{A}_{3y}$, to all velocities in $\dot{\mathbf{x}}'$ was obtainable, it could be expressed in terms of partial derivatives, as follows:

$$\mathbf{P} = \begin{bmatrix} 1 & \frac{\partial A1_y}{\partial A1_x} & 0 & \frac{\partial A2_y}{\partial A1_x} & \frac{\partial A3_x}{\partial A1_x} & 0 \\ 0 & \frac{\partial A1_y}{\partial A2_x} & 1 & \frac{\partial A2_y}{\partial A2_x} & \frac{\partial A3_x}{\partial A2_x} & 0 \\ 0 & \frac{\partial A1_y}{\partial A3_y} & 0 & \frac{\partial A2_y}{\partial A3_y} & \frac{\partial A3_x}{\partial A3_y} & 1 \end{bmatrix}^T \tag{17}$$

The resulting multiplication of $\mathbf{J}'$ in equation (16) with the constraining matrix $\mathbf{P}$ in equation (17) yields:

$$\mathbf{J}'\mathbf{P} = \begin{bmatrix} \frac{\partial\theta_{1,1}}{\partial A1_x} & \frac{\partial\theta_{1,1}}{\partial A2_x} & \frac{\partial\theta_{1,1}}{\partial A3_y} \\ \frac{\partial\theta_{1,2}}{\partial A1_x} & \frac{\partial\theta_{1,2}}{\partial A2_x} & \frac{\partial\theta_{1,2}}{\partial A3_y} \\ \frac{\partial\theta_{1,3}}{\partial A1_x} & \frac{\partial\theta_{1,3}}{\partial A2_x} & \frac{\partial\theta_{1,3}}{\partial A3_y} \end{bmatrix} \tag{18}$$

This matrix $\mathbf{J}'\mathbf{P}$ is square and dimensionally homogeneous. The singular values of this matrix have a clear physical interpretation and therefore may be used in the dexterity analysis of the corresponding mechanism.

2.3.1 Identification of independent parameters

To obtain equation (18), the set $\dot{A}_{1x}, \dot{A}_{2x}, \dot{A}_{3y}$ was chosen as the set of independent Cartesian components. Clearly, six unique sets of independent parameters may be used to define the end effector velocity $\dot{\mathbf{x}}''$. That is, any subset consisting of three elements from the six elements of $\dot{\mathbf{x}}'$ which includes at least one x component and at least one y component may be used. These subsets are:

Case I:	$[\dot{A}_{1_x}\ \dot{A}_{2_x}\ \dot{A}_{3_y}]^T$	Case IV:	$[\dot{A}_{1_y}\ \dot{A}_{2_y}\ \dot{A}_{3_x}]^T$
Case II:	$[\dot{A}_{1_x}\ \dot{A}_{2_y}\ \dot{A}_{3_x}]^T$	Case V:	$[\dot{A}_{1_y}\ \dot{A}_{2_x}\ \dot{A}_{3_y}]^T$
Case III:	$[\dot{A}_{1_y}\ \dot{A}_{2_x}\ \dot{A}_{3_x}]^T$	Case VI:	$[\dot{A}_{1_x}\ \dot{A}_{2_y}\ \dot{A}_{3_y}]^T$

In the following formulation of the constraint equations and alternative inverse displacement solution, the independent end effector parameters will be arbitrarily chosen as Case I (*i.e.*, $\dot{A}_{1x}, \dot{A}_{2x}, \dot{A}_{3y}$). The solutions using any of the potential six cases listed above have a similar form.

2.3.2 Constraint equations

It can in fact be shown that a relationship between $[\dot{A}_{1x}, \dot{A}_{2x}, \dot{A}_{3y}]^T$ to $\dot{\mathbf{x}}'$, *i.e.*, the matrix $\mathbf{P}$ in equation (18), can be obtained. Consider Figure 3 representing the end effector platform. The point D lies on the bisection of the line segment A_1A_2 so:

$$D_x = A_{1_x} + (A_{2_x} - A_{1_x})/2 \tag{19}$$

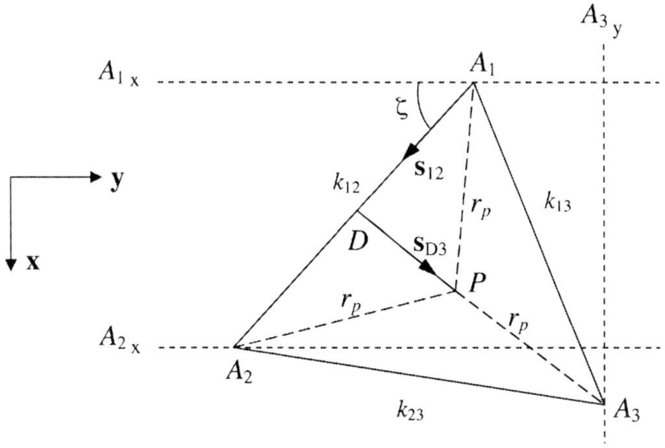

Figure 3: End effector notation for the planar 3-RRR parallel manipulator.

The angle ζ made between line segment A_1A_2 with the negative y-axis is:

$$\zeta = \sin^{-1}\left(\frac{|A_{2_x} - A_{1_x}|}{k_{12}}\right) \tag{20}$$

where k_{12} is the length of the line segment between points A_1 and A_2.

Consider the case where the variables A_{1x}, A_{2x} and A_{3y} are known. Therefore, the vertices of the triangle representing the end effector platform lie somewhere on the three dashed lines shown in Figure 3. When these three dashed lines are used to constrain the vertices of the end effector platform, there are two possible solutions for the unit vector $\mathbf{s}_{12}$:

$$\mathbf{s}_{12} = \begin{bmatrix} s_\zeta & \pm c_\zeta & 0 \end{bmatrix}^T \tag{21}$$

The vector $\mathbf{s}_{D3}$ may be obtained by cross multiplying the vector $\mathbf{s}_{12}$ with $\pm\ \vec{\mathbf{k}}$ (recalling that $\vec{\mathbf{k}} = [0\ \ 0\ \ 1]^T$):

$$\mathbf{s}_{D3} = \mathbf{s}_{12} \times \pm\vec{\mathbf{k}} = \begin{bmatrix} \pm c_\zeta & \pm s_\zeta & 0 \end{bmatrix}^T \tag{22}$$

As a result, there are four possible solutions for vector s_{D3} each corresponding to one of the four unique solutions in Figure 4.

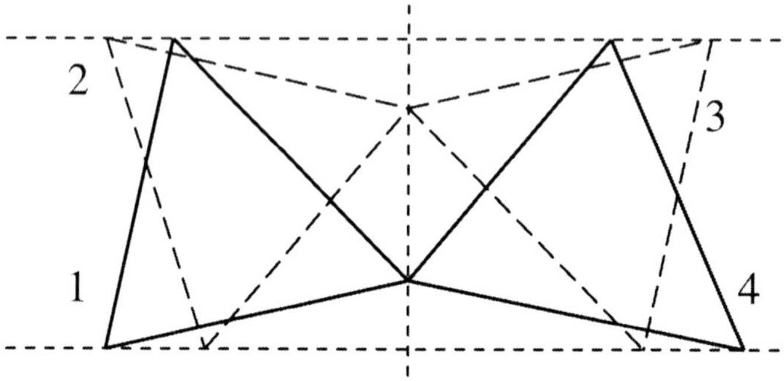

Figure 4: Four possible solutions where a single Cartesian coordinate of each of three points on the end effector platform are known.

Letting e represent the magnitude of the line segment $\overline{DP}$:

$$e = \sqrt{\left(r_p^2 - \frac{k_{12}^2}{4}\right)} \tag{23}$$

where k_{12} can be obtained from the platform radius r_p and the angle between lines $\overline{PA_1}$ and $\overline{PA_2}$. Letting vector $\mathbf{D}$ represent a vector from the origin of the base frame to point D (see Figure 3), a solution for the vector $\mathbf{A}_3$ locating point A_3 with respect to the origin is:

$$\mathbf{A}_3 = \mathbf{D} + (e + r_p) \begin{bmatrix} \pm c_\zeta & \pm s_\zeta & 0 \end{bmatrix}^T \tag{24}$$

From which the first component is

$$A_{3_x} = D_x \pm (e + r_p)\, c_\zeta \tag{25}$$

The same method may then be reversed to find $D_y = A_{3y} \pm (e + r_p)\, s_\zeta$.
Similarly, solutions are found for A_{1y} and A_{2y} as:

$$A_{1_y} = D_y + \left(\frac{k_{12}}{2}\right)(\pm c_\zeta) \tag{26}$$

$$A_{2_y} = D_y - \left(\frac{k_{12}}{2}\right)(\pm c_\zeta) \tag{27}$$

To obtain a single solution for the direction of vector s_{D3} instead of the four possible solutions in equation (22), the true position and orientation of the platform in conventional variables, *i.e.*, x, y and ϕ, are required. Since in workspace volume determination or path planning, these are in fact known, the following decision rules may be used to obtain a unique solution in the coordinates A_{1x}, A_{1y}, A_{2x}, A_{2y}, A_{3x} and A_{3y} . If $x > D_x$, then all terms associated with $\pm c_\zeta$ are in fact $+c_\zeta$ and vice versa. Similarly, if $A_{3y} > y$, then all terms associated with $\pm s_\zeta$ are in fact $+ s_\zeta$ and vice versa.

2.3.3 Alternative inverse displacement solution

In the preceding section, the remaining three Cartesian coordinates of the three points A_i were determined based on one of the Cartesian coordinates being given for each point. This provides full knowledge as to the position of the end effector platform and points A_i. The solution for each limb's pose may be obtained by completing the inverse displacement solution provided in (Tsai, 1999) or (Arsenault & Boudreau, 2004) where points A_i are known. The solution leads to two solutions for each limb. In (Arsenault & Boudreau, 2004), these are referred to as *working modes*. The different solutions correspond to either *elbow up* or *elbow down* configurations of each limb. As there are two solutions for each limb, and three limbs, there are therefore a total of $2^3 = 8$ possible solutions to the inverse displacement problem.

2.3.4 Constraining Jacobian

The first derivative with respect to time of equations (25) through (27) yields the various elements of the matrix $\mathbf{P}$ in equation (17). As previously mentioned, six unique sets of independent end effector variables may be used to obtain the square dimensionally-homogeneous Jacobian matrix.

2.4 Singularity analysis

Singularity analysis of the 3-RRR manipulator has been explored extensively in (Tsai, 1999; Bonev & Gosselin, 2001; Arsenault & Boudreau, 2004). Essentially two singularities exist for this manipulator. An inverse singular configuration occurs whenever one of the three limbs is fully stretched out, or when the distal link overlaps the proximal link of any limb. At such configurations, instantaneous rotations of the actuated revolute joint do not alter the end effector pose.

A direct singular configuration exists whenever the lines collinear with the distal links have a common intersection for all three limbs. In Figure 2, the direction of these lines is represented for limb 2 by vector $\mathbf{c}_2$. At these singular configurations, an instantaneous rotation around the point of intersection of the above mentioned lines, may be obtained without any displacement of the actuators.

Singular configurations are also mathematically introduced by the constraining matrix **P** which do not correspond to physical singular configurations of the manipulator. First, recall the equilateral triangle $A_1A_2A_3$ used to model the end effector (Figure 5). The mechanism's degrees of freedom include a translational ability in x and y and a rotational ability in the plane, *i.e.*, angle ϕ. These three points were used in the formulation of the 3×6 dimensionally homogeneous Jacobian matrix **J**'.

For each of the six sets of potential independent end effector variables for the planar mechanisms described in Section 2.3.1, the poses listed in Table 1 are observed to yield a rank deficient constrained and dimensionally homogeneous Jacobian matrix **J**'**P**.

It is also observed that these singular configurations occur at all x and y positions tested. For the first three cases, where two of the three x-coordinates are considered independent, these singular configurations are introduced when the line made between the two points, whose x-coordinates are independent, is parallel with the x-axis. Similarly, for the last three cases where two of the three y coordinates are considered independent, these singular configurations occur when the line made between the two points, whose y-coordinates are independent, is parallel with the y-axis.

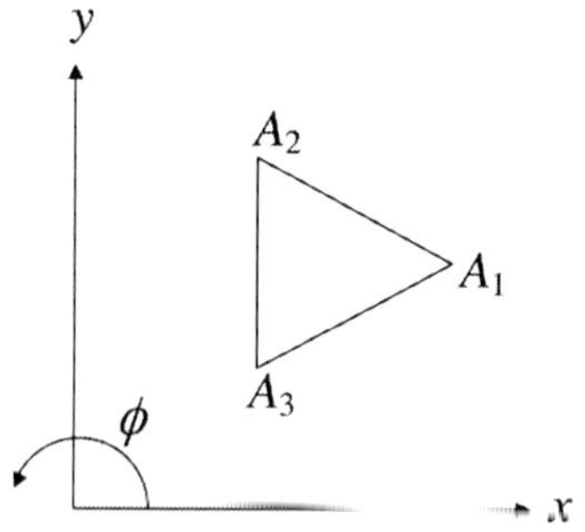

Figure 5: The end effector of a planar mechanism modelled as a triangle. End effector is at a mathematically-introduced singularity if independent variables in Case VI are chosen.

Case No.	Independent Parameters	Mathematically Singular Poses
I	$[\dot{A}_{1_x}\ \dot{A}_{2_x}\ \dot{A}_{3_y}]^T$	$\phi = 30°,\ 210°, -150°, -330°$
II	$[\dot{A}_{1_x}\ \dot{A}_{2_y}\ \dot{A}_{3_x}]^T$	$\phi = -30°, -210°, 150°, 330°$
III	$[\dot{A}_{1_y}\ \dot{A}_{2_x}\ \dot{A}_{3_x}]^T$	$\phi = 90°,\ 270°, -90°, -270°$
IV	$[\dot{A}_{1_y}\ \dot{A}_{2_y}\ \dot{A}_{3_x}]^T$	$\phi = -60°, -240°,\ 120°,\ 300°$
V	$[\dot{A}_{1_y}\ \dot{A}_{2_x}\ \dot{A}_{3_y}]^T$	$\phi = 60°,\ 240°, -120°, -300°$
VI	$[\dot{A}_{1_x}\ \dot{A}_{2_y}\ \dot{A}_{3_y}]^T$	$\phi = 0°, 180°, -180°$

Table 1: Observed mathematically-introduced singularities for the 3-RRR planar parallel manipulator.

The source of this issue is a function of the constraints being imposed by the manipulator's limbs.

The points A_1, A_2 and A_3 are constrained to lie in the xy-plane. Recall that the constraining matrix **P** is formulated based on the implicit constraints imposed on the end effector by the manipulator's limbs, but not explicitly on the architecture itself.
The following is a purely mathematical examination of the terms within the constraining matrix **P** which create the rank deficiencies not inherent to the mechanism.
Consider Case VI as listed in Section 2.3.1, where the independent parameters are identified as A_{1x}, A_{2y} and A_{3y} . The following is a symbolic representation of the resulting constraint matrix:

$$\mathbf{P}_6 = \begin{bmatrix} 1 & 0 & 0 \\ \frac{\partial A_{1_y}}{\partial A_{1_x}} & \frac{\partial A_{1_y}}{\partial A_{2_y}} & \frac{\partial A_{1_y}}{\partial A_{3_y}} \\ \frac{\partial A_{2_x}}{\partial A_{1_x}} & \frac{\partial A_{2_x}}{\partial A_{2_y}} & \frac{\partial A_{2_x}}{\partial A_{3_y}} \\ 0 & 1 & 0 \\ \frac{\partial A_{3_x}}{\partial A_{1_x}} & \frac{\partial A_{3_x}}{\partial A_{2_y}} & \frac{\partial A_{3_x}}{\partial A_{3_y}} \\ 0 & 0 & 1 \end{bmatrix} \tag{28}$$

Given the independent parameters associated with Case VI, the equivalent angle ζ of Figure 3 is defined as:

$$\zeta = \sin^{-1}\left(\frac{|A_{3_y} - A_{2_y}|}{k_{23}}\right) \tag{29}$$

where k_{23} is the length of the line segment between points A_2 and A_3. The angle ζ is defined differently depending on the identified independent parameters. The partial derivative $\partial\zeta$ taken with respect to the various independent parameters, appears in the formulation of many of the entries of equation (28). As a result, when the line between points A_2 and A_3 is parallel with the y-axis (as depicted in Figure 5), the magnitude of the projection of line segment (A_2A_3) onto the y-axis will instantaneously undergo no change for any change in angle ζ. Therefore, the partial derivative $\partial\zeta / \partial|A_{3y} - A_{2y}|$ is equal to infinity. For instance, for a pose where $\phi = 0°$, the constraining matrix **P** may be expressed numerically as:

$$\mathbf{P}_6 = \begin{bmatrix} 1 & 0 & 0 \\ 0 & \infty & -\infty \\ 1 & -\infty & \infty \\ 0 & 1 & 0 \\ 1 & \frac{0}{0} & \frac{0}{0} \\ 0 & 0 & 1 \end{bmatrix} \tag{30}$$

As discussed, Jacobian matrices obtained for the other five cases at the same pose, are not rank deficient and therefore may still be used to obtain a measurement of dexterity.

3. Dexterity measurement

One of the objectives of performing dexterity analyses on parallel manipulators is to obtain an understanding of how sensitive the end effector pose is relative to the actuator displacement. As discussed, for some cases, this has historically been achieved through observation of the Jacobian matrix condition number.

The condition number of the screw based Jacobian matrix **J** and dimensionally homogeneous Jacobian matrix **J**' throughout a chosen path are depicted in Figure 6. Clearly, the planned trajectory either passes through or very near a singular configuration as evidenced by the rapidly increasing condition number of the screw-based Jacobian matrix at approximately t = 0.9 sec. In fact, it can be shown that for the defined path, the manipulator passes through a direct singular configuration where the three vectors $\mathbf{c}_i$ depicted in Figure 2 intersect at a single point.

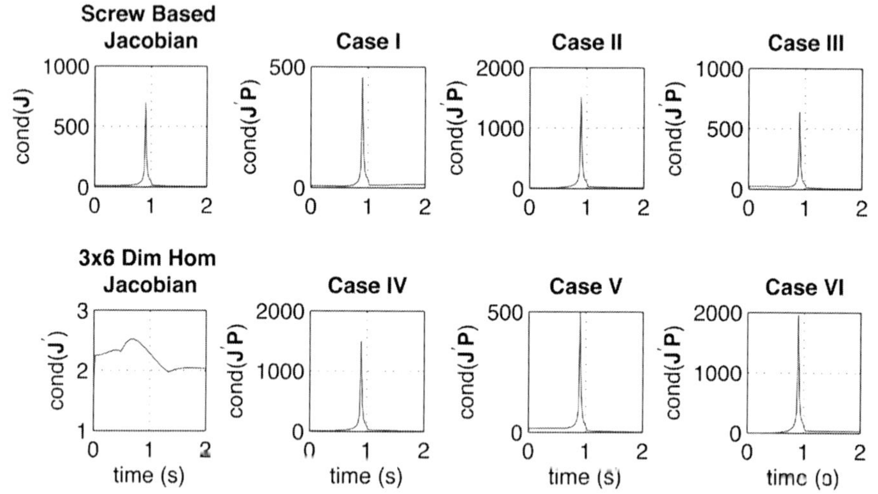

Figure 6: The condition number of each of the formulated Jacobian matrices throughout the planned trajectory.

However, **J**', the 3 × 6 dimensionally homogeneous Jacobian matrix developed by (Kim & Ryu, 2003), does not suggest the same. Instead, its condition number gives the impression that the manipulator is relatively near isotropic condition throughout the defined path. Obviously then, the 3 × 6 dimensionally homogenous Jacobian matrix is not suitable as a dexterity measure. Because three of the six columns of **J**' are dependent on the other three columns, the eigenvalues of **J**' could correspond to velocity directions in the task space which are not obtainable. Therefore, the eigenvalues and singular values of that matrix are essentially meaningless.

Figure 6 also depicts the results obtained by observing the condition number of each of the six constrained dimensionally homogeneous Jacobian matrices. Each of the constrained Jacobian matrices clearly agree that the arbitrarily chosen trajectory has the manipulator passing near a singular configuration. The six matrices **J'P** are constrained based on the manipulator's motion capabilities and therefore accurately predict singular configurations, as shown. Furthermore, their terms are dimensionally homogeneous. Therefore, their condition numbers allow a suitable means of measuring dexterity.

3.1 Reachable workspace

The reachable workspace of the 3-RRR planar parallel manipulator is depicted in Figure 7a). For the workspace plots presented in this section, the values of the architectural parameters are $r_b = 1$, $r_p = 0.4$, $b = 0.5$ and $c = 0.4$. Here, architectural parameter values are arbitrarily chosen such that results obtained in workspace analysis are comparable, in this case, with the serial RRR planar manipulator to be studied later in this chapter.

3.2 Dexterous workspace

In Section 2.3.1, six potential sets of independent end effector velocities were identified to lead to the formulation of six unique constrained and dimensionally-homogeneous Jacobian matrices. Using only one of these matrices as a dexterity measure could lead to potential bias.

To cope with having six constrained and dimensionally-homogeneous Jacobian matrices from which to measure dexterity, and the issues which arise by introducing the artificial singularity conditions discussed in Section 2.4, the minimum condition number of all six Jacobian matrices is proposed as a dexterity measure. This measure is essentially the minimum ratio between the largest actuator effort required to move in a direction in one of the six defined task-space variable sets, with the effort required to move in the easiest direction using the same task space variables. This avoids the issue of introduced singularities by the constraining matrix as the lowest condition number of the six matrices will only be high when the manipulator is near a true singular configuration.

It is also suggested that measures using the singular values also be included. By doing so, both the velocity or accuracy characteristics of the manipulator are obtained, in addition to an indication of how 'near-isotropic' the architecture is at the studied pose. In this section, the singular values of all six Jacobian matrices (provided the corresponding constraining matrix has not introduced a singularity), must lie within imposed limits.

3.2.1 Dexterity defined by the Jacobian matrix condition number

Figure 7b) depicts the dexterous workspace of the 3-RRR manipulator when the condition number of $\mathbf{J}'\mathbf{P}_i$ (where the sub-index i refers to Case i for $i = 1 \ldots 6$), is arbitrarily limited to a maximum of 60.

It can be shown that the region of the workspace removed from that of the manipulator's reachable workspace corresponds to the vicinity of a singular configuration where the three vectors $\mathbf{c}_i$ intersect at a common point, as discussed in Section 2.4.

Figure 8 depicts the cross section of both the reachable workspace in Figure 7a) and dexterous workspace in Figure 7b) at $\phi = 0$. At this value of ϕ, the reachable workspace border at $y = 0$ and $x \approx 0.42$ corresponds to a configuration where both limbs two and three are in the fully stretched position. However, this region of the workspace also corresponds to an architectural pose near the direct singular configuration where the three vectors $\mathbf{c}_i$ intersect. Therefore, in the vicinity of the reachable workspace border at $y = 0$, the manipulator is near both inverse and direct singular configurations. It should be expected that this region of the workspace has poor dexterity which is confirmed by Figure 8.

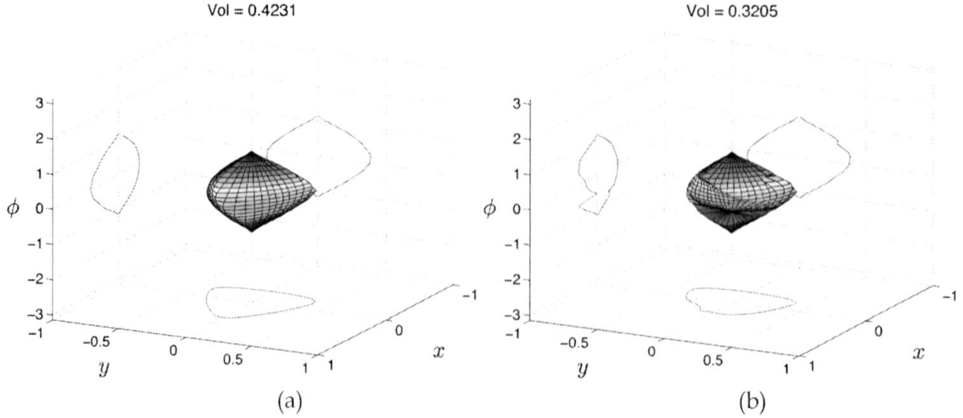

Figure 7: a) Reachable and b) dexterous workspace of the 3-RRR parallel manipulator when defined using a maximum allowable Jacobian matrix condition number of 60. Angle ϕ is expressed in radians.

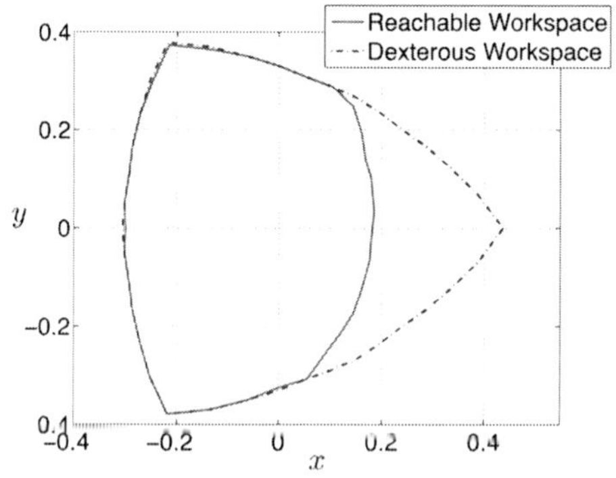

Figure 8: Cross section of both reachable and dexterous workspaces (when defined by a limit on the Jacobian matrix condition number of 60) of the 3-RRR parallel planar manipulator at $\phi = 0$.

3.2.2 Dexterity defined by the Jacobian matrix condition number and minimum singular value

The singular values within the workspace depicted in Figure 7b) vary within the range $0.0056366 \leq \sigma \leq 5.7377$. The dexterous workspace for this manipulator when also restricted to a minimum limit on the singular value of $\sigma \geq 0.1$ for any of the six Jacobian matrices is depicted in Figure 9a). An exception is made for the singular values of any of the six Jacobian matrices should that matrix falsely represent a singular configuration.

The workspace in Figure 7b) has only marginally decreased in volume when compared to the dexterous workspace obtained when limiting only the Jacobian matrix condition number. The necking of the workspace at $\phi \approx -0.65$ occurs because at this pose, the manipulator is near a singular configuration where the three vectors $\mathbf{c}_i$ intersect at a common point.

3.2.3 Dexterity defined by the Jacobian matrix condition number and maximum singular value

Similarly, a limit of $\sigma \leq 2.0$ is imposed on the six Jacobian matrices, with the exception noted earlier, to obtain the dexterous workspace for the 3-RRR manipulator depicted. The resulting workspace obtained using this upper limit is shown in Figure 9b).

Although nearly 10% greater in volume than the dexterous workspace depicted in Figure 9a), both depictions clearly indicate the same singular configuration as discussed earlier when the distal and proximal links of one of the three kinematic branches overlap.

4. The serial RRR planar manipulator

The serial RRR planar manipulator is one of the most trivial of all manipulators. For that reason, it is frequently used as a demonstration example in many texts in robot kinematics, *e.g.*, (Tsai, 1999; Craig, 2003). Through these texts, the majority of necessary work for workspace determination has been presented. Therefore only a brief summary of the required details will be presented here.

4.1 Mechanism architecture

The RRR serial planar architecture is depicted in Figure 10. It consists of three links and three actuated revolute joints. The first actuated revolute joint connects the first limb represented by vector b to the base and may rotate b around point O by angle θ_1. The second actuated revolute joint at B connects the first link to the second, represented by vector **c**. This second joint rotates **c** with respect to **b** by angle θ_2. Finally, the third actuated revolute joint at C may rotate the end effector (vector **d**) by angle θ_3 with respect to **c**. Here, the end effector is represented as triangle $A_1A_2A_3$. Similar to the 3-RRR planar parallel architecture, the serial RRR planar architecture is confined to two translational DOF and one rotational DOF, all in the xy-plane.

4.2 Kinematics

As depicted in Figure 10, there are also two solutions to the inverse displacement problem for this manipulator. These correspond to an *elbow up* and *elbow down* configuration of the manipulator. The inverse displacement solution is provided in (Tsai, 1999; Craig, 2003).

Instead of using an alternative form of the inverse displacement solution to aid in the formulation of a dimensionally-homogeneous Jacobian matrix, it is greatly simplified in the case of serial manipulators, if the forward displacement solution is used instead. First, consider Figure 11, depicting the notation used to relate the three points on the end effector. The lengths of sides a_2 and a_3 may be found by using the cosine law:

$$\begin{aligned} a_2^2 &= r_p^2 + r_p^2 - 2r_p r_p \cos(\pi - \alpha) \\ &= 2r_p^2 - 2r_p^2 \cos(\pi - \alpha) \\ &= 2r_p^2 (1 - \cos(\pi - \alpha)) \end{aligned} \tag{31}$$

$$a_3^2 = 2r_p^2 (1 - \cos(\pi - \beta)) \tag{32}$$

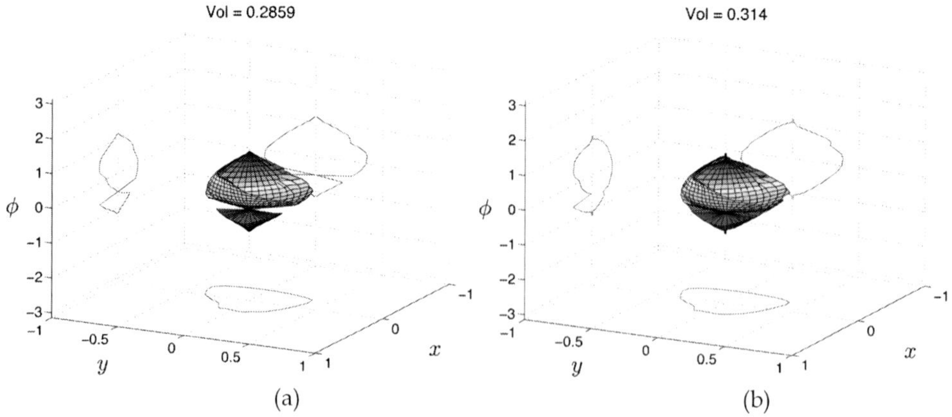

Figure 9: Dexterous workspace of the 3-RRR parallel manipulator when defined using a maximum allowable Jacobian matrix condition number and a) a minimum singular value of 0.1 or b) maximum singular value of 2. Angle ϕ is expressed in radians.

If the joint displacements were known, points B, C, A_1, A_2 and A_3 could be determined as:

$$B = \begin{bmatrix} b\cos\theta_1 \\ b\sin\theta_1 \end{bmatrix} \tag{33}$$

$$C = \begin{bmatrix} b\cos\theta_1 + c\cos(\theta_1+\theta_2) \\ b\sin\theta_1 + c\sin(\theta_1+\theta_2) \end{bmatrix} \tag{34}$$

$$A_1 = \begin{bmatrix} b\cos\theta_1 + c\cos(\theta_1+\theta_2) + d\cos(\theta_1+\theta_2+\theta_3) \\ b\sin\theta_1 + c\sin(\theta_1+\theta_2) + d\sin(\theta_1+\theta_2+\theta_3) \end{bmatrix} \tag{35}$$

$$A_2 = \begin{bmatrix} A_{1_x} + a_2\cos(\theta_1+\theta_2+\theta_3+\frac{\alpha}{2}) \\ A_{1_y} + a_2\sin(\theta_1+\theta_2+\theta_3+\frac{\alpha}{2}) \end{bmatrix} \tag{36}$$

$$A_3 = \begin{bmatrix} A_{1_x} + a_3\cos(\theta_1+\theta_2+\theta_3-\frac{\beta}{2}) \\ A_{1_y} + a_3\sin(\theta_1+\theta_2+\theta_3-\frac{\beta}{2}) \end{bmatrix} \tag{37}$$

where $\theta_1 + \theta_2 + \theta_3 = \phi$. The first derivative of these equations may be used to formulate the various elements of a dimensionally-homogeneous Jacobian matrix.

As discussed in (Tsai, 1999), this manipulator is in a singular configuration whenever the manipulator is either fully extended, *i.e.*, whenever $\theta_2 = \theta_3 = 0°$, or when the second link overlaps the first, *i.e.*, whenever $\theta_2 = 0°$ or $\theta_2 = 180°$.

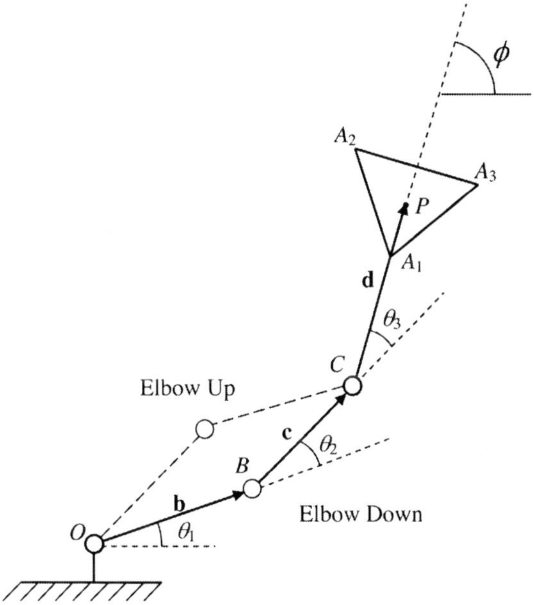

Figure 10: Architecture of the serial RRR planar manipulator.

4.3 Reachable workspace

For the serial manipulator used in the following numerical examples, architectural parameters are arbitrarily chosen to be $b = c = d = 1$. The end effector is represented as an equilateral triangle with vertices A_i. The length of each of the three line segments $\overline{A_iP}$ is equal to 1. Theoretically, infinite rotation of the end effector is obtainable in the plane; however, in order to obtain a result which may be compared to the parallel case (where for the architectural variables used, only a finite rotation was achievable), workspace envelopes obtained in the following sections will be limited to a minimum and maximum rotation of $-\pi \leq \phi \leq \pi$). The reachable workspace for this manipulator, when using the aforementioned limits, is depicted in Figure 12a). The x and y translations refer to the displacement of point P on the end effector platform depicted in Figure 10.
It is immediately clear the tremendous advantage the serial manipulator has over its parallel counterpart in terms of reachable workspace volume.

4.4 Dexterous workspace

As previously discussed, special consideration must be given to the six potential constrained and dimensionally-homogeneous Jacobian matrices that may be used to measure dexterity and the potential singularities introduced by the constraining matrix $\mathbf{P}_i$ (for the parallel case). For the serial case, the six possible Jacobian matrices are denoted by $\mathbf{J}_i$ corresponding to case i as noted in Section 2.3.1.
Similar to the parallel case, never will more than one of the six Jacobian matrices falsely represent a singular configuration at the same pose. However, it can be demonstrated that the condition number of all six matrices simultaneously and rapidly increase in the vicinity

of true singular configurations. Therefore, using the minimum condition number of the six Jacobian matrices remains a plausible index for dexterity.

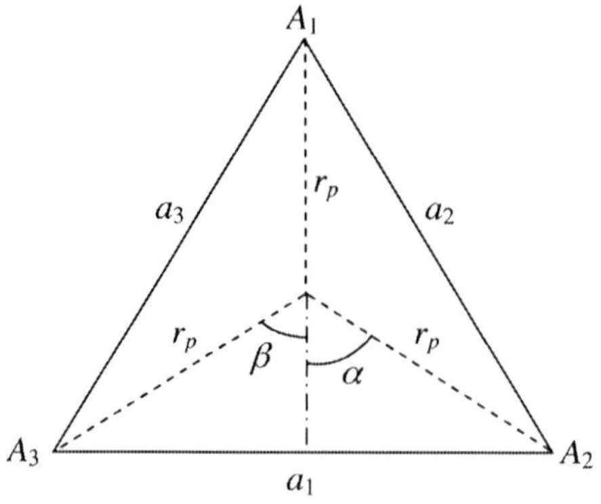

Figure 11: End effector notation for the RRR serial manipulator.

4.4.1 Dexterity measured by the Jacobian matrix condition number

Figure 12b) depicts the RRR serial manipulator's dexterous workspace when restricted to a maximum limit of 60 on the minimum condition number of any of the six Jacobian matrices (with the exception noted earlier for Jacobian matrices which falsely represent singular configurations). The portion of the workspace removed from that of the reachable workspace in Figure 12a) corresponds to the singular configuration where **b** and **c** in Figure 10 are collinear. Therefore, using a limit on the minimum Jacobian matrix condition number remains a potential index for dexterity as it is expected that the manipulator should have poor dexterity in this region. Figure 13 is a cross sectional view of the dexterous workspace depicted in Figure 12b) at $\phi = 0$. For the architectural variables used, at $\phi = 0$, the serial RRR manipulator is in an *interior* singular configuration (Tsai, 1999) at $x = 1$ and $y = 0$. At this pose, vectors **b** and **c** overlap. This is depicted in Figure 14.

4.4.2 Dexterity measured by the Jacobian matrix condition number and maximum singular value

It is important to note that the Jacobian matrix developed for the serial RRR manipulator maps $\dot{\mathbf{q}}$ to $\dot{\mathbf{x}}$ instead of $\dot{\mathbf{x}}$ to $\dot{\mathbf{q}}$ as for the 3-RRR parallel manipulator. Therefore, if a meaningful comparison is to be made, limits on the singular values of $\mathbf{J}^{-1}$ should be imposed, rather than $\mathbf{J}$ for the serial manipulator. This is of no consequence in the comparison of the two manipulators when the condition number limit is imposed as the condition number of $\mathbf{J}^{-1}$ is equal to the condition number of $\mathbf{J}$.

The singular values of $\mathbf{J}^{-1}$ within the workspace depicted in Figure 12b) vary within the range $0.4309 \leq \sigma \leq \infty$. It can be shown that when the singular values $\mathbf{J}^{-1}$ are limited to $\sigma \leq 2.0$, to provide comparison to the corresponding result for the 3-RRR planar parallel manipulator, no workspace volume is obtained. Instead, for illustration purposes, Figure 15a) depicts the workspace volume where singular values are limited to $\sigma \leq 50$. Even at the relatively large allowed value for the singular values, the workspace is significantly reduced from that of Figure 12b) and is highly segmented.

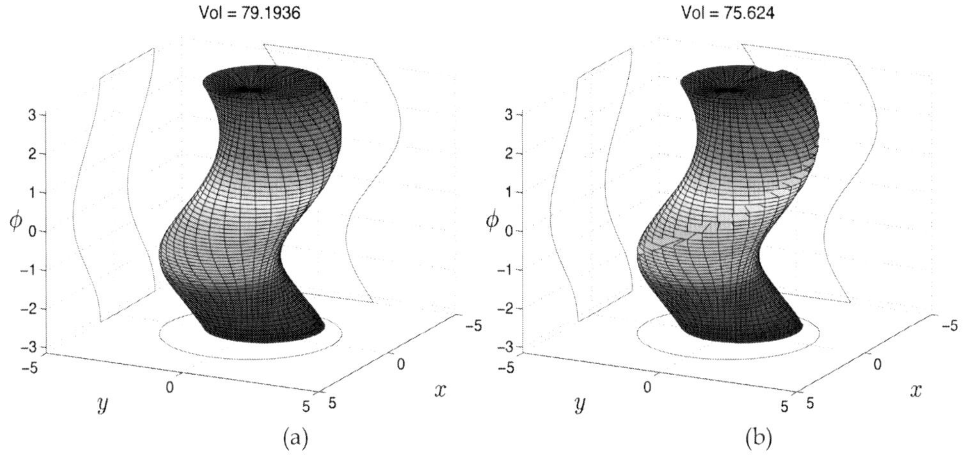

Figure 12: a) Reachable and b) dexterous workspace of the planar <u>RRR</u> serial manipulator when defined using a maximum allowable Jacobian matrix condition number. Angle ϕ is expressed in radians.

4.4.3 Dexterity measured by the Jacobian matrix condition number and minimum singular value

Similarly, a limit may be imposed on the minimum allowable singular value of any of the six Jacobian matrices with the exception noted earlier. When the singular values are limited to $\sigma \geq 0.1$, the dexterous workspace depicted in Figure 15b) is obtained.

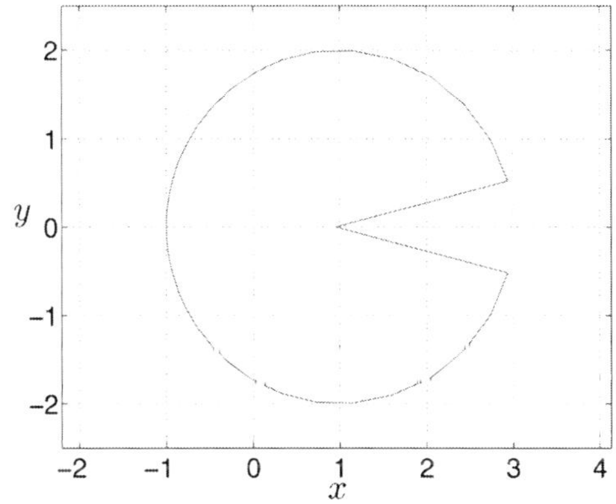

Figure 13: Cross section of the dexterous workspace when defined by a limit on the Jacobian matrix condition number of the serial <u>RRR</u> planar manipulator at $\phi = 0$.

Recall that the workspace corresponding to the parallel manipulator in Figure 9b) had only slightly decreased in volume when compared to that of Figure 7b). However, the workspace of the serial manipulator has not decreased at all.

Again, it should be emphasised that if architectural parameters were optimised to obtain the largest workspace volume possible, different results would be obtained.

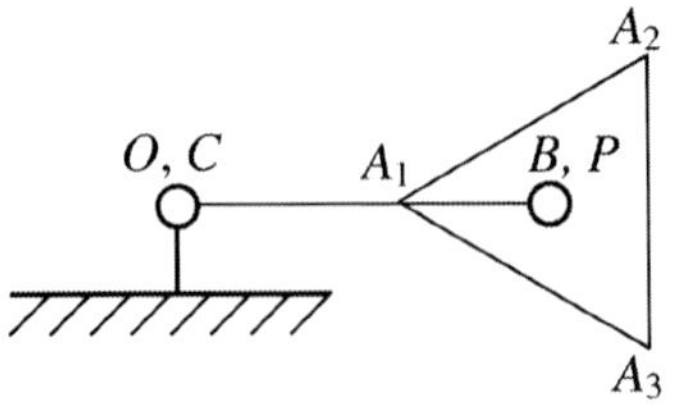

Figure 14: Singular configuration of the serial RRR manipulator.

Recall that the workspace corresponding to the parallel manipulator in Figure 9b) had only slightly decreased in volume when compared to that of Figure 7b). However, the workspace of the serial manipulator has not decreased at all.
Again, it should be emphasised that if architectural parameters were optimised to obtain the largest workspace volume possible, different results would be obtained.

5. Dexterous workspace comparison of parallel and serial planar manipulators

In (Pond, 2006; Pond & Carretero, 2007), different parallel manipulators were quantitatively compared in terms of dexterity using the formulation describer earlier for the dimensionally homogeneous constrained Jacobian matrix. This section will study the effect of the arbitrarily chosen limits on the condition number and singular values on the results obtained for comparison between the serial and parallel manipulators discussed in this chapter. This is the first time such quantitative study has been made for such dissimilar architectures.
For each of the following three subsections, a set of curves will be provided depicting the difference in workspace volume between the serial and parallel manipulators as the limits used to obtain them are varied. In order to better illustrate the changes, the plots are presented on suitable scales.

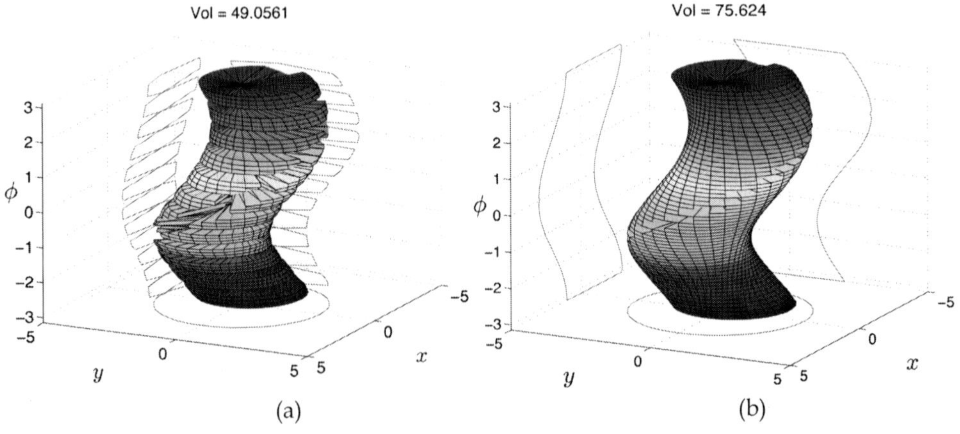

Figure 15: Dexterous workspace of the planar RRR serial manipulator when defined using a maximum allowable Jacobian matrix condition number and a) maximum singular value of 50 or b) minimum allowable singular value of 0.1. Angle ϕ is expressed in radians.

5.1 Dexterity measured by Jacobian matrix condition number

Figure 16a) depicts the dexterous workspace size as a function of the limiting value as the maximum allowable Jacobian matrix condition number. This set of curves emphasises the difference in size between the workspace of the two manipulators at limits of high condition numbers. Note that the y-axis of the graph is on a log scale.

5.2 Dexterity measured by Jacobian matrix condition number and minimum singular value

As noted earlier, the range of singular values within the serial manipulator's workspace is fairly large ($0.4309 \leq \sigma \leq \infty$). However, as Figure 16b) suggests, singular values are far denser in the lower end of this range.

In the previous section, when the singular values were limited to a minimum of $\sigma \geq 0.1$, the serial manipulator had not decreased in volume yet that of the parallel manipulator had. It is important to recall that when the limit is imposed on the lowest allowable singular value, an emphasis is being placed on obtaining high degrees of accuracy and stiffness. Figure 16b) clearly shows, however, that the volume of the serial manipulator's workspace rapidly decreases through the approximate range $0.25 \leq \sigma_{min} \leq 0.4$. Above this range, the parallel manipulator provides the largest workspace volume.

Therefore, these results suggest that, of the two manipulators, for the architectural variables used, the parallel manipulator outperforms the serial manipulator within the range of approximately $\sigma_{min} \geq 0.4$. Naturally, this conclusion can only be made for the specific architectural variables used in this study.

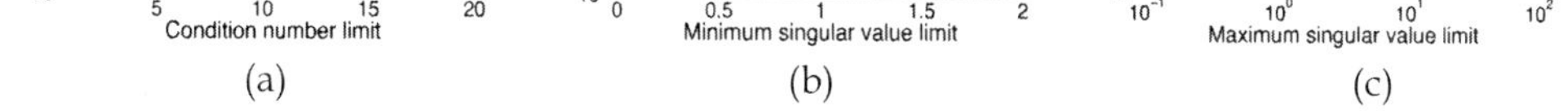

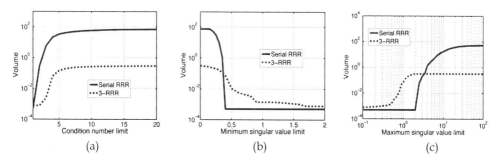

Figure 16: Dexterous workspace comparison based on a) a limit on the condition number, b) a limitation on the minimum allowable singular value, and c) a limitation on the maximum allowable singular value.

5.3 Dexterity measured by Jacobian matrix condition number and maximum singular value

Figure 16c) compares the dexterous workspace volumes of both the serial and parallel planar manipulators when limited by the condition number and a maximum singular value. Recall that the range of singular values within the serial manipulator's workspace is much larger than the corresponding range for the parallel manipulator. The workspace volume of the serial manipulator only begins to significantly increase in volume at a relatively higher limit of approximately $\sigma_{max} \geq 2$. Conversely, at this limit, the workspace corresponding to the parallel manipulator has obtained its full volume as depicted in Figure 16c).

This is an interesting result as lower singular values correspond to higher end effector velocities. This suggests that the parallel architecture studied also provides the largest workspace volume when high end effector velocities are required, to a limit of approximately $\sigma_{max} \geq 4$ where the serial manipulator then provides the largest workspace volume.

6. Conclusions

Through either method of obtaining a constrained dimensionally homogeneous Jacobian matrix (proposed by (Gosselin, 1992) or by (Pond & Carretero, 2006)) for planar mechanisms, a choice exists on which of the potential six Cartesian velocity components on the end effector be used to define the task space velocity variables. The choice has an influence on the resulting Jacobian matrix and therefore its condition number and singular values. Without constraining the Jacobian matrix, the condition number was demonstrated to be essentially meaningless, as in (Kim & Ryu, 2003).

In terms of measuring dexterity, the constrained dimensionally homogeneous Jacobian matrices ($\mathbf{J}'\mathbf{P}$) are superior to the screw based Jacobian matrix ($\mathbf{J}$) in that they are dimensionally consistent. Furthermore, the six matrices ($\mathbf{J}'\mathbf{P}$) are superior to the 3 × 6 dimensionally homogeneous matrix ($\mathbf{J}'$) in that they are constrained, and therefore provide true dexterous information.

The condition number and singular values of each of the six matrices ($\mathbf{J}'\mathbf{P}$) are different for any given pose. Therefore, dexterity measures involving only one of the six ($\mathbf{J}'\mathbf{P}$) matrices are potentially bias. Four potential strategies for dexterity measurement have been proposed based on the condition number and/or singular values of the Jacobian matrices obtained in all six cases. Each measure has a distinct physical meaning, as discussed.

In sum, the Jacobian matrix formulation presented in this chapter allows, for the first time, to *quantitatively* compare different mechanism architectures with complex degrees of freedom in terms of dexterity. Moreover, as illustrated in this chapter, the formulation is not limited to parallel manipulators as it can also be used to quantitatively compare the dexterity of different architectures as long as the end effector is represented by an equivalent set of points. Quantitative dexterity comparisons will allow robot designers to better select proper mechanisms for specific tasks.

7. References

J. Angeles (2003). *Fundamentals of Robotic Mechanical Systems*, Springer-Verlag, ISBN: 978-0-387-95368-7, New York.

M. Arsenault & R. Boudreau (2004). The synthesis of three-degree-of-freedom planar parallel manipulators with revolute joints (3-RRR) for an optimal singularity-free workspace. *Journal of Robotic Systems*, Vol.21, No.5, page numbers (259 274).

M. Badescu & C. Mavroidis (2004). Workspace optimization of 3-legged UPU and UPS parallel platforms with joint constraints. *Journal of Mechanical Design*, Vol.126, No.2, page numbers (291-300), ISSN: 1050-0472.

I. A. Bonev & C. M. Gosselin (2001). Singularity loci of planar manipulators, Proceedings of the 2nd workshop on computational kinematics, Seoul, South Korea, 2001.

J. A. Carretero; R. P. Podhorodeski; M. A. Nahon & C. M. Gosselin (2000). Kinematic analysis and optimization of a new three degree-of-freedom spatial parallel manipulator. *Journal of Mechanical Design*, Vol.122, No.1, page numbers (17-24), ISSN: 1050-0472.

J. J. Craig (2003). *Introduction to Robotics: Mechanics and Control*, Prentice Hall, ISBN: 978- 0-201-54361-2 , New York.

K. L. Doty; C. Melchiorri; E. M. Schwartz & C. Bonivento (1995). Robot manipulability. *IEEE Transactions on Robotics and Automation*, Vol.11, No.3, page numbers (462-268), ISSN: 1042-296X.

C. M. Gosselin (1988). *Kinematic analysis, optimization and programming of parallel robotic manipulators*, Ph.D thesis, Department of Mechanical Engineering, McGill University, ISBN: 0315485043, Montreal, Canada.

C. M. Gosselin & J. Angeles (1989). The optimum kinematic design of a spherical three-degreeof- freedom parallel manipulator. *Journal of Mechanisms, Transmissions, and Automation in Design*, Vol.19, No.4, page numbers (202-207), ISSN: 0738-0666.

C. M. Gosselin (1992). The optimum design of robotic manipulators using dexterity indices. *Journal of Robotics and Autonomous Systems*, Vol.9, No.4, page numbers (213-226).

S. G. Kim & J. Ryu (2003). New dimensionally homogeneous Jacobian matrix formulation by three end-effector points for optimal design of parallel manipulators. *IEEE Transactions on Robotics and Automation*, Vol.19, No.4, page numbers (731-737), ISSN: 1042-296X.

K. Lee & D. Shah (1988). Kinematic analysis of a three-degree-of-freedom in parallel actuated manipulator. *IEEE Journal of Robotics and Automation*, Vol.4, No.3, page numbers (354- 360), ISSN: 0882-4967.

G. Pond (2006). *Dexterity and workspace characteristics of complex degree of freedom parallel manipulators*, Ph.D thesis, Department of Mechanical Engineering, University of New Brunswick, Fredericton, Canada.

G. Pond & J. A. Carretero (2006). Formulating Jacobian matrices for the dexterity analysis of parallel manipulators. *Mechanism and Machine Theory*, Vol.41, No.12, page numbers (1505-1519), ISSN: 0094-114X

G. Pond & J. A. Carretero (2007). Quantitative dexterous workspace comparisons of parallel manipulators. *Mechanism and Machine Theory*, Vol.42, No.10, page numbers (1388-1400), ISSN: 0094-114X.

B. Siciliano (1999). The Tricept robot: inverse kinematics, manipulability analysis and closedloop direct kinematics algorithm. *Robotica*, Vol.17, No.4, page numbers (437-445), ISSN: 0263-5747.

D. Stewart (1965). A platform with six degrees of freedom, Proceedings of the Institute of Mechanical Engineering, pp. 371-386, London, UK, 1965.

G. Strang. (2003). *Introduction to Linear Algebra*, Wellesley Cambridge Pr, ISBN: 978-0-961-40889-3, Wellesley MA.

L.-W. Tsai. (1999). *Robot Analysis: The Mechanics of Serial and Parallel Manipulators*, John Wiley and Sons Inc., ISBN: 978-0-471-32593-2, New York.

L.-W. Tsai & S. Joshi (2000). Kinematics and optimization of a spatial 3-UPU parallel manipulator. *Journal of Mechanical Design*, Vol.122, No.4, page numbers (439-446), ISSN: 1050-0472.

11

Calibration of 3-d.o.f. Translational Parallel Manipulators Using Leg Observations

Anatol Pashkevich[1,2], Damien Chablat[1], Philippe Wenger[1], and Roman Gomolitsky[2]
[1]*Institut de Recherche en Communications et Cybernétique de Nantes*
[2]*Belarusian State University of Informatics and Radioelectronics*
[1]*France*
[2]*Belarus*

1. Introduction

Parallel kinematic machines (PKM) are commonly claimed as appealing solutions in many industrial applications due to their inherent structural rigidity, good payload-to-weight ratio, high dynamic capacities and high accuracy (Tlusty et al., 1999; Tsai, 1999; Merlet, 2000; Wenger et al., 2001). However, while PKM usually exhibit a much better repeatability compared to serial mechanisms, they may not necessarily possess a better accuracy that is limited by manufacturing/assembling errors in numerous links and passive joints (Wang & Masory, 1993). Thus, the PKM accuracy highly relies on an accurate kinematic model, which must be carefully tuned (calibrated) for each manipulator individually.

Similar to serial manipulators, PKM calibration techniques are based on the minimization of a parameter-dependent error function, which incorporates residuals of the kinematic equations (Schröer et al., 1995; Wampler et al., 1995; Fassi et al., 2007; Legnani et al., 2007). For parallel manipulators, the inverse kinematic equations are considered computationally more efficient, contrary to the direct kinematics, which is usually analytically unsolvable for PKM. But the main difficulty with this technique is the full-pose measurement requirement, which is very hard to implement (Innocenti, 1995; Iurascu & Park, 2003; Daney, 2003; Jeong et al., 2004; Huang et al., 2005). Hence, a number of studies have been directed at using the subset of the pose measurement data, which however creates another problem, the identifiability of the model parameters (Khalil & Besnard, 1999; Daney & Emiris, 2001; Besnard & Khalil, 2001; Rauf et al., 2004, 2006).

Popular approaches in parallel robot calibration deal with one-dimensional pose errors using a double-ball-bar system or other measuring devices as well as imposing mechanical constraints on some elements of the manipulator (Zhuang et al., 1999; Thomas et al., 2003; Daney, 1999). However, in spite of hypothetical simplicity, it is hard to implement in practice since an accurate extra mechanism is required to impose these constraints. Additionally, such methods reduce the workspace size and the identification efficiency.

Another category of calibration methods, the self- or autonomous calibration, is implemented by minimizing the residuals between the computed and measured values of the active and/or redundant joint sensors (Hesselbach et al., 2005). Adding extra sensors at

usually unmeasured joints is very attractive from a computational point of view, since it allows getting the data in the whole workspace and potentially reduces impact of the measurement noise. However, only a partial set of the parameters may be identified in this way, since the internal sensing is unable to provide sufficient information for the robot end-effector absolute location (Zhuang, 1997; Williams et al., 2006).
More recently, several hybrid calibration methods were proposed that utilize intrinsic properties of a particular parallel machine allowing extracting the full set of the model parameters (or the most essential of them) from a minimum set of measurements. It worth mentioning an innovative approach developed by Renaud et al. (2004 - 2006) who applied the vision-based measurement technique for the parallel manipulators calibration from the leg observations. In this approach, the source data are extracted from the leg images, without any strict assumptions on the end-effector poses. The only assumption is related to the manipulator architecture (the mechanism is actuated by linear drives located on the base). However, current accuracy of the camera-based measurements is not high enough yet to apply this method in industrial environment.
This chapter summarises the authors' results in the area of parallel robotics (Pashkevich et al., 2005, 2006) and focuses on the calibration of the Orthoglide-type mechanisms, which is also actuated by linear drives located on the manipulator base and admits technique of Renaud et al. (2004, 2005). But, in contrast to the known works, our approach assumes that the leg location is observed for specific manipulator postures, when the tool centre point (TCP) moves along the Cartesian axes. For these postures and for the nominal Orthoglide geometry, the legs are strictly parallel to the corresponding Cartesian planes. So, the deviations of the manipulator geometry influence on the leg parallelism that gives the source data for the parameter identification. The main advantage of this approach is the simplicity of the measuring system that can avoid using computer vision and is composed of standard comparator indicators, which are common in industry.

2. Orthoglide mechanism

2.1 Manipulator architecture

The Orthoglide is a three d.o.f. parallel manipulator actuated by linear drives with mutually orthogonal axes. Its kinematic architecture is presented in Fig. 1 and includes three identical parallel chains that will be further referred to as "legs". Each manipulator leg is formally described as PRPaR - chain, where P, R and Pa denote the prismatic, revolute, and parallelogram joints respectively. The output machinery (with a tool mounting flange) is connected to the legs in such a manner that the tool moves in the Cartesian space with fixed orientation (i.e. restricted to translational motions). The Orthoglide workspace has a regular, quasi-cubic shape. The input/output equations are simple and the velocity transmission factors are equal to one along the x, y and z direction at the isotropic configuration, like in a conventional serial PPP machine. The latter is an essential advantage for machining applications (Wenger & Chablat, 2000; Chablat & Wenger, 2003).
Another specific feature of the Orthoglide mechanism, which will be further used for calibration, is displayed during the end-effector motions along the Cartesian axes. For example, for the x-axis motion, the sides of the x-leg parallelogram must retain strictly parallel to the x-axis. Hence, the observed deviation may be a data source for calibration.

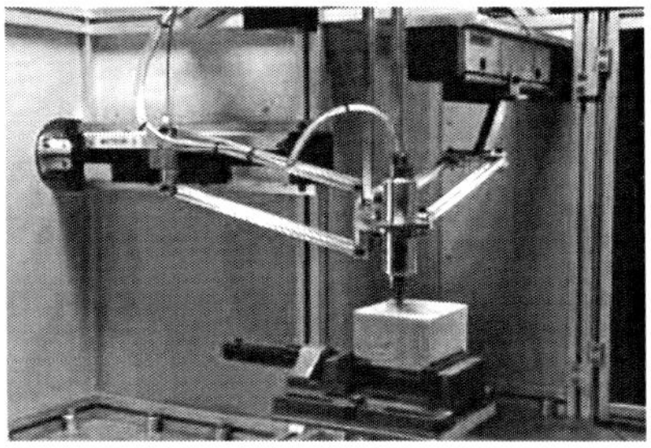

(a)

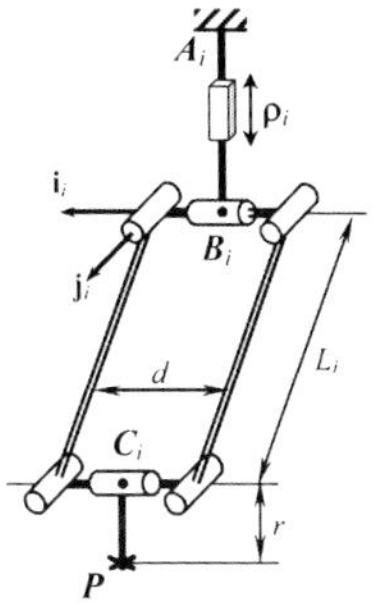

(b)

Fig. 1. The architecture of Orthoglide manipulator (a) and kinematics of its leg (b) (© CNRS Photothèque / CARLSON Leif)

For a small-scale Orthoglide prototype used for the calibration experiments, the workspace size is approximately equal to 200×200×200 mm³ with the velocity transmission factors bounded between 1/2 and 2 (Chablat & Wenger, 2003). The legs nominal geometry is defined by the following parameters: L = 310.25 mm, d = 80 mm, r = 31 mm where L, d are the parallelogram length and width, and r is the distance between the points C_i and the tool centre point P (see Fig. 1b).

2.2 Modelling assumptions

Following previous studies on the PKM accuracy (Wang & Massory, 1993; Renaud et al.; 2004, Caro et al., 2006), the influence of the joint defects is assumed negligible compared to the encoder offsets and the link length deviations. This validates the following modelling assumptions:

i. the manipulator parts are supposed to be rigid bodies connected by perfect joints;
ii. the manipulator legs (composed of a prismatic joint, a parallelogram, and two revolute joints) generate a four degrees-of-freedom motions;
iii. the articulated parallelograms are assumed to be perfect but non-identical;
iv. the linear actuator axes are mutually orthogonal and are intersected in a single point to ensure a translational movement of the end-effector;
v. the actuator encoders are perfect but located with some errors (offsets).

Using these assumptions, calibration equations will be derived based on the observation of the parallel motions of the manipulator legs.

2.3 Kinematic model

Since the kinematic parallelograms are admitted to be non-identical, the kinematic model developed in our previous works (Pashkevich et al., 2005, 2006) should be extended to describe the manipulator with different length leg parameters.

Under the adopted assumptions, similar to the equal-leg case, the articulated parallelograms may be replaced by the kinematically equivalent bar links. Besides, a simple transformation of the Cartesian coordinates (shifted by the vector $(r, r, r)^T$, see Fig. 1b) allows us to eliminate the tool offset. Hence, the Orthoglide geometry can be described by a simplified model, which consists of three rigid links connected by spherical joints to the tool centre point at one side and to the allied prismatic joints at another side (Fig. 2). Corresponding formal definition of each leg can be presented as PSS, where P and S denote the actuated prismatic joint and the passive spherical joint respectively.

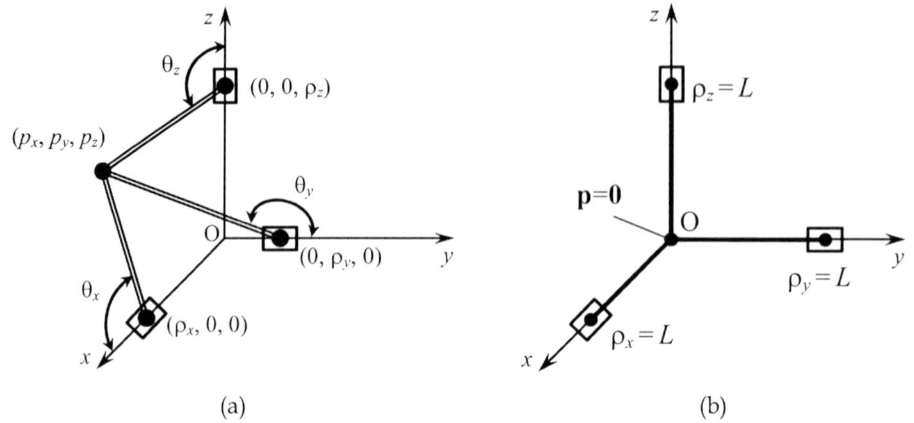

Fig. 2. Orthoglide simplified model (a) and its "zero" configuration (b)

Thus, if the origin of a reference frame is located at the intersection of the prismatic joint axes and the x, y, z-axes are directed along them (see Fig. 2), the manipulator kinematics may be described by the following equations

$$\mathbf{p} = \begin{bmatrix} (\rho_x + \Delta\rho_x) + \cos\theta_x \cos\beta_x L_x + r \\ \sin\theta_x \cos\beta_x L_x \\ -\sin\beta_x L_x \end{bmatrix} \tag{1a}$$

$$\mathbf{p} = \begin{bmatrix} -\sin\beta_y L_y \\ (\rho_y + \Delta\rho_y) + \cos\theta_y \cos\beta_y L_y + r \\ \sin\theta_y \cos\beta_y L_y \end{bmatrix} \tag{1b}$$

$$\mathbf{p} = \begin{bmatrix} \sin\theta_z \cos\beta_z L_z \\ -\sin\beta_z L_z \\ (\rho_z + \Delta\rho_z) + \cos\theta_z \cos\beta_z L_z + r \end{bmatrix} \tag{1c}$$

where $\mathbf{p} = (p_x, p_y, p_z)^T$ is the output vector of the TCP position, $\boldsymbol{\rho} = (\rho_x, \rho_y, \rho_z)^T$ is the input vector of the prismatic joints variables, $\Delta\boldsymbol{\rho} = (\Delta\rho_x, \Delta\rho_y, \Delta\rho_z)^T$ is the encoder offset vector, θ_i, β_i, $i \in \{x, y, z\}$ are the parallelogram orientation angles (internal variables), and L_i are the length of the corresponding leg.

After elimination of the internal variables θ_i , β_i , the kinematic model (1) can be reduced to three equations

$$\left(p_i - (\rho_i + \Delta\rho_i)\right)^2 + p_j^2 + p_k^2 = L_i^2 \tag{2}$$

which includes components of the input and output vectors **p** and **ρ** only. Here, the subscripts $i, j, k \in \{x, y, z\}$, $i \neq j \neq k$ are used in all combinations, and the joint variables ρ_i are obeyed the prescribed limits $\rho_{\min} < \rho_i < \rho_{\max}$ defined in the control software (for the Orthoglide prototype, $\rho_{\min}$ = -100 mm and $\rho_{\max}$ = +60 mm).
It should be noted that, for the case $\Delta\rho_x = \Delta\rho_y = \Delta\rho_z = 0$ and $L_x = L_y = L_z = L$, the nominal "mechanical-zero" posture of the manipulator corresponds to the Cartesian coordinates $\mathbf{p}_0 = (0, 0, 0)^T$ and to the joints variables $\boldsymbol{\rho}_0 = (L, L, L)$. Moreover, in this posture, the *x*-, y- and *z*-legs are oriented strictly parallel to the corresponding Cartesian axes. But the joint offsets and the leg length differences cause the deviation of the "zero" TCP location and corresponding deviation of the leg parallelism, which may be measured and used for the calibration. Hence, six parameters ($\Delta\rho_x, \Delta\rho_y, \Delta\rho_z, L_x, L_y, L_z$) define the manipulator geometry and are in the focus of the proposed calibration technique.

2.4 Inverse and direct kinematics

The inverse kinematic relations are derived from the equations (2) in a straightforward way and only slightly differ from the "nominal" case:

$$\rho_i = p_i + s_i\sqrt{L_i^2 - p_j^2 - p_k^2} - \Delta\rho_i \tag{3}$$

where $s_x, s_y, s_z \in \{\pm 1\}$ are the configuration indices defined for the "nominal" geometry as the signs of $\rho_x - p_x$, $\rho_y - p_y$, $\rho_z - p_z$, respectively. It is obvious that expressions (3) give eight different solutions, however the Orthoglide prototype assembling mode and the joint limits reduce this set to a single case corresponding to $s_x = s_y = s_z = 1$.
For the direct kinematics, equations (2) can be subtracted pair-to-pair that gives linear relations between the unknowns p_x, p_y, p_z, which may be expressed in the parametric form as

$$p_i = \frac{\rho_i + \Delta\rho_i}{2} + \frac{t}{\rho_i + \Delta\rho_i} - \frac{L_i^2}{2(\rho_i + \Delta\rho_i)}, \tag{4}$$

where *t* is an auxiliary scalar variable. This reduces the direct kinematics to the solution of a quadratic equation $At^2 + Bt + C = 0$ with the coefficients

$$A = \sum_{i \neq j} (\rho_i + \Delta\rho_i)^2 (\rho_j + \Delta\rho_j)^2; \qquad B = \prod_i (\rho_i + \Delta\rho_i)^2 - \sum_{i \neq j \neq k} L_i^2 (\rho_j + \Delta\rho_j)^2 (\rho_k + \Delta\rho_k)^2;$$

$$C = \prod_i (\rho_i + \Delta\rho_i)^2 \cdot \left(\sum_i (\rho_i + \Delta\rho_i)^2 / 4 - \sum_i L_i^2 / 2 \right) + \sum_{i \neq j \neq k} L_i^4 (\rho_j + \Delta\rho_j)^2 (\rho_k + \Delta\rho_k)^2 / 4$$

where $i, j, k \in \{x, y, z\}$. From two possible solutions that give the quadratic formula, the Orthoglide prototype (see Fig. 1) admits a single one $t = (-B + \sqrt{B^2 - 4AC}) / (2A)$ corresponding to the selected manipulator assembling mode.

2.4 Differential relations

To obtain the calibration equations, let us derive first the differential relations for the TCP deviation for three types of the Orthoglide postures:

i. "*maximum displacement*" postures for the directions x, y, z (Fig. 3a);
ii. "*mechanical zero*" or the isotropic posture (Fig. 3b);
iii. "*minimum displacement*" postures for the directions x, y, z (Fig. 3c);

These postures are of particular interest for the calibration since, in the "nominal" case, a corresponding leg is parallel to the relevant pair of the Cartesian planes.

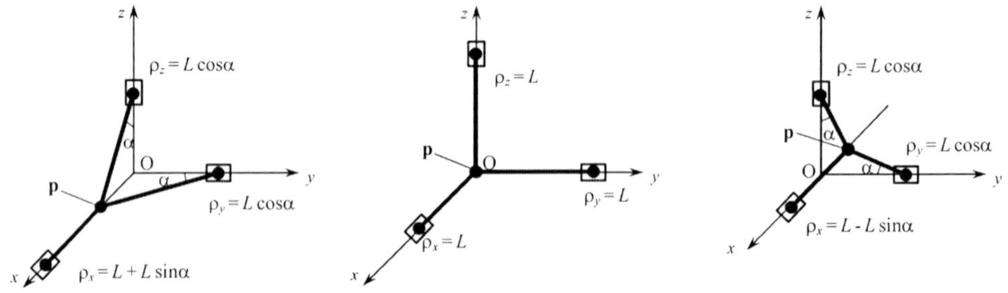

(a) : XMax posture (b) : Zero posture (c) : XMin posture

Fig. 3. Specific postures of the Orthoglide (for the x-leg motion along the Cartesian axis X)

The manipulator Jacobian with respect to the parameters $\Delta\boldsymbol{\rho} = (\Delta\rho_x, \Delta\rho_y, \Delta\rho_z)$ and $\mathbf{L} = (L_x, L_y, L_z)$ can be derived by straightforward differentiating of the kinematic equations (2), which yields

$$\begin{bmatrix} p_x-\rho_x & p_y & p_z \\ p_x & p_y-\rho_y & p_z \\ p_x & p_y & p_z-\rho_z \end{bmatrix} \cdot \frac{\partial \mathbf{p}}{\partial \boldsymbol{\rho}} = \begin{bmatrix} p_x-\rho_x & 0 & 0 \\ 0 & p_y-\rho_y & 0 \\ 0 & 0 & p_z-\rho_z \end{bmatrix}; \quad \begin{bmatrix} p_x-\rho_x & p_y & p_z \\ p_x & p_y-\rho_y & p_z \\ p_x & p_y & p_z-\rho_z \end{bmatrix} \cdot \frac{\partial \mathbf{p}}{\partial \mathbf{L}} = \begin{bmatrix} L_x & 0 & 0 \\ 0 & L_y & 0 \\ 0 & 0 & L_z \end{bmatrix}$$

Thus, after the matrix inversions and multiplications, the desired Jacobian can be written as

$$\mathbf{J}(\mathbf{p}, \boldsymbol{\rho}) = \left[\mathbf{J}_{\rho}(\mathbf{p}, \boldsymbol{\rho});\ \mathbf{J}_{L}(\mathbf{p}, \boldsymbol{\rho})\right] \tag{5}$$

where

$$\mathbf{J}_{\rho}(.) = \begin{bmatrix} 1 & \dfrac{p_y}{p_x-\rho_x} & \dfrac{p_z}{p_x-\rho_x} \\ \dfrac{p_x}{p_y-\rho_y} & 1 & \dfrac{p_z}{p_y-\rho_y} \\ \dfrac{p_x}{p_z-\rho_z} & \dfrac{p_y}{p_z-\rho_z} & 1 \end{bmatrix}^{-1} \qquad \mathbf{J}_{L}(.) = \begin{bmatrix} \dfrac{p_x-\rho_x}{L_x} & \dfrac{p_y}{L_x} & \dfrac{p_z}{L_x} \\ \dfrac{p_x}{L_y} & \dfrac{p_y-\rho_y}{L_y} & \dfrac{p_z}{L_y} \\ \dfrac{p_x}{L_z} & \dfrac{p_y}{L_z} & \dfrac{p_z-\rho_z}{L_z} \end{bmatrix}^{-1}$$

It should be noted that, for the sake of computing convenience, the above expression includes both the Cartesian coordinates p_x, p_y, p_z and the joint coordinates ρ_x, ρ_y, ρ_z, but only one of these sets may be treated as independent taking into account the kinematic equations.

For the "*Zero*" posture, the differential relations are derived in the neighbourhood of the point $\{p_0 = (0, 0, 0)\ ;\ \rho_0 = (L, L, L)\}$, which after substitution to (5) gives the Jacobian matrix

$$\mathbf{J}_0 = \left[\begin{array}{ccc|ccc} 1 & 0 & 0 & -1 & 0 & 0 \\ 0 & 1 & 0 & 0 & -1 & 0 \\ 0 & 0 & 1 & 0 & 0 & -1 \end{array}\right] \quad (6)$$

Hence, in this case, the TCP displacement is related to the joint offsets and the leg lengths variations ΔL_i by trivial equations

$$\Delta p_i = \Delta \rho_i - \Delta L_i;\ \ i \in \{x, y, z\} \quad (7)$$

For the "*XMax*" posture, the Jacobian is computed in the neighbourhood of the point $\{\mathbf{p} = (LS_\alpha, 0, 0)\ ;\ \boldsymbol{\rho} = (L{+}LS_\alpha, LC_\alpha, LC_\alpha)\ \}$, where α is the angle between the *y*-, *z*-legs and the X-axes: α = asin (ρ_{max}/L); $S_\alpha = \sin(\alpha)$, $C_\alpha = \cos(\alpha)$. This gives the Jacobian

$$\mathbf{J}_x^+ = \left[\begin{array}{ccc|ccc} 1 & 0 & 0 & -1 & 0 & 0 \\ T_\alpha & 1 & 0 & -T_\alpha & -C_\alpha^{-1} & 0 \\ T_\alpha & 0 & 1 & -T_\alpha & 0 & -C_\alpha^{-1} \end{array}\right] \quad (8)$$

where $T_\alpha = \tan(\alpha)$. Hence, the differential equations for the TCP displacement may be written as $\Delta p_x = \Delta \rho_x - \Delta L_x$

$$\begin{aligned} \Delta p_y &= T_\alpha \Delta \rho_x + \Delta \rho_y - T_\alpha \Delta L_x - C_\alpha^{-1} \Delta L_y \\ \Delta p_z &= T_\alpha \Delta \rho_x + \Delta \rho_z - T_\alpha \Delta L_x - C_\alpha^{-1} \Delta L_z \end{aligned} \quad (9)$$

It can be proved that similar results are valid for the *YMax* and *ZMax* postures (differing by the indices only), and also for the *XMin*, *YMin*, *ZMin* postures. In the latter case, the angle α should be computed as $\alpha = \text{asin}(\rho_{\text{min}}/L)$.

3. Calibration method

3.1 Measurement technique

To identify the Orthoglide kinematic parameters specified in the previous section, two approaches can be used, which employ different measurement techniques to evaluate the leg-to-surface parallelism. The first of them (Fig. 4a) assumes two measurements for the same leg posture (to assess distances from both leg ends to the base surface). The second technique (Fig. 4b) assumes a fixed location of the measuring device but two distinct leg postures, which are assumed to be parallel to each other in the nominal case.

It is obvious that, for the perfectly calibrated manipulator, both methods give zero differences for each measurement pair. In contrasts, the non-zero differences contain source information for the parameter identification. However, the first method involves absolute measurements that require essential implementation efforts; besides it allows evaluating parallelism only for the X- and Y-legs with respect to the XY-plane. So, the second method will be used here.

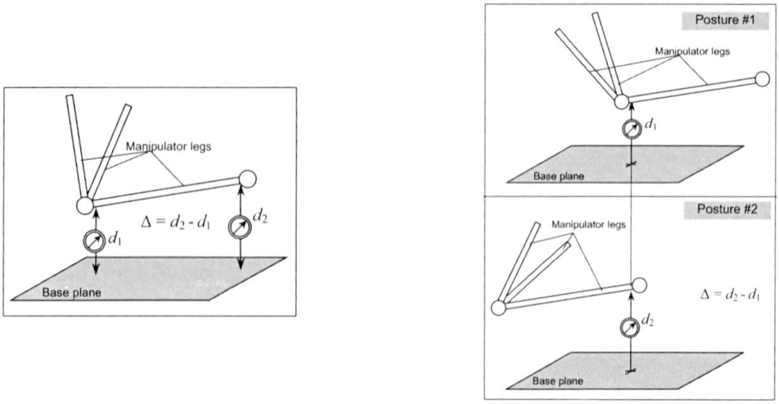

(a) *single-posture / double-sensor method* (b) *double-posture / single-sensor method*

Fig. 4. Measuring the leg parallelism with respect to the base plane

For this method, which employs the relative measurements and allows assessing the leg parallelism with respect to both relevant planes (XY- and XZ-planes for the X-leg, for instance), the calibration experiment may be arranged in the following way:

Step 1. Move the manipulator to the *Zero* posture; locate two gauges in the middle of the X-leg (orthogonal to the leg and parallel to the axes Y and Z); get their readings.

Step 2. Move the manipulator sequentially to the *XMax* and *XMin* postures, get the gauge readings, and compute the differences Δy_x^+, Δz_x^+, Δy_x^-, Δz_x^- with respect to the "Zero" posture values.

Step 3+. Repeat steps 1, 2 for the Y- and Z-legs and compute the differences Δx_y^+, Δz_y^+, Δx_y^-, Δz_y^-, and Δx_z^+, Δy_z^+, Δx_z^-, Δy_z^- corresponding to these legs.

In the above description, the variable following the symbol Δ denotes the measurement direction (x, y or z), the subscript defines the manipulator leg, and the superscript indicates the manipulator posture ('+' for *XMax* and '-' for *XMin*). For example, Δz_x^+ denotes the z-coordinate deviation of the X-leg for the *XMax* posture with respect to *Zero* location.

3.2 Calibration equations

The system of calibration equations can be derived in two steps. First, it is required to define the gauges' initial locations that are assumed to be positioned at the leg middle at the *Zero* posture, i.e. at the points $(\mathbf{p}+\mathbf{r}_i)/2$, $i \in \{x, y, z\}$ where the vectors r_i define the prismatic joints centres: $\mathbf{r}_x = (L+\Delta\rho_x;\ 0;\ 0)^T$; $\mathbf{r}_y = (0;\ L+\Delta\rho_y;\ 0)^T$; $\mathbf{r}_z = (0;\ 0;\ L+\Delta\rho_z)^T$. Hence, using the equation (7), the gauge initial locations can be expressed as

$$\mathbf{g}_x^0 = \left[\ (L - \Delta L_x)/2 + \Delta\rho_x; \quad (\Delta\rho_y - \Delta L_y)/2; \quad (\Delta\rho_z - \Delta L_z)/2 \right]^T$$

$$\mathbf{g}_y^0 = \left[\ (\Delta\rho_x - \Delta L_x)/2; \quad (L - \Delta L_y)/2 + \Delta\rho_y; \quad (\Delta\rho_z - \Delta L_z)/2 \right]^T \tag{10}$$

$$\mathbf{g}_z^0 = \left[\ (\Delta\rho_x - \Delta L_x)/2; \quad (\Delta\rho_y - \Delta L_y)/2; \quad (L - \Delta L_z)/2 + \Delta\rho_z \right]^T$$

Afterwards, for the *XMax*, *YMax*, *ZMax* postures, the leg spatial location is also defined by two points, namely, (i) the tool centre point $\mathbf{p}$, and (ii) the centre of the prismatic joint $\mathbf{r}_i$. For example, for the *XMax* posture, the TCP position is $\mathbf{p}_x^{\max} = (LS_\alpha + \Delta\rho_x - \Delta L_x;\ *;\ *)$, the prismatic joint position is $\mathbf{r}_x^{\max} = (L + LS_\alpha + \Delta\rho_x;\ 0;\ 0)$. So, the leg is located along the line

$$\mathbf{s}_x(\mu) = \mu \cdot \mathbf{p}_x^{\max} + (1-\mu)\cdot \mathbf{r}_x^{\max}; \quad \mu \in [0;1]$$

Since the x-coordinate of the gauge is kept constant (for X-leg measurements), the parameter μ may be obtained from the equation $[\mathbf{s}_x(\mu)]_x = [\mathbf{g}_x^0]_x$, which yields:

$$\mu = 0.5 + S_\alpha - S_\alpha \cdot \Delta L_x / L$$

Hence, after some transformations, the deviations of the X-leg measurements (between the *XMax* and *Zero* postures) may be expressed as

$$\Delta y_x^+ = (0.5 + S_\alpha) T_\alpha \Delta\rho_x + S_\alpha \Delta\rho_y - (0.5 + S_\alpha) T_\alpha \Delta L_x - ((0.5 + S_\alpha) C_\alpha^{-1} - 0.5)\Delta L_y$$

$$\Delta z_x^+ = (0.5 + S_\alpha) T_\alpha \Delta\rho_x + S_\alpha \Delta\rho_z - (0.5 + S_\alpha) T_\alpha \Delta L_x - ((0.5 + S_\alpha) C_\alpha^{-1} - 0.5)\Delta L_z$$

A similar approach may be applied to the *XMin* posture, as well as to the corresponding postures for the Y- and Z-legs. This gives the system of twelve linear equations in six unknowns:

$$\begin{bmatrix} a_1 & b_1 & 0 & -c_1 & -b_1 & 0 \\ b_1 & a_1 & 0 & -b_1 & -c_1 & 0 \\ a_2 & b_2 & 0 & -c_2 & -b_2 & 0 \\ b_2 & a_2 & 0 & -b_2 & -c_2 & 0 \\ \hline 0 & a_1 & b_1 & 0 & -c_1 & -b_1 \\ 0 & b_1 & a_1 & 0 & -b_1 & -c_1 \\ 0 & a_2 & b_2 & 0 & -c_2 & -b_2 \\ 0 & b_2 & a_2 & 0 & -b_2 & -c_2 \\ \hline a_1 & 0 & b_1 & -c_1 & 0 & -b_1 \\ b_1 & 0 & a_1 & -b_1 & 0 & -c_1 \\ a_2 & 0 & b_2 & -c_2 & 0 & -b_2 \\ b_2 & 0 & a_2 & -b_2 & 0 & -c_2 \end{bmatrix} \cdot \begin{bmatrix} \Delta\rho_x \\ \Delta\rho_y \\ \Delta\rho_z \\ \Delta L_x \\ \Delta L_y \\ \Delta L_z \end{bmatrix} = \begin{bmatrix} \Delta x_y^+ \\ \Delta y_x^+ \\ \Delta x_y^- \\ \Delta y_x^- \\ \hline \Delta y_z^+ \\ \Delta z_y^+ \\ \Delta y_z^- \\ \Delta z_y^- \\ \hline \Delta x_z^+ \\ \Delta z_x^+ \\ \Delta x_z^- \\ \Delta z_x^- \end{bmatrix} \tag{11}$$

where $a_i = \sin(\alpha_i)$; $b_i = (0.5 + \sin(\alpha_i))\tan(\alpha_i)$; $c_i = (0.5 + \sin(\alpha_i))/\cos(\alpha_i) - 0.5$; $i \in \{1,2\}$, and $\alpha_1 = \text{asin}(\rho_{max}/L) > 0$; $\alpha_2 = \text{asin}(\rho_{min}/L) < 0$. This system can be solved using the pseudoinverse of Moore-Penrose, which ensures the minimum of the residual square sum for corresponding linear approximation of the kinematic equations that is valid for small values of $\Delta\rho_x, \Delta\rho_y, \ldots \Delta L_y, \Delta L_z$. Otherwise, it is prudent to apply straightforward numerical optimisation, which fits the experimental data to the manipulator kinematic model (1).

3.4 Calibration accuracy

Because of the measurement noise, the developed technique may produce some errors in estimates of the model parameters. Thus, for practical applications, it is worth to evaluate the statistical properties of the calibration errors.

Within the linear calibration equations (11), the impact of the measurement noise may be evaluated using general techniques from the identification theory, under the standard assumptions concerning the primary measurement errors $\xi(.)$ (zero-mean independent and identically distributed Gaussian random variables with the standard deviation σ). For these assumptions, the covariance matrix of the estimated parameters is written as (Ljung, 1999)

$$\mathbf{V}(\Delta\boldsymbol{\rho}, \Delta\mathbf{L}) = (\mathbf{J}^T\mathbf{J})^{-1} \cdot \mathbf{J}^T \cdot \mathbf{E}(\Delta\mathbf{s} \cdot \Delta\mathbf{s}^T) \cdot \mathbf{J} \cdot (\mathbf{J}^T\mathbf{J})^{-1} \tag{12}$$

where **E**(.) denotes the mathematical expectation, **J** is the identification Jacobian, and Δs is the vector of the measurement errors in the right-hand side of the system (11). However, in contrast to the standard technique, the vector $\Delta\mathbf{s}$ includes some statistically-dependent components because the same measurement values, corresponding to the Zero position, are subtracted from those corresponding to the Max and Min postures. In particular,

$$\Delta\mathbf{s} = \left[\xi(x_y^+) - \xi(x_y^0),\ \xi(y_x^+) - \xi(y_x^0), \ldots\ \xi(z_x^+) - \xi(z_x^0)\right]^T \tag{13}$$

where the index sequence strictly corresponds to (11). Thus, the covariance $\mathbf{E}(\Delta\mathbf{s} \cdot \Delta\mathbf{s}^T)$ is the 12×12 non-identity matrix that after relevant transformations may be expressed as

$$\mathbf{E}\left(\Delta\mathbf{s} \cdot \Delta\mathbf{s}^T\right) = \sigma^2 \cdot \begin{bmatrix} \mathbf{G} & \mathbf{0} & \mathbf{0} \\ \mathbf{0} & \mathbf{G} & \mathbf{0} \\ \mathbf{0} & \mathbf{0} & \mathbf{G} \end{bmatrix}_{12\times12}; \quad \mathbf{G} = \begin{bmatrix} 2 & 0 & 1 & 0 \\ 0 & 2 & 0 & 1 \\ 1 & 0 & 2 & 0 \\ 0 & 1 & 0 & 2 \end{bmatrix} \tag{14}$$

Hence, using expressions (12), (14) it is possible to evaluate the identification accuracy (via the covariance matrix (12)) for the set of parameters $\{\Delta\rho_x, \Delta\rho_y, \ldots \Delta L_y, \Delta L_z\}$ provided the measurement error parameter σ is known. For instance, for the Orthoglide prototype described in sub-section 2.1 and the Max/Min posture characteristic angles $\alpha_1 = 11.0°$ and $\alpha_2 = -18.7°$, the measurement noise with $\sigma = 10^{-2}$ mm causes the mean-square errors for the $\Delta\rho_x, \Delta\rho_y, \ldots \Delta L_y, \Delta L_z$ of about 0.07 mm.

4. Experimental results

4.1 Experimental setup

For experimental verification of the developed technique, we used the measuring system composed of standard comparator indicators with resolution of 10.0 μm. The indicators were attached to universal magnetic stands that allow fixing them on the manipulator base. This system is sequentially used for measuring the *X*-, *Y*-, and *Z*-leg parallelism while the manipulator moves between the *Max, Min* and *Zero* postures. (It is obvious that for industrial applications it is worth using more sophisticated digital indicators with the resolution of 1.0 μm or less, which yield more accurate calibration results.)

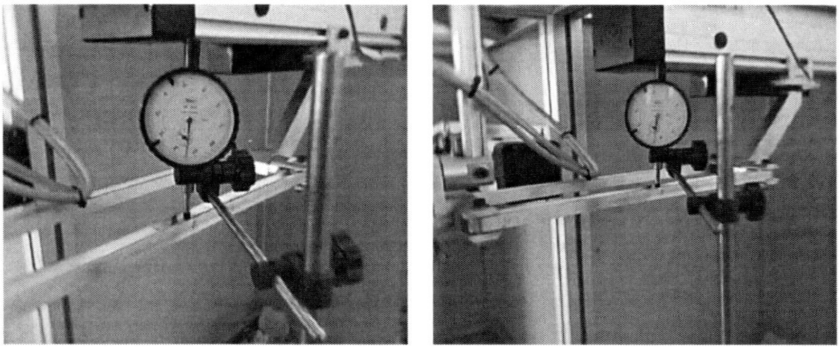

Fig. 5. Experimental setup for calibration experiments

For each measurement, the indicators are located on the mechanism base in such a manner that a corresponding leg is admissible for the gauge contact for all intermediate postures (Fig. 5). The *Min* and *Max* postures are constrained by the software limits and defined as $\rho_{\min} = -100.00$ mm and $\rho_{\max} = +60.00$ mm respectively. Initial position of the indicator corresponds to the leg middle point at the manipulator *Zero* posture.

During experiments, the legs were moved sequentially via the following postures: Zero → Max → Min → Zero→ ... . To reduce the measurement errors, the measurements were repeated three times for each leg. Then, the results were averaged and used for the parameter identification. It should be noted that the measurements demonstrated very high repeatability compared to the encoder resolution (dissimilarity was less than 20.0 μm).

4.2 Calibration results and their analysis

The experimental study included three types of experiments targeted to the following objectives: (#1) validation of modelling assumptions; (#2) obtaining source data for the parameter identification; and (#3) verification of the calibration results.

Experiment #1. The first calibration experiment demonstrated rather high parallelism deviation for the legs at the *Max* and *Min* postures, up to 2.37 mm as shown in Table1. This indicated low accuracy of the nominal kinematic model and motivated necessity of the calibration. On the other hand, the milling accuracy evaluated in separate tests was quite good. However, this is not an indicator of high absolute accuracy but just a proof of the Orthoglide architecture advantages (the milling tests were perfect just because of the high homogeneity of the manipulator workspace in the neighbourhood of the isotropic location).

The straightforward application of the proposed calibration algorithm to the data set #1 was not optimistic: in the frames of the adopted kinematic model the root-mean-square (r.m.s.) deviation for the legs can be reduced down from 1.19 mm to 0.74 mm only (see Table 1 where $\Delta x_y = \Delta x_y^+ - \Delta x_y^-$, $\Delta x_z = \Delta x_z^+ - \Delta x_z^-$, etc.). Besides, the statistical estimation of the measurement noise parameter σ (based on the residual analysis) also yielded unrealistic result compared to the encoder resolutions (0.01 mm). This impelled to conclude that some modelling assumptions are not valid and the manipulator mechanics required more careful tuning, especially orientation of the linear actuator axes (that are assumed to be mutually orthogonal and to intersect in a single point). Thus, the manipulator mechanics was re-tuned, in particular spatial locations of the actuator axes were adjusted.

Data Source	Δx_y	Δx_z	Δy_x	Δy_z	Δz_x	Δz_y	r.m.s.
Experiment #1 (before mechanical tuning and before calibration)							
Measurements #1	+0.52	+1.58	+2.37	-0.25	-0.57	-0.04	1.19
Expected improvement	-0.94	+0.63	+1.07	-0.84	-0.27	+0.35	0.74
Experiment #2 (after mechanical tuning, before calibration)							
Measurements #2	-0.43	-0.37	+0.42	-0.18	-1.14	-0.70	0.62
Expected improvement	-0.28	+0.25	+0.21	-0.14	-0.13	+0.09	0.20
Experiment #3 (after calibration and adjusting of $\Delta\rho$*)*							
Measurements #3	-0.23	+0.27	+0.34	-0.10	-0.09	+0.11	0.21
Expected improvement	-0.29	+0.23	+0.25	-0.17	-0.10	+0.08	0.20

Table 1. Experimental data and expected improvements of accuracy via calibration [mm]

Experiment #2. The second calibration experiment (after mechanical tuning) yielded lower parallelism deviations, less than 0.62 mm in terms of the deviations Δx_y, Δx_z, ... (see Table 1), which is about twice better than in the first experiment. Besides, the expected residual reduction was also essential (0.20 mm) that justified validity of the modelling assumptions. For these data, the developed calibration algorithm was applied for three sets of the model parameters: for the full set $\{\Delta\rho, \Delta L\}$ and for the reduced sets $\{\Delta\rho\}$, $\{\Delta L\}$. As follows from the identification results (Tables 2, 3), the calibration algorithm is able to identify simultaneously both the joint offsets and $\Delta\rho$ and the link lengths ΔL. However, both $\Delta\rho$ and ΔL (separately) demonstrate roughly the same influence on the residual reduction, from 0.32 mm to 0.14 mm (in terms of the deviations $\Delta x_y^+, \Delta x_y^-, \Delta x_z^+, \Delta x_z^-, \ldots$), while the full set $\{\Delta\rho, \Delta L\}$ gives further residual reduction down to 0.12 mm only. This motivates considering $\Delta\rho$ as the most essential parameters to be calibrated. Accordingly, the identified vales of joint offsets $\Delta\rho_x$, $\Delta\rho_y$, $\Delta\rho_z$ were incorporated in the Orthoglide control software.

Experiment #3. The third experiment was targeted to the validation of the calibration results, i.e. assessing the leg parallelism while using the kinematic model with the parameters identified from the data set #2. This experiment demonstrated very good agreement with the expected values of $\Delta x_y^+, \Delta x_y^-, \Delta x_z^+, \Delta x_z^-, \ldots$. In particular, the maximum deviation reduced down from 0.62 mm to 0.24 mm, and the root-mean-square value decreased down from 0.32 mm to 0.15 mm (expected value is 0.14 mm). On the other hand, further fitting of the kinematic model to the third data set gives both negligible improvement in the deviations and very small alteration of the model parameters. It is evident that further reduction of the parallelism deviation is bounded by the manufacturing errors and, by non-geometric reasons.

Residuals	Experiment 2				Experiment 3			
	Exper. data	*Expected improvement*			*Exper. data*	*Expected improvement*		
		$\{\Delta\rho, \Delta L\}$	$\{\Delta\rho\}$	$\{\Delta L\}$		$\{\Delta\rho, \Delta L\}$	$\{\Delta\rho\}$	$\{\Delta L\}$
Δx_y^+	-0.19	-0.09	-0.03	-0.03	-0.07	0.02	0.04	0.04
Δy_x^+	0.08	0.12	0.03	0.04	0.02	0.04	0.02	0.02
Δx_y^-	0.22	0.09	0.13	0.12	0.10	-0.07	-0.06	-0.06
Δy_x^-	-0.34	-0.10	-0.13	-0.13	-0.24	0.01	0.00	0.00
Δx_z^+	-0.29	-0.41	-0.32	-0.33	0.01	-0.02	-0.01	0.00
Δz_x^+	-0.52	-0.45	-0.39	0.42	0.11	-0.02	-0.03	0.04
Δx_z^-	0.08	0.23	0.26	0.26	-0.19	-0.05	-0.04	-0.04
Δz_x^-	0.62	0.55	0.57	0.56	-0.03	0.10	0.09	0.09
Δy_z^+	0.02	-0.04	-0.13	-0.12	0.07	-0.03	-0.05	-0.05
Δz_y^+	-0.24	0.29	-0.26	-0.27	-0.21	-0.05	-0.07	-0.07
Δy_z^-	0.20	-0.03	0.06	-0.06	0.17	0.04	0.03	0.03
Δz_y^-	0.45	0.48	0.51	0.50	0.27	0.11	0.10	0.10
Average	0.32	0.12	0.14	0.14	0.15	0.13	0.14	0.14

Table 2. Residual compensation using different sets of kinematic parameters [mm]

Set of parameters	Identified values [mm]						Residuals
	$\Delta\rho_x$	$\Delta\rho_y$	$\Delta\rho_z$	ΔL_x	ΔL_y	ΔL_z	
$\{\Delta\rho, \Delta L\}$	4.66	-5.36	1.46	5.20	-5.96	3.16	0.12
$\{\Delta\rho\}$	-0.48	0.49	-1.67	–	–	–	0.14
$\{\Delta L\}$	–	–	–	0.50	-0.52	1.69	0.14

Table3. Calibration results for parameters $\Delta\rho$ and ΔL

Resume. Hence, the calibration results confirm validity of the proposed identification technique and its ability to tune the joint offsets and link lengths from observations of the leg parallelism. However, for these partucular experiments, combined influence of the parameters $\{\Delta\rho, \Delta L\}$ may be roughly decribed by the diffrence $\{\Delta\rho - \Delta L\}$ that allows us to simplify modifications of the kinematic model included in the control software. Another conclusion is related to the modelling assumption: for further accuracy improvement it is prudent to generalize the manipulator model by including parameters describing orientation of the prismatic joint axes, which is equavalet to relaxing some modelling assumption.

5. Conclusions

Recent advances in parallel robot architectures encourage related research on kinematic calibration of parallel mechanisms. This paper proposes a novel calibration approach based on observations of manipulator leg parallelism with respect to the Cartesian planes. Presented for the Orthoglide-type mechanisms, this approach may be also applied to other manipulator architectures that admit parallel leg motions (along the Cartesian axes) or, in more general case, allow locating the leg in several postures with a common intersection point.

The proposed calibration technique employs a simple and low-cost measuring system composed of standard comparator indicators attached to the universal magnetic stands. They are sequentially used for measuring the deviation of the relevant leg location while the manipulator moves the tool-center-point in the directions x, y and z. From the measured differences, the calibration algorithm estimates the joint offsets and link lengths that are treated as the most essential parameters that are difficult to identify by other methods.

The presented theoretical derivations deal with the sensitivity analysis of the proposed measurement method and also with the calibration accuracy. The validity of the proposed approach and efficiency of the developed numerical algorithm were confirmed by the calibration experiments with the Orthoglide prototype, which allowed dividing the residual root-mean-square by three.

To increase the calibration precision, future work will focus on the development of the specific assembling fixture ensuring proper location of the linear actuators and also on the expanding the set of the identified model parameters and compensation of the non-geometric errors that are not identified within the frames of the adopted model.

6. References

Besnard, S. & Khalil, W. (2001). Identifiable parameters for parallel robots kinematic calibration. In: *IEEE International Conference on Robotics and Automation*, Vol. 3, pp. 2859-2866, (May 2001), Seoul, Korea.

Caro, S., Wenger, P., Bennis, F. & Chablat, D. (2006). Sensitivity Analysis of the Orthoglide, a 3-DOF Translational Parallel Kinematic Machine. *ASME Journal of Mechanical Design*, Vol. 128, No 2, (March 2006), 392-402.

Chablat, D. & Wenger, P. (2003). Architecture Optimization of a 3-DOF Parallel Mechanism for Machining Applications, the Orthoglide. *IEEE Transactions on Robotics and Automation*, Vol. 19, No 3, (June 2003), 403-410.

Daney, D. (1999). Self calibration of Gough platform using leg mobility constraints. In: *Proceedings of the 10th World Congress on the Theory of Machine and Mechanisms*, pp. 104–109, (June 1999), Oulu, Finland.

Daney, D. (2003). Kinematic Calibration of the Gough platform. *Robotica*, Vol. 21, No 6, (Dec. 2003), 677-690.

Daney, D. & Emiris I.Z. (2001). Robust parallel robot calibration with partial information. In: *IEEE International Conference on Robotics and Automation*, Vol. 4, pp. 3262-3267, May 2001, Seoul, Korea.

Fassi I., Legnani G., Tosi D. & Omodei A. (2007). Calibration of Serial Manipulators: Theory and Applications. In: *Industrial Robotics: Programming, Simulation and Applications*, Proliteratur Verlag, Mammendorf, Germany, pp. 147 - 170.

Hesselbach, J., Bier, C., Pietsch, I., Plitea, N., Büttgenbach, S., Wogersien, A. & Güttler, J. (2005). Passive-joint sensors for parallel robots. *Mechatronics*, Vol. 15, No 1, (Feb. 2005), 43-65.

Huang, T., Chetwynd, D. G., Whitehouse, D. J., & Wang, J. (2005). A general and novel approach for parameter identification of 6-dof parallel kinematic machines. *Mechanism and Machine Theory*, Vol. 40, No 2, (Feb. 2005), 219-239.

Innocenti, C. (1995). Algorithms for kinematic calibration of fully-parallel manipulators In: *Computational Kinematics*, J-P. Merlet and B. Ravani (Eds.), pp. 241-250, Dordrecht: Kluwer Academic Publishers.

Iurascu, C.C. & Park, F.C. (2003). Geometric algorithm for kinematic calibration of robots containing closed loops. *ASME Journal of Mechanical Design*, Vol. 125, No 1, (March 2003), 23-32.

Jeong, J., Kang, D., Cho, Y.M., & Kim, J. (2004). Kinematic calibration of redundantly actuated parallel mechanisms. *ASME Journal of Mechanical Design*, Vol. 126, No 2, (March 2004), 307-318.

Khalil, W. & Besnard, S. (1999). Self calibration of Stewart–Gough parallel robots without extra sensors. *IEEE Transactions on Robotics and Automation*, Vol. 15, No 6, (Dec. 1999), 1116–1112.

Legnani, G., Tosi; D., Adamini, R. & Fassi, I.. (2007). Calibration of Parallel Kinematic Machines: theory and applications. In: *Industrial Robotics: Programming, Simulation and Applications*, Proliteratur Verlag, Mammendorf, Germany, pp. 171 - 194.

Ljung, L. (1999). *System identification : theory for the user* (2nd ed). Prentice Hall, New Jersey.

Merlet, J.-P. (2000). *Parallel Robots*. Kluwer Academic Publishers, Dordrecht.

Pashkevich, A., Wenger, P. & Chablat, D. (2005). Design strategies for the geometric synthesis of Orthoglide-type mechanisms. *Mechanism and Machine Theory*, Vol. 40, No 8, (Aug. 2005), 907-930.

Pashkevich, A., Chablat, D. & Wenger, P. (2006). Kinematics and workspace analysis of a three-axis parallel manipulator: the Orthoglide. *Robotica*, Vol. 4, No 1, (Jan. 2006), 39-49.

Pashkevich, A., Chablat, D. & Wenger, P. (2006) Kinematic Calibration of Orthoglide-Type Mechanisms. Information Control Problems in Manufacturing 2006 (A Proceedings Volume from the 12th IFAC Conference 17-19 May 2006, Saint-Etienne, France), pp. 149-154, 2006

Rauf, A., Kim, S.-G. & Ryu, J. (2004). Complete parameter identification of parallel manipulators with partial pose information using a new measurements device. *Robotica*, Vol. 22, No 6, (Nov. 2004), 689-695.

Rauf, A., Pervez A. & Ryu, J. (2006). Experimental results on kinematic calibration of parallel manipulators using a partial pose measurement device. *IEEE Transactions on Robotics*, Vol. 22, No 2, (Apr. 2006), 379-384.

Renaud, P., Andreff, N., Pierrot, F., & Martinet, P. (2004). Combining end-effector and legs observation for kinematic calibration of parallel mechanisms. In: *IEEE International Conference on Robotics and Automation*, Vol. 4, pp. 4116-4121, (Apr.-May 2004), New Orleans, USA.

Renaud, P., Andreff, N., Gogu, G. & Martinet, P. (2005). Kinematic calibration of parallel mechanisms: a novel approach using legs observation. *IEEE Transactions on Robotics*, Vol. 21, No 4, (Aug. 2005), 529-538.

Renaud, P., Vivas, A., Andreff, N., Poignet, P., Martinet, P., Pierrot, F. & Company, O. (2006). Kinematic and dynamic identification of parallel mechanisms. *Control Engineering Practice*, Vol. 14, No 9, (Sept. 2006), 1099-1109

Schröer, K., Bernhardt, R., Albright, S., Wörn, H., Kyle, S., van Albada, D., Smyth, J. & Meyer, R. (1995). Calibration applied to quality control in robot production. *Control Engineering Practice*, Vol. 3; No 4, (Apr. 1995), 575-580.

Thomas, F., Ottaviano, E., Ros, L., & Ceccarelli, M. (2005). Performance analysis of a 3–2–1 pose estimation device. *IEEE Transactions on Robotics*, Vol. 21, No 3, (June 2005), 88-297.

Tlusty, J., Ziegert, J.C. & Ridgeway, S. . (1999). Fundamental comparison of the use of serial and parallel kinematics for machine tools, *CIRP Annals*, Vol. 48, No 1, 351-356.

Tsai, L. W. (1999). *Robot analysis: the mechanics of serial and parallel manipulators*. John Wiley & Sons, New York.

Wampler, C.W., Hollerbach, T.M. & Arai, T. (1995). An implicit loop method for kinematic calibration and its application to closed chain mechanisms. *IEEE Transactions on Robotics and Automation*, Vol. 11, No 5, (Oct. 1995), 710–724.

Wang, J. & Masory, O. (1993). On the accuracy of a Stewart platform - Part I: The effect of manufacturing tolerances. In: *IEEE International Conference on Robotics and Automation*, Vol. 1, pp. 114–120, (May 1993), Atlanta, USA.

Wenger, P. & Chablat, D. (2000). Kinematic analysis of a new parallel machine-tool: the Orthoglide. In: *7th International Symposium on Advances in Robot Kinematics*, pp. 305-314, (June 2000), Portoroz, Slovenie.

Wenger, P., Gosselin, C. & Chablat, D. (2001). Comparative study of parallel kinematic architectures for machining applications. In: *Workshop on Computational Kinematics*, pp. 249-258, (May 2001), Seoul, Korea.

Williams, I., Hovland, G. & Brogardh, T. (2006). Kinematic error calibration of the gantry-tau parallel manipulator. In: *IEEE International Conference on Robotics and Automation*, pp. 4199-4204, (May 2006), Orlando, USA.

Zhuang, H. (1997). Self-calibration of parallel mechanisms with a case study on Stewart platforms. *IEEE Transactions on Robotics and Automation*, Vol. 13, No 3, (June 1997), 387–397.

Zhuang, H., Motaghedi, S.H. & Roth, Z.S. (1999). Robot calibration with planar constraints. In: *IEEE International Conference of Robotics and Automation*, Vol. 1, pp. 805-810, (May 1999), Detroit, USA.

12

Kinematic Parameters Auto-Calibration of Redundant Planar 2-Dof Parallel Manipulator

Shuang Cong[1)], Chunshi Feng[1)], Yaoxin Zhang[1)],
Zexiang Li[2)] and Shilon Jiang[3)]
[1)] *University of Science and Technology of China*
[2)] *Hong Kong University of Science and Technology*
[3)] *Googol Technology (Shenzhen) Limited*
P. R. China

1. Introduction

Parallel manipulators have the advantage of high speed and high precision in the theory of mechanisms. This has opened up broad possibilities for the use of parallel manipulators in many fields. But in real applications, due to the inevitable manufacturing tolerances and assembling errors, the actual kinematic parameters of parallel manipulators are always unequal to the nominal values and calibration procedures have to be implemented to compensate the kinematic parameter errors between them.

According to the metrology devices adopted, calibration methods of parallel manipulators can be classified into two categories, the external calibration methods and the auto-calibration methods. External calibration methods rely on the precise external 3D measuring devices, such as laser tracking systems (Koseki et al., 1998; Vincze et al., 1994), mechanical devices (Jeong et al., 1999) and camera systems (Zou & Notash, 2001; Renaud et al., 2006). With these external devices, one can measure the end-effector position of parallel manipulators and calibrate the kinematic parameters by minimizing either the errors between the measured end-effector positions and the estimated end-effector positions (Masory et al., 1993), or the errors between the measured joint positions and the estimated joint positions (Zhuang et al., 1995; Zhuang et al., 1998). The auto-calibration methods rely on the redundant joint sensors of parallel manipulators, which can be achieved by adding extra sensors to the uninstrumented joints (Baron & Angeles, 1998; Zhuang, 1997; Wampler et al., 1995; Patel & Ehmann, 2000), or by constraining the motion of end-effector or some joints (Khalil & Besnard, 1999; Wang & Masory, 1993). With the redundant joint sensors, extra information can be obtained for the sampled configurations without employing any external measuring devices (Hollerbach & Wampler, 1996; Yiu et al., 2003c; Chiu & Perng, 2004), and the auto-calibration procedure is usually implemented by minimizing a function of closed-loop constraint errors. Obviously, it is more convenient to measure the sampled configurations by the redundant joint sensors than the external 3D measuring devices. especially for the parallel manipulator with inherent redundant joint sensors. But it is usually difficult to minimize the closed-loop constraint error function, and the calibration results of the auto-calibration methods are usually dependent on the error function adopted in the calibration.

In this chapter, we will calibrate the kinematic parameters of a planar 2-dof parallel manipulator. In the literatures, this type of parallel manipulator has been studied from different aspects. Yiu and Zhang studied the kinematics of the parallel manipulator (Yiu & Li, 2003b; Zhang, 2006). Liu studied the singularities of the parallel manipulator with a geometric method (Liu et al., 2001a; Liu et al., 2003). Furthermore, the dynamics and controller design problem of the parallel manipulator were studied by Liu (Liu et al., 2001a; Liu et al., 2001b; Liu & Li, 2002), Kock (Kock & Schumacher, 2000a; Kock & Schumacher, 2000b), Yiu (Yiu & Li, 2001; Yiu & Li, 2003a), Cheng (Cheng et al., 2003), Shen (Shen et al., 2003) and Zhang (Zhang & Cong, 2005). In this chapter, we will study the calibration and solve three problems. In the second part of the chapter, based on the study of the relationship between the projected tracking error of the joint angles and the error of the sensor zero positions, we propose a projected tracking error function for the calibration of the sensor zero positions of a planar 2-dof parallel manipulator with a redundant joint sensor. With a simple searching strategy for the minimal value of the error function, an auto-calibration procedure is designed, and the validity of the calibration procedure is verified through actual experiments on a real redundant planar 2-dof parallel manipulator. In the third part of the chapter, by eliminating the passive joint positions, we derive another type of error function with only the variables of the active joint positions. Moreover, by decoupling the products items of the kinematic parameters in the error function into the linear combinations of a group of new variables, the error function minimization process is simplified and the calibration precision can be improved further. Based on two error functions proposed in this section, an auto-calibration method and design procedure is given, and the validity of the auto-calibration method is studied with stepwise simulations. Under the assumption that only one coordinate is known accurately forehand, the other 11 kinematic parameters of the parallel manipulator including 3 sensor zero positions, 6 link lengths and 2 base coordinates can be calibrated precisely. In order to obtain the global optimum and auto-calibrate all parameters of the parallel manipulator, in the fourth part of the chapter, three stochastic optimization algorithms including genetic algorithm (Holland, 1975), particle swarm optimization (Kennedy & Eberhart, 1995) and differential evolution (Storn & Price, 1995) are applied to minimize the error functions proposed in the third part of the chapter, respectively. In the applications, the performances of the applied algorithms on the problem are compared under the different methods. Finally, actual calibration is carried out based on differential evolution algorithms, and the results demonstrate that all of the 12 parameters of the parallel manipulator are calibrated with high accuracy. We'll end the chapter with the conclusions.

2. Auto-calibration of sensor zero positions based on the projected tracking Error

2.1 Calibration problem of the sensor zero positions

The structure of the planar 2-dof parallel manipulator to be calibrated is shown in Fig. 1, in which the parallel manipulator consists of 6 links with the same lengths, and 3 active joints located at A_1, A_2, A_3, and 3 passive joints located at B_1, B_2, B_3. The end-effector of the parallel manipulator coincides with O. According to Fig. 1, a reference frame is established in the workspace of the parallel manipulator. The zero positions of the joint angles are all defined as the positive direction of the X axis of the reference frame, and the positive directions of the angles are all defined as the anticlockwise direction.

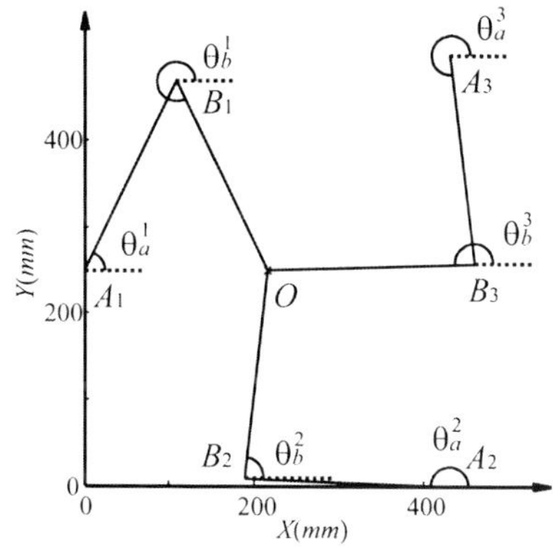

Fig. 1 Structure of the planar 2-dof parallel manipulator

The kinematics of the parallel manipulator has been studied by Yiu and Zhang (Yiu & Li, 2003b; Zhang, 2006), and the end-effector coordinate can be calculated from the active joint angles through the kinematics. For the active joints located at A_1, A_2, A_3, each is attached with a position sensor, through which the active joint angles can be measured. But due to the assembly errors, there is always some bias angle between each sensor zero position and the zero position of the active joint. So active joint angles can be formulated as the sum of the sensors readings and the sensor zero positions.

$$\theta_a^i = \theta_m^i + \theta_z^i,\ i = 1,2,3 \tag{1}$$

in which symbols θ_a^i, $i = 1,2,3$ refer to the active joint angles, symbols θ_m^i, $i = 1,2,3$ refer to the sensor readings and symbols θ_z^i, $i = 1,2,3$ refer to the estimations of the sensor zero positions. Obviously the precision of θ_z^i, $i = 1,2,3$ determines the precision of the parallel manipulator in real applications. But it is difficult to measure the sensor zero positions directly. So calibration procedure has to be implemented to estimate their actual values.

Usually, the sensor zero positions are estimated through the following manual procedure. First move the end-effector to a predefined position in the workspace manually, for which the corresponding active joint angles θ_a^i, $i = 1,2,3$ have been known accurately. Then record the sensor readings θ_m^i, $i = 1,2,3$ and one can get the estimations of the sensor zero positions θ_z^i, $i = 1,2,3$ by subtracting the sensor readings from the active joint angles.

The calibration procedure mentioned above is convenient to be implemented, but the precision of the calibration results of the sensor zero positions is usually limited, since it is difficult to move the end-effector to the predefined position precisely, there is always several millimeter error between the real position of the end-effector and the predefined position. To solve this problem, Yiu proposed two iterative algorithms to calibrate the parallel manipulator (Yiu et al., 2003c), but the robustness of Yiu's method was not proved. Here we propose a new calibration method based on the projected tracking error of the active joint angles of the parallel manipulator.

2. 2 Error function based on the projected tracking error

To formulate the tracking error of the parallel manipulator, we define the tracking error of the end-effector dx, dy and the tracking error of the joint angles $d\theta_a^i$, $d\theta_b^i$, $i = 1,2,3$ as follows:

$$dx = x - \hat{x}, dy = y - \hat{y}$$
$$d\theta_a^i = \theta_a^i - \hat{\theta}_a^i, d\theta_b^i = \theta_b^i - \hat{\theta}_b^i, i = 1,2,3 \quad (2)$$

in which symbols $(\hat{x},\hat{y})$, $\hat{\theta}_a^i, i=1,2,3$ and $\hat{\theta}_b^i, i=1,2,3$ refer to the end-effector coordinate, active joint angles and passive joint angles of the desired path respectively, while symbols (x,y), θ_a^i, $i = 1,2,3$ and θ_b^i, $i = 1,2,3$ are the counterparts of the real path respectively.
According to (1), the tracking error of the active joint angles $d\theta_a^i$, $i = 1,2,3$ can be formulated as the sum of the tracking error of the sensor readings $d\theta_m^i$, $i = 1,2,3$ and the error of the sensor zero positions $d\theta_z^i$, $i = 1,2,3$. So one can have:

$$d\theta_a^i = d\theta_m^i + d\theta_z^i, i = 1, 2, 3 \quad (3)$$

in which symbols $d\theta_m^i$, $d\theta_z^i$, $i = 1, 2, 3$ are defined by

$$\begin{matrix} d\theta_m^i = \theta_m^i - \hat{\theta}_m^i \\ d\theta_z^i = \theta_z^i - \hat{\theta}_z^i \end{matrix}, i = 1,2,3 \quad (4)$$

where symbols $\hat{\theta}_m^i, i=1,2,3$ and θ_m^i, $i = 1, 2,3$ refer to the desired sensor readings and real sensor readings respectively, symbols $\hat{\theta}_z^i, i=1,2,3$ and θ_z^i, $i = 1,2,3$ refer to the actual value and estimated value of the sensor zero positions respectively. According to (4), the tracking error of the sensor readings $d\theta_m^i$ is defined as the difference between the actual value and the desired value of the sensor readings, and the error of the sensor zero positions $d\theta_z^i$ is defined as the difference between the estimated value and the actual value of the sensor zero positions.
Denote the link length of the parallel manipulator by symbol l, and the coordinates of the three bases by (x_a^i, y_a^i), $i = 1,2,3$. Then according to the kinematics of the parallel manipulator, one can express end-effector coordinate of the desired path $(\hat{x},\hat{y})$ through following equations:

$$\begin{matrix} \hat{x} = x_a^i + l \bullet \cos\left(\hat{\theta}_a^i\right) + l \bullet \cos\left(\hat{\theta}_b^i\right) \\ \hat{y} = y_a^i + l \bullet \sin\left(\hat{\theta}_a^i\right) + l \bullet \sin\left(\hat{\theta}_b^i\right) \end{matrix}, i = 1,2,3 \quad (5)$$

As a result of trajectory tracking, the real path of the end-effector (x, y) can be expressed as follows:

$$x = x_a^i + l \bullet cos(\theta_a^i) + l \bullet cos(\theta_b^i)$$

$$y = y_a^i + l \bullet \sin(\theta_a^i) + l \bullet \sin(\theta_b^i) \quad i = 1, 2, 3 \quad (6)$$

Suppose that the estimated values of the sensor zero positions θ_z^i, $i = 1,2,3$ are accurate and equal to the actual values $\hat{\theta}_z^i, i=1,2,3$ exactly, so one can have

$$\theta_z^i = \hat{\theta}_z^i, i = 1,2,3 \tag{7}$$

According to (7) and (1), the active joint angles can be measured accurately, and both of the desired path and the real path lie in the configuration space of the parallel manipulator accurately. By subtracting (5) from (6), one can formulate the tracking error of the parallel manipulator as following equations:

$$\begin{aligned} x - \hat{x} &= l \bullet \cos\left(\theta_a^i\right) + l \bullet \cos\left(\theta_b^i\right) - l \bullet \cos\left(\hat{\theta}_a^i\right) - l \bullet \cos\left(\hat{\theta}_b^i\right) \\ y - \hat{y} &= l \bullet \sin\left(\theta_a^i\right) + l \bullet \sin\left(\theta_b^i\right) - l \bullet \sin\left(\hat{\theta}_a^i\right) - l \bullet \sin\left(\hat{\theta}_b^i\right) \end{aligned}, i = 1,2,3 \tag{8}$$

Implement the Taylor series expansion on (8) and ignore the high-order items of the tracking errors, one can have:

$$\begin{bmatrix} dx \\ dy \end{bmatrix} = \begin{bmatrix} -l \bullet \sin\left(\hat{\theta}_a^i\right) & -l \bullet \sin\left(\hat{\theta}_b^i\right) \\ l \bullet \cos\left(\hat{\theta}_a^i\right) & l \bullet \cos\left(\hat{\theta}_b^i\right) \end{bmatrix} \begin{bmatrix} d\theta_a^i \\ d\theta_b^i \end{bmatrix}, i = 1,2,3 \tag{9}$$

Solve (9), one can express the tracking error of the joint angles $d\theta_a^i$, $d\theta_b^i$, i = 1, 2, 3 by the tracking error of the end-effector dx, dy as following equations:

$$\begin{bmatrix} d\theta_a^i \\ d\theta_b^i \end{bmatrix} = \frac{1}{l \bullet \sin\left(\hat{\theta}_a^i - \hat{\theta}_b^i\right)} \begin{bmatrix} -\cos\left(\hat{\theta}_b^i\right) & -\sin\left(\hat{\theta}_b^i\right) \\ \cos\left(\hat{\theta}_a^i\right) & \sin\left(\hat{\theta}_a^i\right) \end{bmatrix} \begin{bmatrix} dx \\ dy \end{bmatrix}, i = 1,2,3 \tag{10}$$

And the tracking error of the active joint angles $d\theta_a^i$, i = 1,2,3 can be expressed by the tracking error of the end-effector as follows:

$$\begin{bmatrix} d\theta_a^1 \\ d\theta_a^2 \\ d\theta_a^3 \end{bmatrix} = - \begin{bmatrix} \dfrac{\cos\left(\hat{\theta}_b^1\right)}{l \bullet \sin\left(\hat{\theta}_a^1 - \hat{\theta}_b^1\right)} & \dfrac{\sin\left(\hat{\theta}_b^1\right)}{l \bullet \sin\left(\hat{\theta}_a^1 - \hat{\theta}_b^1\right)} \\ \dfrac{\cos\left(\hat{\theta}_b^2\right)}{l \bullet \sin\left(\hat{\theta}_a^2 - \hat{\theta}_b^2\right)} & \dfrac{\sin\left(\hat{\theta}_b^2\right)}{l \bullet \sin\left(\hat{\theta}_a^2 - \hat{\theta}_b^2\right)} \\ \dfrac{\cos\left(\hat{\theta}_b^3\right)}{l \bullet \sin\left(\hat{\theta}_a^3 - \hat{\theta}_b^3\right)} & \dfrac{\sin\left(\hat{\theta}_b^3\right)}{l \bullet \sin\left(\hat{\theta}_a^3 - \hat{\theta}_b^3\right)} \end{bmatrix} \begin{bmatrix} dx \\ dy \end{bmatrix} \tag{11}$$

Define the tracking error vector of the active joint angles by $d\theta_a = [d\theta_a^1 \;\; d\theta_a^2 \;\; d\theta_a^3]^T$ and the tracking error vector of the end-effector by $dxy = [dx\ dy]^T$. Then one can express (11) by:

$$d\theta_a = J \bullet dxy \tag{12}$$

where the symbol J refers to the Jacobian matrix of the active joint angles with the end-effector coordinate.

With the assumption expressed in (7), one can have $d\theta_a^i = d\theta_m^i$, i = 1, 2, 3. So (12) can be formulated as follows:

$$d\theta_m = J \bullet dxy \tag{13}$$

where the symbol $d\theta_m$ is defined by $d\theta_m = [d\theta_m^1\ d\theta_m^2\ d\theta_m^3]^T$.
Define the image space of the Jacobian matrix J as the feasible subspace, and the orthogonal complement of the feasible subspace as the infeasible subspace. According to (13), the tracking error vector of the sensor readings $d\theta_m$ lies in the feasible subspace and the projection of $d\theta_m$ into the infeasible subspace is zero, which can be formulated by

$$Pd\theta_m = 0 \tag{14}$$

where symbol P is the linear projection that can project the vector into the infeasible subspace. The matrix representation of P can be formulated as

$$P = I - J\,(J^T J)^{-1} J^T \tag{15}$$

where symbol I is the identity matrix and symbol J is the Jacobian matrix of the active joint angles with the end-effector coordinate. Obviously, the matrix P is symmetric and idempotent, so one can have following equations:

$$P = P^T$$

$$P = P^2 \tag{16}$$

With the linear projection P, one can project the vector of the tracking error $d\theta_a$ into the infeasible subspace, and define the projected tracking error of the active joint angles by

$$d\theta_e = Pd\theta_a \tag{17}$$

where the symbol $d\theta_e = [d\theta_e^1\ d\theta_e^2\ d\theta_e^3]^T$ refers to the projected tracking error of the active joint angles and equals to the infeasible component of $d\theta_a$.
Let $d\theta_z = [d\theta_z^1\ d\theta_z^2\ d\theta_z^3]^T$ be the vector of the error of the sensor zero positions. With (3), one can have that $d\theta_a$ equals to the sum of $d\theta_m$ and $d\theta_z$. So the infeasible component of $d\theta_a$ equals to the sum of the infeasible component of $d\theta_m$ and $d\theta_z$. According to (14), the tracking error vector of the sensor readings $d\theta_m$ can be viewed as a feasible component of $d\theta_a$, and its infeasible component always vanishes. Then for the error of the sensor zero positions $d\theta_z$, which usually has nonvanishing feasible component and infeasible component simultaneously, its infeasible component equals to the infeasible component of $d\theta_a$. At last through (3) and (14) one can have following equation for the projected tracking error $d\theta_e$:

$$d\theta_e = Pd\theta_a = P(d\theta_m + d\theta_z) = Pd\theta_m + Pd\theta_z = Pd\theta_z \tag{18}$$

According to (18), the projected tracking error of the active joint angles $d\theta_e$ equals to the projection of the vector $d\theta_z$ into the infeasible subspace. So by minimizing the following function of the projected tracking error $d\theta_e$, the value of $d\theta_z$ can be minimized and the sensor zero positions can be calibrated:

$$E = \frac{1}{n}\sum_{j=1}^{n} d\theta_{ej}^T \bullet d\theta_{ej} \tag{19}$$

in which the symbol $d\theta_{ej}$ is the jth sampled projected tracking errors and the symbol n refers to the number of the sampled projected tracking errors.

2.3 Analysis of robustness of the error function

In real application, the readings of sensor are usually inaccurate, so the precision of the calibration results may be limited by the accuracy of the sensor readings. In this section, we will prove that the error function expressed in (19) is robust to the measurement error of the sensor readings, so accurate calibration results can be obtained by minimizing the error function.

Suppose that the number of the sampled sensor readings is n, and denote the measurements error of the jth sensor readings by the symbol $\delta_{mi}=[\delta^1_{mi}\ \delta^2_{mi}\ \delta^3_{mi}]^T$. According to (3), the tracking error of the active joint can be formulated by

$$d\theta_a = \delta_{mj} + d\theta_{mj} + d\theta_z \,,\ j = 1, \dots ,n \tag{20}$$

where symbols $d\theta_{aj}$, $j = 1,\dots,n$ is the tracking error of the active joint angles and $d\theta_{mj}$, $j = 1, \dots ,n$ is the tracking error of the sensor readings. Then following two propositions can be proved for the robustness of the error function to the measurement error δ_{mj}, $j = 1, \dots ,n$.

Proposition 1: Suppose that the measurements error vector δ_{mj}, $j = 1, \dots ,n$ lies in the feasible subspace. The value of the error function E will not be affected by the measurement error and accurate calibration results of the sensor zero positions can be obtained by minimizing the error function E.

Proof: For the assumption that the measurement error δ_{mj}, $j = 1, \dots ,n$ lies in the feasible subspace, one can have

$$P_j\delta_{mj} = 0,\ j = 1, \dots ,n \tag{21}$$

where the symbol P_j refer to the linear projection of the jth desired configuration. Then following equation can be obtained for the projected tracking error of the active joint angles $d\theta_{ej}$, $j = 1, \dots ,n$:

$$d\theta_{ej} = P_j d\theta_{aj} = P_j(\delta_{mj} + d\theta_{mj} + d\theta_z) = P_j\delta_{mj} + P_j\, d\theta_{mj} + P_j\, d\theta_z = P_j\, d\theta_z \,,\ j = 1, \dots ,n \tag{22}$$

According to (22), although the sensor readings are not accurate, the projected tracking error of the active joint angles $d\theta_{ej}$, $j = 1, \dots ,n$ still equals to the projection of the vector $d\theta_z$ into the infeasible subspace. Substitute (22) into (19), one can find that the value of the error function E will not be affected by the measure error δ_{mj}, $j = 1, \dots ,n$. By minimizing the error function, the error vector $d\theta_z$ can be minimized, and accurate calibration results can be obtained for the sensor zero positions. So the robustness of the error function is proved.

Proposition 2: Suppose that the measurement error vector δ_{mj}, $j = 1, \dots ,n$ lies in the infeasible subspace. Denote the mean value of δ_{mj}, $j = 1, \dots ,n$. by $\bar{\delta}_m$, the variance of δ_{mj}, $j = 1, \dots ,n$. By σ^2. If the mean value $\bar{\delta}_m$ equals to zero, then accurate calibration results of the sensor zero positions still can be obtained by minimizing the error function E.

Proof: For the assumption that the measurement error δ_{mj}, $j = 1, \dots ,n$ lies in the infeasible subspace, one can have

$$P_j\delta_{mj} = \delta_{mj} \,,\ j = 1, \dots ,n \tag{23}$$

Then following equations can be obtained for the projected tracking error of the active joint angles $d\theta_{ej}$, $j = 1, \dots ,n$:

$$d\theta_{ej} = P_j d\theta_{aj} = P_j(\delta_{mj} + d\theta_{mj} + d\theta_z) = P_j\delta_{mj} + P_j\, d\theta_{mj} + P_j\, d\theta_z = \delta_{mj} + P_j\, d\theta_z \,,\ j = 1, \dots ,n \tag{24}$$

Substitute (24) into (19), one can calculate the value of the error function E as follows:

$$\begin{aligned} E &= \frac{1}{n}\sum_{j=1}^{n} d\theta_{ej}^{T} \bullet d\theta_{ej} = \frac{1}{n}\sum_{j=1}^{n}\left(\delta_{mj} + P_j d\theta_z\right)^T \bullet \left(\delta_{mj} + P_j d\theta_z\right) \\ &= \frac{1}{n}\sum_{j=1}^{n}\left(\delta_{mj}^T\delta_{mj} + \delta_{mj}P_j d\theta_{zj} + d\theta_{zj}^T P_j^T \delta_{mj} + d\theta_z^T P_j^T P_j d\theta_z\right) \\ &= \frac{1}{n}\sum_{j=1}^{n}\delta_{mj}^T\delta_{mj} + \frac{2}{n}\sum_{j=1}^{n} d\theta_z^T P_j^T \delta_{mj} + \frac{1}{n}\sum_{j=1}^{n} d\theta_z^T P_j^T P_j d\theta_z \end{aligned} \tag{25}$$

With (16) and (23), equation (25) can be simplified as follows:

$$E = \frac{1}{n}\sum_{j=1}^{n}\delta_{mj}^T\delta_{mj} + \frac{2}{n} d\theta_z^T \sum_{j=1}^{n}\delta_{mj} + \frac{1}{n}\sum_{j=1}^{n} d\theta_z^T P_j^T P_j d\theta_z \tag{26}$$

Consider the assumption of the mean value and variance of the measurement error, one can have

$$\begin{aligned} \bar{\delta}_m &= \frac{1}{n}\sum_{j=1}^{n}\delta_{mj} = 0 \\ \sigma^2 &= \frac{1}{n}\sum_{j=1}^{n}\delta_{mj}^T\delta_{mj} \end{aligned} \tag{27}$$

Then with (27), the value of the error function E can be formulated as follows:

$$E = \sigma^2 + \frac{1}{n}\sum_{j=1}^{n} d\theta_z^T P_j^T P_j d\theta_z \tag{28}$$

One can see that, with the measurement error lie in the infeasible subspace, the value of the error function E equals to the norm of the projected tracking error plus the variance of the measurement error, which is a constant. So by minimizing the error function E, accurate calibration results of the sensor zero positions still can be obtained, and the robustness of the error function is proved.

2.4 Auto-calibration procedure based on the error function

In this subsection, with a simple searching strategy for the minimal value of error function proposed above, we will design an auto-calibration procedure for the sensor zero positions of the parallel manipulator. The auto-calibration procedure based on the simple searching strategy is proposed as follows:

Step 1: Move the end-effector manually to a predefined reference point O, for which the active joint angles have been known accurately forehand. Denote the coordinate of O by symbol (x_o, y_o), and denote the actual coordinate of the end-effector position by symbol (x_r, y_r). Here we call (x_r, y_r) as the initial estimation of the reference point O

Step 2: Denote the current real position of the end-effector by symbol R. Take R as the estimation of O, and record the sensor readings corresponding to R. One can get the estimation of the sensor zero positions by subtracting the sensor readings of R from the active joint angles of O.

Step 3: Drive the end-effector of the parallel manipulator to track a predefined circular trajectory, for example, the circular trajectory with center O. Record the sensor readings of the real trajectory corresponding to the interpolation point of the desired trajectory. Then calculate the tracking errors of the active joint angles $d\theta_a$ corresponding to each interpolation point of the desired trajectory by subtracting the estimated sensor zero positions and the sensor readings from the desired active joint angles.

Step 4: Calculate the projected tracking errors $d\theta_e$ through (18) and the value of the error function E through (19).

Step 5: Take the initial estimation (x_r, y_r) of O as the center of the searching region, and the predefined value d as the scope of the searching region. Then drive the end-effector to the points with coordinates (x_r+d, y_r), (x_r+d, y_r+d), (x_r, y_r+d), (x_r-d, y_r+d), (x_r-d, y_r), (x_r-d, y_r-d), (x_r, y_r-d), (x_r+d, y_r-d) in order. Take these points as estimations of the reference point O and repeat the operations described in step 2 to step 4 for each of these points. Then one can get an estimation of the sensor zero positions and a value of the error function E for each of these points.

Step 6: Find out the coordinate of the end-effector corresponding to the minimal E among the 9 estimations of the reference point O, and denote it by symbol (x_m, y_m). If the coordinate (x_m, y_m) equals to the initial estimation (x_r, y_r), then go to step 7. If not, take the coordinate (x_m, y_m) as a new initial estimation of the reference point O, and go to step 2.

Step 7: If the value of d is smaller than the predefined lower limit, then the calibration procedure comes to the end and the sensor zero positions corresponding to the point (x_r, y_r) is the calibration result. If not, divide the variable d by 2 and go to step 2.

2.5 Experiments on a real planar parallel manipulator

With the calibration procedure proposed above, we will calibrate the sensor zero positions of Googol Tech Ltd's GPM2002, which is a planar 2-dof parallel manipulator with a redundant joint sensor. The mechanisms of the parallel manipulator GPM2002 is shown in Fig.2, in which GPM2002 is composed of 6 links and 6 joints. Similar to Fig.1, a reference frame is established in the workspace of GPM2002. Under the reference frame, the coordinate of the 3 bases are (0,250), (433,0) and (433,500) respectively, and the coordinate of the home position of the end-effector is (216.5,250). The lengths of all the 6 links equal to 244mm. The 3 joints located at the bases are actuated by an AC servo motor respectively, while the other 3 joints are unactuated. Each of the AC servo motor is embedded with an internal absolute encoder, with which the active joint angles can be measured.

Based on the auto-calibration procedure proposed in subsection 2.4, an auto-calibration program is realized with VC++. The reference point O of the auto-calibration procedure is defined as home position of the end-effector (216.5,250), and the desired trajectory to be tracked is defined as a circle with center (216.5,250). The initial value of d is defined as 4mm and the lower limit of d 0.125mm. Then calibration experiments are implemented and the bias angels between the sensor zero positions and the active joint zero positions of GPM2002 are calibrated. The calibration experiments were implemented for 3 times, each time with a different initial estimation of the reference point (216.5, 250), and so different initial estimation of the sensor zero positions. Experiment results are shown in Table 1.

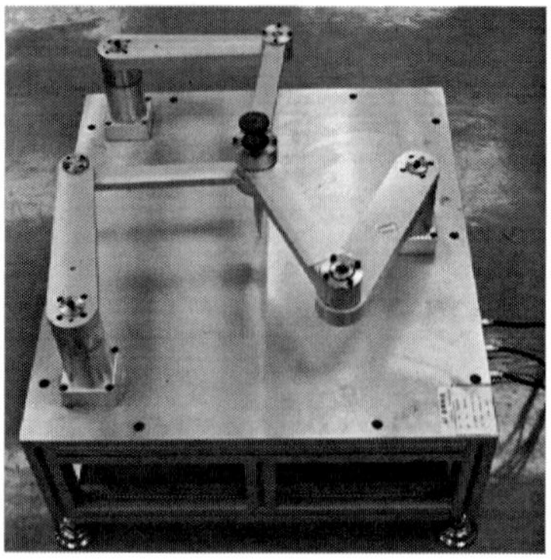

Fig. 2. Structure of GPM2002

	Initial Estimation (*rad*)			Calibration Results (*rad*)		
	Active Joint1	Active Joint2	Active Joint3	Active Joint1	Active Joint2	Active Joint3
Experiment1	1.0325	3.1045	4.8055	1.0661	3.0797	4.7974
Experiment2	1.0362	3.0796	4.8413	1.0652	3.0798	4.7971
Experiment3	1.0888	3.0651	4.7880	1.0663	3.0800	4.7966

Table 1. Calibration results of the sensor zero positions

As shown in Table 1, from different initial estimated values, the calibrated sensor zero positions will converge to an identical value with the decrease of the projected tracking errors in the end.

3. Kinematic parameters calibration of redundant planar 2-dof parallel manipulator with a new error function

In the former section, the three sensor zero positions have been calibrated by optimizing the error function we proposed. In this section, we will further calibrate the other parameters of the parallel manipulator. Based on the minimization of the closed-loop constraint errors, Yiu proposed an auto-calibration procedure for the planar 2-dof parallel manipulator (Yiu et al., 2003c). But for the parallel manipulator, the difficulty of minimizing the closed-loop constraint error function increases as the number of kinematic parameters to be calibrated increases. Among the 12 independent kinematic parameters, only 3 sensor zero positions were calibrated successfully in Yiu's paper and also in the second section of present chapter, while the other 9 parameters were supposed to be known beforehand. By eliminating the passive joint variables of the closed-loop constraint equations, we will simplify the formulation of the closed-loop constraint equations, and propose a new error function to calibrate not only the sensor zero positions but also other kinematic parameters of the parallel manipulator. Compared to Yiu's error function, which involves both active joint positions and passive joint positions as variables, our error function involves only the active joint positions as variables. Besides, by decoupling the product item of the kinematic parameters of the error function into linear combinations of a group of new variables, we simplify the minimization process and improve the calibration precision further.

3.1 A new error function

To formulate the kinematics of GPM2002, we denote the coordinate of the end-effector O by (x, y), the active joint angles by θ_a^i , i = 1,2,3, the passive joint angles by θ_b^i , i = 1,2,3, the lengths of the links connected to the active joints by l_a^i, i = 1,2,3, the lengths of the links connected to the passive joints by l_b^i, i = 1,2,3, and the coordinates of A_1, A_2, A_3 by (x_a^i, y_a^i), i = 1,2,3. While the nominal values of the link lengths l_a^i, l_b^i, i = 1,2,3 are all 244mm, the nominal values of the base coordinates (x_a^i, y_a^i), i = 1,2,3 are (0, 250), (433,0), (433,500) respectively. Then from the joint angles, the coordinate of the end-effector can be calculated through following equations:

$$x = x_a^i + l_a^i \bullet cos(\theta_a^i) + l_b^i \bullet cos(\theta_b^i)$$

$$y = y_a^i + l_a^i \bullet sin(\theta_a^i) + l_b^i \bullet sin(\theta_b^i) \qquad i = 1,2,3 \tag{29}$$

Define x_b^i, y_b^i, i = 1,2,3 as follows:

$$x_b^i = x_a^i + l_a^i \bullet cos(\theta_a^i)$$

$$y_b^i = y_a^i + l_a^i \bullet sin(\theta_a^i) \quad i = 1,2,3 \tag{30}$$

Substitute (30) into (29), one can reformulate (29) into the following quadratic equations:

$$(x- x_b^i)^2+(y- y_b^i)^2 = l_b^{i2}, \ i =1,2,3 \tag{31}$$

From (29), (30) and (31), we can have the following equation:

$$2(x_b^2-x_b^1)x+2(y_b^2-y_b^1)y = d_2-d_1$$

$$2(x_b^3-x_b^1)x+2(y_b^3-y_b^1)y = d_3-d_1 \tag{32}$$

with d_i, i = 1,2,3 defined as $d_i = x_b^{i2}+y_b^{i2}- l_b^{i2}$, i = 1,2,3.

Then the coordinates of the end-effector (x,y) can be solved from (32), and expressed as following equations:

$$x = [d_1(y_b^2-y_b^3)+ d_2(y_b^3-y_b^1)+ d_3(y_b^1-y_b^2)]/2[x_b^1 (y_b^2-y_b^3)+x_b^2 (y_b^3-y_b^1)+x_b^3 (y_b^1-y_b^2)]$$

$$y = [d_1(x_b^3-x_b^2)+ d_2(x_b^1-x_b^3)+ d_3(x_b^2-x_b^1)]/2[x_b^1 (y_b^2-y_b^3)+x_b^2 (y_b^3-y_b^1)+x_b^3 (y_b^1-y_b^2)] \tag{33}$$

Furthermore, the passive joint angles corresponding to the coordinates of the end-effector can be calculated as follows:

$$\theta_b^i = arctg((y-y_b^i)/(x-x_b^i)) \in (-\pi, \pi], \text{ i} = 1,2,3 \tag{34}$$

Besides the 6 link lengths l_a^i, l_b^i, i = 1,2,3 and the 6 base coordinates (x_a^i, y_a^i), i = 1,2,3, 3 more parameters are included in the kinematics to compensate the undetermined bias angles between the actual zero positions of the joint sensors and the predefined zero positions of the joint angles. Here, home position of the end-effector (216.5,250) is defined as zero positions. Denote the bias angles by $\Delta\theta_a^i$, i = 1,2,3, and the readings of the encoders by $\tilde{\theta}_a^i, i = 1,2,3$, then one can express the active joint angles as follows:

$$\theta_a^i = \tilde{\theta}_a^i + \Delta\theta_a^i, i = 1,2,3 \tag{35}$$

Among the 6 base coordinates, 3 of them must be set to their nominal values to establish the coordinate frame before calibration. If not, the manipulator can move freely in the plane, and infinite solutions can be obtained through calibration and from these solutions, it is impossible to tell which solution is the actual one. With 3 coordinates being predefined, there are altogether 12 kinematic parameters to be calibrated. Without losing generality, we would suppose that the base coordinates x_a^1, y_a^1, y_a^2 are equal to their nominal values, and regard them as constants for the calibration of GPM2002. Thus the kinematic parameters to be calibrated for GPM2002 include 3 sensor zero positions $\Delta\theta_a^i$, i = 1,2,3, 6 link lengths l_a^i, l_b^i, i = 1,2,3 and 3 base coordinates x_a^2, x_a^3, y_a^3. For the calibration of the parallel manipulator, the kinematic parameter errors can be represented by the closed-loop constrained equations (Yiu et al., 2003c):

$$E_1 = \begin{bmatrix} x - x_a^1 - l_a^1 \cdot \cos\left(\tilde{\theta}_a^1 + \Delta\theta_a^1\right) - l_b^1 \cdot \cos\left(\theta_b^1\right) \\ y - y_a^1 - l_a^1 \cdot \sin\left(\tilde{\theta}_a^1 + \Delta\theta_a^1\right) - l_b^1 \cdot \sin\left(\theta_b^1\right) \\ x - x_a^2 - l_a^2 \cdot \cos\left(\tilde{\theta}_a^2 + \Delta\theta_a^2\right) - l_b^2 \cdot \cos\left(\theta_b^2\right) \\ y - y_a^2 - l_a^2 \cdot \sin\left(\tilde{\theta}_a^2 + \Delta\theta_a^2\right) - l_b^2 \cdot \sin\left(\theta_b^2\right) \\ x - x_a^3 - l_a^3 \cdot \cos\left(\tilde{\theta}_a^3 + \Delta\theta_a^3\right) - l_b^3 \cdot \cos\left(\theta_b^3\right) \\ y - y_a^3 - l_a^3 \cdot \sin\left(\tilde{\theta}_a^3 + \Delta\theta_a^3\right) - l_b^3 \cdot \sin\left(\theta_b^3\right) \end{bmatrix} = 0 \tag{36}$$

Then for n sampled configurations $\tilde{\theta}_{aj}^1, \tilde{\theta}_{aj}^2, \tilde{\theta}_{aj}^3, j = 1, \cdots, n$, $6n$ equations and $5n$+12 variables can be obtained based on the closed-loop constrained equation (34). Among the variables, $2n$ variables are end-effector coordinates (x^j, y^j), j = 1, ... ,n, $3n$ variables are passive joint angles θ_{bj}^1, θ_{bj}^2, θ_{bj}^3, j = 1, ... ,n, and the remaining 12 variables are kinematic parameters to be calibrated. Obviously, if 12 sampled configurations are chosen, then 72 equations and 72 variables can be obtained, and the kinematic parameters can be calculated by solving the equations. If more configurations are sampled, then the number of the equations will exceeds the number of the variables, and the parallel manipulator can be calibrated by minimizing the norm of the vector E_1 corresponding to the sampled configurations. For example, for n sampled configurations, one can implement the calibration by minimizing the following function J_1 (Yiu et al., 2003c):

$$J_1 = \sum_{j=1}^{n} E_{1j}^T E_{1j} = \sum_{j=1}^{n} \left\| E_{1j} \right\|^2 \tag{37}$$

By eliminating the items involving passive joint angles, the closed-loop constrained equations can also be expressed as follows:

$$E_2 = \begin{bmatrix} \left(x - x_a^1 - l_a^1 \cdot \cos\left(\tilde{\theta}_a^1 + \Delta\theta_a^1\right)\right)^2 + \left(y - y_a^2 - l_a^1 \cdot \sin\left(\tilde{\theta}_a^1 + \Delta\theta_a^1\right)\right)^2 - (l_b^1)^2 \\ \left(x - x_a^2 - l_a^2 \cdot \cos\left(\tilde{\theta}_a^2 + \Delta\theta_a^2\right)\right)^2 + \left(y - y_a^2 - l_a^2 \cdot \sin\left(\tilde{\theta}_a^2 + \Delta\theta_a^2\right)\right)^2 - (l_b^2)^2 \\ \left(x - x_a^3 - l_a^3 \cdot \cos\left(\tilde{\theta}_a^3 + \Delta\theta_a^3\right)\right)^2 + \left(y - y_a^3 - l_a^3 \cdot \sin\left(\tilde{\theta}_a^3 + \Delta\theta_a^3\right)\right)^2 - (l_b^3)^2 \end{bmatrix} = 0 \tag{38}$$

Based on (38), $3n$ equations and $2n$+12 variable can be obtained with n sampled configurations. $2n$ variables are end-effector coordinates (x^j, y^j), j = 1, ... ,n, and the remaining other 12 variables are kinematic parameters of the parallel manipulator. And also, with enough sampled configurations, one can calibrate the unknown kinematic parameters of the parallel manipulator by minimizing the following function J_2:

$$J_2 = \sum_{j=1}^{n} E_{2j}^T E_{2j} = \sum_{j=1}^{n} \left\| E_{2j} \right\|^2 \tag{39}$$

Therefore, the calibration problem can be converted into a minimization problem, in which either error function J_1 in (37) or our new proposed the error function J_2 in (39) can be used as the error function.

3.2 Calibration procedure based on the new error function

The calibration procedure based on the minimization of the error function J_2 proposed in (39) are as follows:

1. Choose the kinematic parameters to be calibrated from the set of the 12 kinematic parameters mentioned above. Evaluate other kinematic parameters by other means and take them as constants for the calibration. Then choose n sampled configurations of the parallel manipulator and record the readings of the encoders $\tilde{\theta}_{aj}^1, \tilde{\theta}_{aj}^2, \tilde{\theta}_{aj}^3, j = 1, \cdots, n$. Make sure that the number of the sampled configurations exceeds the number of calibrated kinematic parameters.
2. Choose J_2 in (39) as error function.
3. Choose suitable initial values for variables. For the geometric parameters, the nominal value of the parameter can be used as the initial value. For the sensor zero positions, suitable initial value can be obtained through following procedures: move the end-effector manually to a reference point, e.g. the home position (216.5, 250), and record the readings of the absolute encoder $\tilde{\theta}_{a0}^i, i = 1,2,3$, then the initial value of the sensor zero positions can be calculated through equation $\Delta\hat{\theta}_a^i = \theta_{a0}^i - \tilde{\theta}_{a0}^i$. Here symbols $\theta_{a0}^i, i = 1,2,3$ refer to the active joint angles of the reference point, which can be calculated through inverse kinematic transformation.
4. Calculate the estimations of joint angles and end-effector coordinates of the sampled configurations. With the estimated value of the sensor zero positions, the estimated active joint angles can be calculated through $\hat{\theta}_{aj}^i = \tilde{\theta}_{aj}^i + \Delta\hat{\theta}_{aj}, i = 1,2,3, j = 1, \cdots, n$. Then the

estimations of the end-effector coordinates $\hat{x}_j, \hat{y}_j, j = 1,\cdots,n$ and the estimations of the passive joint angles $\hat{\theta}^i_{bj}, i = 1,2,3, j = 1,\cdots,n$ can be calculated through (33) and (34) respectively.

5. Solve the minimizing problem with the initial values of the variables obtained in step 3. For the minimization, the estimations of joint angles and end-effector coordinates can be calculated as in step 4, and the values of the variables corresponding to the minimal value of the error function can be regarded as calibration results.

3.3 Experimental results

To verify the validity of the calibration procedure and the error function proposed in subsection 3.1, simulation experiments are implemented in this subsection. In the experiments, a predefined 'actual value' is set for every kinematic parameter, and the encoder readings of the sampled configurations are calculated from the sampled end-effector coordinates with these 'actual values'. Then according to the calibration procedure proposed in 3.2, we can calibrate the parameters involved. We implement the simulation experiments by Matlab™ program, and adopt its optimizing function '*fmincon*' to solve the problems. Furthermore, we adopt a stepwise strategy for the experiments, and in each step we calibrate only a part of the kinematic parameters with the assumption that the remaining other parameters have been known accurately. Then by decreasing the number of parameters supposed to be known and increasing the number of parameters to be calibrated step by step, we try to calibrate as many parameters as possible with the calibration procedure. For the purpose of the calibration accuracy comparson, we do each experiment by using of both error functions J_1 and J_2.

The results of experiments can be examined by two means. The first one is to compare the calibrated results of the kinematic parameters with the predefined 'actual value'. The other one is to compare the calibrated end-effector coordinates $(\hat{x}_j, \hat{y}_j), j = 1,\cdots,n$ with the 'actual end effector coordinates' (x_j, y_j), $j = 1, \ldots, n$ of the sampled configurations, through the following 'kinematics model root mean square error' (Yiu et al., 2003c):

$$kmrmse = \sqrt{\frac{1}{n}\sum_{j=1}^{n}\left(\left(x_j - \hat{x}_j\right)^2 + \left(y_j - \hat{y}_j\right)^2\right)} \tag{40}$$

A. Calibration of Sensor Zero Positions

Suppose that all of the link lengths and the base coordinates have been measured accurately, and take the 3 unknown sensor zero positions $\Delta\theta^i_a$, $i = 1,2,3$ as variables, we can calibrate the variables by solving these problems:

$$\min J_1(\Delta\theta^1_a, \Delta\theta^2_a, \Delta\theta^3_a)$$

$$\min J_2(\Delta\theta^1_a, \Delta\theta^2_a, \Delta\theta^3_a)$$

The point (210,245) is taken as the estimation of the home point, and the initial estimations of $\Delta\theta^1_a$, $\Delta\theta^2_a$, $\Delta\theta^3_a$ are calculated by subtracting the encoder readings corresponding to point (210,245) from the active joint angles corresponding to the home point (216.5,250). 3 sampled configurations and 9 sampled configurations are chosen respectively for the simulations. Results are shown in Table 2, from which we can see that the precision of calibration results can be improved by increasing the number of the sampled configurations, as well as that more precise calibration results can be obtained by adopting J_2 as the error function.

Calibrated variable	Actual value	Initial estimation	Results calibrated with J_1		Results calibrated with J_2	
			3 samples	9 samples	3 samples	9 samples
$\Delta\theta_a^1$ (*rad*)	1.00000	1.00913	1.00078	1.00000	0.99999	1.00000
$\Delta\theta_a^2$ (*rad*)	1.00000	0.97689	0.99948	1.00000	1.00001	1.00000
$\Delta\theta_a^3$ (*rad*)	1.00000	1.02748	0.99976	1.00000	1.00000	1.00000
kmrmse of samples(*mm*)			1.47188e-1	7.85886e-4	1.49488e-3	1.26624e-5
Number of iterations			12	9	14	13

Table 2. Calibration results of sensor zero positions

B. Calibration of Sensor Zero Positions and Link Lengths

In this subsection, only the base coordinates are supposed to be known, while all of the sensor zero positions and the link lengths are chosen as variables. Then the calibration problem can be converted into the following problems:

$$\min J_1(\Delta\theta_a^1, \Delta\theta_a^2, \Delta\theta_a^3, l_a^1, l_b^1, l_a^2, l_b^2, l_a^3, l_b^3)$$

$$\min J_2(\Delta\theta_a^1, \Delta\theta_a^2, \Delta\theta_a^3, l_a^1, l_b^1, l_a^2, l_b^2, l_a^3, l_b^3)$$

As we'll demonstrate that error function J_2 is easier than J_1 and more precise results can be obtained through minimizing J_2. Before we move on, take a look at J_2, one can find that there are 2 product items of the variables in J_2, which can be expressed as follows:

$$\begin{aligned} l_a^i \bullet \cos\left(\tilde{\theta}_{aj}^i + \Delta\theta_a^i\right) &= \left(l_a^i \bullet \cos\left(\Delta\theta_a^i\right)\right) \bullet \cos\left(\tilde{\theta}_{aj}^i\right) - \left(l_a^i \bullet \sin\left(\Delta\theta_a^i\right)\right) \bullet \sin\left(\tilde{\theta}_{aj}^i\right) \\ l_a^i \bullet \sin\left(\tilde{\theta}_{ai}^j + \Delta\theta_a^i\right) &= \left(l_a^i \bullet \sin\left(\Delta\theta_a^i\right)\right) \bullet \cos\left(\tilde{\theta}_{aj}^i\right) + \left(l_a^i \bullet \cos\left(\Delta\theta_a^i\right)\right) \bullet \sin\left(\tilde{\theta}_{aj}^i\right) \end{aligned} \quad (41)$$

in which the product items $l_a^i \bullet \cos\left(\tilde{\theta}_{aj}^i + \Delta\theta_a^i\right)$ and $l_a^i \bullet \sin\left(\tilde{\theta}_{aj}^i + \Delta\theta_a^i\right)$ are decoupled into linear combinations of $l_a^i \bullet cos(\Delta\theta_a^i)$ and $l_a^i \bullet sin(\Delta\theta_a^i)$. So the product items in the error function can be eliminated by choosing $l_a^i \bullet cos(\Delta\theta_a^i)$ and $l_a^i \bullet sin(\Delta\theta_a^i)$ as calibrated variables. Let $l_{ac}^i = l_a^i \bullet cos(\Delta\theta_a^i)$, $l_{as}^i = l_a^i \bullet sin(\Delta\theta_a^i)$, $i = 1,2,3$, then we have:

$$E_3 = \begin{bmatrix} \left(x - x_a^1 - \left(l_{ac}^1 \cdot \cos\left(\tilde{\theta}_a^1\right) - l_{as}^1 \cdot \sin\left(\tilde{\theta}_a^1\right)\right)\right)^2 + \ldots \left(y - y_a^1 - \left(l_{ac}^1 \cdot \sin\left(\tilde{\theta}_a^1\right) + l_{as}^1 \cdot \cos\left(\tilde{\theta}_a^1\right)\right)\right)^2 - (l_b^1)^2 \\ \left(x - x_a^2 - \left(l_{ac}^2 \cdot \cos\left(\tilde{\theta}_a^2\right) - l_{as}^2 \cdot \sin\left(\tilde{\theta}_a^2\right)\right)\right)^2 + \ldots \left(y - y_a^2 - \left(l_{ac}^2 \cdot \sin\left(\tilde{\theta}_a^2\right) + l_{as}^2 \cdot \cos\left(\tilde{\theta}_a^2\right)\right)\right)^2 - (l_b^2)^2 \\ \left(x - x_a^3 - \left(l_{ac}^3 \cdot \cos\left(\tilde{\theta}_a^3\right) - l_{as}^3 \cdot \sin\left(\tilde{\theta}_a^3\right)\right)\right)^2 + \ldots \left(y - y_a^3 - \left(l_{ac}^3 \cdot \sin\left(\tilde{\theta}_a^3\right) + l_{as}^3 \cdot \cos\left(\tilde{\theta}_a^3\right)\right)\right)^2 - (l_b^3)^2 \end{bmatrix} = 0 \quad (42)$$

$$J_3 = \sum_{j=1}^{n} E_{3j}^T E_{3j} = \sum_{j=1}^{n} \left\| E_{3j} \right\|^2 \quad (43)$$

Thus we can also calibrate through another new error function that:

$$\min J3(l_{ac}^1, l_{as}^1, l_{ac}^2, l_{as}^2, l_{ac}^3, l_{as}^3, l_b^1, l_b^2, l_b^3)$$

When the calibration results are obtained by minimizing J_3, we can calculate the parameters of the manipulator by following equations:

$$l_a^i = (l_{ac}^i{}^2 + l_{as}^i{}^2)^{0.5}$$

$$\Delta\theta_a^i = arctg(l_{as}^i / l_{ac}^i) \quad i = 1,2,3 \tag{44}$$

Now we have three error functions. We take the same initial estimations of $\Delta\theta_a^1$, $\Delta\theta_a^2$, $\Delta\theta_a^3$ as used in the calibration of the sensor zero positions. The initial estimations for l_a^i, l_b^i, $i = 1,2,3$ are taken as the nominal values. 21 sampled configurations are taken. The sensor zero positions and the link lengths of the parallel manipulator are calibrated by solving the minimizing problems. Results are shown in Table 3, from which one can find that, with the same sampled configurations and the same initial estimations of the kinematic parameters, the precision of the calibration results are improved by adopting l_{ac}^i, l_{as}^i $i = 1,2,3$ as calibrated variables. And also, with J_3 as the error function, iterations cost gets lower to find the solution.

Calibrated variable	Actual value	Initial estimation	Results calibrated with function J_1	Results calibrated with function J_2	Results calibrated with function J_3
$\Delta\theta_a^1$ (*rad*)	1.00000	1.01052	0.99996	1.00001	1.00000
$\Delta\theta_a^2$ (*rad*)	1.00000	0.97614	1.00004	1.00000	1.00000
$\Delta\theta_a^3$ (*rad*)	1.00000	1.02659	1.00008	1.00001	1.00000
l_a^1 (*mm*)	244.1	244	244.08300	244.10012	244.09994
l_a^2 (*mm*)	244.2	244	244.19647	244.19818	244.19999
l_a^3 (*mm*)	243.5	244	244.49562	244.50038	244.49998
l_b^1 (*mm*)	243.8	244	243.78108	243.80393	243.79992
l_b^2 (*mm*)	244.2	244	244.21085	244.19883	244.20004
l_b^3 (*mm*)	244.2	244	244.22065	244.20262	244.20006
kmrmse of samples(*mm*)			3.21669e-3	1.56383e-3	1.33563e-5
Number of iterations			46	47	33

Table 3. Calibration results of sensor zero positions and link lengths

C. Calibration of Sensor Zero Positions, Link Lengths and Base Coordinates

The sensor zero positions and link lengths have been calibrated successfully in last subsection, next we will calibrate all of the kinematic parameters. Since the superiority of the error function J_3 has been demonstrated in the last subsection, we use only J_3 as the error function to calibrate the parameters here. Choose l_{ac}^i, l_{as}^i, l_b^i, $i = 1,2,3$ and the base coordinates x_a^2, x_a^3, y_a^3 as variables. Minimize the following function J_3:

$$min\ J3(l^1_{ac}, l^1_{as}, l^2_{ac}, l^2_{as}, l^3_{ac}, l^3_{as}, l^1_b, l^2_b, l^3_b, x^2_a, x^3_a, y^3_a)$$

31 sample configurations are taken and calibration results are shown in Table 4, from which one can see that there are several millimeter errors between the calibrated results and the predefined 'actual values' of the kinematic parameters. Further research reveals that, if we assume that the base coordinate x^2_a is known accurately before calibration, precise calibration results still can be obtained for the other 11 kinematic parameters by solving the problem:

$$min\ J_3(l^1_{ac}, l^1_{as}, l^2_{ac}, l^2_{as}, l^3_{ac}, l^3_{as}, l^1_b, l^2_b, l^3_b, x^3_a, y^3_a)$$

The same sampled configurations and the same initial estimations of the kinematic parameters are employed with that of the 12 parameter experiment, except that the base coordinates x^2_a is supposed to be known beforehand. Calibration results are shown in last column of the Table 4, from which one can see that all 11 kinematic parameters, including 3 sensor zero positions and 8 geometric parameters, can be calibrated accurately.

Calibrated variable	Actual value	Initial estimation	12 values calibrated with J_3	11values calibrated with J_3
$\Delta\theta^1_a$ (rad)	1.00000	1.01052	1.0072	1.00000
$\Delta\theta^2_a$ (rad)	1.00000	0.97612	1.0072	1.00000
$\Delta\theta^3_a$ (rad)	1.00000	1.02678	1.0072	1.00000
l^1_a (mm)	244.1	244	244.0998	244.0999
l^2_a (mm)	244.2	244	247.1998	244.2000
l^3_a (mm)	243.5	244	247.3005	244.4999
l^1_b (mm)	243.8	244	246.8959	243.7999
l^2_b (mm)	244.2	244	244.3009	244.1999
l^3_b (mm)	244.2	244	244.3005	244.2001
x^3_a (mm)	433.05	433	436.7084	433.0502
y^3_a (mm)	499.96	500	506.2960	499.9598
x^2_a (mm)	433.04	433	440.3557	
$kinrmse$ of samples(mm)		3.3521		1.0211e-4
Number of iterations		34		70

Table 4. Calibration results of 12 parameters

4. Complete kinematic parameters auto-calibration using stochastic optimization algorithms

We have already selected and defined some error functions in the former section and optimized them using Matlab™ function. But, as we know, the Matlab™ function we used is a local optimization method. As a real-world optimization problem, the corresponding error functions are very complex both on the number of variables and their multimodal features, thus it is very hard to converge to a global minimum using non-global optimization methods. According to No Free Lunch Theorems (NFLT, Wolper & Macready, 1997), no algorithm is perfect on all the problems. Although there have been a lot of tests on benchmark functions to evaluate different algorithms on different performances, for a specific optimization problem, an optimization algorithm has to be chosen appropriately to the structure of the problem itself.

There are mainly two classes of optimization algorithms. The first ones are the deterministic algorithms, including gradient-based algorithms *etc.* In this class, one must have some information about the objective function, such as gradients, and this information is used to determine the search direction in a deterministic manner at every step of the algorithm. If a problem is linear or nonlinear but convex, deterministic algorithms are readily applied to solve the problem and can perform very well. But, generally speaking, real-world problems are hardly such easy class. Most real-world problems are nonlinear, non-convex, multi-dimensional and have a lot of local minima. For these real-world problems, deterministic algorithms are inappropriate or bear very poor performances, because the objective function information is not available in many cases or the algorithms run very big risk to be trapped in local minimum and cannot escape. Due to these drawbacks, the use of deterministic algorithms in real-world applications is very limited. To address the problem of convergence to local optima, stochastic optimization algorithms are proposed and have been playing a rapidly growing role in the past few decades. Different from deterministic optimization algorithms, stochastic optimization algorithms deliberately introduce randomness into the search process and inherently accept weak candidate solutions, thus the search propcess could escape from local optima to local the global optimum. Moreover, the algorithms are less sensitive to noises and modeling errors. These algorithms mainly include genetic algorithms (GA, Holland, 1975; Goldberg, 1989), differential evolution (DE, Storn & Price, 1995) and particle swarm optimization (PSO, Kennedy & Eberhart, 1995) *etc.* Most of these algorithms are also inherently parallel, which makes the algorithms more efficient in searching for global solutions.

As for our calibration problem, it is not hard to find that the error functions in (37), (39) and (43) mentioned above are in continuous spaces, nonlinear, non-convex, multi-dimensional and have a lot of local minima. GA, PSO and DE are then the very natural choices. We designed auto-calibration based on GA, PSO and DE for simulation experiments and actual system calibration.

4.1 Auto-calibration based on GA

GA is a population-based optimization algorithm, in which a candidate solution is called an individual and individuals constitute population. The quality of the individuals is termed as their fitness, the higher quality an individual has, the higher fitness it owns. The individuals evolve mainly through reproduction, crossover and mutation operations. In our auto-calibration work based on GA, individuals are represented as binary string. If the search range of a parameter is $[X_{min}, X_{max}]$ and the precision requirement is p, then the length of the string to present this parameter is $L = \lceil \log 2((X_{max} - X_{min})/p) \rceil$. The individual length is the sum of the 12 parameters' string length. The population size is N. The initial population is initialized randomly. Then by decoding, the real values of the parameters to be calibrated are obtained. Use the expression of the base coordinates (30) and the equations (32) and (33) to estimate the positions of the end-effector. When the positions of the end-effector are estimated, using error functions J_1, J_2 or J_3 given in (37), (39) or (43) as the fitness functions, respectively, we can evaluate the fitness of the individuals. In the reproduction stage, we use tournament selection (Goldberg & Deb, 1991), which can make a lower selection pressure compared with roulette wheel selection. Copy the winner of the tournaments into the population. After reproduction, crossover operation is to be carried out with probability *PCrossover*. In our experiments, we apply single bit crossover, in which one bit is selected as the crossover point. With probability *PMutation*, we employ two bits mutation, in which two bits are mutated to produce a new individual. The process of reproduction, crossover and mutation iterates until the stop criteria that is defined later. In order to maintain the diversity of the population, half of the individuals are reinitialized when the population fitness does not improve for a number of generations *NR*. Both in the reproduction and reinitializing, the best individual is maintained.

4.2 Auto-calibration based on PSO

PSO is a new population-based optimization algorithm. Different from GA, searching is carried out straightforward in the search space of PSO and no genetic operation is needed. Every solution in the PSO algorithm is called a particle. A particle has its location and flying velocity. Define a particle's location and velocity of the kth iteration as X^k and V^k, and the state of this particle in the next iteration can be calculated as follows:

$$V^{k+1}=w^k V^k+C_1 R_1(P_{best}-X^k)+C_2 R_2(G_{best}-X^k)$$

$$X^{k+1}=X^k+V^{k+1} \tag{45}$$

in which w is inertia weight (Shi & Eberhart, 1998), C_1 and C_2 are predefined acceleration constants, R_1 and R_2 are random numbers generated in the range of [0,1]. P_{best} is the best location obtained ever by the particle itself, and G_{best} is the best location ever detected by the whole population.

In auto-calibration experiments based on PSO, suppose the population size is N and the N particles fly in a 12 dimensional search space. The location of the particle is represented as $X=(x_1, x_2, ..., x_{12})$, corresponding to the solution of the 12 parameters. The velocity of the particles is represented as $V=(v_1, v_2, ..., v_{12})$, corresponding to the flying over distance of the particle. The locations and velocities of the particles are initialized randomly. In the iterations, estimate the positions of the end-effector and evaluate the quality of the particles in the same way as it does in the auto-calibration based on GA. If a particle beats its P_{best}, update its P_{best}. And if it beats G_{best}, then update G_{best} correspondingly. Then according to (45), update every particle's location and velocity. In order to level up the search efficiency, search space $[X_{min}, X_{max}]$ is set to constrain the particles' movement. If a particle outpaces the border, it is placed on the border of the space. Also, in order to maintain the diversity of the population, half of the particles are reinitialized except the best when the population does not improve for NR generations.

4.3 Auto-calibration based on DE

DE is a new population-based optimization algorithm. Its operations are the same as GA in name, but very different in nature. The concepts of individual and population are the same as those of GA. In an n variable problem, the individual in DE is represented by a vector $X=(x_1, x_2, \ldots, x_n)$. And the population for each generation G can be represented as $X_{i,G}$, $i = 1,2,\ldots, N$, in which N is the population size. In mutation,

$$V_{i,G+1} = X_{r1,G} + F \bullet (X_{r2,G} - X_{r3,G}) \tag{46}$$

where there are random indexes r1, r2, r3$\in\{j \mid j\neq i, j\in[1,N]\}$ and $F\in[0,2]$. And in crossover,

$$\begin{aligned} U_{i,G+1} &= (U_{1i,G+1}, U_{2i,G+1}, \cdots, U_{ni,G+1}) \\ U_{ji,G+1} &= \begin{cases} V_{ji,G+1} & if(randb(j) \le CR) or\ j = rnbr(i) \\ X_{ji,G} & if(randb(j) > CR) or\ j \ne rnbr(i) \end{cases} \quad j=1,2,\cdots,n \quad i=1,2,\cdots N \end{aligned} \tag{47}$$

in which $randb(j)\in[0,1]$ is the jth evaluation of a norm random number. $CR\in[0,1]$ is the crossover constant set by the user. $Rnbr(i)$ is a randomly chosen index from n dimensions to ensure that at least one dimension parameter from $V_{i,G+1}$ can be attained by $U_{i,G+1}$. In selection stage,

$$X_{i,G+1} = \begin{cases} U_{i,G+1}, & if\ U_{i,G+1}\ is\ better\ than\ X_{i,G} \\ X_{i,G}, & otherwise \end{cases} \tag{48}$$

In our simulation based on DE, n=12.The initial population is randomly generated. In every generation, according to (46), we employ mutation. And then according to (47), crossover operation is implemented. After mutation and crossover, we check the individuals whether they are still in the search range of $[X_{min}, X_{max}]$. If any parameter is out of this range, this parameter is randomly regenerated. When this checking is finished, in the same way as the evaluation of individuals in GA and PSO, we evaluate $X_{i,G}$ and $U_{i,G+1}$. Based on the quality of $X_{i,G}$ and $U_{i,G+1}$, selection operation is carried out according to (48) and the individuals that are selected constitute the new population of next generation.

4.4 Simulation experiments

In the simulation experiments, we supposed that the base coordinates x_a^1, y_a^1, x_a^2 are equal to their nominal values. We sampled 50 configurations arbitrarily in the workspace of the manipulator, and recorded the encoder readings as $\tilde{\theta}_{aj}^i, j=1,2,3\ i=1,2,\cdots 50$.

The values of the kinematic parameters, including their nominal values provided by the producer, 'actual values' we predefined, and their ranges in the search space are set as in Table.5.

The control variable settings for each algorithm are described as follows.

In GA, we define population N=100, precision requirement p=1.0e-4, crossover probability $PCrossover$=0.85 and mutation probability $PMutation$=0.15, tournament scale is 4. Reinitialize half individuals when the error function value keeps still over NR=50 iterations. For PSO, we define population N=100, C_1=C_2=2 and the inertia weight w decreased from 0.9 to 0.1

with the iterations. Reinitialize half particles when the error function value keeps still over NR=50 iterations. In DE, N=150 , F=0.5 and CR=0.8 .
Our main task in simulation experiments is to test the convergence performances of each algorithm under error functions J_1, J_2 and J_3. Since the population of DE is not equal to that of GA and PSO, and also there are differences in the nature of algorithms, it is meaningless to compare the convergence performances with the generations. Instead, we can compare their convergences with the evaluations of the error functions. Thus, in order to make this uniform comparing criterion, we define that search process of each algorithm stops when the error functions evaluation times reach 5.0e7. The simulation experimental results are shown in Fig.3, Fig.4 and Fig.5, respectively. For clarity, the errors are presented as $\log_{10}(J_i)$, i = 1,2,3, and this error is defined as the convergence performance. One can see from Fig.3 that under function J_1, GA performs worst. It converges to 1.3477e3. DE performs best. It converges to 2.5351e-3 and its converging speed is also very fast. Between them is PSO, which converges to 2.6869e2. From Fig.4 one can see that due to using error function J_2, the errors get lower compared with J_1. DE is still the best one. It converges to 5.5626e-14 at a fast speed. With the product items decoupled, the search process would become easy, which is also verified by simulation results shown in Fig. 5, from which one can see that under J_3, the search results improve to different extents for different algorithms. Among them, DE improves most. By means of DE, J_3 reaches 3.5946e-20 in Fig. 4, while in Fig. 4, J_2 only gets to 5.5626e-14.

	y_a^2 (mm)	x_a^3 (mm)	y_a^3 (mm)	l_a^1 (mm)	l_a^2 (mm)	l_a^3 (mm)
Nominal Values	0	433	500	244	244	244
Actual Values	0.3	433.5	499.4	244.1	244.2	243.5
X_{min}	-5	428	495	239	239	239
X_{max}	5	438	505	249	249	249
	l_b^1 (mm)	l_b^2 (mm)	l_b^3 (mm)	$\Delta\theta_a^1$ (rad)	$\Delta\theta_a^2$ (rad)	$\Delta\theta_a^3$ (rad)
Nominal Values	244	244	244	0	0	0
Actual Values	243.8	244.2	244.6	0.01	-0.01	0.01
X_{min}	239	239	239	-0.2	-0.2	-0.2
X_{max}	249	249	249	0.2	0.2	0.2

Table 5. Value settings for the simulation experiments

All the results of the auto-calibration simulation experiments can also been seen in Table 6.

	GA	PSO	DE
J_1	1.3447e3	2.6869e2	2.5351e-3
J_2	8.9451e2	1.0079e2	5.5626e-14
J_3	1.0344e2	8.9561e0	3.5946e-20

Table 6. All the results of the simulation experiments

From the simulation experimental results given above, we can obtain the conclusions that: 1) Error function J_2 is simpler than J_1 by eliminating the items involving passive joint angles, and Error function J_3 is simpler than J_2 owing to decoupling the products items; 2) The DE has the best performances under all the three error functions both in convergence accuracy and speed. The algorithms on this calibration problem go from worst to best is: GA, PSO and DE, and 3) Since DE converges to 3.5946e-20 under J_3, very close to zero, the solution might be very close to the 'actual values'.

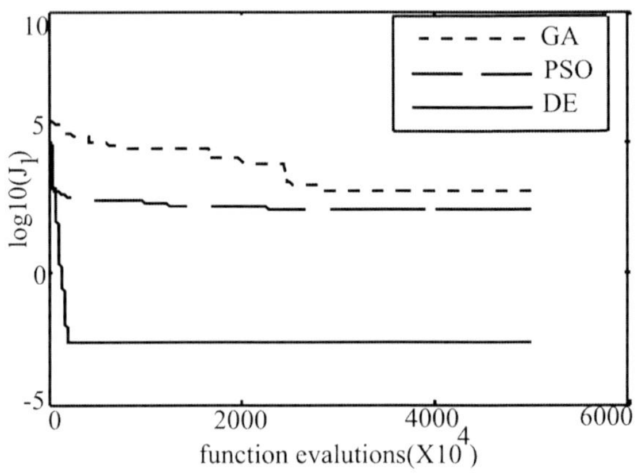

Fig. 3. The performances of the algorithms under function J_1

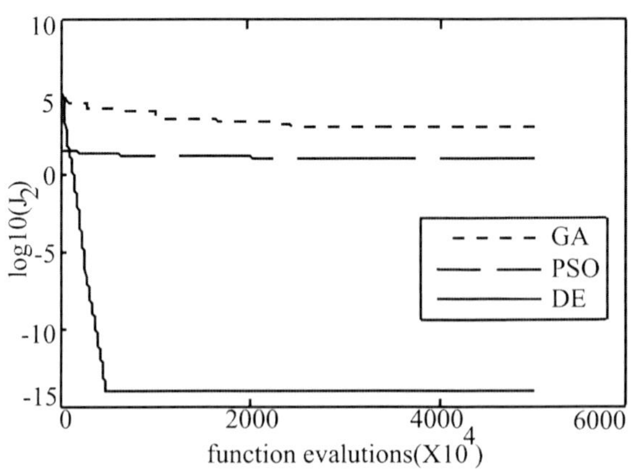

Fig. 4. The performances of the algorithms under function J_2

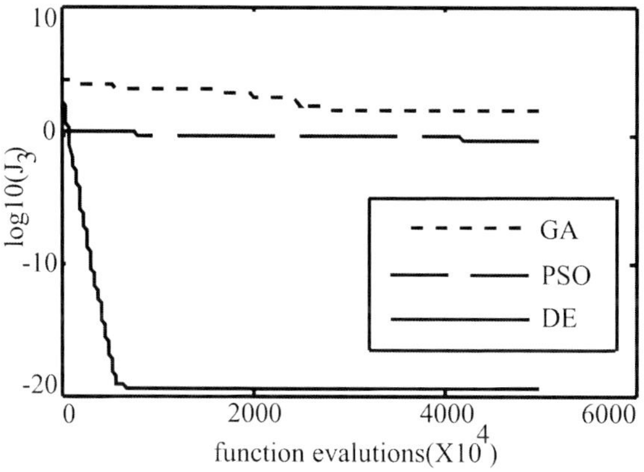

Fig.5 The performances of the algorithms under function J_3

In order to find whether DE's calibration solution under J_3 is really close to the 'actual values', we compare the solution's values to the 'actual values' and calculate the *kmrmse* of the solution, which is presented in Table 7.

parameters	actual values	DE calibration solution
y_a^2 (mm)	0.3000	0.3076
x_a^3 (mm)	433.5000	433.1470
y_a^3 (mm)	499.4000	499.6045
l_a^1 (mm)	244.1000	244.0004
l_a^2 (mm)	244.2000	244.1004
l_a^3 (mm)	243.5000	243.4006
l_b^1 (mm)	243.8000	243.7005
l_b^2 (mm)	244.2000	244.1004
l_b^3 (mm)	244.6000	244.5002
$\Delta\theta_a^1$ (rad)	0.01000	0.0107
$\Delta\theta_a^2$ (rad)	-0.01000	-0.0092
$\Delta\theta_a^3$ (rad)	0.01000	0.0107
kmrmse of samples(mm)		0.0371

Table 7. Calibration results of DE under J_3

One can see from Table 7 that the largest error between DE calibration solution and 'actual values' is 0.3530mm, most of the errors are less than 0.1mm, and the errors of the three sensor zero positions are all less than 0.001rad. Also, the *kmrmse* of the solution is only 0.0371mm, demonstrating that the calibration solution is very close to the 'actual values' indeed.

4.5 Actual calibration

In the simulation experiments above, the results demonstrate that J_3 is the best error function and DE is the most efficient algorithm for our calibration problem. Based on this conclusion, we applied DE calibration with error function J_3 on actual parallel manipulator. The nominal parameter values of the manipulator have been mentioned before. The search ranges are the same as presented in Table 5.

First we sampled 50 configurations all in a line in the work space of the manipulator. But we came across great difficulties in calibrating a rational result. A rather long time had passed when we found that the problem was in the sampled configurations. If the sampled configurations are in a line, the movements of the links are not sufficient, some of the links move widely, but others move little. Thus not all the components of the manipulator are excited sufficiently. Besides, due to the inherent error in sensors, if the sampled configurations are in a line, the errors in the three sensors are not balanced. On considering these factors, we sampled 50 configurations in a circle around the geometrical center of the work space. By doing so, all the components of the manipulator can be excited sufficiently and the sensor errors can be balanced completely. The calibration results based on these sampled configurations proved that our analysis was right.

Using the 50 sampled configurations in a circle around the geometrical center of the work space and through DE method, whose parameter settings are the same as those in the simulation experiments, we obtained the all the 12 parameters. Because of the inevitable inaccuracy of the sensors, the error J_3 in actual experiment cannot get down to the level of the simulation. After about 1e6 evaluations of the error function of J_3, the error does not improve any more. We stopped the optimization procedure then, and regarded the best solution found ever as the calibration results. The results are reported in Table 8.

parameters	nominal values	values through calibration
y_a^2 (mm)	0.0000	2.7571
x_a^3 (mm)	433.0000	436.2436
y_a^3 (mm)	500.0000	501.9123
l_a^1 (mm)	244.0000	243.6527
l_a^2 (mm)	244.0000	242.5634
l_a^3 (mm)	244.0000	242.4579
l_b^1 (mm)	244.0000	243.8194
l_b^2 (mm)	244.0000	243.4168
l_b^3 (mm)	244.0000	246.8952
$\Delta\theta_a^1$ (rad)	0.0000	6.2385e-3
$\Delta\theta_a^2$ (rad)	0.0000	1.0623e-2
$\Delta\theta_a^3$ (rad)	0.0000	2.8643e-3

Table 8. The results of the actual calibration

For the purpose of comparison, we calculated the J_3 error under the two sets of parameters. Under the nominal values, the J_3 error is 2.1517e5, while under the calibrated values, J_3 gets down to 3.1841e2, which demonstrates that the accuracy of calibrated parameters is much higher than that of the nominal values.

5. Conclusion

In this chapter, we implemented the kinematic auto-calibration of a redundant planar 2-dof parallel manipulator. In this process, we first calibrated the error of the sensor zero positions by optimizing an projected tracking error function, and also the robustness of this method has been proved. Furthermore, in order to calibrate the other parameters of this parallel manipulator, we gave another error function based on the closed-loop constraint equations. By decoupling the product items in the error function, we simplified the optimization and more precise result was obtained. But, at most 11 out 12 parameters could be calibrated using only local optimization method. In order to calibrate all of the parameters, global optimization methods including GA, PSO and DE were applied. In simulation experiments, differential evolution was proved to be the most approriate algorithm for the calibration problem. Finally, all the parameters of an real-world redundant planar 2-dof parallel manipulator were calibrated successfully by applying differential evolution to optimize the decoupled error function.

6. Acknowledgements

This work was supported by the National Natural Science Fundation of China under Grant No. 50375148.

7. References

Baron, L. & Angeles, J. (1998). The on-line direct kinematics of parallel manipulators under joint-sensor redundancy. *Proceedings of international symposium on advances in robot kinematics*, pp 126-137, Strobl, Austria, 1998.

Cheng, H.; Yiu, Y.K. & Li, Z.X. (2003) Dynamics and Control of Redundantly Actuated Parallel Manipulators. *IEEE/ASME Transactions on Mechatronics*. Vol.8, No.4, pp.483-491.

Chiu, Y.J. & Perng, M.H. (2004). M.H. Self-calibration of a general hexapod manipulator with enhanced precision in 5-DOF motions. *Mechanism and Machine Theory* Vol.39, No.1, pp.1 23.

Glodberg, D.E.(1989). *Genetic Algorithms in Search. Optimization and Machine Learning.* Addison-Wesley Press.

Goldberg, D. E. & Deb, K. (1991). A comparative analysis of selection schemes used in genetic algorithms. *Foundations of Genetic Algorithms.* pp. 69-93, San Mateo, CA: Morgan Kaufmann.

Holland,J. (1975). Adaptation in Nature and Artificial System, The University of Michigan Press.

Hollerbach, J.M. & Wampler, C.W. (1996). The calibration index and taxonomy for robot kinematic calibration methods. *International Journal of Robotics Research* Vol.15, No.6, pp.573-591.

Jeong, W.J.; Kim, H.S. & Kwak, K.Y. (1999). Kinematics and workspace analysis of a parallel wire mechanism for measuring a robot pose. *Mechanism and Machine Theory*, Vol. 34, No. 6, pp. 825-841.

Kennedy, J. & Eberhart, R. (1995). Particle swarm optimization. *Proceedings of IEEE International Conference of Neural Networks*, pp.1942-1948, Perth, Australia, 1995.

Koseki, Y.; Arai, T.; Sugimoto, K.; Takatuji, T. & Goto, M. (1998). Design and accuracy evaluation of high-speed and high-precision parallel mechanism. *Proceedings of the IEEE International Conference on Robotics and Automation*, pp. 1340-1345, Leuven, May, 1998.

Khalil, W. & Besnard, S. (1999). Self calibration of stewart-gough parallel robots without extra sensors. *IEEE Transactions on Robotics and Automation* Vol.15, No.6, pp.1116-1121.

Kock, S. & Schumacher, W. (2000a). Mixed elastic and rigid-body dynamic model of an actuation redundant parallel robot with high-reduction gears. *Proceedings of the International Conference on Robotics and Automation*, pp. 1918-1923, 2000.

Kock, S. & Schumacher, W. (2000b) Control of a fast parallel robot with a redundant chain and gearboxes: experimental results. *Proceedings of the International Conference on Robotics and Automation*, pp. 1924-1929, 2000.

Liu, G.F.; Cheng, H.; Xiong, Z.H.; Wu, X.Z.; Wu, Y.L. & Li, Z.X. (2001a). Distribution of singularity and optimal control of redundant parallel manipulators. *Proceedings of the International Conferentce on Intelligent Robots and Systems*, pp. 177-182, Maui, 2001.

Liu, G.F.; Wu, Y.L.; Wu, X.Z.; Yiu, Y.K. & Li, Z.X. (2001b). Analysis and control of redundant parallel manipulators. *Proceedings of the International Conference on Robotics and Automation*, pp. 3748-3754, 2001

Liu, G.F. & Li, Z.X. (2002) A unified geometric approach to modeling and control of constrained mechanical systems. *IEEE Transactions on Robotics and Automation*. Vol.18, No.4, pp.574-587.

Liu, G.F.; Lou, Y. & Li, Z. (2003). Singularities of parallel manipulators: A geometric treatment. *IEEE Transactions on Robotics and Automation*. Vol.19, No.4, pp.579-594.

Masory, O.; Wang, J. & Zhuang, H. (1993). On the accuracy of a stewart platform-Part II: kinematic calibration and compensation. *Proceedings of the IEEE international conference on robotics and automation*, pp 725-731, Atlanta, USA, 1993.

Patel, A.J. & Ehmann, K.F. (2000). Calibration of a hexapod machine tool using a redundant leg. *International Journal of Maching Tools and Manufacture* Vol.40, No.4, pp.489-512.

Renaud, P.; Andreff, N.; Lavest, J,M. & Dhome, M. (2006). Simplifying the kinematic calibration of parallel mechanisms using vision-based metrology. *IEEE Transaction on Robotics*, Vol. 22, No. 1, pp. 12-22.

Shen, H.; Wu, X.Z; Liu, G.F & Li, Z.X. (2003). Hybrid position/force adaptive control of redundantly actuated parallel manipulators. *Zidonghua Xuebao/Acta Automatica Sinica*, Vol.29, No.4, pp.567-572.

Shi, Y. & Eberhart, R. (1998). A Modified Particle Swarm Optimize. *Evolutionary Computation Proceedings 1998. IEEE World Congress on Computational Intelligence, the 1998 IEEE International Conference on* , pp.69-73.

Storn.R. & Price.K. (1995). Differential Evolution—A Simple and Efficient Adaptive Scheme for Global Optimization over Continuous Spaces. *Technical Report TR-95-012*, ICSI.

Vincze, M.; Prenninger, J.P. & Gander, H. (1994). A laser tracking system to measure position and orientation of robot end-effectors under motion. *International Journal of Robotics Research*, Vol. 13, No. 4, pp. 305-314.

Wampler, C.W.; Hollerbach, J.M. & Arai, T. (1995). An implicit loop method for kinematic calibration and its application to closed-chain mechanisms. *IEEE Transactions on Robotics and Automation* Vol.11, No.5, pp.710-724.

Wang, J. & Masory, O. (1993). On the accuracy of a stewart platform-Part I, the effect of manufacturing tolerances. *Proceedings of the IEEE international conference on robotics and automation*, Atlanta, USA, pp.114 - 120, 1993.

Wolpert, D.H. & Macready, W.G.. (1997). No Free Lunch Theorems for Optimization. *IEEE Transaction on Evolutionary Compuatation*, Vol,1, No.1, pp.67-82.

Yiu, Y.K. & Li, Z.X. (2001). Dynamics of a Planar 2-dof Redundant Parallel Robot. *International Conference on Mechatronics Technology*, pp. 339-244, 2001

Yiu, Y.K. & Li, Z.X. (2003a). PID and adaptive robust control of a 2-dof over-actuated parallel manipulator for tracking different trajectory. *International Symposium on Computational Intelligence in Robotics and Automation*, pp. 1052-1057, 2003.

Yiu,Y.K. & Li, Z.X. (2003b). Optimal forward kinematics map for a parallel manipulator with sensor redundancy. Proceedings of *International Symposium on Computational Intelligence in Robotics and Automation*, pp. 354-359 ,2003.

Yiu, Y.K.; Meng, J. & Li, Z.X. (2003c). Auto-calibration for a parallel manipulator with sensor redundancy. *Proceedings of the IEEE international conference on robotics and automation*, pp 3660-3665, Taipei, Taiwan, 2003.

Zhang, Y.X. & Cong, S. (2005). Optimal motion control and simulation of redundantly actuated 2-dof planar parallel manipulator. *Journal of System Simulation*, Vol.17, No.10, pp.2450-2454.

Zhang, L.J. (2006). Performance analysis and dimension optimization of 2-dof parallel manipulators, *Dissertation for the Doctoral Degree in Engineering*, Yanshan university.

Zhuang, H.; Masory, O. & Yan, J. (1995). Kinematic calibration of a stewart platform using pose measurements obtained by a single theodolite. *Proceedings of the IEEE international conference on intelligent robots and systems*, pp 329-334, Pittsburgh, USA, (1995)

Zhuang, H. (1997). Self-calibration of parallel mechanisms with a case study on stewart platforms. *IEEE Transaction on Robotics and Automation* Vol.13, No.3, pp.387-397.

Zhuang, H.; Yan, J. & Masory, O. (1998). Calibration of stewart platforms and other parallel manipulators by minimizing inverse kinematic residuals. *Journal of Robotics and System* Vol.15, No.7, pp.395-405.

Zou, H. & Notash, L. (2001). Discussions on the camera-aided calibration of parallel manipulators. *Proceedings of CCToMM Symposium on Mechanisms, Machines, and Mechatronics,* Montreal, Canada, 2001.

13

Error Modeling and Accuracy of TAU Robot

Hongliang Cui, Zhenqi Zhu,
Zhongxue Gan and
Torgny Brogardh

1. Introduction

The TAU parallel configuration is rooted in a series of inventions and was masterminded by Torgny Brogardh [1][2][3][4]. The configuration of the robot simulates the shape of " τ " like the name of the Delta after the " ∇ " shape configuration of another parallel robot.

As shown in Fig. 1, the basic TAU configuration consists of 3 driving axes, 3 arms, 6 linkages, 12 joints and a moving (tool) plate. There are 6 chains connecting the main column to the end-effector in TAU configuration. The TAU robot is a typical 3/2/1 configuration. There are 3 parallel and identical links and another 2 parallel and identical links. Six chains will be used to derive all kinematic equations. Table 1 highlights the features of the TAU configuration.

On the subject of D-H modeling, Tasi [5], Raghavan [6], Abderrahim and Whittaker[12] have applied the method and studied the limitations of various modeling methods. On the subject of forward kinematics, focus has been on finding closed form solutions based on various robotic configurations, and numerical solutions for difficult configurations of robots. It can be found in the work done by Dhingra [8], Shi [14], Didrit [16], Zhang [17], Nanua [18], Sreenivasan [19], Griffis and Duffy [20], Lin [21]. On the subject of error analysis, Wang and Masory [7], Gong [11], Patel and Ehmann [13] used forward solutions to obtain errors. Jacobian matrix was also used in obtaining errors. On the subject of the variation of parallel configurations, from the work done by Dhingra [9][10], Geng and Haynes [15], the influence of the configurations on the methods of finding closed form solutions can be found.

In this paper, the D-H model is used to define the TAU robot, a complete set of parameters is included in the modeling process. Kinematic modeling and error modeling are established with all errors using Jacobian matrix method for the TAU robot. Meanwhile, a very effective Jacobian Approximation Method is introduced to calculate the forward kinematic problem instead of Newton-Raphson method. It denotes that a closed form solution can be obtained instead of a numerical solution. A full size Jacobian matrix is used in carrying out error analysis, error budget, and model parameter estimation and identification. Simulation results indicate that both Jacobian matrix and Jacobian Approximation Method are correct and have an accuracy of micron meters. ADAMS simulation results are used in verifying the established models.

	Serial Robot	Stewart Platform	Tau configuration
Stiffness	Low	High	High (simulation)
Accuracy	Low	High	High (simulation)
Workspace	Large	Small	Large
Footprint	Small	Large	Small
Inverse solution in general	Easy	Easy	Difficult
Analytical inverse solution	Easy	Easy	Difficult
Forward solution in general	Easy	Difficult	Easy
Analytical forward solution	Easy	Difficult	Easy

Table 1. Comparison of kinematic properties of TAU and other robots.

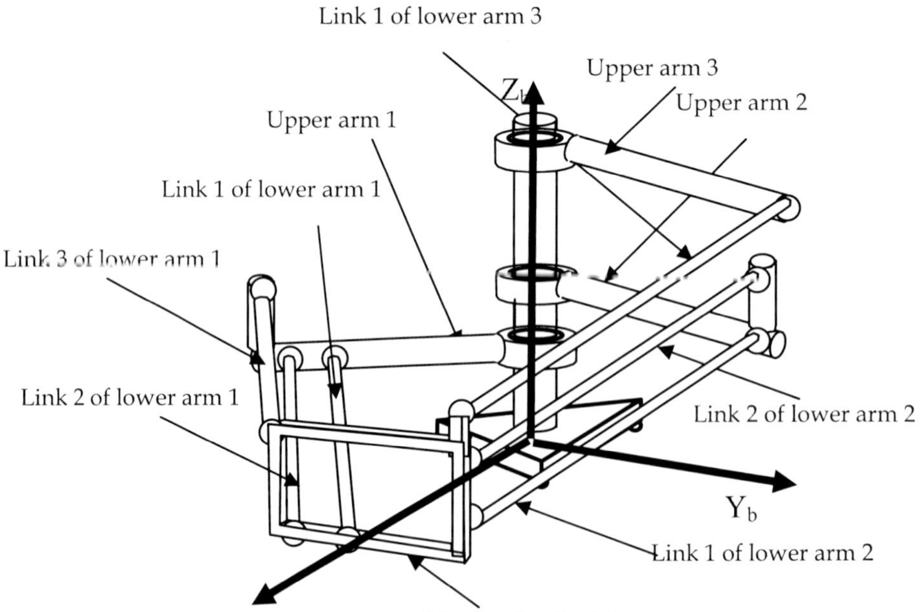

Fig. 1 TAU robot configuration

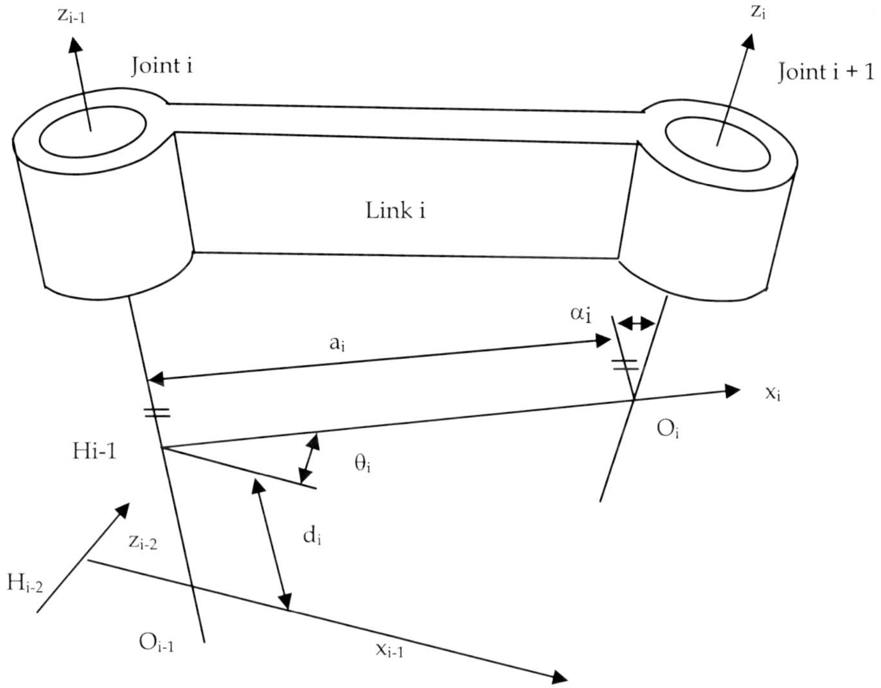

Fig. 2 Parameter definition of D-H model

2. Kinematic modeling

2.2 The D-H model of TAU robot

For the TAU robot, the D-H model is used for the following purposes:

(1) Fully describing the kinematic positional relationship among all the links and joints.

(2) Accurately and easily integrating the error model into a full parameter model.

(3) Standardizing and parameterizing the TAU model to establish dynamic coupling control model.

With the parameters defined in Fig. 2, the D-H model transformation matrix can be obtained as follows

$$A = \begin{bmatrix} \cos\theta_i & \sin\theta_i & 0 & -a_i \\ -\cos\alpha_i \sin\theta_i & -\cos\alpha_i \sin\theta_i & \sin\alpha_i & -d_i \sin\alpha_i \\ \sin\alpha_i \sin\theta_i & -\cos\alpha_i \sin\theta_i & \cos\alpha_i & -d_i \cos\alpha_i \\ 0 & 0 & 0 & 1 \end{bmatrix}$$

2.3 Inverse kinematics and forward kinematics

For the TAU robot, the inverse kinematic and forward kinematic are relatively simple. The six equations of kinematic chains remain 3, as shown in Fig. 3, based on the condition of parallel and identical links.

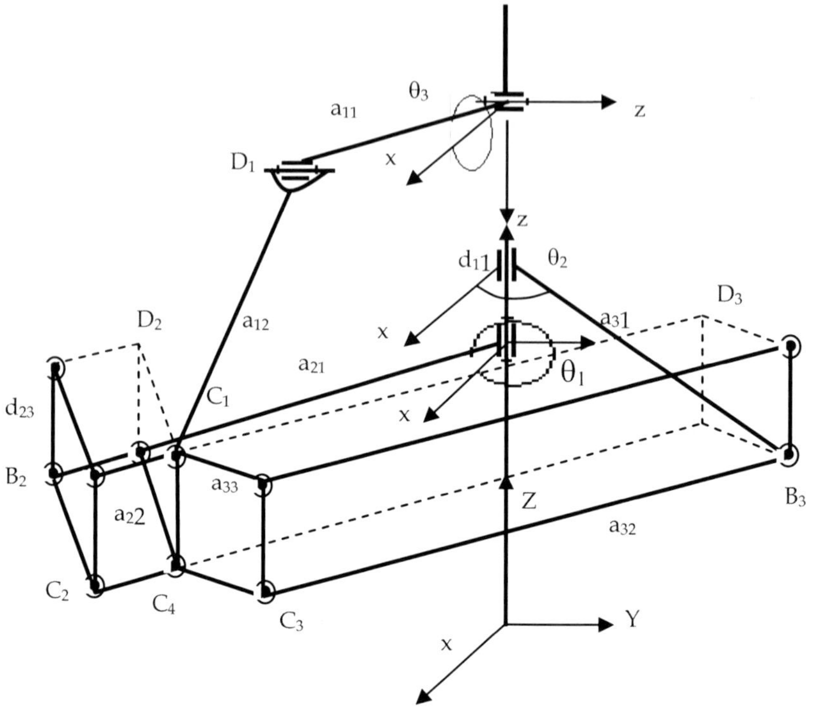

Fig. 3 Tau parallel mechanism

Coordinates of D1 are obtained as,

$$d_{1x} = a_{11}\cos((\theta_1+\theta_2)/2)\cos\theta_3$$
$$d_{1y} = a_{11}\cos((\theta_1+\theta_2)/2)\sin\theta_3$$
$$d_{1z} = -a_{11}\sin\theta_1 + d_{11}$$

$$c_{1x} = p_x$$
$$c_{1y} = p_y$$
$$c_{1z} = p_z$$

Where P_x, P_y, and P_z are the coordinates of C1.

$$dist(d_1 - c_1) = a_{12} \tag{1}$$

Coordinates of D2 are obtained as,

$$d_{2x} = a_{21}\cos(\theta_1)$$
$$d_{2y} = a_{21}\sin(\theta_1)$$
$$d_{2z} = d_{21} + d_{23}$$

$$c_{2x} = p_x$$
$$c_{2y} = p_y$$
$$c_{2z} = p_z - d_{23}$$

$$dist(d_2 - c_1) = a_{22} \tag{2}$$

Coordinates of D3 are obtained as,

$$d_{3x} = a_{31}\cos(\theta_2) - a_{33}\cos(120 + \theta_1)$$
$$d_{3y} = a_{31}\sin(\theta_2) - a_{33}\sin(120 + \theta_1)$$
$$d_{3z} = d_{31}$$

$$dist(d_3 - c_1) = a_{32} \tag{3}$$

For inverse kinematics, simplify the Equation 2 and assume next expressions,

$$\cos(\delta) = \frac{p_x}{\sqrt{p_x^2 + p_y^2}},\ \sin(\delta) = \frac{p_y}{\sqrt{p_x^2 + p_y^2}} \tag{4a}$$

The new equation 5a can be obtained from Equation 2.

$$2a_{21}\sqrt{p_x^2 + p_y^2}\left(\frac{p_x}{\sqrt{p_x^2 + p_y^2}}\cos\theta_1 + \frac{p_y}{\sqrt{p_x^2 + p_y^2}}\sin\theta_1\right) = a_{21}^2 + (p_x^2 + p_y^2 + p_z^2) - a_{22}^2 \tag{5a}$$

Then substitute the equation 4a into equation 5a to get

$$\cos(\theta_1 - \delta) = \frac{a_{21}^2 + (p_x^2 + p_y^2 + p_z^2) - a_{22}^2}{2a_{21}\sqrt{p_x^2 + p_y^2}}$$

Thus,

$$\theta_1 = \cos^{-1}\left[\frac{a_{21}^2 + (p_x^2 + p_y^2 + p_z^2) - a_{22}^2}{2a_{21}\sqrt{p_x^2 + p_y^2}}\right] + \delta \tag{6a}$$

where $\delta = tg^{-1}(\frac{p_y}{p_x})$

Assume next expressions as,

$$p'_x = p_x - a_{33}\cos(\theta_1 + 120) \qquad \cos(\gamma) = \frac{p'_x}{\sqrt{p'^2_x + p'^2_y}}$$
$$\text{and} \tag{7a}$$
$$p'_y = p_y - a_{33}\sin(\theta_1 + 120) \qquad \sin(\gamma) = \frac{p'_y}{\sqrt{p'^2_x + p'^2_y}}$$

Substitute the Equation 7a into Equation 3, the equation 8a can be obtained as,

$$\theta_2 = \cos^{-1}[\frac{a_{31}^2 - a_{32}^2 + (p'^2_x + p'^2_y + p'^2_z) - a_{22}^2}{2a_{31}\sqrt{p'^2_x + p'^2_y}}] + \gamma \tag{8a}$$

where $\gamma = tg^{-1}(\frac{p'_x}{p'_y})$

Also the Equation 9a can be obtained by substituting the equation 6a, 8a into equation 1.

$$\theta_3 = \cos^{-1}[\frac{a_{11}^2 + p_x^2 + p_y^2 + (p_z - d_{11})^2 - a_{12}^2}{2\sqrt{[a_{11}\cos(\frac{\theta_1 + \theta_2}{2}) + a_{11}\sin(\frac{\theta_1 + \theta_2}{2})]^2 + (p_z - d_{11})^2}}] - \phi \tag{9a}$$

where $\phi = tg^{-1}[\frac{p_z - d_{11}}{a_{11}\cos(\frac{\theta_1 + \theta_2}{2}) + a_{11}\sin(\frac{\theta_1 + \theta_2}{2})}]$

For forward kinematics, it is relatively easy. Subtractig equation 2 from Equation 1 for eliminating the square items (p_x^2, p_y^2, p_z^2), then do the same procedure to Equation 2 and 3, finally three linear equations can be obtained. The three length equations are applied to solve inverse and forward problems. A closed form solution can be obtained from the three equations for both inverse and forward problems.

3. Jacobian matrix of TAU robot with all error parameters

In error analysis, error sensitivity is represented by the Jacobian matrix. Derivation of the Jacobian matrix can be carried out after all the D-H models are established. For the TAU robot, the 3-DOF kinematic problem will become a 6-DOF kinematic problem. The kinematic problem becomes more complicated.

In fact, the error sensitivity is formulated through $\frac{\partial x}{\partial g_i}$, $\frac{\partial y}{\partial g_i}$, $\frac{\partial z}{\partial g_i}$ where x, y, z represent the position of the tool plate and dg_i is the error source for each component. So the following equations can be obtained:

$$dx = \sum_1^N \frac{\partial x}{\partial l_i} dg_i \qquad dy = \sum_1^N \frac{\partial y}{\partial l_i} dg_i \qquad dz = \sum_1^N \frac{\partial z}{\partial l_i} dg_i$$

The error model is actually a 6-DOF model since all error sources have been considered. It includes both the position variables X, Y, Z and also rotational angles α, β, γ.
From the six kinematic chains, equations established based on D-H models are

$$\begin{aligned} f_1 &= f_1(x, y, z, \alpha, \beta, \gamma, g) = 0 \\ f_2 &= f_2(x, y, z, \alpha, \beta, \gamma, g) = 0 \\ &\cdots\cdots\cdots\cdots\cdots\cdots\cdots\cdots \\ f_6 &= f_6(x, y, z, \alpha, \beta, \gamma, g) = 0 \end{aligned}$$

Differentiating all the equations against all the variables $x, y, z, \alpha, \beta, \gamma$ and g, where g is a vector including all geometric parameters:

$$\frac{\partial f_i}{\partial x}\cdot dx + \frac{\partial f_i}{\partial y}\cdot dy + \frac{\partial f_i}{\partial z}\cdot dz + \frac{\partial f_i}{\partial \alpha}\cdot d\alpha + \frac{\partial f_i}{\partial \beta}\cdot d\beta + \frac{\partial f_i}{\partial \gamma}\cdot d\gamma + \sum_j \frac{\partial f_i}{\partial g_j}\cdot dg_j = 0 \quad (4)$$

Rewrite it in matrix as

$$\begin{bmatrix} \frac{\partial f_1}{\partial x} & \frac{\partial f_1}{\partial y} & \frac{\partial f_1}{\partial z} & \frac{\partial f_1}{\partial \alpha} & \frac{\partial f_1}{\partial \beta} & \frac{\partial f_1}{\partial \gamma} \\ \frac{\partial f_2}{\partial x} & \frac{\partial f_2}{\partial y} & \frac{\partial f_2}{\partial z} & \frac{\partial f_2}{\partial \alpha} & \frac{\partial f_2}{\partial \beta} & \frac{\partial f_2}{\partial \gamma} \\ \frac{\partial f_3}{\partial x} & \frac{\partial f_3}{\partial y} & \frac{\partial f_3}{\partial z} & \frac{\partial f_3}{\partial \alpha} & \frac{\partial f_3}{\partial \beta} & \frac{\partial f_3}{\partial \gamma} \\ \frac{\partial f_4}{\partial x} & \frac{\partial f_4}{\partial y} & \frac{\partial f_4}{\partial z} & \frac{\partial f_4}{\partial \alpha} & \frac{\partial f_4}{\partial \beta} & \frac{\partial f_4}{\partial \gamma} \\ \frac{\partial f_5}{\partial x} & \frac{\partial f_5}{\partial y} & \frac{\partial f_5}{\partial z} & \frac{\partial f_5}{\partial \alpha} & \frac{\partial f_5}{\partial \beta} & \frac{\partial f_5}{\partial \gamma} \\ \frac{\partial f_6}{\partial x} & \frac{\partial f_6}{\partial y} & \frac{\partial f_6}{\partial z} & \frac{\partial f_6}{\partial \alpha} & \frac{\partial f_6}{\partial \beta} & \frac{\partial f_6}{\partial \gamma} \end{bmatrix} \bullet \begin{bmatrix} dx \\ dy \\ dz \\ d\alpha \\ d\beta \\ d\gamma \end{bmatrix} = \begin{bmatrix} \sum_j \frac{-\partial f_1}{\partial g_j} dg_j \\ \sum_j \frac{-\partial f_2}{\partial g_j} dg_j \\ \sum_j \frac{-\partial f_3}{\partial g_j} dg_j \\ \sum_j \frac{-\partial f_4}{\partial g_j} dg_j \\ \sum_j \frac{-\partial f_5}{\partial g_j} dg_j \\ \sum_j \frac{-\partial f_6}{\partial g_j} dg_j \end{bmatrix} \quad (5)$$

In a compact form, it becomes

$$J_1 dX = dG \tag{6}$$

Where

$$dG=\begin{bmatrix} \sum_j \frac{-\partial f_1}{\partial g_j} dg_j \\ \sum_j \frac{-\partial f_2}{\partial g_j} dg_j \\ \sum_j \frac{-\partial f_3}{\partial g_j} dg_j \\ \sum_j \frac{-\partial f_4}{\partial g_j} dg_j \\ \sum_j \frac{-\partial f_5}{\partial g_j} dg_j \\ \sum_j \frac{-\partial f_6}{\partial g_j} dg_j \end{bmatrix} = -\begin{bmatrix} \frac{\partial f_1}{\partial g_1} & \frac{\partial f_1}{\partial g_2} & \cdots & \frac{\partial f_1}{\partial g_N} \\ \vdots & \vdots & \cdots & \vdots \\ \frac{\partial f_6}{\partial g_1} & \frac{\partial f_6}{\partial g_2} & \cdots & \frac{\partial f_6}{\partial g_N} \end{bmatrix}_{6\times N} \bullet \begin{bmatrix} dg_1 \\ dg_2 \\ \vdots \\ dg_N \end{bmatrix}_{N\times 1} \tag{7}$$

From Eq. (7) above, we have

$$dG = J_2 dg \tag{8}$$

Substitute Eq.(6) into Eq.(8) to obtain

$$J_1 dX = J_2 dg \tag{9}$$

$$dX = (J_1^{-1} J_2) dg \tag{10}$$

The Jacobian matrix is obtained as $J_1^{-1} \cdot J_2$

$$J = J_1^{-1} \cdot J_2 = \begin{bmatrix} \frac{\partial f_1}{\partial x} & \frac{\partial f_1}{\partial y} & \frac{\partial f_1}{\partial z} & \frac{\partial f_1}{\partial \alpha} & \frac{\partial f_1}{\partial \beta} & \frac{\partial f_1}{\partial \gamma} \\ \frac{\partial f_2}{\partial x} & \frac{\partial f_2}{\partial y} & \frac{\partial f_2}{\partial z} & \frac{\partial f_2}{\partial \alpha} & \frac{\partial f_2}{\partial \beta} & \frac{\partial f_2}{\partial \gamma} \\ \frac{\partial f_3}{\partial x} & \frac{\partial f_3}{\partial y} & \frac{\partial f_3}{\partial z} & \frac{\partial f_3}{\partial \alpha} & \frac{\partial f_3}{\partial \beta} & \frac{\partial f_3}{\partial \gamma} \\ \frac{\partial f_4}{\partial x} & \frac{\partial f_4}{\partial y} & \frac{\partial f_4}{\partial z} & \frac{\partial f_4}{\partial \alpha} & \frac{\partial f_4}{\partial \beta} & \frac{\partial f_4}{\partial \gamma} \\ \frac{\partial f_5}{\partial x} & \frac{\partial f_5}{\partial y} & \frac{\partial f_5}{\partial z} & \frac{\partial f_5}{\partial \alpha} & \frac{\partial f_5}{\partial \beta} & \frac{\partial f_5}{\partial \gamma} \\ \frac{\partial f_6}{\partial x} & \frac{\partial f_6}{\partial y} & \frac{\partial f_6}{\partial z} & \frac{\partial f_6}{\partial \alpha} & \frac{\partial f_6}{\partial \beta} & \frac{\partial f_6}{\partial \gamma} \end{bmatrix}^{-1} \begin{bmatrix} -\frac{\partial f_1}{\partial g_1} & -\frac{\partial f_1}{\partial g_2} & \cdots & -\frac{\partial f_1}{\partial g_N} \\ \vdots & \vdots & \cdots & \vdots \\ -\frac{\partial f_6}{\partial g_1} & -\frac{\partial f_6}{\partial g_2} & \cdots & -\frac{\partial f_6}{\partial g_N} \end{bmatrix} \tag{11}$$

For a prototype of the TAU robotic design, the dimension of the Jacobian matrix is 6 by 71. An analytical solution can be obtained and is used in our analysis.

4. Kinematic modeling with all error parameters (application 1 of the Jacobian matrix)

4.1 Newton-Raphson numerical method

Because of the number of parameters involved as well as the number of error sources involved, the kinematic problem becomes very complicated. No analytical solution can be obtained but numerical solution. The TAU configuration, as a special case of parallel robots, its forward kinematic problem is, therefore, very complicated. The Newton-Raphson method as an effective numerical method can be applied to calculate the forward problem of the TAU robot, with an accurate Jacobian matrix obtained.
Newton-Raphson method is represented by

$$X_{n+1} = X_n - [F'(X_n)]^{-1} \cdot F(X_n)$$

With the six chain equations obtained before, the following can be obtained

$$[F'(X_n)]^{-1} = \mathrm{Inv}\begin{bmatrix} \frac{\partial f_1}{\partial x} & \frac{\partial f_1}{\partial y} & \frac{\partial f_1}{\partial z} & \frac{\partial f_1}{\partial \alpha} & \frac{\partial f_1}{\partial \beta} & \frac{\partial f_1}{\partial \gamma} \\ \frac{\partial f_2}{\partial x} & \frac{\partial f_2}{\partial y} & \frac{\partial f_2}{\partial z} & \frac{\partial f_2}{\partial \alpha} & \frac{\partial f_2}{\partial \beta} & \frac{\partial f_2}{\partial \gamma} \\ \frac{\partial f_3}{\partial x} & \frac{\partial f_3}{\partial y} & \frac{\partial f_3}{\partial z} & \frac{\partial f_3}{\partial \alpha} & \frac{\partial f_3}{\partial \beta} & \frac{\partial f_3}{\partial \gamma} \\ \frac{\partial f_4}{\partial x} & \frac{\partial f_4}{\partial y} & \frac{\partial f_4}{\partial z} & \frac{\partial f_4}{\partial \alpha} & \frac{\partial f_4}{\partial \beta} & \frac{\partial f_4}{\partial \gamma} \\ \frac{\partial f_5}{\partial x} & \frac{\partial f_5}{\partial y} & \frac{\partial f_5}{\partial z} & \frac{\partial f_5}{\partial \alpha} & \frac{\partial f_5}{\partial \beta} & \frac{\partial f_5}{\partial \gamma} \\ \frac{\partial f_6}{\partial x} & \frac{\partial f_6}{\partial y} & \frac{\partial f_6}{\partial z} & \frac{\partial f_6}{\partial \alpha} & \frac{\partial f_6}{\partial \beta} & \frac{\partial f_6}{\partial \gamma} \end{bmatrix}$$

This equation is used later to calculate the forward kinematic problem, and it is also compared with the method described in the next section.

4.2 Jacobian approximation method

A quick and efficient analytical solution is still necessary even though an accurate result has been obtained by the N-R method. The N-R result is produced based on iteration of numerical calculation, instead of from an analytical closed form solution. The N-R method is too slow in calculation to be used in on-line real time control. No certain solution is guaranteed in the N-R method. So a Jacobian approximation method is needed.
The Jacobian approximation method is established. Using this method, error analysis, calibration, compensation, and on-line control model can be established. As the TAU robot is based on a 3-DOF configuration, instead of a general Stewart platform, the Jacobian

approximate modification can be obtained based the 3-DOF analytical solution without any errors. The mathematical description of the Jacobian approximation method can be described as follows.

For forward kinematics,

$$X = F(\theta, \varepsilon)$$
$$X = F(\theta, 0) + J_{FORWARD} \cdot d\varepsilon$$

Where $J_{FORWARD} = F'(\theta, \varepsilon)$ and ε represents error.

Thus, the analytical solution $F(\theta, 0)$ and $F(X, 0)$, is obtained. Therefore, the Jacobian Approximation as an analytical solution is obtained and solving nonlinear equations using N-R method is not necessary in this case.

5. Determination of independent design variables using SVD method (application 2 of Jacobian matrix)

With the reality that all the parts of a robot have manufacturing errors and misalignment errors as well as thermal errors, errors should be considered for any of the components in order to accurately model the accuracy of the robot. Error budget is carried out in the study and error sensitivity of robot kinematics with respect to any of the parameters can be obtained from the error modeling. This is realized through the established Jacobian matrix.

To find those parameters in the error model that are linearly dependent and those parameters that are difficult to observe, the Jacobian matrix is analyzed. SVD method (Singular Value Decomposition) is used in such an analysis.

A methodical way of determining which parameters are redundant is to investigate the singular vectors. An investigation of the last column of the V vector will reveal that some elements are dominant in order of magnitude. This implies that corresponding columns in the Jacobian matrix are linearly dependent. The work of reducing the number of error parameters must continue until no singularities exist and the condition number has reached an acceptable value.

A total of 40 redundant design variables of the 71 design parameters are eliminated by observing the numerical Jacobian matrix obtained. Table 2 in Appendix A lists the remaining calibration parameters.

6. Error budget and results (application 3 of Jacobian matrix)

When the SVD is completed and a linearly independent set of error model parameters determined, the Error Budget can be determined. The mathematical description of the error budget is as follows:

$$J = U \bullet S \bullet V^T$$
$$dX = J \bullet dg = U \bullet S \bullet V^T \bullet dg$$
$$U^T \bullet dX = S \bullet V^T \bullet dg$$

Assume $U^T \bullet dX = d\overline{X}$ and $V^T \bullet dg = d\overline{g}$. So we have $d\overline{g} = d\overline{X} / S_{ii}$, finally,

$$dg = (V \bullet U^T \bullet dX) / S_{ii} \tag{12}$$

Thus if the dX is given as the accuracy of the Tau robot, the error budget dg can be determined. Given the D-H parameters for all three upper arms and the main column, the locations of the joints located at each of the three upper arms are known accurately. The six chain equations are created for the six link lengths, as follows:

$$F = \left\{ \begin{bmatrix} f1(upperarm\_po\text{int}\,s, TCP\_po\text{int}\,s) \\ f2(upperarm\_po\text{int}\,s, TCP\_po\text{int}\,s) \\ f3(upperarm\_po\text{int}\,s, TCP\_po\text{int}\,s) \\ f4(upperarm\_po\text{int}\,s, TCP\_po\text{int}\,s) \\ f5(upperarm\_po\text{int}\,s, TCP\_po\text{int}\,s) \\ f6(upperarm\_po\text{int}\,s, TCP\_po\text{int}\,s) \end{bmatrix} \right.$$

Where $TCP\_po\text{int} = f(px, py, pz, \alpha, \beta, \gamma)$

$$Upperarm\_po\text{int} = f(\varepsilon)$$

and ε is a collection of all the design parameters. Thus,

$$F = \left\{ \begin{bmatrix} F1(\varepsilon, px, py, pz, \alpha, \beta, \gamma) \\ F2(\varepsilon, px, py, pz, \alpha, \beta, \gamma) \\ F3(\varepsilon, px, py, pz, \alpha, \beta, \gamma) \\ F4(\varepsilon, px, py, pz, \alpha, \beta, \gamma) \\ F5(\varepsilon, px, py, pz, \alpha, \beta, \gamma) \\ F6(\varepsilon, px, py, pz, \alpha, \beta, \gamma) \end{bmatrix} \right.$$

An error model is developed based on the system of equations as described above. A total of 71 parameters are defined to represent the entire system, the 71 parameters include all the D-H parameters for the 3 upper arms, as well as the coordinates (x, y, z) of the 6 points at both ends of the 6 links, respectively. Appendix B (Table 3) presents the error budget.

7. Simulation results

The Jacobian approximation method is verified by the following two different approaches: (1) 6-DOF forward kinematic analysis (Newton-Raphson method), and (2) ADAMS simulation results.
Fig. 4 shows the error between Jacobian approximation method and ADAMS simulation results, and Fig. 5 gives the error between the N-R method and ADAMS simulation results.

In Fig. 4, the maximum error is 1.53um with an input error of 1 mm. The Jacobian approximation method has a very high accuracy compared with simulation results.

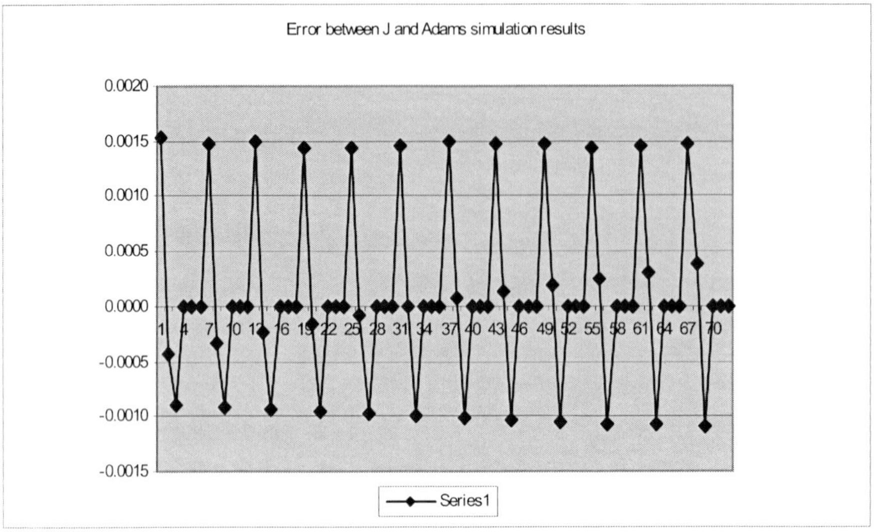

Fig. 4 Error between Jacobian approximation method and ADAMS simulation results

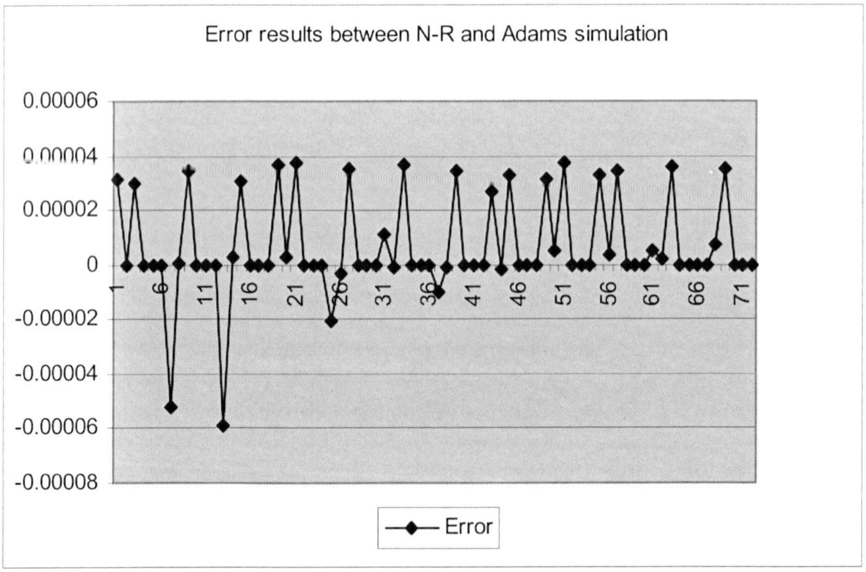

Fig. 5 Error between N-R method and ADAMS simulation results

Based on the D-H model of TAU with all error parameters, inverse and forward kinematic models have been established. From the point of view of mathematics, the TAU kinematic problem is to solve 6 nonlinear equations using Newton-Raphson method with Jacobian matrix as the searching direction and accurate results have been obtained up to 0.06 um compared with ADAMS simulation results as shown in Fig. 5. Appendix C (Table 4) gives the comparison between Jacobian Matrix and N-R method.

8. Conclusions

It can be observed from the results, that Jacobian Matrix is effective with an accuracy up to 1.53 um with an input error of 1 mm (Link 1 of lower arm 1). This was verified using ADAMS simulation results. Results from N-R method match very well with ADAMS simulation with a difference of only 0.06 um.

Based on the D-H model and an accurate Jacobian matrix, a series of results have been presented including error analysis, forward kinematic, redundant variable determination, error budget, and Jacobian approximation method. The Jacobian approximation method can be used in on-line control of the robot. For the TAU robot, a closed form solution of a forward kinematic problem is reached with a high accuracy instead of N-R numerical solution. The simulation results are almost perfect compared with that from ADAMS.

9. Acknowledgement

Authors from Stevens Institute of Technology are grateful to the ABB Corporate Research Center for the use of its research facilities and successful collaboration. ABB Corporate Research Center deeply appreciates the solid work done by Stevens's participants.

10. References

Brogangrdh, "Design of high performance parallel arm robots for industrial applications," Proceedings of A symposium Commemorating the Legacy, Works, and Life of Sir Robert Stawell Ball Upon the 100th Anniversary of A Treatise on the Theory of Screws , University of Cambridge, Trinity College, July 9-11, 2000.

Brogangrdh, et al, "Device for relative movement of two elements," United States Patent 6425303, July 30, 2002.

Brogangrdh, et al, "Device for relative movement of two elements," United States Patent 6336374, January 8, 2002.

Brogangrdh, et al, "Device for relative movement of two elements," United States Patent 6301988, October 16, 2001.

Lung-wen Tsai "Robot Analysis-The Mechanics of Serial and Parallel Manipulators", John Wiley & Sons, Inc.

M. Raghavan, "The Stewart Platform of General Geometry Has 40 Configurations," ASME Journal of Mechanical Design, June 1993, Vol. 115, pp 277-282.

Jian Wang and Oren Masory , "On the Accuracy of A Stewart Platform- Part I The Effect of Manufacturing Tolerances," IEEE, 1050-4729/ 1993, pp 114-120.

A. K. Dhingra, A. N. Almadi, D. Kohli, "Closed-Form Displacement Analysis of 8, 9 and 10-Link Mechanisms," Mechanism and Machine Theory, 35, 2000, pp 821-850.

A. K. Dhingra, A. N. Almadi, D. Kohli, "A Grobner-Sylvester Hybrid Method For Closed-Form Displacement Analysis of Mechanisms," 1998 ASME Design Engineering Technical Conference, Atlanta, GA.

A. K. Dhingra, A. N. Almadi, D. Kohli, "Closed-Form Displacement Analysis of 8, 9 and 10-Link Mechanisms, Part II" Mechanism and Machine Theory, 35, 2000, pp 851-869.

Chunhe Gong, Jingxia Yuan, Jun Ni, "Nongeometric Error Identification and Compensation for Robotic System by Inverse Calibration," International Journal of Machine Tools & Manufacture 40 (2000) 2119-2137.

M. Abderrahim, A. R. Whittaker, "Kinematic Model Identification of Industrial manipulators," Robotics and Computer Integrated Manufacturing 16 (2000), 1-8.

Amit J. Patel, Kornel F. Ehmann, "Calibration of a Hexapod Machine Tool Using a Redundant Leg," International Journal of Machine Tools & Manufacture 40 (2000) 489-512.

Xiaolun Shi, R. G. Fenton, "A Complete and General Solution to the Forward Kinematics Problem of Platform-Type Robotic Manipulators," IEEE, 1050-4729/ 1994, pp 3055-3062.

Z. Jason Geng and Leonard S. Haynes, "A 3-2-1 Kinematic Configuration of a Stewart Platform and its Application to Six Degree of Freedom Pose Measurements," Robotics & Computer-Integrated Manufacture, Vol. 11, No. 1, pp23-34, 1994.

Olivier Didrit, Michel Petitot, and Eric Walter "Guaranteed Solution of Direct Kinematic Problems for General Configurations of Parallel Manipulators," IEEE Transactions on Robotics and Automation, Vol. 14, No.2, April 1998.

Chang-de Zhang, Shin-Min Song, "Forward Kinematics of a Class of Parallel (Stewart) Platform with Closed-Form Solutions," Proceedings of the 1991 IEEE International Conference on Robotics and Automation, Sacramento, California-April 1991.

Prabjot Nanua, Kenneth J. Waldron, and Vasudeva Murthy, "Direct kinematic Solution of a Stewart platform," IEEE Transactions on Robotics and Automation, Vol. 6, No.4, August 1990.

S. V. Sreenivasan and K. J. Waldron, P. Nanua"Closed-Form Direct Displacement Analysis of a 6-6 Stewart Platform," Mech. Mach. Theory Vol. 29. No. 6, pp 855-864, 1994.

M. Griffis and J. Duffy, "A Forward Displacement Analysis of a Class of Stewart Platform," Journal of Robotic System 6 (6), 703 720 (1989) by John Wiley & Sons, Inc.

W. Lin, M. Griffis, J. Duffy, "Forward Displacement Analyses of the 4-4 Stewart Platforms," Transaction of the ASME Vol. 114, September 1992, pp444-450.

Appendix A

Parameter Number	Parameter Definition	Parameter
16	height of the TCP	a
22	joint 3	a6
23	arm3	a7
24	joint 1 & arm 1	d1
25	short arm 1	d3
28	joint3	d6
31	joint_link11_arm1	y1
34	joint_link21_arm1	y2
37	joint_link31_arm1	y3
40	joint_link12_arm2	y4
43	joint_link22_arm2	y5
46	joint_link13_arm3	y6
48	joint_link11p	x11
49	joint_link11p	y11
51	joint_link31p	x22
52	joint_link31p	y22
54	joint_link21p	x33
55	joint_link21p	y33
56	joint_link21p	z33
57	joint_link12p	x44
58	joint_link12p	y44
59	joint_link12p	z44
60	joint_link22p	x55
61	joint_link22p	y55
62	joint_link22p	z55
63	joint_link13p	x66
64	joint_link13p	y66
67	link11	L1
68	link31	L2
69	link21	L3
70	link22	L4

Table. 2 List of the independent design variables

Appendix B

Error Budget			
Variable No.	Description	Name	Budget
1	drive 1	Joint 1	32 arcsec
2	drive 2	Joint 2	1.17 arcsec
3	drive 3	Joint 3	1.2 arcsec
17	joint 1 and arm 1	a1	1.62 um

24		d1	363 um
4		sit1	10.4 arcsec
10		afa1	110 arcsec
18	joint_link11_arm 1	a2	373 um
19	short arm 1	a3	174 um
25		d3	449 um
5		sit3	9.24 arcsec
11		afa3	9.45 arcsec
20	joint 2 and arm 2	a4	1.9 mm
26		d4	485 um
6		sit4	1.22 arcsec
12		afa4	38.5 arcsec
21	short arm 2	a5	430 um
27		d5	D
7		sit5	11.2 arcsec
13		afa5	D
22	joint 3	a6	0
28		d6	D
8		sit6	4.64 arcsec
14		afa6	D
23	arm 3	a7	0
29		d7	D
9		sit7	6.14 arcsec
15		afa7	D
30	joint_link11_arm1	x1	D
31		y1	43 um
32		z1	123 um
33	joint_link21_arm1	x2	D
34		y2	49.4 um
35		z2	D
36	joint_link31_arm1	x3	115 um
37		y3	108 um
38		z3	D
39	joint_link12_arm2	x4	D
40		y4	1.28 mm
41		z4	D
42	joint_link22_arm2	x5	2.6 mm
43		y5	68.2 um
44		z5	D
45	joint_link13_arm3	x6	D
46		y6	21.6 um
47		z6	213 um
48	joint_link11_platform	x11	50 um
49		y11	50 um

50		z11	D
51	joint_link31_platform	x22	50 um
52		y22	50 um
53		z22	D
54	joint_link21_platform	x33	50 um
55		y33	50 um
56		z33	13.3 um
57	joint_link12_platform	x44	50 um
58		y44	50 um
59		z44	37.9 um
60	joint_link22_platform	x55	50 um
61		y55	50 um
62		z55	398 um
63	joint-link13_platform	x66	50 um
64		y66	50 um
65		z66	50 um
16	height of the TCP	a	436 um
66	link 13	L0	0
67	link 11	L1	88 um
68	link 31	L2	151 um
69	link 21	L3	54.3 um
70	link 22	L4	213 um
71	link 12	L5	1.47 mm

Table 3 Error budget

Appendix C

Drive Angles	TCP Pose	Jacobian	Newton_raphson	Error between J and N
joint1=0 joint2=0 joint3=0	X	0.00E+00	1.53E-03	0.001531339
	Y	-1.81E+00	-1.81E+00	-0.0049559
	Z	-1.61E-16	-9.20E-04	-0.000919889
	afa	5.01E-03	5.01E-03	2.634E-07
	bta	-9.32E-19	-9.33E-19	-1.00679E-21
	gma	-9.32E-19	-9.32E-19	-1.5976E-22
joint1=3.75 joint2=3.75 joint3=2	X	1.19E-01	1.20E-01	0.00119916
	Y	-1.81E+00	1.81E+00	-0.0009736
	Z	-2.09E-16	-9.45E-04	-0.000945018
	afa	5.01E-03	5.01E-03	2.7566E-06
	bta	0.00E+00	9.46E-16	9.45683E-16
	gma	0.00E+00	-4.84E-16	-4.84153E-16
joint1=7.5 joint2=7.5 joint3=4	X	2.37E-01	2.38E-01	0.00135537
	Y	-1.80E+00	-1.80E+00	0.0007562
	Z	-1.79E-16	-9.69E-04	-0.000968876
	afa	5.02E-03	5.02E-03	3.547E-07
	bta	0.00E+00	3.15E-16	3.14853E-16
	gma	0.00E+00	-4.82E-16	-4.82129E-16

joint1=11.25 joint2=11.25 joint3=6	X	3.54E-01	3.55E-01	0.00149511
	Y	-1.78E+00	-1.78E+00	0.0001837
	Z	-1.79E-16	-9.91E-04	-0.000991397
	afa	5.03E-03	5.03E-03	3.263E-06
	bta	0.00E+00	-3.10E-18	-3.10077E-18
	gma	-9.32E-19	1.15E-18	2.0782E-18
joint1=15 joint2=15 joint3=8	X	4.70E-01	4.71E-01	0.00111796
	Y	-1.75E+00	-1.75E+00	-0.0027737
	Z	-5.96E-17	-1.01E-03	-0.001012624
	afa	5.05E-03	5.05E-03	1.7286E-06
	bta	0.00E+00	0.00E+00	0
	gma	0.00E+00	0.00E+00	0
joint1=18.75 joint2=18.75 joint3=10	X	5.83E-01	5.85E-01	0.00173003
	Y	-1.72E+00	-1.72E+00	0.0017688
	Z	-5.96E-17	-1.03E-03	-0.001032565
	afa	5.07E-03	5.08E-03	6.0465E-06
	bta	4.66E-19	-6.39E-16	-6.39425E-16
	gma	-9.32E-19	9.59E-16	9.6015E-16
joint1=22.5 joint2=22.5 joint3=12	X	6.94E-01	6.96E-01	0.00184612
	Y	-1.68E+00	-1.68E+00	0.0036642
	Z	2.09E-16	-1.05E-03	-0.00105122
	afa	5.11E-03	5.11E-03	-3.4323E-06
	bta	0.00E+00	-8.47E-22	-8.47033E-22
	gma	0.00E+00	8.47E-22	8.47033E-22
joint1=26.25 joint2=26.25 joint3=14	X	8.03E-01	8.04E-01	0.00099179
	Y	-1.63E+00	-1.63E+00	0.002734
	Z	0.00E+00	-1.07E-03	-0.001068582
	afa	5.14E-03	5.14E-03	3.7091E-06
	bta	0.00E+00	3.26E-16	3.25672E-16
	gma	0.00E+00	-4.78E-16	-4.77901E-16
joint1=30 joint2=30 joint3=16	X	9.07E-01	9.09E-01	0.00170544
	Y	-1.57E+00	-1.57E+00	-0.0012306
	Z	-2.09E-16	-1.08E-03	-0.001084643
	afa	5.19E-03	5.19E-03	-2.0346E-06
	bta	0.00E+00	8.47E-22	8.47033E-22
	gma	0.00E+00	0.00E+00	0
joint1=33.75 joint2=33.75 joint3=18	X	1.01E+00	1.01E+00	-0.0004597
	Y	-1.51E+00	-1.51E+00	0.0015319
	Z	1.49E-16	-1.10E-03	-0.001099391
	afa	5.24E-03	5.24E-03	-7.54E-08
	bta	0.00E+00	-6.75E-16	-6.74923E-16
	gma	0.00E+00	4.55E-18	4.54772E-18
joint1=37.5 joint2=37.5 joint3=18	X	1.10E+00	1.11E+00	0.0060663
	Y	-1.44E+00	-1.44E+00	0.0007547
	Z	2.98E-17	-1.11E-03	-0.001112819
	afa	5.30E-03	5.30E-03	2.869E-07
	bta	0.00E+00	0.00E+00	0
	gma	0.00E+00	0.00E+00	0
joint1=41.25 joint2=41.25 joint3=22	X	1.20E+00	1.20E+00	-0.002128
	Y	-1.36E+00	-1.36E+00	-0.0038563
	Z	-2.98E-17	-1.12E-03	-0.001124931
	afa	5.37E-03	5.37E-03	-1.1E-07
	bta	0.00E+00	0.00E+00	0
	gma	0.00E+00	0.00E+00	0

Table 4 Results of the comparison between Jacobian Matrix and N-R method

Specific Parameters of the Perturbation Profile Differentially Influence the Vertical and Horizontal Head Accelerations During Human Whiplash Testing

Loriann M. Hynes, Natalie S. Sacher and James P. Dickey
Human Health and Nutr. Sci., University of Guelph
Canada

1. Introduction

Whiplash experiments using human subjects are important tools for evaluating biological response during collisions and can provide key insights into mechanisms of injury (Muhlbauer et al., 1999). Human experimentation, including whiplash-like perturbation testing, is essential to evaluate the kinematic, kinetic and electromyographic responses to enable mathematical models to predict the loads within neck structures (Choi & Vanderby, 1999). Although some experimentation has been performed with staged collisions between actual automobiles (Welcher & Szabo, 2001; Severy et al., 1955; Castro et al., 2001; Brault et al., 1998), more typically, human testing has been performed using experimental sleds (Dehner et al., 2007; Kumar et al., 2005b; Kumar et al., 2002; Muhlbauer et al., 1999; Siegmund et al., 2003; Kaneoka et al., 1999).

One limitation of sled testing is that the perturbation pulse varies from trial to trial due to the varying inertia between subjects, and the varying effective inertia as subjects respond differently to the perturbation. This is an important issue since certain properties of the crash pulse influence the risk of injury (Kullgren et al., 2000; Hynes & Dickey, 2008). In order to address this limitation, some researchers have developed advanced test sleds that incorporate feedback-controlled linear motors to enable them to precisely control the properties of the perturbation pulse (Siegmund et al., 2005; Siegmund et al., 2004). This approach offers considerable advantage over spring (Magnusson et al., 1999), gravity (Kaneoka et al., 1999) and pneumatic (Kumar et al., 2002; Hernandez et al., 2005) driven sleds, yet it is still limited to simulating the anterior-posterior motion of vehicles. In contrast, research from experimental automobile collision studies that have used real vehicles has shown that the vehicle accelerations include substantial vertical accelerations (Severy et al., 1955; Welcher & Szabo, 2001) due to factors such as the vehicle suspension and bumper height mismatch (Siegel et al., 2001).

Peak head acceleration is thought to be a key variable in whiplash, and typical and reproducible head/neck motion patterns have been observed in experiments with male (Muhlbauer et al., 1999) and female (Dehner et al., 2007) volunteers; however, various published studies have observed different responses which cloud the interpretation of this body of research. For example, Kumar and colleagues (2005b; 2002; 2005a) consistently report that the peak magnitude of the head accelerations is less than the sled peak acceleration while others (Severy et al., 1955; Magnusson et al., 1999; Siegmund et al., 2003; Muhlbauer et al., 1999) report peak head accelerations that exceed the sled acceleration. Many human studies of whiplash-like perturbations have evaluated the horizontal component of the head acceleration response and have not fully appreciated the vertical component, although these two loading directions have different biological implications (Nightingale et al., 2000; Siegmund et al., 2001) and presumably different thresholds for injury. As notable exceptions, Siegmund et al. (2004) performed an in-depth analysis of the vertical and horizontal head accelerations in response to a series of increasing intensity perturbations, Welcher et al. (2001) evaluated the vertical and horizontal head accelerations of a single female subject exposed to 5 in-car collisions of differing severity, and Hernandez et al (2005) evaluated the displacement and acceleration responses of both female and male subjects exposed to two levels of perturbation. Siegmund et al. (2004) observed that high acceleration, high velocity perturbations consistently produced the largest muscle activations, head horizontal and vertical accelerations, head angular accelerations and velocities, and head angles compared to low acceleration, low velocity perturbations. Siegmund et al., (2004) and Welcher et al., (2001) both reported that the horizontal and vertical accelerations were highly correlated. Hernandez et al., (2005) observed that the angular head displacements as well as the rearward and forward angular head accelerations were somewhat increased in the fast case compared to the slow case. In addition, they noted that the males presented two times higher upward linear head acceleration than females in the unexpected condition. The purpose of this study was to investigate the relationship between the vertical and horizontal head accelerations during low velocity horizontal whiplash-like perturbations. Information about the relationship between the vertical and horizontal head accelerations during low velocity perturbations is essential to enable extrapolation of research findings from low velocity (non-injurious) perturbations to higher severity situations. We have used a commercial parallel robot as a feedback-controlled motion platform to provide the different perturbation pulses.

2. Methods

Permission for this study was obtained from the University of Guelph Research Ethics Board and written consent was obtained from all subjects. Seventeen subjects underwent a cervical spine orthopaedic examination performed by a Certified Athletic Therapist, as well as questionnaire screening, to ensure they were free of any neck pain and/or obvious neck pathology. We excluded subjects who reported being involved in a car accident in the previous five years.

A robotic platform (R2000, PRSCo, New Hampshire, USA) was used to apply the low-velocity whiplash-like perturbations. The accuracy of the robot is ±0.001mm ±0.001 degrees in Cartesian space, permitting precisely controlled and repeatable perturbations. All robot motion was restricted to the posterior-anterior direction. Two specific robotic displacement profiles were generated (Figure 1). One profile reflected the kinematics of a spring-powered experimental sled with a peak velocity of 2.14 kph and peak acceleration of 0.41g termed "mild", which was generated by integrating the published acceleration pulse (Figure 3, Magnusson et al., 1999). The second profile had a higher peak acceleration (0.94g), peak velocity (3.06 kph), and a shorter time to peak acceleration, and was termed "moderate".

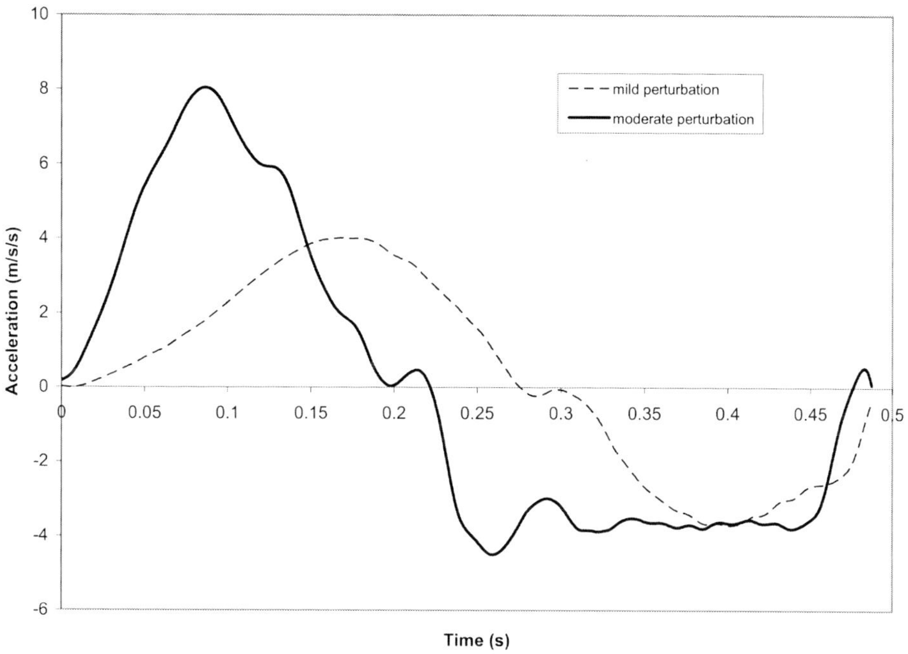

Fig. 1. Time course of platform accelerations for the mild and moderate rear-end perturbations. The mild profile reflects the perturbation from Magnusson et al., 1999; the moderate profile has a larger magnitude and shorter time-to-peak acceleration.

Subjects were seated in a fully functional 1991 Honda Accord front passenger car seat mounted to the robot's platform (Fig. 2). Two triaxial accelerometers (Crossbow CSL04LP3±4g Module) were used; one was fixed to the subjects' foreheads to measure head accelerations, as in previous studies (Kumar et al., 2002; Kumar et al., 2004a) while the second accelerometer was mounted on the robotic platform to determine the initial onset of the platform movement. Each subject was exposed to 10 perturbation trials; 5 moderate and 5 mild, presented in random order. Subjects were provided notice of the impending perturbation using a countdown; however, subjects were unaware of the magnitude of the oncoming perturbation.

Accelerometer data were collected at a sampling frequency of 1000Hz and processed using LabVIEW 7.0 (National Instruments). Data collection was initiated one second before each perturbation.

The peak vertical and anterior-posterior (A-P) head accelerations were extracted for the first moderate and mild perturbations. Statistical analysis was performed using Graphpad Software Inc., Version 4.03. Non-parametric statistics were used since the variance was not homogenous. As there is no non-parametric equivalent to two-way ANOVAs, pair-wise comparisons with paired Wilcoxon t tests were used and significance was set at the 0.05 level.

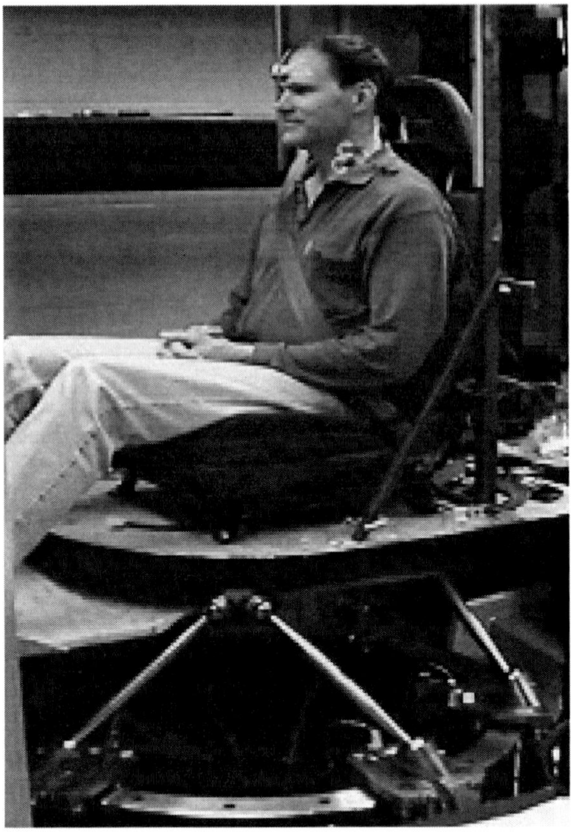

Fig. 2. Experimental setup showing a subject sitting on the automobile seat mounted to the top platform of the parallel robot. Surface EMG and head accelerometer data was collected during the perturbations with the subject secured in the car seat using the standard safety seat belt and head restraint.

3. Results

All subjects successfully completed the entire experimental protocol; no subjects complained of neck pain during or after the experiment. The magnitude of the head accelerations were significantly greater in the moderate perturbations compared to the mild perturbations ($p=0.0003$ for vertical, effect size $=0.92$, $p=0.0003$ for horizontal head accelerations, effect size $= 1.13$; Figure 2). In the mild perturbations, the magnitude of the vertical and horizontal head accelerations were not significantly different ($p>0.05$). However, in the moderate perturbations, the magnitude of the horizontal head accelerations were significantly larger than the vertical accelerations ($p=0.0007$; effect size $= 0.445$). Both the horizontal and vertical head acceleration magnitudes were larger than the platform accelerations in all cases (refer to the horizontal lines in Figure 3).

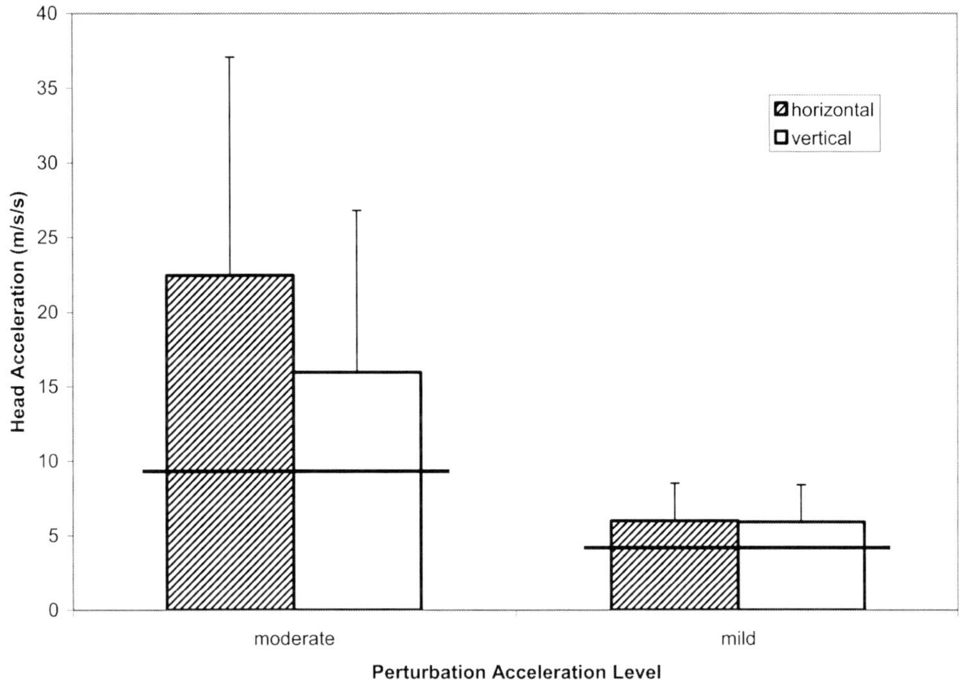

Fig. 3. Bar graph demonstrating the peak head accelerations (mean and one standard deviation) in the vertical and horizontal directions during the mild (0.41 g) and moderate (0.94 g) rear-end whiplash-like horizontal perturbations. The horizontal lines reflect the magnitude of the horizontal robotic platform accelerations.

4. Discussion

This study directly compared two different perturbation profiles, using repeated measures, to evaluate vertical and horizontal head accelerations during whiplash-like perturbations. We observed that the magnitude of the vertical head accelerations depended on the specific perturbation parameters; the horizontal and vertical acceleration magnitudes were not significantly different in the mild perturbation, but the horizontal head accelerations were significantly larger than the vertical accelerations during the moderate perturbations. These findings illustrate that human subjects have different responses to whiplash-like perturbations depending on the specific acceleration profile parameters, including peak acceleration. This finding is in contrast to one study that found that the vertical and horizontal head accelerations were highly correlated for seven different perturbation profiles (Siegmund et al., 2004), but somewhat supported by a different study that observed differences in the magnitude of the vertical head acceleration between female and male subjects (Hernandez et al., 2005). Our finding supports a recent in vitro experiment that observed that the crash pulse shape influences the peak loading and the injury tolerance levels of the neck in simulated low-speed side-collisions (Kettler et al., 2006).

Several recent studies have reported typical and reproducible head/neck motion and acceleration patterns during perturbation testing (Dehner et al., 2007; Muhlbauer et al., 1999). It is essential to appreciate that these patterns are modulated by the specific perturbation profile. Parameters such as the time to peak acceleration, in addition to the magnitude of the acceleration and velocity, appear to influence the resulting head/neck motion. We document that the relationship between the vertical and horizontal head accelerations depend on the specific perturbation pulse; we recommend that all studies should publish their perturbation pulses to aid in comparisons between studies.

We observed that horizontal platform perturbations led to both vertical and horizontal head accelerations. However, our accelerometer measurements were influenced by the location of the accelerometer (forehead in this experiment, similar to other research studies c.f. Kumar et al. (2002) and (2004a)). We have subsequently performed testing to evaluate the differences in accelerometer measurements between mounting the accelerometer on the forehead and temple, since the temple location is closer to the center of mass of the head (Muhlbauer et al., 1999). These tests revealed that the peak horizontal forehead accelerations were approximately 16% less, and the vertical forehead accelerations 38% greater, than the peak temple accelerations. These differences arise since the forehead accelerations are also sensitive to rotational accelerations of the head, and are similar to the 16% changes in peak acceleration between mounting accelerometers on the top of the head compared to the forehead (Mills & Carty, 2004). Nevertheless, the fact that we observed systematic differences in forehead accelerations with different perturbation profiles remains and indicates that differences would also be present for temple or head center of mass linear and/or angular accelerations; the specific features of the perturbation profile, such as the peak acceleration, influence the head acceleration responses. Another limitation of this study was that the peak acceleration of the perturbation profile was comparatively quite low. However, it is important to note that these perturbation profiles produced head accelerations and neck muscle activation patterns similar to previous experiments investigating human responses to whiplash-like perturbations (Severy et al., 1955; Magnusson et al., 1999; Siegmund et al., 2003) and that the use of a parallel robot permitted more precise control over the motion patterns than alternative testing approaches.

Clearly there is additional potential for parallel robots in this area; although some researchers have used linear sleds to simulate offset collisions by orienting the subject at an angle to the direction of sled travel (Kumar et al., 2004b), as 6 df mechanisms, parallel robots could be programmed to move in three-dimensional space to reflect offset collisions more realistically. We are currently undertaking research projects in which we are applying concurrent vertical and horizontal perturbations, and a second study in which we are evaluating different perturbation directions.

5. Conclusions

The level of the perturbation acceleration influences the resulting acceleration of the head, in both the vertical and horizontal directions. A parallel robotic platform facilitated this research by enabling feedback-controlled motion for the perturbations.

6. Acknowledgements

The parallel robot was purchased by a grant from the Canadian Foundation for Innovation, and funding for this study was provided by a grant from the AUTO21, one of the Canadian Networks of Centres of Excellence.

7. References

Brault, J.R., Wheeler, J.B., Siegmund, G.P., & Brault, E.J. (1998). Clinical response of human subjects to rear-end automobile collisions. *Archives of Physical Medicine and Rehabilitation*, Vol. 79, pp. 72-80, ISSN 0003-9993.

Castro, W.H.M., Meyer, S.J., Becke, M.E.R., Nentwig, C.G., Hein, M.F., Ercan, B.I. et al. (2001). No stress - no whiplash? Prevalence of "whiplash" symptoms following exposure to a placebo rear-end collision. *International Journal of Legal Medicine*, 114, pp. 316-322, ISSN 0937-9827.

Choi, H. & Vanderby, R. (1999). Comparison of biomechanical human neck models: Muscle forces and spinal loads at C4/5 level. *Journal of Applied Biomechanics*, Vol. 15, pp. 120-138, ISSN 1065-8483.

Dehner, C., Elbel, M., Schick, S., Walz, F., Hell, W., & Kramer, M. (2007). Risk of injury of the cervical spine in sled tests in female volunteers. *Clinical Biomechanics*, Vol. 22, pp. 615-622, ISSN 0268-0033.

Hernandez, I.A., Fyfe, K.R., Heo, G., & Major, P.W. (2005). Kinematics of head movement in simulated low velocity rear-end impacts. *Clinical Biomechanics*, Vol. 20, pp. 1011-1018, ISSN 0268-0033.

Hynes, L.M. & Dickey, J.P. (2008). The rate of change of acceleration: Implications to head kinematics during rear-end impacts. *Accident Analysis and Prevention*, In Press, ISSN 0001-4575.

Kaneoka, K., Ono, K., Inami, S., & Hayashi, K. (1999). Motion analysis of cervical vertebrae during whiplash loading. *Spine*, Vol. 24, pp. 763-769, ISSN 0362-2436.

Kettler, A., Fruth, K., Claes, L., & Wilke, H.J. (2006). Influence of the crash pulse shape on the peak loading and the injury tolerance levels of the neck in in vitro low-speed side-collisions. *Journal of Biomechanics*, Vol. 39, pp. 323-329, ISSN 0021-9290.

Kullgren, A., Krafft, M., Nygren, A., & Tingvall, C. (2000). Neck injuries in frontal impacts: influence of crash pulse characteristics on injury risk. *Accident Analysis and Prevention*, Vol. 32, pp. 197-205, ISSN 0001-4575.

Kumar, S., Ferrari, R., & Narayan, Y (2004a). Electromyographic and kinematic exploration of whiplash-type neck perturbations in left lateral collisions. *Spine*, Vol. 29, pp. 650-659, ISSN 0362-2436.

Kumar, S., Ferrari, R., & Narayan, Y. (2004b). Electromyographic and kinematic exploration of whiplash-type rear impacts: effect of left offset impact. *The Spine* Journal, Vol. 4, pp. 656-665, ISSN 1529-9430.

Kumar, S., Ferrari, R., & Narayan, Y. (2005a). Kinematic and electromyographic response to whiplash loading in low-velocity whiplash impacts--a review. *Clinical Biomechanics*, Vol. 20, 343-356, ISSN 0268-0033.

Kumar, S., Ferrari, R., & Narayan, Y. (2005b). Turning away from whiplash. An EMG study of head rotation in whiplash impact. *Journal of Orthopaedic Research*, Vol. 23, pp. 224-230, ISSN 0736-0266.

Kumar, S., Narayan, Y., & Amell, T. (2002). An electromyographic study of low-velocity rear-end impacts. *Spine*, 27, pp. 1044-1055, ISSN 0362-2436.

Magnusson, M.L., Pope, M.H., Hasselquist, L., Bolte, K.M., Ross, M., Goel, V.K. et al. (1999). Cervical electromyographic activity during low-speed rear impact. *European Spine Journal*, 8, pp. 118-125, ISSN 0940-6719.

Mills, D. & Carty, G. (2004). Comparative Analysis of Low Speed Live Occupant Crash Test Results to Current Literature. Proceedings of the Canadian Multidisciplinary Road Safety Conference XIV, pp. 1-14, June 2004, Ottawa, Ontario.

Muhlbauer, M., Eichberger, A., Geigl, B.C., & Steffan, H. (1999). Analysis of kinematics and acceleration behavior of the head and neck in experimental rear-impact collisions. *Neuro-Orthopedics*, Vol. 25, pp. 1-17, ISSN 0177-7955.

Nightingale, R.W., Camacho, D.L., Armstrong, A.J., Robinette, J.J., & Myers, B.S. (2000). Inertial properties and loading rates affect buckling modes and injury mechanisms in the cervical spine. *Journal of Biomechanics*, Vol. 33, pp. 191-197, ISSN 0021-9290.

Severy, D.M., Mathewson, J.H., & Bechtol, C.O. (1955). Controlled automobile rearend collisions, an investigation of related engineering and medical phenomena. *Canadian Services Medical Journal*, Vol. 11, pp. 727-759.

Siegel, J.H., Loo, G., Dischinger, P.C., Burgess, A.R., Wang, S.C., Schneider, L.W. et al. (2001). Factors influencing the patterns of injuries and outcomes in car versus car crashes compared to sport utility, van, or pick-up truck versus car crashes: Crash Injury Research Engineering Network Study. *The Journal of Trauma*, Vol. 51, pp. 975-990, ISSN 0022-5282.

Siegmund, G.P., Heinrichs, B.E., Chimich, D.D., DeMarco, A.L., & Brault, J.R. (2005). The effect of collision pulse properties on seven proposed whiplash injury criteria. *Accident Analysis and Prevention*, Vol. 37, pp. 275-285, ISSN 0001-4575.

Siegmund, G.P., Myers, B.S., Davis, M.B., Bohnet, H.F., & Winkelstein, B.A. (2001). Mechanical evidence of cervical facet capsule injury during whiplash: a cadaveric study using combined shear, compression, and extension loading. *Spine*, Vol. 26, pp. 2095-2101, ISSN 0362-2436.

Siegmund, G.P., Sanderson, D.J., & Inglis, J.T. (2004). Gradation of Neck Muscle Responses and Head/Neck Kinematics to Acceleration and Speed Change in Rear-End Collisions. *STAPP Car Crash Journal*, Vol. 48, pp. 419-430, ISSN 1532-8546.

Siegmund, G.P., Sanderson, D.J., Myers, B.S., & Inglis, J.T. (2003). Awareness affects the response of human subjects exposed to a single whiplash-like perturbation. *Spine*, Vol. 28, pp. 671-679, ISSN 0362-2436.

Welcher, J.B. & Szabo, T.J. (2001). Relationships between seat properties and human subject kinematics in rear impact tests. *Accident Analysis and Prevention*, Vol. 33, pp. 289-304, ISSN 0001-4575.

Welcher, J.B., Szabo, T.J., & Voss, D.P. (2001). Human Occupant Motion in Rear-End Impacts: Effects of Incremental Increases in Velocity Change. SAE 2001 World Congress, pp. 241-249, Warrendale, PA, Society of Automotive Engineers, Inc, April 2001.

15

Neural Network Solutions for Forward Kinematics Problem of HEXA Parallel Robot

M. Dehghani, M. Eghtesad, A. A. Safavi, A. Khayatian, and M. Ahmadi
Shiraz University
I.R. Iran

1. Introduction

Forward kinematics problem of parallel robots is a very difficult problem to solve in comparison to the serial manipulators due to their highly nonlinear relations between joint variables and position and orientation of the end effector. This problem is almost impossible to be solved analytically. Numerical methods are the most common approaches to solve this problem. Nevertheless, the possible lack of convergence of these methods is the main drawback. In this chapter, two types of neural networks - multilayer perceptron (MLP) and wavelet based neural network (wave-net) - are used to solve the forward kinematics problem of the HEXA parallel manipulator. This problem is solved in a typical workspace of this robot. Simulation results show the advantages of employing neural networks, and in particular wavelet based neural networks, to solve this problem.

2. Review of forward kinematics problem of parallel robot

The idea of designing parallel robots started in 1947 when D. Stewart constructed a flight simulator based on his parallel design (Stewart, 1965). Then, other types of parallel robots were introduced (Merlet, 1996). Parallel manipulators have received increasing attention because of their high stiffness, high speed, high accuracy and high carrying capability (Merlet, 2002). However, parallel manipulators are structurally more complex, and also require a more complicated control scheme; in addition, they have a limited workspace in compare to serial robots. Therefore, parallel manipulators are the best alternative of serial robots for tasks that require high load capacity in a limited workspace.

A parallel robot is made up of an end-effector that is placed on a mobile platform, with n degrees of freedom, and a fixed base linked together by at least two independent kinematic chains (Tsai, 1999). Actuation takes place through m simple actuators, (see Fig. 1).

Similar to serial robots, kinematic analysis of parallel manipulators contains two problems: forward kinematics problem (FKP) and inverse kinematics problem (IKP). In parallel robots unlike serial robots, solution to IKP is usually straightforward but their FKP is complicated. FKP involves a system of nonlinear equations that usually has no closed form solution (Merlet, 2001).

Traditional methods to solve FKP of parallel robots have focused on using algebraic formulations to generate a high degree polynomial or a set of nonlinear equations. Then, methods such as interval analysis Merlet, 2004), algebraic elimination (Lee, 2002), Groebner basis approach Merlet, 2004) and continuation (Raghavan, 1991) are used to find the roots of the polynomials or to solve nonlinear equations. The FKP is not fully solved just by finding all the possible solutions. Further schemes are needed to find a unique actual position of the platform among all the possible solutions. Use of iterative numerical procedures (Merlet, 2007), (Wang, 2007) and auxiliary sensors (Baronet *et al.*, 2000) are the two commonly adopted schemes to further lead to a unique solution. Numerical iteration is usually sensitive to the choice of initial values and nature of the resulting constraint equations. The auxiliary sensors approach has practical limitations, such as cost and measurement errors. No matter how the forward kinematics problem may be solved, direct determination of a unique solution is still a challenging problem.

Artificial neural networks (ANNs) are computational models comprising numerous nonlinear processing elements arranged in patterns similar to biological neural networks. These computational models have now become exciting alternatives to conventional approaches in solving a variety of engineering and scientific problems. Traditional neural networks are back propagation networks that are trained with supervision, using gradient-descent training technique which minimizes the squared error between the actual outputs of the network and the desired outputs. Two common types of them are multilayer perceptron (MLP) and radial basis function (RBF) are used in modeling of different problems. Recently wavelet neural networks have been presented by Zhang *et al.* in 1992 based on wavelet decomposition (Zhang *et al.*, 1992). The proposed wavelet neural network (WNN) inspired by feed forward neural networks and wavelet decompositions is an efficient alternative to multilayer perceptron (MLP) and redial basis function (RBF) neural networks for process modeling and classifying problems. The structure of proposed WNN is similar to that of the radial basis function (RBF) networks, except that their main activation function is replaced by orthogonal basis functions with simple network topology (Zhang, 1995). The WNN can further result in a convex cost index to which simple iterative solutions such as gradient descent rules are justifiable and are not in danger of being trapped in local minima when choosing the orthogonal wavelets as the activation functions in the nodes (Zhang *et al.*, 1992). Wave-nets are a class of wavelet-based neural networks with hierarchical multiresolution learning. Wave-nets were introduced by Bakshi and Stephanopoulos (Bakshi & Sephanopolus, 1993). Then, their nature and applications were thoroughly investigated by Safavi (Safavi & Romagnoli, 1997). There have also been other attempts at using wavelets for NNs, with the learning algorithms that are different from wave-nets (Szu *et al.*, 1992).

Some researchers have tried using neural networks for solving the FKP of parallel robots (Geng *et al.*, 1992), (Yee, 1997). Almost all of prior researches have focused on using ANNs approach to solve FKP of Stewart platform. Few of them have also applied this method to solve FKP of other parallel robot (Ghobakhlo *et al.*, 2005), (Sadjadian *et al.*, 2005). In this chapter, we focus on HEXA parallel robot, first presented by Pierrot (Pierrrot *et al.*,1990), whose platform is coupled to the base by 6 RUS-limbs, where R stands for revolute joint, U stands for universal joint and S stands for spherical joint (see Fig. 2). Complete description of HEXA robot is presented in Section 2.

The solution of IKP of HEXA was first presented in (Pierrrot *et al.*, 1990) by F. Pierrrot who solved the system of nonlinear equations and obtained a unique solution for the problem. A numerical solution for FKP of HEXA parallel robot was presented by J.P. Merlet in (Merlet, 2001). FKP of this robot has no closed form solution and at most 40 assembly modes (assembly modes are different configurations of the end-effecter with given values of joint variables) exist for this problem. He suggested iterative methods for solving HEXA FKP. But, these methods have some drawbacks, such as being lengthy procedures and giving incorrect answers (Merlet, 2001). Utilization of the passive joint sensors; however, enables one to find closed form solutions. In (Last *et al.*, 2005) it has been shown that a minimum number of three passive joint sensors are needed for solving the FKP analytically.

In this chapter, two neural network approaches are used to solve FKP of HEXA robot. To carry out this task, we first estimate the IKP in some positions and orientations -posses- of the workspace of the robot. Then a multilayer perceptron (MLP) network and a wave-net are trained with data obtained by solving IKP. We test the networks in the other positions and orientations of the workspace. Finally the simulation results will be presented and these two networks will be compared.

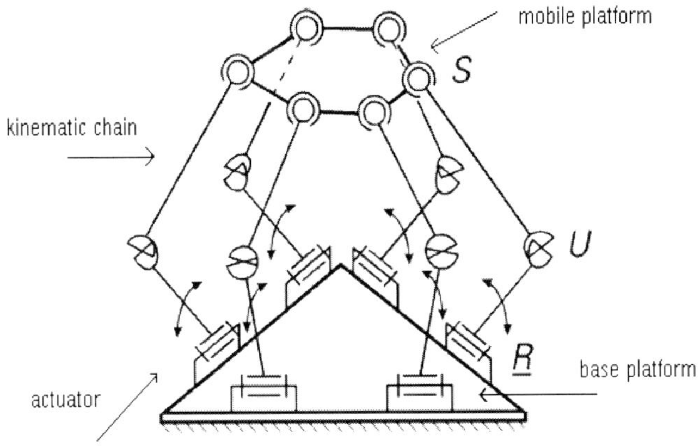

Fig. 1. A typical RUS parallel robot (Bonev *et al.*, 2000)

The rest of the chapter is organized as follows: Section 2 contains HEXA mechanism description. Kinematic modeling of the manipulator is discussed in Section 3 where inverse and forward kinematics are studied and the need for appropriate method to solve forward kinematics is justified. MLP network and wave-net method to solve FKP are discussed in section 4. In section 5 the results of solving FKP for HEXA parallel manipulator robot by these networks are presented. Comparison of these networks and conclusion are discussed in section 6.

3. Mechanism description

There are different classes of parallel robots. Undoubtedly, the most popular member of the 6-RUS class is the HEXA robot (Pierrrot *et al.*, 1990), of which an improved version is already available. The first to propose this architecture, however, was Hunt in 1983 (Hunt, 1983). Some other prototypes have been constructed by Sarkissian in 1990 (Sarkissian *et al.*, 1990), by Zamanov (Zamanov *et al.* 1992) and by Mimura in 1995 (Mimura, 1995). The latter has even performed a detailed set of analyses on this type of manipulator. Two other designs are also commercially available by Servos & Simulation Inc. as motion simulation systems (Merlet, 2001). Finally, a more recent and more peculiar design has been introduced by

Hexel Corp., dubbed as the "Rotary Hexapod" (Merlet, 2001). Among these different versions, Pierrrot's HEXA robot is considered in this chapter (see Fig. 2).

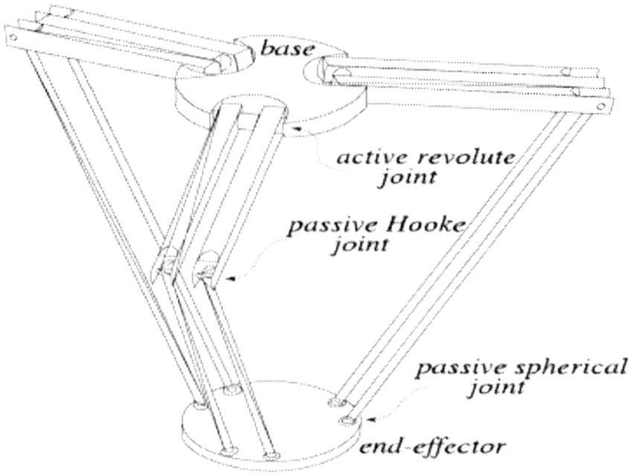

Fig. 2. Pierrrot's HEXA robot (Pierrrot *et al.*, 1990)

All types of HEXA robots are 6-DOF parallel manipulators that have the following characteristics:

a) With multiple closed chains, it can realize a greater structural stiffness.

b) To prevent the angular error of each motor from accumulating, it can realize a higher accuracy of the end-effecter position.

c) As all the actuators can be placed collectively on the base, it can realize a very light mechanism.

Consequently, HEXA enjoys the advantages of faster motions, better accuracy, higher stiffness and greater loading capacities over the serial manipulators (Uchiyama *et al.*, 1992).

4. Kinematic modeling

As in the case of conventional serial robots, kinematics analysis of parallel manipulators is also performed in two phases. In forward or direct kinematics the position and orientation of the mobile platform is determined given the leg lengths. This is done with respect to a base reference frame. In inverse kinematics we use position and orientation of the mobile platform to determine actuator lengths. For all types of parallel robots, IKP is easily solved. For HEXA parallel robot this problem was solved by Pierrrot (Pierrrot *et al.*, 1990). Brief solution of IKP is presented by Bruyninckx in (Bruyninckx, 1997). Fig. 3 shows one mechanical chain in HEXA design. In each chain, M specifies the length of the crank which is the mechanical link between the revolute and universal joints, and L gives the length of the rod which connects universal and spherical joints. Other parameters, H, h and a, are introduced as shown in Fig. 4 The relationship between the joint angles $\theta_{i,j}$ (i=1,2,3 and j=1,2), robot parameters and position and orientation of the end-effector can be obtained from the following procedure. The joint angle $\theta_{i,j}$ moves the end point of crank of ith leg to the position p_i given by

$$p_i = b_i + R_{ib}^i R(X, \theta_{i,j})[0 \quad 0 \quad M]^T \tag{1}$$

In this equation, the joint angle $\theta_{i,j}$ is the only unknown variable. The positions p_i are connected to a mobile platform pivot point t_i by links of known length L. Matrix R_{ib}^i is the rotation matrix between the base frame {bs} and a reference frame constructed in the actuated R joint, with X-axis along the joint axis and the Z-axis along the direction of the first link corresponding to a zero joint angle $\theta_{i,j}$ (see Fig. 3). Matrix $R(X, \theta_{,j})$ is the rotation matrix corresponding to a rotation about the X axis by the angle $\theta_{i,j}$:

$$R(X, \theta_{i,j}) = \begin{bmatrix} 1 & 0 & 0 \\ 0 & \cos(\theta_{i,j}) & -\sin(\theta_{i,j}) \\ 0 & \sin(\theta_{i,j}) & \cos(\theta_{i,j}) \end{bmatrix} \tag{2}$$

In each chain, a loop closure formulation can be adopted as follows (see Fig. 3):

$$\overrightarrow{t_i b_i} = \overrightarrow{t_i p_i} + \overrightarrow{p_i b_i} \tag{3}$$

with

$$\left|\overrightarrow{b_i p_i}\right| = M \tag{4}$$

$$\left|\overrightarrow{t_i p_i}\right| = L \tag{5}$$

It is possible to solve (3), (4), (5), for $\theta_{i,j}$:

$$\theta_{i,j} = 2 * \tan^{-1}\left(\frac{V_{i,j} \pm \sqrt{V_{i,j}^{\ 2} - W_{i,j}^{\ 2} + U_{i,j}^{\ 2}}}{U_{i,j} + W_{i,j}}\right) \tag{6}$$

where

$$V_{i,j} = -\mu_{i,j} \tag{7}$$

$$U_{i,j} = \lambda_{i,j} - H \tag{8}$$

$$W_{i,j} = \frac{L^2 - M^2 + (\lambda_{i,j} - H)^2 + (\rho_{i,j} - (-1)^j a)^2 + \mu_{i,j}^{\ 2}}{2L} \tag{9}$$

And $[\lambda_{i,j} \quad \rho_{i,j} \quad \mu_{i,j} \quad 1]^T$ is the position vector of the pivot point t_i in the reference frame constructed in the actuated R joint (Pierrrot *et al.*, 1990). The same equations can be used to derive the HEXA forward kinematic model, but the closed form solution to FKP can not be found. So, we propose to use numerical schemes by neural network approach for solving FKP in the workspace of the robot.

5. Artificial neural networks

The inspiration for neural networks comes from researches in biological neural networks of the human brains. Artificial neural network (ANN) is one of those approaches that permit imitating of the mechanisms of learning and problem solving functions of the human brain which are flexible, highly parallel, robust, and fault tolerant. In artificial neural networks implementation, knowledge is represented as numeric weights, which are used to gather the relationships between data that are difficult to realise analytically, and this iteratively adjusts the network parameters to minimize the sum of the squared approximation errors using a gradient descent method. Neural networks can be used to model complex relationship without using simplifying assumptions, which are commonly used in linear approaches. One category of the neural networks is the back propagation network which is trained with supervision, using gradient-descent training technique and minimizes the squared error between the actual outputs of the network and the desired outputs.

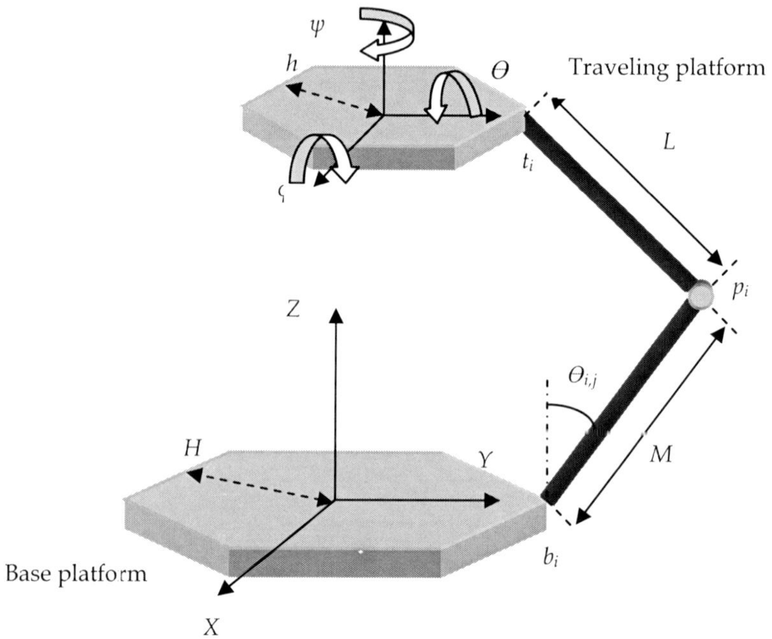

Fig. 3. A typical chain of the HEXA design. The joint angle $\theta_{i,j}$ is variable and measured; the lengths L and M of the "base" and "top" limbs of each chain are constant; the angles of all other joints are variable but not measured. Note that the joint between L and M is two degrees of freedom universal joint, so that the link L does not necessarily lie in the plane of the figure.

5.1 Multilayer perceptron (MLP)

The MLP is one of the typical back propagation ANNs and consists of an input layer, some hidden layers and an output layer, as shown in Fig. 5.

MLP is trained by back propagation of errors between desired values and outputs of the network using gradient descent or conjugate gradient algorithms. The network starts training after the weight factors are initialized randomly. Valid data consisting of the input vector and the corresponding desired output vector is fed to the network and the difference between the output layer result and the corresponding desired output result is used to adjust the weights by back propagation of the errors. This procedure continues until errors are small enough or no more weight changes occur. A first challenge in training the back propagation neural network is the choice of the appropriate network architecture, i.e. number of hidden layers and number of nodes of each layer. There is no available theoretical result which such choice may rely on. This can only be determined by user's experience (Medsker *et al.*, 1994).

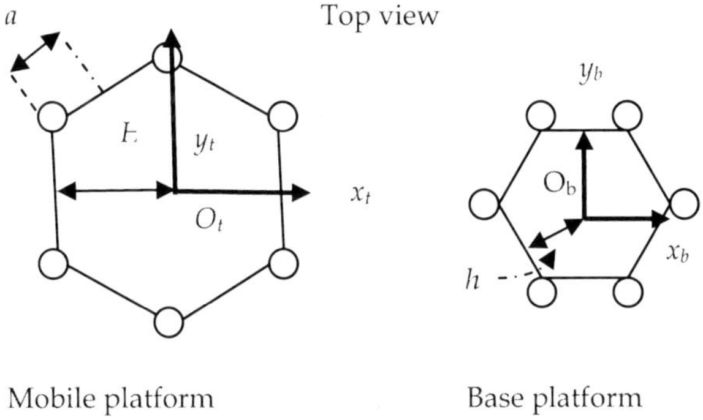

Fig. 4. Top views of the base and mobile platforms

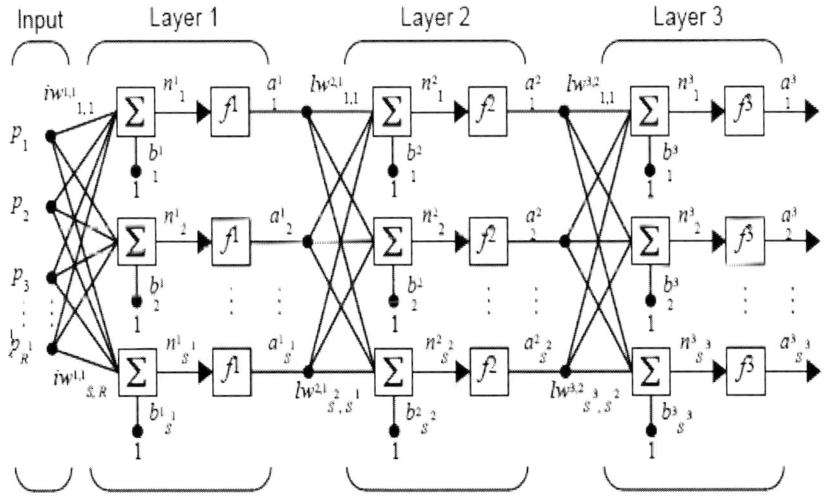

Fig. 5. Schematic of the MLP network (Geng *et al.*, 1992)

5.2 Wavelet based neural network (wave-net)

The hierarchical multiresolution wavelet based network, namely wave-net, was first introduced by Bakhshi (Bakshi and Sephanopolus, 1993) and was further investigated by Safavi (Safavi and Romagnoli, 1997). There has been another approach to develop wavelet based neural network with almost an MLP structure presented by Zhang (Zhang *et al.*, 1992). However, the latter type of neural network lacks an efficient use of the capabilities of wavelets and multiresolution analysis and therefore is not considered in this chapter.

5.2.1 Wavelets and multiresolution analysis (MRA)

Wavelets are a new family of localized basis functions and have found many applications in quite a large area of science and engineering (Daubechies, 1992). These basis functions can be used to express and approximate other functions. They are functions with a combination of powerful features, such as orthonormality, locality in time and frequency domains, different degrees of smoothness, fast implementations, and in some cases compact support. Wavelets are usually introduced in a multiresolution framework developed by Mallat (Mallat, 1989). These are shortly explained in the following. Consider a function F(X) in L2(R), where L2(R) denotes the vector space of all measurable, square integrable one-dimensional functions. The function can be expressed as

$$F(X) = F_0(X) + \sum_{m=-\infty}^{m=0} \sum_{k=-\infty}^{k=+\infty} d_{m,k}\psi_{m,k}(X) \tag{11}$$

where

$$F_0(X) = \sum_{k=-\infty}^{k=+\infty} a_{0,k}\varphi_{0,k}(X) \tag{12}$$

Here, the function $\varphi_{m,k}$ (not to be confused with the orientation angle φ)is called a scaling function of the multiresolution analysis (MRA) and a family of scaling functions of the MRA is expressed as;

$$\varphi_{m,k}(X) = 2^{-m/2}\varphi(2^{-m}X - k) \qquad m,k \in Z \tag{13}$$

Where 2^{-m} and k correspond respectively to the dilation and translation factors of the scaling function, and $2^{-m/2}$ is an energy normalization factor. The wavelets, denoted by $\psi_{m,k}$ (not to be confused with the orientation angle ψ), can easily be obtained from $\varphi_{m,k}$. A family of wavelets may be represented as:

$$\psi_{m,k}(X) = 2^{-m/2}\psi(2^{-m}X - k) \qquad m,k \in Z \tag{14}$$

To gain a thorough understanding of the role of scaling functions and wavelets within the multiresolution approximation framework see (Daubechies, 1992).

5.2.2 Wave-net learning

Equation (11) describes the basic framework of a wave-net in that it explains how each wavelet co-operates in the whole approximation scheme. It also shows that the scaling functions are only used at the earliest stage of approximation to produce F_0, after which the approximation scheme uses only wavelets. Fig. 6 depicts a typical wave-net structure. The hierarchical nature of the scheme is also obvious. Once the first approximation to a function F is obtained, that is F_0, one can get a better approximation, namely F_{-1}, by including wavelets of the same dilation factor as the scaling function, here m=0. Adding wavelets of the next highest resolution, here m= -1, leads to an approximation F_{-2} , finer than the previous one F_{-1}. This process is continued until the original function is reconstructed or an arbitrary degree of accuracy for the approximation is obtained.
In the above hierarchical approach, wavelets with different dilations and translations are incorporated.
The approaches to find the network coefficients, $a_{m,k}$ and $d_{m,k}$ are presented by Safavi (Safavi and Romagnoli, 1997).

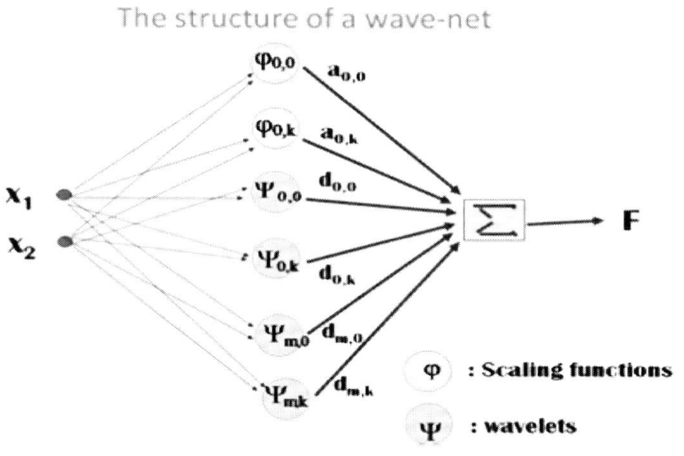

Fig. 6. The wave-net structure

6. Neural network solution for FKP

In order to model HEXA FKP with neural networks, first, a typical workspace for the robot is determined. Then, IKP is solved in some points of the workspace and finally the MLP and wave-net are trained with the data of IK solution in the typical robot workspace.

6.1 The workspace analysis

It is well known that parallel manipulators have a rather limited and complex workspace. Six parameters consisting of three coordinates of position of center of mass for mobile platform in the base frame (X, Y, Z) and three RPY orientation angles of mobile platform with respect to the base frame (three angles of mobile platform orientation in space consist of φ, ψ and θ angles, see Fig. 3) vary in the HEXA workspace.

Complete analysis of HEXA workspace is presented in (Bonev *et al.*, 2000) by A. Bonev. We use a typical workspace shown in Fig. 7. In this workspace, end-effector can move 300 millimeters in both directions of X and Y axes; also it can move 600 millimeters in positive Z direction. In all positions of the workspace, mobile platform can rotate in the range of [-π/3, π/3] for φ, ψ and θ angles. Fig. 7 shows the typical workspace which is used in this chapter. The geometric parameters of the robot are given in Table 1.

H	h	M	L	a
360mm	51mm	220mm	280mm	51mm

Table1. Geometric parameters of HEXA parallel robot

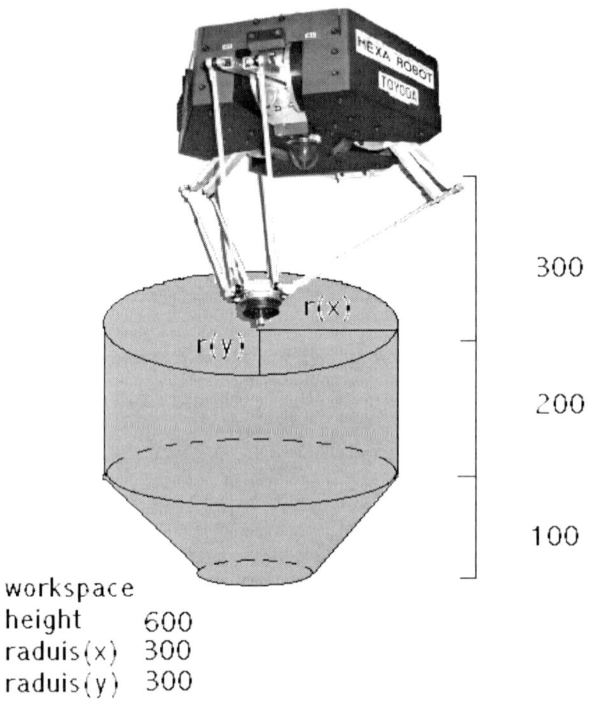

Fig. 7. A typical workspace for the HEXA parallel robot

6.2 Neural network solution for FKP

Now a MLP network can be trained with the data generated by the solution of IKP. In order to model the FKP in terms of 6 variables of positions and orientations of the mobile platform, a MLP network with a configuration of 6×13×13×13×13×13×6 has been developed with the smallest error and has been used to model FKP. In other words, the

ANN model has 6 inputs consisting of 6 joint angles, 5 hidden layers with 13 neurons in each layer, and 6 neurons in the output layer. The activation functions used in the hidden layers and the output layer are logarithmic and pure linear, respectively. The number of patterns used for training and test are 17500 and 35000, respectively. The network is trained over 1200 epochs with error back propagation training. Each network is evaluated by comparing the predictions and the true outputs, resulting in a prediction error for each orientation angle. The autocorrelation coefficients are also computed for the predicted error of each orientation angle.

6.3 Wave-net solution for FKP

In order to model the FKP with wave-net, MRA framework is used to approximate this process in different resolutions. Inputs, outputs and the number of patterns used for training and test are similar to the MLP network. The network is trained in resolutions m=0,-1 and -2 and the best results of modeling are reached at resolution -2. Figure 10 shows the training results for the successive resolutions zero, -1 and -2 for the X, Y, Z positions. For φ, ψ and θ angles the results are not represented due to the similarity and also to save space.

6.4 Modeling results

In this section the result of modeling FKP are presented. Error parameters in the tables are:
mse ; maximum squared error performance function
mae ; maximum absolute error performance function
nrmse ; normalized root minimum square error
Figures 8-11 show the modelling error and the correlations between the outputs of networks and the target outputs.

6.4.1 Modeling results with the MLP network

Table 2 and Figs. 8 and 9 show the results of FKP solution by MLP; Table 2 shows the resulted errors of FKP modeling.
It is apparent from Table 2 that *mse* , *mae* and *nrmse* in all joints are less than $2*10^{-5}$, 0.01 and 0.01 respectively, in test data. *mae* indicates maximum absolute error of modeling; therefore, maximum error of position and orientation of mobile platform is not bigger than 1 millimeter in position and 0.1 degrees in orientation in the worst case. *mse* shows the maximum of the average of errors in all points and so the average error of FKP solution in the typical workspace is less than $2*10^{-5}$. R in Table 2 indicates linear regression between output of the network and the target data. The closer regression to 1, the better the modeling is. The linear regression of all joints is more than 0.99 which shows very good quality modeling results. Fig. 9 shows the error of modeling in 1000 sample test points of typical workspace. For these sample posses the errors of modeling in position and orientation are very small and can be neglected.

6.4.2 Modeling results with wave-net

Figures 10 and 11 show the results of FKP solution by wave-net. Table 3 shows the resulted errors of FKP modeling. In Table 3 *mse* and *mae* in all joints are less than 10^{-6}, 10^{-2}, respectively, for test data. Therefore, maximum error orientation of mobile platform is not greater than 10^{-2} degrees in orientation for the worst case. Besides, the average error of FKP solution in the typical workspace is less than e-6. *R* (linear regression) in Table 3 of all joints is more than 0.999 which shows good modeling results. So, comparing the results of the MLP network and wave-net, wave-net model has smaller prediction error for FKP modeling of HEXA robot.

7. Comparison of MLP and wave-net results

In section 6 two approaches were used to model the FKP of HEXA robot - MLP network and wavelet based neural network. Though both neural network approaches showed great potential for this study, some comparison between these two approached are presented here. It is apparent from the results that errors of modelling by wave-net is less than MLP network, also the required time for modeling by wave-net is smaller than MLP; therefore, the wave-net modeling shows superior results in comparison to the MLP. Table 4 shows the results of modeling with these networks.

Figure 11 shows the linear regression between target X and Y positions and wave-net outputs. The same regressions can be obtained for φ, θ and ψ angles and Z position which are omitted here because of the similarity.

Variable	*mse*	*mae*	*nrmse*	*R*
X	1.3232e-005	0.0089	0.01	0.999
Y	5.76992e-006	0.0076	0.0094	0.999
Z	1.79034e-005	0.0091	0.0045	0.999
φ	5.77768e-006	0.01	0.0073	0.988
θ	1.20364e-006	0.009	0.0034	0.988
ψ	2.1676e-006	0.0087	0.0045	0.999

Table 2. The resulted errors of FKP modeling by test data with MLP network

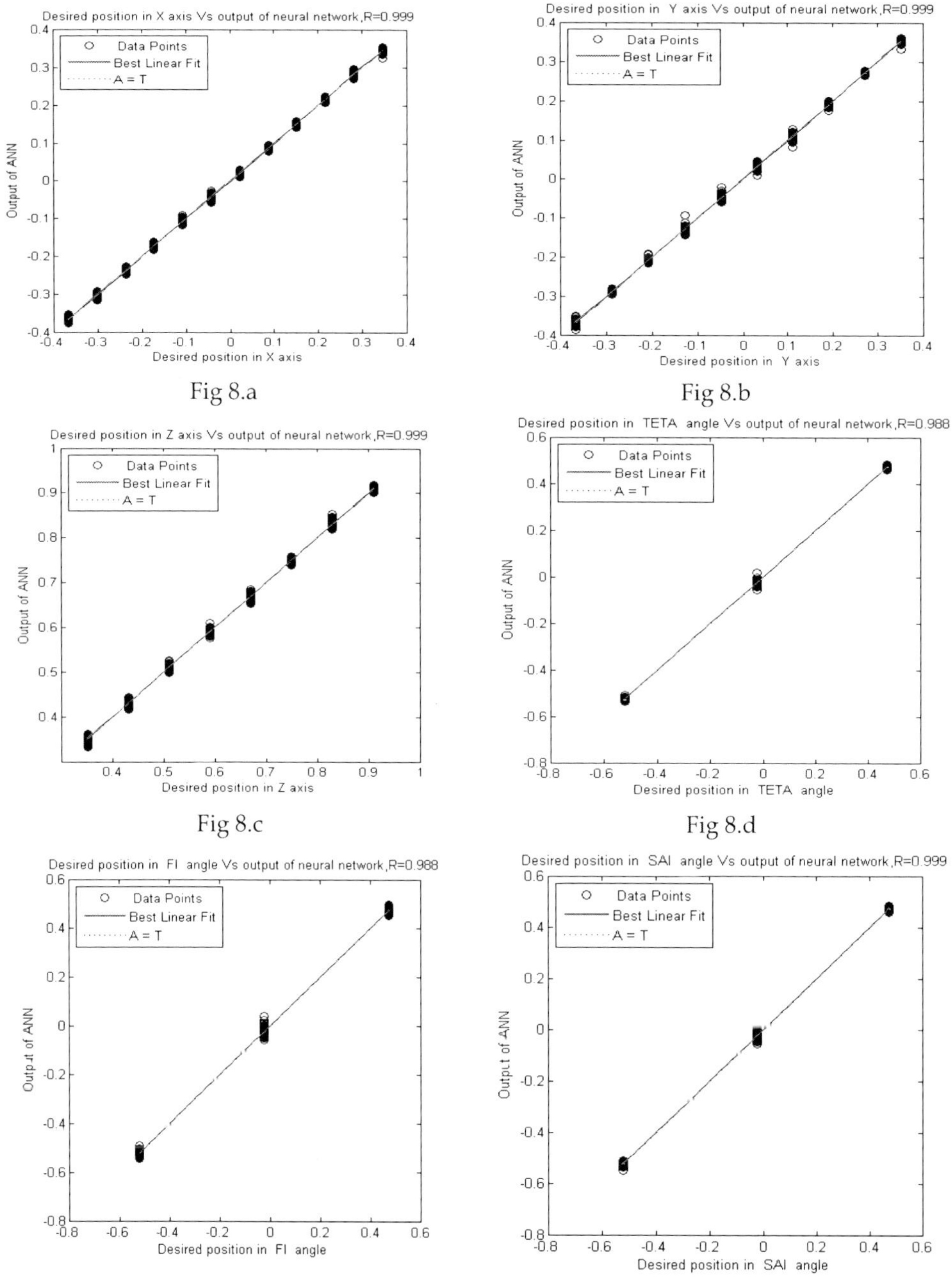

Fig. 8. The results of HEXA parallel robot modeling with ANN for X,Y,Z axes and φ, ψ , θ angles, from 8-a to 8-f, respectively.

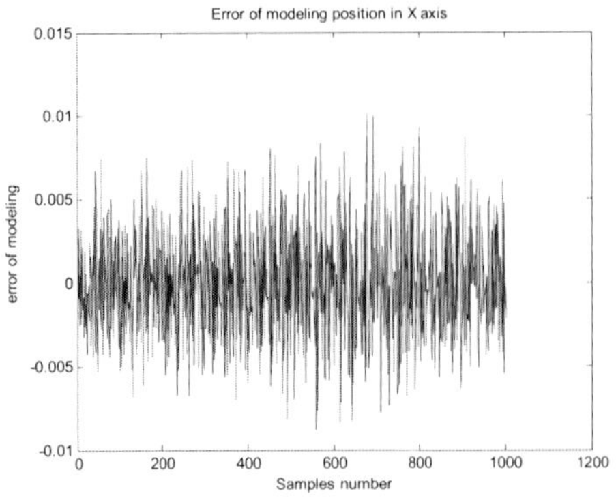

Fig 9.a

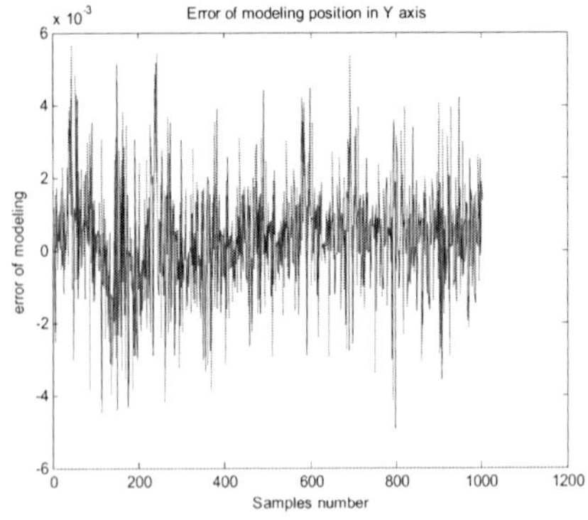

Fig 9.b

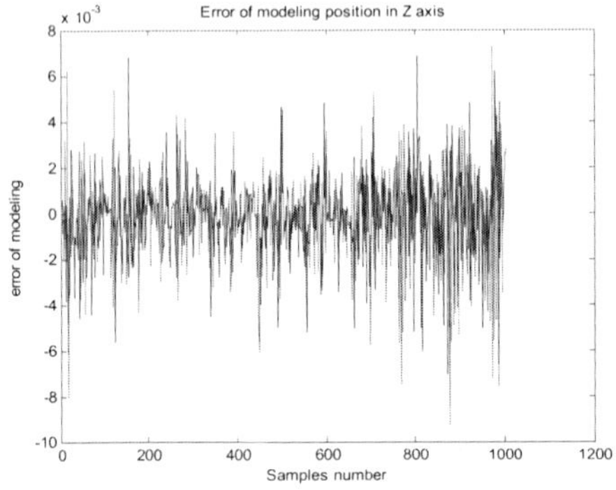

Fig 9.c

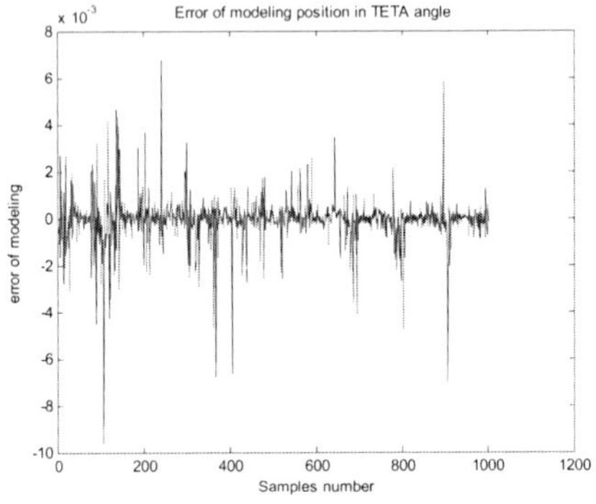

Fig 9.d

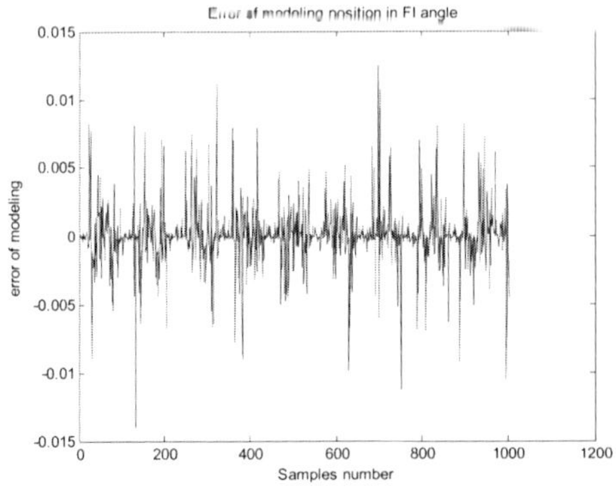

Fig 9.e

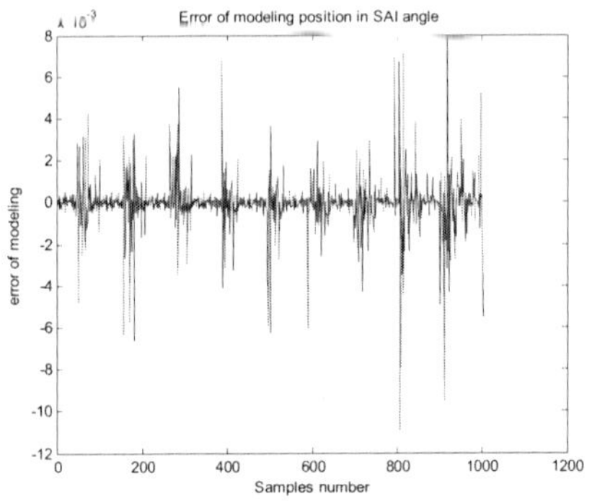

Fig 9.f

Fig. 9. The error of HEXA parallel robot modeling with ANN for X,Y,Z axes and φ, ψ, θ angles, from 9-a to 9-f, respectively.

Resolution 0

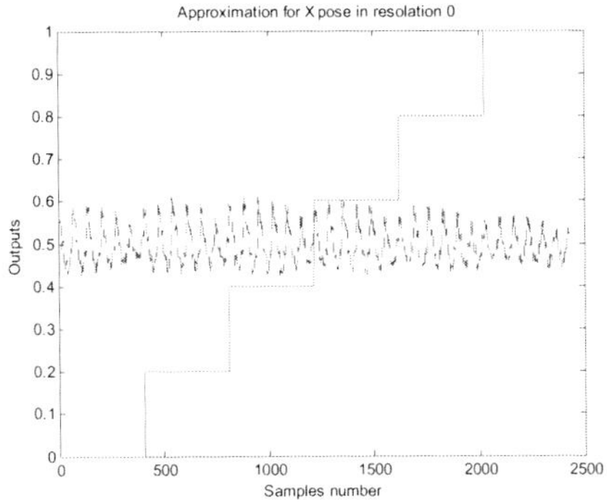

(Fig. 10.1) X

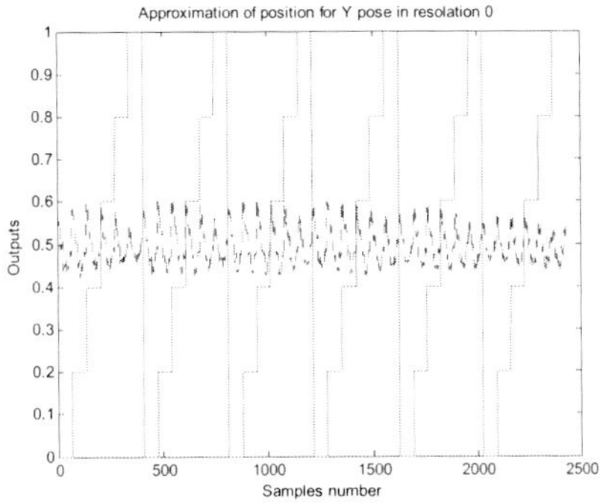

(Fig. 10.2) Y

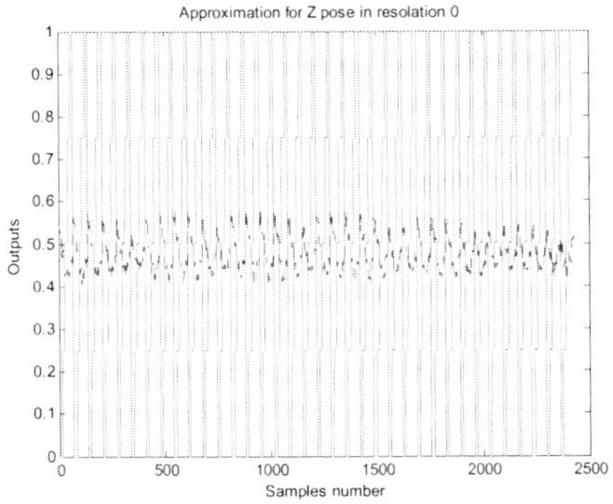

(Fig. 10.3) Z

Resolution -1

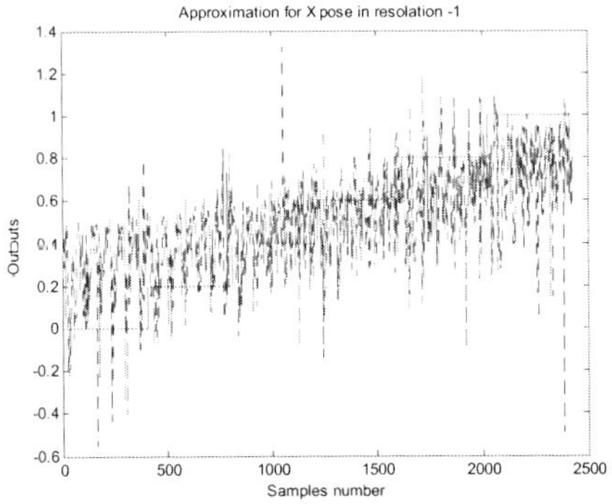

(Fig. 10.4) X

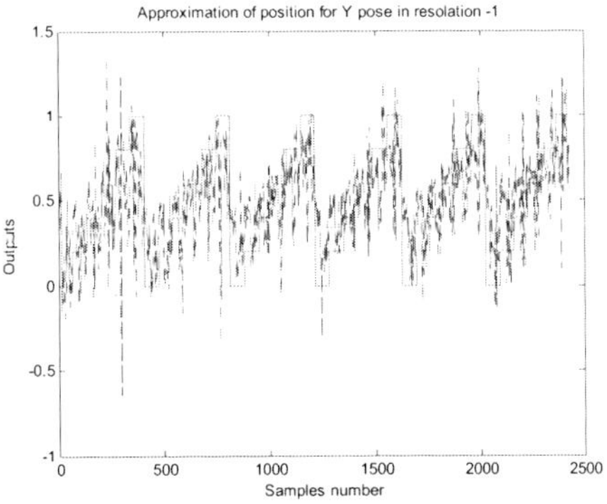

(Fig. 10.5) Y

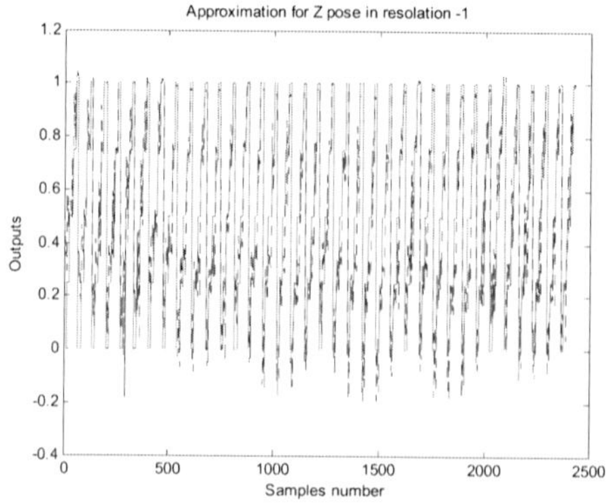

(Fig. 10.6) Z

Resolution -2

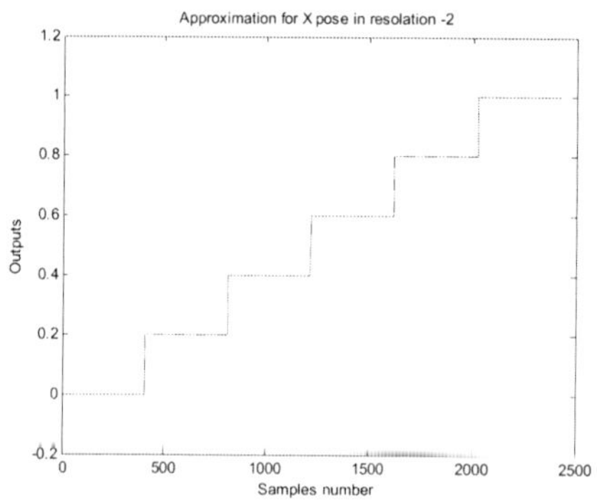

(Fig. 10.7) X

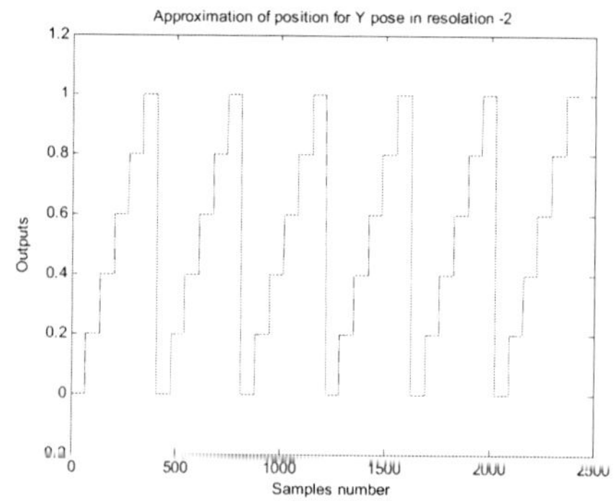

(Fig. 10.8) Y

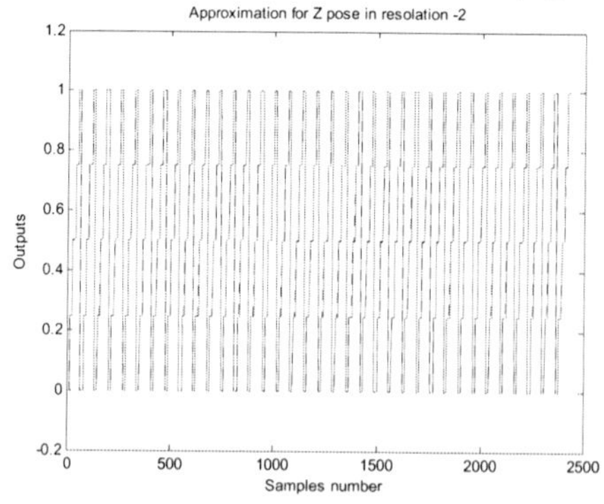

(Fig. 10.9) Z

Fig. 10 – Modeling results of X, Y, Z positions in resolution 0,-1 and -2 by the trained data, respectively

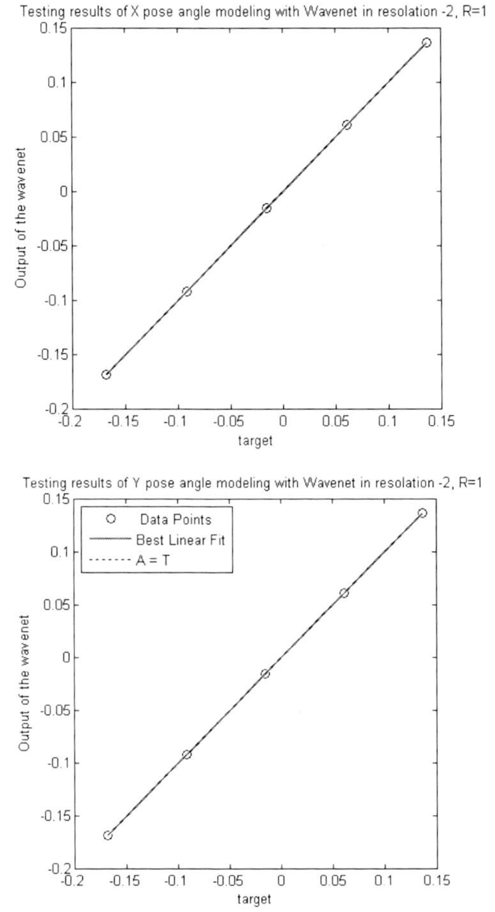

a b

Fig. 11 - Modeling results of X and Y positions with the wave-net, a is X model and b is Y model.

Variable	mse	mae	R
Ψ	8.2568e-010	2.5947e-004	1
Y	2.6346e-013	4.6090e-006	1
Z	1.2103e-006	4.7103e-002	0.9999
Φ	1.1402e-09	2.9911e-004	0.9999
θ	8.2568e-09	2.5947e-003	1
X	1.8501e-015	3.1252e-008	1

Table 3. The resulted errors of FKP modeling by test data with wave-net

	Wave-net			*MLP*		
	Training time 33 min			*Training time 123 min*		
Variable	*mse*	*mae*	*R*	*mse*	*mae*	*R*
Ψ	8.26e-010	2.60e-004	1	1.33e-005	0.0089	1
Y	2.64e-013	4.61e-006	1	5.77e-006	0.0076	1
Z	1.21e-006	4.71e-002	0.999	1.79e-005	0.0091	0.999
Φ	1.15e-09	2.99e-004	0.999	5.78e-006	0.01	0.999
θ	8.26e-09	2.60e-003	1	1.20e-006	0.009	1
X	1.85e-015	3.13e-008	1	1.85e-015	3.13e-008	1

Table 4. The comparison between results of modeling by wave-net and MLP

8. Conclusion

In this chapter, we proposed to use neural networks for FK solution of HEXA robot, which can be elaborated to generate the best estimation of forward kinematics of the robot. The research results in this chapter are quite important as they solve a problem for which there is no known closed form solution. Besides, the presented solution in this research has the better prediction and obtains smaller error in compare to the other works which have studied FKP of HEXA robot to the best of our knowledge.

9. References

Bakshi, B.R. and G. Sephanopoulos (1993). Wave-net: a multiresolution, hierarchical neural network with localised learning, AIChE Journal Vol. 39 No. 1.

Baron, L. and J. Angeles (2000). The direct kinematics of parallel manipulators under joint-sensor redundancy. IEEE conf. Trans. Robot. Autom.

Bonev, A. and M. Gosselin (2000). A Geometric Algorithm for the computation of the constant orientation workspace of 6 RUS parallel manipulator. Proc. of DETC'00 ASME Design Engineering Technical Conferences and Computers and Information in Engineering Conf. Baltimore.

Bruyninckx, H. (1997). The HEXA: A fully-parallel manipulator with closed form position and velocity kinematic, In Int. Conf. Robotics and Automation, pages 2657-2662, Albuquerque.

Daubechies, I. (1992). Ten Lectures on Wavelets, SIAM, Philadelphia, PA.

Ghobakhlo, A. and M. Eghtesad (2005). Neural network solution for the forward kinematics problem of a redundant hydraulic shoulder, Aicon.

Geng, Z. and L. Haynes (1992). Neural network solution for the forward kinematics problem of a Stewart platform. Robotic, Computer and Integrat. Manuf. Vol. 9, No. 6, 485-495.

Hunt, K. H. (1983). Structural kinematic in parallel actuated robot arms, Journal of Mechanisms, Transmissions and Automations in Design.

Last, P. C. Budde, C. Bier and J. Hesselbach (2005) HEXA-parallel-structure calibration by means of angular passive joint sensors. Proc. of the IEEE Int. Conf. on Mechatronics & Automation Niagara Falls, Canada.

Lee, T. Y and J. K. Shim, Forward kinematics for the general 6- 6 Stewart platform using algebraic elimination, Mech.Theory 36, 1073–1085.

Mallat, S.G. (1989) A theory for multi-resolution signal decomposition: the wavelet representation, IEE Trans. Pattern Analysis Mach. Int. vol. 11, no. 7.

Medsker, L. and J. Liebowitz (1994). Design and development of expert systems and neural networks. Macmillan, New York.

Merlet, J. P. (2007). Direct kinematics of parallel manipulators. J. of Robotica vol. 25.

Merlet, J. P. (2004). Solving the forward kinematics of a Gough-Type parallel manipulator with interval analysis, The Int. Journal of Robotics Research Vol. 23, No. 3.

Merlet, J. P. (2002). Still a long way to go on the road for parallel mechanisms. ASME Conf. Montreal, Canada.

Merlet, J. P. (2001). Parallel Robots (Solid Mechanics and Its Applications), Kluwer Academic Publishers.

Merlet, J. P. (1996). Direct kinematic of planer parallel manipulator. Proc. In IEEE Int. Conf. on Robotic and Automation, Minneapolis.

Mimura, N. and Y. Funahashi (1995) A new analytical system applying 6 dof parallel link manipulator for evaluating motion sensation. In IEEE Int.Conf on Robotic and Automation, Nagoya.

Pierrrot, F. M. Uchiyama, P. Dauchez and A. Fournier (1990). A new design of a 6-DOF parallel robot, in J. Robotics and Mechatronics, Vol. 2, No. 4.

Raghavan, M. (1991). The Stewart platform of general geometry has 40 configurations, Proc. the ASME Design and Automation Conf. Chicago, Vol. 32.

Sadjadian, H., H. M. Taghirad and A. Fatehi (2005) Neural Network Solution for Computing the Forward Kinematic of a Redundant Parallel Manipulator. Int. Journal of Computational intelligence, Vol 2, No 1.

Sarkissian, Y. L. and T.F. Parikyan (1990) manipulator , Russion, Patent, n1585144.

Safavi, A. A. and J. A. Romagnoli (1997). Application of Wavelet-based Neural Networks to the Modeling and Optimisation of an Experimental Distillation Column, EngngApplic. Artif. Intell. Vol. 10, No. 3, 301-313, Elsevier Science Ltd.

Siciliano, B and L. Sciavico (1995). Modeling and Control of Robot Manipulators, McGraw-Hill.

Stewart, D. (1965). Platform with Six Degrees of Freedom, Proc. of the Inst. Mech. Eng.

Szu, H. H. and S. Kadambe (1992). Neural network adaptive wavelets for signal representation, journal of Optical engineering Vol. 31 No. 9.

Tsai, L. W. (1999) Robot Analysis (The Mechanics of Serial and Parallel Manipulators). Wiley-Interscience Publication (John Wiley & Sons, Inc).

Uchiyama, M., K. Iimura, F. Pierrrot, K. Unno, and O. Toyama (1992.). Design and control of a very fast 6-DOF parallel robot, Proc. of the IMACS/SICE Int. Symp. on Robotics, Mechatronics and Manufacturing Systems.

Wang, Y. (2007). Direct numerical solution to forward kinematics of general Stewart–Gough platforms, J. of Robotica Vol. 25

Yee, C. S. (1997). Forward kinematics solution of Stewart platform using neural networks. Neurocomputing Vol. 16, No. 4, 333–349.

Zamanov, V. B. and Z. M. Sotirov (1992), Parallel manipulators in robotics, In int. Symp. on Robotic Mechatronic and manufacturing systems, Kobe.

Zhang, J. G., G. Walter, Y. Miao and W. Lee (1995). Wavelet neural networks for function learning, IEEE Transactions on Signal Processing, Vol. 43, pp. 1485-1497.

Zhang, Q. and A. Benveniste (1992). Wavelet networks, IEEE Transactions on Neural Networks, Vol. 3, No. 6.

16

Acceleration Analysis of 3-RPS Parallel Manipulators by Means of Screw Theory

J. Gallardo, H. Orozco, J.M. Rico, C.R. Aguilar and L. Pérez
Department of Mechanical Engineering, Instituto Tecnológico de Celaya, FIMEE, Universidad de Guanajuato, México

1. Introduction

According to the notation proposed by the International Federation for the Theory of Mechanisms and Machines IFToMM (Ionescu, 2003); a parallel manipulator is a mechanism where the motion of the end-effector, namely the moving or movable platform, is controlled by means of at least two kinematic chains. If each kinematic chain, also known popularly as limb or leg, has a single active joint, then the mechanism is called a fully-parallel mechanism, in which clearly the nominal degree of freedom equates the number of limbs. Tire-testing machines (Gough & Whitehall, 1962) and flight simulators (Stewart, 1965), appear to be the first transcendental applications of these complex mechanisms. Parallel manipulators, and in general mechanisms with parallel kinematic architectures, due to benefits --over their serial counterparts-- such as higher stiffness and accuracy, have found interesting applications such as walking machines, pointing devices, multi-axis machine tools, micro manipulators, and so on. The pioneering contributions of Gough and Stewart, mainly the theoretical paper of Stewart (1965), influenced strongly the development of parallel manipulators giving birth to an intensive research field. In that way, recently several parallel mechanisms for industrial purposes have been constructed using the, now, classical hexapod as a base mechanism: Octahedral Hexapod HOH-600 (Ingersoll), HEXAPODE CMW 300 (CMW), Cosmo Center PM-600 (Okuma), F-200i (FANUC) and so on. On the other hand one cannot ignore that this kind of parallel kinematic structures have a limited and complex-shaped workspace. Furthermore, their rotation and position capabilities are highly coupled and therefore the control and calibration of them are rather complicated.
It is well known that many industrial applications do not require the six degrees of freedom of a parallel manipulator. Thus in order to simplify the kinematics, mechanical assembly and control of parallel manipulators, an interesting trend is the development of the so called defective parallel manipulators, in other words, spatial parallel manipulators with fewer than six degrees of freedom. Special mention deserves the Delta robot, invented by Clavel (1991); which proved that parallel robotic manipulators are an excellent option for industrial applications where the accuracy and stiffness are fundamental characteristics. Consider for instance that the Adept Quattro robot, an application of the Delta robot, developed by Francois Pierrot in collaboration with Fatronik (Int. patent appl. WO/2006/087399), has a

2.0 kilograms payload capacity and can execute 4 cycles per second. The Adept Quattro robot is considered at this moment the industry's fastest pick-and-place robot.
Defective parallel manipulators can be classified in two main groups: Purely translational (Romdhane et al, 2002; Parenti-Castelli et al, 2000; Carricato & Parenti-Castelli, 2003; Di Gregorio & Parenti-Castelli, 2002; Ji & Wu, 2003; Kong & Gosselin, 2004a; Kong & Gosselin, 2002) or purely spherical (Alizade et al, 1994; Di Gregorio, 2002; Gosselin & Angeles, 1989; Kong & Gosselin, 2004b; Liu & Gao 2000). A third class is composed by parallel manipulators in which the moving platform can undergo mixed motions (Parenti-Castelli & Innocenti, 1992; Gallardo-Alvarado et al, 2006; Gallardo-Alvarado et al, 2007). The 3-RPS, Revolute + Prismatic +Spherical, parallel manipulator belongs to the last class and is perhaps the most studied type of defective parallel manipulator.
The 3-RPS parallel manipulator was introduced by Hunt (1983) and has been the motive of an exhaustive research field where a great number of contributions, approaching a wide range of topics, kinematic and dynamic analyses, synthesis, singularity analysis, extensions to hyper-redundant manipulators, etc; have been reported in the literature, see for instance Lee & Shah (1987), Kim & Tsai (2003), Liu & Cheng (2004), Lu & Leinonen (2005). In particular, screw theory has been proved to be an efficient mathematical resource for determining the kinematic characteristics of 3-RPS parallel manipulators, see for instance Fang & Huang (1997), Huang and his co-workers (1996, 2000, 2001, 2002); including the instantaneous motion analysis of the mechanism at the level of velocity analysis (Agrawal, 1991).
This paper addresses the kinematics of 3-RPS parallel manipulators, including position, velocity and acceleration analyses. Firstly the forward position analysis is carried out in analytic form solution using the Sylvester dialytic elimination method. Secondly the velocity and acceleration analyses are approached by means of the theory of screws. To this end, the velocity and reduced acceleration states of the moving platform, with respect to the fixed platform, are written in screw form through each one of the limbs of the mechanism. Finally, the systematic application of the Klein form to these expressions allows obtaining simple and compact expressions for computing the velocity and acceleration analyses. A case study is included.

2. Description of the mechanism

A 3-RPS parallel manipulator, see Fig. 1, is a mechanism where the moving platform is connected to the fixed platform by means of three extendible limbs. Each limb is composed by a lower body and an upper body connected each other by means of an active prismatic joint. The moving platform is connected at the upper bodies via three distinct spherical joints while the lower bodies are connected to the fixed platform by means of three distinct revolute joints.
An effective general formula for determining the degrees of freedom of closed chains still in our days is an open problem. An exhaustive review of formulae addressing this topic is reported in Gogu (2005). Regarding to the existing methods of computation, these formulae are valid under specific conducted considerations. For the parallel manipulator at hand, the mobility is determined using the well-known Kutzbach-Grübler formula

$$F = 6(n - j - 1) + \sum_{i=1}^{j} f_i \tag{1}$$

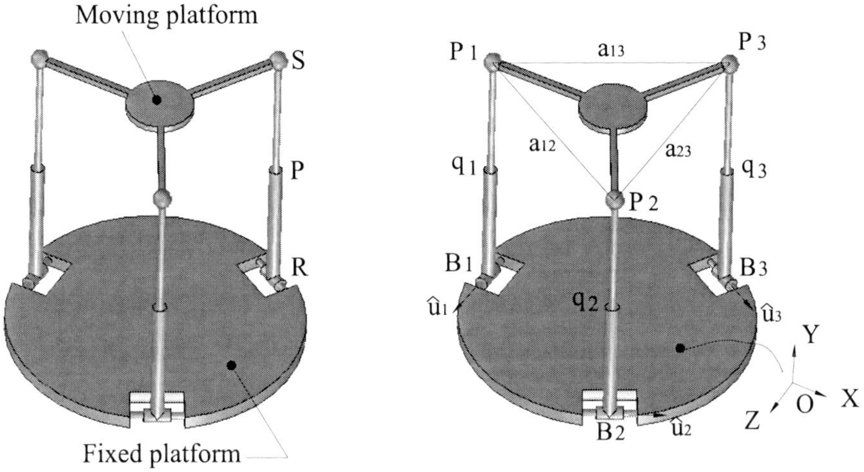

Fig. 1. The 3-RPS parallel manipulator and its geometric scheme.

Where n is the number of links, j is the number of kinematic pairs and f_i is the number of freedoms of the i-th pair. Thus, taking into account that for the mechanism at hand n=8, j=9 and $\sum_{i=1}^{j} f_i = 15$; then the degrees of freedom of it are equal to 3, an expected result.

2. Position analysis

In this section the forward finite kinematics of the 3-RPS parallel manipulator is approached using analytic procedures. The inverse position analysis is considered here a trivial task and therefore it is omitted.

The geometric scheme of the spatial mechanism is shown in the right side of Fig. 1. Accordingly with this figure; B_i, q_i and P_i denotes, respectively, the nominal position of the revolute joint, the length of the limb and the center of the spherical joint in the same limb. While u_i denotes the direction of the axis associated to the revolute joint. On the other hand a_{mn} represents the distance between the centers of two spherical joints.

In this work, the forward position analysis of the 3-RPS parallel manipulator consists of finding the pose, position and orientation, of the moving platform with respect to the fixed platform given the three limb lengths or generalized coordinates q_i of the parallel manipulator. To this end, it is necessary to compute the coordinates of the three spherical joints expressed in the reference frame XYZ.

When the limbs of the parallel manipulator are locked, the mechanism becomes a 3-RS structure. In order to simplify the analysis, the reference frame XYZ, attached at the fixed platform, is chosen in such a way that the points B_i lie on the XZ plane. Under this consideration the axes of the revolute joints are coplanar and three constraints are imposed by these joints as follows

$$(P_i - B_i) \bullet u_i = 0 \qquad i \in \{1,2,3\} \tag{2}$$

where the dot denotes the usual inner product operation of the three dimensional vectorial algebra. It is worth to mention that expressions (2) were not considered, in the form derived, by Tsai (1999), and therefore the analysis reported in that contribution requires a particular arrangement of the positions of the revolute joints over the fixed platform accordingly to the reference frame XYZ. Furthermore, clearly expressions (2) are applicable not only to tangential 3-RPS parallel manipulators, like the mechanism of Fig. 1, but also to the so-called concurrent 3-RPS parallel manipulators.
On the other hand, clearly the limb lengths are restricted to

$$(P_i - B_i) \bullet (P_i - B_i) = q_i^2 \qquad i \in \{1,2,3\} \tag{3}$$

Finally, three compatibility constraints can be obtained as follows

$$\begin{aligned} (P_2 - P_3) \bullet (P_2 - P_3) &= a_{23}^2 \\ (P_1 - P_3) \bullet (P_1 - P_3) &= a_{13}^2 \\ (P_1 - P_2) \bullet (P_1 - P_2) &= a_{12}^2 \end{aligned} \tag{4}$$

Expressions (2)-(4) form a system of nine equations in the nine unknowns $\{X_1, Y_1, Z_1, X_2, Y_2, Z_2, X_3, Y_3, Z_3\}$. In what follows, expressions (2-4) are reduced systematically into a highly non linear system of three equations in three unknowns. Afterwards, a sixteenth-order polynomial in one unknown is derived using the Sylvester dialytic elimination method.
It follows from Eqs. (2) that

$$X_1 = f(Z_i) \qquad i \in \{1,2,3\} \tag{5}$$

On the other hand with the substitution of (5) into expressions (3), the reduction of terms leads to

$$Y_i^2 = p_i \qquad i \in \{1,2,3\} \tag{6}$$

where p_i are second-degree polynomials in Z_i . Finally, the substitution of Eqs. (6) into Eqs. (4) results in the following highly non-linear system of three equations in the three unknowns Z_1, Z_2 and Z_3

$$\begin{aligned} c_1Z_2^2 + c_2Z_3^2 + c_3Z_2^2Z_3 + c_4Z_2Z_3^2 + c_5Z_2Z_3 + c_6Z_2 + c_7Z_3 + c_8 &= 0 \\ d_1Z_1^2 + d_2Z_3^2 + d_3Z_1^2Z_3 + d_4Z_1Z_3^2 + d_5Z_1Z_3 + d_6Z_1 + d_7Z_3 + d_8 &= 0 \\ e_1Z_1^2 + e_2Z_2^2 + e_3Z_1^2Z_2 + e_4Z_1Z_2^2 + e_5Z_1Z_2 + e_6Z_1 + e_7Z_2 + e_8 &= 0 \end{aligned} \tag{7}$$

therein c, d and e are coefficients that are calculated accordingly to the parameters and generalized coordinates, namely the length limbs of the parallel manipulator.

Expressions (7) are similar to those introduced in Tsai (1999); however their derivation is simpler due to the inclusion, in this contribution, of Eqs. (2).
Please note that only two of the unknowns are present in each one of Eqs. (7) and therefore their solutions appear to be an easy task. For example, Z_2 and Z_3 can be obtained as functions of Z_1 from the last two quadratic equations; afterwards the substitution of these variables into the first quadratic yields a highly non-linear equation in Z_1. The handling of such an expression is a formidable an unpractical task. Thus, an appropriated strategy is required for solving the system of equations at hand. Some options are

- A numerical technique such as the Newton-Raphson method. It is an effective option, however only one and imperfect solution can be computed, and there are not guarantee that all the solutions will be calculated.
- Using computer algebra like Maple©. An absolutely viable option that guarantee the computation of all the possible solutions.
- The application of the Sylvester dialytic elimination method. An elegant option that allows to compute all the possible solutions.

In this contribution the last option was selected and in what follows the results will be presented.
With the purpose to eliminate Z_3, the first two quadratics of (7) are rewritten as follows

$$\begin{aligned} p_1 Z_3^2 + p_2 Z_3 + p_3 &= 0 \\ p_4 Z_3^2 + p_5 Z_3 + p_6 &= 0 \end{aligned} \tag{8}$$

where p_1, p_2 and p_3 are second-degree polynomials in Z_2 while p_4, p_5 and p_6 are second-degree polynomials in Z_1. After a few operations, the term Z_3 is eliminated from (8). With this action, two linear equations in two unknowns, the variable Z_3 and the scalar 1, are obtained. Casting in matrix form such expressions it follows that

$$M_1 \begin{bmatrix} Z_3 \\ 1 \end{bmatrix} = \begin{bmatrix} 0 \\ 0 \end{bmatrix} \tag{9}$$

where

$$M_1 = \begin{bmatrix} p_1 p_5 - p_2 p_4 & p_1 p_6 - p_3 p_4 \\ p_3 p_4 - p_1 p_6 & p_3 p_5 - p_2 p_6 \end{bmatrix}$$

It is evident that expression (9) is valid if, and only if, $\det(M_1) = 0$. Thus clearly one can obtain

$$p_7 Z_2^4 + p_8 Z_2^3 + p_9 Z_2^2 + p_{10} Z_2 + p_{11} = 0 \tag{10}$$

where p_7, p_8, p_9, p_{10} and p_{11} are fourth-degree polynomials in Z_1; and the first step of the Sylvester dialytic elimination method finishes with the computation of this eliminant.

Please note that Eq. (10) and the last quadratic of Eqs. (7) represents a non-linear system of two equations in the unknowns Z_1 and Z_2, and in what follows it is reduced into an univariate polynomial equation. As an initial step, that last quadratic of (7) is rewritten as

$$p_{12}Z_2^2 + p_{13}Z_2 + p_{14} = 0, \tag{11}$$

where p_{12}, p_{13} and p_{14} are second-degree polynomials in Z_1. It is very tempting to assume that the non-linear system of two equations formed by (10) and (11) can be easily solved obtaining first Z_2 in terms of Z_1 from Eq. (11) and later substituting it into Eq. (10). However, when one realize this apparent evident action with the aid of computer algebra, an excessively long expression is derived, and its handling is a hazardous task. Thus, the application of the Sylvester dialytic elimination method is a more viable option.

In order to avoid extraneous roots, it is strongly advisable the deduction of a minimum of linear equations. For example, the term Z_2^4 is eliminated multiplying Eq. (10) by p_{12} and Eq. (11) by $p_7Z_2^2$. The substraction of the obtained expressions leads to

$$(p_{13}p_7 - p_{12}p_8)Z_2^3 + (p_{14}p_7 - p_{12}p_9)Z_2^2 - p_{12}p_{10}Z_2 - p_{12}p_{11} = 0. \tag{12}$$

Expressions (11) and (12) can be considered as a linear system of two equations in the four unknowns Z_2^3, Z_2^2, Z_2 and 1. Therefore it is necessary the search of two additional linear equations.

An equation is easily obtained multiplying Eq. (11) by Z_2

$$p_{12}Z_2^3 + p_{13}Z_2^2 + p_{14}Z_2 = 0. \tag{13}$$

The search of the fourth equation is more elusive, for details the reader is referred to Tsai (1999). To this end, multiplicate Eq. (10) by $(p_{12}Z_2 + p_{13})$ and Eq. (11) by $(p_7Z_2^3 + p_8Z_2^2)$. The subtraction of the resulting expressions leads to

$$\begin{aligned}(p_{12}p_9 - p_7p_{14})Z_2^3 + (p_{12}p_{10} + p_{13}p_9 - p_9p_{14})Z_2^2 \\ + (p_{12}p_{11} + p_{13}p_{10})Z_2 + p_{13}p_{11} = 0\end{aligned} \tag{14}$$

Casting in matrix form expressions (11)-(14) it follows that

$$M_2\begin{bmatrix} Z_2^3 \\ Z_2^2 \\ Z_2 \\ 1 \end{bmatrix} = \begin{bmatrix} 0 \\ 0 \\ 0 \\ 0 \end{bmatrix}, \tag{15}$$

where

$$M_2 = \begin{bmatrix} 0 & p_{12} & p_{13} & p_{14} \\ p_{13}p_7 - p_{12}p_8 & p_{14}p_7 - p_{12}p_9 & -p_{12}p_{10} & -p_{12}p_{11} \\ p_{12} & p_{13} & p_{14} & 0 \\ p_{12}p_9 - p_7p_{14} & p_{12}p_{10} + p_{13}p_9 - p_8p_{14} & p_{12}p_{11} + p_{13}p_{10} & p_{13}p_{11} \end{bmatrix}$$

Clearly expression (15) is valid if, and only if, $\det(M_2) = 0$. Therefore, this eliminant yields a sixteenth-order polynomial in the unknown Z_1.

It is worth to mention that expressions (10) and (11) have the same structure of those derived by Innocenti & Parenti-Castelli (1990) for solving the forward position analysis of the Stewart platform mechanism. However, this work differs from that contribution in that, while in this contribution the application of the Sylvester Dialytic elimination method finishes with the computation of the determinant of a 4x4 matrix, the contribution of Innocenti & Parenti-Castelli (1990), a more general method than the presented in this section, finishes with the computation of the determinant of a 6x6 matrix.

Once Z_1 is calculated, Z_2 and Z_3 are calculated, respectively, from expressions (11) and the second quadratic of (8) while the remaining components of the coordinates, X_i and Y_i, are computed directly from expressions (5) and (6), respectively. It is important to mention that in order to determine the feasible values of the coordinates of the points P_i, the signs of the corresponding discriminants of Z_2, Z_3 and Y_i must be taken into proper account. Of course, one can ignore this last recommendation if the non-linear system (3) is solved by means of computer algebra like Maple©.

Finally, once the coordinates of the centers of the spherical joints are calculated, the well-known 4×4 transformation matrix T results in

$$T = \begin{bmatrix} R & r_{C/O} \\ 0_{1\times 3} & 1 \end{bmatrix}, \tag{16}$$

where, $r_{C/O} = (P_1 + P_2 + P_3)/3$ is the geometric center of the moving platform, and R is the rotation matrix.

3. Velocity analysis

In this section the velocity analysis of the 3-RPS parallel manipulator is carried out using the theory of screws which is isomorphic to the Lie algebra e(3). This section applies well known screw theory; for readers unfamiliar with this mathematical resource, some appropriated references are provided at the end of this work (Sugimoto, 1987; Rico and Duffy, 1996; Rico et al, 1999).

The mechanism under study is a spatial mechanism, and therefore the kinematic analysis requires a six-dimensional Lie algebra. In order to satisfy the dimension of the subspace spanned by the screw system generated in each limb, the 3-RPS parallel manipulator can be modelled as a 3-R*RPS parallel manipulator, see Huang and Wang (2000), in which the revolute joints R* are fictitious kinematic pairs. In this contribution, see Fig. 2, each limb is modelled as a Cylindrical + Prismatic + Spherical kinematic chain, CPS for brevity. It is straightforward to demonstrate that this option is simpler than the proposed in Huang and Wang (2000). Naturally, this model requires that the joint rate associated to the translational displacement of the cylindrical joint be equal to zero.

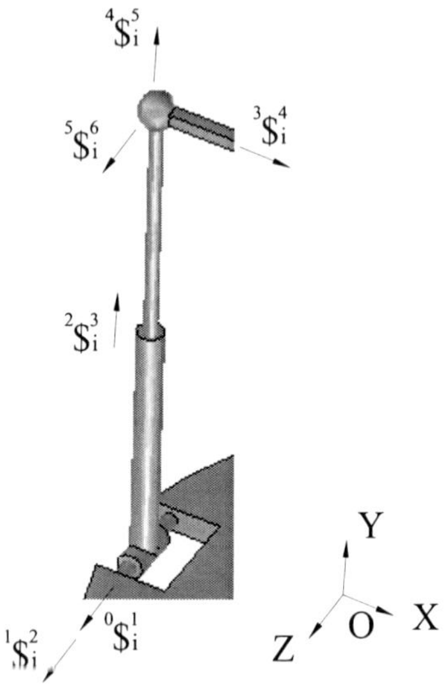

Fig. 2. A limb with its infinitesimal screws

Let $\omega = (\omega_X, \omega_Y, \omega_Z)$ be the angular velocity of the moving platform, with respect to the fixed platform, and let $\mathbf{v}_O = (v_{OX}, v_{OY}, v_{OZ})$ be the translational velocity of the point O, see Fig. 2; where both three-dimensional vectors are expressed in the reference frame XYZ. Then, the velocity state $V_O = [\omega \quad \mathbf{v}_O]$, also known as the twist about a screw, of the moving platform with respect to the fixed platform, can be written, see Sugimoto (1987), through the j-th limb as follows

$$\sum_{i=0}^{5} {}_i\omega_{i+1}^j \, {}^i\$_j^{i+1} = V_O \qquad j \in \{1,2,3\}, \tag{17}$$

where, the joint rate ${}_2\omega_3^j = \dot{q}_j$ is the active joint associated to the prismatic joint in the j-th limb, while ${}_0\omega_1^j = 0$ is the joint rate of the prismatic joint associated to the cylindrical joint. With these considerations in mind, the inverse and forward velocity analyses of the mechanism under study are easily solved using the theory of screws.
The inverse velocity analysis consists of finding the joint rate velocities of the parallel manipulator, given the velocity state of the moving platform with respect to the fixed platform. Accordingly to expression (17), this analysis is solved by means of the expression

$$\Omega_j = J_j^{-1} V_O. \tag{18}$$

Therein

- $J_j = \begin{bmatrix} {}^0\$_j^1 & {}^1\$_j^2 & {}^2\$_j^3 & {}^3\$_j^4 & {}^4\$_j^5 & {}^5\$_j^6 \end{bmatrix}$ is the Jacobian of the j-th limb, and
- $\Omega_j = \begin{bmatrix} {}_0\omega_1^j & {}_1\omega_2^j & {}_2\omega_3^j & {}_3\omega_4^j & {}_4\omega_5^j & {}_5\omega_6^j \end{bmatrix}^T$ is the matrix of joint velocity rates of the j-th limb.

On the other hand, the forward velocity analysis consists of finding the velocity state of the moving platform, with respect to the fixed platform, given the active joint rates $\dot{q}_j$. In this analysis the Klein form of the Lie algebra e (3) plays a central role.
Given two elements $\$_1 = \begin{bmatrix} s_1 & s_{O1} \end{bmatrix}$ and $\$_2 = \begin{bmatrix} s_2 & s_{O2} \end{bmatrix}$ of the Lie algebra e (3), the Klein form, $\{*,*\}$, is defined as follows

$$\{\$_1, \$_2\} = s_1 \bullet s_{O2} + s_2 \bullet s_{O1}. \tag{19}$$

Furthermore, it is said that the screws $\$_1$ and $\$_2$ are reciprocal if $\{\$_1, \$_2\} = 0$.
Please note that the screw ${}^4\$_i^5$ is reciprocal to all the screws associated to the revolute joints in the same limb. Thus, applying the Klein form of the screw ${}^4\$_i^5$ to both sides of expression (17), the reduction of terms leads to

$$\{V_O, {}^4\$_i^5\} = \dot{q}_i \qquad i \subset \{1,2,3\}. \tag{20}$$

Following this trend, choosing the screw ${}^5\$_i^6$ as the *cancellator screw* it follows that

$$\{V_O, {}^5\$_i^6\} = 0 \qquad i \in \{1,2,3\}. \tag{21}$$

Casting in a matrix-vector form expression (20) and (21), the velocity state of the moving platform is calculated from the expression

$$J^T \Delta\, V_O = \dot{Q}, \tag{22}$$

wherein

- $J = \left[{}^4\$_1^5 \quad {}^4\$_2^5 \quad {}^4\$_3^5 \quad {}^5\$_1^6 \quad {}^5\$_2^6 \quad {}^5\$_3^6 \right]$ is the Jacobian of the parallel manipulator,
- $\Delta = \begin{bmatrix} 0_{3\times3} & I_3 \\ I_3 & 0_{3\times3} \end{bmatrix}$ is an operator of polarity, and
- $\dot{Q} = \left[\dot{q}_1 \quad \dot{q}_2 \quad \dot{q}_3 \quad 0 \quad 0 \quad 0 \right]^T$.

Finally, once the angular velocity of the moving platform and the translational velocity of the point O are obtained, respectively, as the primal part and the dual part of the velocity state $V_O = \left[\omega \quad v_O \right]$, the translational velocity of the center of the moving platform, vector v_C, is calculated using classical kinematics. Indeed

$$v_C = v_O + \omega \times r_{C/O} . \tag{23}$$

Naturally, in order to apply Eq. (22) it is imperative that the Jacobian J be invertible. Otherwise, the parallel manipulator is at a singular configuration, with regards to Eq. (18).

4. Acceleration analysis

Following the trend of Section 3, in this section the acceleration analysis of the parallel manipulator is carried out by means of the theory of screws.

Let $\dot{\omega} = (\dot{\omega}_X, \dot{\omega}_Y, \dot{\omega}_Z)$ be the angular acceleration of the moving platform, with respect to the fixed platform, and let $a_O = (a_{OX}, a_{OY}, a_{OZ})$ be the translational acceleration of the point O; where both three-dimensional vectors are expressed in the reference frame XYZ. Then the reduced acceleration state $A_O = \left[\dot{\omega} \quad a_O - \omega \times v_O \right]$, or accelerator for brevity, of the moving platform with respect to the fixed platform can be written, for details see Rico & Duffy (1996), through each one of the limbs as follows

$$\sum_{i=0}^{5} {}_i\dot{\omega}_{i+1}^j \, {}^i\$_j^{i+1} + \$_{Lie\text{-}j} = A_O \qquad j \in \{1,2,3\}, \tag{24}$$

where $\$_{Lie-j}$ is the *Lie screw* of the j-th limb, which is calculated as follows

$$\$_{Lie\text{-}j} = \sum_{k=0}^{4} \left[{}_k\omega_{k+1}^j \, {}^k\$_j^{k+1} \quad \sum_{r=k+1}^{5} {}_r\omega_{r+1}^j \, {}^r\$_j^{r+1} \right],$$

and the brackets $\left[* \quad * \right]$ denote the Lie product.

Equation (24) is the basis of the inverse and forward acceleration analyses.

The inverse acceleration analysis, or in other words the computation of the joint acceleration rates of the parallel manipulator given the accelerator of the moving platform with respect to the fixed platform, can be calculated, accordingly to expression (24), as follows

$$\dot{\Omega}_j = J_j^{-1}(A_O - \$_{Lie-j}), \tag{25}$$

where $\dot{\Omega}_j = \begin{bmatrix} {}_0\dot{\omega}_1^j & {}_1\dot{\omega}_2^j & {}_2\dot{\omega}_3^j & {}_3\dot{\omega}_4^j & {}_4\dot{\omega}_5^j & {}_5\dot{\omega}_6^j \end{bmatrix}^T$ is the matrix of joint acceleration rates. On the other hand, the forward acceleration analysis, or in other words the computation of the accelerator of the moving platform with respect to the fixed platform given the active joint rate accelerations $\ddot{q}_j$ of the parallel manipulator; is carried out, applying the Klein form of the reciprocal screws to Eq. (24), using the expression

$$J^T \Delta\, A_O = \ddot{Q}, \tag{26}$$

where

$$\ddot{Q} = \begin{bmatrix} \ddot{q}_1 + \left\{ {}^4\$_1^5, \$_{Lie-1} \right\} \\ \ddot{q}_2 + \left\{ {}^4\$_2^5, \$_{Lie-2} \right\} \\ \ddot{q}_3 + \left\{ {}^4\$_3^5, \$_{Lie-3} \right\} \\ \left\{ {}^5\$_1^6, \$_{Lie-1} \right\} \\ \left\{ {}^5\$_2^6, \$_{Lie-2} \right\} \\ \left\{ {}^5\$_3^6, \$_{Lie-3} \right\} \end{bmatrix}$$

Once the accelerator $A_O = \begin{bmatrix} \dot{\omega} & a_O - \omega \times v_O \end{bmatrix}$ is calculated, the angular acceleration of the moving platform is obtained as the primal part of A_O, whereas the translational acceleration of the point O is calculated upon the dual part of the accelerator. With these vectors, the translational acceleration of the center of the moving platform, vector a_C, is computed using classical kinematics. Indeed

$$a_C = a_O + \dot{\omega} \times r_{C/O} + \omega \times (\omega \times r_{C/O}). \tag{27}$$

Finally, it is interesting to mention that Eq. (26) does not require the values of the passive joint acceleration rates of the parallel manipulator.

5. Case study. Numerical example

In order to exemplify the proposed methodology of kinematic analysis, in this section a numerical example, using SI units, is solved with the aid of computer codes.
The parameters and generalized coordinates of the example are provided in Table 1.

$$
\begin{aligned}
B_1 &= (.1246762518, 0, .4842063942)\\
B_2 &= (.3569969122, 0, -.3500759985)\\
B_3 &= (-.4816731640, 0, -.1341303959)\\
u_1 &= (.9684127885, 0, -.2493525036)\\
u_2 &= (-.7001519970, 0, -.7139938243)\\
u_3 &= (-.2682607918, 0, .9633463279)\\
a_{12} &= a_{13} = a_{23} = \sqrt{3}/2\\
q_1 &= -0.5\sin^2(t)\cos(t)\\
q_2 &= 0.35\sin[t\sin(t)\cos(t)]\\
q_3 &= -0.35\sin(t)\cos[t\sin(t/2)]\\
0 &\le t \le 2\pi
\end{aligned}
$$

Table 1. Parameters and instantaneous length of each limb of the parallel manipulator

According with the data provided in Table 1, at the time t=0 the sixteenth polynomial in Z_1 results in

$$
\begin{aligned}
&.490873788e09 + .627748325e10 Z_1 + .246379238e11 Z_1^2 - .82281001e10 Z_1^3 -\\
&.281160758e12 Z_1^4 - .444113311e12 Z_1^5 + .964036155e12 Z_1^6 + .2739680775e13 Z_1^7 -\\
&.108993550e13 Z_1^8 - .672039554e13 Z_1^9 + .3921344e11 Z_1^{10} + .786657045e13 Z_1^{11} -\\
&.64783709e12 Z_1^{12} - .373666459e13 Z_1^{13} + .195532604e13 Z_1^{14} - .3787349072e12 Z_1^{15} +\\
&.261153294e11 Z_1^{16} = 0.
\end{aligned}
$$

The solution of this univariate polynomial equation, in combination with expressions (5) and (6), yields the 16 solutions of the forward position analysis, which are listed in Table 2. Taking solution 3 of Table 2 as the initial configuration of the parallel manipulator, the most representative numerical results obtained for the forward velocity and acceleration analyses are shown in Fig. 3.

Solution	P_1	P_2	P_3
1,2	(-.086,± 0.307, - .335)	(.432,±.994, - .424)	(-.364,±1.093, - .101)
3,4	(.121,± .899,.471)	(.361, ± .999, - .354)	(-.468,±1.099,-.130)
5,6	(.161, ± .888,.625)	(.236,±.985, - .231)	(.544,±.273,.151)
7,8	(-.099,±.054, - .385)	(-.091,±.778,.089)	(.558,±.209,.155)
9,10	(.193,±.857,.749)	(-.321,±.312,.314)	(.528,±.333,.147)
11,12	(.182,±.869,.709)	(-.326,±.287,.320)	(-.185,±1.056,-.051)
13,14	(-.104,±.194i,-.407)	(-.628,±.950i,.615)	(.578,±.004i,.160)
15,16	(-.104,±.195i,-.407)	(-.657,±1.009i,.644)	(.578,±.004i,.160)

Table 2. The sixteen solution of the forward position analysis

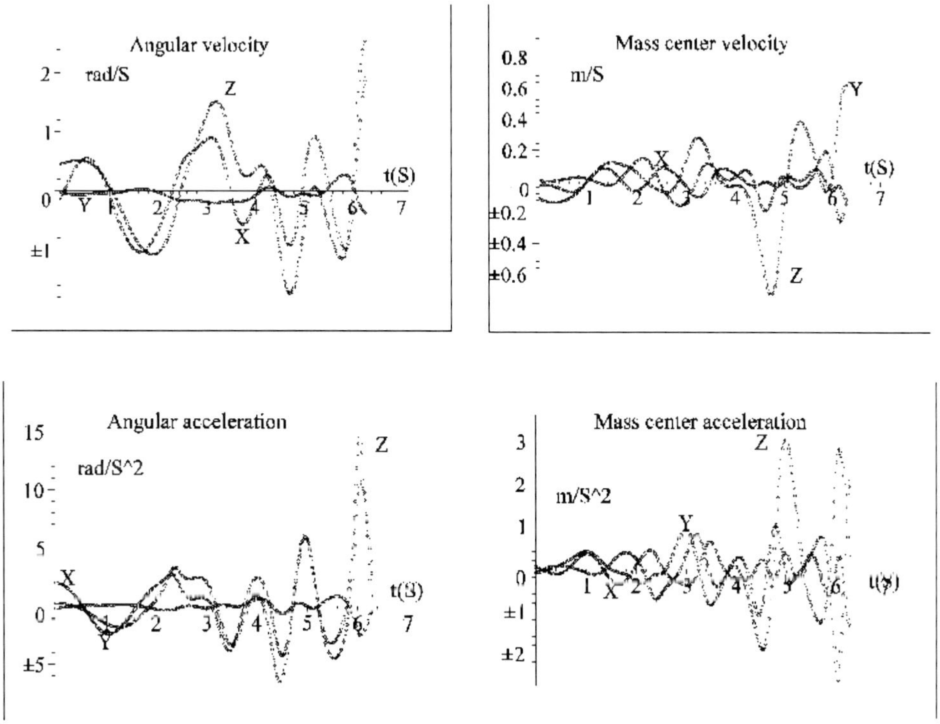

Fig. 3. Forward kinematics of the numerical example using screw theory

Furthermore, the numerical results obtained via screw theory are verified with the help of special software like ADAMS©. A summary of these numerical results is reported in Fig. 4.

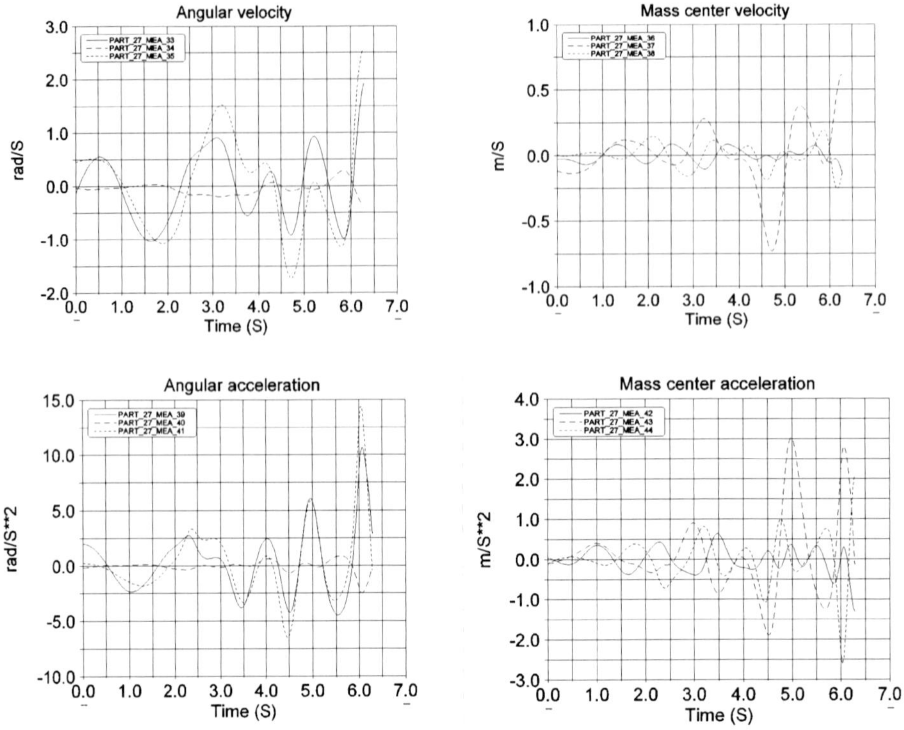

Fig. 4. Forward kinematics of the numerical example using ADAMS©

Finally, please note how the results obtained via the theory of screws are in excellent agreement with those obtained using ADAMS©.

6. Conclusions

In this work the kinematics, including the acceleration analysis, of 3-RPS parallel manipulators has been successfully approached by means of screw theory. Firstly, the forward position analysis was carried out using recursively the Sylvester dialytic elimination method, such a procedure yields a 16-th polynomial expression in one unknown, and therefore all the possible solutions of this initial analysis are systematically calculated. Afterwards, the velocity and acceleration analyses are addressed using screw theory. To this end, the velocity and reduced acceleration states of the moving platform, with respect to the fixed platform are written in screw form through each one of the three limbs of the manipulator. Simple and compact expressions were derived in this contribution for solving the forward kinematics of the spatial mechanism by taking advantage of the concept of reciprocal screws via the Klein form of the Lie algebra e (3). The obtained expressions are simple, compact and can be easily translated into computer codes. Finally, in order to exemplify the versatility of the chosen methodology, a case study was included in this work.

7. Acknowledgements

This work has been supported by Dirección General de Educación Superior Tecnológica, DGEST, of México

8. References

Agrawal, S.K. (1991). Study of an in-parallel mechanism using reciprocal screws. *Proceedings of 8th World Congress on TMM*, 405-408.

Alizade, R.I., Tagiyev, N.R. & Duffy, J. (1994). A forward and reverse displacement analysis of an in-parallel spherical manipulator. *Mechanism and Machine Theory*, Vol. 29, No. 1, 125-137.

Carricato, M. & Parenti-Castelli, V. (2003). A family of 3-DOF translational parallel manipulators. *ASME journal of Mechanical Design*, Vol. 125, No. 2, 302-307.

Clavel, R. (1991). Conception d'un robot parallèle rapide à 4 degrés de liberté. Ph.D. Thesis, EPFL, Lausanne, Switzerland.

Di Gregorio, R. (2002). A new family of spherical parallel manipulators. *Robotica*, Vol. 20, 353-358.

Di Gregorio, R. & Parenti-Castelli, V. (2002). Mobility analysis of the 3-UPU parallel mechanism assembled for a pure translational motion. *ASME Journal of Mechanical Design*, Vol. 124, No. 2, 259-264.

Fang, Y. & Huang, Z. (1997). Kinematics of a three-degree-of-freedom in-parallel actuated manipulator mechanism. *Mechanism and Machine Theory*, Vol. 32, 789-796.

Gallardo-Alvarado, J., Rico-Martínez, J.M. & Alici, G. (2006). Kinematics and singularity analyses of 4-dof parallel manipulator using screw theory. *Mechanism and Machine Theory*, Vol. 41, No. 9, 1048-1061.

Gallardo-Alvarado, J., Orozco-Mendoza, H. & Maeda-Sánchez, A. (2007). Acceleration and singularity analyses of a parallel manipulator with a particular topology. *Meccanica*, Vol. 42, No. 3, 223-238.

Gogu, G. (2005). Mobility of mechanisms: a critical review. *Mechanism and Machine Theory*, Vol. 40, No. 9, 1068-1097.

Gosselin, C.M. & Angeles, J. (1989). The optimum kinematic design of a spherical three-degree-of-freedom parallel manipulator. *ASME Journal of Mechanical Design*, Vol. 111, No. 2, 202-207.

Gough, V.E. & Whitehall, S.G. (1962). Universal tyre test machine, Proceedings of the FISITA Ninth International Technical Congress, pp. 117-137, May.

Huang, Z. & Fang, Y.F. (1996). Kinematic characteristics analysis of 3 DOF in-parallel actuated pyramid mechanism. *Mechanism and Machine Theory*, Vol. 31, No. 8, 1009-1018.

Huang, Z. & Wang, J. (2000) Instantaneous motion analysis of deficient-rank 3-DOF parallel manipulator by means of principal screws. In: *Proceedings of A Symposium Commemorating the Legacy, Works, and Life of Sir Robert Stawell Ball Upon the 100th Anniversary of a Treatise on the Theory of Screws*. Cambridge.

Huang, Z. & Wang, J. (2001). Identification of principal screws of 3-DOF parallel manipulators by quadric degeneration. *Mechanism and Machine Theory*, Vol. 36, No. 8, 893-911.

Huang, Z., Wang, J. & Fang, Y. (2002) Analysis of instantaneous motions of deficient-rank 3-RPS parallel manipulators. *Mechanism and Machine Theory*, Vol. 37, No. 2, 229-240.

Hunt, K.H. (1983). Structural kinematics of in-parallel-actuated robot arms. *ASME Journal of Mechanisms, Transmissions, and Automation in Design*, Vol. 105, 705-712.

Innocenti, C. & Parenti-Castelli, V. (1990). Direct position analysis of the Stewart platform mechanism. *Mechanism and Machine Theory*, Vol. 35, No. 6, 611--621.

Ionescu, T. (2003). Standardization of terminology. *Mechanism and Machine Theory*, Vol. 38, No. 7-10, 597-1111.

Ji, P. & Wu, H. (2003). Kinematics analysis of an offset 3-UPU translational parallel robotic manipulator. *Robotics and Autonomous Systems*, Vol. 42, No. 2, 117-123.

Kim, H.S. & Tsai, L.-W. (2003). Kinematic synthesis of a spatial 3-RPS parallel manipulator. *ASME Journal of Mechanical Design*, Vol. 125, 92-97.

Kong, X. & Gosselin, C.M. (2004a). Type synthesis of 3-DOF translational parallel manipulators based on screw theory. *ASME Journal of Mechanical Design*, Vol. 126, No. 1, 83-92.

Kong, X. & Gosselin, C.M. (2004b). Type synthesis of 3-DOF spherical parallel manipulators based on screw theory. *ASME Journal of Mechanical Design*, Vol. 126, No. 1, 101-108.

Kong, X.-W & Gosselin, C.M. (2002). Kinematics and singularity analysis of a novel type of 3-CRR 3-DOF translational parallel manipulator. *International Journal of Robotics Research*, Vol. 21, No. 9, 791-798.

Lee, K.M. & Shah, D.K. (1987). Kinematic analysis of a three-degree-of-freedom in-parallel actuated manipulator, *Proceedings IEEE International Conference on Robotics and Automation*, Vol. 1, 345-350.

Liu, C.H. & Cheng, S. (2004). Direct singular positions of 3RPS parallel manipulators. *ASME Journal of Mechanical Design*, Vol. 126, 1006-1016.

Liu, X.-J. & Gao, F., (2000). Optimum design of 3-DOF spherical parallel manipulators with respect to the conditioning and stiffness indices. *Mechanism and Machine Theory*, Vol. 35, No. 9, 1257-1267.

Lu Y. & Leinonen, T. (2005). Solution and simulation of position-orientation for multi-spatial 3-RPS Parallel mechanisms in series connection. *Multibody System Dynamics*, Vol. 14, 47-60.

Parenti-Castelli, V., Di Gregorio, R. & Bubani, F. (2000). Workspace and optimal design of a pure translation parallel manipulator. *Meccanica*, Vol. 35, No. 3, 203-214.

Parenti-Castelli, V. & Innocenti, C. (1992). Forward displacement analysis of parallel mechanisms: Closed form solution of PRR-3S and PPR-3S structures. *ASME Journal of Mechanical Design*, Vol. 114, No. 1, 68-73.

Rico, J.M. & Duffy, J. (1996). An application of screw algebra to the acceleration analysis of serial chains. *Mechanism and Machine Theory*, Vol. 31, 445-457.

Rico, J.M., Gallardo, J. & Duffy J. (1999). Screw theory and higher order kinematic analysis of open serial and closed chains. *Mechanism and Machine Theory*, Vol. 34, No. 4, 559-586.

Romdhane, L., Affi, Z. & Fayet, M. (2002). Design and singularity analysis of a 3-translational-dof in-parallel manipulator. *ASME Journal of Mechanical Design*, Vol. 124, No. 3, 419-426.

Stewart, D. (1965). A platform with six degrees of freedom. *Journal Proceedings IMECHE Part I*, Vol. 180, No. 15, 371-386.

Sugimoto, K. (1987). Kinematic and dynamic analysis of parallel manipulators by means of motor algebra. *ASME Journal of Mechanisms, Transmissions, and Automation in Design*, Vol. 109, 3-7.

Tsai, L.-W. (1999) *Robot analysis*. John Wiley & Sons.

17

Multiscale Manipulations with Multiple Parallel Mechanism Manipulators

Gilgueng Hwang and Hideki Hashimoto
The University of Tokyo
Japan

1. Introduction

While recent years have brought an explosive growth in new microelectromechanical system (MEMS) devices ranging from accelerometers, oscillators, micro optical components, to micro-fluidic and biomedical devices, our concern is now moving towards complex microsystems that combine sensors, actuators, computation and communication in a single micro device. It is widely expected that these devices will lead to dramatic developments and a huge market, analogous to microelectronics.

However, several problems (e.g., sticking effect) exist which are preventing fully autonomous manipulation with dextrous skills at the micro-scale. Dextrous manipulation, requiring precise control of forces and motions, cannot be accomplished with a conventional robotic gripper; any slave should be more anthropomorphic in design. These kinds of dextrous manipulations still need human operator assistance.

Task-based autonomous manipulation has been explored by many researchers to overcome micromanipulation problems. Fearing et al. developed an automated microassembly system with ortho-tweezers and force sensing (Fearing et al., 2001). However, the workspace was really small and the range of the target object was pretty much limited.

Inspection, prototyping or repairing of a miniaturized system which require human's flexible intelligence relies on the human's operation under a microscope. These tasks are very stressful and cannot be done by an autonomous system such as Fearing's prototyping machine. As a human-centred manipulation, teleoperation has been researched for the ability to display the expanded micro environment to the human operator. By combining with haptic interfaces which provide force feedback, a human operator could operate a micro scale object with enough telepresence and reality, increasing the human's operability and operational efficiency. There are many applications based on bilateral teleoperation control (Kosuge et al., 1995; Hannaford & Anderson, 1988). A chopstick-like micromanipulation system having a two-fingered micro-hand as a slave was developed (Tanikawa & Arai, 1999). However, this system did not support force feedback to the operator, which does not provide enough telepresence. A surgical system such as the da Vinci System (Dosis et al., 2003) is a good application since humans and robots collaborate with each other for the purpose of high performance with safety. In our previous work, a tele-micromanipulation system was proposed to enable micro tasks without stress (Ando et al., 2001). The above teleoperated systems had enough functionality in a specified application target instead of losing dexterity during operation.

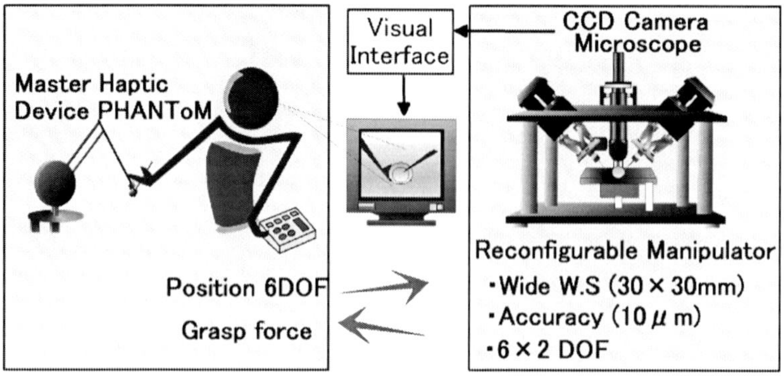

Fig. 1. Concept of micromanipulation system using multiple parallel mechanism micromanipulators

There are still many applications which require human intelligence to overcome the limitation of artificial intelligence (AI) technology and the limited dexterity of the slave manipulator. For these reasons, humans should intervene in the manipulation process. Dexterous manipulation, requiring precise control of forces and motions, cannot be accomplished with a conventional robotic gripper; a slave should be more anthropomorphic in design. These kinds of dextrous manipulations still need a human operator's assistance. In the mean while, the single-master multi-slave (SMMS) system was developed using the virtual internal model (Kosuge et al., 1990). It enables better dexterous teleoperation system without increasing the human's operational d.o.f. However, their work relied on the reference position distribution without considering the force feedback which made the comparative analysis with other systems. The communication delay issue of SMMS teleoperation by decomposing the dynamics of multiple slaves has been studied showing several simulated results (Lee & Spong, 2005). However, it was not implemented to the real time system for the practical systems different dynamics than the theory.

These concepts of multiple robot configurations from macro scale can be the break through of conventional micro/nanomanipulation.In our research, we implement dexterous micromanipulation system based on the SMMS concept (Hwang et al., 2007). A single PHANToM haptic device as a master device and dual 6-d.o.f. parallel micromanipulators as slave devices are adopted as shown in Fig. 1.

Using dual-slave manipulators is expected to enhance the performance or dexterity of the total system compared to our previous work, which had a single slave manipulator (Ando et al., 2001).

This chapter continues with the system structure of tele-micromanipulation systems and the parallel manipulator as a slave device is briefly introduced. The singular position and manipulability analyses are done in the Section 3. Section 4 covers the overall strategy and the mapping method between the PHANToM master device and the dual slave manipulators. Finally, several experimental results (e.g., accuracy evaluation, master–slave position/force-mapping method) are shown. Notation is based on the reference (Paul, 1981).

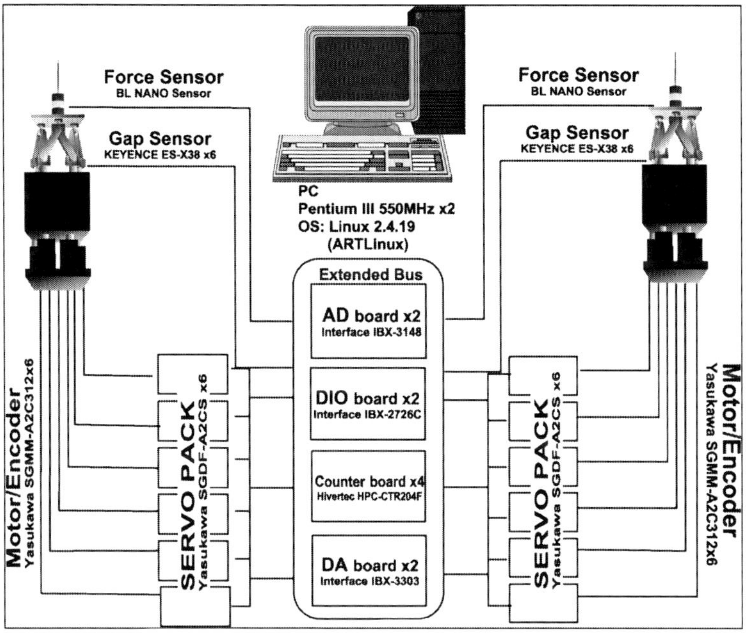

Fig. 2. Electronics and control box for dual parallel mechanism micromanipulators connected to the pc

2. SMMS system overview

In this section, a tele-micromanipulation system is introduced. Figure 2 shows the configuration of our tele-micromanipulation system. A parallel manipulator having an

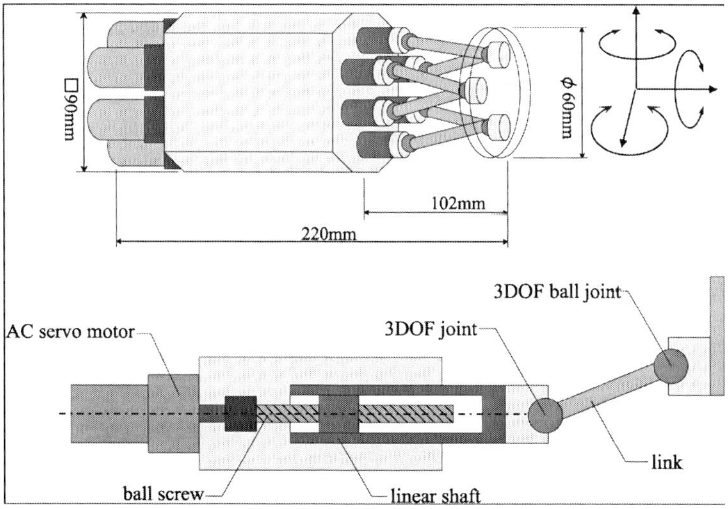

Fig. 3. Parallel mechanism micromanipulator

original mechanism is used as a slave manipulator in the teleoperation system. The slave manipulator and master system are connected using an Ethernet and they are used to perform teleoperations through the network. A bilateral control system was adopted to realize the overall teleoperation system.

2.1 SMMS architecture

The proposed SMMS tele-micromanipulation system is an extension of our previous system (Ando et al., 2001) that supported 1:1 master–slave tele-micromanipulation.
We proposed a prototype of the SMMS tele-micromanipulation system which applies dual-parallel micromanipulators in a 'V' configuration which is similar to the natural human hand position while assembling or manipulating an object.
Figure 1 show the overall system configuration including the master haptic device, slave manipulators and the visual interface connected on RT-Middleware (Ando et al., 2005). RT-Middleware technology has several advantages for constructing SMMS system architecture such as providing a flexible network connection environment which is required for multi-modal displaying or human–robot shared control. The shared controller plays a major role in generating the multi-slave's reference position and internal force by human–robot cooperation. However, in this paper, we do not go further with the shared control. Each slave controller drives six link shafts to position the parallel mechanism end-effector within about 10 μm positioning accuracy. There are three RT components with a position/force IO interface through the network. Both slave manipulators are controlled by a PC (Pentium III 500 MHz × 2). A real-time extension for Linux (ART-Linux) is used as the operating system to perform motion control at 1-kHz sampling rate.
The slave manipulator with a parallel link mechanism as shown in Fig. 3 is used as a slave manipulator. Compared to the serial mechanisms used in normal robot arms, the parallel mechanism has merits such as high stiffness, high speed and high precision.
A serial link mechanism PHANToM haptic device is adopted as our master device. This device allows the 6-d.o.f. measurement including 3 d.o.f. of each translational and rotational motion. Also, the 3-d.o.f. translational force feedback is supported by this master device. Real Time Linux was used as the operating system to control the master with 1-kHz sampling frequency.

3. Kinematics analysis of slave device

It has already been mentioned in Section 2 that there exist several characteristics of a novel parallel mechanism slave micromanipulator being used in this research. However, in the cooperative manipulation of a multi-slave system, several problems of parallel manipulators such as a singular position possibly become more serious than in independent manipulation. Therefore, singular position and the manipulability of the parallel manipulator used in this research are discussed in this section.
The most feasible arrangement of both manipulators considering the result of manipulability is also described in the latter part in this section.

3.1 Inverse kinematics of parallel micromanipulator

It is already mentioned in chapter.3 that there exist several characteristics of a novel parallel mechanism slave micromanipulator being used in this research. However, in the cooperative manipulation of multi-slave system, several problems of parallel manipulators such as singular position possibly become more serious than in the independent manipulation. Therefore, analyses on the kinematics, singular position, and manipulability of the parallel manipulator which is used in this research should be given in this section. The most feasible arrangement of both manipulators is also described in the latter part in this section. An inverse kinematics analysis on the parallel mechanism micromanipulator is described in this section. Figure 4 shows coordinate system adopted for kinematics analysis. Point O is the origin and the criteria coordinate is the base coordinate system.

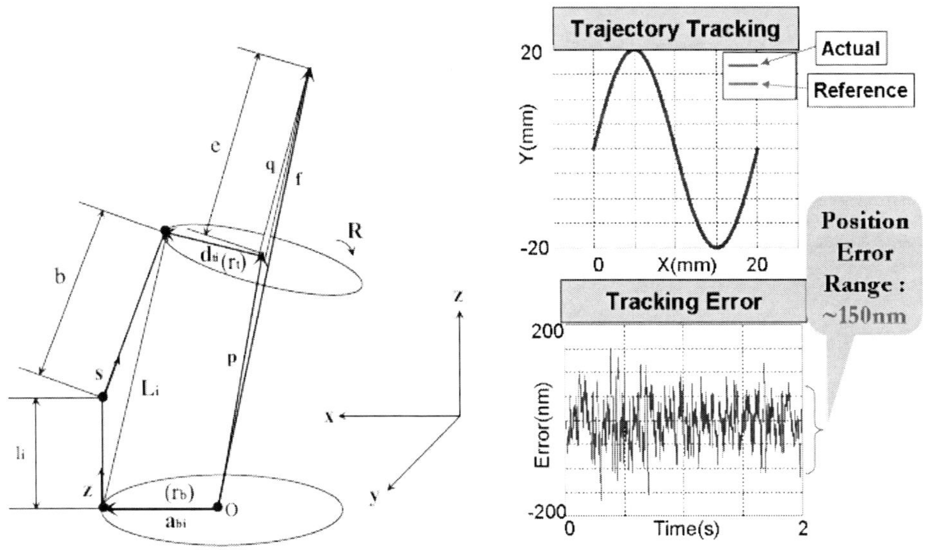

(a) (b)

Fig. 4. Coordinate system of parallel mechanism manipulator (a), and position tracking (b)

We define each variables and constants in the following manner.

(ϕ, θ, φ) : posture of end-effector,

R : Rotation matrix which represents the posture of the end-effector,

i : Chain number,

a_b : Vector from centre of base to base joints,

r_b : Length of a_b,

d_t : Vector from centre of end-plate to end-table joints,

r_t : Length of d_t,

z : Unit vector from base joint datum point to actuator joints datum point,

l_i : Length of chain of prismatic joints,

s : Unit vector from actuator joints datum point to end-plate joints datum point,

From relations between base joints and end-effector joints, we get

$$p + Rd_{ti} - a_{bi} = l_i z + bs, \tag{1}$$

Where L_i is used for $p + Rd_{ti} - a_{bi}$.

$$L_i - l_i z = bs\,. \tag{2}$$

Both sides of equation are squared and because $z^2 = 1, s^2 = 1$, and we get the following equation.

$$l_i^2 - 2(L_i \cdot z)l_i + L_i^2 - b^2 = 0. \tag{3}$$

This equation is solved for l_i, and we get,

$$l_i = (L_i \cdot z) \pm \sqrt{b^2 - L_i^2 + (L_i \cdot z)^2}\,. \tag{4}$$

Where, $L_i = (L_x, L_y, L_z), z = (0,0,1)$ and using the constraints that the chains sign of square root is negative, we get the following equation.

$$l_i = L_{zi} - \sqrt{b^2 - L_{xi}^2 - L_{yi}^2}\,. \tag{5}$$

Thus, the length of the link l_i is determined by the tip position of the end-effector $F(x, y, z, \varphi, \theta, \phi)$. In other words, we can compute the reference link length l_{1-6} from the position and posture of end-effector.

A simulation was conducted to verify the validity of the calculated inverse kinematics for our purpose. As y-directional reference sinusoidal input for simulation is set as following.

$$Y - 20 \sin 2\pi t \tag{6}$$

The resolution of encoder attached to each linear driving link is given as the 0.122μm. Figure 4b shows the result of the simulation. It is quite well being tracked to the given sinusoidal trajectory with the maximum tracking error 0.150μm. This result proves the validity of the inverse kinematics calculation conducted here, also almost of the error is generated from the quantization process with the encoder resolution.

3.1 Singular position analysis

There exists a singular position in the manipulator workspace. The d.o.f., decreases around this position or point. Therefore, it should be avoided in the control of the manipulator. Here, we need to analyze this singular position of our system by some calculations.

First of all, the Jacobian matrix can be mathematically defined as follows. Assuming that an n-d.o.f. manipulator is working in an m-dimensional task space, where $m < n$, we have:

$$v = J_v(x)i \tag{7}$$

Where v and i indicate the Cartesian and joint velocity vectors defined in the task space R^n and the joint space R^m, respectively, and J_v represents a $m \times n$ Jacobian matrix. J_v is also considered as a Jacobian matrix which contains ϕ as an Euler angle.

In the case that we put the column vector of J_v as M_j (j = 1 ~ 6), the relation between v and i can be described as:

$$v = \sum_{j=1}^{6} i_j M_j , \tag{8}$$

Therefore, the singular position can be obtained by the calculation of the determinant of the Jacobian matrix J_v :

$$\det J_v = 0 \tag{9}$$

The singular position may cause some problems which can be measured by the manipulability. Therefore, we need to visualize the manipulability of dual manipulators on the overall workspace.

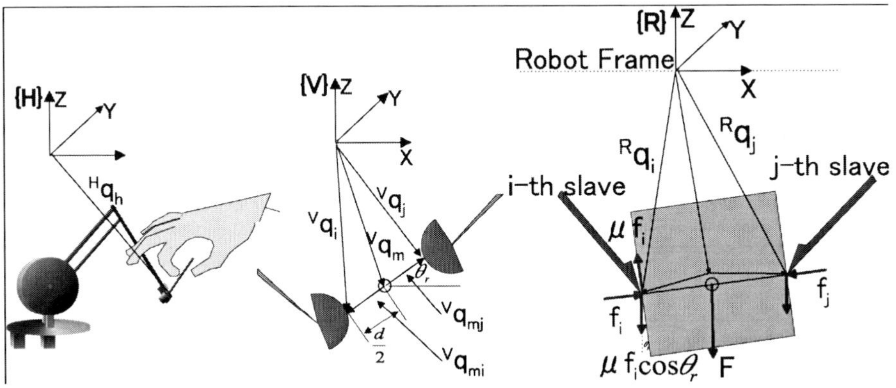

Fig. 5. Kinematical model of the cooperative system. The universe, and left and right manipulator's coordinate frames are described as {A}, {B}, {C}, respectively.

3.2 Multi-micromanipulator analysis

In the case of the dual micromanipulator system, a proper coordinate system for each manipulator needs to be analyzed. Figure 5 shows the proposed model of the cooperative system.

There are several advantages of this design which can be summarized as high flexibility to random target objects, high dexterity, etc.

The angle between two manipulators is 90◦ and the initial points of the end-effector are set as (5, 0, 5) mm and (−5, 0, 5) mm. Frame B is described by:

$${}^A_B C = Rot(y,45) \cdot Trans(5,0,-5), \tag{10}$$

$$ {}^{A}_{C}C = Rot(y,-45) \cdot Trans(-5,0,-5). \tag{11} $$

Equations (10) and (11) show the conversion matrix for arranging the coordinate system of each manipulator.

3.3 Manipulability measure

The manipulability measure w is proposed to measure quantitatively the ability to change the position and orientation of the end-effector from the view point of the kinematics (Yoshikawa, 1985). There is also a trade-off between accuracy and manipulability. As the proposed parallel manipulator has no redundant d.o.f., w is represented as follows:

$$ w = \left|\det(J_v)\right|. \tag{12} $$

The kinematical manipulability in the dual-micromanipulator's workspace which is parallel to the xy plane can be referred from the reference (Hwang et al., 2007). The human operator is possibly able to avoid the singular position which is located around the position with zero manipulability. It is easily verified that mapping the centre position of both manipulators with the haptic reference position gives consistency of manipulability, because manipulability is plotted as symmetrical to the plane x = 0.

4. Manipulation strategy

4.1 Master/slave mapping

Figure 5 depicts the kinematical model of the proposed system including mapping method. The reference position of the haptic device is mapped into the centric of both end-effectors' tip positions. At the same time, the internal force generated by both end-effectors conflicts with the target object is fed into the user through the haptic device. In this section, a novel mapping method between the SMMS system arranged as shown in sections 2 and 3 and finally a valid manipulation strategy are discussed. Basically, the reference position of the haptic device should be scaled into the center of both manipulators' tip positions during the free motion. The reference position to each slave manipulator is calculated from the current position and posture of the haptic device. It enables to assure more workspace for both of manipulators. Also, the same manipulability of both manipulators can be obtained on z axis (xy=0), because the manipulability of both manipulators is symmetrically plotted to the plane x = 0. It can be the most feasible interface to the human operator with only two-dimensional visual information. Several experimental results to verify the proposed mapping method will further be shown in section 5.

4.2 Virtual mapping method

This section introduces a novel mapping method between SMMS devices arranged as shown in the section 2 and 3 (see Figure 1) and an improved manipulation strategy is proposed in order to emphasize the human manipulation in the micro world.

Since the master device and the slave devices have very different kinematical structures in this system, the usual direct mapping method (e.g., joint-to-joint mapping) cannot be useful in our system. A virtual mapping method was used to connect between the human's hand and the non-anthrophormous slave device, where their feedback strategy was based on the fingertip-level force feedback (Griffin et al., 2003). Our system has the serial link master device but the parallel link slave devices that define a new manipulation approach. This paper introduces a novel approach to realize more dexterous control with the simplified master device to control multiple slave devices, discussed in relation to the object based roll and yaw angle control to adjust the grasp force control. Figure 5 and 6 show this new manipulation approach.

The roll and yaw angle of the phantom haptic interface is measured and used to provide the roll and yaw orientation of the manipulated object. Then the reference grasping force is proportional to the measured pitch angle of PHANToM haptic interface. This reference object size is mapped into the width of both the slave's tip positions.

As shown in Fig. 5, the virtual mapping parameters are described in the virtual robot coordinate which is denoted as V. Given the reference centre position of the virtual object q_m, we need to calculate the reference tip position of each end-effector, q_i, q_j. The centre position of the virtual object is described as ${}^V q_m$. ${}^V q_{mi}$ and ${}^V q_{mj}$ describe the vector from the virtual object centre to each of slave tip position. This term can be obtained from the roll and yaw angles (θ_r, θ_y) commanded from the master haptic device. The reference grasping force F_g^r is generated by the pitch angle (θ_p). Then, ${}^V q_{mi}$ and ${}^V q_{mj}$ can be calculated by the following rotational matrix:

$$ {}^V q_{mi} = \begin{bmatrix} \cos\theta_r \cos\theta_y & -\sin\theta_r & \cos\theta_r \sin\theta_y \\ \sin\theta_r \cos\theta_y & \cos\theta_r & \sin\theta_r \sin\theta_y \\ -\sin\theta_y & 0 & \cos\theta_y \end{bmatrix} \begin{bmatrix} -d/2 \\ 0 \\ 0 \end{bmatrix}. \tag{7} $$

where d is the distance between two virtual tip positions which is calculated to assign virtual dynamics (spring constant: s_v) to the virtual object as follows:

$$ F_g^r = s_v \frac{1}{d}, \quad d = \Delta\theta_p = \left| {}^V q_i - {}^V q_j \right|. \tag{8} $$

Then, each slave's tip position based on the virtual object is:

$$ {}^V q_i = {}^V q_m + {}^V q_{mi}, \quad {}^V q_j = {}^V q_m + {}^V q_{mj}. \tag{9} $$

4.3 Force feedback stratogy

Here the question "what kind of haptic feedback is efficient?" arises. To answer this question, we need to consider the several factors including the contact type between the end-effector tip and the object. A haptic feedback contributes to increase the operability of human resulting in the improved overall performance of manipulation. Considering that our concern is to develop the single-master multi-slave telemicromanipulation framework, the conventional master, slave direct force feedback is not feasible. As shown in the Figure.4.8, in the case of 1:1 master slave system, a direct feedback from the force sensor is an effective for recognizing the slave side by human. If an 1:N master slave system as the grasping an object by human's hand with five fingertips, still human recognizes the grasping condition from multiple fingertip sensing. To assure the stable grasping in 1:N system, each end-effector of N slave devices should cooperate with each other to regulate the grasping force by transferring the grasping force to the human operator through the haptic interface. This strategy helps to regulate the grasping force by both the human's control and robot's control. Looking it in coordinate system, the exerted force by the object to the slave and the feedback forces in case of 1:1 feedback strategy are denoted as ${}^{R}f_i$ and ${}^{U}f_i$ respectively. These are scaled between the master and the slave to amplify the force in the micro environment. However, in the case of 1:N feedback strategy, it is assumed that the virtual thread exerting the feedback force (${}^{U}f_i$) to the human operator by the slave's applied forces (${}^{R}f_i$) to the object.

Then, the force diagram between the multi-slaves and the object should be analyzed to calculate the proper feedback to the user.

μ denotes the friction coefficient between the object and the slave. f_i is the internal force term at the ith slave's contact position. μf_i describes the frictional force upward caused by the internal force (f_i at a ith contact position. θ_r denotes the commanded roll angle of the manipulated object which forms the orthogonal downward force ($\mu f_i \cos\theta_r$). If N slaves are handling an object, the resisting force (F) downward is,

$$F = \mu N f_i \cos\theta_r \tag{10}$$

Assuming that the scaling factor between the master and the slave for micromanipulation tasks is given as,

$$A_p = diag(A_p, A_p, A_p) \tag{11}$$

Then the feedback force to the human operator which is defined here as the ${}^{U}f_i$ is obtained from,

$${}^{u}f_i = A_p F \tag{12}$$

A question is now arising mainly about how to obtain f_i in real time and deliver it to user. The simple calculation to obtain f_i is implemented to the SMMS system in section 4.4.

4.4 Internal force decomposition

When multiple manipulators grasp an object, the force applied by multiple robots can be decomposed into motion-inducing force and internal force. Especially, the internal force should be kept in a certain range to assure the stable grasping and safety of object. In the proposed cooperative master–slave manipulation system, the internal force to squeeze the grasped object is fed into the human operator through the master haptic device. Figure 5 depicts two manipulators cooperating to grasp a single object.

Force decomposition using the theory of metric spaces and generalized inverses was attempted (Bonitz & Hsia, 1994). It is assumed that each manipulator grasps the object rigidly, exerting both forces and moments on the object. The net force at the object frame is related to the forces applied by the manipulators by:

$$f_{obj} = J_o^T f \tag{13}$$

where $f_{obj} = \begin{bmatrix} f_o^T & m_o^T \end{bmatrix}^T$ is the net force and moment at the object frame, $J_o^T = \begin{bmatrix} J_{o1}^T & J_{o2}^T \end{bmatrix}$. J_{oi} is the Jacobian from the object frame to the ith end-effector frame, $f = \begin{bmatrix} f_1^T & m_1^T & f_2^T & m_2^T \end{bmatrix}^T$. f_i is the force and m_i is the moment applied by the ith end-effector. $p_i = \begin{bmatrix} p_{ix} & p_{iy} & p_{iz} \end{bmatrix}^T$ is the vector from the ith end-effector to the object frame:

$$J_{oi}^T = \begin{bmatrix} I_3 & O_3 \\ -P_i & I_3 \end{bmatrix}^T \tag{14}$$

$$P_i = \begin{bmatrix} 0 & -p_{iz} & p_{iy} \\ p_{iz} & 0 & -p_{ix} \\ -p_{iy} & p_{ix} & 0 \end{bmatrix} \tag{15}$$

The applied force f is decomposed into motion-inducing, f_M, and internal, f_I, with:

$$f_I = (1 - J_o^{T\Phi} J_o^T) f, \tag{16}$$

Where, $J_o^{T\Phi} J_o^T$ is a generalized inverse of J_o^T. The projections $P_M = J_o^{T\Phi} J_o^T$ and $P_I = I - J_o^{T\Phi} J_o^T$ project the applied force onto the motion inducing force subspace, F_M, and the internal force inducing force subspace, F_I, respectively. If $rank(J_o^T M J_o) = rank(J_o^T)$, the weighting matrix M can be used to compute a generalized inverse of J_o^T. Therefore:

$$J_o^{T\Phi} = M J_o (J_o^T M J_o)^{-1} = \frac{1}{2} \begin{bmatrix} \overline{I_3} & \overline{O_3} \\ P_1 & I_3 \\ I_3 & O_3 \\ P_2 & I_3 \end{bmatrix}. \tag{17}$$

Using (13) and (14), the decomposition of f is:

$$f_I = \begin{bmatrix} f_1 - f_2 \\ m_1 + (P_2 - P_1) f_2 - m_2 \\ -(f_1 - f_2) \\ (P_1 - P_2) f_1 - m_1 + m_2 \end{bmatrix}. \tag{18}$$

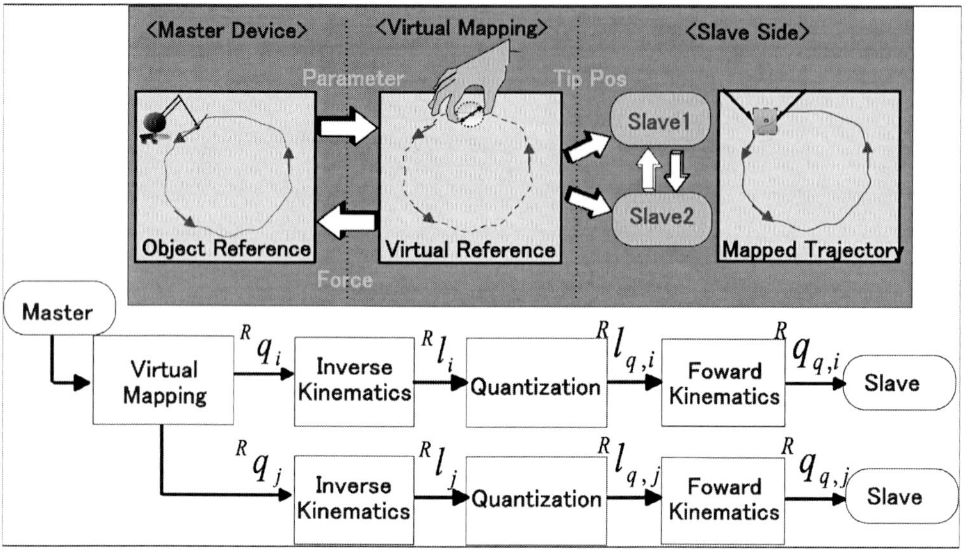

Fig. 6. Virtual mapping experiment: ${}^{R}q_{i,j}$ are robot's Cartesian references, ${}^{R}l_{i,j}$ are robot's link coordinates, ${}^{R}l_{q,i}$ and ${}^{R}l_{q,j}$ are quantized robot link configuration and ${}^{R}q_{q,i}$ and ${}^{R}q_{q,j}$ are resultant robot control configuration.

4.5 Manipulation strategy

In much previous master-slave manipulation research, the whole task was done by only the human operator.

However, in the case of the SMMS system, there exist d.o.f. differences between the master and the slave. Therefore, it is a good idea to build a whole manipulation process with several task phases.

In this paper, a human machine cooperative manipulation strategy is proposed as shown in Fig. 6. User is assumed to monitor the target object under the two-dimensional (2-D) visual information on the task space.

As a first phase, both slave manipulators approach to the target object following the reference position scaled from the haptic device operated by the human. Once one of both end-effectors is contacted with object, the grasping phase is started. An autonomous grasping algorithm leads to a stable grasp. Then, the user makes a movement of the target object grasped between both end-effectors. To release the object, operator needs to lead the object to be touched on the ground. The problem caused by the d.o.f. difference between the master and the slave is overcome through the proposed manipulation strategy.

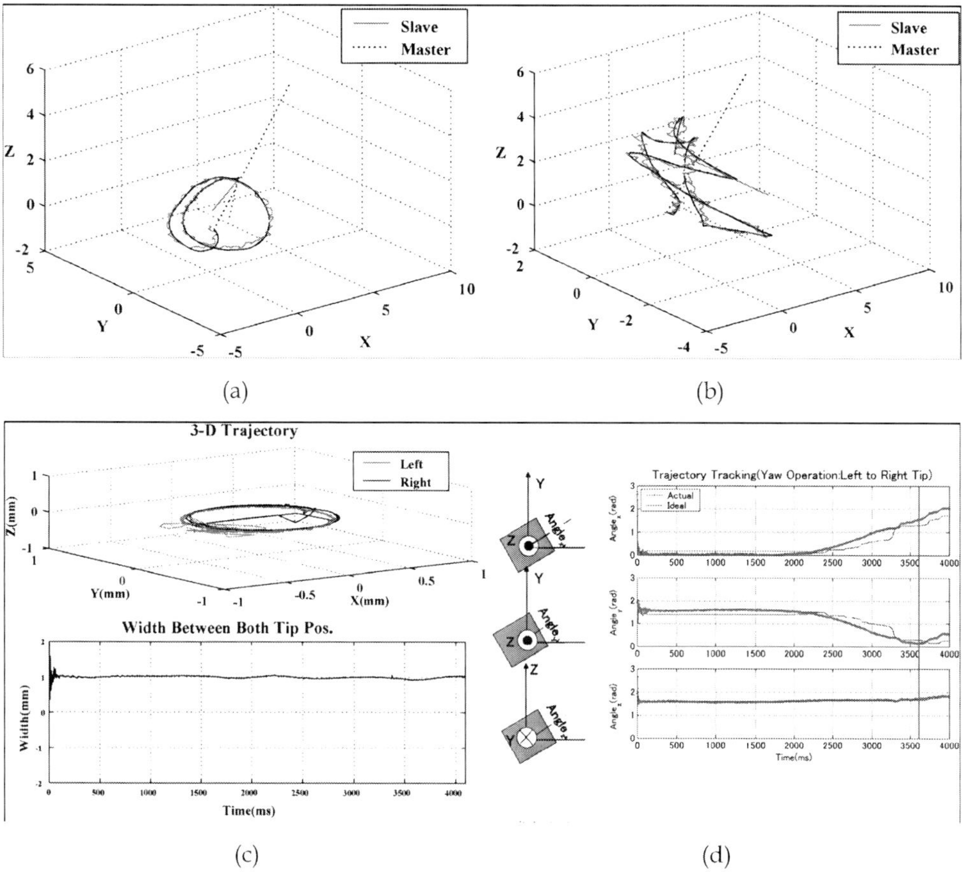

Fig. 7. Dual-micromanipulator's manipulability (w). Upper and lower panels show the left and right manipulators, respectively.

5. Experiments

5.1 Positioning accuracy experiment

First of all, a rectilinear movement experiment was performed to evaluate the system's overall positioning accuracy, which is a trade-off with manipulability. Figure 6 shows the parallel line drawing. Intervals between lines of 10 µm were used. Observed intervals of parallel lines are about 11 µm. In this case intervals are not the same as the reference intervals; however, almost regular intervals were observed. Movement error is caused by an encoder resolution which was set as 0.122 µm.

5.2 Trajectory tracking experiment

Figure 6 describes the controller of an overall system to show the mapping. An experiment on micro-positional synchronization using our micromanipulation system was conducted.

The user operates the PHANToM device with a 3-D random or circular trajectory. This PHANToM reference trajectory generated by the human operator contains the scale factor between the master and the slave. The trajectory of the PHANToM reference and the resulting trajectory of both end-effectors centre position is shown in Fig. 7a,b. We generated two different trajectories which are circular and random motions with a single master device. The following trajectory of both slave end-effectors centre position is depicted as a solid line. The validity of our proposed positional mapping method in real-time operation is verified with this result. Through the result shown in Fig. 7a,b, both slave end-effectors are kept at a certain width during the experiment. The reference trajectory of the single master device is controlling two slave devices without coupling them to smaller scale applications such as micro- and nanomanipulations, which require higher positioning accuracy. These results strengthen the feasibility of the single-master controlled multi-slave system.
Other experiments were conducted to verify the proposed object mapping method. A human operator is handling a 1 cm^3 cubic object with the PHANToM haptic interface. In Fig. 5, the vector from the left to right contact position is defined as:

$$ {}^{R}q_{12} = {}^{R}q_{2} - {}^{R}q_{1}, \tag{17} $$

Where ${}^{R}q_{i}$ depicts the ith contact position in the robot coordinate system. The yaw angle of the virtual object is controlled by a sinusoidal wave or the human operator's arbitrary operation. First, the sinusoidal yaw angle input is given to prove the proposed mapping method. Figure 7c,d show the 3-D trajectory of both end-effectors' tip positions and the width between them. The reference position of the virtual centre is constrained to be static. It is verified that the width is kept static during the operation. The reference roll, pitch and yaw angles are converted between the user's frame and the robot's frame as follows:

$$ {}^{R}\theta_{y,ref} = \frac{1}{36}\,{}^{U}\theta_{y,ref} + 1.0, \tag{18} $$

$$ {}^{R}\theta_{p,ref} = 0.05\,{}^{U}\theta_{p,ref}, \tag{19} $$

$$ {}^{R}\theta_{r,ref} = 0.1\,{}^{U}\theta_{r,ref}, \tag{20} $$

These reference inputs are decided empirically by considering both the workspace and object scale. The experimental results are shown here. In Fig. 7d, the object's orientation is shown following the yaw operation by the human operator. Anglex , Angley and Anglez are angles between ${}^{R}q_{12}$ and each axis. These results show that the developed SMMS system can realize low-cost, object based 6 d.o.f., and even compact-sized moment control by both slaves' translational movement. It is promising to the development of dexterous micro/nanomanipulation system with ultra-high precision positioning accuracy.

5.3 Force mapping experiment

Several experiments were further performed to verify how feasible it is for us to adopt the decomposed internal force derived in section 4. Another purpose of this experiment is to obtain the desired internal force for stable grasping during several primitive tasks. A square stylen block 10 mm in width, length and height was used in this experiment. Figure 16 shows the internal force plotted for several grasping phases such as keeping the grasped object, parallel movement and 10 times iterative grasping/releasing. Transferring force to the user gives the teleoperation more transparency which is closely correlated to the user operability and the task performance. In the case of used stylen bock experiment, it is found that a reasonable choice of the internal force during the grasp phase should be made around 5 mN in each direction. Also, the Figure 6 describes the controller of a phase transition should be done by the event which is the change of internal force around the transition phase. Especially, in the iterative grasping/releasing phase, it is possibly able to be used for compensating the deficiency of the master's d.o.f.

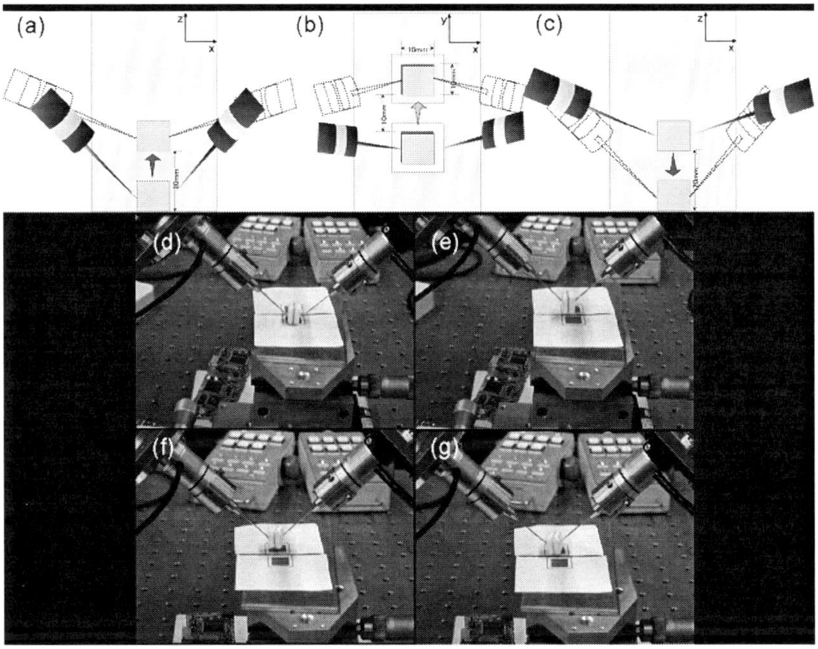

Fig. 8. Styren block pick-and-place experiment: schematic (a) pick up 10 mm (b) move 10 mm (c) release onto substrate. Operation: (d) grasp (e) pick up (f) move (g) release.

5.4 Pick-and-place of styrene block

We conducted several experiments using a 10 mm cubic styrene block, roe (4-6 mm) to demonstrate the feasibility of our proposed system for deformable objects in different scale and mechanical properties.

First, pick-and-place of a stylen cubic block (1 cm^3) was performed to show validity of the proposed dexterous manipulation strategy. Figure 8 shows the demo experiment. The distance between the initial and the target point is 1 cm. The human operator recognizes the target object from the visual display of the microscope and operates the single haptic interface to the contact position of the object. Then, the pitch angle of the haptic interface is controlled by the user to give enough grasp force with the internal grasp force display.

Then, the grasped object is moved to the target goal. To release the grasped object, the pitch angle is controlled again by the user with the feeling of internal force.
Force sensors at each end-effector measure real-time 6 axial force data and implement low-level internal grasping force to the user through the haptic interface and network between the master and slave (Fig. 9). The user`s work speed is recorded to clearly define mode transition time. Styrene block handling assumes stable contact between the probe and styrene block, which causes higher friction at the probe that could damage membranes in real cell handling. But it is sufficient to show the feasibility of our proposal in biotweezing even as small as on the smaller scale. It also demonstrates amplified motion without being limited by microscope depth-of-focus problems during pick-and-place tasks.

5.5 Salmon roe pick-and-place

After confirming feasibility in preliminary experiments, we conducted cell handling and injection using roe from 4 to 6 mm in diameter, which is easy to obtain and provides visual feedback without the need for a microscope (Fig. 10). To handle smaller cells, it is necessary to solve autofocusing problems.

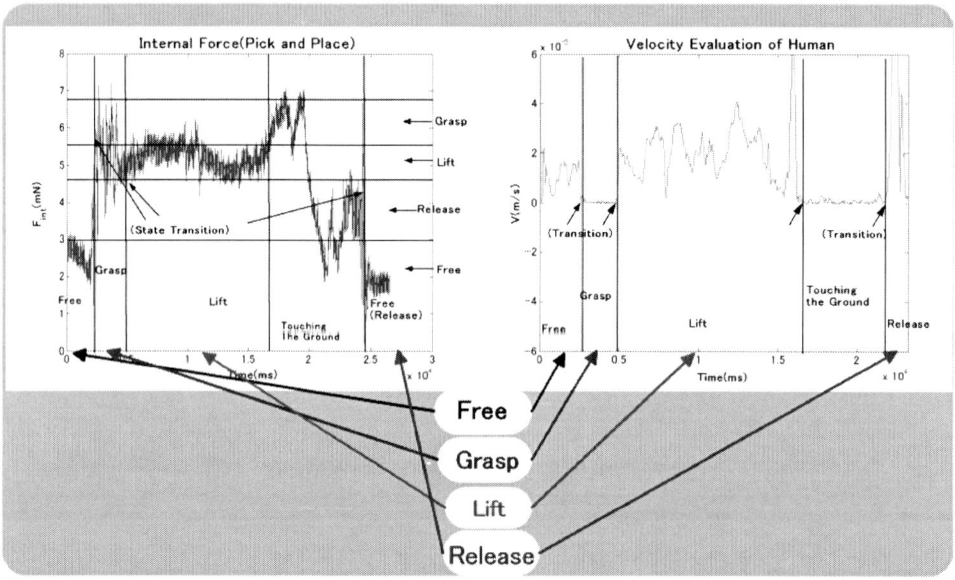

Fig. 9. Internal force during the styrene block pick-and-place operation and user operation speed evaluation.

6. Conclusion & outlook

We have shown a dexterous micromanipulation system based on single-master and multi-slave device configuration and its deformable object micromanipulation such as styrene block and roe. The system enables 6 d.o.f. object position control using 2 parallel mechanism micromanipulators. It has potential applications for manipulating, characterization, complicated device assembly, and biological cell manipulation. In addition to the manipulation dexterity, an object based 6 d.o.f. control was achieved using compact size low-cost moment control which has enough accuracy by using multiple slaves translation motion. First, we analyzed parallel mechanism and multiple manipulators kinematics and further evaluated the singular position and manipulability of the system. Second, one of the most serious problems, i.e., such as the mapping method between the master and the slave, was discussed. A novel micromanipulation strategy which is the most feasible for the SMMS system was given. Experimental results including styrene block and salmon roe manipulation were shown to verify the feasibility of the proposed method to wide variety of applications. In parallel, we have already developed a human-robot shared internal grasp force control via a network using a SMMS system. The stability analysis on the controller passivity over delayed communication is being performed. Also, virtual fixture-based haptic guidance tele-micromanipulation to improve human operability is implemented on the SMMS system and evaluated from the viewpoint of human operability and task performance, which can be challenging research. Finally the proposed system will further be implemented in smaller scale manipulations by improving the conventional sensing and actuation limitation.

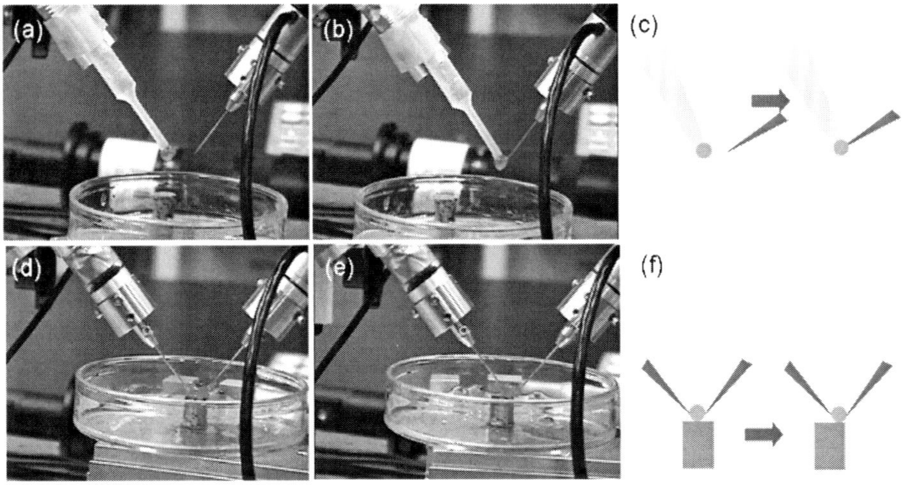

Fig. 10. Screenshots of salmon roe tweezing, indentation (a,b,c) and pick-and-place experiment (d,e,f).

7. References

Ando, N.; Korondi, P & Hashimoto, H. (2001). Development of micromanipulator and haptic interface for networked micromanipulation, *IEEE/ASME Transactions on Mechatronics*, Vol.6, No.4, pp.417–427, ISSN 1083-4435.

Ando, N.; Suehiro, T.; Kitagaki, K.; Kotoku, T. & Yoon, W. (2005). Implementation of RT composit components and a component manager, *Proceedings of the 2005 IEEE/RSJ International Conference on Intelligent Robots and Systems*, Vol.1, pp.3933–3938, ISBN 0-7803-8912-3, Edmonton, Canada, Aug. 2005.

Bonitz. R. & Hsia, T. (1994). Force decomposition in cooperating manipulators using theory of metric spaces and generalized inverses, *Proceedings of 1994 IEEE International*

Conference on Robotics and Automation, Vol.2, pp.1521–1527, ISBN 0-8186-5330-2, San Diego, USA, May. 1994.

Cleary, K. & Arai, T. (1991). A prototype parallel manipulator: kinematics, construction, software, workspace results, and singularity analysis, *Proceedings of the 1991 IEEE International Conference on Robotics and Automation*, Vol.1, pp.566–57, ISBN 0-8186-2163-X, Sacramento, USA, Apr. 1991.

Dosis, A.; Bello, F.; Rockall, T.; Munz, Y.; Moorthy, K.; Martin, S. & Darzi, A. (2003). ROVIMAS: a software package for assessing surgical skills using the da Vinci telemanipulator system, *Proceedings of 4th International IEEE EMBS Special Topic Conference on Information Technology Applications in Biomedicine*, Vol.1, pp. 326-329, ISBN 0-7803-7667-6, Birmingham, UK, Apr. 2003.

Fichter, E. F. (1986). A Stewart-Platform based manipulator: general theory and practical construction, *International Journal of Robotics Research*, Vol.5, No.2, pp.157–182, ISSN 0278-3649.

Griffin, W. B.; Provancher W. R. & Cutkosky, M. R. (2003). Feedback strategies for shared control in dexterous telemanipulation, *Proceedings of the 2003 IEEE/RSJ International Conference on Intelligent Robots and Systems*, Vol.3, pp.2791–2796, ISBN 0-7803-7860-1, Las Vegas, USA, Oct. 2003.

Hannoford, B. & Anderson, R. (1988). Experimental and simulation studies of hard contact in force reflecting teleoperation, *Proceedings of the 1995 IEEE International Conference on Robotics and Automation*, Vol.1, pp.584–589, ISBN 0-8186-0852-8, Pittsburgh, USA, Apr. 1988.

Hwang, G. & Hashimoto, H. (2007). Development of a single-master multi-slave tele-micromanipulation system, *Advanced Robotics*, Vol.21, No.3-4, pp.329–349, ISSN 0169-1864.

Kosuge, K.; Ishikawa, J.; Furuta, K. & Sakai, M. (1990). Control of single-master multi-slave manipulator system using VIM, *Proceedings of the 1990 IEEE International Conference on Robotics and Automation* , Vol.1, pp.1172–1177, ISBN 0-8186-9061-5, Cincinnati, USA, May. 1990.

Kosuge, K.; Itoh T.; Fukuda T. & Otsuka, M. (1995).Tele-manipulation system based on task-oriented virtual tool, *Proceedings of the 1995 IEEE International Conference on Robotics and Automation*, Vol.1, pp.351–356, ISBN 0-7803-1965-6, Nagoya, Japan, May. 1995.

Lee D. J. & Spong, M. W. (2005). Bilateral teleoperation of multiple cooperative robots over delayed communication networks: theory, *Proceedings of 2005 IEEE International Conference on Robotics and Automation*, Vol.1, pp.360-365, ISBN 0-7803-8914-X, Barcelona, Spain, Apr. 2005.

Paul, R. P. (1981). *Robot Manipulators: Mathematics, Programming, and Control*, ISBN 0-2621-6082X, MIT Press, Cambridge, USA.

Tanikawa, T. & Arai, T. (1999). Development of a micro-manipulation system having a two-fingered micro-hand, *IEEE Transactions on Robotics and Automation*, Vol.15, No.1, pp.152–162, ISSN 1042-296X.

Thompson J. A. & Fearing, R. S. (2001). Automating microassembly with ortho-tweezers and force sensing, *Proceedings of the 2001 IEEE/RSJ International Conference on Intelligent Robots and Systems*, Vol.1, pp.1327–1334, ISBN 0-7803-6612-3, Hawaii, USA, Nov. 2001.

Yoshikawa, T. (1985). Manipulability of robotic mechanisms, *International Journal of Robotics Research*, Vol.4, No.2, pp.3–9, ISSN 0278-3649.

18

Principal Screws and Full-Scale Feasible Instantaneous Motions of Some 3-DOF Parallel Manipulators

Z. Huang, J. Wang and S. H. Li
Robotics Research Center, Yanshan University
P. R. China

1. Introduction

With the development of parallel robot, various lower-mobility parallel mechanisms are proposed, especially, the 3-DOF parallel mechanisms have interested many researchers. Hunt (1983) proposed the first 3-DOF 3-RPS parallel mechanism. Lee and Shah (1988) addressed various possible applications of the mechanism. Waldron et al. (1989) studied an ARTISAN manipulator. Clavel (1988) proposed the DELTA. Gosselin and Angeles (1988, 1989) proposed an optimum kinematic design for planar and spherical 3-DOF parallel manipulators. Song (1995) studied a force-compensating device based on 3-RPS mechanism. Huang and Fang (1996) proposed some novel 3-DOF parallel mechanisms. Di Gregorio (1999) discussed the influence of flexibility of a 3-DOF parallel mechanism on the platform motion.

We know that a primary and basic step towards understanding a mechanism is to find all the feasible instantaneous motions or twists it can produce. In other words, it needs to determine both the range of the twist pitches and the distribution of the twist axes. It is important to correctly use a robot manipulator and plan its trajectory.

In robotics practice, for a six-DOF manipulator its end-effector has infinite moving possibility and can undergo any given twist in 3D space. For a lower-mobility manipulator it also has infinite moving possibility, however, it is clear that there are many motions impossible to realize. To find all the feasible instantaneous motions of a lower-mobility serial robot is easier than that of the 6 DOF one. The possible twists of a lower-mobility serial robot are obtained only by the linear combination of its joint screws. Nevertheless, for lower-mobility parallel mechanism, it is extraordinary difficult. To solve the problem we use the screw theory.

One hundred years ago Ball (1900) published his classical work on screw theory. Hunt (1978) further developed screw theory. He discussed all the screw systems. The screw systems were distinguished as general and special cases basically according to the pitches of principal screws. Gibson & Hunt (1990 a & b) classified first-order, second-order and third-order screw system further by means of projective geometry and gave the planar representation of the general three-system. Any screw motion of a 3-DOF rigid body can be expressed by a linear combination of its three principal screws in the three-system. The spatial distribution of axes of all the screws of the screw system in three-dimension space is regular. For example, all the screws of the second-order screw system lie on a cylindroid. For a third-order screw system or three-system, all the screws with the same pitch may lie on a hyperboloid of one sheet. The cylindroid or hyperboloid depicts the distributions of the positions and orientations of all screw axes of that screw system.

The key to determine both the range of the twist pitches and the distribution of the twist axes in three-dimension space is to get principal screws of the screw system. The principal screw is a very important concept in screw theory. After the principal screws are obtained, it is easy to know all possible motions of the mechanisms at a the given instant. In a second-order screw system there are two principal screws. For a third-order screw system, there exist three principal screws. The third-order screw system is the most important and complicated one.

The twist screw system of a 3-DOF parallel mechanism is a third-order screw system. To study the mechanism, we need to determine its three principal screws. Parkin (1990) specified the principal screws of the three-system from three given screws by adopting the mutual moment operation. Tsai & Lee (1993) studied the principal screws from three known screws by means of eigenvector. Zhang & Xu (1998) constructed the principal screws from three known screws by using algebraic method.

From another point of view, we put forward a directly and simply analytical method for identifying the principal screws for a 3-DOF parallel mechanism and obtained a full-scale feasible instantaneous motion of that mechanism. Fang & Huang (1998) firstly established the important relationship between the principal screws and Jacobian matrix of the 3-DOF mechanism, and identified the principal screws of the third-order screw system using the quadratic equation degenerating theory. After that, based on the relationship between the pitch/axis and the Jacobian matrix of the mechanism two equations are obtained. Then, another simpler and more effective principle (Huang & Wang 2001), the quadric degenerating theory, was further proposed for identifying the principal screws. For applying the principle to lower-mobility parallel mechanisms, corresponding two 3×3 Jacobian matrices are needed to establish firstly. This can be realized by using the imaginary-mechanism method proposed by Yan & Huang (1985) and developed by Huang & Wang (1992).

In this Chapter two typical examples are also discussed. One is a 3-DOF 3-RPS parallel manipulator (Huang & Wang, 2002). Another example is a special 3-UPU parallel mechanism (Huang, Li & Zuo, 2004). The analysis discovers this mechanism has some interesting and exceptional characteristics. All above analysis are important for enriching the mechanism theory and beneficial the mechanical design of the similar mechanical system.

2. Principal screws

The principal screw principle may be used to study the feasible instant motion. Its important merit is that it can illustrate a full-scale feasible instantaneous motion of a mechanism at any given configuration.

2.1 The screw representation

In screw theory (Ball, 1900; Hunt, 1978 & Huang, et al, 2006), a straight line in 3D space can be expressed by two vectors, $\mathbf{S}$ and S_0. Their dual combination is called a line vector, (S, S_0). The line vector can be expressed as

$$\$ = (\mathbf{S};\ \mathbf{S}_0) = (\mathbf{S};\ \mathbf{r}\times\mathbf{S}) = (l\ \ m\ \ n;\ p\ \ q\ \ r)$$

where S is a unit vector along the straight line; l, m, and n are three direction cosines of S; p, q, and r are the three elements of the cross product of $\boldsymbol{r}$ and S; $\boldsymbol{r}$ is a position vector of any point on the line or the line vector. $(S;\ S_0)$ is also called Plücker coordinates of the line vector and it consists of six components in total. For a line vector we have $\mathbf{S}\cdot\mathbf{S}_0 = 0$. When $S\cdot S = 1$, it is a unit line vector. When $\mathbf{S}_0 = \mathbf{0}$, the line vector, $(\mathbf{S};\ \mathbf{0})$, passes through the origin point.

When $\mathbf{S}\cdot\mathbf{S}_0 \neq 0$, it is defined as a screw

$$\$ = (\mathbf{S};\ \mathbf{S}_0) = (\mathbf{S};\ \mathbf{r}\times\mathbf{S} + h\mathbf{S}) \tag{1}$$

When $\mathbf{S}\cdot\mathbf{S} = 1$, it is a unit screw. The pitch of the screw is

$$h = \mathbf{S}\cdot\mathbf{S}_0 / \mathbf{S}\cdot\mathbf{S}$$

If the pitch of a screw is equal to zero, the screw degenerates into a line vector. In other words, a unit screw with zero-pitch (h = 0) is a line vector. The line vector can be used to express a revolute motion or a revolute pair in kinematics, or a unit force along the line in statics. If the pitch of a screw goes to infinity, $h = \infty$, the screw is expressed as

$$\$ = (0;\ \mathbf{S}) = (0\ \ 0\ \ 0;\ l\ \ m\ \ n)$$

and called a couple in screw theory. That means a unit screw with infinity-pitch, $h = \infty$, is a couple. The couple can be used to express a translation motion or a prismatic pair in kinematics, or a couple in statics. $\mathbf{S}$ is its direction cosine.

Both the revolute pair and prismatic pair are the single-DOF kinematic pair. The multi-DOF kinematic pair, such as cylindrical pair, universal joint or spherical pair, can be considered as the combination of some single-DOF kinematic pairs, and represented by a group of screws.

The twist motion of a robot end-effector can be described by a screw. The linear velocity $\boldsymbol{v}_P$ of a selected reference point P on the end-effector and the angular velocity $\boldsymbol{\omega}$ of the end-effector are given according to the task requirements. Therefore, the screw of the end-effector can be expressed by the given kinematic parameters $\boldsymbol{v}_P$ and $\boldsymbol{\omega}$

$$\$_i = \boldsymbol{\omega} + \in \mathbf{v}^o = \boldsymbol{\omega} + \in (\mathbf{v}_P + \mathbf{r}_P \times \boldsymbol{\omega})$$

where $\in$ is a dual sign; $\boldsymbol{v}^o$ is the velocity of the point coincident with the original point in the body; r_P is a positional vector indicating the reference point on the end-effector of the manipulator. When the original point of the coordinate system is coincident with point P, the pitch and axis can be determined by the following two equations

$$h = \frac{\boldsymbol{\omega}\cdot\mathbf{v}_P}{\boldsymbol{\omega}\cdot\boldsymbol{\omega}} \tag{2}$$

$$\mathbf{r}\times\boldsymbol{\omega}=\mathbf{v}_P-h\boldsymbol{\omega} \tag{3}$$

If a mechanism has three DOF, the order of the screw system is three. The motion of the three-order mechanism can be determined by three independent generalized coordinates. These independent generalized coordinates are often selected as three input-pair rates. The $\boldsymbol{v}_P$ and $\boldsymbol{\omega}$ of a robot can then be determined by these three input joint rates

$$\begin{aligned}\mathbf{v}_P&=[G]\dot{\mathbf{q}}\\ \boldsymbol{\omega}&=[G']\dot{\mathbf{q}}\end{aligned}\qquad \dot{\mathbf{q}}=\{\dot{q}_1\quad \dot{q}_2\quad \dot{q}_3\}^T \tag{4}$$

where [G] and [G′] are 3×3 first-order influence coefficient matrices (Thomas & Tesar, 1983). Substituting Eq. (4) into Eqs. (2, 3), the screw can also be described as the function of the joint rates

$$h=\frac{\dot{\mathbf{q}}^T[G']^T[G]\dot{\mathbf{q}}}{\dot{\mathbf{q}}^T[G']^T[G']\dot{\mathbf{q}}} \tag{5}$$

$$[\mathbf{r}][G']\dot{\mathbf{q}}=([G]-h[G'])\dot{\mathbf{q}} \tag{6}$$

where $[r]$ is a skew-symmetrical matrix of vector $\mathbf{r}=(x\quad y\quad z)^T$. Suppose we give the following expressions

$$\mathbf{u}=\dot{q}_1/\dot{q}_3\quad;\quad \mathbf{w}=\dot{q}_2/\dot{q}_3 \tag{7}$$

and then

$$\dot{\mathbf{q}}=(u\quad w\quad 1)\dot{q}_3$$

In this case, the pitch and the axis equations are given by

$$h=\frac{\{u\quad w\quad 1\}[G']^T[G]\{u\quad w\quad 1\}^T}{\{u\quad w\quad 1\}[G']^T[G']\{u\quad w\quad 1\}^T} \tag{8}$$

$$[\mathbf{r}][G']^T\{u\quad w\quad 1\}^T=([G]+[\mathbf{r}_P][G']-h[G'])\{u\quad w\quad 1\}^T \tag{9}$$

where $[r_P]$ is a skew-symmetrical matrix of coordinate of the point P.

2.2 Principal screws of three-order screw system

A third-order screw system has three principal screws. The three principal screws are mutually perpendicular and intersecting at a common point generally. Any screw in the screw system is the linear combination of the three principal screws. In the third-order screw system, two pitches of three principal screws are extremum, and the pitches of all other screws lie between the maximum pitch and the minimum pitch. Therefore to get the three principal screws is the key step to analyze the full-scale instantaneous motion of any 3-DOF mechanism. For obtaining the three principal screws there are two useful principles, the quadratic curve degenerating theory and quadric degenerating theory.

2.2.1 Quadratic curve degenerating theory

Let h_α, h_β and h_γ be pitches of the three principal screws and suppose $h_\gamma < h < h_\alpha$. Ball (1900) gave a graph illustrating the full-scale plane representation of a third-order system with quadratic curves, and each quadratic curve has identical pitch. If the pitch of any screw in the system is equal to h_α, h_β or h_γ, the quadratic equation will degenerate. When $h = h_\alpha$ or $h = h_\gamma$, the quadratic equation collapses into two virtual straight lines intersecting at a real point; when $h = h_\beta$, the quadratic equation collapses into two real straight lines (Hunt 1978).
Expanding Eq. (8), we have

$$a_{11}u^2 + 2a_{12}uw + a_{22}w^2 + 2a_{13}u + 2a_{23}w + a_{33} = 0 \tag{10}$$

where the coefficient a_{ij}, $(i, j = 1 \sim 3)$, is a function of pitch h and the elements of the matrices $[G]$ and $[G']$. From the quadratic equation degenerating principle, the determinant of the coefficient matrix should be zero, that is

$$\mathbf{D} = \begin{vmatrix} a_{11} & a_{12} & a_{13} \\ a_{21} & a_{22} & a_{23} \\ a_{31} & a_{32} & a_{33} \end{vmatrix} = 0, \quad (a_{ij} = a_{ji}) \tag{11}$$

Expanding the Eq. (11) we have

$$c_1h^3 + c_2h^2 + c_3h + c_4 = 0 \tag{12}$$

where c_i, $(i = 1 \sim 4)$, is a function of the elements of $[G]$ and $[G']$. Three roots of the Eq. (12) are pitches, h_α, h_β and h_γ, of the three principal screws. Substituting the pitch of principal screw into Eq. (10), the above quadratic equation degenerates into two straight lines, the root, $(u_i \quad w_i)$, of the two equations is

$$\begin{aligned} u_i &= \frac{a_{22}a_{13} - a_{12}a_{23}}{a_{12}^2 - a_{11}a_{22}} \\ w_i &= -\frac{a_{23}}{a_{22}} - \frac{a_{12}}{a_{22}}u_i \end{aligned} \qquad i = 1, 2, 3 \tag{13}$$

Each set of $(u_i \quad w_i)$ corresponds three inputs $(u_i \quad w_i \quad 1)$. Three sets of $(u_i \quad w_i)$, $i = \alpha, \ \beta, \ \gamma$, correspond three output twists, i.e., three principal screws.

When the pitches of three principal screws are obtained, substituting the three values into Eq. (9), the axis equations of three principal screws can also be obtained.

2.2.2 Quadric degenerating theory

The quadric degenerating theory is an easier method for calculating the principal screws. Eq. (6) can be further simplified as

$$[A]\ \dot{\mathbf{q}} = 0 \tag{14}$$

where

$$[A] = [\mathbf{r}][G'] - [G] + h[G']$$

is a 3×3 matrix. $[G]$ and $[G']$ are also 3×3 first-order kinematic influence coefficient matrices, which are functions of the structure parameters of the mechanism. Since not all the components of vector $\dot{q}$ are zeros in general, the necessary and sufficient condition that ensures the solutions of Eq. (14) being non-zero is that the determinant of the matrix $[A]$ is equal to zero. Namely (Huang & Wang 2001)

$$Det[\mathbf{A}] = 0 \tag{15}$$

Expanding Eq. (15), we obtain the position equation describing all the screw axes

$$c_{11}x^2 + c_{22}y^2 + c_{33}z^2 + 2c_{12}xy + 2c_{23}yz + 2c_{13}xz + 2c_{14}x + 2c_{24}y + 2c_{34}z + c_{44} = 0 \tag{16}$$

where the coefficients, c_{ij} (i=1, 2, 3, 4, j=1, 2, 3, 4), are the function of pitch h as well as coefficients g_{ij}, b_{ij}, the latter are relative with the elements of matrices [G] and [G′] in Appendix (Huang & Wang 2001). The Eq. (16) is a quadratic equation with three elements, x, y and z. It expresses a quadratic surface in space. The spatial distribution of all the screw axes in 3D is quite complex. Generally, all the screw axes lie on a hyperboloid of one sheet if every coefficient in Eq. (16) contains the same pitch h.

2.2.2.1 Pitches of three principal screws

For a third-order screw system there exist three principal screws α, β and γ. Let h_α, h_β and h_γ be the pitches of the three principal screws, and also suppose $h_\alpha > h_\beta > h_\gamma$.

We know that the quadric surface, Eq. (16), collapses into a straight line where the principal screws α or γ lies, when $h = h_\alpha$ or $h = h_\gamma$. The quadric surface degenerates into two intersecting planes, when $h = h_\beta$, and the intersecting line is just the axis of principal screw β (Hunt 1978). According to this nature, we can identify the three principal screws of the three-system.

The quadric has four invariants, I, J, D and Δ, and they are

$$I = c_{11} + c_{22} + c_{33}$$

$$\Delta = \begin{vmatrix} c_{11} & c_{12} & c_{13} & c_{14} \\ c_{21} & c_{22} & c_{23} & c_{24} \\ c_{31} & c_{32} & c_{33} & c_{34} \\ c_{41} & c_{42} & c_{43} & c_{44} \end{vmatrix} \quad ; \quad D = \begin{vmatrix} c_{11} & c_{12} & c_{13} \\ c_{21} & c_{22} & c_{23} \\ c_{31} & c_{32} & c_{33} \end{vmatrix} \tag{17}$$

$$J = c_{11}c_{22} + c_{22}c_{33} + c_{11}c_{33} - c_{12}^2 - c_{23}^2 - c_{13}^2 \qquad \left(c_{ij} = c_{ji}\right)$$

Expanding D, and let it equal to zero, $D = 0$, we have the expression

$$a_1h^3 + a_2h^2 + a_3h + a_4 = 0 \tag{18}$$

where the coefficients a_i (i=1, ..., 4) are also the function of g_{ij} , b_{ij} and h. Three possible roots can be obtained by solving Eq. (18), and these three roots correspond to pitches of the three principal screws. When the pitch in the system is equal to one of the three principal screw pitches, the invariant Δ is zero as well. It satisfies the condition that the quadric degenerates into a line or two intersecting planes. Therefore, the key to identify the principal screws in the third-order system is that the quadric, Eq. (16), degenerates into a line or a pair of intersecting planes.

2.2.2.2 The axes of principal screws and principal coordinate system

The coordinate system that consists of three principal screws is named the principal coordinate system. We know that the most concise equation of a hyperboloid is under its principal coordinate system. Now, we look for the principal coordinate system of the hyperboloid.

Equation (16) represented in the base coordinate system can be transformed into the normal form of the hyperboloid of one sheet in the principal coordinate system. After the pitches of the three principal screws are obtained, the pitch of any screw in the system is certainly within the range of $h_\gamma < h < h_\alpha$. The general three-system (Hunt 1978) appears only when three pitches of the three principal screws all are finite and also satisfy $h_\gamma \neq h_\beta \neq h_\alpha$. The axes of all the screws with the same pitch in the range from h_γ to h_β or from h_β to h_α form a hyperboloid of one sheet. In this case the invariant D is not equal to zero, and the quadrics are the concentric hyperboloids. By solving Eq. (19)

$$\begin{cases} c_{11}x + c_{12}y + c_{13}z + c_{14} = 0 \\ c_{21}x + c_{22}y + c_{23}z + c_{24} = 0 \\ c_{31}x + c_{32}y + c_{33}z + c_{34} = 0 \end{cases} \tag{19}$$

the root of Eq. (19) is just the center point o′ $\left(x_0 \quad y_0 \quad z_0\right)$ of the hyperboloid. It is clear that the point o' is also the origin of the principal coordinate system. The coordinate translation is

$$\begin{cases} x = x' + x_0 \\ y = y' + y_0 \\ z = z' + z_0 \end{cases} \tag{20}$$

The eigenequation of the quadric is

$$\mathrm{k}^3 - \mathrm{I}\mathrm{k}^2 + \mathrm{J}\mathrm{k} - \mathrm{D} = 0 \tag{21}$$

Its three real roots k_1, k_2, k_3 are the three eigenvalues, and not all the roots are zeros. In general, $k_1 \neq k_2 \neq k_3$. The corresponding three unit eigenvectors $(\lambda_1 \;\; \mu_1 \;\; \nu_1)$, $(\lambda_2 \;\; \mu_2 \;\; \nu_2)$ and $(\lambda_3 \;\; \mu_3 \;\; \nu_3)$ are perpendicular each other, and corresponding three principal screws, α, β *and* γ, form the coordinate system (o'-$x'y'z'$). The principal coordinate system (o'-$\alpha\beta\lambda$) can then be constructed by a following coordinate rotation

$$\begin{cases} x = \lambda_1 x' + \lambda_2 y' + \lambda_3 z' \\ y = \mu_1 x' + \mu_2 y' + \mu_3 z' \\ z = \nu_1 x' + \nu_2 y' + \nu_3 z' \end{cases} \tag{22}$$

After the coordinate transformation, the normal form of the hyperboloid is

$$k_1 x^2 + k_2 y^2 + k_3 z^2 + \frac{\Delta}{D} = 0 \tag{23}$$

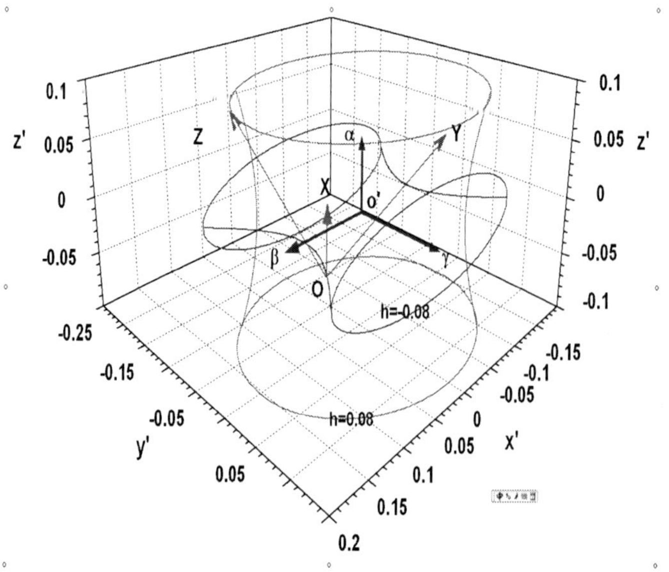

Fig.1. Hyperboloid of one sheet

Hunt (1978) gave that when h lies within the range $h_\beta < h < h_\alpha$, the central symmetrical axis of the hyperboloid is α, and the semi-major axis of its central elliptical section in the $\beta\gamma$-plane always lies along β. For $h_\gamma < h < h_\beta$, the central symmetrical axis of the hyperboloid is γ, and the semi-major axis of its central elliptical section in the $\beta\gamma$-plane is also along β, Fig.1. Therefore, we may easily determine the three axes of the principal coordinate system.

3. Imaginary mechanism and Jacobian matrix

In order to determine the pitches and axes using Eqs. (4-9), the key step is to determine 3×3 Jacobian matrices [G] and [G′]. For a 3-DOF parallel mechanism to determine the [G] and [G′] is difficult. Here the imaginary-mechanism principle (Yan & Huang, 1985; Huang & Wang, 1992) can solve the issue easily.

Note that, the imaginary-mechanism principle with unified formulas is a general method, and can be applied for kinematic analysis of any lower-mobility mechanism. An example is taken to introduce how to set the matrices [G] and [G′].

Fig. 2(a) shows a 3-DOF 3-RPS mechanism consisting of an upper platform, a base platform, and three kinematic branches. Each of its three branches is comprise of a revolute joint R, a prismatic pair P and a spherical pair S, which is a RPS serial chain. The axes of three revolute joints are tangential to the circumcircle of the lower triangle.

The mechanism has three linear inputs, $\dot{L}_1$, $\dot{L}_2$, $\dot{L}_3$.

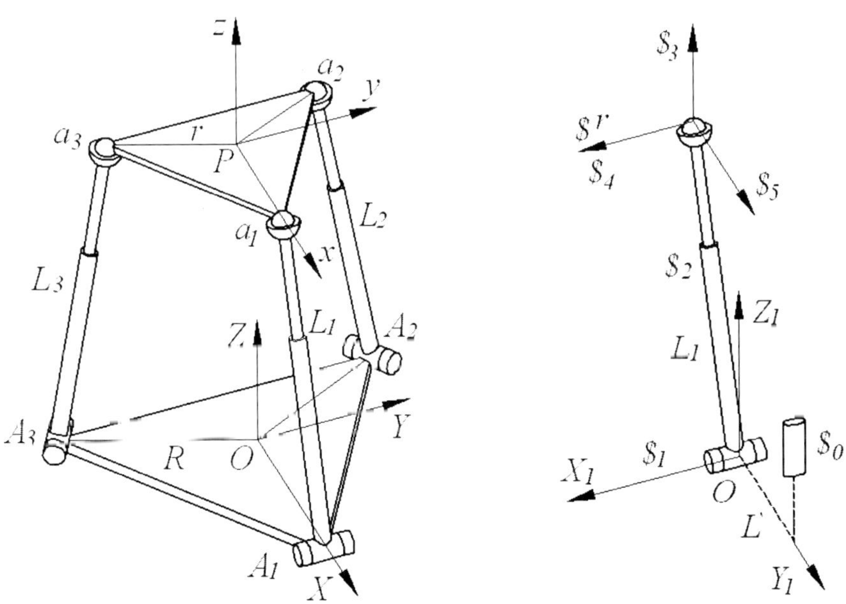

a) Mechanism sketch b) Imaginary branch

Fig.2. 3-DOF 3-RPS parallel mechanism

3.1 Imaginary twist screws of branches

Each kinematic branch of the 3-RPS mechanism may be represented by five single-DOF kinematic pairs as RPRRR. In order to get the Jacobian matrix by means of the method of kinematic influence coefficient of a 6-DOF parallel mechanism (Huang 1985), we may transform this 3-DOF mechanism into an imaginary 6-DOF one in terms of the kinematic equivalent principle. An imaginary link and an imaginary revolute pair, $\$_0$, with single-DOF, are added to each branch of the mechanism. Then each branch becomes an imaginary 6-DOF serial chain. In order to keep a kinematic equivalent effect, let the amplitude ω_0 of the imaginary screw $\$_0$ of each branch always be zero; and let each screw system formed by imaginary $\$_0$ and the other five screws of the primary branch RPRRR be linearly independent.

Considering the imaginary pair $\$_0$, the Plücker coordinates of all six screws shown in Fig. 3b with respect to local o-$X_1Y_1Z_1$ coordinate system are

$$\begin{array}{ll} \$_1 = \{1 \;\; 0 \;\; 0 ; 0 \;\; 0 \;\; 0\} & \$_4 = \{1 \;\; 0 \;\; 0 ; 0 \;\; L_0 \zeta \;\; -L_0 \psi\} \\ \$_2 = \{0 \;\; 0 \;\; 0 ; 0 \;\; \psi \;\; \zeta\} & \$_5 = \{0 \;\; \zeta \;\; -\psi ; -L_0 \;\; 0 \;\; 0\} \\ \$_3 = \{0 \;\; \psi \;\; \zeta ; 0 \;\; 0 \;\; 0\} & \$_0 = \{0 \;\; 0 \;\; 1 ; L' \;\; 0 \;\; 0\} \end{array} \tag{24}$$

where ψ and ζ are directional cosines of the screw axes $\$_2$ and $\$_3$. The screw matrix of each branch with respect to the local coordinate system is $[Gg] = \{\$_0 , \$_1 , \$_2 , \$_3 , \$_4 , \$_5\}$, and we have $[G_i^0] = [A_i^0][Gg]$.

3.2 Imaginary Jacobian matrix

For each serial branch, the motion of the end-effector of the 3-RPS mechanism can be represented by the following expression

$$\mathbf{V}_H = [G_i^0]\, \dot{\boldsymbol{\varphi}}^{(i)} \qquad i = 1,2,3 \tag{25}$$

where $\mathbf{V}_H = \{\boldsymbol{\omega} \;\; \mathbf{v}_P\}^T$ is a six dimension vector; $\boldsymbol{\omega}$ is the angular velocity of the moving platform; v_P is the linear velocity of the reference point P in the moving platform; and $\dot{\boldsymbol{\varphi}}^{(i)} = (\dot{\varphi}_0^{(i)} \;\; \dot{\varphi}_1^{(i)} \;\; \dot{\varphi}_2^{(i)} \;\; \dot{\varphi}_3^{(i)} \;\; \dot{\varphi}_4^{(i)} \;\; \dot{\varphi}_5^{(i)})$ is a vector of joint rates. If $[G_i^0]$ is non-singular

$$\dot{\boldsymbol{\varphi}}^{(i)} = [G_0^i]\, \mathbf{V}_H \qquad i = 1 , 2 , 3 \tag{26}$$

where $[G_0^i] = [G_i^0]^{-1}$

The input rates $\dot{L}_1 , \dot{L}_2 , \dot{L}_3$ of the mechanism are known and the rate of each imaginary link is zero, which is equal to known. Then for each branch we have

$$\dot{\boldsymbol{\varphi}}^{(i)} = (\dot{\varphi}_0 \;\; \dot{\varphi}_1 \;\; \dot{\varphi}_2 \;\; \dot{\varphi}_3 \;\; \dot{\varphi}_4 \;\; \dot{\varphi}_5)^{(i)} = (0 \;\; \dot{\varphi}_1 \;\; \dot{L}_1 \;\; \dot{\varphi}_3 \;\; \dot{\varphi}_4 \;\; \dot{\varphi}_5)^{(i)} \quad i = 1,2,3 \tag{27}$$

Taking the first row and third row from the matrix $\left[G_0^i\right]$ in Eq. (26) of each branch, there are six linear equations. A new matrix equation can be established

$$\dot{\mathbf{q}} = \left[G_H^q\right]\mathbf{V}_H \quad \dot{\mathbf{q}} = \left\{\dot{L}_1 \ \ \dot{L}_2 \ \ \dot{L}_3 \ \ 0 \ \ 0 \ \ 0\right\} \tag{28}$$

where

$$\left[G_H^q\right] = \left[\left[G_0^1\right]_{3:} \ \left[G_0^2\right]_{3:} \ \left[G_0^3\right]_{3:} \ \left[G_0^1\right]_{1:} \ \left[G_0^2\right]_{1:} \ \left[G_0^3\right]_{1:}\right]^T \in R^{6\times 6}$$

where $\left[G_0^i\right]_{i:}$ represents the ith row of matrix $\left[G_0^i\right]$. If the matrix $\left[G_H^q\right]$ is non-singular, from Eq. (28)

$$\mathbf{V}_H = \left[G_q^H\right]\dot{\mathbf{q}} \tag{29}$$

where

$$\left[G_q^H\right] = \left[G_H^q\right]^{-1} \tag{30}$$

Since the 3-RPS mechanism has three freedoms, it needs three inputs. The matrix $\left[G_L^H\right]$ formed by taking the first three columns of the matrix $\left[G_q^H\right]$ is a 6×3 Jacobian matrix. Therefore

$$\mathbf{V}_H = \left[G_L^H\right]\dot{\mathbf{L}} \tag{31}$$

As $\mathbf{V}_H = \left\{\boldsymbol{\omega} \ \ \mathbf{v}_P\right\}^T$, Eq. (31) can be separated into two equations

$$\mathbf{v}_p = [\mathrm{G}]\dot{\mathbf{L}}\ ; \quad \boldsymbol{\omega} = [\mathrm{G}']\dot{\mathbf{L}} \tag{32}$$

where $\left[G'\right]$ is the first three rows of $\left[G_L^H\right]$; $\left[G\right]$ is the last three rows of $\left[G_L^H\right]$. Then we obtain the 3×3 matrices [G] and [G′]. From the analysis process we know that the matrices $\left[G\right]$ and $\left[G'\right]$ are independent of the chosen of these imaginary pairs.

4. Full-scale feasible instantaneous screws of 3-RPS mechanism

Now, we continue to study the 3-RPS mechanism, Fig. 2, to get the full-scale feasible instantaneous motion. The parameters of the mechanism are : R=0.05 m; r=0.05 m; L_0=0.2 m; L'=0.04 m. Three configurations will be discussed.

4.1 Upper platform is parallel to the base

Substituting given geometrical parameters and expanding Eq. (8), we have Eq. (10)

$$a_{11}u^2 + 2a_{12}uw + a_{22}w^2 + 2a_{13}u + 2a_{23}w + a_{33} = 0 \tag{33}$$

Eq. (33) is a quadratic equation with two variables, u and w. It will degenerate, if Equation (11) is satisfied. Expanding Eq. (11) we have the Eq. (12)

$$ah^3 + bh^2 + ch + d = 0 \tag{34}$$

The three roots of Eq. (34) are just three pitches of the three principal screws. Substituting each root h into Eq. (33) the quadratic equation degenerates into two linear equations expressing two straight lines. The intersecting point (u, w) of the two lines can be obtained. Then, the axis of the principal screw can also be obtained by using Eq. (9).

When the moving platform is parallel to the fixed one, it follows that: $a = b = c = d = 0$; i.e., all the coefficients of Eq. (34) are zeroes. From algebra, the three roots, h, can be any constant. For some reasons, which we will present below, however, the three roots of Eq. (34) should be $(\infty \ \ 0 \ \ 0)$. When $h \to \infty$, we have $u = 1$, $w = 1$, then the inputs are $\dot{\mathrm{L}} = \{u \ \ w \ \ 1\} = \{1 \ \ 1 \ \ 1\}$. The output motion is a pure translation, namely $\$_{Z1} = \{0 \ \ 0 \ \ 0 \ \ ; \ \ 0 \ \ 0 \ \ 1\}$. When the pitch of the principal screw is zero, $h = 0$, $u = 0/0$; $w = 0/0$. Mathematically, u and w both can be any value except one. All other roots of Eq. (34) will not be considered, as they are algebraically redundant. Then, the corresponding three principal screws can be written as

$$\begin{aligned} \$_{z1} &= \{0 \ \ 0 \ \ 0; \ \ 0 \ \ 0 \ \ 1\} \\ \$_{z2} &= \{0 \ \ 1 \ \ 0; \ \ -P_x \ \ 0 \ \ 0\} \\ \$_{z3} &= \{1 \ \ 0 \ \ 0; \ \ 0 \ \ P_z \ \ 0\} \end{aligned} \tag{35}$$

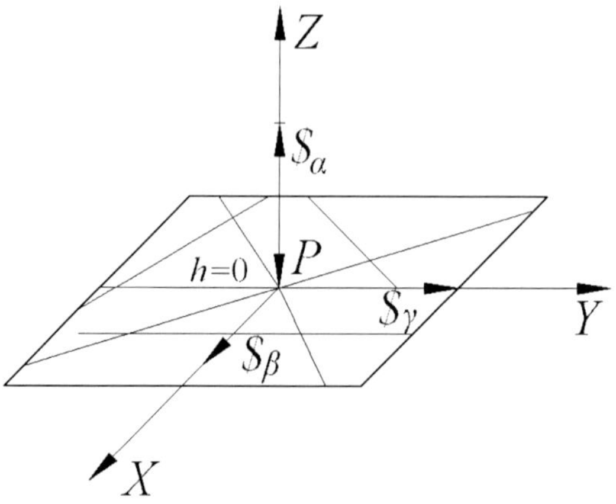

Fig. 3. The spatial distribution of the screws when the upper parallel to the base

Any output motion may be considered as a linear combination of the three principal screws. The full-scale distribution result, Fig.3, of all screws obtained by linear combinations of three principal screws can also be verified by using another method presented in Huang et al.,

(1996), and is identical with the actual mechanism model in our laboratory. The three principal screws belong to the fourth special three-system presented by Hunt (1978).
When the upper platform is parallel to the fixed platform, all possible output twists of the upper platform except the translation along the Z direction are rotations corresponding screws with zero pitch. Their axes all lie in the moving platform and in all the directions. Fig. 3 shows the full-scale possible twist screws with zero-pitch. Therefore from this figure you don't attempt to make the moving platform rotate round any axis not on the plane shown in Fig. 3. That is impossible.

4.2 The upper platform rotates by an angle α about line a_2a_3

When the upper platform continually rotates by an angle α about line a_2a_3, namely the mechanism is in the configuration that the lengths of the two input links are the same. Note that, for this kind of mechanisms the platform cannot continually rotate about axes lying in the plane shown in Fig.3 except some three axes including a_2a_3. In other words, it is very often impossible that the platform can continually rotated about an axis lying in the plane, as shown in Fig.3, (Zhao et al, 1999).
The coordinates of point a_1 on the upper platform and point A_1 on the base have the following values

$$a_1 = \{r(3\cos\alpha - 1)/2 \quad 0 \quad L_0 + 3r\sin\alpha/2\} \quad \mathbf{A}_1 = \{R \quad 0 \quad 0\} \tag{36}$$

In this configuration, the screw system including the imaginary pair of the first chain corresponding to $\left[G_1^0\right]$ with respect to the fixed coordinate system is

$$\begin{aligned}
\$_1 &= \{\mathbf{S}_1;\ \mathbf{S}_{01}\} = \{\mathbf{S}_l;\ \mathbf{A}_1 \times \mathbf{S}_l\} \\
\$_2 &= \{\mathbf{S}_2;\ \mathbf{S}_{02}\} = \{\mathbf{0};\ \mathbf{L}_l\}/\|\mathbf{L}_l\| \\
\$_3 &= \{\mathbf{S}_3;\ \mathbf{S}_{03}\} = \{\mathbf{L}_l;\ \mathbf{a}_1 \times \mathbf{L}_l\}/\|\mathbf{L}_l\| \\
\$_4 &= \{\mathbf{S}_4;\ \mathbf{S}_{04}\} = \{\mathbf{S}_l;\ \mathbf{a}_1 \times \mathbf{S}_l\} \\
\$_5 &= \{\mathbf{S}_5;\ \mathbf{S}_{05}\} = \{\mathbf{L}_l \times \mathbf{S}_4;\ \mathbf{a}_1 \times \mathbf{L}_l \times \mathbf{S}_4\}/\|\mathbf{L}_l \times \mathbf{S}_4\| \\
\$_0 &= \{0 \quad 0 \quad 1;\ 0 \quad -L' \quad 0\}
\end{aligned} \tag{37}$$

where $\mathbf{S}_l = \mathbf{S}_4 = \{0 \quad -1 \quad 0\}$, $\mathbf{L}_l = \mathbf{a}_1 - \mathbf{A}_1$.
The twist screw systems of the other two chains corresponding $\left[G_2^0\right]$ and $\left[G_3^0\right]$ are the same as the case that the upper platform is parallel to the base. Establishing matrices $[G]$ and $[G']$, we can solve principal screws by using the previous method.
Suppose $\alpha = 30^\circ$, the pitches of three principal screws can be obtained by solving Eq. (34). They are $h_\alpha = 5.13\times10^5; h_\beta = 0; h_\gamma = -5.13\times10^5$. When $I_2 = 0$, where I_2 is the two-order determinant of coefficients of the quadratic equation, its two roots are $h_1 = -0.0057$, $h_2 = 0.0165$. There are six types of the quadratic curve for the same configuration of the mechanism, as shown in Table 1. The pitch h varies between h_α and h_γ.

Each point in Fig. 4 denotes a pitch h of a twist screw of the moving platform relative the three inputs (u, w, 1). You can get the output pitch of the instantaneous twist when three inputs are given. Fig.4 also shows the relation between inputs and the six types of quadratic curves with different pitches in this configuration of the mechanism.

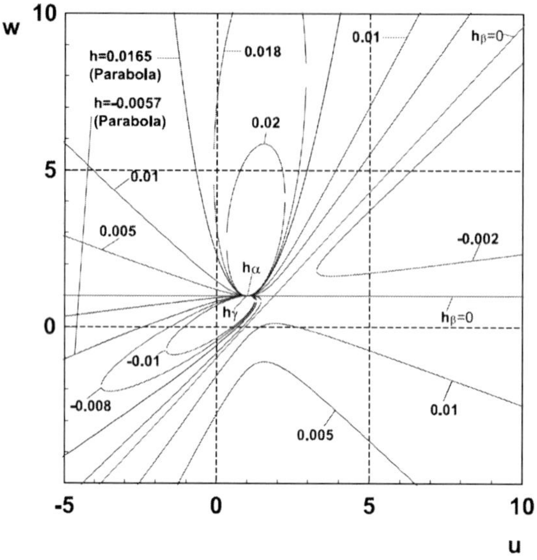

Fig. 4. When the upper platform rotates 30° about a_2a_3

The range of the value of h		Type of conics
In 30° configuration	In general configuration	
$0.0165256 < h < 5.13\times10^5$ or $-5.13\times10^5 < h < -0.0057003$	$0.0131215 < h < 4.28\times10^5$ or $-4.28\times10^5 < h < -0.0160208$	Real ellipse
$h > 5.13\times10^5$ or $h < -5.13\times10^5$	$h > 4.28\times10^5$ or $h < -4.28\times10^5$	Imaginary ellipse
$h_\alpha = 5.13\times10^5$ or $h_\gamma = -5.13\times10^5$	$h_\alpha = 4.28\times10^5$ or $h_\gamma = -4.28\times10^5$	Dot ellipse
$-0.0057003 < h < 0.0165256$	$-0.0160208 < h < 0.0131215$	Hyperbola
$h_\beta = 0$	$h_\beta = 0.0079$	A pair of intersecting real lines
$h = 0.0165256$ or $h = -0.0057003$	$h = 0.0131215$ or $h = -0.0160208$	Parabola

Table 1. Six types of the quadratic curves

The twist screws with the same pitch, h, form a quadratic curve. The pure rotations with zero pitch are illustrated as a pair of intersecting real straight lines in the figure.

The two straight lines can also be obtained and proved by using another method proposed by Huang & Fang (1996). The three principal screws are

$$\$^1_m = \{0 \quad -1 \quad 0 \quad ; \quad 0.2 \quad 0 \quad 0.1\}$$

$$\$^2_m = \{0.966 \quad 0 \quad 0.259 \quad ; \quad 0 \quad 0.22 \quad 3.96\times10^6\} \tag{38}$$

$$\$^3_m = \{-0.966 \quad 0 \quad -0.259 \quad ; \quad 0 \quad -0.22 \quad 3.96\times10^6\}$$

The screw $\$^m$ with infinite pitch can be obtained by a linear combination of $\$^2_m$ and $\$^3_m$

$$\$^m = \{0 \quad 0 \quad 0 \quad ; \quad 0 \quad 0 \quad 1\}$$

It expresses a pure translation along the Z-direction. $\$^1_m$ with zero pitch is a pure rotation about an axis parallel to the Y-axis. $\$^2_m$ is a twist screw with $h \neq 0$ and deviates from the normal direction of $\$^m$. The three screws, $\$^m$, $\$^1_m$ and $\$^2_m$

$$\$^m = \{0 \quad 0 \quad 0 \quad ; \quad 0 \quad 0 \quad 1\}$$

$$\$^1_m = \{0 \quad -1 \quad 0 \quad ; \quad 0.2 \quad 0 \quad 0.1\} \tag{39}$$

$$\$^2_m = \{0.966 \quad 0 \quad 0.259 \quad ; \quad 0 \quad 0.22 \quad 3.96\times10^6\}$$

form a set of new principal screws, which is just the seventh special three-system screws presented by Hunt (1978), Tsai and Lee (1993).

4.3 General configuration of the 3-RPS mechanism

In any general configuration, the lengths of three legs of the parallel manipulator are different. The coordinates of the points a_1, a_2 and a_3 with respect to the coordinate system *P-xyz* are

$$\begin{aligned}\mathbf{a}_1 &= \{r \quad 0 \quad 0\}^T \\ \mathbf{a}_2 &= \{-r/2 \quad \sqrt{3}r/2 \quad 0\}^T \\ \mathbf{a}_3 &= \{-r/2 \quad -\sqrt{3}r/2 \quad 0\}^T\end{aligned} \tag{40}$$

Since the transformation matrix from the system *P-xyz* to the fixed system *O-XYZ* is [T]. The coordinates of the points with respect to the fixed coordinate system *O-XYZ* are

$$\{\mathbf{P}_i \quad 1\}^T = [T]\{\mathbf{a}_i \quad 1\}^T \qquad i = 1,2,3 \tag{41}$$

The unit vectors u_1, u_2 and u_3 representing revolute axes with respect to the fixed system are

$$\begin{aligned} \mathbf{u}_1 &= \{0 \;\; 1 \;\; 0\}^T \\ \mathbf{u}_2 &= \{-\sqrt{3}/2 \;\; -1/2 \;\; 0\}^T \\ \mathbf{u}_3 &= \{\sqrt{3}/2 \;\; -1/2 \;\; 0\}^T \end{aligned} \tag{42}$$

The screw systems of the three serial chains in the fixed system can be expressed as following

$$\begin{aligned} \$_1^i &= \{\mathbf{S}_1 ;\;\; \mathbf{S}_{01}\} = \{\mathbf{u}_1 ;\;\; \mathbf{A}_i \times \mathbf{u}_i\} \\ \$_2^i &= \{\mathbf{S}_2 ;\;\; \mathbf{S}_{02}\} = \{\mathbf{0} ;\;\; \mathbf{L}_i\} / \|\mathbf{L}_i\| \\ \$_3^i &= \{\mathbf{S}_3 ;\;\; \mathbf{S}_{03}\} = \{\mathbf{L}_i ;\;\; \mathbf{P}_i \times \mathbf{L}_i\} / \|\mathbf{L}_i\| \qquad i = 1,2,3 \\ \$_4^i &= \{\mathbf{S}_4 ;\;\; \mathbf{S}_{04}\} = \{\mathbf{u}_i ;\;\; \mathbf{P}_i \times \mathbf{u}_i\} \\ \$_5^i &= \{\mathbf{S}_5 ;\;\; \mathbf{S}_{05}\} = \{\mathbf{L}_i \times \mathbf{u}_i ;\;\; \mathbf{P}_i \times \mathbf{L}_i \times \mathbf{u}_i\} / \|\mathbf{L}_i \times \mathbf{u}_i\| \end{aligned} \tag{43}$$

Three imaginary revolute pairs added to three branches are supposed all in Z-direction and passing through points k_1, k_2 and k_3, respectively. They are on the lines from original point O to the points A_1, A_2 and A_3, respectively. All lengths are L', then the coordinates of the points k_1, k_2 and k_3 are expressed as three vectors

$$\begin{aligned} \mathbf{k}_1 &= \{L' \;\; 0 \;\; 0\} \\ \mathbf{k}_2 &= \{-L'/2 \;\; \sqrt{3}L'/2 \;\; 0\} \\ \mathbf{k}_3 &= \{-L'/2 \;\; -\sqrt{3}L'/2 \;\; 0\} \end{aligned} \tag{44}$$

The three corresponding imaginary twist screws are

$$\$_0^i = \{\mathbf{S}_0 ;\;\; \mathbf{k}_i \times \mathbf{S}_0\} \qquad i = 1,2,3 \tag{45}$$

where $\mathbf{S}_0 = \{0 \;\; 0 \;\; 1\}$.

The matrices $\left[G_i^0\right]$ corresponding screw systems of the three branches with respect to the fixed coordinate system are

$$\left[G_i^0\right] = \{\$_0^i \;\; \$_1^i \;\; \$_2^i \;\; \$_3^i \;\; \$_4^i \;\; \$_5^i\} \qquad i = 1,2,3 \tag{46}$$

When the coordinates of center point of the upper platform with respect to the fixed system are given as

$$X = 0.002 \;\; m, \quad Y = 0.001 \;\; m, \quad Z = 0.22 \;\; m$$

the pitches of the three principal screws can be obtained as:

$$h_\alpha = 4.28 \times 10^5 \,; h_\beta = 0.0079 \,; h_\gamma = -4.28 \times 10^5 \,.$$

When $I_2 = 0$, two possible roots of the pitch are $h_1 = -0.016$, $h_2 = 0.013$. There are also six types of conics in this configuration, Table 1. Fig.5 illustrates a planar representation of pitches of all possible twist screws in this case.

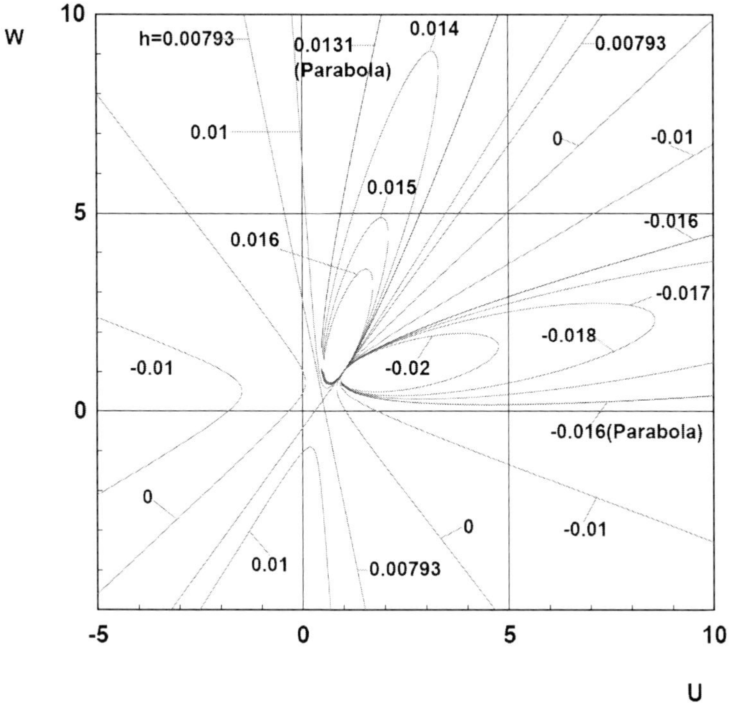

Fig. 5. The planar representations of the twist screws in any general configuration

The coordinates (u, w) of the principal screw with h_α are (1.0004133965, 1.000387461). The (u, w) corresponding h_γ are (1.0004134267, 1.000387451). They both are too close to be distinguished by naked eye in the figure. The three principal screws can be obtained as

$$
\begin{aligned}
\$^1_m &= \{-0.97 \quad 0.23 \quad 0 \quad ; \quad -0.06 \quad -0.22 \quad 0.06\} \\
\$^2_m &= \{0.22 \quad 0.95 \quad 0.21 \quad ; \quad -0.204 \quad 0.395 \quad 4.1\times 10^6\} \qquad (47) \\
\$^3_m &= \{-0.22 \quad -0.95 \quad -0.21 \quad ; \quad 0.204 \quad -0.395 \quad 4.1\times 10^6\}
\end{aligned}
$$

The screw $\$^m = \{0 \quad 0 \quad 0 \quad ; \quad 0 \quad 0 \quad 1\}$ with infinite pitch, $h^m = \infty$, can be obtained by the linear combination of $\$^2_m$ and $\$^3_m$. $\$^m$ expresses a pure translation along the Z direction.

$\$_m^1$ with $h_m^1 = 0$ is perpendicular to Z-axis. $\$_m^2$ with $h_m^2 \neq 0$ deviates from the normal direction of $\$^m$. Therefore, the three principal screws, $\$^m$, $\$_m^1$ and $\$_m^2$, also form a seventh special three-system. Therefore, the formation of all linear combinations of $\$^m$, $\$_m^1$ and $\$_m^2$ in three-dimensional space, as shown in Fig.6, is a hyperbolic paraboloid.

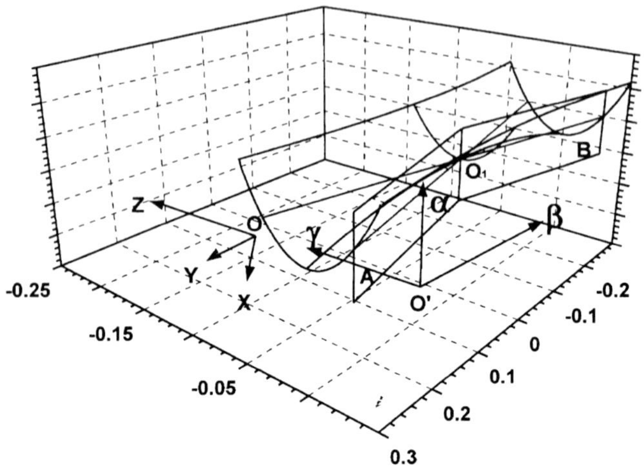

Fig. 6. The spatial distribution of the screws in General configuration

5. Full-scale feasible instantaneous screw of a 3-UPU mechanism

In this section we discuss an interesting 3-DOF special 3-UPU mechanism. It has some special inconceivable characteristics.

5.1 First-order influence matrices and kinematic analysis

The 3-UPU mechanism, as shown in Fig. 7a, consists of a fixed pyramid $A_1A_2A_3$, a moving pyramid $a_1a_2a_3$ and three UPU kinematic chains. Three centrelines of the three prismatic pairs in the initial position are mutually perpendicular. The middle two revolute pairs, $\$_2$ and $\$_4$, Fig. 7b, adjacent to the prismatic pair in every branch, are mutually perpendicular, moreover they both are perpendicular to the prismatic pair. This is different with general 3-D translational 3-UPU parallel mechanism (Tsai & Stamper, 1996). The base coordinate system is *O-XYZ*. The length of each side of the cubic mechanism is *m*.
For this special 3-UPU mechanism, each branch of the mechanism has equivalent five single-DOF kinematic pairs. According to the imaginary-mechanism method mentioned in Section 3, an imaginary link and an imaginary revolute pair denoted by a screw with zero pitch, $\$_{0i}$, are added to each branch, as shown in Fig. 7c. Then, each branch has six single-DOF kinematic pairs. Note that it is necessary to let the angular velocity amplitude of $\$_0$ for each branch always be zero.

For each six-DOF serial branch, the motion of the end-effector of the 3-UPU mechanism can be represented as

$$\mathbf{V}_H = \left[G_i^0\right]\dot{\phi}^i \quad (i = 1, 2, 3) \tag{48}$$

Based on the Eq. (48) and Section 3, the matrix equation as well as $[G']$ and $[G]$ can be obtained

$$\mathbf{V}_H = \left[G_L^H\right]\dot{\mathbf{q}} \tag{49}$$

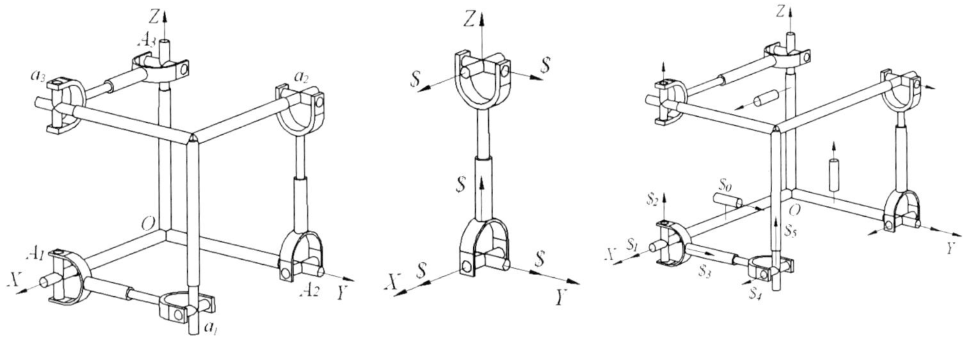

a. 3-UPU mechanism b. A limb c. An imaginary linkage

Fig. 7. Initial Position Mechanism Sketch

$$\mathbf{V}_p = [G]\dot{\mathbf{q}} \quad \boldsymbol{\omega} = [G']\dot{\mathbf{q}} \tag{50}$$

where $[G']$ is the first three rows of $\left[G_L^H\right]$; $[G]$ is the last three rows of $\left[G_L^H\right]$. They both are 3×3 matrices.

5.2. Initial configuration

Fig. 7a shows the initial configuration of the mechanism, m = 1.0 m, l = 0.3 m, and $d_1 = d_2 = d_3$. For each branch of the mechanism $\dot{\phi}_0^i$, $\left(\dot{\phi}_0^i = 0\right)$, and $\dot{q}_i$, $(i = 1, 2, 3)$, are denoted as inputs.

Assume the three lengths from the origin O to the centers of three imaginary pairs all to be $l = m - d_i$, which lie on the X-axis, Y-axis and Z-axis, respectively. d_i is the distance between the first two kinematic pairs including the imaginary pair.

The first-order influence coefficient matrices of the three branches are

$$\left[G_i^0\right] = (\$_0 \quad \$_1 \quad \$_3 \quad \$_4 \quad \$_5) \ , (i = 1, 2, 3)$$

According to Eq. (32), we obtain the two matrices

$$[G'] = \begin{bmatrix} 0 & 0 & 0 \\ 0 & 0 & 0 \\ 0 & 0 & 0 \end{bmatrix} \qquad [G] = \begin{bmatrix} 0 & 0 & 1 \\ 1 & 0 & 0 \\ 0 & 1 & 0 \end{bmatrix} \tag{51}$$

From Eq.s (51) and (12), we get the coefficients of the Eq. (12) as

$$c_1 = c_2 = c_3 = c_4 = 0 \tag{52}$$

The result is very special and implies that the roots of Eq. (12) can be any values. For this special situation to determine the three values we should consider other conditions. From section 2.2 of the References (Huang et al., 2004; and Huang & Fang, 1996) the three roots should all be infinite. That means the three roots, h_α, h_β and h_γ, all are ∞. The three principal screws belong to the sixth special third-order system presented by Hunt (1978). The three mutually perpendicular screws correspond with three independent translational motions. Obviously, along any direction in space there also exists an instant translational motion by linear combination of the three screws.

However, by further analysis we find that only three feasible translational motions can continue along the three coordinate axes, respectively. The feasible translational motions along all other directions in 3-D space are only instantaneous. It is easy to recognize, that when a small finite translation occurs not along the coordinate axis from the initial mechanism configuration, all three UPU chains are not the same as the configuration shown in Fig. 7b, and the three constraint screws will change and not similar that in the first configuration. Not all constrained motions are rotational. Therefore, the finite translation can occur only independently along each one of the three coordinate axes. In other words, three twists with ∞ pitch cannot be linearly combined at this initial position and the mechanism is not the same as the general 3D translational parallel mechanism proposed by Tsai & Stamper, (1996). The mechanism has such a very unusual characteristic.

5.3. The second configuration

The parameters of the mechanism are assumed as: $m = 1.0$ m, $l = 0.3$ m; and $a = 0.2$ m is the displacement of the moving pyramid along the X-axis, Fig. 8. In this case we have

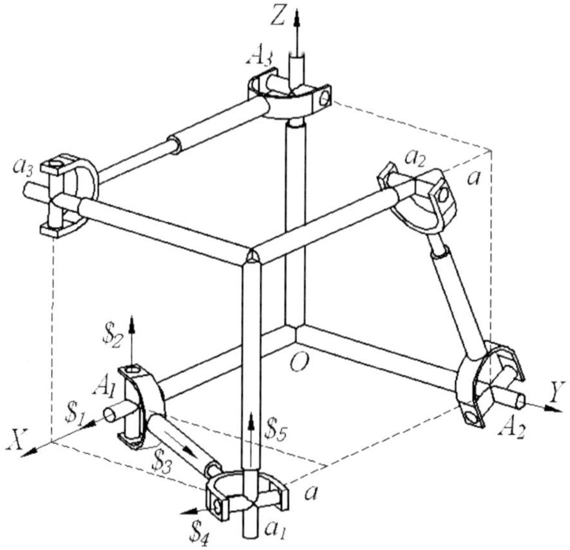

Fig.8. UPU branch after moving along the X-axis

$$[G'] = \begin{bmatrix} 0 & 0 & 0 \\ 0.00567188 & 0.170156 & -0.0344828 \\ 0.170156 & 0.0567188 & 1.03448 \end{bmatrix}$$

$$[G] = \begin{bmatrix} -0.0567188 & -0.170156 & 1.03448 \\ 0.850782 & 0.0283594 & -0.172414 \\ 0.0351657 & 1.05497 & -0.213793 \end{bmatrix} \tag{53}$$

Substituting $[G]$ and $[G']$ into the Eq. (8) and according to the Eq. (10), we have

$$D = \begin{vmatrix} a_{11} & a_{12} & a_{13} \\ a_{21} & a_{22} & a_{23} \\ a_{31} & a_{32} & a_{33} \end{vmatrix} = 0 \tag{54}$$

Expanding and solving the equation, we have

$$\begin{aligned} h_\alpha &= 5.6 \\ h_\beta &= -5.6 \\ h_\gamma &= -2.16318 \times 10^{17} \end{aligned} \tag{55}$$

where one is infinite, the other two are finite values with opposite signs. Therefore any screw in the screw system is the linear combination of the three principal screws and its pitch is inside the scope, $-5.6 \le h \le 5.6$. Three principal kinematic screws are

$$\begin{aligned} \$_\alpha &= (0 \quad 1.0 \quad 1.0; 88.4053 \quad 5.0 \quad 6.2)/\sqrt{2} \\ \$_\beta &= (0 \quad -1.0 \quad 1.0; 71.7085 \quad 5.0 \quad -6.2)/\sqrt{2} \\ \$_\gamma &= (0 \quad 0 \quad 0; 1 \quad 0 \quad 0) \end{aligned} \tag{56}$$

and the vector equations of three axes are

$$\begin{aligned} \mathbf{r} \times \mathbf{S}_\alpha &= (88.4053 \quad 5 \quad 6.2)^T / \sqrt{2} \\ \mathbf{r} \times \mathbf{S}_\beta &= (71.7085 \quad 5 \quad 6.2)^T / \sqrt{2} \\ \mathbf{r} \times \mathbf{S}_\gamma &= (0 \quad 0 \quad 0)^T \end{aligned} \tag{57}$$

where $\mathbf{S}_\alpha$, $\mathbf{S}_\beta$ and $\mathbf{S}_\gamma$ are three direction vectors of the three principal screws. Comparing with Eq. (15) in Reference Huang et al., (2004), the three screws in that Eq.(15) are just the linear combination of the three principle screws in Eq. (56). That means the result is correct and proved mutually. This system belongs to a third special three-system screw.

When different h value is substituted into Eq. (10), we may obtain different quadratic equation. Giving one set of input $(u \quad w \quad 1)$, the corresponding pitch of output motion is shown in Fig.9. Figure 9 illustrates the full-scale feasible instantaneous motion at that moment.

We know that each pitch of the screw determines a quadratic equation, Eq. (10). Here all quadratic equations degenerate into a pair of intersecting straight lines, when h lies within the range $-5.6 < h < 5.6$; It is because that two invariants of all the quadratic equations, Eq. (10), satisfy D = 0 and δ < 0. Similarly, when h = 5.6 and h = -5.6 both quadratic equations collapse into two pairs of superposed straight lines, they are respectively

$$u = w$$

$$0.173127u + 0.173127w = 0.0679061 \tag{58}$$

The quadratic equation collapses into a point which is just the intersecting point of all the straight lines, as shown in Fig. 9, when $h = h_\gamma = -2.16381 \times 10^{17} = -\infty$.

Fig. 9 illustrates the finite-and-infinite pitch graph of the third special three-system screw including the finite pitches in the scope from -5.6 to 5.6 and an infinite pitch.

Each point in the figure indicates the relation between the input $\begin{pmatrix} u & w & 1 \end{pmatrix}$ and the output pitch, h. It is necessary to point out that, for a six-DOF mechanism, infinite pitches of its infinite feasible instant motions distribute in an infinite scope $\begin{pmatrix} -\infty & \infty \end{pmatrix}$, but for this 3-UPU mechanism its infinite possibility is only in a limited scope (-5.6, 5.6) plus a point with infinite pitch value.

From Fig. 9, we can find that all the straight lines pass through a common point, which is a very special point. The pitch values of all the straight lines are finite, but at the special point, the pitch suddenly becomes infinite

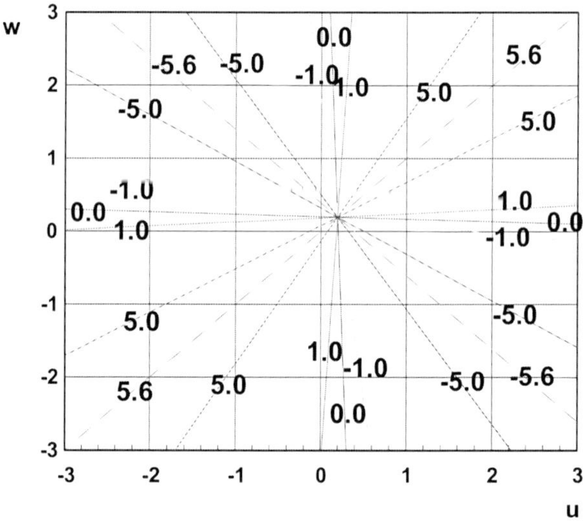

Fig. 9. The Pitch of the Twist at the Second Configuration of 3-TPT Mechanism.

6. Future research

Based on this principle many three-degrees of freedom parallel mechanisms need to be further analyzed.

7. Conclusions

This chapter presents a study on the full-scale instant twists motions of 3-DOF parallel manipulators. The study is of extremely benefit to understand and correctly apply a mechanism. It is based on principal screws of the screw system. The key problem is to derive three principal screws from a given 3-DOF mechanism. It needs to set the relation between the pitches of the principal screws and the three linear inputs of the mechanism.

In this chapter, the effective method to identify the principal screws of a third-order screw system of 3-DOF mechanisms is presented. For obtaining the principal screws there introduce two methods, the quadratic curve degenerating theory and quadric degenerating theory. Besides, the imaginary-mechanism influence coefficient principle is also used.

In the following sections two mechanisms are discussed using the principle. Analyzing the full-scale screws the planar representations of pitches and the spatial distributions of the axes are illustrated.

It is necessary to conclude that the special 3-UPU mechanism has some exceptional interesting characteristics. At the initial configuration, the moving pyramid can continually translate along the *X*- or *Y*- or *Z*-axis, however, for all other directions the translational freedom is only instantaneous. At a general configuration, all the straight lines with different pitch pass through a common point, a very special point. The pitch values of all the straight lines are finite, at the intersecting point, however, the pitch is infinite.

8. Acknowledgement

The research work reported here is supported by NSFC under Grant No. 59575043 and 50275129.

9. References

Ball, R.S. (1900). *The Theory of Screws*. England: Cambridge University Press.

Clavel, R. (1988). DELTA, A fast robot with parallel geometry, *Proc. of the Int. Symp. on Industrial Rob. Switzerland*, pp. 91-100.

Di Gregorio, R. et al.(1999). Influence of leg flexibility on the kinetostatic behaviour of a 3-DOF fully-parallel manipulator, *Proceedings of 10th World Congress on the Theory of Machine and Mechanisms*, June .20-24, Oulu, Finland. 3, pp. 1091-1098.

Fang, Y. F. & Huang, Z. (1998). Analytical Identification of the Principal Screw of the Third Order Screw System. *Mech.& Mach. Theory*, 33(7), 987-992.

Gibson, C. G. & Hunt, K. H. (1990 a). Geometry of Screw Systems-1 Screws Genesis and Geometry. *Mech.& Mach. Theory*. 25(1) 1-10.

Gibson, C. G. & Hunt, K. H. (1990 b). Geometry of Screw Systems-2 Classification of Screw Systems. *Mech.& Mach. Theory*. 25(1), 11-27.

Gosselin, C. M. & Angeles, J. (1988). The optimum kinematic design of a planar three-DOH parallel manipulator, *Transactions of the ASME Journal Mech Trans. Autom. Des.*, 110 (1), 35-41.

Gosselin, C. M. & Angeles, J.(1989). The optimum kinematic design of a spherical three-degree-of-freedom parallel manipulator, *Transactions of the ASME Journal Mech. Trans. Autom. Des.*, 111 (2), 202-207.

Huang, Z. (1985). Modeling Formulation of 6-DOF multi-loop Parallel Manipulators, Part-1: Kinematic Influence Coefficients, *Proc. of the 4th IFToMM International Symposium on Linkage and Computer Aided Design Methods*, Bucharest, Romania, Vol. II-1, 155-162

Huang, Z. & Fang, Y.F. (1996). Kinematic Characteristics Analysis of 3-DOF In-Parallel Actuated Pyramid Mechanisms. *Mech. & Mach. Theory*, 31(8), 1009-1018.

Huang, Z.; Li, S.H. & Zuo, R.G. (2004). Feasible instantaneous motions and kinematic characteristics of a special 3-DOF 3-UPU parallel manipulator. *Mechanism and Machine Theory*, 2004, 39(9), 957-970

Huang, Z.; Tao, W. S. & Fang, Y. F. (1996). Study on the Kinematic Characteristics of 3-DOF Parallel Actuated Platform Mechanisms. *Mech. & Mach. Theory*, 31(8), 999-1007.

Huang, Z; Zhao, Y.S. & Zhao, T.S., *Advanced Spatial Mechanism*, Beijing, Higher Education Press, 2006 (in Chinese)

Huang, Z. & Wang, H. B. (1992). Dynamic Force Analysis of n-DOF Multi-Loop Complex Spatial Mechanism. *Mechanism and Machine Theory*, 27(1), 97-105.

Huang, Z. & Wang, J. (2001). Identification of principal screws of 3-DOF parallel manipulators by quadric degeneration. *Mechanism and Machine Theory*, Vol 36(8), 893-911

Huang, Z. & Wang, J. (2002). Huang Z, Wang J, Analysis of Instantaneous Motions of Deficient-Rank 3-RPS Parallel Manipulators. *Mechanism and Machine Theory*, 37(2):229-240

Hunt, K. H. (1978). Kinematic Geometry of Mechanisms. Oxford University Press,

Hunt, K. H. (1983). Structural Kinematics of In-Parallel-Actuated Robot Arms. *Trans. ASME J. Mech. Trans. Auto. Des.*, 105(4), 705-712.

Lee, K.M. & Shah, D.K. (1988). Kinematic analysis of a three-degree-of-freedom parallel actuated manipulator. *IEEE Trans. Robotics Autom.*, 4 (3), 354-360.

Parkin, J. A. (1990). Co-ordinate Transformations of Screws with Application to Screw Systems and Finite Twists. *Mech. & Mach. Theory*, 25(6), 689-699.

Song, S.M. & Zhang, M.D. (1995). A Study of Reactional Force Compensation Based on Three-Degree-of Freedom Parallel Platforms. *J. Robotic System*, 12 (12), 783-794,

Thomas, M. & Tesar, D.(1983). Dynamic modeling of serial manipulator arms. *J. Dyn. Sys. Meas. Cont.*, 104(9), 218-227.

Tsai, M. J. & Lee, H. W. (1993). On the Special Bases of Two-and-Three Screw Systems, *Trans. of the ASME J Mech Design*, 115 540-546.

Tsai, L. W. Stamper, R. (1996). A parallel manipulator with only translational degrees of freedom. ASME 96-DETC-MECH-112. Irvine(CA), USA,.

Waldron, K. J.; Raghavan, M. & Roth, B. (1989). Kinematics of a hybird series–parallel manipulation system. *Transactions of the ASME Journal Mech Trans. Autom.* Des., 111, 211-221.

Yan, J. & Huang, Z. (1985). Kinematic Analysis of Multi-Loop Spatial Mechanism, *Proc. Of the 4th IFToMM International Symposium on Linkage and Computer Aided Design Method*, Bucharest, Vol.2-2, 439-446

Zhang, W. X. & Xu, Z. C. (1998). Algebraic Construction of the Three-System of Screws. *Mech. Mach. Theory*, 33 (7), 925-930.

Zhao, T.S.; Zhao, Y.S. & Huang, Z. (1999). Physical and Mathematical Conditions of Existence of Axes about Which Platform of Deficient-Rank Parallel Robots Can Rotate continuously. *Robot*, 21(5), 347-351 (in Chinese)

19

Singularity Robust Inverse Dynamics of Parallel Manipulators

S. Kemal Ider
Middle East Technical University Ankara, Turkey

1. Introduction

Parallel manipulators have received wide attention in recent years. Their parallel structures offer better load carrying capacity and more precise positioning capability of the end-effector compared to open chain manipulators. In addition, since the actuators can be placed closer to the base or on the base itself the structure can be built lightweight leading to faster systems (Gunawardana & Ghorbel, 1997; Merlet, 1999; Gao et al., 2002).
It is known that at *kinematic singular positions* of serial manipulators and parallel manipulators, arbitrarily assigned end-effector motion cannot in general be reached by the manipulator and consequently at those configurations the manipulator loses one or more degrees of freedom. In addition, the closed loop structure of parallel manipulators gives rise to another type of degeneracy, which can be called *drive singularity,* where the actuators cannot influence the end-effector accelerations instantaneously in certain directions and the actuators lose the control of one or more degrees of freedom. The necessary actuator forces become unboundedly large unless consistency of the dynamic equations are guaranteed by the specified trajectory.
The previous studies related to the drive singularities mostly aim at finding only the locations of the singular positions for the purpose of avoiding them in the motion planning stage (Sefrioui & Gosselin, 1995; Daniali et al, 1995; Alici, 2000; Ji, 2003; DiGregorio, 2001; St-Onge & Gosselin, 2000). However unlike the kinematic singularities that occur at workspace boundaries, drive singularities occur inside the workspace and avoiding them limits the motion in the workspace. Therefore, methods by which the manipulator can move through the drive singular positions in a stable fashion are necessary.
This chapter deals with developing a methodology for the inverse dynamics of parallel manipulators in the presence of drive singularities. To this end, the conditions that should be satisfied for the consistency of the dynamic equations at the singular positions are derived. For the trajectory of the end-effector to be realizable by the actuators it should be designed to satisfy the *consistency conditions*. Furthermore, for finding the appropriate actuator forces when drive singularities take place, the dynamic equations are modified by using higher order derivative information. The linearly dependent equations are replaced by the *modified equations* in the neighborhoods of the singularities. Since the locations of the drive singularities and the corresponding modified equations are known (as derived in Section 3), in a practical scenario the actuator forces are found using the modified equations

in the vicinity of the singular positions and using the regular inverse dynamic equations elsewhere. Deployment motions of 2 and 3 dof planar manipulators are analyzed to illustrate the proposed approach (Ider, 2004; Ider, 2005).

2. Inverse dynamics and singular positions

Consider an n degree of freedom parallel robot. Let the system be converted into an open-tree structure by disconnecting a sufficient number of unactuated joints. Let the degree of freedom of the open-tree system be m, i.e. the number of the independent loop closure constraints in the parallel manipulator be m-n. Let $\boldsymbol{\eta} = [\eta_1, ..., \eta_m]^T$ denote the joint variables of the open-tree system and $\mathbf{q} = [q_1, ..., q_n]^T$ the joint variables of the actuated joints. The m-n loop closure equations, obtained by reconnecting the disconnected joints, can be written as

$$\phi_i(\eta_1, ..., \eta_m) = 0 \qquad i=1,...,m\text{-}n \tag{1}$$

and can be expressed at velocity level as

$$\Gamma_{ij}^G \dot{\eta}_j = 0 \qquad i=1,...,m\text{-}n \qquad j=1,...,m \tag{2}$$

where $\Gamma_{ij}^G = \dfrac{\partial \phi_i}{\partial \eta_j}$. A repeated subscript index in a term implies summation over its range. The prescribed end-effector Cartesian variables $x_i(t)$, $i=1,...,n$ represent the tasks of the non-redundant manipulator. The relations between the joint variables due to the tasks are

$$f_i(\eta_1, ..., \eta_m) = x_i \qquad i=1,...,n \tag{3}$$

Equation (3) can be written at velocity level as

$$\Gamma_{ij}^P \dot{\eta}_j = \dot{x}_i \qquad i=1,...,n \qquad j=1,...,m \tag{4}$$

where $\Gamma_{ij}^P = \dfrac{\partial f_i}{\partial \eta_j}$. Equations (2) and (4) can be written in combined form,

$$\boldsymbol{\Gamma} \dot{\boldsymbol{\eta}} = \mathbf{h} \tag{5}$$

where $\boldsymbol{\Gamma}^{\mathrm{T}} = \left[\boldsymbol{\Gamma}^{G^{\mathrm{T}}} \quad \boldsymbol{\Gamma}^{P^{\mathrm{T}}} \right]$ which is an $m \times m$ matrix and $\mathbf{h}^{\mathrm{T}} = \left[\mathbf{0} \quad \dot{\mathbf{x}}^{\mathrm{T}} \right]$. The derivative of equation (5) gives the acceleration level relations,

$$\boldsymbol{\Gamma} \ddot{\boldsymbol{\eta}} = -\dot{\boldsymbol{\Gamma}} \dot{\boldsymbol{\eta}} + \dot{\mathbf{h}} \tag{6}$$

The dynamic equations of the parallel manipulator can be written as

$$\mathbf{M} \ddot{\boldsymbol{\eta}} - \boldsymbol{\Gamma}^{G^{\mathrm{T}}} \boldsymbol{\lambda} - \mathbf{Z}^{\mathrm{T}} \mathbf{T} = \mathbf{R} \tag{7}$$

where $\mathbf{M}$ is the $m \times m$ generalized mass matrix and $\mathbf{R}$ is the vector of the generalized Coriolis, centrifugal and gravity forces of the open-tree system, $\boldsymbol{\lambda}$ is the $(m-n)\times 1$ vector of the joint forces at the loop closure joints, $\mathbf{T}$ is the $n \times 1$ vector of the actuator forces, and each row of $\mathbf{Z}$ is the direction of one actuator force in the generalized space. If the variable of the joint which is actuated by the i th actuator is η_k, then for the i th row of $\mathbf{Z}$, $Z_{ik}=1$ and $Z_{ij}=0$ for j=1,...,m ($j \neq k$).

Combining the terms involving the unknown forces $\boldsymbol{\lambda}$ and $\mathbf{T}$, one can write equation (7) as

$$\mathbf{A}^{\mathrm{T}}\boldsymbol{\tau} = \mathbf{M}\ddot{\boldsymbol{\eta}} - \mathbf{R} \tag{8}$$

where the m × m matrix $\mathbf{A}^{\mathrm{T}}$ and the $m \times 1$ vector $\boldsymbol{\tau}$ are

$$\mathbf{A}^{\mathrm{T}} = \begin{bmatrix} \boldsymbol{\Gamma}^{G^{\mathrm{T}}} & \mathbf{Z}^{\mathrm{T}} \end{bmatrix} \tag{9}$$

and

$$\boldsymbol{\tau}^{\mathrm{T}} = \begin{bmatrix} \boldsymbol{\lambda}^{\mathrm{T}} & \mathbf{T}^{\mathrm{T}} \end{bmatrix} \tag{10}$$

The inverse dynamic solution of the system involves first finding $\ddot{\boldsymbol{\eta}}$, $\dot{\boldsymbol{\eta}}$ and $\boldsymbol{\eta}$ from the kinematic equations and then finding $\boldsymbol{\tau}$ (and hence $\mathbf{T}$) from equation (8).

For the prescribed $\mathbf{x}$(t), $\ddot{\boldsymbol{\eta}}$ can be found from equation (6), $\dot{\boldsymbol{\eta}}$ from equation (5) and $\boldsymbol{\eta}$ can be found either from the position equations (1,3) or by numerical integration. However during the inverse kinematic solution, singularities occur when $|\boldsymbol{\Gamma}|=0$. At these configurations, the assigned $\dot{\mathbf{x}}$ cannot in general be reached by the manipulator since, in equation (3), a vector $\mathbf{h}$ lying outside the space spanned by the columns of $\boldsymbol{\Gamma}$ cannot be produced and consequently the manipulator loses one or more degrees of freedom.

Singularities may also occur while solving for the actuator forces in the dynamic equation (8), when $|\mathbf{A}|=0$. For each different set of actuators, $\mathbf{Z}$ hence the singular positions are different. Because this type of singularity is associated with the locations of the actuators, it is called *drive singularity* (or *actuation singularity*). At a drive singularity the assigned $\ddot{\boldsymbol{\eta}}$ cannot in general be realized by the actuators since, in equation (8), a right hand side vector lying outside the space spanned by the columns of $\mathbf{A}^{\mathrm{T}}$ cannot be produced, i.e. the actuators cannot influence the end-effector accelerations instantaneously in certain directions and the actuators lose the control of one or more degrees of freedom. (The system cannot resist forces or moments in certain directions even if all actuators are locked.) The actuator forces become unboundedly large unless consistency of the dynamic equations are guaranteed by the specified trajectory.

Let $\boldsymbol{\Gamma}^{Gu}$ be the $(m-n)\times(m-n)$ matrix which is composed of the columns of $\boldsymbol{\Gamma}^{G}$ that correspond to the variables of the unactuated joints. Since $Z_{ik}=1$ and $Z_{ij}=0$ for $j \neq k$, the drive singularity condition $|\mathbf{A}|=0$ can be equivalently written as $|\boldsymbol{\Gamma}^{\mathrm{Gu}}|=0$.

In the literature the singular positions of parallel manipulators are mostly determined using the kinematic expression between $\dot{\mathbf{q}}$ and $\dot{\mathbf{x}}$ which is obtained by eliminating the variables of the unactuated joints (Sefrioui & Gosselin, 1995; Daniali et al, 1995; Alici, 2000; Ji, 2003; DiGregorio, 2001; St-Onge & Gosselin, 2000),

$$\mathbf{J}\dot{\mathbf{q}} + \mathbf{K}\dot{\mathbf{x}} = \mathbf{0} \tag{11}$$

References (Sefrioui & Gosselin, 1995 ; Daniali et al, 1995; Ji, 2003) name the condition $|\mathbf{J}| = 0$ as "Type I singularity" and the condition $|\mathbf{K}| = 0$ "Type II singularity". And in reference (DiGregorio, 2001) they are called "inverse problem singularity" and "direct problem singularity", respectively. Since it shows the lost Cartesian degrees of freedom, the condition $|\mathbf{\Gamma}| = 0$ shown above corresponds to $|\mathbf{J}| = 0$. For the drive singularity, equation (2) can be written as

$$\mathbf{\Gamma}^{Gu}\,\dot{\boldsymbol{\eta}}^{u} = -\mathbf{\Gamma}^{Ga}\,\dot{\mathbf{q}} \tag{12}$$

where $\boldsymbol{\eta}^{u}$ is the vector of the joint variables of the unactuated joints and $\mathbf{\Gamma}^{Ga}$ is the matrix composed of the columns of $\mathbf{\Gamma}^{G}$ associated with the actuated joints. Since after finding $\dot{\boldsymbol{\eta}}^{u}$ from eqn (12) one can find $\mathbf{h}$ and hence $\dot{\mathbf{x}}$ from eqn (5) directly, the drive singularity condition $|\mathbf{A}| = 0$ (i.e. $|\mathbf{\Gamma}^{Gu}| = 0$) given above is equivalent to $|\mathbf{K}| = 0$. It should be noted that the identification of the singular configurations as shown here is easier since elimination of the variables of the passive joints is not necessary.

3. Consistency conditions and modified equations

At the motion planning stage one usually tries to avoid singular positions. This is not difficult as far as inverse kinematic singularities are concerned because they usually occur at the workspace boundaries (DiGregorio, 2001). In this paper it is assumed that $\mathbf{\Gamma}$ always has full rank, i.e. the desired motion is chosen such that the system never comes to an inverse kinematic singular position. On the other hand, drive singularities usually occur inside the workspace and avoiding them restricts the functional workspace. It is therefore important to devise techniques for passing through the singular positions while the stability of the control forces is maintained. To this end, equation (8) must be made consistent at the singular position. In other words, since the rows of $\mathbf{A}^{\mathrm{T}}$ become linearly dependent, the same relation must also be present between the rows of the right hand side vector ($\mathbf{M}\ddot{\boldsymbol{\eta}} - \mathbf{R}$), so that it lies in the vector space spanned by the columns of $\mathbf{A}^{\mathrm{T}}$.

3.1 Consistency conditions and modified equations when rank(A) becomes m-1

At a drive singularity, usually rank of $\mathbf{A}$ becomes m-1. Let at the singular position the s th row of $\mathbf{A}^{\mathrm{T}}$ become a linear combination of the other rows of $\mathbf{A}^{\mathrm{T}}$.

$$A_{sj}^{\mathrm{T}} = \alpha_p A_{pj}^{\mathrm{T}} \qquad p = 1,...,m \quad (p \neq s)\,, \;\; j = 1,...,m \tag{13}$$

where α_p are the linear combination coefficients (which may depend also on η_i). Notice that only those rows of $\mathbf{A}^{\mathrm{T}}$ which are associated with the unactuated joints can become linearly dependent, hence α_p corresponding to the actuated joints are zero. Then for the rows of equation (6) one must have

$$A_{sj}^{\mathrm{T}}\,\tau_j - \alpha_p\,A_{pj}^{\mathrm{T}}\,\tau_j = M_{sj}\,\ddot{\eta}_j - R_s - \alpha_p(M_{pj}\,\ddot{\eta}_j - R_p) \tag{14}$$

Substitution of equation (13) into equation (14) yields

$$M_{sj}\,\ddot{\eta}_j - R_s = \alpha_p(M_{pj}\,\ddot{\eta}_j - R_p) \tag{15}$$

Equation (15) represents the *consistency condition* that $\ddot{\eta}_j$ should satisfy at the singular position. Since $\ddot{\eta}_j$ are obtained from the inverse kinematic equations (6), the trajectory $\ddot{\mathbf{x}}$ must be planned in such a way to satisfy equation (15) at the drive singularity. Otherwise an inconsistent trajectory cannot be realized and the actuator forces grow without bounds as the drive singularity is approached. Time derivative of equation (14) is

$$(A_{sj}^{\mathrm{T}} - \alpha_p\,A_{pj}^{\mathrm{T}})\,\dot{\tau}_j + (\dot{A}_{sj}^{\mathrm{T}} - \alpha_p\,\dot{A}_{pj}^{\mathrm{T}} - \dot{\alpha}_p\,A_{pj}^{\mathrm{T}})\,\tau_j = (M_{sj} - \alpha_p M_{pj})\,\dddot{\eta}_j$$

$$+(\dot{M}_{sj} - \alpha_p\dot{M}_{pj} - \dot{\alpha}_p M_{pj})\,\ddot{\eta}_j - \dot{R}_s + \alpha_p\dot{R}_p + \dot{\alpha}_p R_p \tag{16}$$

Now, because equation (13) holds at the singular position, there exists a neighborhood in which the first term in equation (16) is negligible compared to the other terms. Therefore in that neighborhood this term can be dropped to yield

$$(\dot{A}_{sj}^{\mathrm{T}} - \alpha_p\,\dot{A}_{pj}^{\mathrm{T}} - \dot{\alpha}_p\,A_{pj}^{\mathrm{T}})\,\tau_j = (M_{sj} - \alpha_p M_{pj})\,\dddot{\eta}_j + (\dot{M}_{sj} - \alpha_p\dot{M}_{pj} - \dot{\alpha}_p M_{pj})\,\ddot{\eta}_j - \dot{R}_s + \alpha_p\dot{R}_p + \dot{\alpha}_p R_p \tag{17}$$

Equation (17) is the *modified equation* that can be used to replace the s th row of equation (8) or any other equation in the linearly dependent set.

3.2 Consistency conditions and modified equations when rank(A) becomes r<m

In the general case where the rank of $\mathbf{A}^{\mathrm{T}}$ becomes $r < m$ at the singular position, let rows s_k, $k = 1,...,m\text{-}r$ of $\mathbf{A}^{\mathrm{T}}$ become linear combinations of the other r rows of $\mathbf{A}^{\mathrm{T}}$,

$$A_{s_k j}^{\mathrm{T}} = \alpha_{kp}\,A_{pj}^{\mathrm{T}} \qquad p = 1,...,m \quad (p \neq s_k),\ \ j = 1,...,m\,,\ \ k = 1,...,m\text{-}r \tag{18}$$

where α_{kp} are the linear combination coefficients. Then the following relations must be present among the rows of equation (8)

$$A_{s_k j}^{\mathrm{T}}\,\tau_j - \alpha_{kp}\,A_{pj}^{\mathrm{T}}\,\tau_j = M_{s_k j}\,\ddot{\eta}_j - R_{s_k} - \alpha_{kp}(M_{pj}\,\ddot{\eta}_j - R_p) \qquad k = 1,...,m\text{-}r \tag{19}$$

The consistency relations are obtained as below

$$M_{s_k j}\,\ddot{\eta}_j - R_{s_k} = \alpha_{kp}(M_{pj}\,\ddot{\eta}_j - R_p) \qquad k = 1,...,m\text{-}r \tag{20}$$

Substitution of equation (18) into the derivative of equation (19) yields the modified equations,

$$(\dot{A}^{\mathrm{T}}_{s_k j} - \alpha_{kp}\dot{A}^{\mathrm{T}}_{pj} - \dot{\alpha}_{kp}A^{\mathrm{T}}_{pj})\,\tau_j = (M_{s_k j} - \alpha_{kp}M_{pj})\,\dddot{\eta}_j + (\dot{M}_{s_k j} - \alpha_{kp}\dot{M}_{pj} - \dot{\alpha}_{kp}M_{pj})\,\ddot{\eta}_j$$

$$-\dot{R}_{s_k} + \alpha_{kp}\dot{R}_p + \dot{\alpha}_{kp}R_p \qquad k = 1,\ldots,m\text{-}r \tag{21}$$

3.3 Inverse dynamics algorithm in the presence of drive singularities

When the linearly dependent dynamic equations in equation (8) are replaced by the modified equations, equation (8) takes the following form, which is valid in the vicinity of the singular configurations.

$$\mathbf{D}^{\mathrm{T}}\boldsymbol{\tau} = \mathbf{S} \tag{22}$$

where in the case the s th row of $\mathbf{A}^{\mathrm{T}}$ becomes a linear combination of the other rows,

$$D^{\mathrm{T}}_{ij} = \begin{cases} A^{\mathrm{T}}_{ij} & i \neq s \\ \dot{A}^{\mathrm{T}}_{ij} - \alpha_p\dot{A}^{\mathrm{T}}_{pj} - \alpha_p\dot{A}^{\mathrm{T}}_{pj} & i = s \end{cases} \tag{23}$$

and

$$S_i = \begin{cases} M_{ij}\ddot{\eta}_j - R_i & i \neq s \\ (M_{ij} - \alpha_p M_{pj})\,\dddot{\eta}_j + (\dot{M}_{ij} - \alpha_p\dot{M}_{pj} - \dot{\alpha}_p M_{pj})\,\ddot{\eta}_j - \dot{R}_i + \alpha_p\dot{R}_p + \dot{\alpha}_p R_p & i = s \end{cases} \tag{24}$$

In the general case when the rank of $\mathbf{A}^{\mathrm{T}}$ becomes r, $\mathbf{D}^{\mathrm{T}}$ and $\mathbf{S}$ take the following form.

$$D^{\mathrm{T}}_{ij} = \begin{cases} A^{\mathrm{T}}_{ij} & i \neq s_k,\ k = 1,\ldots,m-r \\ \dot{A}^{\mathrm{T}}_{ij} - \alpha_{kp}\dot{A}^{\mathrm{T}}_{pj} - \dot{\alpha}_{kp}A^{\mathrm{T}}_{pj} & i = s_k,\ k = 1,\ldots,m-r \end{cases} \tag{25}$$

and

$$S_i = \begin{cases} M_{ij}\,\ddot{\eta}_j - R_i & i \neq s_k,\ k = 1,\ldots,m-r \\ (M_{ij} - \alpha_{kp}M_{pj})\,\dddot{\eta}_j + (\dot{M}_{ij} - \alpha_{kp}\dot{M}_{pj} - \dot{\alpha}_{kp}M_{pj})\,\ddot{\eta}_j - \dot{R}_i + \alpha_{kp}\dot{R}_p + \dot{\alpha}_{kp}R_p & i = s_k,\ k = 1,\ldots,m-r \end{cases} \tag{26}$$

Notice that $\dddot{\boldsymbol{\eta}}$ in the modified equation should be found from the derivative of equation (6),

$$\boldsymbol{\Gamma}\,\dddot{\boldsymbol{\eta}} = -2\dot{\boldsymbol{\Gamma}}\,\ddot{\boldsymbol{\eta}} - \ddot{\boldsymbol{\Gamma}}\,\dot{\boldsymbol{\eta}} + \ddot{\mathbf{h}} \tag{27}$$

$\dddot{\boldsymbol{\eta}}$ obtained from equation (27) corresponds to the prescribed end-effector jerks $\dddot{\mathbf{x}}$ (in $\ddot{\mathbf{h}}$). Also the coefficients of the forces in the modified equations (17,21) depend on velocities. Therefore, if at the singularity the system is in motion, then by the modified equations the driving forces affect the end-effector jerk instantaneously in the singular directions.

The *inverse dynamics algorithm* in the presence of drive singularities is given below.

1. Find the loci of the positions where the actuation singularities occur and find the linear dependency coefficients associated with the singular positions.
2. If the assigned path of the end-effector passes through singular positions, design the trajectory so as to satisfy the consistency conditions at the singular positions.

3. Set time $t = 0$.
4. Calculate $\boldsymbol{\eta}$, $\dot{\boldsymbol{\eta}}$ and $\ddot{\boldsymbol{\eta}}$ from kinematic equations.
5. If the manipulator is in the vicinity of a singular position, i.e. $|g(\eta_1,...,\eta_m)| < \varepsilon$ where $g(\eta_1,...,\eta_m) = 0$ is the singularity condition and ε is a specified small number, calculate $\ddot{\boldsymbol{\eta}}$ from eqn (27) and then find $\boldsymbol{\tau}$ (hence **T**) from equation (22).
6. If the manipulator is not in the vicinity of a singular position, i.e. $|g(\eta_1,...,\eta_m)| > \varepsilon$, find $\boldsymbol{\tau}$ (hence **T**) from equation (8).
7. Set $t = t + \Delta t$. If the final time is reached, stop. Otherwise continue from step 3.

4. Case studies

4.1 Two degree of freedom 2-RRR planar parallel manipulator

The planar parallel manipulator shown in Figure 1 has 2 degrees of freedom ($n = 2$). Considering disconnection of the revolute joint at *P*, the joint variable vector of the open-chain system is $\boldsymbol{\eta} = [\theta_1 \quad \theta_2 \quad \theta_3 \quad \theta_4]^{\mathrm{T}}$. The joints at *A* and *C* are actuated, i.e. $\mathbf{q} = [\theta_1 \quad \theta_2]^{\mathrm{T}}$. The end point P is desired to make a deployment motion *s*(*t*) along a straight line whose angle with x-axis is $\gamma = 330^o$, starting from initial position $x_{P_o} = -0.431$ m, $y_{P_o} = 1.385$ m. The time of the motion is $T = 1$ s and its length is $L = 2.3$ m in the positive s sense.

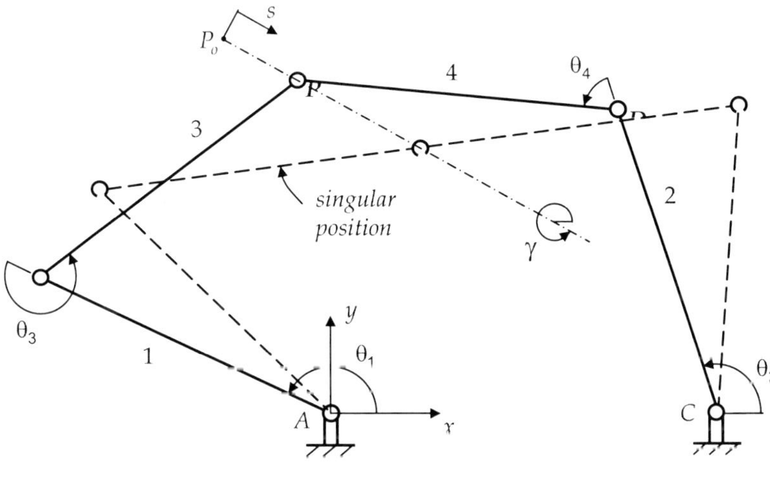

Figure 1. Two degree of freedom 2-RRR planar parallel manipulator.

The moving links are uniform bars. The fixed dimensions are labelled as $r_o = AC$, $r_1 = AB$, $r_2 = CD$, $r_3 = BP$ and $r_4 = DP$. The numerical data are $r_o = 1.75$m, $r_1 = r_2 = r_3 = r_4 = 1.4$m, $m_1 = m_2 = 6$ kg and $m_3 = m_4 = 4$kg.

The loop closure constraint equations at velocity level are $\boldsymbol{\Gamma}^G \dot{\boldsymbol{\eta}} = \mathbf{0}$ where

$$\mathbf{\Gamma}^G = \begin{bmatrix} -r_1 s_1 - r_3 s_{13} & r_2 s_2 + r_4 s_{24} & -r_3 s_{13} & r_4 s_{24} \\ r_1 c_1 + r_3 c_{13} & -r_2 c_2 - r_4 c_{24} & r_3 c_{13} & -r_4 c_{24} \end{bmatrix} \tag{28}$$

Here $s_i = \sin\theta_i$, $c_i = \cos\theta_i$, $s_{ij} = \sin(\theta_i + \theta_j)$, $c_{ij} = \cos(\theta_i + \theta_j)$. The prescribed Cartesian motion of the end point P, $\mathbf{x}$ can be written as

$$\mathbf{x} = \begin{bmatrix} x_P(t) \\ y_P(t) \end{bmatrix} = \begin{bmatrix} x_{P_o} + s(t)\sin\gamma \\ y_{P_o} + s(t)\cos\gamma \end{bmatrix} \tag{29}$$

Then the task equations at velocity level are $\mathbf{\Gamma}^P \dot{\boldsymbol{\eta}} = \dot{\mathbf{x}}$, where

$$\mathbf{\Gamma}^P = \begin{bmatrix} -r_1 s_1 - r_3 s_{13} & 0 & -r_3 s_{13} & 0 \\ r_1 c_1 + r_3 c_{13} & 0 & r_3 c_{13} & 0 \end{bmatrix} \tag{30}$$

The mass matrix $\mathbf{M}$ and the vector of the Coriolis, centrifugal and gravitational forces $\mathbf{R}$ are

$$\mathbf{M} = \begin{bmatrix} M_{11} & 0 & M_{13} & 0 \\ 0 & M_{22} & 0 & M_{24} \\ M_{13} & 0 & M_{33} & 0 \\ 0 & M_{24} & 0 & M_{44} \end{bmatrix} \tag{31}$$

where

$$M_{11} = m_1 \frac{r_1^2}{3} + m_3 (r_1^2 + \frac{r_3^2}{3} + r_1 r_3 c_3),\quad M_{13} = m_3 (\frac{r_3^2}{3} + \frac{r_1 r_3 c_3}{2}),\quad M_{33} = m_3 \frac{r_3^2}{3}$$

$$M_{22} = m_2 \frac{r_2^2}{3} + m_4 (r_2^2 + \frac{r_4^2}{3} + r_2 r_4 c_4),\quad M_{24} = m_4 (\frac{r_4^2}{3} + \frac{r_2 r_4 c_4}{2}),\quad M_{44} = m_4 \frac{r_4^2}{3} \tag{32}$$

and

$$\mathbf{R} = \begin{bmatrix} R_1 \\ R_2 \\ R_3 \\ R_4 \end{bmatrix} = \begin{bmatrix} -\ m_3 r_1 r_3 s_3 \dot{\theta}_3 (\dot{\theta}_1 - \frac{1}{2}\dot{\theta}_3) + \frac{1}{2} m_1 g\ r_1 c_1 + m_3 g\ (r_1 c_1 + \frac{1}{2} r_3 c_{13}) \\ \frac{1}{2} m_3 r_1 r_3 s_3 \dot{\theta}_1^2 + \frac{1}{2} m_3 g\ r_3 c_{13} \\ -\ m_4 r_2 r_4 s_4 \dot{\theta}_4 (\dot{\theta}_2 - \frac{1}{2}\dot{\theta}_4) + \frac{1}{2} m_2 g\ r_2 c_2 + m_4 g\ (r_2 c_2 + \frac{1}{2} r_4 c_{24}) \\ \frac{1}{2} m_4 r_2 r_4 s_4 \dot{\theta}_2^2 + \frac{1}{2} m_4 g\ r_4 c_{24} \end{bmatrix} \tag{33}$$

Since the variables of the actuated joints are θ_1 and θ_2, the matrix $\mathbf{Z}$ composed of the actuator direction vectors is

$$\mathbf{Z} = \begin{bmatrix} 1 & 0 & 0 & 0 \\ 0 & 1 & 0 & 0 \end{bmatrix} \tag{34}$$

Then the coefficient matrix of the constraint and actuator forces, $\mathbf{A}^{\mathrm{T}}$ is

$$\mathbf{A}^{\mathrm{T}} = \begin{bmatrix} -r_1 s_1 - r_3 s_{13} & r_1 c_1 + r_3 c_{13} & 1 & 0 \\ r_2 s_2 + r_4 s_{24} & -r_2 c_2 - r_4 c_{24} & 0 & 1 \\ -r_3 s_{13} & r_3 c_{13} & 0 & 0 \\ r_4 s_{24} & -r_4 c_{24} & 0 & 0 \end{bmatrix} \tag{35}$$

The drive singularities are found from $|\mathbf{A}| = 0$ as $\sin(\theta_1 + \theta_3 - \theta_2 - \theta_4) = 0$, i.e. as the positions when points A, B and D become collinear. Hence, drive singularities occur inside the workspace and avoiding them limits the motion in the workspace. Defining a path for the operational point P which does not involve a singular position would restrict the motion to a portion of the workspace where point D remains on one side of the line joining A and D. In fact, in order to reach the rest of the workspace (corresponding to the other closure of the closed chain system) the manipulator has to pass through a singular position.

When the end point comes to $s = L_d = 0.80\,\mathrm{m}$, $\theta_1 + \theta_3$ becomes equal to $\pi + \theta_2 + \theta_4$, hence a drive singularity occurs. At this position the third row of $\mathbf{A}^{\mathrm{T}}$ becomes r_3 / r_4 times the fourth row. Then, for consistency of equation (8), the third row of the right hand side of equation (8) should also be r_3 / r_4 times the fourth row. The resulting consistency condition that the generalized accelerations must satisfy is obtained from equation (15) as

$$M_{31}\ddot{\theta}_1 - \frac{r_3}{r_4} M_{24}\ddot{\theta}_2 + M_{33}\ddot{\theta}_3 - \frac{r_3}{r_4} M_{44}\ddot{\theta}_4 = R_3 - \frac{r_3}{r_4} R_4 \tag{36}$$

Hence the time trajectory $s(t)$ of the deployment motion should be selected such that at the drive singularity the generalized accelerations satisfy equation (36).

An arbitrary trajectory that does not satisfy the consistency condition is not realizable. This is illustrated by considering an arbitrary third order polynomial for $s(t)$ having zero initial and final velocities, i.e. $s(t) = \frac{3Lt^2}{T^2} - \frac{2Lt^3}{T^3}$. The singularity position is reached when $t = 0.48\,\mathrm{s}$. The actuator torques are shown in Figure 2. The torques grow without bounds as the singularity is approached and become infinitely large at the singular position. (In Figure 2 the torques are out of range around the singular position.)

For the time function s(t), a polynomial is chosen which satisfies the consistency condition at the drive singularity in addition to having zero initial and final velocities. The time T_d when the singular position is reached and the velocity of the end point P at T_d, $v_P(T_d)$ can be arbitrarily chosen. The loop closure relations, the specified angle of the acceleration of P and the consistency condition constitute four independent equations for a unique solution of $\ddot{\theta}_i$, $i = 1,\ldots,4$ at the singular position. Hence, using θ_i and $\dot{\theta}_i$ at T_d, the acceleration of P at T_d, $a_P(T_d)$ is uniquely determined. Consequently a sixth order polynomial is selected where $s(0) = 0$, $\dot{s}(0) = 0$, $s(T) = L$, $\dot{s}(T) = 0$, $s(T_d) = L_d$, $\dot{s}(T_d) = v_P(T_d)$ and $\ddot{s}(T_d) = a_P(T_d)$. T_d and $v_P(T_d)$ are chosen by trial and error to prevent any overshoot in s or $\dot{s}$. The values used are $T_d = 0.55\,\mathrm{s}$ and $v_P(T_d) = 3.0\,\mathrm{m/s}$, yielding $a_P(T_d) = 18.2\,\mathrm{m/s^2}$. $s(t)$ so obtained is given by equation (37) and shown in Figure 3.

$$s(t) = 30.496\,t^2 - 154.909\,t^3 + 311.148\,t^4 - 265.753\,t^5 + 81.318\,t^6 \tag{37}$$

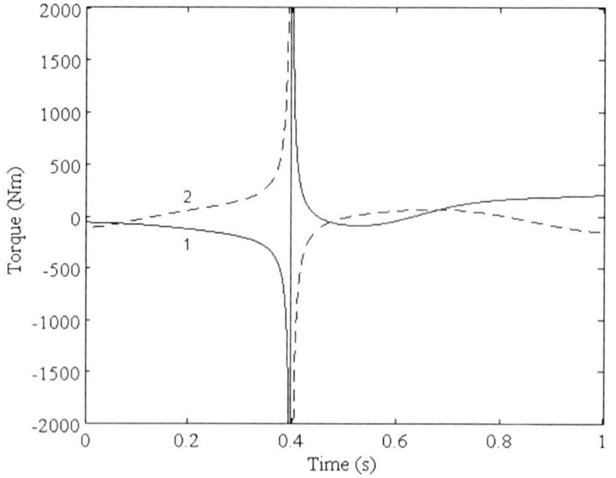

Figure 2. Motor torques for the trajectory not satisfying the consistency condition: 1. T_1, 2. T_2

Furthermore, even when the consistency condition is satisfied, $\mathbf{A}^{\mathrm{T}}$ is ill-conditioned in the vicinity of the singular position, hence $\boldsymbol{\tau}$ cannot be found correctly from equation (8). Deletion of a linearly dependent equation in that neighborhood would cause task violations due to the removal of a task. For this reason the modified equation (17) is used to replace the dependent equation in the neighborhood of the singular position. The modified equation, which relates the actuator forces to the system jerks, takes the following form.

$$(\dot{A}_{31}^{\mathrm{T}} - \frac{r_3}{r_4}\dot{A}_{41}^{\mathrm{T}})\,\tau_1 + (\dot{A}_{32}^{\mathrm{T}} - \frac{r_3}{r_4}\dot{A}_{42}^{\mathrm{T}})\,\tau_2 = M_{31}\dddot{\theta}_1 - \frac{r_3}{r_4}M_{24}\dddot{\theta}_2 + M_{33}\dddot{\theta}_3 - \frac{r_3}{r_4}M_{44}\dddot{\theta}_4$$

$$+\dot{M}_{31}\ddot{\theta}_1 - \frac{r_3}{r_4}\dot{M}_{24}\ddot{\theta}_2 + \dot{M}_{33}\ddot{\theta}_3 - \frac{r_3}{r_4}\dot{M}_{44}\ddot{\theta}_4 - \dot{R}_3 + \frac{r_3}{r_4}\dot{R}_4 \tag{38}$$

The coefficients of the constraint forces in eqn (38) are

$$\dot{A}_{31}^{\mathrm{T}} - \frac{r_3}{r_4}\dot{A}_{41}^{\mathrm{T}} = -r_3(\dot{\theta}_1 + \dot{\theta}_3)c_{13} - r_3(\dot{\theta}_2 + \dot{\theta}_4)c_{24} \tag{39a}$$

$$\dot{A}_{32}^{\mathrm{T}} - \frac{r_3}{r_4}\dot{A}_{42}^{\mathrm{T}} = -r_3(\dot{\theta}_1 + \dot{\theta}_3)s_{13} - r_3(\dot{\theta}_2 + \dot{\theta}_4)s_{24} \tag{39b}$$

which in general do not vanish at the singular position if the system is in motion.
Once the trajectory is chosen as above such that it renders the dynamic equations to be consistent at the singular position, the corresponding θ_i, $\dot{\theta}_i$ and $\ddot{\theta}_i$ are obtained from inverse kinematics, and when there is no actuation singularity, the actuator torques T_1 and

T_2 (along with the constraint forces λ_1 and λ_2) are obtained from equation (8). However in the neighborhood of the singular position, equation (22) is used in which the third row of equation (8) is replaced by the modified equation (38). The neighborhood of the singularity where equation (22) is utilized is taken as $\left|\theta_1 + \theta_3 - \theta_2 - \theta_4 - 180^o\right| < \varepsilon = 1^o$. The motor torques necessary to realize the task are shown in Figure 4. At the singular position the motor torques are found as $T_1 = -138.07\,\text{Nm}$ and $T_2 = -30.66\,\text{Nm}$. To test the validity of the modified equations, when the simulations are repeated with $\varepsilon = 0.5^o$ and $\varepsilon = 1.5^o$, no significant changes occur and the task violations remain less than 10^{-4} m.

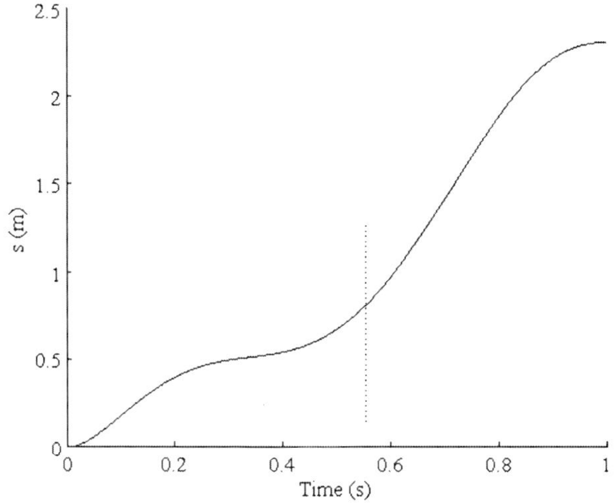

Figure 3. Time function satisfying the consistency condition.

4.2 Three degree of freedom 2-RPR planar parallel manipulator

The 2-*RPR* manipulator shown in Figure 5 has 3 degrees of freedom (n=3). Choosing the revolute joint at D for disconnection (among the passive joints) the joint variable vector of the open chain system is $\boldsymbol{\eta} = [\theta_1 \;\; \zeta_1 \;\; \theta_2 \;\; \zeta_2 \;\; \theta_3]^{\mathrm{T}}$, where $\zeta_1 = AB$ and $\zeta_2 = CD$. The link dimensions of the manipulator are labelled as $a = AC$, $b = BD$, $c = BP$ and $\alpha = \angle PBD$. The position and orientation of the moving platform is $\mathbf{x} = [x_P \;\; y_P \;\; \theta_3]^{\mathrm{T}}$ where x_P, y_P are the coordinates of the operational point of interest P in the moving platform.

The velocity level loop closure constraint equations are $\boldsymbol{\Gamma}^G \dot{\boldsymbol{\eta}} = \mathbf{0}$, where

$$\boldsymbol{\Gamma}^G = \begin{bmatrix} -\zeta_1 \sin\theta_1 & \cos\theta_1 & \zeta_2 \sin\theta_2 & -\cos\theta_2 & -b\sin\theta_3 \\ \zeta_1 \cos\theta_1 & \sin\theta_1 & -\zeta_2 \cos\theta_2 & -\sin\theta_2 & b\cos\theta_3 \end{bmatrix} \tag{40}$$

The prescribed position and orientation of the moving platform, $\mathbf{x}(t)$ represent the tasks of the manipulator. The task equations at velocity level are $\boldsymbol{\Gamma}^P \dot{\boldsymbol{\eta}} = \dot{\mathbf{x}}$ where

$$\mathbf{\Gamma}^P = \begin{bmatrix} -\zeta_1 \sin\theta_1 & \cos\theta_1 & 0 & 0 & -c\ \sin(\theta_3+\alpha) \\ \zeta_1 \cos\theta_1 & \sin\theta_1 & 0 & 0 & c\ \cos(\theta_3+\alpha) \\ 0 & 0 & 0 & 0 & 1 \end{bmatrix} \tag{41}$$

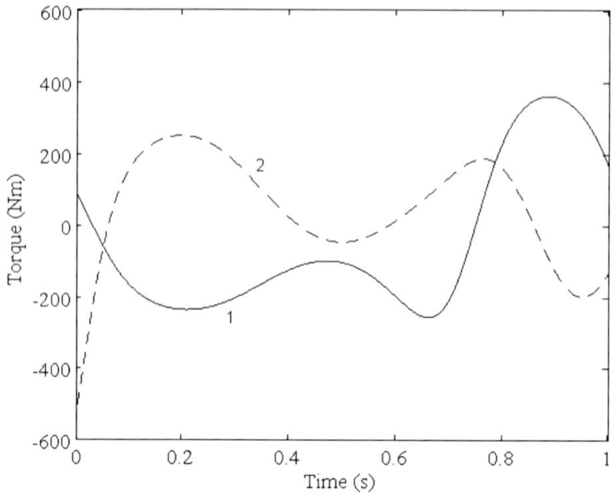

Figure 4. Motor torques for the trajectory satisfying the consistency condition: 1. T_1, 2. T_2.

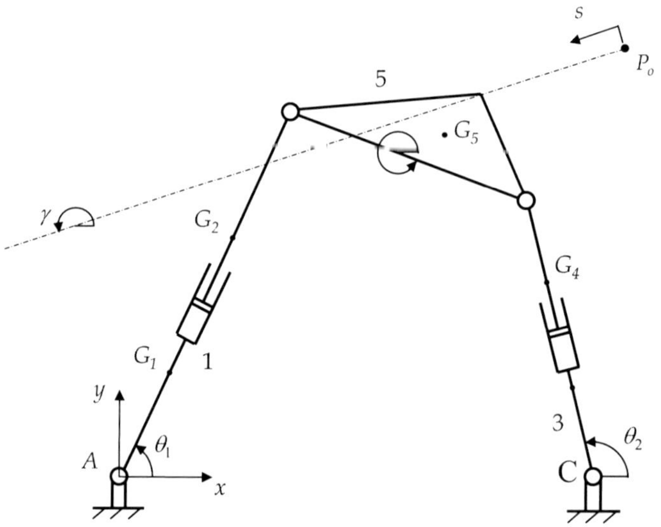

Figure 5. 2-RPR planar parallel manipulator.

Let the joints whose variables are θ_1, ζ_1 and ζ_2 be the actuated joints. The actuator force vector can be written as $\mathbf{T} = [T_1 \quad F_1 \quad F_2]^{\mathrm{T}}$ where T_1 is the motor torque corresponding to

θ_1, and F_1 and F_2 are the translational actuator forces corresponding to ζ_1 and ζ_2, respectively. Consider a deployment motion where the platform moves with a constant orientation given as $\theta_3 = 320^\circ$ and with point P having a trajectory $s(t)$ along a straight line whose angle with x-axis is $\gamma = 200^\circ$, starting from initial position $x_{P_o} = 0.800\text{ m}$, $y_{P_o} = 0.916\text{ m}$ (Figure 5). The time of the deployment motion is $T = 1\text{s}$ and its length is $L = 1.5\text{ m}$. Hence the prescribed Cartesian motion of the platform can be written as

$$\mathbf{x} = \begin{bmatrix} x_P(t) \\ y_P(t) \\ \theta_3(t) \end{bmatrix} = \begin{bmatrix} x_{P_o} + s(t)\sin\gamma \\ y_{P_o} + s(t)\cos\gamma \\ 320^\circ \end{bmatrix} \tag{42}$$

The link dimensions and mass properties are arbitrarily chosen as follows. The link lengths are $AC = a = 1.0\text{m}$, $BD = b = 0.4\text{m}$, $BP = c = 0.2\text{m}$, $\angle PBD = \alpha = 0$. The masses and the centroidal moments of inertia are $m_1 = 2\text{kg}$, $m_2 = 1.5\text{kg}$, $m_3 = 2\text{kg}$, $m_4 = 1.5\text{kg}$, $m_5 = 1.0\text{kg}$, $I_1 = 0.05\text{kgm}^2$, $I_2 = 0.03\text{kgm}^2$, $I_3 = 0.05\text{kgm}^2$, $I_4 = 0.03\text{kgm}^2$ and $I_5 = 0.02\text{kgm}^2$. The mass center locations are given by $AG_1 = g_1 = 0.15\text{m}$, $BG_2 = g_2 = 0.15\text{m}$, $CG_3 = g_3 = 0.15\text{m}$, $DG_4 = g_4 = 0.15\text{m}$, $BG_5 = g_5 = 0.2\text{m}$ and $\angle G_5BD = \beta = 0$.

The generalized mass matrix $\mathbf{M}$ and the generalized inertia forces involving the second order velocity terms $\mathbf{R}$ are

$$\mathbf{M} = \begin{bmatrix} M_{11} & 0 & 0 & 0 & M_{15} \\ 0 & M_{22} & 0 & 0 & M_{25} \\ 0 & 0 & M_{33} & 0 & 0 \\ 0 & 0 & 0 & M_{44} & 0 \\ M_{51} & M_{52} & 0 & 0 & M_{55} \end{bmatrix}, \quad \mathbf{R} = \begin{bmatrix} R_1 \\ R_2 \\ R_3 \\ R_4 \\ R_5 \end{bmatrix} \tag{43}$$

where M_{ij} and R_i are given in the Appendix.

For the set of actuators considered, the actuator direction matrix $\mathbf{Z}$ is

$$\mathbf{Z} = \begin{bmatrix} 1 & 0 & 0 & 0 & 0 \\ 0 & 1 & 0 & 0 & 0 \\ 0 & 0 & 0 & 1 & 0 \end{bmatrix} \tag{44}$$

Hence, $\mathbf{A}^{\mathrm{T}}$ becomes

$$\mathbf{A}^{\mathrm{T}} = \begin{bmatrix} -\zeta_1\sin\theta_1 & \zeta_1\cos\theta_1 & 1 & 0 & 0 \\ \cos\theta_1 & \sin\theta_1 & 0 & 1 & 0 \\ \zeta_2\sin\theta_2 & -\zeta_2\cos\theta_2 & 0 & 0 & 0 \\ -\cos\theta_2 & -\sin\theta_2 & 0 & 0 & 1 \\ -b\sin\theta_3 & b\cos\theta_3 & 0 & 0 & 0 \end{bmatrix} \tag{45}$$

Since $|\mathbf{A}| = b\,\zeta_2 \sin(\theta_2 - \theta_3)$, drive singularities occur when $\zeta_2 = 0$ or $\sin(\theta_2 - \theta_3) = 0$. Noting that ζ_2 does not become zero in practice, the singular positions are those positions where points B, D and C become collinear.

Hence, drive singularities occur inside the workspace and avoiding them limits the motion in the workspace. Avoiding singular positions where $\theta_2 - \theta_3 = \pm n\pi$ $(n = 0,1,2,...)$ would restrict the motion to a portion of the workspace where point D is always on the same side of the line BC. This means that in order to reach the rest of the workspace (corresponding to the other closure of the closed chain system) the manipulator has to pass through a singular position.

When point P comes to $s = L_d = 0.662\,\text{m}$, a drive singularity occurs since θ_2 becomes equal to $\theta_3 + \pi$. At this position the third and fifth rows of $\mathbf{A}^{\mathrm{T}}$ become linearly dependent as $A_{3j}^T - \frac{\zeta_2}{b} A_{5j}^T = 0$, $j = 1,...,5$. The consistency condition is obtained as below

$$M_{33}\ddot{\theta}_2 - \frac{\zeta_2}{b}(M_{51}\ddot{\theta}_1 + M_{52}\ddot{\zeta}_1 + M_{55}\ddot{\theta}_3) = R_3 - \frac{\zeta_2}{b} R_5 \tag{46}$$

The desired trajectory should be chosen in such a way that at the singular position the generalized accelerations should satisfy the consistency condition.

If an arbitrary trajectory that does not satisfy the consistency condition is specified, then such a trajectory is not realizable. The actuator forces grow without bounds as the singular position is approached and become infinitely large at the singular position. This is illustrated by using an arbitrary third order polynomial for $s(t)$ having zero initial and final velocities, i.e. $s(t) = \frac{3\,L\,t^2}{T^2} - \frac{2\,L\,t^3}{T^3}$. The singularity occurs when $t = 0.46\,\text{s}$. The actuator forces are shown in Figures 6 and 7. (In the figures the forces are out of range around the singular position.)

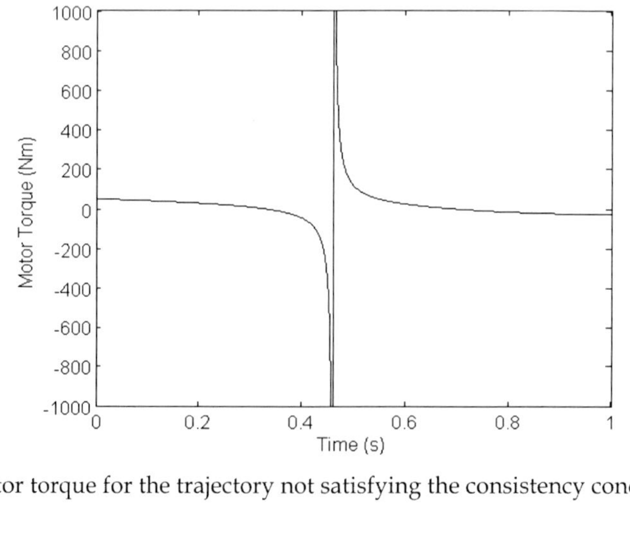

Figure 6. Motor torque for the trajectory not satisfying the consistency condition.

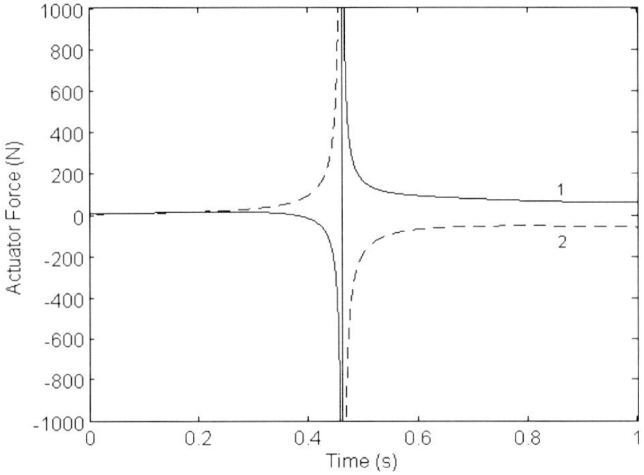

Figure 7. Actuator forces for the trajectory not satisfying the consistency cond.: 1. F_1, 2. F_2.

For the time function $s(t)$ a polynomial is chosen that renders the dynamic equations to be consistent at the singular position in addition to having zero initial and final velocities. The time T_d when singularity occurs and the velocity of the end point when $t = T_d$, $v_P(T_d)$ can be arbitrarily chosen. The acceleration level loop closure relations, the specified angle of the acceleration of P ($\gamma = 200°$), the specified angular acceleration of the platform ($\ddot{\theta}_3 = 0$) and the consistency condition constitute five independent equations for a unique solution of $\ddot{\eta}_i$, $i = 1,...,5$ at the singular position. Hence, using $\boldsymbol{\eta}$ and $\dot{\boldsymbol{\eta}}$ at T_d, the acceleration of P at T_d, $a_P(T_d)$ is uniquely determined. Consequently a sixth order polynomial is selected where $s(0) = 0$, $\dot{s}(0) = 0$, $s(T) = L$, $\dot{s}(T) = 0$, $s(T_d) = L_d$, $\dot{s}(T_d) = v_P(T_d)$ and $\ddot{s}(T_d) = a_P(T_d)$. The values used for T_d and $v_P(T_d)$ are $0.62\,\text{s}$ and $1.7\,\text{m/s}$ respectively, yielding $a_P(T_d) = 10.6\,\text{m/s}^2$. $s(t)$ so obtained is shown in Figure 8 and given by equation (47).

$$s(t) = 20.733\,t^2 - 87.818\,t^3 + 146.596\,t^4 - 103.669\,t^5 + 25.658\,t^6 \tag{47}$$

Bad choices for T_d and $v_P(T_d)$ would cause local peaks in $s(t)$ implying back and forth motion of point P during deployment along its straight line path.

However, even when the equations are consistent, in the neighborhood of the singular positions $\mathbf{A}^{\mathrm{T}}$ is ill-conditioned, hence $\boldsymbol{\tau}$ cannot be found correctly from equation (8). This problem is eliminated by utilizing the modified equation valid in the neighborhood of the singular position. The modified equation (17) takes the following form

$$B_j\,\tau_j = Q \qquad j = 1,2 \tag{48}$$

where

$$B_1 = \dot{A}^T_{31} - \frac{\zeta_2}{b}\dot{A}^T_{51} - \frac{\dot{\zeta}_2}{b}A^T_{51}, \qquad B_2 = \dot{A}^T_{32} - \frac{\zeta_2}{b}\dot{A}^T_{52} - \frac{\dot{\zeta}_2}{b}A^T_{52} \tag{49a}$$

$$Q = M_{33}\dddot{\theta}_2 - \frac{\zeta_2}{b}(M_{51}\dddot{\theta}_1 + M_{52}\dddot{\zeta}_1 + M_{55}\dddot{\theta}_3) + \dot{M}_{33}\ddot{\theta}_2 - \frac{\zeta_2}{b}(\dot{M}_{51}\ddot{\theta}_1 + \dot{M}_{52}\ddot{\zeta}_1$$

$$+\dot{M}_{55}\ddot{\theta}_3) - \frac{\dot{\zeta}_2}{b}(M_{51}\ddot{\theta}_1 + M_{52}\ddot{\zeta}_1 + M_{55}\ddot{\theta}_3) - \dot{R}_3 + \frac{\zeta_2}{b}\dot{R}_5 + \frac{\dot{\zeta}_2}{b}R_5 \tag{49b}$$

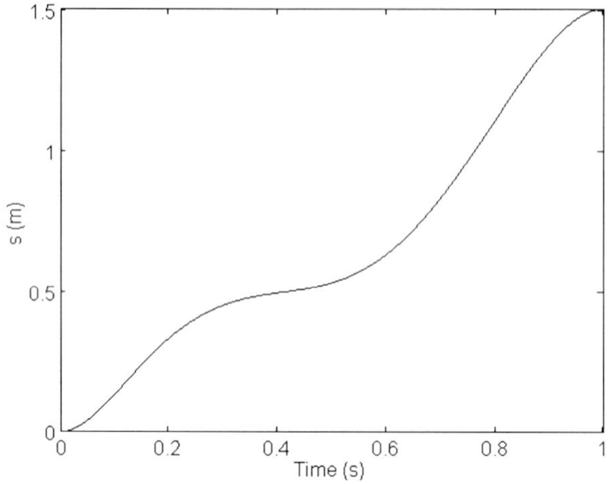

Figure 8. A time function that satisfies the consistency condition.

Once the trajectory is specified, the corresponding $\boldsymbol{\eta}$, $\dot{\boldsymbol{\eta}}$ and $\ddot{\boldsymbol{\eta}}$ are obtained from inverse kinematics, and when there is no actuation singularity, the actuator forces T_1, F_1 and F_2 (and the constraint forces λ_1 and λ_2) are obtained from equation (8). However in the neighborhood of the singularity, **A** is ill-conditioned. So the unknown forces are obtained from equation (22) which is obtained by replacing the third row of equation (8) by the modified equation (48). The neighborhood of the singular position where equation (22) is utilized is taken as $\left|\theta_2 - \theta_3 + 180^{\circ}\right| < \varepsilon = 0.5^{\circ}$. The motor torques and the translational actuator forces necessary to realize the task are shown in Figures 9 and 10, respectively. At the singular position the actuator forces are $T_1 = 30.31$ Nm, $F_1 = 26.3$ N and $F_2 = 1.61$ N. The joint displacements under the effects of the actuator forces are given in Figures 11 and 12. To test the validity of the modified equations in a larger neighborhood, when the simulations are repeated with $\varepsilon = 1^{\circ}$, no significant changes are observed, the task violations remaining less than 10^{-5} m.

5. Conclusions

A general method for the inverse dynamic solution of parallel manipulators in the presence of drive singularities is developed. It is shown that at the drive singularities, the actuator forces cannot influence the end-effector accelerations instantaneously in certain directions. Hence the end-effector trajectory should be chosen to satisfy the consistency of the dynamic equations when the coefficient matrix of the drive and constraint forces, **A** becomes singular. The satisfaction of the consistency conditions makes the trajectory to be realizable by the actuators of the manipulator, hence avoids the divergence of the actuator forces.

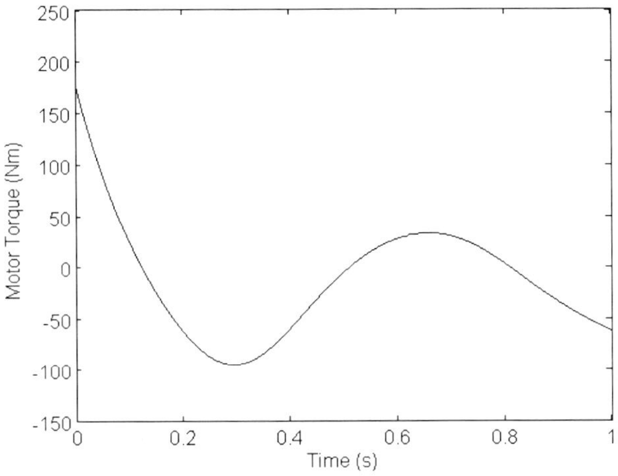

Figure 9. Motor torque for the trajectory satisfying the consistency condition

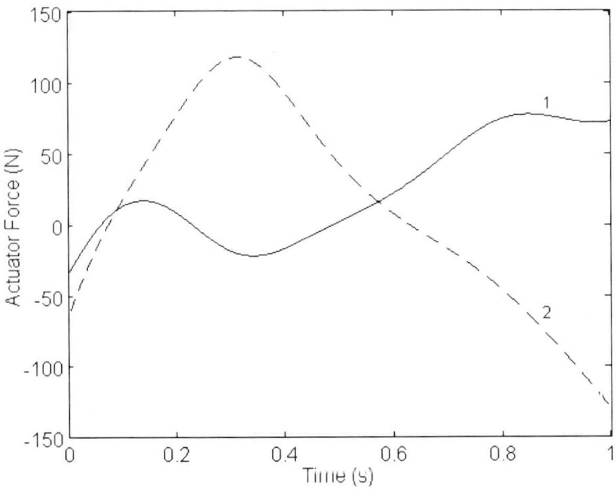

Figure 10. Actuator forces for the trajectory satisfying the consistency condition: 1. F_1 , 2. F_2

To avoid the problems related to the ill-condition of the force coefficient matrix, **A** in the neighborhood of the drive singularities, a modification of the dynamic equations is made using higher order derivative information. Deletion of the linearly dependent equation in that neighborhood would cause task violations due to the removal of a task. For this reason the modified equation is used to replace the dependent equation yielding a full rank force coefficient matrix.

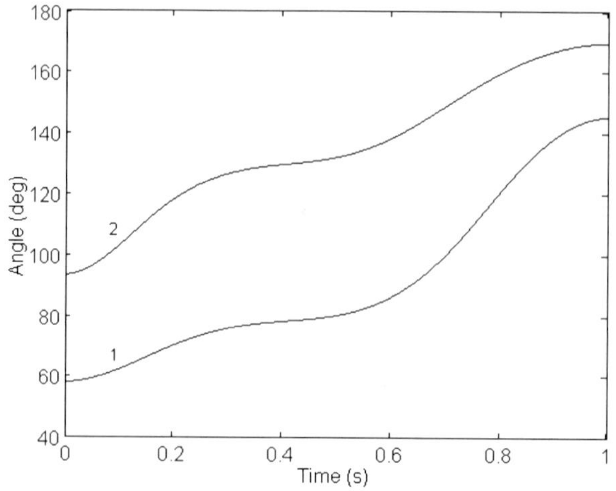

Figure 11. Rotational joint displacements: 1. θ_1 , 2. θ_2 .

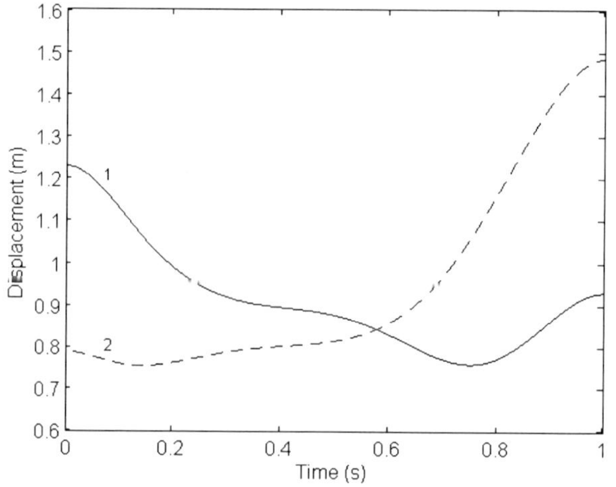

Figure 12. Translational joint displacements: 1. ζ_1 , 2. ζ_2 .

6. References

Alıcı, G. (2000). Determination of singularity contours for five-bar planar parallel manipulators, *Robotica,* Vol. 18, No. 5, (September 2000) 569-575.

Daniali, H.R.M.; Zsombor-Murray, P.J. & Angeles, J. (1995). Singularity analysis of planar parallel manipulators, *Mechanism and Machine Theory,* Vol. 30, No. 5, (July 1995) 665-678.

Di Gregorio, R. (2001). Analytic formulation of the 6-3 fully-parallel manipulator's singularity determination, *Robotica,* Vol. 19, No. 6, (September 2001) 663-667.
Gao, F.; Li, W.; Zhao, X.; Jin, Z. & Zhao, H. (2002). New kinematic structures for 2-, 3-, 4-, and 5-DOF parallel manipulator designs, *Mechanism and Machine Theory,* Vol. 37, No. 11, (November 2002) 1395-1411.
Gunawardana, R. & Ghorbel, F. (1997). PD control of closed-chain mechanical systems: an experimental study, *Proceedings of the Fifth IFAC Symposium on Robot Control,* Vol. 1, 79-84, Nantes, France, September 1997, Cambridge University Press, New York.
Ider, S.K. (2004). Singularity robust inverse dynamics of planar 2-RPR parallel manipulators, *Proceedings of the Institution of Mechanical Engineers, Part C: Journal of Mechanical Engineering Science,* Vol. 218, No. 7, (July 2004) 721-730.
Ider, S.K. (2005). Inverse dynamics of parallel manipulators in the presence of drive singularities, *Mechanism and Machine Theory,* Vol. 40, No. 1, (January 2005) 33-44.
Ji, Z. (2003) Study of planar three-degree-of-freedom 2-RRR parallel manipulators, *Mechanism and Machine Theory,* Vol. 38, No. 5, (May 2003) 409-416.
Kong, X. & Gosselin, C.M. (2001). Forward displacement analysis of third-class analytic 3-RPR planar parallel manipulators, *Mechanism and Machine Theory,* Vol. 36, No. 9, (September 2001) 1009-1018.
Merlet, J.-P. (1999). Parallel Robotics: Open Problems, *Proceedings of Ninth International Symposium of Robotics Research,* 27-32, Snowbird, Utah, October 1999, Springer-Verlag, London.
Sefrioui, J. & Gosselin, C.M. (1995). On the quadratic nature of the singularity curves of planar three-degree-of-freedom parallel manipulators, *Mechanism and Machine Theory,* Vol. 30, No. 4, (May 1995) 533-551.

Appendix

The elements of **M** and $\boldsymbol{R}$ of the 2-RPR parallel manipulator shown in equation (41) are given below, where m_i, $i = 1,...,5$ are the masses of the links, I_i, $i = 1,...,5$ are the centroidal moments of inertia of the links and the locations of the mass centers G_i, $i = 1,...,5$ are indicated by $g_1 = AG_1$, $g_2 = BG_2$, $g_3 = CG_3$, $g_4 = DG_4$, $g_5 = BG_5$ and $\beta = \angle G_5BD$.

$$M_{11} = m_1 g_1^2 + I_1 + m_2(\zeta_1 - g_2)^2 + I_2 + m_5\zeta_1^2 \tag{A1}$$

$$M_{15} = m_5\zeta_1 g_5 \cos(\theta_1 - \theta_3 - \beta) \tag{A2}$$

$$M_{22} = m_2 + m_3 \tag{A3}$$

$$M_{25} = m_5 g_5 \sin(\theta_1 - \theta_3 - \beta) \tag{A4}$$

$$M_{33} = m_3 g_3^2 + I_3 + m_4(\zeta_2 - g_4)^2 + I_4 \tag{A5}$$

$$M_{44} = m_4 \tag{A6}$$

$$M_{51} = m_5\zeta_1 g_5 \cos(\theta_1 - \theta_3 - \beta) \tag{A7}$$

$$M_{52} = m_5 g_5 \sin(\theta_1 - \theta_3 - \beta) \tag{A8}$$

$$M_{55} = m_5 g_5^2 + I_5 \tag{A9}$$

$$R_1 = 2m_2(\zeta_1 - g_2)\dot{\zeta}_1\dot{\theta}_1 + m_5\zeta_1 g_5\dot{\theta}_3^2 \sin(\theta_1 - \theta_3 - \beta) + [m_1 g_1 + m_2(\zeta_1 - g_2) + m_5\zeta_1]g\cos\theta_1 \tag{A10}$$

$$R_2 = -m_5 g_5\dot{\theta}_3^2 \cos(\theta_1 - \theta_3 - \beta) - m_2(\zeta_1 - g_2)\dot{\theta}_1^2 - m_5\zeta_1\dot{\theta}_1^2 + (m_2 + m_5)g\sin\theta_1 \tag{A11}$$

$$R_3 = 2m_4(\zeta_2 - g_4)\dot{\zeta}_2\dot{\theta}_2 + [m_3 g_3 + m_4(\zeta_2 - g_4)]g\cos\theta_2 \tag{A12}$$

$$R_4 = -m_4(\zeta_2 - g_4)\dot{\theta}_2^2 + m_4 g\sin\theta_2 \tag{A13}$$

$$R_5 = m_5 g_5[2\dot{\zeta}_1\dot{\theta}_1 \cos(\theta_1 - \theta_3 - \beta) - \zeta_1\dot{\theta}_1^2 \sin(\theta_1 - \theta_3 - \beta) + g\cos(\theta_3 + \beta)] \tag{A14}$$

20

Control of a Flexible Manipulator with Noncollocated Feedback: Time Domain Passivity Approach

Jee-Hwan Ryu[1], Dong-Soo Kwon[2] and Blake Hannaford[3]
Korea University of Technology and Education[1],
Korea Advanced Institute of Science and Technology[2],
University of Washington[3]
R. of Korea[1,2]
USA[3]

1. Introduction

Flexible manipulators are finding their way in industrial and space robotics applications due to their lighter weight and faster response time compared to rigid manipulators. Control of flexible manipulators has been studied extensively for more than a decade by several researchers (Book 1993, Cannon and Schmitz 1984, De Luca and Siciliano 1989, Siciliano and Book 1988, Vidyasagar and Anderson 1989, and Wang and etc. 1989). Despite their applications, control of flexible manipulators has proven to be rather complicated.

It is well known that stabilization of a flexible manipulator can be greatly simplified by collocating the sensors and the actuator, in which the input-output mapping is passive (Wang and Vidyasagar 1990), and a stable controller can be easily devised independent of the structure details. However, the performance of this collocated feedback turns out to be not satisfactory due to a week control of the vibrations of the link (Chudavarapu and Spong 1996). This initiated finding other noncollocated output measurements like the position of the end-point of the link to increase the control performance (Cannon and Schmitz 1984). However, if the end-point is chosen as the output and the joint torque is chosen as the input, the system becomes a nonminimum phase one, hence possibly behave actively. As a result, the small increment of the output feedback controller gains can easily make the closed-loop system unstable. This had led many researchers to seek other outputs for which the passivity property is enjoyed.

Wang and Vidyasagar (1990) proposed the so-called reflected tip position as such an output. This corresponds to the rigid body deflection minus the deflection at the tip of the flexible manipulator. Pota and Vidyasagar (1991) used the same output to show that in the limit, for a non uniform link, the transfer function from the input torque to the derivative of the reflected tip position is passive whenever the ratio of the link inertia to the hub inertia is sufficiently small. Chodavarapu and Spong (1996) considered the virtual angle of rotation, which consists of the hub angle of rotation augmented with a weighted value of the slope of the link at its tip. They showed that the transfer function with this output is minimum phase and that the zero dynamics are stable.

Despite the fact that these previous efforts have succeeded in numerous kinds of applications, the critical drawback was that these are model-based approaches requiring the system parameters or the dynamic structure information at the least. However, interesting systems are uncertain and it is usually hard to obtain the exact dynamic parameters and structure information.

In this paper, we introduce a different way of treating noncollocated control systems without any model information. Recently developed stability guaranteed control method based on time-domain passivity control (Hannaford and Ryu 2002, Ryu, Kwon, and Hannaford 2002) is applied.

2. Review of stability guaranteed control with time domain passivity approach

2.1 Network model

In our previous paper (Ryu, Kwon, and Hannaford 2002), the traditional control system view could be analyzed in terms of energy flow by representing it in a network point of view. Energy here was defined as the integral of the inner product between the conjugate input and output, which may or may not correspond to a physical energy. We partition the traditional control system into three elements, the trajectory generator (consisting of the trajectory generator), the control element (consisting of the controller, actuator and sensors) and the plant (consisting of the plant). The connection between the controller element and the plant is a physical interface at which, suitable conjugate variables define the physical energy flow between controller and plant. The connection between trajectory generator and controller, which traditionally consists of a one-way command information flow, is modified by the addition of a virtual feedback of the conjugate variable. For a motion control system, the trajectory generator output would be a desired velocity (v_d), and the virtual feedback would be equal to the controller output (τ) (Fig. 1).

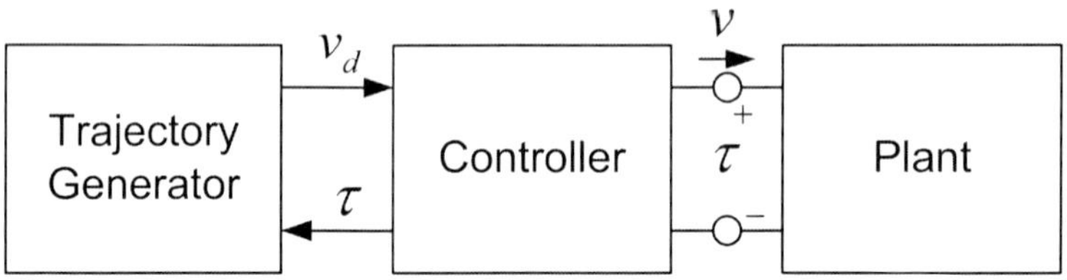

Fig. 1. Network view of a motion control system

To show that this consideration is generally possible for motion control systems, we physically interpret these energy flows. We consider a general tracking control system with a position PID and feed forward controller for moving a mass (M) on the floor with a desired velocity (v_d). The control system can be described by a physical analogy with Fig. 2. The position PD controller is physically equivalent to a virtual spring and damper whose reference position is moving with a desired velocity (v_d). In addition Integral Controller (u_I) and the feed forward controller (u_{FF}) can be regarded as internal force sources. Since the mass and the reference position are connected with the virtual spring and damper, we can obtain the desired motion of the mass by moving the reference position with the desired velocity. The important point is that if we want to move the reference position with the desired velocity (v_d), force is required. This force is determined by the impedance of the controller and the plant. Physically this force is equivalent to the controller (PID and feed forward) output (τ). As a result, the conjugate pair $(v_d \text{ and } \tau)$ simulates the flow of virtual input energy from the trajectory generator, and the conjugate pair $(v \text{ and } \tau)$ simulates the flow of real output energy to the plant. Through the above physical interpretation, we can construct a network model for general tracking control systems (Fig. 1), and this network model is equivalently described with Fig. 3 whose trajectory generator is a current (or velocity) source with electrical-mechanical analogy. Note that electrical-mechanical analog networks enforce equivalent relationships between effort and flow. For the mechanical systems, forces replace voltages in representing effort, while velocities representing currents in representing flow.

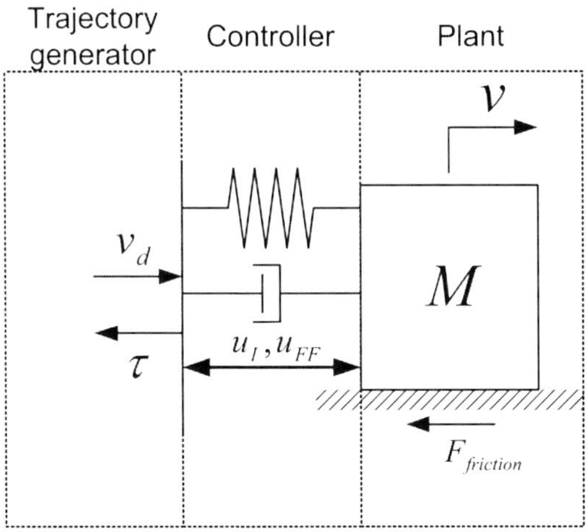

Fig. 2. Physical analogy of a motion control system

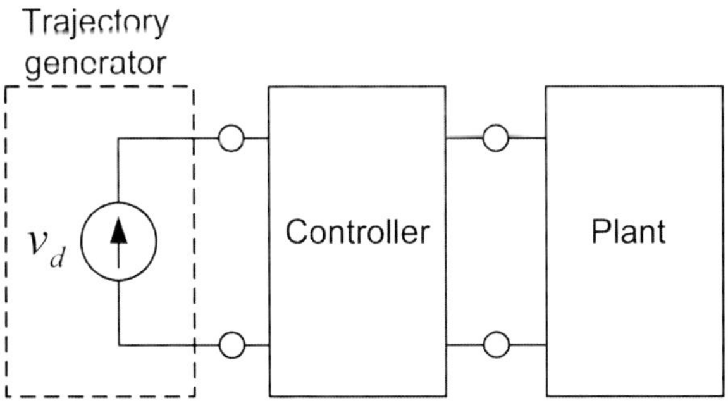

Fig. 3. Network view of general motion control systems

2.2 Stability concept

From the circuit representation (Fig. 4), we find that the virtual input energy from the trajectory generator depends on the impedance of the connected network. If the connected network (controller and plant) with the trajectory generator is passive, the control system can remain passive (Desoer and Vidyasagar 1975) since the trajectory generator creates just the amount of energy necessary to make up for the energy losses of the connected passive network. This is just like a normal electric circuit. Thus we have to make the connected network passive to guarantee the stability of the control system since passivity is a sufficient condition for stability.

In addition, the plant is uncertain and has a wide variation range of impedance or admittance (from zero to infinite). Thus, the controller 2-port should be passive to guarantee stability with any passive plant.

2.3 Time domain passivity approach

A new, energy-based method has been presented for making large classes of control systems passive by making the controller 2-port passive based on the time-domain passivity concept. In this section, we briefly review time-domain passivity control.

First, we define the sign convention for all forces and velocities so that their product is positive when power enters the system port (Fig. 4). Also, the system is assumed to have initial stored energy at $t = 0$ of $E(0)$. The following widely known definition of passivity is used.

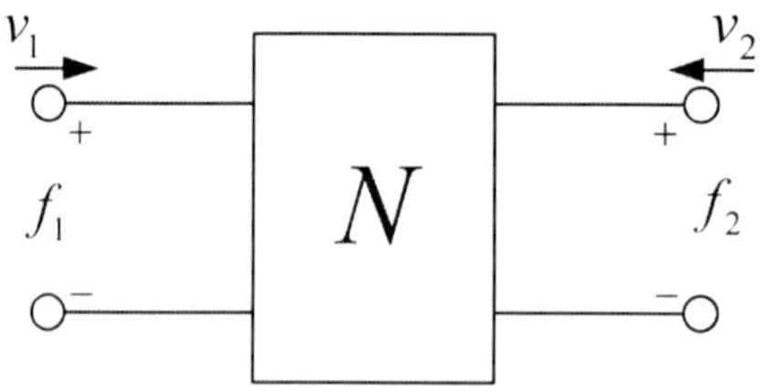

Fig. 4. Two-port network

Definition 1: *The two-port network, N, with initial energy storage $E(0)$ is passive if and only if,*

$$\int_0^t \left(f_1(\tau)v_1(\tau) + f_2(\tau)v_2(\tau)\right)d\tau + E(0) \geq 0, \qquad \forall t \geq 0 \tag{1}$$

for admissible forces (f_1, f_2) and velocities (v_1, v_2).

Equation (1) states that the energy supplied to a passive network must be greater than negative $E(0)$ for all time (van der Schaft 2000, Adams and Hannaford 1999, Desoer and Vidyasagar 1975, Willems 1972).

The conjugate variables that define power flow in such a computer system are discrete-time values, and the analysis is confined to systems having a sampling rate substantially faster than the dynamics of the system so that the change in force and velocity with each sample is small. Thus, we can easily "instrument" one or more blocks in the system with the following "Passivity Observer," (PO) for a two-port network to check the passivity (1).

$$E_{obsv}(n) = \Delta T \sum_{k=0}^{n} \left(f_1(k)v_1(k) + f_2(k)v_2(k)\right) + E(0) \tag{2}$$

where ΔT is the sampling period.
If $E_{obsv}(n) \geq 0$ for every n, this means the system dissipates energy. If there is an instance when $E_{obsv}(n) < 0$, this means the system generates energy and the amount of generated energy is $-E_{obsv}(n)$.
Consider a two-port system which may be active. Depending on operating conditions and the specifics of the two-port element's dynamics, the PO may or may not be negative at a particular time. However, if it is negative at any time, we know that the two-port may then be contributing to instability. Moreover, we know the exact amount of energy generated and we can design a time-varying element to dissipate only the required amount of energy. We call this element a "Passivity Controller" (PC). The PC takes the form of a dissipative element in a series or parallel configuration for the input causality (Hannaford and Ryu 2002).

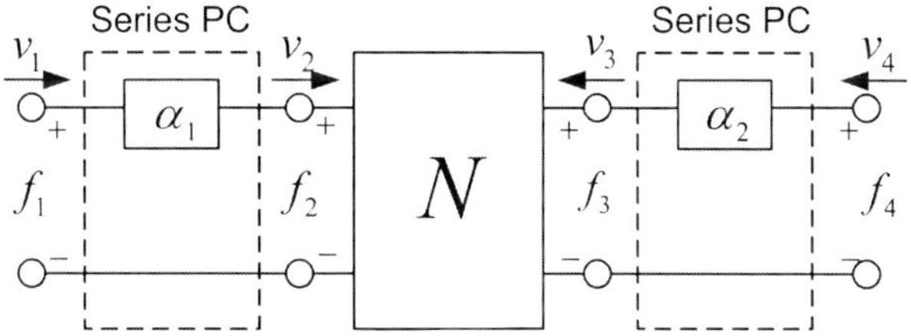

Fig. 5. Configuration of series PC for 2-port network

For a 2-port network with impedence causality at each port, we can design two series PCs (Fig. 5) in real time as follows:

1) $v_1(n) = v_2(n)$ and $v_3(n) = v_4(n)$ are inputs
2) $f_2(n)$ and $f_3(n)$ are the outputs of the system.
3) $W(n) = W(n-1) + f_2(n)v_2(n) + f_3(n)v_3(n) + \alpha_1(n-1)v_2(n-1)^2 + \alpha_2(n-1)v_3(n-1)^2$ is the PO

Two series PCs can be designed for several cases

4)
$$\alpha_1(n) = \begin{cases} -W(n)/v_2(n)^2 & \text{if case } 2, 4.2 \\ \dfrac{-f_2(n)v_2(n)}{v_2(n)^2} & \text{if case } 4.1 \\ 0 & \text{if case } 1, 3 \end{cases} \tag{3}$$

5)
$$\alpha_2(n) = \begin{cases} -W(n)/v_3(n)^2 & \text{if case } 3 \\ \dfrac{-(W(n-1) + f_3(n)v_3(n))}{v_3(n)^2} & \text{if case } 4.1 \\ 0 & \text{if case } 1, 2, 4.2 \end{cases} \tag{4}$$

where each case is as follows:
Case 1: energy does not flow out

$$W(n) \geq 0$$

Case 2: energy flows out from the left port

$$W(n) < 0,\ f_2(n)v_2(n) < 0,\ f_3(n)v_3(n) \geq 0$$

Case 3: energy flows out from the right port

$$W(n) < 0,\ f_2(n)v_2(n) \geq 0,\ f_3(n)v_3(n) < 0$$

Case 4: energy flows out from the both ports: as we mentioned above, in this paper, we divide it into two cases. The first case is when the produced energy from the right port is greater than the previously dissipated energy:

$$4.1\ W(n) < 0,\ f_2(n)v_2(n) < 0,\ f_3(n)v_3(n) < 0, W(n-1) + f_3(n)v_3(n) < 0$$

in this case, we only have to dissipate the net generation energy of the right port as the second line in Eq. (4). The second case is when the produced energy from the right port is less than the previously dissipated energy:

$$4.2\ W(n) < 0,\ f_2(n)v_2(n) < 0,\ f_3(n)v_3(n) < 0, W(n-1) + f_3(n)v_3(n) \geq 0$$

in this case we don't need to activate the right port PC, and also reduce the conservatism of the left port PC as the fist line of Eq. (3).

6) $f_1(n) = f_2(n) + \alpha_1(n)\, v_2(n) \Rightarrow$ output

7) $f_4(n) = f_3(n) + \alpha_2(n)\, v_3(n) \Rightarrow$ output

Please see (Ryu, Kwon and Hannaford 2002a, 2002b) for more detail about two-port time domain passivity control approach.

3. Implementation issues

This section addresses how to implement the time domain passivity control approach to flexible manipulator with noncollocated feedback. Consider a single link flexible manipulator having a planar motion, as detailed in Fig. 6. v_e is the end-point velocity, v_a is the velocity of the actuating position, and τ is the control torque at the joint.

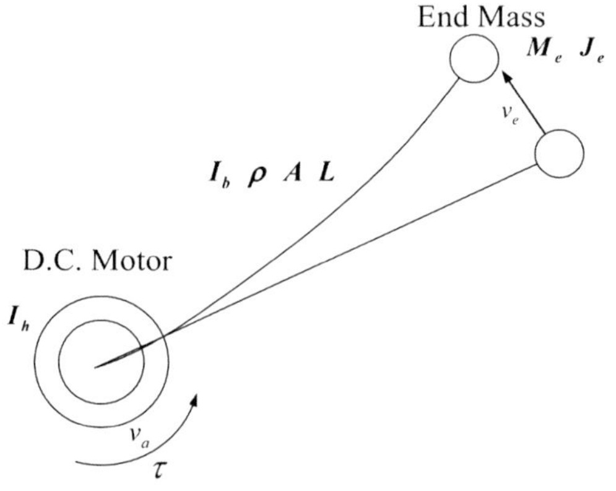

Fig. 6. A Single-link flexible manipulator

3.1 Network modeling

When we feedback end-point position to control the motion of the flexible manipulator, a network model (including causality) of the overall control system is depicted as in Fig. 7. v_{ed} means a desired velocity of the end-point. In this case, we have to consider one important thing. If the input-output relation of the plant is active, the time domain passivity control scheme cannot be applied, since the time domain passivity control scheme has been developed in the framework that the input-output relation of the plant is passive. If the end-point position is a plant output and joint torque is a plant input, the input-output relation of the plant is possibly active. Thus the overall control system may not be passive even though the controller remains passive.

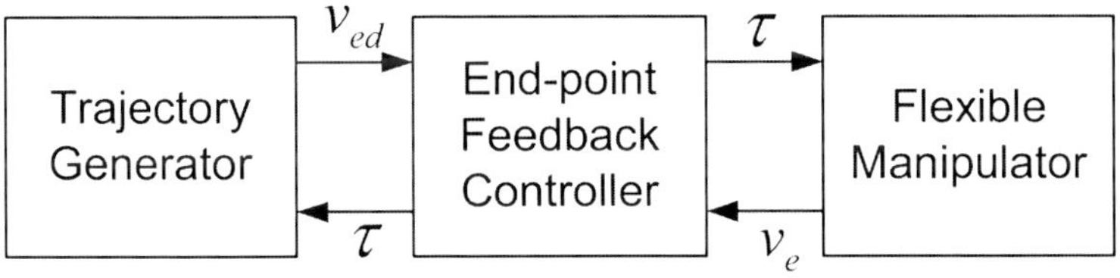

Fig. 7. A network model of flexible manipulator with end-point feedback

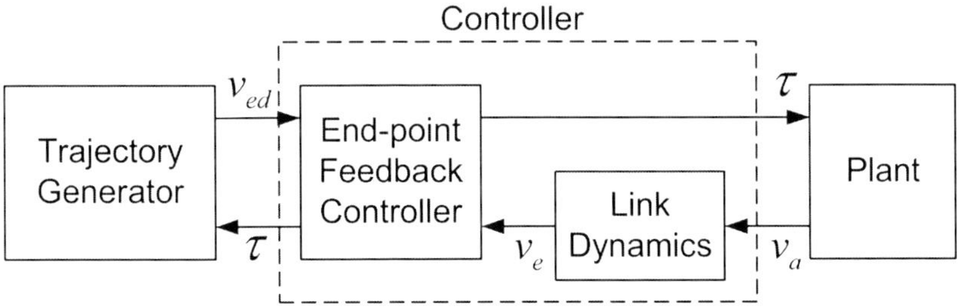

Fig. 8. A network model of flexible manipulator with end-point feedback

To solve this problem, we make the above network model suitable to our framework. The important physical fact is that the conjugate input-output pair (v_e, τ) is not simulating physical output energy from the controller to the flexible manipulator, the energy flows into the flexible manipulator through only the place where the actuator is attached. Even the controller use non-collocated sensor information to generate controller output, the actual physical energy that is transmitted to the flexible manipulator is determined by the conjugate pair at the actuating position. Based on the above inspection, it is clear that the controlled result about the joint torque is joint velocity, and the joint velocity cause end-point motion. Therefore, we can extract a link dynamics from the joint velocity to the end-point velocity from the flexible manipulator (which has noncollocated feedback) as in Fig. 8. The noncollocated (possibly active) system is then separated into the collocated (passive) system and a dynamics from the collocated output (joint velocity) to the noncollocated output (end-point velocity). As a result, if it is possible, and it generally is, to use the velocity information of the actuating position, we can construct the network model (controller and passive plant) that is suitable to our framework as in Fig. 8 by including the link dynamics that cause the noncollocation problem into the controller.

3.2 Designing the PO/PC

First, for designing the PO, it is necessary to check the real-time availability of the conjugate signal pairs at each port of the controller. The conjugate pair at the port that is connected with the trajectory generator is usually available since the desired trajectory (v_{ed}) is given and the controller output (τ) is calculated in real-time. Furthermore, the conjugate pair is generally available for the other port that is connected to the plant since the same controller output (τ) is used, and the output velocity of the actuating position (v_a) is measured in real-time. Thus, the PO is designed as

$$E_{obsv}(n) = \Delta T \sum_{k=0}^{n} \left(\tau(k)v_{ed}(k) - \tau(k)v_a(k)\right) + E(0) = \Delta T \cdot W(n).$$

After designing the PO, the causality of each port of the controller should be determined in order to choose the type of PC for implementation. In a noncollocated flexible manipulator control system, the output of the trajectory generator is the desired velocity (v_{ed}) of the end-point, and the controller output (τ) is feedback to the trajectory generator. Thus, the port that is connected with the trajectory generator has impedance causality. Also, the other port of the controller has impedance causality because a motion controlled flexible manipulator usually has admittance causality (force input (τ) and joint velocity output (v_a)). Thus, two series PCs have to be placed at each port to guarantee the passivity of the controller (Fig. 9). From the result in (Ryu, Kwon and Hannaford 2002b), the initial energy of the controller is as follows:

$$E(0) = \frac{1}{2} K_p e(0)^2$$

where K_p is a proportional gain and $e(0)$ is the position error of the end-point at the starting time.

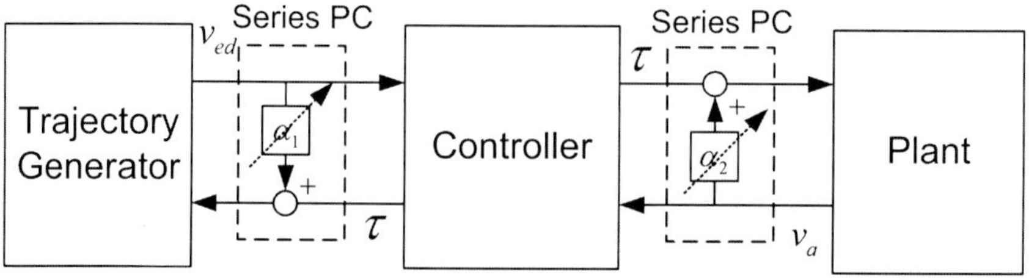

Fig. 9. Configuration of PC for a flexible manipulator with end-point feedback

4. Simulation examples

Many researchers have used a flexible manipulator for testing newly developed control methods due to its significant control challenges. In this section, the proposed stability guaranteed control scheme for noncollocated control systems is tested for feasibility with a simulated flexible link manipulator.

The experimentally verified single link flexible manipulator model (Kwon and Book 1994) is employed in this paper. A single link flexible manipulator having a planar motion is detailed in Fig. 6. The rotational inertia of the servo motor, the tachometer, and the clamping hub are modeled as a single hub inertia I_h. The payload is modeled as an end mass M_e and a rotational inertia J_e. The joint friction is included in the damping matrix. The system parameters in Fig. 6 are given in Table 1. The closed form dynamic equation is derived using the assumed mode method. For the system dynamic model, the flexible mode is modeled up to the third mode, that is, an 8th order system is considered.

In this section, a stable tip position feedback control is achieved for a flexible manipulator by using the PO/PC.

The following PD controller gain is used

$$K_P = 30\,,\ \ K_D = 0.8$$

In this noncollocated feedback systems, the hub angle can be considered as a conjugate pair with joint torque to calculate physical energy output flow into the flexible manipulator (see Section 3.A).

Without the PC turned on, tip-position tracking control is simulated (Fig. 10). The desired tip-position trajectory is as follows:

$$x_d(t) = 0.1\sin(t)$$

Link	EI : stiffness (Nm²)	11.85	H : thickness (m)	47.63E-4
	ρ A : unit length mass (kg/m)	0.2457	L : length	1.1938
Tip mass	M_e : mass (kg)	0.5867	Je : rot. Inertia (kgm²)	0.2787
Hub	I_h : rot. Inertia (kgm²)		0.016	

Table 1. Physical properties of a single-link flexible manipulator

The tip-position can not follow the desired trajectory, tip position and control input have oscillation which increases with time (Fig. 10a,b), the PO (Fig. 10c) grow to more and more negative values.

Stable tip-position tracking is achieved with the PC turned on. Tip-position tracks the desired trajectory very well (Fig. 11a), and the PO is constrained to positive values (Fig. 11c). The PC at the both side is active only when these are required, and dissipate the just amount of energy generation (Fig. 11d)

5. Conclusions

In this paper, we propose a stability guaranteed control scheme of noncollocated feedback control systems without any model information. The main contribution of this research is proposing a method to implement the PO/PC for a possibility active plant due to the noncollocated feedback. We separate the active plant into passive one and a transfer function from the collocated output to the noncollocated output. Therefore, the control system can be fit to our PO/PC framework. As a result, we can achieve stable control even for the noncollocated control system.

6. References

R. J. Adams and B. Hannaford, "Stable Haptic Interaction with Virtual Environments," *IEEE Trans. Robot. Automat.*, vol. 15, no. 3, pp. 465-474, 1999.

W. J. Book, "Controlled Motion in an elastic world," *ASME Journal of Dynamic Systems Measurement and Control*, 115(2B), pp. 252-261, 1993.

R. H. Cannon and E. Schmitz, "Initial Experiments on the End-point Control of a Flexible One-link Robot," *Int. Journal of Robotics Research*, vol. 3, no. 3, pp. 62-75, 1984.

P. A. Chudavarapu and M. W. Spong, "On Noncollocated Control of a Single Flexible Link," *IEEE Int. Conf. Robotics and Automation*, MN, April, 1996.

C. A. Desoer and M. Vidyasagar, *Feedback Systems: Input-Output Properties*, New York: Academic, 1975.

B. Hannaford and J. H. Ryu, "Time Domain Passivity Control of Haptic Interfaces," *REGULAR paper in the IEEE Trans. On Robotics and Automation*, Vol. 18, No. 1, pp. 1-10, 2002.

A. De Luca and B. Siciliano, "Trajectory Control of a Non-linear One-link Flexible Arm," *Int. Journal of Control*, vol. 50, no. 5, pp. 1699-1715, 1989.

H. R. Pota and M. Vidyasagar, "Passivity of Flexible Beam Transfer Function with Modified Outputs," *IEEE Int. Conf. Robotics and Automation*, pp. 2826-2831, 1991.

J. H. Ryu, D. S. Kwon, and B. Hannaford, "Stable Teleoperation with Time Domain Passivity Control," *Proc. of the 2002 IEEE Int. Conf. on Robotics & Automation*, Washington DC, USA, pp1863-1869, 2002a.

J. H. Ryu, D. S. Kwon, and B. Hannaford, "Stability Guaranteed Control: Time Domain Passivity Approach," will be published in IROS2002, 2002b.

B. Sciliano and W. J. Book, "A Singular Perturbation Approach to Control of Lightweight Flexible Manipulators," *Int. Journal of Robotics Research*, vol. 7, no. 4, pp. 79-90, 1988.

A.J. van der Schaft, "L2-Gain and Passivity Techniques in Nonlinear Control," Springer, Communications and Control Engineering Series, 2000.

M. Vidyasagar and B. D. O. Anderson, "Approximation and Stabilization of Distributed Systems by Lumped Systems," *Systems and Control Letters*, vol. 12, no. 2, pp. 95-101, 1989.

J. C. Willems, "Dissipative Dynamical Systems, Part I: General Theory," *Arch. Rat. Mech. An.*, vol. 45, pp. 321-351, 1972.

W. J. Wang, S. S. Lu and C. F. Hsu, "Experiments on the Position Control of a One-link Flexible robot arm," *IEEE Trans. on Robotics and Automation*, vol. 5, no. 3, pp. 373-377, 1089.

D. Wang and M. Vidyasagar, "Passivity Control of a Single Flexible Link," *IEEE Int. Conf. Robotics and Automation*, pp. 1432-1437, 1990.

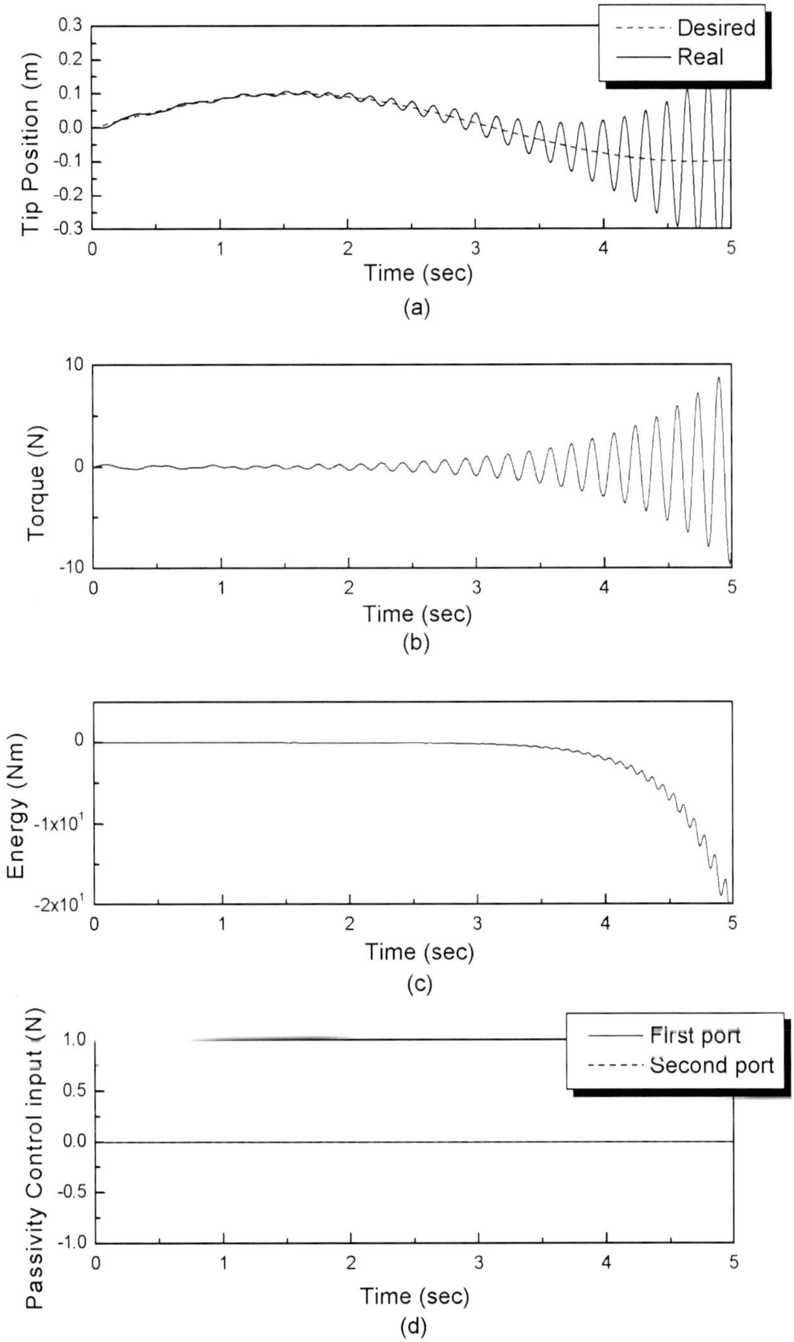

Fig. 10. Tip-position feedback without the PC

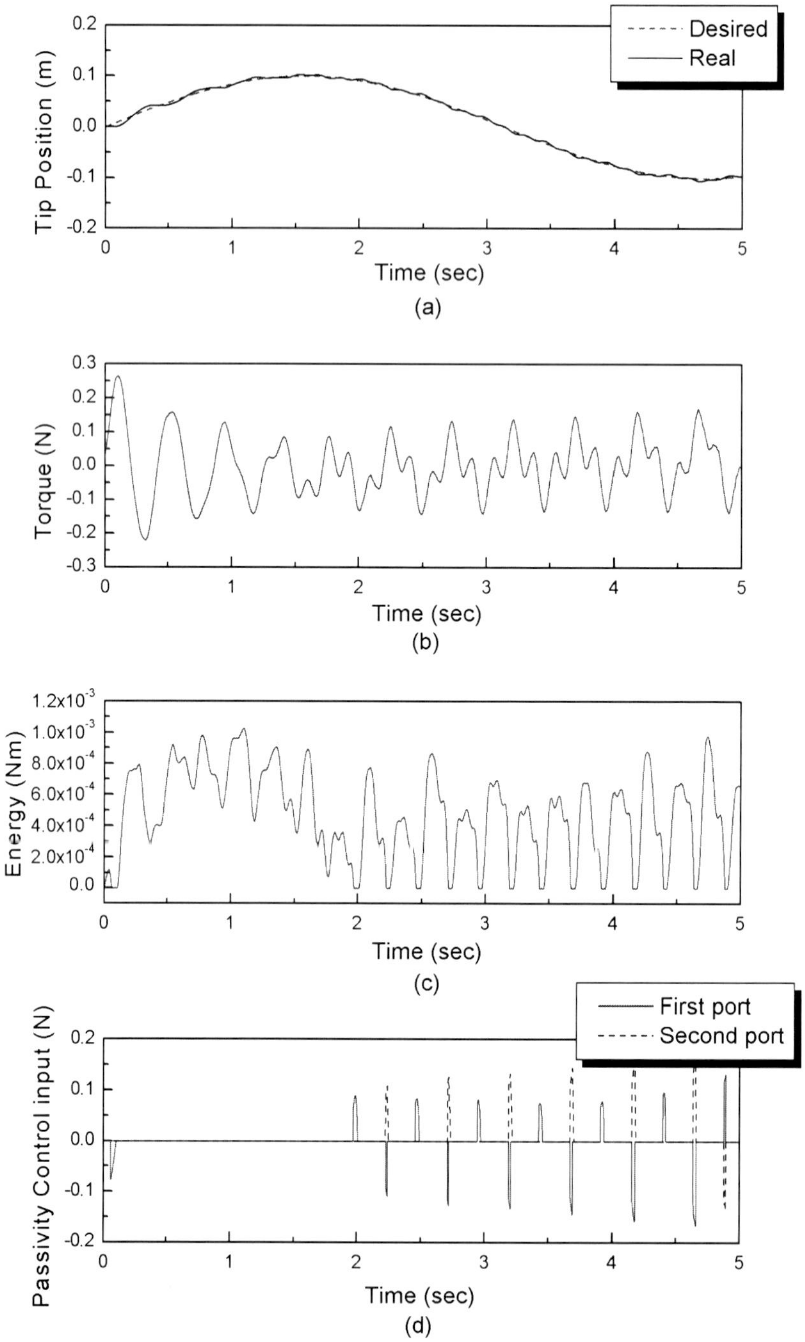

Fig. 11. Tip-position feedback with the PC

21

Dynamic Modelling and Vibration Control of a Planar Parallel Manipulator with Structurally Flexible Linkages

Bongsoo Kang[1] and James K. Mills[2]
Hannam University[1], University of Toronto[2]
South Korea[1]
Canada[2]

1. Introduction

A parallel manipulator provides an alternative design to serial manipulators, and can be found in many applications such as mining machines (Arai et al., 1991). Through the design of active joints such that actuators are fixed to the manipulator base, the mass of moving components of the parallel manipulator is greatly reduced, and high speed and high acceleration performance may be achieved. Parallel manipulators, comprised of closed-loop chains due to multiple linkages of the parallel structure, also provide high mechanical rigidity, but adversely exhibit smaller workspace and associated singularities. Considerable research has focused on kinematic analysis and singularity characterization of these devices (Gosselin & Angeles, 1990, Merlet, 1996). Planar parallel manipulators typically consist of three closed chains and a moving platform. According to the arrangement of their joints in a chain, these mechanisms are classified as PRR, RRR etc. where P denotes a prismatic joint and R denotes a revolute joint respectively.

An assembly industry, such as the electronic fabrication, demands high-speed, high acceleration placement manipulators, with corresponding lightweight linkages, hence these linkages deform under high inertia forces leading to unwanted vibrations. Moreover, such multiple flexible linkages of a parallel manipulator propagate their oscillatory motions to the moving platform where a working gripper is located. Therefore, such vibration must be damped quickly to reduce settling time of the manipulator platform position and orientation. A number of approaches to develop the dynamic model of parallel manipulators with structural flexibility have been presented in the literature(Fattah et al., 1995, Toyama et al. 2001), but relatively few works related to vibration reduction of a parallel manipulator have been published. Kozak (Kozak et al., 2004) linearized the dynamic equations of a two-degree-of-freedom parallel manipulator locally, and applied an input shaping technique to reduce residual vibrations through modification of the reference command given to the system. Kang (Kang et al., 2002) modeled a planar parallel manipulator using the assumed modes method, and presented a two-time scale controller for linkage vibration attenuation of the planar parallel manipulator. Since both the input shaping technique and the two-time scale control scheme, applied to parallel manipulators,

can only control command inputs to joint actuators of a manipulator, their performance on vibration reduction of flexible linkage are limited.
This chapter introduces a methodology for the dynamic analysis of a planar parallel mechanism beginning with the rigid-body model of a planar parallel manipulator. Flexible deformations of each linkage are expressed by the product of time-dependant functions and position-dependant functions, i.e. an assumed modes model. Overall dynamic equations of the motion for a planar parallel manipulator are formulated by Lagrangian equations. Then, an active damping approach using piezoelectric material actuators is presented to damp out oscillation of linkages of a planar parallel manipulator. Also an integrated control scheme is designed to permit the platform of the parallel manipulator to follow a given trajectory while simultaneously damping structural vibration of flexible linkages. Attached directly to the surface of flexible linkages, piezoelectric materials deform under an applied control voltage, producing shear forces, which counteract shear stresses that occur due to deformation of the linkages. Transducers are often developed from either of two types of material: polyvinylidene fluoride (PVDF) or lead zirconium titanate (PZT). PVDF is lightweight and is mainly used as a measurement device for detecting vibration, although the PVDF can also be used as an actuator. PZT has been used as actuators for micro mechanisms and has a higher strain constant than PVDF.
In simulations, both the PZT and PVDF piezoelectric actuators are applied to a planar parallel manipulator with flexible linkages and the respective performance of two actuators are compared. The dynamics of the planar parallel platform are selected such that the linkages have considerable flexibility, to better exhibit the effects of the vibration damping control system proposed. Simulation results show that the PZT actuator can give better performance in vibration attenuation than the PVDF layer.

2. Rigid-body analysis of a planar parallel manipulator

2.1 Architecture of a planar parallel manipulator

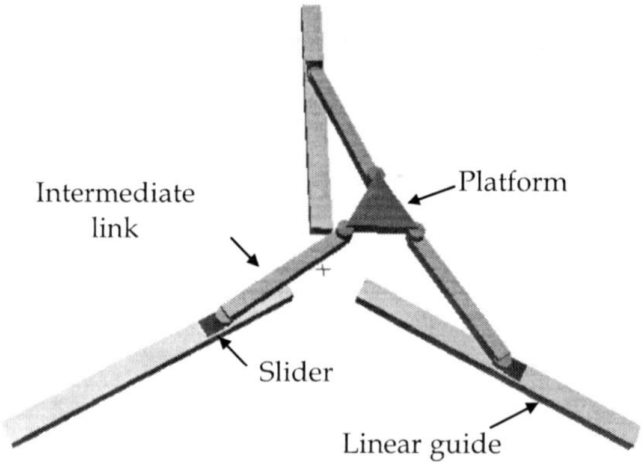

Fig. 1. Configuration of a $\underline{P}RR$ type planar parallel manipulator

The architecture of a $\underline{P}RR$ type parallel mechanism is illustrated in Fig. 1. The underline of the first character of $\underline{P}RR$ means that the first Prismatic joint is an active joint derived by actuators. The moving platform, a regular triangular shape, exhibits translation and rotational motion in a plane. Three intermediate links between the moving platform and sliders play a role to convert the actuating force into movements of the platform. Both ends of the intermediate link are composed of non-actuated revolute joints. The sliders move along the linear guide and their motions can be achieved by a ball-screw mechanism. The proposed planar manipulator is categorized as a PRR type, because a closed-loop chain consists of a prismatic joint and two consecutive revolute joints. In contrast to well-known RPR type parallel manipulators, actuators of the proposed PRR configuration remain stationary that results in low inertia of moving parts. Workspace analysis of a planar parallel manipulator has been addressed (Gosselin et al., 1996, Heerah et al., 2002) and singularity analysis of a planar parallel manipulator has been studied (Gosselin & Angeles, 1990, Merlet, 1996).

2.2 Kinematics

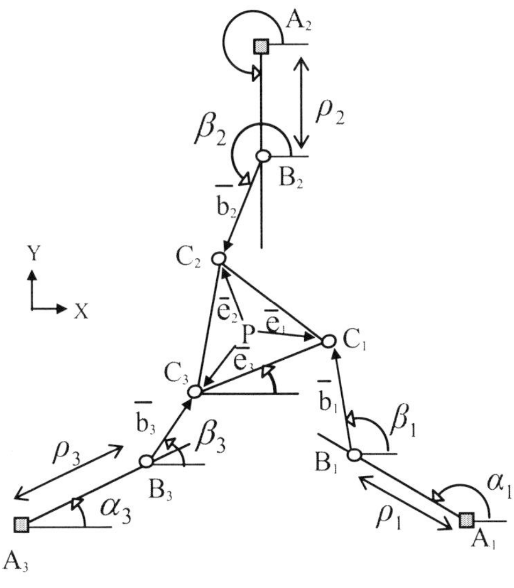

Fig. 2. Coordinate system of rigid-body model

We will begin with formulation of the rigid body model of the proposed planar parallel manipulator, and then include structural flexibility of the linkage in Section 3. Prior to derivation of dynamics of the parallel manipulator, the inverse kinematic solution of the manipulator is formulated to define kinematic relations between active joints, and position and orientation of the platform. Then, based on these kinematic relations, equations of the motion for a parallel manipulator are formulated later.

Generalized coordinates for the PRR manipulator are defined, as shown in Fig. 2. The position of the reference X-Y frame is arbitrary. A_i is the origin of the ith linear guide and B_i is the position of the ith slider. C_i is the position of the revolute joint of the platform facing with the ith linkage and P is the position of the platform at its mass center. Three linkages, including associated coordinates, are numbered with a subscript, *i*, starting from the lower right link in a counterclockwise direction. The pose of the moving platform at its mass center can be written with respect to the reference X-Y frame as

$$\overline{X}_P := [x_P \quad y_P \quad \phi]^T \tag{1}$$

The displacement of the sliders from their origin, A_i, to B_i, are expressed as

$$\overline{\rho} := [\rho_1 \quad \rho_2 \quad \rho_3]^T \tag{2}$$

β_i is defined as the angle at B_i between the X-axis of the fixed frame and the ith intermediate link and α_i is the angle at A_i between the X-axis of the fixed frame and the ith linear guide. The position and orientation of the ith link at its mass center can also be written as

$$\overline{X}_i := [x_i \quad y_i \quad \beta_i]^T \tag{3}$$

For each chain, a loop close equation can be written using position vectors defined as

$$\overline{A_iP} + \overline{PC_i} = \overline{A_iB_i} + \overline{B_iC_i} \qquad i=1,2,3 \tag{4}$$

Fig. 3. shows the diagram of the loop close equation when the first link is considered. The right-hand side of equation (4), the coordinates of C_i, is written as

$$x_{ci} = x_{ai} + \rho_i \cos\alpha_i + l\cos\beta_i \tag{5}$$

$$y_{ci} = y_{ai} + \rho_i \sin\alpha_i + l\sin\beta_i \tag{6}$$

where x_{ai} and y_{ai} are coordinates of point A_i respectively and l is length of the linkage.

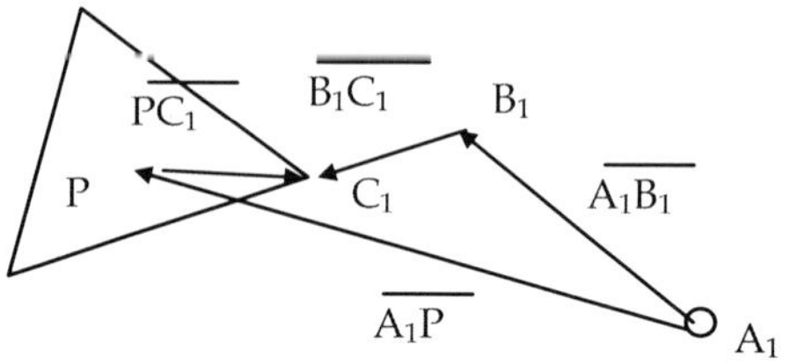

Fig. 3. Schematic diagram of loop close equation

Also, the x and y coordinates of point C_i, i.e., the left-hand side of equations (4), can be formulated using platform coordinates as

$$x_{ci} = x_p + x'_{ci}\cos\phi - y'_{ci}\sin\phi \tag{7}$$

$$y_{ci} = y_p + x'_{ci}\sin\phi + y'_{ci}\cos\phi \tag{8}$$

x'_{ci} and y'_{ci} are x and y coordinates of Ci respectively measured from the mass center of the platform, P, when ϕ is zero, as shown in Fig. 4.

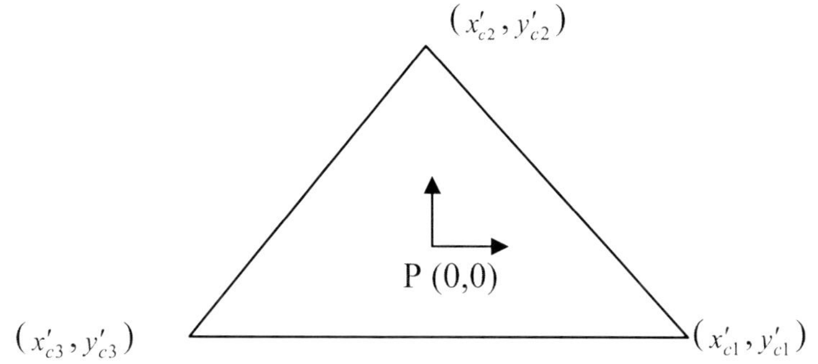

Fig. 4. Platform coordinates

From equations (5-8), a closed-form solution is calculated as

$$\rho_i = M_i \pm \sqrt{l^2 - S_i^2} \qquad i=1,2,3 \tag{9}$$

where:

$$M_i = (x_{ci} - x_{ai})\cos\alpha_i + (y_{ci} - y_{ai})\sin\alpha_i$$

$$S_i = (x_{ci} - x_{ai})\sin\alpha_i - (y_{ci} - y_{ai})\cos\alpha_i$$

Since there are two possible solutions for each chain, this manipulator can take on a maximum of eight configurations for a set of given coordinates of the platform. Note, only if the argument of the square root in equation (9) becomes zero, dose equation (9) have a unique solution. If the argument turns out to be negative, there is no solution to satisfy given kinematic requirements. Compared with the inverse kinematic solution described above, forward kinematic solutions of a planar parallel manipulator are much more difficult to solve (Merlet, 1996).

3. Dynamic analysis of flexible linkages

As industry demands high-speed machines, and hence lightweight linkages which deform under high inertial forces, we must consider structural flexibility of linkages in modeling a parallel manipulator. A single flexible link has been modeled in (Bellezza et all., 1990) and a serial type manipulator with both rigid and flexible links has been presented in (Low & Vidyasagar, 1988). Flexible models of a parallel manipulator have been studied in (Fattah et al., 1995).

A coordinate system for the flexible model is identical with the rigid-body model shown in Fig. 2. Only difference is the existence of the lateral deformation, $w_i(l)$, at the distal end of the ith linkage, C_i, due to flexibility of the linkage, as shown in Fig. 5. Out of X-Y plane deformations are not considered here. This is the subject of another analysis. If length of the linkage, l, is much longer than thickness of the linkage, the linkage can be treated as an Euler-Bernoulli beam (Genta, 1993).

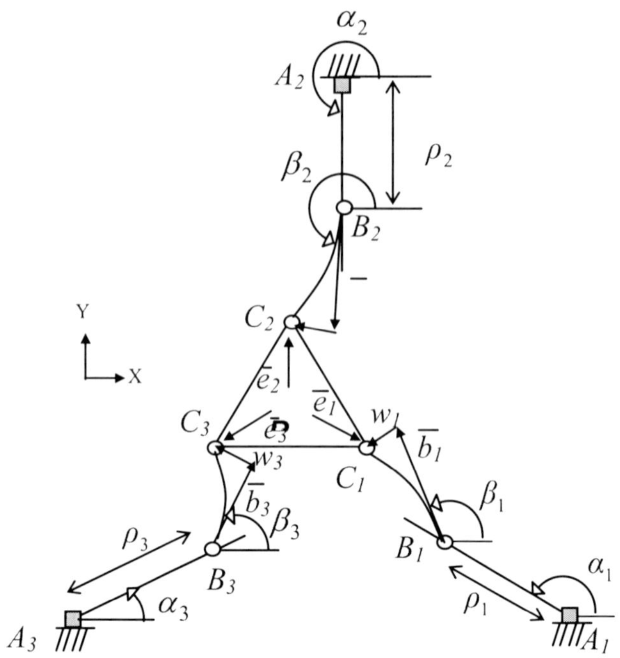

Fig. 5. Coordinate system of structurally-flexible model

Coupled with flexible deformations, the kinematic equations (5-6) are hence modified as follows;

$$x_{ci} = x_{ai} + \rho_i \cos\alpha_i + l\cos\beta_i - w_i(l)\sin\beta_i \tag{10}$$

$$y_{ci} = y_{ai} + \rho_i \sin\alpha_i + l\sin\beta_i + w_i(l)\cos\beta_i \tag{11}$$

Since the left-hand side of equation (4) remains valid for a flexible model,

$$x_{ci} = x_p + x'_{ci}\cos\phi - y'_{ci}\sin\phi \tag{12}$$

$$y_{ci} = y_p + x'_{ci}\sin\phi + y'_{ci}\cos\phi \tag{13}$$

Through Equations (10-13), the inverse kinematic solution of a structurally flexible manipulator is formulated as

$$\rho_i = M_i \pm \sqrt{l^2 + w_i(l)^2 - S_i^2} \qquad i=1,2,3 \tag{14}$$

where:

$$M_i = (x_{ci} - x_{ai})\cos\alpha_i + (y_{ci} - y_{ai})\sin\alpha_i$$

$$S_i = (x_{ci} - x_{ai})\sin\alpha_i - (y_{ci} - y_{ai})\cos\alpha_i$$

Comparing with the inverse kinematic solution of the rigid-body model, equation(9), the linkage deformation is added in the right-hand side of equation (14). In addition, large linkage deformation may lead to no solution to ρ_i, *i=1,2,3*, because the argument of the square root in equation (14) has a negative value.
Evaluation of the derivative of equations (10-13), with respect to time, gives

$$(\dot{x}_P\bar{i} + \dot{y}_P\bar{j}) + \dot{\phi}(\bar{k}\times\bar{e}_i) = \dot{\rho}_i\bar{a}_i + (\dot{\beta}_i + \dot{w}_i(l)/l)(\bar{k}\times\bar{b}_i) \tag{15}$$

where $\bar{i}$, $\bar{j}$ are unit vectors along the reference X-Y frame respectively and $\bar{e}_i$ is shown in Fig. 5. Dot-multiplication of equation (15) by $\bar{b}_i$ leads to

$$\dot{\rho}_i = \frac{1}{\bar{a}_i\cdot\bar{b}_i}\begin{bmatrix} b_{ix} & b_{iy} & e_{ix}b_{iy} - e_{iy}b_{ix}\end{bmatrix}\begin{bmatrix}\dot{x}_P & \dot{y}_P & \dot{\phi}\end{bmatrix}^T := J_{Pi}\,\dot{\bar{X}}_P \tag{16}$$

Subscripts of a position vector, named as x and y, represent X-directional and Y-directional components of the corresponding vector respectively. Cross-multiplication of equation (15) by $\bar{b}_i$ gives

$$\dot{\beta}_i = \frac{1}{l^2}\{\begin{bmatrix}-b_{iy} & b_{ix} & e_{ix}b_{ix} + e_{iy}b_{iy}\end{bmatrix} - (\bar{b}_i\times\bar{a}_i)J_{Pi}\}\dot{\bar{X}}_P - \dot{w}_i(l)/l \tag{17}$$

Accelerations of the sliders and the links are given respectively by

$$\ddot{\rho}_i = \frac{1}{\bar{a}_i\cdot\bar{b}_i}\{\begin{bmatrix} b_{ix} & b_{iy} & e_{ix}b_{iy} - e_{iy}b_{ix}\end{bmatrix}\ddot{\bar{X}}_P - \bar{b}_i\cdot\bar{e}_i\dot{\phi}^2 + \dot{\beta}_i^2 l^2 + \dot{\beta}_i\dot{w}_i(l)l\} \tag{18}$$

$$\ddot{\beta}_i = \frac{1}{l^2}\{\begin{bmatrix}-b_{iy} & b_{ix} & e_{ix}b_{ix} + e_{iy}b_{iy}\end{bmatrix}\ddot{\bar{X}}_P - (\bar{b}_i\times\bar{e}_i)\dot{\phi}^2 - (\bar{b}_i\times\bar{a}_i)\ddot{\rho}_i\} - \ddot{w}_i(l)/l \tag{19}$$

Since three linkages in this analysis are assumed to have structural flexibility, the linkage deforms under high acceleration, as shown in Fig. 5. Flexible deformations can be expressed by the product of time-dependant functions and position-dependant functions, i.e. an assumed modes model (Genta, 1993);

$$w_i(x,t) := \sum_{j=1}^{r}\eta(t)_{ij}\psi_j(\xi) \qquad i=1,2,3 \tag{20}$$

where:
$\xi := x/l$, r :=the number of assumed modes.

Functions η(t) can be considered the generalized coordinates expressing the deformation of the linkage and functions $\psi(\xi)$ are referred to as assumed modes.

Considering boundary conditions of the linkage on B_i and C_i, their behavior is close to a pin (B_i)-free (C_i) motion. Normalized shape functions, satisfying this boundary condition, are selected as:

$$\psi_j(\xi) := \frac{1}{2\sin(\gamma_j)}[\sin(\gamma_j \xi) + \frac{\sin(\gamma_j)}{\sinh(\gamma_j)}\sinh(\gamma_j \xi)] \tag{21}$$

where:

$0 \le \xi \le 1$ and $\gamma_j := (j + 0.25)\pi l$ $j=1,2,\ldots,r$

The first four shape functions are shown in Fig. 6 where the left end (B_i) exhibits zero deformation and the right end (C_i) exhibits a maximum deformation, as expected.
All generalized coordinates are collected to form of a single vector X defined as:

$$X := \begin{bmatrix} \overline{\rho} & \overline{\beta} & \overline{X}_P & \overline{\eta} \end{bmatrix}^T \in R^{9+3r} \tag{22}$$

where:

$\overline{\beta} := \begin{bmatrix} \beta_1 & \beta_2 & \beta_3 \end{bmatrix}^T$

$\overline{\eta} := \begin{bmatrix} \eta_{11} & \cdots & \eta_{1r} & \eta_{21} & \cdots & \eta_{2r} & \eta_{31} & \cdots & \eta_{3r} \end{bmatrix}^T$

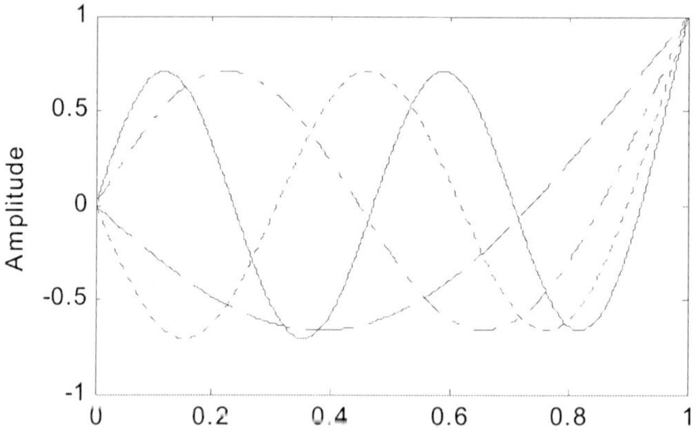

Fig. 6. Amplitude of first four mode shapes of straight beam vs. location along beam ξ (dashed line: first mode, dash-dot line: second mode, dotted line: third mode, solid line: fourth mode)

Using inertia parameters of the manipulator and generalized coordinates, the kinetic energy of three sliders is written as

$$T_S = \sum_{i=1}^{3} \frac{1}{2} m_s \dot{\rho}_i^2 \tag{23}$$

m_s is mass of the sliders.
The kinetic energy of the three links is expressed as

$$T_L = \sum_{i=1}^{3} \frac{1}{2} \int \rho_A [\dot{\rho}_i^2 + (x\dot{\beta}_i + \dot{w}_i)^2 + 2\dot{\rho}_i (x\dot{\beta}_i + \dot{w}_i)\sin(\alpha_i - \beta_i)]dx \tag{24}$$

The kinetic energy of the platform is expressed as

$$T_P = \frac{1}{2} m_P (\dot{x}_P{}^2 + \dot{y}_P{}^2) + \frac{1}{2} I_P \dot{\phi}^2 \tag{25}$$

m_P, I_P are mass and mass moment of the platform respectively.
Therefore, collecting all kinetic energies, equations (23-25), the total kinetic energy of the system is

$$T = \sum_{i=1}^{3} \frac{1}{2} m_s \dot{\rho}_i{}^2 + \sum_{i=1}^{3} \frac{1}{2} \int \rho_A [\dot{\rho}_i{}^2 + (x\dot{\beta}_i + \dot{w}_i)^2 + 2\dot{\rho}_i (x\dot{\beta}_i + \dot{w}_i) \sin(\alpha_i - \beta_i)] dx$$

$$+ \frac{1}{2} m_P (\dot{x}_P{}^2 + \dot{y}_P{}^2) + \frac{1}{2} I_P \dot{\phi}^2 \tag{26}$$

Since gravitational force is applied along Z-direction, perpendicular to the X-Y plane, potential energy due to gravitational force does not changed at all during any in-plane motions of the manipulator. Considering potential energy due to deformation of the linkage, total potential energy of the system is given as

$$V = \sum_{i=1}^{3} \int EI (w_i'')^2 dx \tag{27}$$

where: ρ_A := mass per length of the linkage
E := elastic modulus of the linkage
I := area moment of inertia of the linkage

Evaluating Lagrangian equations of the first type given by

$$\frac{d}{dt}\left(\frac{\partial T}{\partial \dot{X}_i}\right) - \frac{\partial (T - V)}{\partial X_i} = Q_i + \sum_{k=1}^{m} \lambda_k \frac{\partial \Gamma_k}{\partial X_i}, \quad i=1,2,\ldots, 9+3r \tag{28}$$

where: Q_i := generalized force
λ_k := k^{th} Lagrange multiplier
Γ_k := k^{th} constrained equation

the left-hand side of equation (28) is formulated as follows:

$$\frac{d}{dt}\left(\frac{\partial T}{\partial \dot{\rho}_i}\right) - \frac{\partial (T - V)}{\partial \rho_i} = (m_s + m)\ddot{\rho}_i + 0.5ml \sin(\alpha_i - \beta_i)\ddot{\beta}_i + \sum_{j=1}^{r} \ddot{\eta}_{ij} \sin(\alpha_i - \beta_i) \int \rho_A \psi_j dx$$

$$- 0.5ml \cos(\alpha_i - \beta_i)\dot{\beta}_i{}^2 - \sum_{j=1}^{r} \dot{\eta}_{ij} \dot{\beta}_i \cos(\alpha_i - \beta_i) \int \rho_A \psi_j dx \quad i=1,2,3 \tag{29}$$

$$\frac{d}{dt}\left(\frac{\partial T}{\partial \dot{\beta}_i}\right) - \frac{\partial (T - V)}{\partial \beta_i} = 0.5ml \sin(\alpha_i - \beta_i)\ddot{\rho}_i + ml^2 \ddot{\beta}_i / 3 + \sum_{j=1}^{r} \ddot{\eta}_{ij} \int \rho_A x \psi_j dx$$

$$+ \sum_{j=1}^{r} \dot{\eta}_{ij} \dot{\rho}_i \cos(\alpha_i - \beta_i) \int \rho_A \psi_j dx \qquad i=1,2,3 \tag{30}$$

$$\frac{d}{dt}\left(\frac{\partial T}{\partial \dot{X}_P}\right) - \frac{\partial (T-V)}{\partial X_P} = \begin{bmatrix} m_P & 0 & 0 \\ 0 & m_P & 0 \\ 0 & 0 & I_P \end{bmatrix} \begin{bmatrix} \ddot{x}_P \\ \ddot{y}_P \\ \ddot{\phi} \end{bmatrix} \tag{31}$$

$$\frac{d}{dt}\left(\frac{\partial T}{\partial \dot{\eta}_{ij}}\right) - \frac{\partial (T-V)}{\partial \eta_{ij}} = \sin(\alpha_i - \beta_i)\ddot{\rho}_i \int \rho_A \psi_j dx + \ddot{\beta}_i \int \rho_A x \psi_j dx + \ddot{\eta}_{ij} \int \rho_A \psi_j^{\,2} dx$$

$$- \int \rho_A \psi_j dx \cos(\alpha_i - \beta_i)\dot{\beta}_i \dot{\rho}_i + \int EI(\psi_j'')^2 dx \qquad i=1,2,3 \text{ and } j=1,2,..,r \tag{32}$$

m is mass of the linkage.
Since the number of generalized coordinates excluding vibration modes is nine, greater than the number of the degrees-of-freedom of the manipulator, three, six constraint equations should be considered in equations of the motion. From the geometry of three closed-loop chains, equation (4), a fundamental constrained equation is given by

$$\overline{A_i P} + \overline{P C_i} - \overline{A_i B_i} - \overline{B_i C_i} = 0 \qquad i=1,2,3 \tag{33}$$

Dividing equations (33) into an X-axis's component and a Y-axis's component, six constraint equations are given by

$$\Gamma_{2i-1} := \rho_i \cos\alpha_i + l\cos\beta_i - \sum_{j=1}^{r} \eta_{ij} \sin\beta_i - x_P - r\cos(\phi_i + \phi) = 0 \tag{34}$$

$$\Gamma_{2i} := \rho_i \sin\alpha_i + l\sin\beta_i + \sum_{i=1}^{r} \eta_{ij} \cos\beta_i - y_P - r\sin(\phi_i + \phi) = 0 \tag{35}$$

where:
$r\cos(\phi_i) := x'_{ci}$, $r\sin(\phi_i) := y'_{ci}$ $\qquad i=1,2,3$
From equation (34) and (35), the right-hand side of equation (28) is

$$F_{ai} + \sum_{k=1}^{6} \lambda_k \frac{\partial \Gamma_k}{\partial \rho_i} = F_{ai} + \lambda_{2i-1}\cos\alpha_i + \lambda_{2i}\sin\alpha_i \qquad i=1,2,3 \tag{36}$$

$$\sum_{k=1}^{6} \lambda_k \frac{\partial \Gamma_k}{\partial \beta_i} = \lambda_{2i-1}(-l\sin\beta_i - \sum_{j=1}^{r} \eta_{ij}\cos\beta_i) + \lambda_{2i}(l\cos\beta_i - \sum_{j=1}^{r} \eta_{ij}\sin\beta_i) \qquad i=1,2,3 \tag{37}$$

$$F_{ext} + \sum_{k=1}^{6} \lambda_k \frac{\partial \Gamma_k}{\partial X_P} = F_{ext} + \begin{bmatrix} -1 & 0 & -1 & 0 & -1 & 0 \\ 0 & -1 & 0 & -1 & 0 & -1 \\ s3_1 & c3_1 & s3_2 & c3_2 & s3_3 & c3_3 \end{bmatrix} \begin{bmatrix} \lambda_1 \\ \vdots \\ \lambda_6 \end{bmatrix} \tag{38}$$

where:
$s3_i := e_{ix}$, $c3_i := e_{iy}$
F_{ext} is an external force and F_{ai} is an actuating force.

$$\sum_{k=1}^{6}\lambda_k \frac{\partial \Gamma_k}{\partial \eta_{1j}} = -\lambda_1 \sin\beta_1 + \lambda_2 \cos\beta_1 \qquad j=1,2,..,r \tag{39}$$

$$\sum_{k=1}^{6}\lambda_k \frac{\partial \Gamma_k}{\partial \eta_{2j}} = -\lambda_3 \sin\beta_2 + \lambda_4 \cos\beta_2 \qquad j=1,2,..,r \tag{40}$$

$$\sum_{k=1}^{6}\lambda_k \frac{\partial \Gamma_k}{\partial \eta_{3j}} = -\lambda_5 \sin\beta_3 + \lambda_6 \cos\beta_3 \qquad j=1,2,..,r \tag{41}$$

Putting equations (29-32) and equations (36-41) together, the equations of motion for the planar parallel manipulator are complete with a total of $9+3\times r$ equations;

$$\begin{bmatrix} M_{11} & M_{12} & 0 & M_{14} \\ M_{12}^T & M_{22} & 0 & M_{24} \\ 0 & 0 & M_{33} & 0 \\ M_{14}^T & M_{24}^T & 0 & M_{44} \end{bmatrix}\begin{bmatrix} \ddot{\bar{\rho}} \\ \ddot{\bar{\beta}} \\ \ddot{\bar{X}}_P \\ \ddot{\bar{\eta}} \end{bmatrix} + \begin{bmatrix} V_1 \\ V_2 \\ 0 \\ V_4 \end{bmatrix} + \begin{bmatrix} 0 & 0 & 0 & 0 \\ 0 & 0 & 0 & 0 \\ 0 & 0 & 0 & 0 \\ 0 & 0 & 0 & K \end{bmatrix}\begin{bmatrix} \bar{\rho} \\ \bar{\beta} \\ \bar{X}_P \\ \bar{\eta} \end{bmatrix} = \begin{bmatrix} F_a \\ 0 \\ F_{ext} \\ 0 \end{bmatrix} + \begin{bmatrix} J_{\Gamma 1} \\ J_{\Gamma 2} \\ J_{\Gamma 3} \\ J_{\Gamma 4} \end{bmatrix}\begin{bmatrix} \lambda_1 \\ \lambda_2 \\ \lambda_3 \\ \lambda_4 \\ \lambda_5 \\ \lambda_6 \end{bmatrix} \tag{42}$$

where:

$$M_{11} = (m_s + m)\begin{bmatrix} 1 & 0 & 0 \\ 0 & 1 & 0 \\ 0 & 0 & 1 \end{bmatrix} \in R^{3\times 3} \qquad M_{12} = \frac{ml}{2}\begin{bmatrix} s_1 & 0 & 0 \\ 0 & s_2 & 0 \\ 0 & 0 & s_3 \end{bmatrix} \in R^{3\times 3}$$

$$M_{14} = m\begin{bmatrix} s_1\int\psi_1 d\xi & \cdots & s_1\int\psi_r d\xi & 0 & \cdots & 0 & 0 & \cdots & 0 \\ 0 & \cdots & 0 & s_2\int\psi_1 d\xi & \cdots & s_2\int\psi_r d\xi & 0 & \cdots & 0 \\ 0 & \cdots & 0 & 0 & \cdots & 0 & s_3\int\psi_1 d\xi & \cdots & s_3\int\psi_r d\xi \end{bmatrix} \in R^{3\times 3r}$$

$$M_{22} = \frac{ml^2}{3}\begin{bmatrix} 1 & 0 & 0 \\ 0 & 1 & 0 \\ 0 & 0 & 1 \end{bmatrix} \in R^{3\times 3} \qquad M_{33} = \begin{bmatrix} m_p & 0 & 0 \\ 0 & m_p & 0 \\ 0 & 0 & I_P \end{bmatrix} \in R^{3\times 3}$$

$$M_{24} = ml\begin{bmatrix} \int\psi_1\xi\, d\xi & \cdots & \int\psi_r\xi\, d\xi & 0 & \cdots & 0 & 0 & \cdots & 0 \\ 0 & \cdots & 0 & \int\psi_1\xi\, d\xi & \cdots & \int\psi_r\xi\, d\xi & 0 & \cdots & 0 \\ 0 & \cdots & 0 & 0 & \cdots & 0 & \int\psi_1\xi\, d\xi & \cdots & \int\psi_r\xi\, d\xi \end{bmatrix} \in R^{3\times 3r}$$

$$M_{44} = m\begin{bmatrix} \hat{M} & 0 & 0 \\ 0 & \hat{M} & 0 \\ 0 & 0 & \hat{M} \end{bmatrix} \in R^{3r\times 3r} \qquad \hat{M} = \begin{bmatrix} \int \psi_1^2 d\xi & \cdots & 0 \\ \vdots & \cdots & \vdots \\ 0 & \cdots & \int \psi_r^2 d\xi \end{bmatrix} \in R^{r\times r}$$

$$K = \frac{EI}{l^3}\begin{bmatrix} \hat{K} & 0 & 0 \\ 0 & \hat{K} & 0 \\ 0 & 0 & \hat{K} \end{bmatrix} \in R^{3r\times 3r} \quad \hat{K} = \begin{bmatrix} \int \psi_1''^2 d\xi & \cdots & 0 \\ \vdots & \cdots & \vdots \\ 0 & \cdots & \int \psi_r''^2 d\xi \end{bmatrix} \in R^{r\times r}$$

$$V_1 = -\begin{bmatrix} 0.5mlc_1\dot{\beta}_1^2 + \sum_{j=1}^{r} m\dot{\eta}_{1j}\dot{\beta}_1 c_1 \int \psi_j d\xi \\ 0.5mlc_2\dot{\beta}_2^2 + \sum_{j=1}^{r} m\dot{\eta}_{2j}\dot{\beta}_2 c_2 \int \psi_j d\xi \\ 0.5mlc_3\dot{\beta}_3^2 + \sum_{j=1}^{r} m\dot{\eta}_{3j}\dot{\beta}_3 c_3 \int \psi_j d\xi \end{bmatrix} \in R^3 \qquad V_2 = m\begin{bmatrix} \sum_{j=1}^{r} \dot{\eta}_{1j}\dot{\rho}_1 c_1 \int \psi_j d\xi \\ \sum_{j=1}^{r} \dot{\eta}_{2j}\dot{\rho}_2 c_2 \int \psi_j d\xi \\ \sum_{j=1}^{r} \dot{\eta}_{3j}\dot{\rho}_3 c_3 \int \psi_j d\xi \end{bmatrix} \in R^3$$

$$V_4 = -m\begin{bmatrix} c_1\dot{\rho}_1\dot{\beta}_1 \int \psi_1 d\xi \\ \vdots \\ c_1\dot{\rho}_1\dot{\beta}_1 \int \psi_r d\xi \\ c_2\dot{\rho}_2\dot{\beta}_2 \int \psi_1 d\xi \\ \vdots \\ c_2\dot{\rho}_2\dot{\beta}_2 \int \psi_r d\xi \\ c_3\dot{\rho}_3\dot{\beta}_3 \int \psi_1 d\xi \\ \vdots \\ c_3\dot{\rho}_3\dot{\beta}_3 \int \psi_r d\xi \end{bmatrix} \in R^{3r}$$

where : $s_i := \sin(\alpha_i - \beta_i)$ and $c_i := \cos(\alpha_i - \beta_i)$

$$J_{\Gamma 1} = \begin{bmatrix} \cos\alpha_1 & \sin\alpha_1 & 0 & 0 & 0 & 0 \\ 0 & 0 & \cos\alpha_2 & \sin\alpha_2 & 0 & 0 \\ 0 & 0 & 0 & 0 & \cos\alpha_3 & \sin\alpha_3 \end{bmatrix} \in R^{3\times 6}$$

$$J_{\Gamma 2} = \begin{bmatrix} s2_1 & c2_1 & 0 & 0 & 0 & 0 \\ 0 & 0 & s2_2 & c2_2 & 0 & 0 \\ 0 & 0 & 0 & 0 & s2_3 & c2_3 \end{bmatrix} \in R^{3\times 6}$$

where: $s2_i := -l\sin\beta_i - \cos\beta_i \sum_{j=1}^{r} \eta_{ij}$ and $c2_i := l\cos\beta_i - \sin\beta_i \sum_{j=1}^{r} \eta_{ij}$

$$J_{\Gamma 3} = \begin{bmatrix} -1 & 0 & -1 & 0 & -1 & 0 \\ 0 & -1 & 0 & -1 & 0 & -1 \\ s3_1 & -c3_1 & s3_2 & -c3_2 & s3_3 & -c3_3 \end{bmatrix} \in R^{3\times 6}$$

where: $s3_i := e_{ix}$, $c3_i := e_{iy}$

$$J_{\Gamma 4} = \begin{bmatrix} -\sin\beta_1 & \cos\beta_1 & & & & \\ \vdots & \vdots & 0 & & 0 & \\ -\sin\beta_1 & \cos\beta_1 & & & & \\ & & -\sin\beta_2 & \cos\beta_2 & & \\ 0 & & \vdots & \vdots & 0 & \\ & & -\sin\beta_2 & \cos\beta_2 & & \\ & & & & -\sin\beta_3 & \cos\beta_3 \\ 0 & & 0 & & \vdots & \vdots \\ & & & & -\sin\beta_3 & \cos\beta_3 \end{bmatrix} \in R^{3r\times 6}$$

4. Active vibration control

If the intermediate linkages of the planar parallel manipulator are very stiff, an appropriate rigid body model based controller, such as a computed torque controller (Craig, 2003), can yield good trajectory tracking of the manipulator. However, structural flexibility of the linkages transfers unwanted vibration to the platform, and may even lead to instability of the whole system. Since control of linear motions of the sliders alone can not result in both precise tracking of the platform and vibration attenuation of the linkages simultaneously, an additional active damping method is proposed through the use of smart material. As discussed, the vibration damping controller proposed here is applied separately to a PVDF layer and PZT segments, and the performance of each actuator is then compared. Attached to the surface of the linkage, both of these piezoelectric materials generate shear force under applied control voltages, opposing shear stresses which arise due to elastic deformation of the linkages.

The integrated control system for the planar parallel manipulator proposed here consists of two components. The first component is a proportional and derivative (PD) feedback control scheme for the rigid body tracking of the platform as given below:

$$u_i(t) = -k_p(\rho_{di} - \rho_i) - k_d(\dot{\rho}_{di} - \dot{\rho}_i), \quad i=1,2,3 \tag{43}$$

where k_p and k_d are a proportional and a derivative feedback gain respectively. ρ_{di} and $\dot{\rho}_{di}$ are desired displacement and velocity of the ith slider respectively. This signal is used as an input to electrical motors actuating ball-screw mechanisms for sliding motions. In the following, we introduce the second component of the integrated control system separately, for each of the piezoelectric materials examined, a PVDF layer and PZT segments, shown respectively in Fig. 7 and 8.

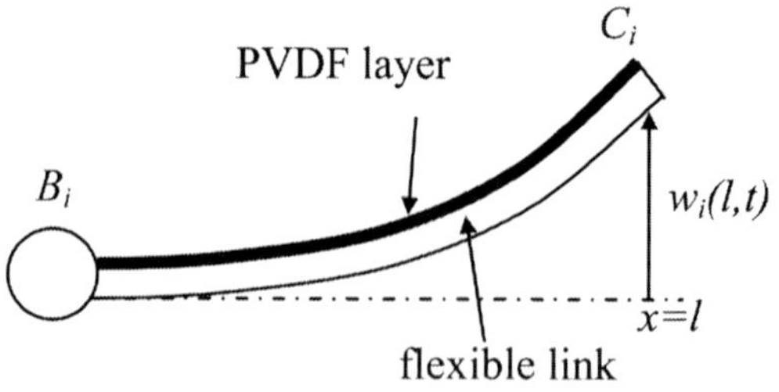

Fig. 7. Intermediate link with PVDF layer

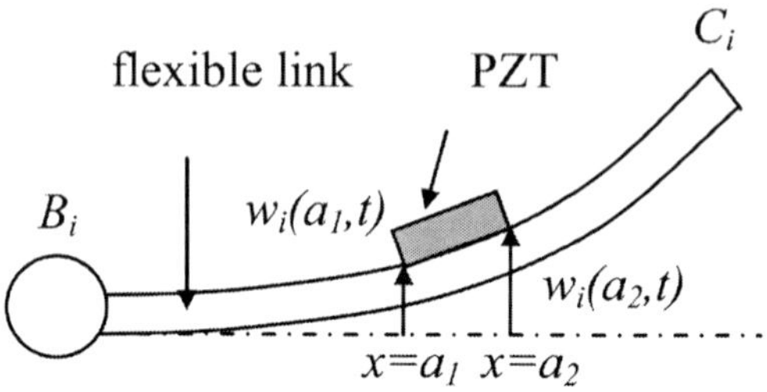

Fig. 8. Intermediate link with PZT actuator

4.1 PVDF actuator control formulation

A PVDF layer can be bonded uniformly on the one side of the linkages of the planar parallel manipulator, as shown in Fig. 7. When a control voltage, v_i, is applied to the PVDF layer, the virtual work done by the PVDF layer is expressed as

$$\delta W_{PVDF} = cv_i(t)\sum_{j=1}^{r}\psi'_j(l)\delta\eta_{ij} \tag{44}$$

where c is a constant representing the bending moment per volt (Bailey & Hubbard, 1985) and l is the link length. $(\cdot)'$ implies differentiation with respect to x. If the control voltage applied to the PVDF layer, v_i, is formulated as

$$v_i(t) = -k_l\dot{w}'_i(l,t) \qquad i=1,2,3 \tag{45}$$

the slope velocity of the linkages, $\dot{w}'(l,t)$, converges to zero, assuming no exogenous disturbances applied to the manipulator, hence vibration of the linkages is damped out. Since the slope velocity, $\dot{w}'(l,t)$, is not easily measured or estimated by a conventional sensor system, an alternative scheme, referred to as the L-type method (Sun & Mills, 1999), is proposed as follows:

$$v_i(t) = -k_l\dot{w}_i(l,t)\, i=1,2,3 \tag{46}$$

Instead of the slope velocity, $\dot{w}'(l,t)$, the linear velocity, $\dot{w}(l,t)$, is employed in formulation of the control law. The linear velocity, $\dot{w}(l,t)$, can be calculated through the integration of the linear acceleration measured by an accelerometer installed at the distal end of the linkages, C_i. The shape function, $\psi_j(\xi=1)$, and its derivative, $\psi'_j(\xi=1)$, have same trend of variation at the distal end of the linkages, C_i, in all vibration modes, as shown in Fig. 6 ;

$$\psi_j(\xi=1)\psi'_j(\xi=1)\geq 0 \quad j=1,2,---,r \tag{47}$$

Therefore, the control system maintains stability when employing the L-type method to formulate the control voltages, v_i .

4.2 PZT actuator control formulation

PZT actuators are manufactured in relatively small sizes, hence several PZT segments can be bonded together to a flexible linkage to damp unwanted vibrations. Assuming that only one PZT segment is attached to each intermediate linkage of the planar parallel manipulator, as shown in Fig. 8, the virtual work done by the PZT actuator is expressed as

$$\delta W_{PZT}=cv_i(t)\sum_{j=1}^{r}[\psi'_j(a_2)-\psi'_j(a_1)]\delta\eta_{ij} \tag{48}$$

a_1 and a_2 denote the positions of the two ends of the PZT actuator measured from B_i along the intermediate linkage, as shown in Fig. 8. As the PVDF layer is, the PZT actuator is controlled using the L-type method as

$$v_i(t)=-k_l[\dot{w}_i(a_2,t)-\dot{w}_i(a_1,t)] \quad i=1,2,3 \tag{49}$$

In contrast to the PVDF layer bonded uniformly to the manipulator linkages, the performance of the L-type scheme for the PZT actuator depends on the location of the PZT actuator. In order to achieve stable control performance, the PZT actuator should be placed in a region along the length of the linkage i.e. $x\in[a_1,a_2]$ as discussed in (Sun & Mills, 1999), where $\psi_j(x)$ and $\psi'_j(x)$ have the same trend of variation,

$$(\psi_j(a_2)-\psi_j(a_1))(\psi'_j(a_2)-\psi'_j(a_1))\geq 0 \tag{50}$$

As the number of vibration modes increases, it is difficult to satisfy the stability condition, given in equation(50), for higher vibration modes, since the physical length of a PZT actuator is not sufficiently small.

5. Simulation results

Simulations are performed to investigate vibrations of the planar parallel manipulator linkages and damping performance of both piezoelectric actuators used in the manipulator with structurally-flexible linkages. Specifications of the manipulator for simulations are listed in Table 1. The first three modes are considered in the dynamic model, i.e. r=3. A sinusoidal function with smooth acceleration and deceleration is chosen as the desired input trajectory of the platform;

$$x_P=\frac{x_f}{t_f}t-\frac{x_f}{2\pi}\sin(\frac{2\pi}{t_f}t) \tag{51}$$

Considering the target-performance in an electrical assembly process, such as wire bonding in integrated circuit fabrication, the goal for the platform is designed to move linearly 2 *mm* (x_f) within 10 *msec* (t_f). Feedback gains of the control system for the slider actuators are listed in Table 2. The feedback gain for piezoelectric actuators, k_I, is selected so that the control voltage, applied to the PVDF layer, does not exceed 600 Volts. A fourth order Runge-Kutta method was used to integrate the ordinary differential equations, given by Equation (42) at a control update rate of 1 *msec*, using MATLAB™ software. Parameters of piezoelectric materials, currently manufactured, are listed in Table 3. The placement position of the PZT actuator is adjusted to a_1=0.66, a_2=0.91, so that the first two vibration modes satisfy the stability condition given in equation (50).
Results of the PVDF layer are shown in Figures 9-12. Figure 9 shows that the error profile of the manipulator platform exhibits large oscillation at the initial acceleration, but continuously decreases due to the damping effect of the PVDF layer applied to the flexible linkages. The error profile of the platform without either of PVDF or PZT, labeled as "no damping" in Figure 9, shows typical characteristics of an undamped system. With Figure 10 showing deformation of the linkages on C_i, it reveals that the PVDF layer can damp structural vibration of the linkages in a gradual way. The first three vibration modes are illustrated in Figure 11. The first mode has twenty times the amplitude than the second mode, and one hundred times the amplitude than the third mode. The control output for the first slider actuator is shown in the upper plot of Figure 12, and control voltage for the first PVDF layer is shown in the lower plot of Figure 12. The control voltage, applied to the PVDF layer, decreases as the amplitude of vibration does.
Results of the PZT actuator are shown in Figures 13-17. Comparing Figure 13 with Figure 9, the PZT actuator exhibits better damping performance than the PVDF layer. The error profile of the platform, with the PZT actuator activated, enters steady state quickly and does not exhibit any vibration in steady state. The structural vibrations of the linkages, illustrated in Figure 14, are completely damped after 60 msec. The first three vibration modes are shown in Figure 15. The first mode has ten times the amplitudes than the other modes. Since the PZT actuator has higher strain constant than the PVDF, the PZT actuator can generate large shear force with relatively small voltage applied. The maximum voltage of the lower plot of Figure 16 is about 200 Volts, while that of the Figure 12 reaches 600 Volts. Due to the length of the linkage and the PZT actuator applied to the linkage, only the first two modes satisfy the stability condition, given by equation (50). However, this has little effect on damping performance, as shown in Figure 14 since the first two modes play dominant roles in vibration. If the placement of the PZT actuator change to a_1=0.4, a_2=0.65, only the first mode satisfies the stability condition, which leads to divergence of vibration modes, as shown in Figure 17.

6. Conclusion

In this chapter, the equations of motion for the planar parallel manipulator are formulated by applying the Lagrangian equation of the first type. Introducing Lagrangian multipliers simplifies the complexities due to multiple closed loop chains of the parallel mechanism and the structurally flexible linkages. An active damping approach applied to two different piezoelectric materials, which are used as actuators to damp unwanted vibrations of flexible linkages of a planar parallel manipulator. The proposed control is applied to PVDF layer and PZT segments. An integrated control system, consisting of a PD feedback controller, applied to electrical motors for rigid body motion control of the manipulator platform, and a L-type controller applied to piezoelectric actuators to damp unwanted linkage vibrations, is developed to permit the manipulator platform to follow a given trajectory while damping vibration of the manipulator linkages. With an L-type control scheme determining a control voltage applied, the piezoelectric materials have been shown to provide good damping performance, and eventually reduce settling time of the platform of the planar parallel manipulator. Simulation results show that the planar parallel manipulator, with the lightweight linkages, during rigid body motion, undergoes persistent vibration due to high acceleration and deceleration. Additionally, the PZT actuator yields better performance in vibration attenuation than the PVDF layer, but may enter an unstable state if the position of the PZT actuator on the linkage violates the stability condition for the dominant vibration modes. In the near future, we will perform vibration experiments with a prototype planar parallel manipulator based on presented simulation results.

Platform	side length mass	0.1 m 0.2 kg
Slider	mass	0.2 kg
Linear guide (Ball-screw)	stroke incline angle	0.4 m 150°, 270°, 30°
Link	length density modulus cross-section	0.2 m 2770 kg/m^3 73 GPa 0.025 m(W) * 0.015 m(H)

Table 1. Specification of the planar parallel manipulator

k_p	10,000 N/m
k_d	500 N-sec/m
k_I	4,000 V-sec/m for PVDF 1,500 V-sec/m for PZT

Table 2. Feedback control gains

	PVDF	PZT
modulus	2 GPa	63 GPa
length	0.2 m	0.05 m
thickness	0.28 mm	0.75 mm
width	0.025 m	0.025 m
density	1800 kg/m^3	7600 kg/m^3
d_{31}	22 * 10^{-12} m/V	110 * 10^{-12} m/V

Table 3. Parameters of piezoelectric materials

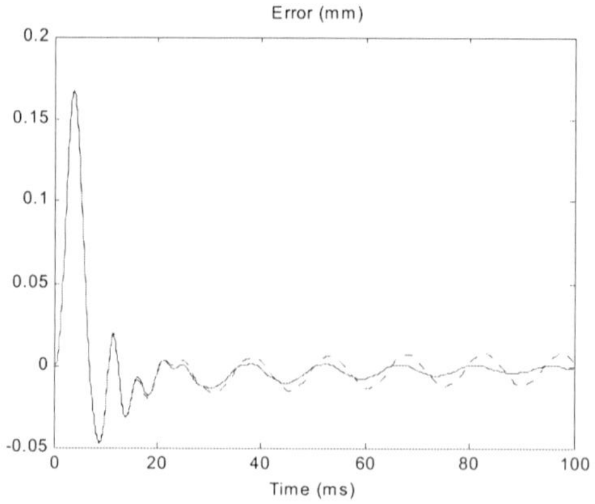

Fig. 9. Error profile of the platform (dotted: no damping, solid: with PVDF layer)

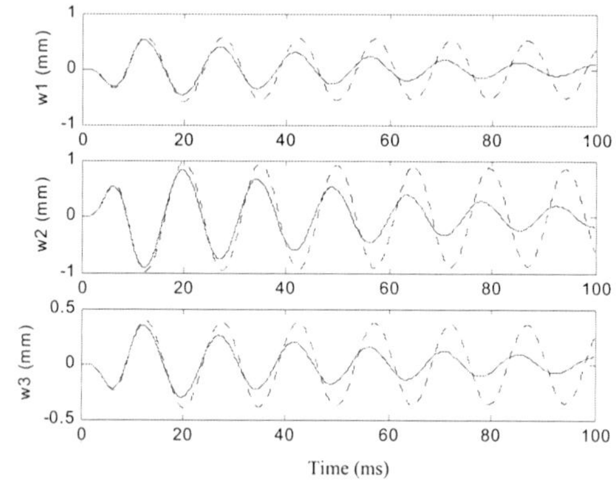

Fig. 10. Flexible deformation of each link (dotted: no damping, solid: with PVDF layer)

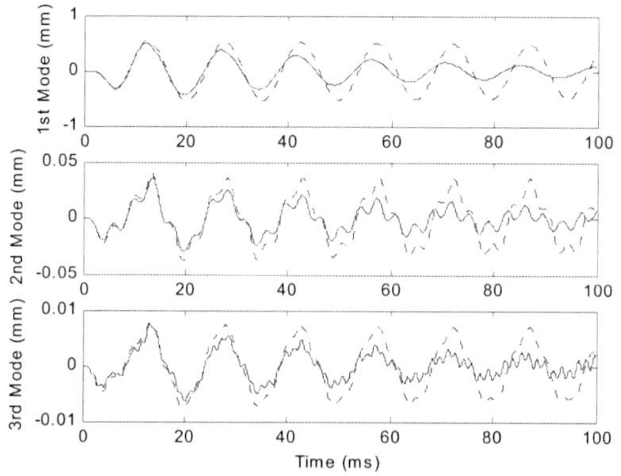

Fig. 11. The first three vibration modes of the first link (dotted: no damping, solid: with PVDF layer)

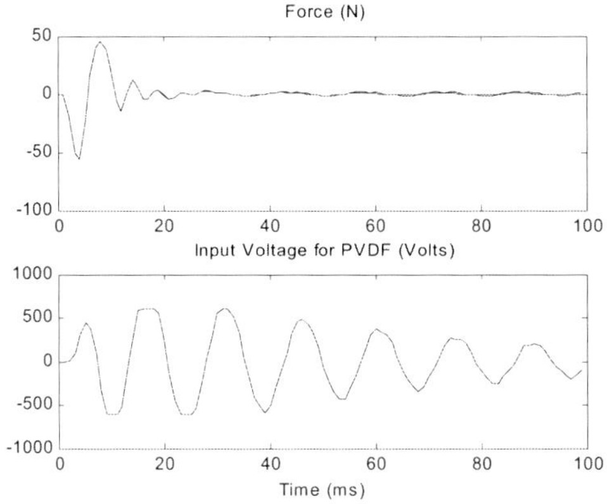

Fig. 12. Control output for the first link

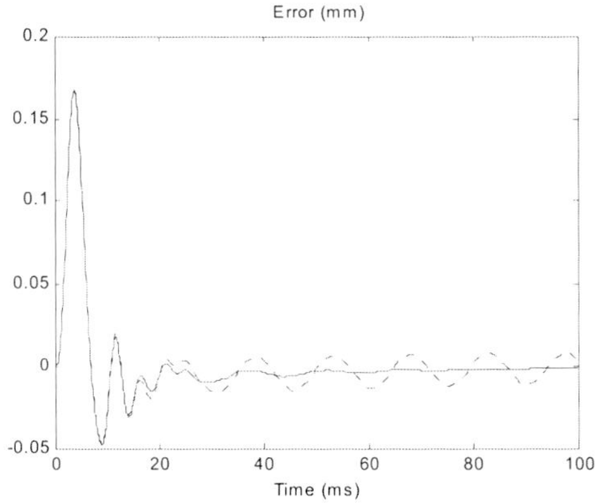

Fig. 13. Error profile of platform (dotted: no damping, solid: with PZT actuator)

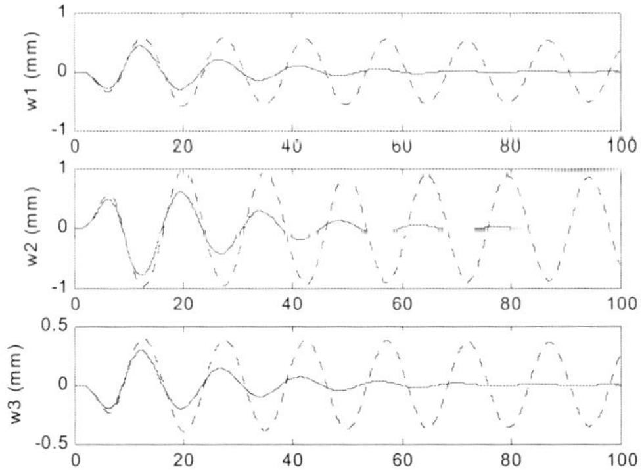

Fig. 14. Flexible deformation of each link (dotted: no damping, solid: with PZT actuator)

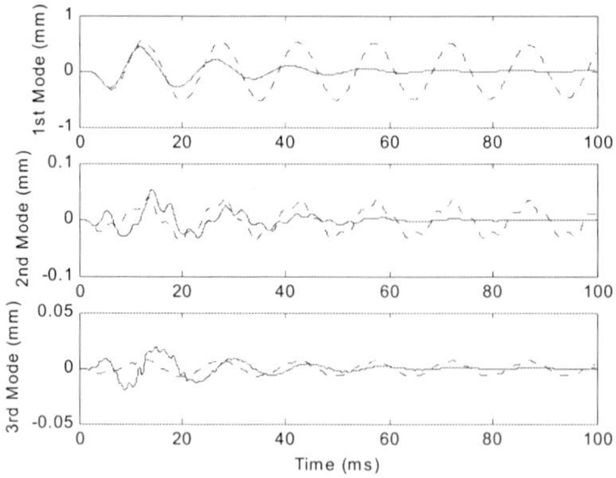

Fig. 15. The first three vibration modes of the first link(dotted: no damping, solid: with PZT actuator)

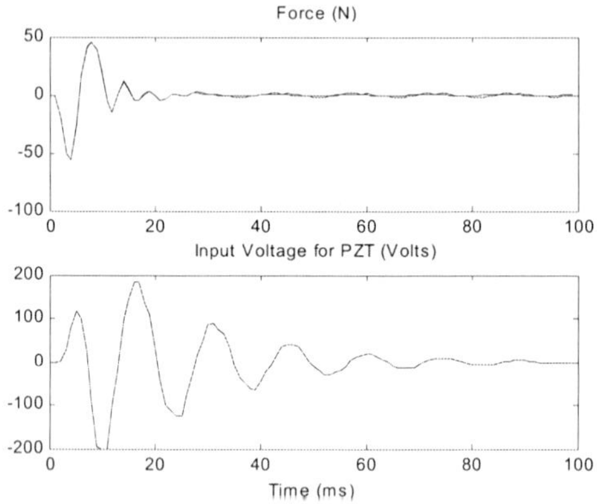

Fig. 16. Control output for the first link

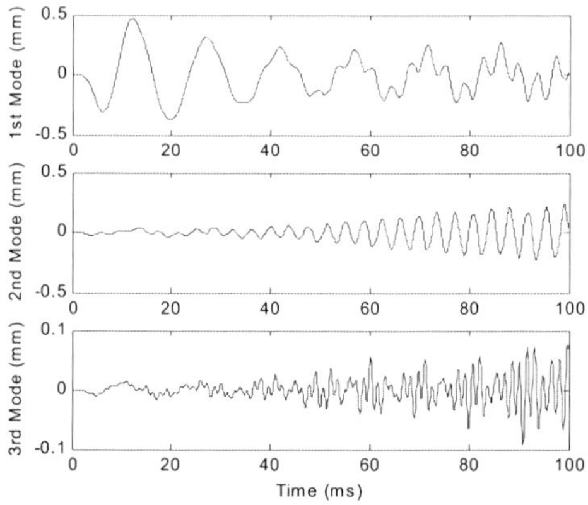

Fig. 17. The first three vibration modes of the first link with PZT actuator located on inappropriate place

7. References

Arai, T.; Cleary, K., Homma, K., Adachi, H., & Nakamura, T. (1991), Development of parallel link manipulator for underground excavation task, *Proceedings of International Symposium on Advanced Robot Technology*, pp. 541-548, Tokyo, Japan, March 1991

Bailey, T. & Hubbard, J. (1985), Distributed piezoelectric-polymer active vibration control of a cantilever beam, *Journal of Guidance, Control and Dynamics*, Vol. 8, No. 5, pp. 605-611, 0731-5090

Bellezza, F.; Lanari, L. & Ulivi, G. (1990), Exact modeling of the flexible slewing link, *Proceedings of IEEE International Conference on Robotics and Automation*, pp. 734-739, 0-8186-9061-5, Cincinnati, USA, May 1990, IEEE

Craig, J. J. (2003), *Introduction to Robotics*, Prentice-Hall, ISBN-10: 0387985069

Fattah, A.; Angeles, J. & Misra, K. A. (1995), Dynamic of a 3-DOF spatial parallel manipulator with flexible links, *Proceedings of IEEE International Conference on Robotics and Automation*, pp. 627-632, 0-7803-1965-6, Nagoya, Japan, May 1995, IEEE

Genta, G. (1993), *Vibration of structures and machines*, Springer, ISBN-10: 0387985069

Gosselin, C. & Angeles, J. (1990), Singularity analysis of closed-loop kinematic chains, *IEEE Transactions on Robotics and Automation*, Vol. 6, No. 3, pp. 281-290, 1552-3098

Gosselin C.; Lemieux, S., & Merlet, J.-P. (1996), A new architecture of planar three- degree-of-freedom parallel manipulator, *Proceedings of IEEE International Conference on Robotics and Automation*, pp. 3738-3743, 0-7803-2988, Minneapolis, USA, April 1996, IEEE

Heerah, I.; Kang, B., Mills, J. K., & Benhabib, B. (2002), Architecture selection and singularity analysis of a 3-degree-of-freedom planar parallel manipulator, *Proceedings of ASME 2002 Design Engineering Technical Conferences and Computers and Information in Engineering Conference*, pp. 1-6, 791836037, Montreal, Canada, October 2002, ASME

Kang, B.; Yeung, B., and Mills, J. K. (2002), Two-time scale controller design for a high speed planar parallel manipulator with structural flexibility, *Robotica*, Vol. 20, No. 5, pp. 519-528, 0263-5747

Kozak, K.; Ebert-Uphoff, I., & Singhose, W. E. (2004), Locally linearized dynamic analysis of parallel manipulators and application of input shaping to reduce vibrations, *ASME Journal of Mechanical Design*, Vol. 126, No. 1, pp. 156-168, 1050-0472

Low, K. H. & Vidyasagar, M. (1988), A Lagrangian formulation of the dynamic model for flexible manipulator systems, *ASME Journal of Dynamic Systems, Measurement, and Control*, Vol. 110, pp. 175-181, 0022-0434

Merlet, J.-P. (1996), Direct kinematics of planar parallel manipulators, *Proceedings of IEEE International Conference on Robotics and Automation*, pp. 3744-3749, 0-7803-2988, Minneapolis, USA, April 1996, IEEE

Sun, D. & Mills, J. K. (1999), PZT actuator placement for structural vibration damping of high speed manufacturing equipment, *Proceedings of the American Control Conference*, pp. 1107-1111, 0780349903, San Diego, USA, June 1999, IEEE

Toyama, T.; Shibukawa, T., Hattori, K., Otubo, K., and Tsutsumi, M. (2001), Vibration analysis of parallel mechanism platform with tilting linear motion actuators, *Journal of the Japan Society of Precision Engineering*, Vol. 67, No. 9, pp. 458-1462, 0916-78X

22

Task Space Approach of Robust Nonlinear Control for a 6 DOF Parallel Manipulator

Hag Seong Kim
Agency for Defense Development
Korea

1. Introduction

The dynamics and kinematics of a parallel manipulator has been widely researched by virtue of its a high force-to-weight ratio and widespread applications ranging from vehicle or flight simulator to machine tool despite a smaller workspace than a serial robot system (Merlet, 2000). Such a parallel system has been paid special attention as a typical multi-input multi-output nonlinear system to retain a high control performance. A control scheme for a 6 DOF parallel manipulator can be classified into two groups: a joint space based control scheme (Honegger et al., 2000; Kang et al., 1996; Kim et al., 2000; Nguyen et al., 1993; Sirouspour & Salcudean, 2001) and a task space based control strategy (Kang et al., 1996; Park, 1999; Ting et al., 1999). It is easy to realize the joint space based control scheme to a parallel manipulator as if the decoupled single-input single-output (SISO) control systems activate for a parallel mechanism. The simplicity has let many research activities pursue more specific approaches. As a result, the novel joint space based control approaches have been studied to improve the control performance by rejecting the nonlinear effects in the equations of motion (Honegger et al., 2000; Kang et al., 1996; Kim et al., 2000; Nguyen et al., 1993; Sirouspour & Salcudean, 2001). Particularly, for a parallel system driven by a hydraulic-servo system, joint space based robust nonlinear control scheme (Kim et al., 2000) has proposed. However, the research has dealt with excessively conservative uncertainties including gravity and known dynamic characteristics even though the friction effect can be neglected by the hydrostatic bearing. On the other hand, a task space based control for a 6 DOF parallel manipulator has a potential to meet excellent control performances under system uncertainties: inertia, modeling error, friction, etc. However, its scheme may be realized by the obtained the 6 DOF system state through a costly sensor or a novel nonlinear state estimation methodology. H_∞ robust control strategy (Park, 1999) and the adaptive control scheme (Ting et al., 1999) have been studied as the examples of task space based control. However, there have been still some weak points in the previous researches; the linearized model based approach and a simulation study only, respectively. Another task space based nonlinear control scheme has been proposed to a Stewart platform (Kang et al., 1996). However, it has also shown the computational simulation results only on the assumption to the system uncertainties that seems excessive. Furthermore, its treatment on stick-slip friction is minimal, which may give rise to serious deterioration of control performance in a real system where the frictional property is not negligible.

This paper focus on a theoretical and experimental study to develop a task space based robust nonlinear controller for a 6 DOF parallel system. This study starts from the indirect estimation of the system state essential to the task space based control instead of the application of a costly 6 DOF sensor. The 6 DOF system state is obtained by a numerical forward kinematic solution based on the Newton-Raphson method (Dieudonne et al., 1972; Nguyen et al., 1993) and an alpha-beta tracker (Friedland, 1973; Lewis, 1986). The feasibility of the indirect state estimation method is confirmed by the comparison of the results from the alpha-beta tracker with forward kinematic solution and the measured gyro signals, respectively. Then, the Friedland-Park friction estimator (Friedland & Park, 1992) is employed to attenuate the frictional disturbances in the actuators. The friction estimates are also compared to independently measured friction values (Park, 1999), which show reasonable agreement. Finally, the task space based robust nonlinear control scheme with the proposed estimation methods for system state and friction is proposed and theoretically proved by the representation of the practical stability for a 6 DOF parallel manipulator with uncertainties such as inertia, modeling error, friction, and measurement errors, etc. The proposed controller law exhibits remarkable regulation and tracking control performances to given several inputs. It is also shown that the proposed robust nonlinear control law with task space approach outperforms the task space based nonlinear control without the additional input for a robust control and a PID controller with the two independent estimators for the system state and friction in joint space.

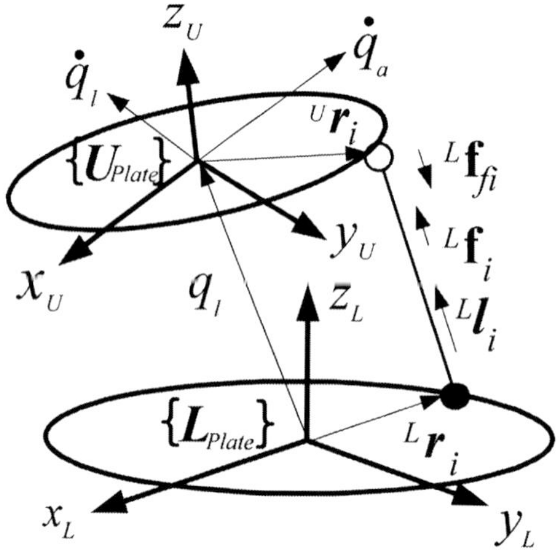

Fig. 1. The vector definitions for the mathematical model of the 6 DOF parallel manipulator

2. Dynamic model of a 6 DOF parallel manipulator

This section briefly describes the dynamic model of a 6 DOF parallel manipulator that has been extensively studied (Dasgupta & Mruthyunjaya, 1998; Kang et al., 1996; Kim et al., 2000). Fig. 1 describes the $\{L_{plate}\}$ and $\{U_{plate}\}$ coordinate systems as the base coordinate system for the inertial frame and the moving coordinate system for the body-fixed frame, respectively. Linear motions along the $x_L - y_L - z_L$ axis are surge (x), sway (y), and heave

(z), respectively. Rotational motions corresponding to the $x_U - y_U - z_U$ Euler angles are roll (θ_r), pitch (θ_p), and yaw (θ_y), respectively. The definition of each vector required to derive the kinematic and dynamic equations of the parallel mechanism are depicted in Fig. 1 ($i = 1$ to 6) as well.

For the angular and linear motions of the parallel manipulator, the following dynamic model can be derived by the Euler-Lagrangian method (Kang et al., 1996; Kim et al., 2000)

$$\mathbf{M}(\mathbf{q}_U,\xi)\ddot{\mathbf{q}}_U + \mathbf{C}(\mathbf{q}_U,\dot{\mathbf{q}}_U,\xi)\dot{\mathbf{q}}_U + \mathbf{G}(\mathbf{q}_U,\xi) = \mathbf{J}^T(\mathbf{q}_U)(\mathbf{f} - \mathbf{f}_f), \tag{1}$$

where $\mathbf{M}(\cdot) \in \mathbf{R}^{6\times6}$ is inertia, $\mathbf{C}(\cdot) \in \mathbf{R}^{6\times6}$ is Coriolis and centrifugal force, $\mathbf{G}(\cdot) \in \mathbf{R}^6$ is gravitational force, $\mathbf{J}(\cdot) \in \mathbf{R}^{6\times6}$ is Jacobian, ξ denotes uncertainties, $\mathbf{q}_U = [\mathbf{q}_l^T \;\; \mathbf{q}_a^T]^T \in \mathbf{R}^6$, $\mathbf{q}_l = [x \;\; y \;\; z]^T$, $\mathbf{q}_a = [\theta_r \;\; \theta_p \;\; \theta_y]^T$, $\mathbf{f} \in \mathbf{R}^6$ denotes the actuator forces, and $\mathbf{f}_f \in \mathbf{R}^6$ is an equivalent friction vector for actuators and joints.

In (1), it is assumed that $\dot{\mathbf{M}} - 2\mathbf{C}$ satisfies the skew symmetric property (Spong & Vidyasagar, 1989) and the parallel system is mechanically designed to avoid singularity of Jacobian matrix in the workspace. It is further assumed that system uncertainties are closed and bounded. The above assumptions are summarized as

Assumption 1. *The Jacobian is not singular.*

Assumption 2. *If ξ (constant or time-varying) represents uncertainties that include inertia, modeling error, and measurement noise, $\xi \in \Xi$, where Ξ is compact set.*

The actuator dynamics (both electrical and mechanical) may be neglected in this system to simplify the system model and apply the robust nonlinear control theory with ease. Nevertheless, in this paper, the friction of each actuator is considered since the friction may be the primary cause that deteriorates a control performance.

3. Estimations of system state and friction

This section briefly describes both the indirect system state estimation methodology and friction estimator. The length of each cylinder may be readily measured by relatively cheap sensor and directly applied to the joint space based control scheme, while the task space based control scheme requires 6 DOF data information which may be extracted by costly sensor. Alternatively, nonlinear observer (Kang, et al., 1998) may be implemented to acquire the 6 DOF system state. However, the idea is not adopted since the overall system stability and control performance may not be guaranteed on the observed state in a short time that can be appeared by undesirable condition called "peaking phenomenon" in a nonlinear system (Khalil, 1996). Furthermore, the mathematical relation between the angular velocities of the upper plate and the linear velocities of actuators (Dieudonne et al., 1972; Honegger et al., 2000; Kim et al., 2000; Nguyen et al., 1993) may be applied to calculate the angular velocities of the rigid upper plate. However, the mathematical relation is somewhat complicated, and can levy much computational time on the control system. Therefore, the following indirect state estimation methodology is presented to surmount such an adverse circumstance. The 6 DOF system state is estimated with the Newton-Raphson method and an alpha-beta tracker. The Newton-Raphson method performs well with a proper choice of the initial condition (Dieudonne et al., 1972). Furthermore, the derivatives of the system states are easily calculated via an alpha-beta tracker even though the tracker is applicable to a system with acceleration of zero mean process noise (Friedland, 1973; Lewis, 1986). In addition to that, the proposed control scheme in this paper also needs the first/second order derivatives of the arbitrary, continuous, desired position information, which cannot be pre-computed. Therefore, the indirect method is available to yield derivatives of arbitrary inputs as well.

This section also considers the equivalent friction estimator in order to reject the undesirable friction property. In general, the frictional property is changeable in the various conditions like lubrication, load, and even time, which means there may be uncertainties in friction. Furthermore, the uncertain and excessive feed-forward compensation may result in a phenomenon like limit cycle or undesirable control performance. Therefore, the Friedland-Park friction observer is pursued as a framework to compensate stick-slip friction among the previous approaches (Amstrong-Hélouvry et al., 1994; Panteley et al., 1998) since the Friedland-Park friction observer has shown excellent observer performance against general friction properties in spite of consideration on ideal Coulomb friction model (Friedland & Park, 1992). Unfortunately, the friction observer cannot be directly applied to the parallel mechanism due to its highly coupled nonlinear dynamics since it is targeted to SISO system. Therefore, the equivalent friction estimator is applied only in the context of friction estimator and control design with the assumption that each actuator system of the parallel manipulator could be modeled as an equivalent SISO system. The uncertainties in the friction estimates are regarded as an element of system uncertainties, which will be discussed later in the control design section. The observer with a readily implemented structure is briefly described for a decoupled parallel system as in a equivalent SISO system in the following.

$$\ddot{x}_f = \dot{v}_f = \frac{1}{m_{eq}}\left(w_f - f_f\right),\ f_f = c_f \cdot \mathrm{sgn}(v_f), \tag{2}$$

where x_f and v_f are the estimated linear displacement, velocity of each actuator, respectively, m_{eq} is the equivalent load of each cylinder, f_f is the actuator friction of each cylinder, c_f is the friction parameter, and w_f is the control force that includes additional robust control and estimated friction terms of each actuator. Then, the parameter $\hat{c}_f$ can be updated by

$$\hat{c}_f = z_f - k_f \left|v_f\right|^{\mu_f}, \tag{3}$$

$$\dot{z}_f = \frac{1}{m_{eq}} k_f \mu_f \left|v_f\right|^{\mu_f - 1}\left(w_f - \hat{f}_f\right)\mathrm{sgn}(v_f), \tag{4}$$

where $\hat{c}_f$ is estimated friction parameter, z_f is variable, $k_f > 0$ and $\mu_f \geq 1$ are constant gains. It should be noted that $\mu_f \geq 1$ since the dynamics of the variable z_f in (4) cannot be defined at the zero velocity in the case of $\mu_f < 1$.

It should be noted that the indirect state estimation scheme and the friction estimator have not been widely applied to a 6 DOF parallel system even though these may be often used independently in practice.

4. Robust nonlinear control design

This section presents the design of the robust nonlinear controller and the accompanying stability analysis for the 6 DOF parallel manipulator control system equipped with the aforementioned estimators in the previous section. The robust nonlinear control theory has been researched widely to guarantee practical stability for a nonlinear system with the detailed definitions (practical stability, uniform ultimate boundedness, etc.) and the assumptions described in Barmish et al.(1983), Corless & Leitmann (1981), and Khalil (1996). However, it may not be straightforward applied to a nonlinear system with stick-slip friction that does not satisfy the Caratheodory condition (Corless & Leitmann, 1981) at zero velocity where the friction represented by set-value map (*Caratheodory condition is mathematically required to guarantee the existence and the continuity of the solution*). Nevertheless, the previous studies (Hahn, 1967; Radcliffe & Southward, 1990) have shown that there still exists a continuous solution under stick-slip friction from a practical viewpoint. Therefore, it may be not too impudent to suppose the existence of a solution to a mathematical model for a real system with friction, which makes it possible to apply a robust nonlinear control theory into a parallel system with stick-slip friction. The following assumptions 3 is additionally made for a robust control design.

Assumption 3. There exist positive constant $\sigma_{\bar{M}}$, $\sigma_{\underline{M}} > 0$ such that

$$\sigma_{\underline{M}}\,\mathbf{I} < \mathbf{M}(\mathbf{q}_U, \xi) < \sigma_{\bar{M}}\mathbf{I}, \tag{5}$$

where $\forall \mathbf{q}_U \in \mathbf{D}_{qr}$, $\mathbf{D}_{qr} = \{\mathbf{q}_U \mid \|\mathbf{q}_U\| \le r,\ r \in [0,\infty)\}$, $\xi \in \Xi$, and Ξ is compact.

If the measurements or estimates of 6 DOF positional data contain uncertainties, the control function with the inverse of $\mathbf{J}^T(\cdot)$ is no longer valid. In this case, the proposed robust nonlinear control strategy requires additional assumptions:

Assumption 4. There exist a constant k_1 such that

$$\|\delta\mathbf{J}^T\mathbf{J}_c^{-T}\| \le k_1 < 1, \tag{6}$$

where $\mathbf{J}_c^T(\mathbf{q}'_U, \xi) = \mathbf{J}^T(\mathbf{q}_U, 0) + \delta\mathbf{J}^T(\delta\mathbf{q}_U, \xi)$, $\mathbf{q}'_U = \mathbf{q}_U - \delta\mathbf{q}_U$, $\mathbf{q}'_U$ is a vector of 6 DOF estimated system state , and $\delta\mathbf{q}_U$ is a vector of uncertainties in the 6 DOF positional estimated values.

Assumption 5. There exist a constant k_2 such that

$$\|\delta\dot{\mathbf{q}}_U + \mathbf{S}\delta\mathbf{q}_U\| \le k_2, \tag{7}$$

where $\delta\dot{\mathbf{q}}_U$ is a vector of uncertainties in the measured or estimated 6 DOF velocities, and $\mathbf{S} = \mathrm{diag}(\mathbf{S}_i) \in \mathbf{R}^{6\times 6}$, $\mathbf{S}_i > 0$.

Assumption 6. *It is assumed that each matrix in (1) can be represented as nominal plus deviation:*

$$\mathbf{M}(\mathbf{q}_U, \xi) = \mathbf{M}_0(\mathbf{q}'_U, 0) + \delta\mathbf{M}(\mathbf{q}_U, \xi), \quad \mathbf{C}(\mathbf{q}_U, \dot{\mathbf{q}}_U, \xi) = \mathbf{C}_0(\mathbf{q}'_U, \dot{\mathbf{q}}'_U, 0) + \delta\mathbf{C}(\mathbf{q}_U, \dot{\mathbf{q}}_U, \xi),$$

$$\mathbf{G}(\mathbf{q}_U, \xi) = \mathbf{G}_0(\mathbf{q}'_U, 0) + \delta\mathbf{G}(\mathbf{q}_U, \xi), \text{ and } \mathbf{f}_f(\mathbf{q}_U, \dot{\mathbf{q}}_U, \xi) = \hat{\mathbf{f}}_f(\mathbf{q}'_U, \dot{\mathbf{q}}'_U, 0) + \delta\mathbf{f}_f(\mathbf{q}_U, \dot{\mathbf{q}}_U, \xi).$$

The excessive uncertainties in the control design (Kang et al., 1996; Kim et al., 2000) including the nominal values of gravitational force and Coriolis force may result in undesirable control performance. Therefore, in the proposed control strategy of this paper, the uncertainties are minimized by directly compensating for the nominal gravitational force, Coriolis force, etc. on the assumption 6. Furthermore, the uncertainties in friction estimates are also considered as the element of system uncertainties.

Remark 7. The existence of the constant k_1 and k_2 in assumptions 4 and 5 seem to be restrictive. If the uncertainties or errors in the measurements or estimates of 6 DOF data are large or cannot be bounded, it is impossible to apply a MIMO robust control scheme. If the 6 DOF positional data are made directly available via a 6 DOF sensor, $\delta\mathbf{q}_U$ is negligible in the assumptions 4-6. The experimental results based on the indirect state estimation methodology show the reasonable agreement of the estimates later in Section 5.1.

Theorem 8. *Suppose that the system (1) satisfies the assumptions 1-6 with the definition of tracking error* $\tilde{\mathbf{q}}_U = \mathbf{q}_U - \mathbf{q}_{U_d}$, *where* $\mathbf{q}_{U_d} \in \mathbf{R}^6$ *is the desired trajectory . In addition, suppose that there exist the bounding functions* $\rho_1(\cdot)$ *and* $\rho_2(\cdot)$ *that satisfy the condition (9). Then, the system (1) is practically stable in the domain* $\mathbf{D}_r = \{\mathbf{e} \in \mathbf{R}^{12} \mid \|\mathbf{e}\| \le r,\ \mathbf{e} \equiv [\tilde{\mathbf{q}}'_U\ \dot{\tilde{\mathbf{q}}}'_U]^T,\ r \in [0,\infty)\}$ *for a given* ε *with the robust nonlinear control law (10).*

$$\mathbf{f}_{eq} = \mathbf{J}_e^{-T}\left\{\mathbf{G}_0(\mathbf{q}'_U) + \mathbf{M}_0(\mathbf{q}'_U)(\ddot{\mathbf{q}}_{U_d} - \mathbf{S}\dot{\tilde{\mathbf{q}}}'_U) + \mathbf{C}_0(\mathbf{q}'_U,\dot{\mathbf{q}}'_U)(\dot{\mathbf{q}}_{U_d} - \mathbf{S}\tilde{\mathbf{q}}'_U) - \mathbf{K}_P\tilde{\mathbf{q}}'_U - \mathbf{K}_V\dot{\tilde{\mathbf{q}}}'_U + \mathbf{J}_e^T(\cdot)\hat{\mathbf{f}}_f\right\}, \quad (8)$$

where $\mathbf{K}_P,\ \mathbf{K}_V \in \mathbf{R}^{6\times 6}$, $\mathbf{K}_P,\ \mathbf{K}_V$ *are symmetric positive definite matrices,*

$$\mathbf{S} = \mathrm{diag}(\mathbf{S}_i) \in \mathbf{R}^{6\times 6},\ \mathbf{S}_i > 0,\ \mathbf{K}_P + \mathbf{S}\,\mathbf{K}_V > 0,\ \begin{bmatrix} \mathbf{SK}_P & \mathbf{0} \\ \mathbf{0} & \mathbf{K}_V \end{bmatrix} > 0,\ \tilde{\mathbf{q}}'_U = \tilde{\mathbf{q}}_U - \delta\mathbf{q}_U,\ \dot{\tilde{\mathbf{q}}}'_U = \dot{\tilde{\mathbf{q}}}_U - \delta\dot{\mathbf{q}}_U,$$

and $\delta\mathbf{q}_U$, $\delta\dot{\mathbf{q}}_U$ *are the uncertainty vectors due to measured or estimated 6 DOF position and velocity errors, respectively.*

$$\boldsymbol{\varphi}(\cdot) = \mathbf{h}_1(\cdot) + \mathbf{h}_2(\cdot) + \mathbf{h}_3(\cdot) - \delta\mathbf{J}^T\mathbf{J}_e^{-T}\mathbf{v},\ \text{and}\ \|\boldsymbol{\varphi}(\cdot)\| \le \rho_1 + k_1\|\mathbf{v}\| \quad (9)$$

where

$$\mathbf{h}_1(\cdot) = -\delta\mathbf{M}(\cdot)\ddot{\mathbf{q}}_{U_d} - \delta\mathbf{C}(\cdot)\dot{\mathbf{q}}_{U_d} - \delta\mathbf{G}(\cdot) - \mathbf{J}_e^T(\cdot)\delta\mathbf{f}_f - \mathbf{M}(\cdot)\delta\ddot{\mathbf{q}}_U - \mathbf{C}(\cdot)\delta\dot{\mathbf{q}}_U,$$

$$\mathbf{h}_2(\cdot) = \delta\mathbf{M}(\cdot)\mathbf{S}\dot{\tilde{\mathbf{q}}}_U + \delta\mathbf{C}(\cdot)\mathbf{S}\tilde{\mathbf{q}}_U,\ \mathbf{h}_3(\cdot) = -\delta\mathbf{J}^T\mathbf{f}_{eq}(\cdot) + \delta\mathbf{J}^T\mathbf{f}_f,\ \text{and}\ \|\mathbf{h}_1(\cdot) + \mathbf{h}_2(\cdot) + \mathbf{h}_3(\cdot)\| \le \rho_1(\cdot) < \rho_2(\cdot).$$

$$\mathbf{f} = \mathbf{f}_{eq} + \mathbf{J}_e^{-T}\mathbf{v},\ \textit{and}\ \ \mathbf{v} = \begin{cases} -\dfrac{\rho_2(\mathbf{e})}{1-k_1}\dfrac{\mathbf{w}}{\|\mathbf{w}\|}, & \textit{if}\ \|\mathbf{w}\| \ge \varepsilon \\ -\dfrac{\rho_2(\mathbf{e})}{1-k_1}\dfrac{\mathbf{w}}{\varepsilon}, & \textit{if}\ \|\mathbf{w}\| < \varepsilon \end{cases}, \quad (10)$$

where $\mathbf{w}(\cdot) = (\dot{\tilde{\mathbf{q}}}'_U + \mathbf{S}\tilde{\mathbf{q}}'_U)\rho_2(\cdot)$.

Proof. Consider the following Lyapunov candidate function:

$$V = \frac{1}{2}(\dot{\tilde{\mathbf{q}}}'_U + \mathbf{S}\tilde{\mathbf{q}}'_U)^T\mathbf{M}(\dot{\tilde{\mathbf{q}}}'_U + \mathbf{S}\tilde{\mathbf{q}}'_U) + \frac{1}{2}\tilde{\mathbf{q}}'^T_U(\mathbf{K}_P + \mathbf{SK}_V)\tilde{\mathbf{q}}'_U. \quad (11)$$

If the Lyapunov candidate function is chosen with real 6 DOF state (not estimated 6 DOF system state), then the practical stability cannot be rigorously proved since there exists unmatched condition caused by the bounded value in the assumption 5. Therefore, the Lyapunov candidate function is selected as (11) on the assumption 5. The positive definite and decrescent property of this candidate function was presented in Kim et al. (2000). As a consequence, there exist constants γ_1, $\gamma_2 > 0$ such that $\gamma_1 \|\mathbf{e}\|^2 \le V \le \gamma_2 \|\mathbf{e}\|^2$. With additional measurement or estimation error and the assumption 6, the system dynamics (1) can be rearranged into

$$\mathbf{M}(\cdot)\ddot{\tilde{\mathbf{q}}}'_U + \mathbf{C}(\cdot)\dot{\tilde{\mathbf{q}}}'_U = -\mathbf{M}_0\ddot{\mathbf{q}}_{U_d} - \mathbf{C}_0\dot{\mathbf{q}}_{U_d} - \mathbf{G}_0 + \mathbf{J}_e^{\prime T}\mathbf{f} - \mathbf{J}_e^{\prime T}\hat{\mathbf{f}}_f - \delta\mathbf{J}^T\mathbf{f} + \delta\mathbf{J}^T\mathbf{f}_f + \mathbf{h}_1(\cdot). \tag{12}$$

If the assumptions 1-4, and 6, the skew symmetric property of $\dot{\mathbf{M}} - 2\mathbf{C}$, and control input (10) are considered, then the derivative of the Lyapunov candidate function (11) becomes after mathematical manipulations

$$\dot{V} \le -\gamma_3\|\mathbf{e}\|^2 + (\dot{\tilde{\mathbf{q}}}'_U + \mathbf{S}\tilde{\mathbf{q}}'_U)^T\mathbf{v} + \left\|\dot{\tilde{\mathbf{q}}}'_U + \mathbf{S}\tilde{\mathbf{q}}'_U\right\|\rho_1 + k_1\left\|\dot{\tilde{\mathbf{q}}}'_U + \mathbf{S}\tilde{\mathbf{q}}'_U\right\| \cdot \|\mathbf{v}\|, \tag{13}$$

where $\gamma_3 = \lambda_{\min}\left(\begin{bmatrix} \mathbf{SK}_P & 0 \\ 0 & \mathbf{K}_V \end{bmatrix}\right)$.

In the case $\|\mathbf{w}\| \ge \varepsilon$, (13) with (9) and (10) can be further reduced to yield

$$\dot{V} \le -\gamma_3\|\mathbf{e}\|^2. \tag{14}$$

In the case $\|\mathbf{w}\| < \varepsilon_2$, (13) can be simplified to

$$\dot{V} \le -\gamma_3\|\mathbf{e}\|^2 + \frac{\varepsilon}{4}. \tag{15}$$

The details of the derivation of (14) and (15) are shown in Khalil (1996).
Subsequently, ε and $\mu(\varepsilon)$ are chosen such that $\varepsilon < 2\gamma_3 \cdot \gamma_2^{-1} \cdot \gamma_1 \cdot r^2$ and $\mu(\varepsilon) = \sqrt{\gamma_3^{-1}(\varepsilon/4) + h}$ for $h > 0$,

$$\dot{V} \le -\gamma_3\|\mathbf{e}\|^2 + \frac{\varepsilon}{4} < 0, \ \forall \mu(\varepsilon) \le \|\mathbf{e}\| < r. \tag{16}$$

Therefore, for any given ε, if $\mu(\varepsilon) < \|\mathbf{e}(t_0)\| < \gamma_2^{-1}(\gamma_1(r))$, then $\dot{V}$ is strictly negative, which implies that there exists a finite time t_1 such that

$$t_1 \le t_0 + \frac{1}{\gamma_3 h}\left\{\gamma_2\|\mathbf{e}(t_0)\|^2 - \gamma_1\left(\frac{\varepsilon}{4\gamma_3} + h\right)\right\},$$

and the state stays in the set $\mathbf{D}_\mu = \{\mathbf{e} : \|\mathbf{e}\| \le \mu(\varepsilon)\}$ after time t_1 (Canudus de Wit, Siciliano, & Bastin, 1996). As a result, the system response is uniformly ultimately bounded, which implies practical stability via the controller (10).

5. Experiments

In this section, the proposed task space based robust nonlinear control strategy is experimentally investigated for a 6 DOF parallel manipulator, which compares to the nonlinear control with the estimators of the system state and friction in a task space and a PID control with the system state and friction estimators in a joint space. In Fig. 2, the control block diagrams displays the implementation of the task space based robust nonlinear control strategy proposed in the previous section and another two control laws, namely, task space based nonlinear control and joint space based PID control which both treat the estimates of the system state and friction.

Fig. 3 shows the experimental apparatus of motion control system with the embedded control structure in Fig. 2, which consists of 1) Six electrical cylinders (ETS32-B08PZ20-CMA150-A, Parker Inc.), 2) Control systems (Pentium III 800 PC-based system), 3) Motor amplifiers (OEM-570T, Compumotor Inc.), 4) D/A board for actuators (AT-A0-6/10, NI Corp.), 5) Encoder board (AT6450, Parker Inc.), and 6) 12bit A/D and D/A converter (LAB-PC+, NI Corp.). The sampling time for the control system is 3msec. Rate transducer (RT02-0820-1, Humphrey Inc.) is applied to investigate the indirect state estimation performance as well. Table 1 describes the parameter values of the parallel manipulator. The experimental results are evaluated in the following procedure. First, the indirect method for the acquisition of the estimated 6 DOF data is examined through the results from the alpha-beta tracker with forward kinematic solution and the measured gyro signals, respectively. Second, the performance of the equivalent friction estimator is evaluated by comparison between the independently measured data (Park, 1999) and the estimates. Finally, the control performance of the proposed robust nonlinear control law (10) in task space is compared to the task space based nonlinear control law (8) and the joint space based PID control with the estimators for the system state and friction.

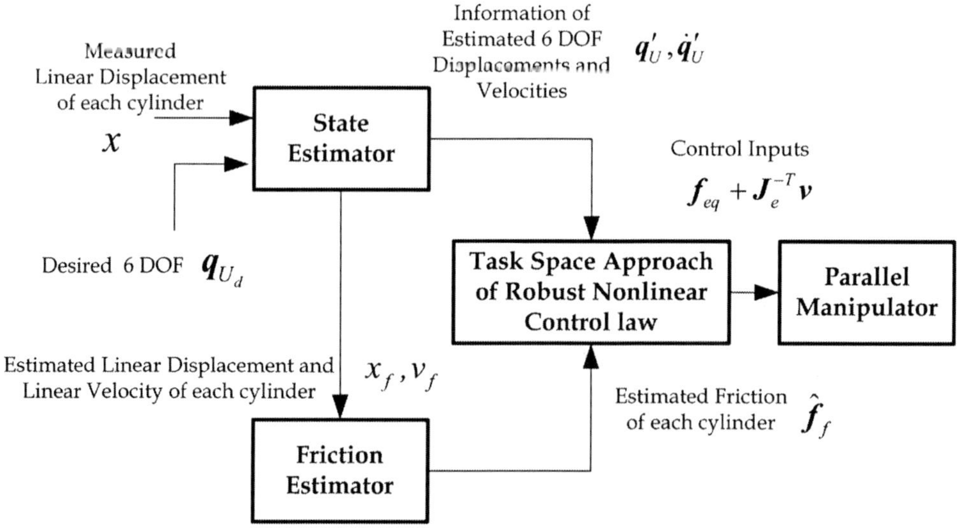

Fig. 2. Control block diagrams (a) Task space based robust nonlinear control scheme with the system state and friction estimators

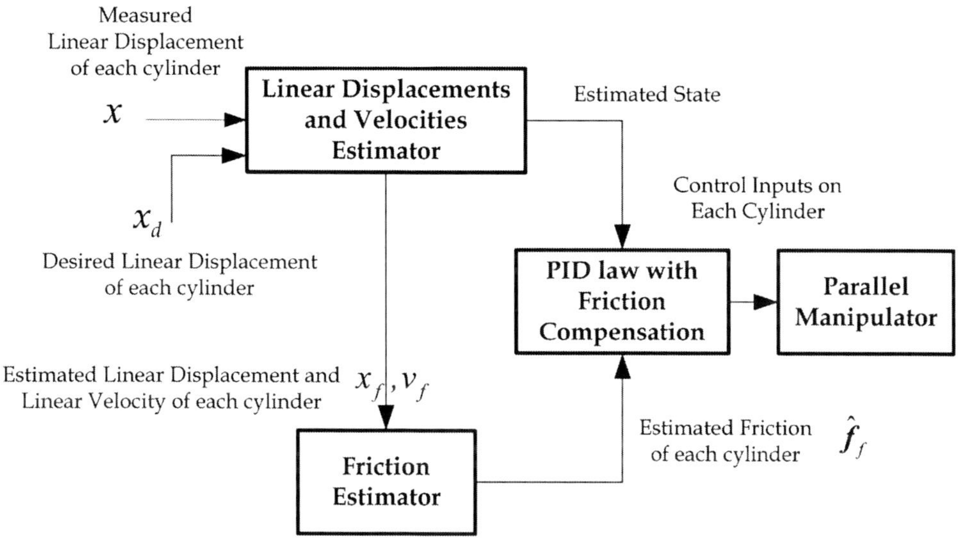

Fig. 2. Control block diagrams (b) Joint space based PID control with the estimators of the system state and friction

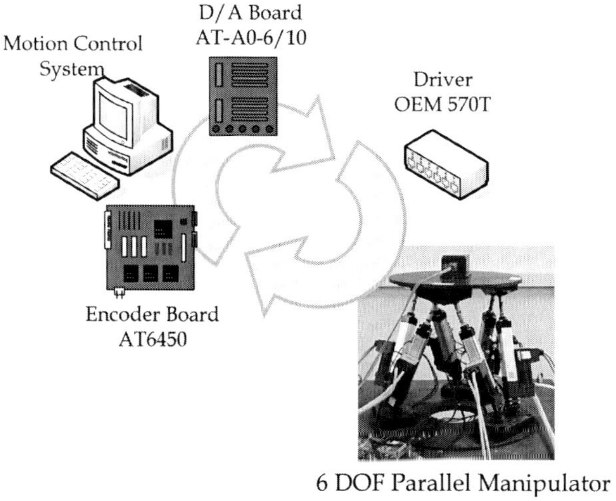

Fig. 3. Experimental apparatus of motion control system

Parameter	Description	Value	Unit
l_{min}, l_{max}	Min./Max. Stroke of Cylinder	0.365/0.51	$[m]$
m_U	Mass of Upper Plate	24.0	$[Kg]$
I_{xx}, I_{yy}, I_{zz}	Moment of Inertia of Upper Plate	0.4315, 0.4316, 0.6111	$[Kg \cdot m^2]$
r_L, r_U	Radius of Lower Plate/Upper Plate	0.24/0.16	$[m]$

Table 1. Parameter values of a 6 DOF parallel manipulator

5.1 State estimator

In this sub-section, the performance of the numerical method and the alpha-beta tracker are investigated to confirm the estimated system state to be feasible prior to the application for the task space based control approach. The sensing and estimation procedure is enumerated as follows;

1) Measure the length of each cylinder.
2) Use the alpha-beta tracker for each length signal
3) Apply the numerical method to obtain a forward kinematic solution
4) Use alpha-beta tracker to acquire the derivatives of 6 DOF positional state from the forward kinematic solution

Firstly, the measured cylinder lengths are compared to the inverse kinematic solution based on the 6 DOF estimates from the Newton-Raphson numerical method (tolerance 10^{-7}) as in Fig. 4. The result shows less than 0.1% errors (normal length 435mm) to a multi-directional sinusoidal inputs (roll (2.0°/1.0Hz), pitch (5.0°/0.5Hz), yaw (2.5°/1.0Hz) and heave (5.0mm/0.5Hz)), which verifies that the assumption 4 is satisfied since k_1 is less than 0.1 in the case of the intended ±5% uncertainty in each cylinder length. The installed rate transducer (RT02-0820-1, Humphrey Inc.) as a sensor providing a base line has checked the fidelity of the 6 DOF estimator through the comparison between the estimated and measured rotational velocities of angular motions.

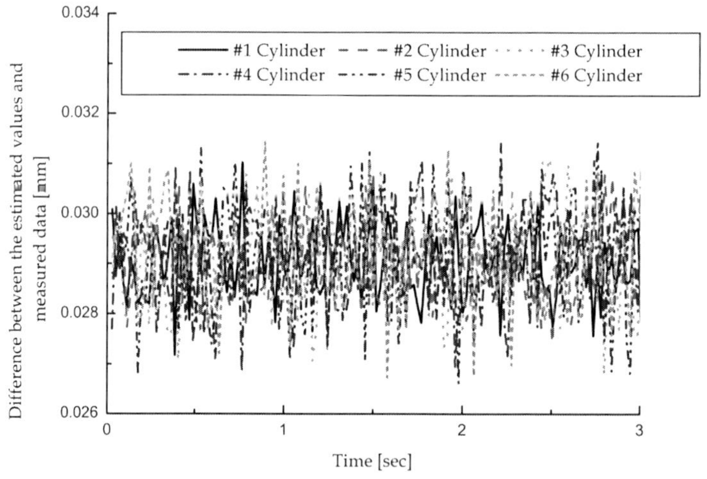

Fig. 4. Errors between measured lengths and inverse kinematic solutions from the estimated 6 DOF positional data based on the numerical method and the alpha-beta tracker

In Fig. 5, the comparisons between the rate transducer readings and the estimated angular velocities to sinusoidal position inputs (Roll: 2.0°/1.0Hz, Pitch: 5.0°/0.5Hz), which gives fidelity that the estimation scheme truly yields the derivatives of motion signals without complicated calculation and the assumption 5 is feasible.

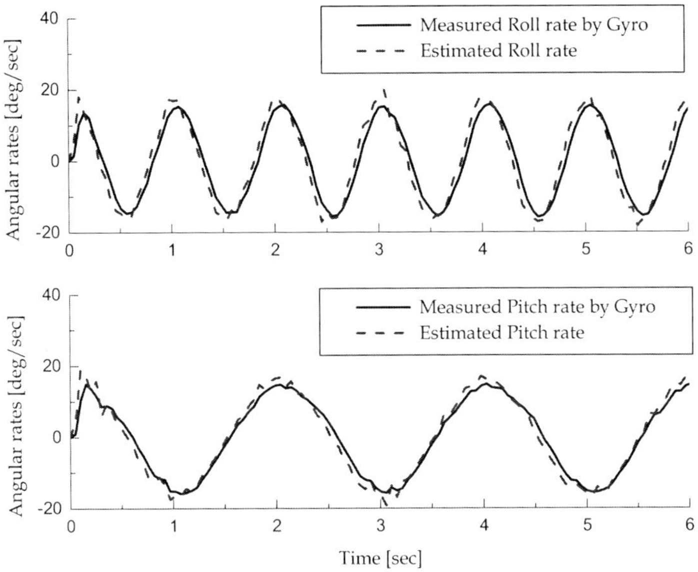

Fig. 5. The measured data and the estimated signals by the alpha-beta tracker to sinusoidal inputs (Roll: 2.0°/1.0 Hz, Pitch: 5.0°/0.5 Hz)

5.2 Friction estimator

This subsection describes the friction estimator proposed to reject the frictional disturbance for the enhanced control performance. As mentioned in Section 3, the excessive or deficient feed-forward friction compensation to step input under uncertain frictional disturbance makes an oscillatory or a sticking steady state, respectively. Fig. 6 shows the example of the phenomenon when incongruent feed-forward friction compensation with independently measured value (Park, 1999) is applied to the same system (the result can be compare to that of Fig. 8 (a) in subsection 5.3 later).

Fig. 7 presents that comparison of errors between PID control with friction estimator and PID control without friction compensator to a roll input (Sine: 5°/0.5Hz), which explains that the friction compensator is truly required. The figure also shows that the good friction estimation result of the 3rd cylinder (other cylinders have similar results) through PID control with the friction estimator and control performance becoming better as the time increases. The gains k_f and μ_f are 10.0 and 1.5, respectively in this estimator. It should be noted that the independently measured friction property of this parallel system (Park, 1999) may depends on load condition, lubrication condition, temperature, even time, etc. In the proposed robust nonlinear control, the difference between the bounded real friction and the estimated friction is considered as the element of the system uncertainty in (9).

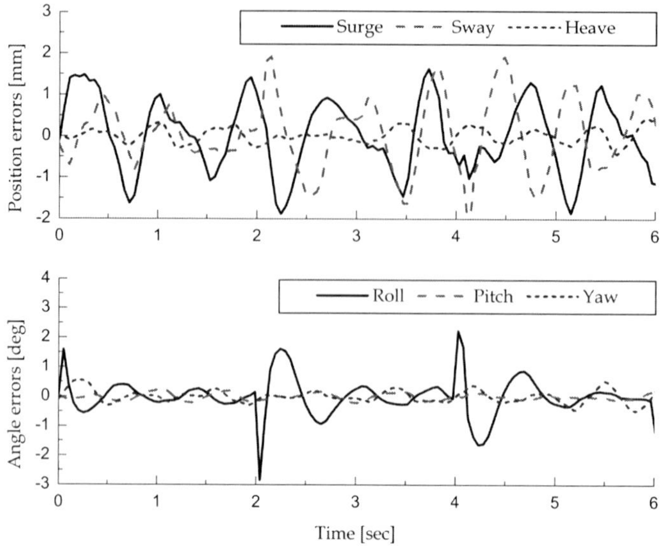

Fig. 6. Influence of the feed-forward friction compensation with PID control to a step input (Roll: 5.0°/0.25 Hz)

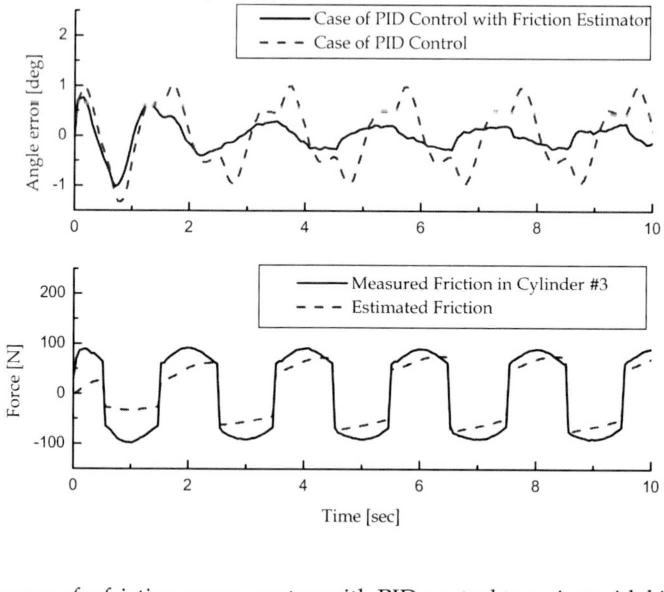

Fig. 7. Effectiveness of a friction compensator with PID control to a sinusoidal input (Roll: 5.0°/0.5 Hz)

5.3 Regulating and tracking performances

In this subsection, the control performance of the proposed task space based robust nonlinear control law with the estimators for the system state and friction (*hereafter this control law is named* TRNCE) is presented. As unbiased benchmarking controllers, task space based nonlinear control law with the system state and friction estimation method (*hereafter the control law is called* TNCE) and joint space based PID control law with the estimators for the system state and friction (*hereafter the control law is* PIDE) are employed. It should be noted that TNCE (8) is similar to a task space based PD controller and handles the perfectly known nonlinearities. On the other hand, the TRNCE (10) deals with the uncertainties of system parameters and frictions additionally. The experimentally tuned PID control gains K_{P_gain}, K_{I_gain}, K_{D_gain} are 100, 800, and 20, respectively, which result in smaller steady state errors than those by the gains in Park (1999). The control gains for the TRNCE and TNCE are:

$$\mathbf{K}_P = 1\times10^5 \cdot \begin{bmatrix} 0.456 & 0 & 0 & 0 & 0.0234 & 0 \\ 0 & 0.456 & 0 & -0.01404 & 0 & 0 \\ 0 & 0 & 3.75 & 0 & 0 & 0 \\ 0 & -0.01404 & 0 & 0.0312 & 0 & 0 \\ 0.0234 & 0 & 0 & 0 & 0.052 & 0 \\ 0 & 0 & 0 & 0 & 0 & 0.0208 \end{bmatrix},$$

$$\mathbf{K}_V = 1\times10^4 \cdot \begin{bmatrix} 0.06 & 0 & 0 & 0 & 0.009 & 0 \\ 0 & 0.12 & 0 & -0.009 & 0 & 0 \\ 0 & 0 & 1.16 & 0 & 0 & 0 \\ 0 & -0.009 & 0 & 0.02 & 0 & 0 \\ 0.009 & 0 & 0 & 0 & 0.02 & 0 \\ 0 & 0 & 0 & 0 & 0 & 0.0027 \end{bmatrix},$$

$\mathbf{S} = 10 \cdot \mathbf{I}$, $\varepsilon = 2.0$, and $k_1 = 0.1$.

It is further assumed that there exist such system uncertainties as 5% in inertia, 5% in gravity force and 1% in Jacobian. The gain matrices chosen above can be easily confirmed the positive definiteness condition in Theorem 8. The TRNCE gains seem much higher than those of PIDE. However, it comes from that the TRNCE calculates the desired force from the gain matrices and Jacobian, while the PIDE produces just control input calculated by the position errors and estimated friction.

Firstly, the regulation performance is investigated. Fig. 8 shows that the nonlinear approaches (TRNCE and TNCE) have superior overall regulating performance to a step input (Roll: 5°/0.25Hz) than the PIDE. With a view point of pseudo- steady-state error to roll motion input, TRNCE shows ±0.3° of error bound; on the other hand, the PIDE shows ±0.1° of superior error bound even though there exists 16% overshoot in the transient response. However, large and oscillatory errors by the PIDE are observed in the other motions; the other motion errors by PIDE are twice or more those by the TRNCE. The regulating performances by TNCE show the similar to those by TRNCE. The above outcomes stem from the fact that the PIDE does not consider the sensitivity of 6 DOF displacements on length variation, that is, Jacobian, which results in overall performance in task space to be inferior to those of TRNCE and TNCE. Reducing the magnitude of ε may give further enhancement of regulation performance. However, such an approach may degrade the control performance by chattering effect as described in Khalil (1996) due to a fast switching control input that may excite high frequency modes in the system.

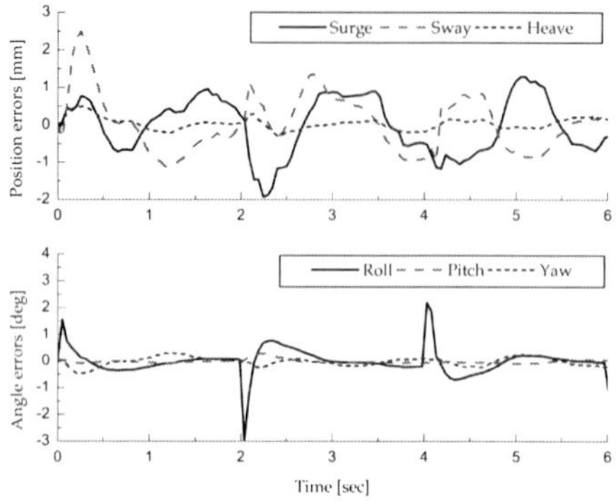

(a) PIDE

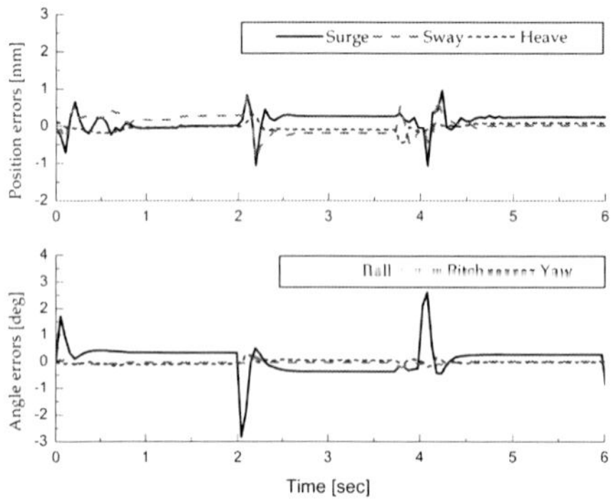

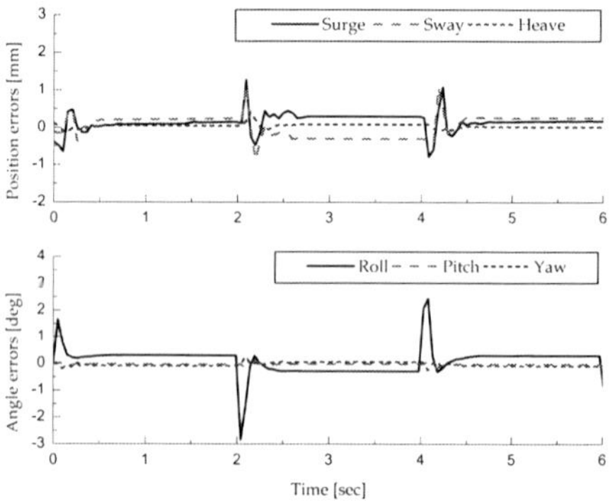

(b) TNCE (c) TRNCE

Fig. 8. Regulating errors of 6DOF motions to a step input (Roll: 5.0°/0.25 Hz)

Tracking errors to a sinusoidal input of roll motion (5.0°/0.5 Hz) are examined as well (not in this paper). In a steady state, the translation error bounds of the TRNCE are smaller than +0.41/-0.4 mm, those of the TNCE are smaller than +0.45/-0.5 mm, while those of the PIDE are larger than +0.8/-1.1 mm. All the rotational errors of the TRNCE are bounded below $+0.28^\circ/-0.31^\circ$, while the maximum errors of the TNCE are stayed at $+0.34^\circ/-0.49^\circ$ and maximum errors of the PIDE are smaller than $\pm 0.29^\circ$. With a viewpoint in only comparison of the min/max steady state error in a roll direction, the PIDE shows the slightly better performance. However, the simple comparisons of maximum and minimum error values cannot represent the overall tracking performance. Therefore, the RMS (root mean square) values in the errors are investigated to confirm the comprehensive tracking performance. If each RMS value of 6 DOF motion errors by PIDE is defined as 100%, then each RMS value of motion errors along six directions (surge, sway, heave, roll, pitch, and yaw) is 40%, 34%, 39%, 94%, 91%, and 62% for TNCE, and 31%, 34%, 37%, 72%, 90%, and 35% for TRNCE, respectively. The RMS values of errors show that nonlinear control laws designed in task space are superior to the PIDE. Furthermore, the TRNCE exhibits the more excellent control performance than the TNCE by the RMS values of errors and the comparison of each maximum value, which result from the reflection of the system uncertainties.

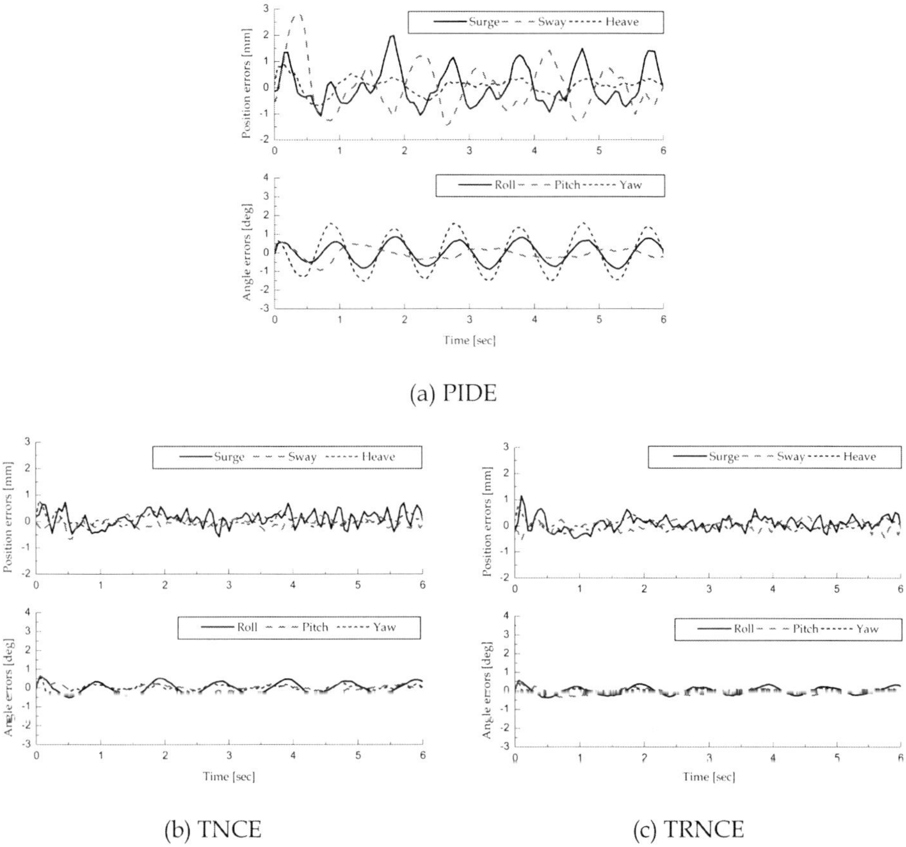

Fig. 9. Tracking errors of 6DOF motions to multi-directional sinusoidal inputs (Roll: 2.0°/1.0 Hz, Pitch: 5.0°/0.5 Hz, Yaw: 2.5°/1.0 Hz, and Heave: 5.0 mm/0.5 Hz)

Fig. 9 presents tracking errors to multi-directional sinusoidal inputs (Roll: 2.0°/1.0Hz, Pitch: 5.0°/0.5Hz, Yaw: 2.5°/1.0Hz, and Heave: 5.0mm/0.5Hz). The TRNCE and TNCE show the remarkable tracking performances superior to those of the PIDE in all 6 DOF directions which is similar in performance tendency to the previous case. The superb performances through the TRNCE and TNCE result from the task space based designs and cancellation of nonlinearities (the inertia force for a given acceleration, the gravitational force, the Coriolis and centrifugal forces). The translation errors of the TRNCE are bounded between +0.77mm and -0.48mm, those of the TNCE lie between +0.76mm and -0.52mm, while those of the PIDE exceed ±1.5mm in a steady state. All the rotational error bounds of the TRNCE lie within ±0.35°, maximum error of the TNCE are bounded below ±0.45°, while those of the PIDE exceeds ±1.5°. The RMS (root mean square) values in the errors are also investigated to confirm the comprehensive tracking performance. In the case that each RMS value of the 6 DOF motion errors is also defined as 100 % by PIDE, each RMS (root mean square) value of the motion errors along six directions (surge, sway, heave, roll, pitch, and yaw) is 45%, 23%, 58%, 51%, 66%, and 13% for TNCE and 38%, 23%, 56%, 36%, 57%, and 9% for TRNCE, respectively. There exists the difference in control performance between the TRNCE and the TNCE, which stems from the additional robust control input considering the system uncertainties. Consequently, it is shown that the TRNCE excels the TNCE and the PIDE in terms of control performances to the multi-directional sinusoidal inputs with high frequency component.

6. Conclusion

This paper proposes and implements the task space approach of a robust nonlinear control with the system state and friction estimation methodologies for the parallel manipulator which is a representative multi-input & multi-output nonlinear system with uncertainties. In order to implement the proposed robust nonlinear control law, the indirect 6 DOF system state estimator is firstly employed and confirmed the outstanding effects experimentally. The indirect system state estimation scheme consists of Newton-Raphson method and the alpha-beta tracker algorithm, which is simple route and readily applicable to a real system instead of a costly 6 DOF sensor or a model-based nonlinear state observer with the actuator length measurements. Secondly, the Friedland Park friction observer is applied as the equivalent friction estimator in joint space which provides the friction estimates to attenuate uncertain frictional disturbance. The suitability of this friction estimation approach is experimentally confirmed as well. Finally, the control performances of the proposed task space based robust nonlinear control law equipped with the estimators of system state and the friction are experimentally evaluated. With viewpoints of regulating and tracking, the remarkable control results to several inputs are shown under system nonlinearity, parameter uncertainties, uncertain friction property, etc. In addition to those, the experimental results shows that the proposed robust nonlinear control scheme in task space surpasses the nonlinear task space control with the estimators and the joint space based PID control with the estimators, which reveal its availability to the practical applications like a robotic system or machine-tool required the task space based control scheme for a precision control performance.

7. References

Amstrong-Hélouvry; B., Dupont, P. & Canudas de Wit, C. (1994). A Survey of Models, Analysis Tools and Compensation Methods for the Control of Machines with Friction. *Automatica*, Vol. 30, No. 7, pp. 1083-1138.

Barmish, B. R.; Corless, M. J. & Leitmann, G. (1983). A New Class of Stabilizing Controllers for Uncertain Dynamical Systems. *SIAM Journal of Control and Optimization*, Vol. 21, pp. 246-255.

Canudus de Wit, C.; Siciliano, B. & Bastin, G. (1996). *Theory of Robot Control*, Springer, Berlin.

Corless, M. J. & Leitmann, G. (1981). Continuous State Feedback Guaranteeing Uniform Ultimate Boundedness for Uncertain Dynamic Systems. *IEEE Transactions on Automatic Control*, Vol. 26, pp. 153-158.

Dasgupta, B. & Mruthyunjaya, T. S. (1998). Closed-Form Dynamic Equations of the General Stewart Platform through the Newton-Euler Approach. *Mechanism and Machine Theory*, Vol. 33, pp. 993-1012.

Dieudonne, J. E.; Parrish, R. V. & Bardusch, R. E. (1972). An Actuator Extension Transformation for a Motion Simulator and an Inverse Transformation applying Newton-Raphson Method. NASA Technical Report D-7067.

Friedland, B. (1973). Optimum Steady-State Position and Velocity Estimation Using Sampled Position Data, *IEEE Transactions on Aerospace and Electronic Systems*, AES-Vol. 9, No. 6, pp. 906-911.

Friedland, B. & Park, Y. J. (1992). On Adaptive Friction Compensation. *IEEE Transactions on Automatic Control*, Vol. 37, No. 10, pp. 1609-1612.

Hahn, W. (1967). *Stability of Motion*, Springer, New York.

Honegger, M.; Brega, R. & Schweitzer, G. (2000). Application of a Nonlinear Adaptive Controller to a 6 dof Parallel Manipulator. *In Proceeding of the 2000 IEEE International Conference on Robotics and Automation*, pp. 1930-1935, San Francisco, April, 2000, CA., USA.

Kang, J. Y.; Kim, D. H. & Lee, K. I. (1996) Robust Tracking Control of Stewart Platform. In *Proceedings of the 35th Conference of Decision and Control*, pp. 3014-3019, Kobe, December, 1996, Japan.

Kang, J. Y.; Kim, D. H. & Lee, K. I. (1998). Robust Estimator Design for Forward Kinematics Solution of a Stewart Platform. *Journal of Robotic Systems*, Vol. 15, Issue 1, pp. 30-42.

Khalil, H. K. (1996). *Nonlinear Systems*, 2nd ed.,Prentice-Hall, New Jersey.

Kim, D. H.; Kang, J. Y. & Lee, K. I. (2000). Robust Tracking Control Design for a 6 DOF Parallel Manipulator. *Journal of Robotic Systems*, Vol. 17, Issue 10, pp. 527-547.

Lewis, F. (1986). *Optimal Estimation with an Introduction to Stochastic Control Theory*, John Wiley and Sons, Inc, USA.

Merlet, J. P. (2000). *Parallel Robots*, Kluwer Academic Publisher, Netherlands.

Nguyen, C. C.; Antrazi, S., Zhou, Z. L. & Campbell, C. (1993). Adaptive Control of a Stewart Platform-Based Manipulator. *Journal of Robotic Systems*, Vol. 10, No. 5, pp.657-687

Panteley, E.; Ortega, R. & Gafvert, M. (1998). An Adaptive friction compensator for global tracking in robot manipulators, *Systems & Control Letters*, Vol. 33, Issue 5, pp. 307-313.

Park, C. G. (1999). Analysis of Dynamics including Leg Inertia and Robust Controller Design for a Stewart Platform, Ph. D. thesis, Seoul National University, Korea.

Radcliffe, C. J. & Southward, S. C. (1990). A Property of Stick-Slip Friction Models which Promotes Limit Cycle Generation. *In Proceedings on American Control Conference*, pp. 1198-1203, May, 1990, San Diego, USA.

Sirouspour, M. R. & Salcudean, S. E. (2001). Nonlinear Control of Hydraulic Robots, *IEEE Transactions on Robotics and Automation*, Vol. 17, No. 2, pp. 173-182.

Spong, M. W. & Vidyasagar, M. (1989). *Robot Dynamics and Control*, John Wiley & Sons, Inc.

Ting, Y.; Chen, Y. S. & Wang, S. M. (1999). Task-space Control Algorithm for Stewart Platform. *In Proceedings of the 38th Conference on Decision and Control*, pp. 3857-3862, December, 1999, Phoenix, Arizona, USA.

23

Tactile Displays with Parallel Mechanism

Ki-Uk Kyung and Dong-Soo Kwon*
Electronics and Telecommunications Research Institute(ETRI)
**Korea Advanced Institute of Science and Technology(KAIST)*
Republic of Korea

1. Introduction

Since more intuitive and realistic interaction between human and computer/robot has been requested, haptics has emerged as a promising element in the field of user interfaces. Particularly for tasks like real manipulation and exploration, the demand for interaction enhanced by haptic information is on the rise.
Researchers have proposed a diverse range of haptic devices. Force feedback type haptic devices with robotic link mechanisms have been applied to teleoperation system, game interfaces, medical simulators, training simulators, and interactive design software, among other domains. However, compared to force feedback interfaces, tactile displays, haptic devices providing skin sense, have not been deeply studied. This is at least partly due to the fact that the miniaturization and the arrangement necessary to construct such systems require more advanced mechanical and electronic components.
A number of researchers have proposed tactile display systems. In order to provide tactile sensation to the skin, work has looked at mechanical, electrical and thermal stimulation. Most mechanical methods involve an array of pins driven by linear actuation mechanisms with plural number of solenoids, piezoelectric actuators, or pneumatic actuators. In order to realize such compact arrangement of stimulators, parallel mechanisms have been commonly adopted.
This chapter deals with parallel mechanisms for tactile displays and their specialized designs for miniaturization and feasibility. In addition, the chapter also covers application of tactile displays for human-computer/robot interfaces.

2. Tactile display research review

Researchers have proposed a diverse range of haptic interfaces for more realistic communication methods with computers. Force feedback devices, which have attracted the most attention with their capacity to physically push and pull a user's body, have been applied to game interfaces, medical simulators, training simulators, and interactive design software, among other domains (Burdea, 1996). However, compared to force feedback interfaces, tactile displays have not been deeply studied. It is clear that haptic applications for mobile devices such as PDAs, mobile computers and mobile phones will have to rely on tactile devices. Such a handheld haptic system will only be achieved through the development of a fast, strong, small, silent, safe tactile display module, with low heat dissipation and power consumption. Furthermore, stimulation methods reflecting human tactile perception characteristics should be suggested together with a device.

A number of researchers have proposed tactile display systems. In order to provide tactile sensation to the skin, work has looked at mechanical, electrical and thermal stimulation. Most mechanical methods involve an array of pins driven by linear actuation mechanisms such as a solenoids, piezoelectric actuators, or pneumatic actuators. Particularly, their mechanisms are focused on miniaturized parallel arrangement of actuators. In 1995, a tactile display composed of solenoids has been investigated and it was applied to an endoscopic surgery simulator (Fisher et al., 1997). One of well known tactile displays is composed of RC servomotors. The servomotor occur linear motion of tactor and the parallel arrangement of tactors form a tactor array of the tactile display (Wagner et al., 2002). Another example is the "Texture Explorer", developed by Ikei's group (Ikei & Shiratory, 2002). This 2×5 flat pin array is composed of piezoelectric actuators and operates at a fixed frequency (~250Hz) with maximum amplitude of 22μm. Summers et al. developed a broadband tactile array using piezoelectric bimorphs, and reported empirical results for stimulation frequencies of 40Hz and 320Hz, with the maximum displacement of 50μm (Summers & Chanter, 2002). Since the tactile displays mentioned above may not result in sufficiently deep skin indentation, Kyung et al. (2006a) developed a 5x6 pin-array tactile display which has a small size, long travel and high bandwidth. However, this system requires a high input voltage and a high power controller. As an alternative to providing normal indentation, Hayward et al. have focused on the tactile sensation of lateral skin stretch and designed a tactile display device which operates by displaying distributed lateral skin stretch at frequencies of up to several kilohertz (Hayward & Cruz-hernandez, 2000; Luk et al., 2006). However, it is arguable that the device remains too large (and high voltage) to be realistically integrated into a mobile device. Furthermore, despite work investigating user performance on cues delivered by lateral skin stretch, it remains unclear whether this method is capable of displaying the full range of stimuli achievable by presenting an array of normal forces. More recently, a miniaturized tactile display adopting parallel and woven arrangement of ultrasonic linear actuators have been proposed (Kyung & Lee, 2008). The display was embedded into a pen-like case and the assembly realized haptic stylus applicable to a touchscreen of mobile communication device.

Konyo et al. (2000) used an electro-active polymer as an actuator for mechanical stimulation. Poletto and Doren (1997) developed a high voltage electro-cutaneous stimulator with small electrodes. Kajimoto et al. (1999) developed a nerve axon model based on the properties of human skin and proposed an electro-cutaneous display using anodic and cathodic current stimulation. Unfortunately, these tactile display devices sometimes involve user discomfort and even pain.

We can imagine a haptic device providing both force and tactile feedback simultaneously. Since Kontarinis et al. applied vibration feedback to a teleoperation (Kontrarinis & Howe, 1995), some research works have had interests in combination of force and tactile feedback. Akamatsu and MacKenzie (1996) suggested a computer mouse with tactile and force feedback increased usability. However, the work dealt with haptic effects rather than precisely controlled force and tactile stimuli. Kammermeier et al. (2004) combined a tactile actuator array providing spatially distributed tactile shape display on a single fingertip with a single-fingered kinesthetic display and verified its usability. However, the size of the tactile display was not small enough to practically use the suggested mechanism. As more practical design, Okamura and her colleagues design a 2D tactile slip display and installed it into the handle of a force feedback device (Webster et al., 2005). Recently, in order to provide texture sensation with precisely controlled force feedback, a mouse fixed on 2DOF mechanism was suggested (Kyung et al., 2006b). A small pin-array tactile display was embedded into a mouse body and it realized texture display with force feedback. More recently, Allerkamp et al. (2007) developed a compact pin-array and they tried to realize the combination of force feedback and tactile display based on the display and vibrations. However, in previous works, the tactile display itself is quite small but its power controller is too big to be used practically.

This chapter focuses on design and evaluation of two tactile displays developed by authors. The tactile displays are based on miniaturized parallel arrangement of actuators. In the section 3, 5x6 pin array based on piezoelectric bimorphs are introduced. The performance of tactile display has been verified by pattern display and the tactile unit is installed in a conventional mouse to provide tactile feedback while using the mouse. In the section 4, a compact tactile display with 3x3 pin array is described. The tactile display unit is embedded into a stylus-like body and the performance of the haptic stylus is introduced.

3. Texture display mouse

3.1 Planar distributed tactile display

Fig. 1 shows the side view of the tactile display assembly (Kyung et al. 2006a). Each step of the stair-like bimorph support holds six bimorphs arranged in two rows. The lower and upper rows are laterally offset by 1.8 mm. Each step is longitudinally offset 1.8mm from the next. 10 tiers of 3 piezoelectric bimorphs are interwoven to address 5 rows and 6 columns of pins (tactors) on 1.8 mm centers. The maximum deflection is greater than 700μm and the bandwidth is about 350Hz. The blocking force is 0.06N. The specifications of the tactile stimulator with piezoelectric bimorphs were verified to ensure that it deforms the user's skin within 32dBSL (sensation level in decibels above threshold). Each bimorph is 35 mm × 2.5 mm with a thickness of 0.6 mm. The size of the cover case is 40 mm × 20 mm × 23 mm. Efforts to minimize the weight of the materials and wiring produced a finished design with a weight of only ~11 grams. The contact area is 9.7mm×7.9mm—a previous study showed this area is sufficient to discern difference in textures.

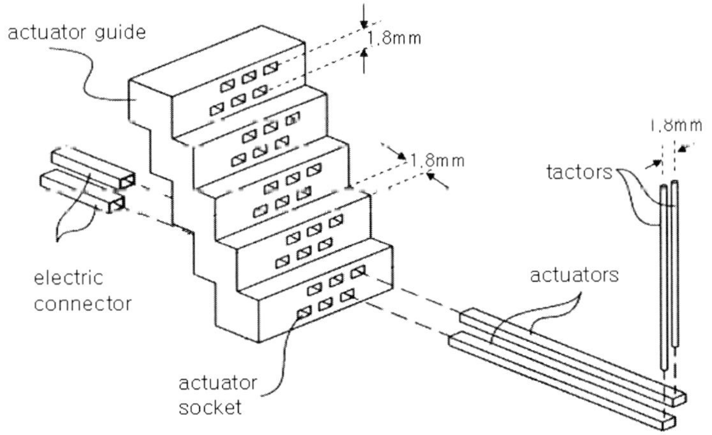

Fig. 1. Profile of the tactile display

Fig. 2 shows the contact interface of our tactile display. The frame is 40mm × 20mm × 23mm. The 30 stacked actuators are piezoelectric bimorphs driven by 150 VDC bias. Since the tactile display unit, which is described in Section 3.1, is small enough to be embedded into a computer mouse, we developed a new texture display mouse that has a tactile display function as well as all functions of a conventional mouse. Fig. 3 shows a prototype of the tactile display mouse. The pin array part of the tactile display is located between two click buttons of the mouse and it does not provide any interference during mouse movement (Kyung et al., 2007).

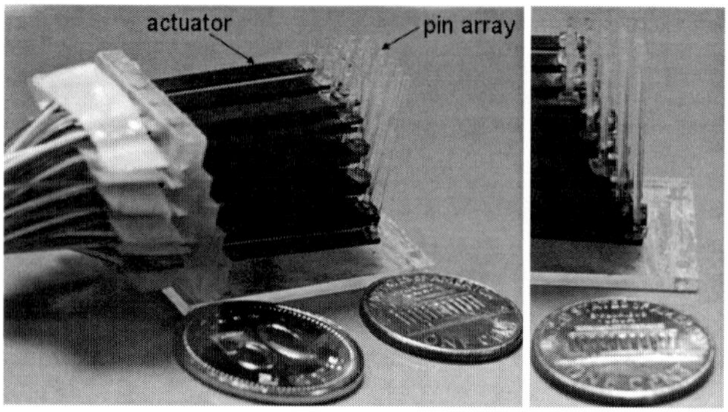

Fig. 2. The texture display unit

Fig. 3. A prototype of the texture display mouse

3.2 Static pattern display

In order to use the proposed haptic mouse as a computer interface, the system should provide some kinds of symbols, icons, or texts in a haptic manner. Therefore, in this set of experiments, the performance of the tactile display was evaluated by asking subjects to discriminate between plain and textured polygons, round figures, and gratings. In these experiments, the actuator voltages were adjusted to set the desired shape, which was then held constant. Subjects were allowed to actively stroke the tactile array with their finger pad. Thus, the experiments were conducted under the condition of active touch with static display.

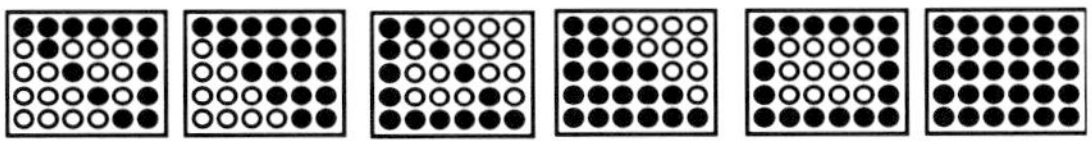

Fig. 4. Planar polygon samples

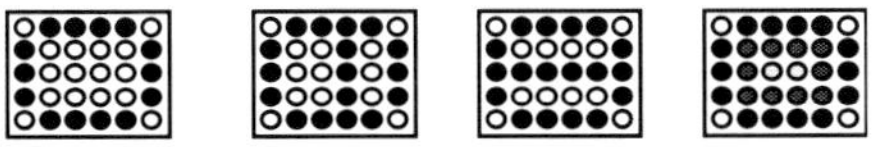

Fig. 5. Rounded shaped samples

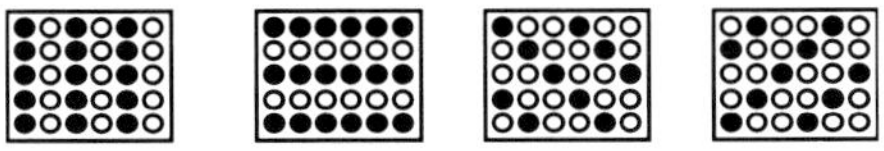

Fig. 6. Grating samples

Experiment I. Polygon discernment: In the first experiment, subjects were asked to ascertain the performance of a tactile display that presented 6 polygons created by the static normal deflections of the pin array. Fig. 4 shows the 6 test samples consisting of blank and filled polygonal outlines. After the presentation of the stimulus, subjects were free to explore it with their finger and were required to make a determination within ten seconds. Each sample was displayed 5 times randomly. Twenty-two naïve subjects (13 men and 9 women), all in their twenties, performed the task (Table 1). The proportion of correct answers (90-99%, depending on the stimulus) far exceeded chance (10%), indicating that the display provides a satisfactory representation of polygons, and that fine features such as fill type and polygon orientation are readily perceived.

Experiment II. Rounded shapes: The purpose of this experiment was to verify that the system could simulate the differences between shapes that were similar and those that had identical boundaries. Four round shapes with distinctive features were presented to the same subjects who participated in Experiment 1. The other conditions, such as response time and active touch, were the same. Three of the samples in this experiment (Fig. 5, the three leftmost shapes) were simple planar outlines. The fourth was a three dimensional half ellipsoid. It is reasonable to suppose that the conspicuous difference of the fourth sample caused the 100% correct answer rate (Table 1). Results for the other shapes are comparable to those found in the polygon discrimination task, indicating that the display does a satisfactory job of rendering round shapes.

Experiment III. Gratings: The same experiment as above was performed using the four grating samples shown in Fig. 6. The interval between each convex line was 3.6 mm. The purpose of this experiment was to verify that the developed system can present gratings and their directions. Table 1 shows the proportion of correct answers for the different gratings.

Sample No.		1	2	3	4	5	6
Percentage of Correct Answers	Experiment I	90.8	98.7	93.3	93.2	97.3	95.9
	Experiment II	97.3	100	91.5	100		
	Experiment III	93.3	95.9	100	95.9		

Table 1. Experimental results

3.3 Vibrotactile pattern display

In this section, we investigate how vibrotaction, particularly at low frequencies with identical thresholds, affects the identification of forms in which only passive touch, and not rubbing, is used. Craig (2002) has already compared the sensitivity of the scanned mode and static mode in discerning tactile patterns, but here we compare the sensitivity of the static mode and synchronized vibrating mode. In these experiments, subjects were not allowed to rub the surface of the tactile display. In order to set the other conditions identical to those in the experiment of section 3.2, except for the vibrotaction, the same texture groups used in section 3.2 were deployed with three different low frequencies: static, 1Hz, and 3Hz. The frequencies were selected based on identical sensation levels, since the magnitudes of the threshold value in the frequency band of 0~5Hz are almost the same.

Table 2 shows that the proportion of correct answers generally increases as the frequency rises from static to 1 Hz to 3Hz. The proportion of correct answers is similar for stimuli presented at 3 Hz and for active touching (Table 2). This suggests that passive touch with low frequency vibration may be a viable alternative to active touch. From a psychophysical and physiological point of view, it seems likely that a 3Hz vibration can effectively stimulate the Merkel cells and that the associated SA I afferent provides the fine spatial resolution necessary for the subject to make the required discriminations. From these results, we expect that the haptic mouse is capable of displaying virtual patterns and characters in real time while the user simply grasps and translates the mouse while exploring the virtual environment.

Sample No.			1	2	3	4	5	6
Percentage of Correct Answers	Polygonal Samples	0Hz	51.4	72.9	55.7	82.9	60.0	45.7
		1 Hz	55.4	90.8	67.1	94.7	90.5	94.7
		3 Hz	70.7	90.5	81.3	86.5	86.8	93.3
	Rounded Samples	0Hz	71.4	72.9	73.2	100		
		1 Hz	89.2	73.0	63.3	94.7		
		3 Hz	81.6	80.3	88.5	94.7		
	Grated Samples	0Hz	56.6	74.3	66.7	59.2		
		1 Hz	93.3	90.8	81.3	81.6		
		3 Hz	83.8	93.2	94.7	85.9		

Table 2. Experimental results

4. Tactile feedback stylus

4.1 Compact tactile display module

This section describes another type of tactile display composed of 3x3 pin array for embedding into a portable device. In order to make a tactile display module, actuator selection is the first and dominant step. The actuator should be small, light, safe, silent, fast, powerful, consume modest amounts of power and emit little heat. Recently, we developed a small tactile display using a small ultrasonic linear motor. We here briefly describe its operation principle and mechanism.

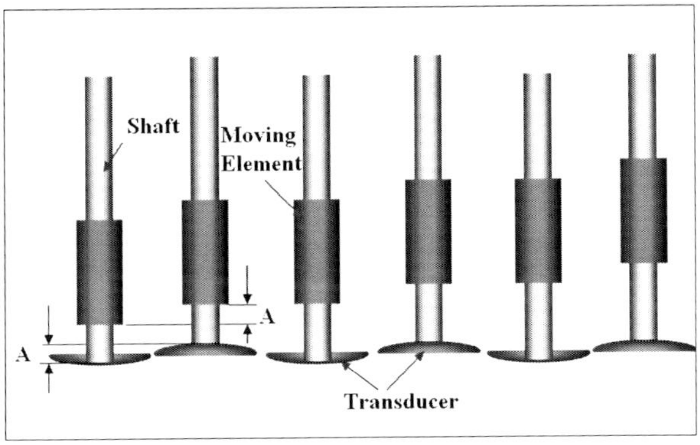

Fig. 7. Operation principle of an actuator

The basic structure and driving principle of the actuator are described in Fig. 7. The actuator is composed of a transducer, shaft and a moving element. The transducer is composed of two piezoelectric ceramic disks and elastic material membranes. The convex motion of the membranes causes lift in the shaft of the motor. The fast restoring concave motion overcomes the static frictional force between the moving element and the shaft and makes the moving element maintain its position. The displacement 'A' of one cycle is sub-micrometer scale and rapid vibration of the membrane at a frequency of 45 kHz (ultrasonic range) causes rapid movement of the moving element. The diameter of the transducer is 4mm and its thickness is 0.5mm. The thrusting force of the actuator is greater than 0.2N and the maximum speed of the moving element is around 30mm/sec. In order to minimize the size of the tactile display module, the actuators were arranged as shown in Fig. 8. Essentially, this figure shows the arrangement of two variations on the actuators - each with different shaft lengths. This design minimizes the gap between actuators. Another feature is that the elements previously described as "moving" are now stationary and fixed together, causing the shafts to become the elements which move when the actuators are turned on. This minimizes the size of the contact point with a user's skin (to the 1mm diameter of the shaft), while maintaining the mechanical simplicity of the system. Fig. 9 shows the implemented tactile display.

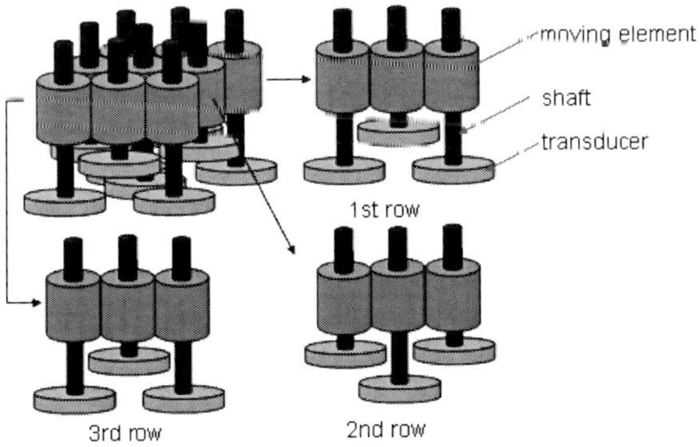

Fig. 8. Design drawing of a tactile display module

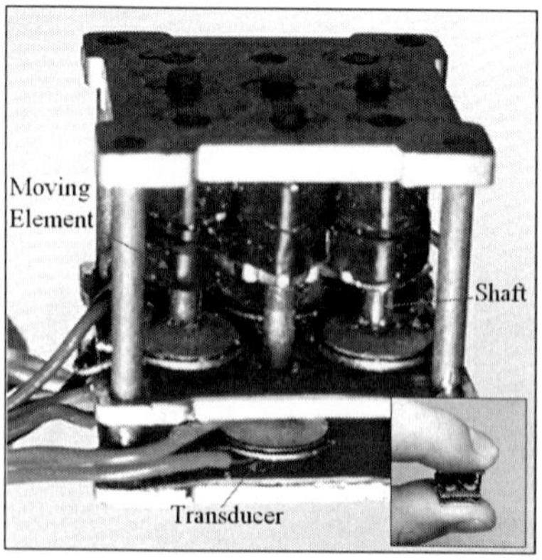

Fig. 9. Implemented tactile display

From the design specification described above, the prototype of the tactile display module has been implemented as shown in Fig. 9. In order to embed the module in a pen, we constructed only a 3x3 pin array. However, it should be noted that the basic design concept is fully extensible; additional columns and rows can be added without electrical interference or changes in pin density. The shaft itself plays the role of tactor and has a travel of 1mm. The distance between two tactors is 3.0mm. Since the actuators operate in the ultrasonic range, they produce little audible noise. The average thrusting force of each actuator exceeds 0.2N, sufficient to deform the skin with an indentation of 1 mm. The total size of the module is 12x12x12 mm and its weight is 2.5grams. Since the maximum speed of a pin is around 30mm/sec the bandwidth of the tactile display is approximately 20Hz when used with a maximum normal displacement of 1mm. If the normal displacement is lower than 1mm, the bandwidth could be increased.

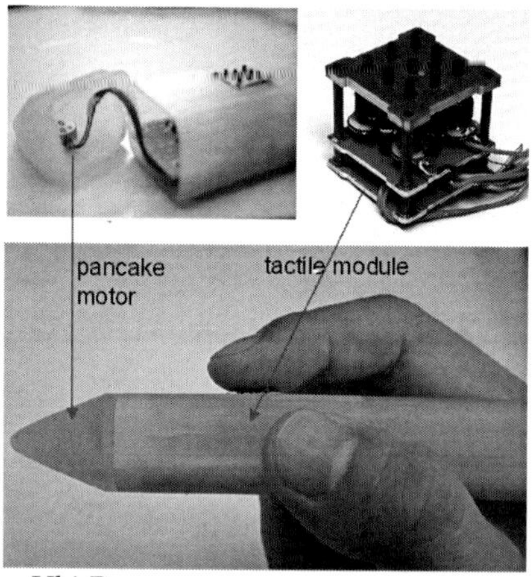

Fig. 10. The prototype of the Ubi-Pen

4.2 Implementation of pen-like tactile display

The pen is a familiar device and interface. Since they are small, portable and easy to handle, styli have become common tools for interacting with mobile communication devices. In order to support richer stylus based tactile cues, we embedded our tactile display module into a pen-like prototype. In addition, as shown in Fig. 10, we installed a pancake-type(coin-type) vibrating motor in the tip of the pen to provide a sense of contact (Kyung & Lee, 2008). The housing of the pen was manufactured by rapid prototyping, and it has a length of 12cm and a weight of 15 grams. Currently, its controller is not embedded. We named this device the Ubi-Pen and intend it for use as an interface to VR, for the blind, to represent textures, and as a symbolic secure communication device. We also suggest it could be used generally as the stylus of a mobile communication device.

4.3 Pattern display of the tactile display module

A common method to evaluate the performance of tactile displays is to test user's performance at recognizing specific patterns. We use Braille as a stimulus set to conduct such a test. Specifically, we conducted a study involving the presentation of the Braille numbers 0~9 on the Ubi-Pen.

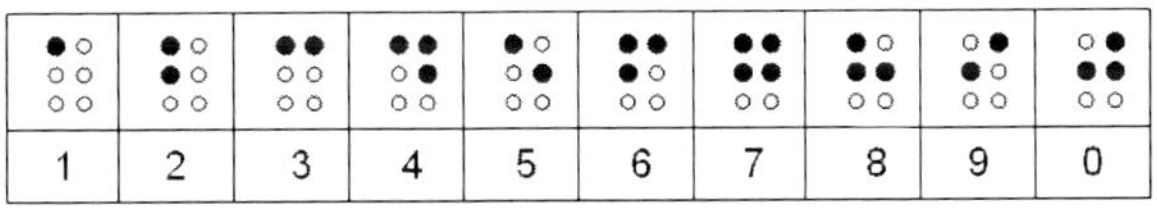

Fig. 11. Braille Patterns for the Experiment

Fig. 11 shows the experimental Braille patterns. Subjects were required to hold the pen such that the tip of their index finger rested over the pin-array part of tactile display module. In this experiment, the Braille display test bas been conducted for the normal and the blind. After setup stage, we conducted a study on recognition rate of the 10 numeric digits in the Braille character set. As these can be displayed on only four pins, we mapped them to the corner pins on our tactile display module. We chose to do this as our user-base was composed of sighted Braille novices. We used three different stimulation frequencies: 0, 2 and 5Hz. (Pins move up and maintain static position at the 0Hz). Pins movement was synchronized. We presented 60 trials in total, each number at each frequency, twice. All presentations were in a random order, and subjects were not advised about the correctness of their responses. 10 subjects participated in the experiment. The Braille stimuli were generated continuously and changed as soon as the subject respond using the graphic user interface. There were 2 minutes breaks after every 20 trials.

Two blind people have participated in the same experiment and the visual guidance in the experiment has been replaced by the speech guidance of experimenter. For all stimuli, they responded exactly and quickly. The Braille expert usually read more than 100 characters, and the blind subjects respond they don't feel any difficulties to read the Braille numbers. Since the duration of each trial was shorter than 1~2 seconds and they answer in the form of speech, we could not measure the duration exactly. Moreover, 4 neighborhood pins have been presented again with identical procedure for the blind people. And they responded more quickly since the gap of pins was more familiar with them. Duration of each trial was always shorter than 1 second.

	Normal subjects	Blind Subjects
Average Percentage of Correct Answers	80.83	100
Average Duration of Each Trial (sec)	5.24	1~2

Table 3. Experimental Results

Table 3 shows the summary of experimental results. Although normal subjects were novice in using the tactile display, the average percentage of correct answers exceeded 80 percent.

The confusions come from the relatively low tactile sensitivity of the novices compared with the sensitivity of the blind. Since the various analysis of the tactile display for the blind is another interesting topic, this will be investigated in our future work

4.4 Image display on touch screen

The Ubi-pen mouse enables tactile pattern display. This program provides a symbolic pointer in the shape of a square, with a size of 15x15 pixels. A user can load any grayscale image. As shown in Fig. 12, when the user touches an image on the touch screen with the Ubi-Pen, the area of the cursor is divided into 9(=3x3) sub-cells and the average gray value of each cell is calculated. Then, this averaged gray value is converted to the intensity of the stimuli displayed on each pin of the tactile display. Figure 13 shows the methodology of the pattern display.

In order to verify texture display performance of the Ubi-Pen, 3 kinds of texture sample groups have been chosen. As described above, every sample is gray images. And we prepared three image groups classified by their feature characteristics. This experiment is to test user's performance at recognizing specific patterns. One of five images in a group is displayed on the screen, but a participant is not able to see the image. He/she sees only a blank square covering the image. The size of the box is same as the image's one and the actual gray values of the image is obtained although the users rubs the blank square. While the user contacts a touch screen, he/she is required discriminating surfaces from scratch-like feeling. The experimental results show in Table 4 and the data verify that the Ubi-Pen and image display scheme works well.

Fig. 12.(a) shows 5 image samples of group I, those are characterized by directions of gratings. The size of each image is 300x270 pixels. The percentage of correct answers in Table 4 clearly shows that the pen type tactile display works very well in discriminating gratings. Average duration of a trial is about only 10 seconds. Fig. 12.(b) shows 5 image samples of group II, those are characterized by groove width. A user feels horizontal gratings during rubbing surfaces, in this experiment however, he/she should detect the variation of gap distance. In order to discriminate these patterns, the stimuli in accordance with movement on the plane should be detected. As shown in Table 4, sample 1, 2 and 4 are easily recognized, and the results for sample 3 and 5 are also acceptable. Users feel a bit more difficult than group I, but the performance of the device is still acceptable. Figure 12.(c) shows 5 image samples of group III, those are characterized by shapes. Since average percentage of correct answers in this group is 77.5, we can accept that we can recognize various patterns by rubbing surface using the proposed device. However, as shown in Table 4, participants have been a bit confused among the image samples except sample 5 whose geometric connection is different. And it takes twice time to give an answer compared to group I. In case of complex pattern, it is reasonable that it takes a long time and error increases. However, improvement of the device is necessary since device itself can cause confusion such as low reality, inconveniency or low density.

	Percentage of Correct Answers					Duration of a Trial (sec)	
	S1	S2	S3	S4	S5	Ave.	Std.
Group1	97.5	92.5	85.0	95.0	92.5	10.7	2.9
Group2	92.5	100	77.5	97.5	75	13.4	4.0
Group3	62.5	77.5	80.0	72.5	95.0	20.6	10.7

Table 4. Experimental Results.

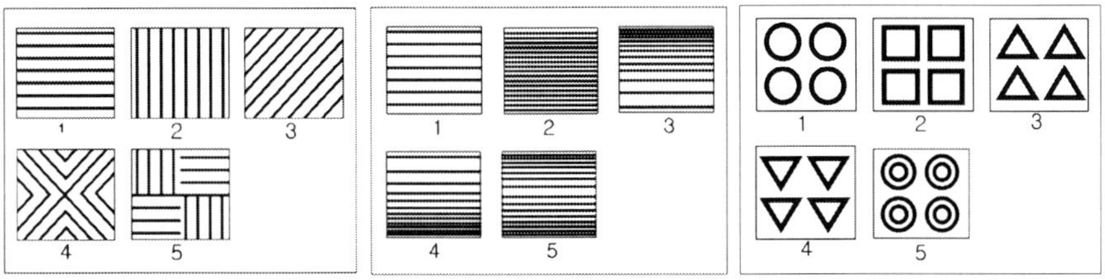

(a) Group I (b) Group II (c) Group III

Fig. 12. Braille Patterns for the Experiment

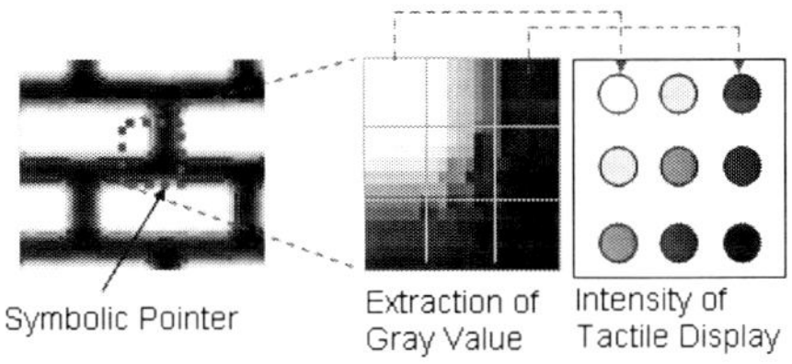

Fig. 13. Methodology of pattern display

5. Summary

This chapter deals with tactile displays and their mechanisms. We briefly reviewed research history of mechanical type tactile displays and their parallel arrangement. And this chapter mainly describes two systems including tactile displays.

The 5x6 pin arrayed tactile display with parallel arrangement of piezoelectric bimorphs has been described in the section 3. The tactile display has been embedded into a mouse device and the performance of the device has been verified from pattern display experiment.

Another focus of this chapter is describing a compact tactile display module and verifying its performance in a pen-like form factor. As described in section 4, a small, safe, low power consuming, silent and light tactile display module with parallel and woven arrangement of ultrasonic linear motors has been built. Using the tactile display, we propose the Ubi-Pen which can provide texture and vibration stimuli. This system shows satisfactory preliminary performance in representing tactile patterns. We also evaluate its capacity to support GUI operations by providing scratching sensation when a user rubs surface displayed on a touch screen.

There have been various trials to develop tactile displays for simulating surface gratings, patterns, roughness and etc. However, so far, the best candidate in designing tactile display has been a pin-array. In order to provide enough indenting stimulation in a pin-array, parallel arrangement of linear mechanism has been necessarily required. In the future, invention of new materials will suggest compacter and more effective design. In this chapter, we have focused on two technologies suggesting examples of miniaturized design concepts of tactile displays adopting parallel mechanisms.

6. References

Akamatsu, M. & MacKenzie, I. S. (1996), Movement characteristics using a mouse with tactile and force feedback, *International Journal of Human-Computer Studies*, 45, 483-493.

Allerkamp, D.; Böttcher, G.; Wolter, F. E.; Brady, A. C.; Qu, J. & Summers, I. R. (2007), A vibrotactile approach to tactile rendering, *The Visual Computer*, Springer, 23, 2, 97-108

Burdea, G. C. (1996), *Force and Touch Feedback for Virtual Reality*. Wiley-Interscience.

Craig, J. C. (2002), Identification of scanned and static tactile patterns, *Perception & Psychophysics*, 64, 1,107-120.

Fischer, H., Neisius, B. & Trapp, R. (1995), Tactile feedback for endoscopic surgery, *Interactive Technology and the New Paradigm for Health Care*, Washington, DC: IOS.

Hayward, V. & Cruz-Hernandez, M. (2000), Tactile display device using distributed lateral skin stretch, *Proc. ASME Vol. DSC-69-2*, 1309-1314.

Ikei, Y. & Shiratori, M. (2002), TextureExplorer : A tactile and force display for visual textures, *Proceedings of HAPTICS 2002*, Orlando, FL, pp. 327-334.

Kajimoto, H.; Kawakami, N.; Maeda, T.; & Tachi, S. (1999), Tactile feeling display using functional electrical stimulation. Proc. of ICAT 99, pp.107-114.

Kammermeier, P.; Kron, A.; Hoogen, J. & Schmidt, G. (2004), Display of holistic haptic sensations by combined tactile and kinesthetic feedback, *Presence-Teleoperators and Virtual Environments*, 13, 1–15.

Kontarinis, D. A. & Howe, R. D. (1995), Tactile display of vibratory information in teleoperation and virtual environments, *Presence: Teleoperators and Virtual Environments*, 4, 4, 387-402.

Konyo, M.; Tadokoro, S. & Takamori, T. (2000), Artificial tactile feel display using soft gel actuators, *Proc. of IEEE ICRA 2000*, pp. 3416-3421.

Kyung, K. U.; Ahn, M. S.; Kwon, D. S. & Srinivasan, M.A. (2006), A compact planar distributed tactile display and effects of frequency on texture judgment, *Advanced Robotics*, 20, 5, 563-580.

Kyung, K. U.; Kwon, D. S. & Yang, G. H. (2006), A novel interactive mouse system for holistic haptic display in a human-computer interface, *International Journal of Human Computer Interaction*, 20, 3, 247–270.

Kyung, K. U.; Kim, S.C. & Kwon, D.S. (2007) Texture Display Mouse: Vibrotactile Pattern and Roughness Display, *IEEE/ASME Transaction on Mechatronics*, 12, 3, 356-360.

Kyung, K. U. & Lee, J. Y. (2008), Design and applications of a pen-like haptic interface with texture and vibrotactile display, *IEEE Computer Graphics and Applications*, In Press.

Luk, J.; Pasquero, J.; Little, S.; MacLean, K. E.; Levesque, V. & Hayward, V. (2006), A Role for Haptics in Mobile Interaction: Initial Design Using a Handheld Tactile Display Prototype. *Proc. of the ACM 2006 Conference on Human Factors in Computing Systems(CHI 2006)*. pp.171-180

Poletto, C. J. & Doren, C. V. (1997), A high voltage stimulator for small electrode electrocutaneous stimulation, *Proc. of the 19th IEEE Int. Conf. on Eng. Med. & Bio. Soc.*, pp.2415-2418.

Summers, I. R. & Chanter, C. M. (2002), A broadband tactile array on the fingertip, *Journal of the Acoustical Society America*, 112, 2118-2126.

Wagner, C. R. Lederman, S. J. Howe, R .D. (2002), A tactile shape display using RC servomotors, *Proceedings. 10th Symposium on Haptic Interfaces for Virtual Environment and Teleoperator Systems*, ISBN: 0-7695-1489-8, pp.354-355

Webster RJ, Murphy TE, Verner LN and Okamura AM (2005), A novel two-dimensional tactile slip display: design, kinematics and perceptual experiments, *Transactions on Applied Perception (TAP)*, 2, 2, 150-165.

24

Design, Analysis and Applications of a Class of New 3-DOF Translational Parallel Manipulators

Yangmin Li and Qingsong Xu
University of Macau,
P. R. China

1. Introduction

In recent years, the progress in the development of parallel manipulators has been accelerated since parallel manipulators possess many advantages over their serial counterparts in terms of high accuracy, velocity, stiffness, and payload capacity, therefore allowing their wide range of applications as industrial robots, flight simulators, parallel machine tools, and micro-manipulators, etc. Generally, a parallel manipulator consists of a mobile platform that is connected to a fixed base by several limbs or legs in parallel as its name implies (Merlet, 2000). Up to now, most 6-DOF parallel manipulators are based on the Gough-Stewart platform architecture due to the aforementioned advantages. However, six DOF is not always required in many situations. Besides, a general 6-DOF parallel manipulator has such additional disadvantages as complicated forward kinematics and excessive singularities within a relatively small size of workspace.

On the contrary, limited-DOF parallel manipulators with fewer than six DOF which not only maintain the inherent advantages of parallel mechanisms, but also possess several other advantages including the reduction of total cost of the device and control, are attracting attentions of various researchers. Many parallel manipulators with two to five DOF have been designed and investigated for pertinent applications. According to the properties of their output motion, the limited-DOF parallel manipulators can be classified into three categories in terms of translational, spherical, and mixed parallel manipulators. The first type allows the mobile platform a purely translational motion, which is useful as a machine tool, a positioner of an automatic assembly line, and so on. The second one enables the output platform only perform a rotational motion around a fixed point, and can be used in such situations as a telescope, an antenna, an end-effector of a robot, etc. And the last one allows the platform to both translate and rotate, and can be employed as a motion simulator, a mixed orientating/positioning tool, and others.

Particularly, due to the application requirements of translational motion, translational parallel manipulators (TPMs) become the focus of a great number of researches. The most well-known TPM is the Delta robot (Clavel, 1988) whose concept then has been realized in several different configurations (Tsai et al., 1996; Li & Xu, 2005), and many other structures have been also proposed in the literature. For example, the 3-UPU, 3-RUU and 3-PUU mechanisms (Tsai & Joshi, 2002), 3-RRC structure (Zhao & Huang, 2000), 3-RPC architecture (Callegari & Tarantini, 2003), 3-CRR manipulator (Kong & Gosselin, 2002; Kim & Tsai, 2003), the Orthoglide (Chablat & Wenger, 2003), etc. Here the notation of R, P, U, and C denotes the revolute joint, prismatic joint, universal joint, and cylindrical joint, respectively. In addition, the recent advances in the systematic type synthesis of TPMs could be found in the literature (Kim & Chung, 2003; Kong & Gosselin, 2004).

It has been seen that most existing TPMs have a complex structure. Especially, some TPMs contain the U and S (spherical) joints which are not easy to manufacture and hence expensive although they are commercially available. From the economic point of view, the simpler of the architecture of a TPM is, the lower cost it will be spent. In previous works of the authors, two novel TPMs with the 3-PRC architecture have been proposed in (Li & Xu, 2006; Xu & Li, 2007). As an overconstrained mechanism, the 3-PRC TPM possesses a simpler structure than expected. However, the mobile platform has a relatively large dimension since the long C joints are mounted on it, which may prevent the TPM's applications in the situations where the mobile platform with a small size is preferred such as a pick-and-place operation over a limited space. In the current research, a new type of parallel mechanism called a 3-PCR TPM is proposed and investigated for various applications. With comparison to a 3-PRC TPM, the mobile platform of a 3-PCR TPM only contains the passive R joints, which allows the generation of a small size output platform accordingly.

The remainder of this chapter is organized in the following way. In section 2, the 3-PCR TPM architecture is described and the mobility is determined by resorting to the screw theory. Ant then, both the inverse and forward kinematics problems have been solved in Section 3, and the velocity equations are derived in Section 4. Next, four types of singular configurations are checked in Section 5, where the mechanism design rules to eliminate them are also given, and the isotropic configurations are derived in Section 6. Afterwards, the manipulator workspace has been obtained by both analytical and numerical approaches in Section 7, and the dexterity evaluations in terms of manipulability and global dexterity index have been carried out in Section 8. Then, in Section 9, the application of a 3-PCR TPM as a CPR medical robot has been proposed in detail, and several variation structures of the 3-PCR TPM have been presented in Section 10. Finally, some concluding remarks are given in Section 11.

2. Description and mobility analysis of the manipulator

2.1 Kinematical architecture

The CAD model of a 3-PCR TPM with intersecting guide ways is graphically shown in Fig. 1 and the schematic diagram is illustrated in Fig. 2, respectively. The TPM consists of a mobile platform, a fixed base, and three limbs with identical kinematical structure. Each limb connects the fixed base to the mobile platform through a P joint, a C joint, and an R joint in sequence, where the P joint is driven by a linear actuator mounted on the fixed base. Thus, the mobile platform is attached to the base by three identical PCR linkages. The following mobility analysis shows that in order to keep the mobile platform from changing its orientation, it is sufficient for the axes of passive joints within limbs to satisfy some certain geometric conditions. That is, the axes of the C and R joints within the same limb are parallel to each other.

The geometry of one typical kinematic chain is depicted in Fig. 3. To facilitate the analysis, as shown in Figs. 2 and 3, we assign a fixed Cartesian frame $O\{x, y, z\}$ at the centered point O of the fixed base, and a moving frame $P\{u, v, w\}$ on the triangle mobile platform at centered point P, with the z- and w-axes perpendicular to the platform, x- and y-axes parallel to u- and v-axes, respectively.

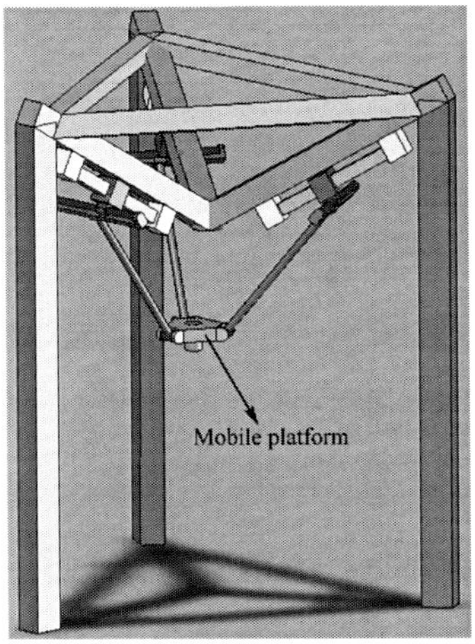

Fig. 1. A 3-PCR TPM with intersecting guide ways.

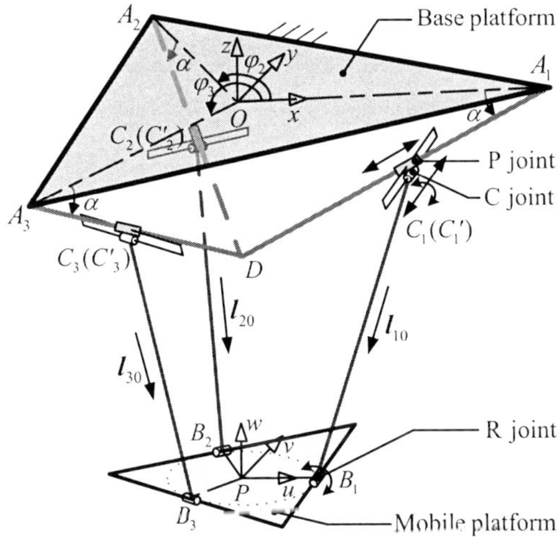

Fig. 2. Schematic diagram of a 3-PCR TPM.

The i-th limb C_iB_i (i = 1, 2, 3) with the length of l is connected to the passive C joint at C_i and connected to the mobile platform as point B_i. Q_i denotes the point on the C joint that is coincident with the initial position of C_i. And the three points B_i lie on a circle of radius b. In addition, the three rails M_iN_i intersect each other at point D and intersect the x-y plane at points A_1, A_2 and A_3 respectively, that lie on a circle of radius a. The sliders of prismatic joints Q_i are restricted to move along the rails between M_i and N_i. Angle α is measured from the fixed base to rails M_iN_i and defined as the actuators layout angle. Without loss of generality, let the x-axis point along OA_1 and the u-axis direct along PB_1. Angle φ_i is defined from the x-axis to OA_1 in the fixed frame, and also from the u-axis to PB_1 in the moving frame. For simplicity, we assign that $\varphi_i = (i-1)\times 120^{\circ}$, which results in a symmetric workspace of the manipulator. Additionally, let $d_{\max}$ and $s_{\max}$ denote the maximum stroke of linear actuators and C joints, respectively, i.e.,

$$-\frac{d_{\max}}{2} \le d_i \le \frac{d_{\max}}{2} \tag{1}$$

$$-\frac{s_{\max}}{2} \le s_i \le \frac{s_{\max}}{2} \tag{2}$$

for i=1, 2, and 3.

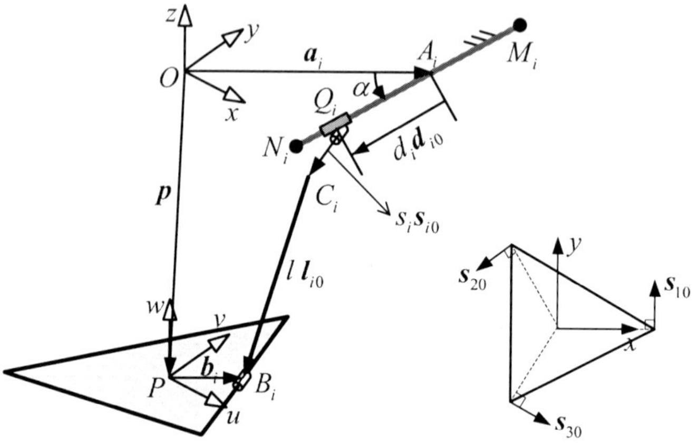

Fig. 3. Representation of direction vectors.

2.1 Mobility analysis of the manipulator

The mobility determination, i.e., the DOF identification, is the first and foremost issue in designing a parallel manipulator. The general Grubler-Kutzbach criterion is useful in mobility analysis for many parallel manipulators; however it is difficult to directly apply this criterion to mobility analysis of some kinds of limited-DOF parallel manipulators. For example, the number of DOF of a 3-PCR TPM given by the general Grubler-Kutzbach criterion is

$$F = \lambda(n-j-1) + \sum_{i=1}^{j} f_i = 6\times(8-9-1)+12 = 0 \tag{3}$$

where λ represents the dimension of task space, n is the number of links, j is the number of joints, and f_i denotes the degrees of freedom of joint i.

The zero number of DOF of a 3-PCR TPM given by the general Grubler-Kutzbach criterion reveals that the 3-PCR TPM is an overconstrained parallel manipulator. Another drawback of the general Grubler-Kutzbach criterion is that it can only derive the number of DOF of

some mechanisms but can not obtain the properties of the DOF, i.e., whether they are translational or rotational DOF.
On the contrary, we can effectively determine the mobility of a 3-PCR TPM by resorting to the screw theory (Hunt, 1990).

2.1.1 Overview of screw and reciprocal screw systems

In screw theory, a unit (normalized) screw is defined by a pair of vectors:

$$\hat{\$} = \begin{bmatrix} \mathbf{s} \\ \mathbf{r} \times \mathbf{s} + h\mathbf{s} \end{bmatrix} \tag{4}$$

where s is a unit vector directing along the screw axis, r denotes the position vector pointing from an arbitrary point on the screw axis to the origin of the reference frame, the vector $r \times s$ defines the moment of the screw axis with respect to the origin of the reference frame, and h represents the pitch of the screw. If the pitch equals to zero, the screw becomes:

$$\hat{\$} = \begin{bmatrix} \mathbf{s} \\ \mathbf{r} \times \mathbf{s} \end{bmatrix} \tag{5}$$

While in case of infinite pitch, the screw reduces to:

$$\hat{\$} = \begin{bmatrix} \mathbf{0} \\ \mathbf{s} \end{bmatrix} \tag{6}$$

A screw can be used to represent a twist or a wrench. With $\$_F$ and $\$_L$ respectively denoting the vectors of the first and last three components of a screw \$, then $\$_F$ and $\$_L$ respectively represent the angular and linear velocities when \$ refers to a twist, and the force and couple vectors when \$ refers to a wrench.
Two screws, namely, $\$_r$ and \$, are said to be reciprocal if they satisfy the following condition.

$$\$_r \circ \$ = [\tilde{\Delta}\$_r]^T \$ = 0 \tag{7}$$

where " $\circ$ " represents the reciprocal product operator, and the matrix $\tilde{\Delta}$, which is used to interchange the first and last three components of a screw ($\$_r$), is defined by:

$$\tilde{\Delta} \equiv \begin{bmatrix} \mathbf{0} & \mathbf{I} \\ \mathbf{I} & \mathbf{0} \end{bmatrix} \tag{8}$$

where **0** and **I** denote a zero matrix and an identity matrix in 3×3, respectively. The physical meaning of reciprocal screws is that the wrench $\$_r$ produces no work along the twist of \$.
Concerning an n-DOF spatial serial kinematic chain with n 1-DOF joints ($n \le 6$), the joint screws (twists) associated with all the joints form an n-order twist system or n-system provided that the n joint screws are linearly independent. The instantaneous twists of the end-effector can be described as follows.

$$\$=\sum_{i=1}^{n}\dot{q}_i\hat{\$}_i \tag{9}$$

where $\dot{q}_i$ is the intensity and $\hat{\$}_i$ is the unit screw associated with the i-th joint.

The reciprocal screw system of the twist system consists of 6-n linearly independent reciprocal screws (wrenches) and is called a (6-n)-order wrench system or (6-n)-system. In what follows, the relevant results of screw theory are utilized for the mobility investigation of a 3-PCR TPM.

2.1.2 Mobility determination of a 3-PCR TPM

For a 3-PCR parallel manipulator presented here, the motion of each limb that can be treated as a twist system is guaranteed under some exerted structural constraints which are termed as a wrench system. The wrench system is a reciprocal screw system of the twist system for the limb. The mobility of the manipulator is then determined by the effect of linear combination of the wrench systems for all limbs

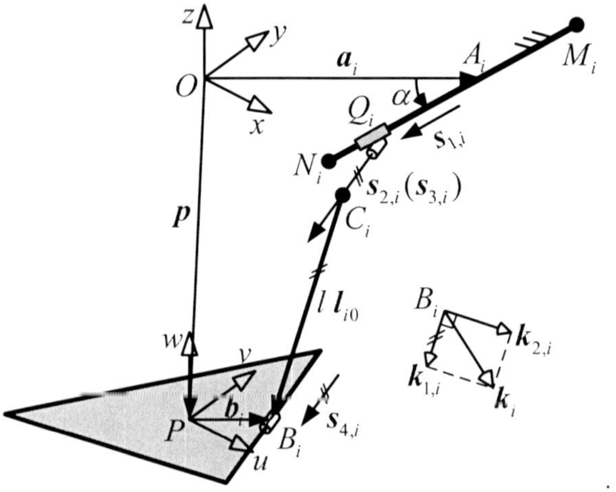

Fig. 4. Representation of screw vectors.

With $\omega=[\omega_x\ \omega_y\ \omega_z]^T$ and $\upsilon=[\upsilon_x\ \upsilon_y\ \upsilon_z]^T$ respectively denoting the vectors for the angular and linear velocities, then the twist of the mobile platform can be defined as $\$_p=[\omega^T\ \upsilon^T]^T$. Considering that a C joint is equivalent to the combination of a P joint with a coaxial R joint, the connectivity of each limb for a 3-PCR TPM is equal to four since each limb consists of four 1-DOF joints. Hence, with reference to Fig. 4, the instantaneous twist $\$_p$ of the mobile platform can be expressed as a linear combination of the four instantaneous twists, i.e.,

$$\$_p=\dot{d}_i\hat{\$}_{1,i}+\dot{\theta}_{2,i}\hat{\$}_{2,i}+\dot{\theta}_{3,i}\hat{\$}_{3,i}+\dot{s}_i\hat{\$}_{4,i} \tag{10}$$

for i=1, 2, 3, where $\dot{\theta}_{j,i}$ is the intensity and $\hat{\$}_{j,i}$ denotes a unit screw associated with the j-th joint of the i-th limb with respect to the instantaneous reference frame P, and

$$\hat{\$}_{1,i} = \begin{bmatrix} \mathbf{0} \\ \mathbf{s}_{1,i} \end{bmatrix} \tag{11}$$

$$\hat{\$}_{2,i} = \begin{bmatrix} \mathbf{0} \\ \mathbf{s}_{2,i} \end{bmatrix} \tag{12}$$

$$\hat{\$}_{3,i} = \begin{bmatrix} \mathbf{s}_{3,i} \\ \mathbf{c}_i \times \mathbf{s}_{3,i} \end{bmatrix} \tag{13}$$

$$\hat{\$}_{4,i} = \begin{bmatrix} \mathbf{s}_{4,i} \\ \mathbf{b}_i \times \mathbf{s}_{4,i} \end{bmatrix} \tag{14}$$

can be identified, where $\mathbf{s}_{j,i}$ represents a unit vector along the *j*-th joint axis of the *i*-th limb, $\mathbf{0}$ denotes a 3×1 zero vector, $\mathbf{b}_i = \overrightarrow{PB}$, $\mathbf{c}_i = \overrightarrow{PC} = \mathbf{b}_i - l\mathbf{l}_{i0}$, and $\mathbf{s}_{2,i} = \mathbf{s}_{3,i} = \mathbf{s}_{4,i} = \mathbf{s}_{i0}$, since the R and C joint axes are parallel to each other.

The screws that are reciprocal to all the joint screws of one limb of a 3-PCR TPM form a 2-system. Hence, two reciprocal screws of the *i*-th limb can be identified as two infinite-pitch wrench screws as follows.

$$\hat{\$}_{r,1,i} = \begin{bmatrix} \mathbf{0} \\ \mathbf{h}_{1,i} \end{bmatrix} \tag{15}$$

$$\hat{\$}_{r,2,i} = \begin{bmatrix} \mathbf{0} \\ \mathbf{h}_{2,i} \end{bmatrix} \tag{16}$$

where $\mathbf{h}_{1,i}$ and $\mathbf{h}_{2,i}$ are two different arbitrary vectors perpendicular to $\mathbf{s}_{i0}$ of the *i*-th limb. $\hat{\$}_{r,1,i}$ and $\hat{\$}_{r,2,i}$ denote two unit couples of constraints imposed by the joints of the *i*-th limb, and are exerted on the mobile platform.

For simplicity, let $\mathbf{h}_{1,i}$ lie in the *u*-*v* plane and $\mathbf{h}_{2,i}$ be vertical to the *u*-*v* plane, respectively, i.e.,

$$\mathbf{h}_{1,1} = [1 \quad 0 \quad 0]^T$$

$$\mathbf{h}_{1,2} = [-\frac{1}{2} \quad \frac{\sqrt{3}}{2} \quad 0]^T$$

$$\mathbf{h}_{1,3} = [-\frac{1}{2} \quad -\frac{\sqrt{3}}{2} \quad 0]^T$$

$$\mathbf{h}_{2,1} = \mathbf{h}_{2,2} = \mathbf{h}_{2,3} = [0 \quad 0 \quad 1]$$

It is observed that the six wrench screws are linearly dependent and form a screw system of order 3, namely a 3-order wrench system of the mobile platform. Since the directions of each

C and R joint axis satisfy the conditions described earlier, i.e., they are invariable, the wrench system restricts three rotations of the mobile platform with respect to the x-, y- and z-axes of the fixed frame at any instant. Thus leads to a TPM with three translational DOF along the x-, y- and z-axes of the fixed frame.
It should be noted that the mobility of a 3-PCR TPM can also be determined by adopting other methods, such as a recent theory of degrees of freedom for complex spatial mechanisms proposed by Zhao (2004), or a group-theoretic approach recommended by Angeles (2005), etc.

3. Kinematics modeling

3.1 Inverse kinematics modeling mobility

The inverse kinematics problem solves the actuated variables from a given position of the mobile platform.
Due to the mobile platform of a 3-PCR TPM delivers only a translational motion, the position of the mobile platform with respect to the fixed frame can be described by a position vector $\mathbf{p} = [p_x \quad p_y \quad p_z]^T = \overrightarrow{OP}$. Besides, the position vectors of points A_i and B_i with respect to frames O and P respectively, can be written as $\boldsymbol{a}_i$ and $\boldsymbol{b}_i$ in the fixed frame as represented in Fig. 3. Then, a vector-loop equation can be written for i-th limb as follows:

$$l\mathbf{l}_{i0} = \mathbf{L}_i - d_i\mathbf{d}_{i0} \tag{17}$$

with

$$\mathbf{L}_i = \mathbf{p} + \mathbf{b}_i - \mathbf{a}_i + s_i\mathbf{s}_{i0} \tag{18}$$

where $\mathbf{l}_{i0}$ is the unit vector along $\overline{C_iB_i}$, d_i represents the linear displacement of i-th actuated joint, $\mathbf{d}_{i0}$ is the unit vector directing along rail M_iN_i, s_i denotes the stroke of i-th C joint, and $\mathbf{s}_{i0}$ is the unit vector parallel to the axes of the C and R joints of limb i, which is denoted in Fig. 3 and can be calculated as:

$$\mathbf{s}_{i0} = \begin{bmatrix} -s\varphi_i & c\varphi_i & 0 \end{bmatrix}^T \tag{19}$$

where c stands for the cosine and s stands for the sine functions.
Substituting (18) into (17) and dot-multiplying both sides of the expression by $\mathbf{s}_{i0}$ allows the derivation of s_i, i.e.,

$$s_i = -\mathbf{s}_{i0}^T\mathbf{p} \tag{20}$$

which lies within the range of $-s_{\max}/2 \le s_i \le s_{\max}/2$.
Dot-multiplying (17) with itself and rearranging the items, yields

$$d_i^2 - 2d_i\mathbf{d}_{i0}^T\mathbf{L}_i + \mathbf{L}_i^T\mathbf{L}_i - l^2 = 0 \tag{21}$$

Then, solving (21) leads to solutions for the inverse kinematics problem:

$$d_i = \mathbf{d}_{i0}^T \mathbf{L}_i \pm \sqrt{(\mathbf{d}_{i0}^T \mathbf{L}_i)^2 - \mathbf{L}_i^T \mathbf{L}_i + l^2} \tag{22}$$

We can observe that there exist two solutions for each actuated variable, hence there are totally eight possible solutions for a given mobile platform position. To enhance the stiffness of the manipulator, only the negative square root in (22) is selected to yield a solution where the three legs are inclined inward from top to bottom.

3.2 Forward kinematics modeling

Given a set of the actuated inputs, the position of the mobile platform is resolved by the forward kinematics.

From (17) and (18), we can derive that

$$l \mathbf{l}_{i0} = \mathbf{p} + s_i \mathbf{s}_{i0} - \mathbf{e}_i \tag{23}$$

where

$$\mathbf{e}_i = \mathbf{a}_i + d_i \mathbf{d}_{i0} - \mathbf{b}_i = [e_{ix} \quad e_{iy} \quad e_{iz}]^T \tag{24}$$

Dot-multiplying (23) with itself and considering (19), (20) and (24), yields

$$(p_x c^2\varphi_i + p_y c\varphi_i s\varphi_i - e_{ix})^2 + (p_x c\varphi_i s\varphi_i + p_y s^2\varphi_i - e_{iy})^2 + (p_z - e_{iz})^2 = l^2 \tag{25}$$

which is a system of three second-degree algebraic equations in the unknowns of p_x, p_y, and p_z.

3.2.1 Forward kinematics solutions

The analytical forward kinematics solution can be obtained by solving (25) via the Sylvester dialytic elimination method, which allows the generation of an eighth-degree polynomial in only one variable as follows.

Firstly, in order to eliminate p_y, writing (25) for i=2 and 3 respectively into a second-degree polynomial in p_y:

$$Ap_y^2 + Bp_y + C = 0 \tag{26}$$

$$Dp_y^2 + Ep_y + F = 0 \tag{27}$$

where A, B, C, D, E, and F are all second-degree polynomials in p_x and p_z.

Taking (27)×A−(26)×D and (27)×C (26)×F respectively, and rewriting the two equations into the matrix form as

$$\begin{bmatrix} AE - BD & AF - CD \\ CD - AF & CE - BF \end{bmatrix} \begin{bmatrix} p_y \\ 1 \end{bmatrix} = \begin{bmatrix} 0 \\ 0 \end{bmatrix} \tag{28}$$

Equation (28) represents a system of two linear equations in p_y and 1. The following equation can be obtained by equating the determinant of the coefficient matrix to zero:

$$(AE - BD)(CE - BF) + (AF - CD)^2 = 0 \tag{29}$$

Secondly, for the purpose of eliminating p_x, we write (29) in the form of

$$Lp_x^4 + Mp_x^3 + Np_x^2 + Pp_x + Q = 0 \tag{30}$$

where L, M, N, P, and Q can be shown to be second-degree polynomials in p_z.
Substituting $\varphi_1 = 0^\circ$ into (25) for i=1, yields

$$(p_x - e_{1x})^2 + e_{1y}^2 + (p_z - e_{1z})^2 = l^2 \tag{31}$$

which can be rewritten as:

$$Gp_x^2 + Hp_x + I = 0 \tag{32}$$

where G, H, and I are all second-degree polynomials in p_z.
Now we can eliminate the unknown p_x from (30) and (32) as follows.
Taking (32)× Lp_x^2 -(30)× G, we can obtain

$$(HL - GM)p_x^3 + (IL - GN)p_x^2 - GPp_x - GQ = 0 \tag{33}$$

Taking (32)× $(Lp_x^3 + Mp_x^2)$ -(30)× $(Gp_x + H)$, yields

$$(GN - LI)p_x^3 + (GP + HN - MI)p_x^2 + (GQ + HP)p_x + HQ = 0 \tag{34}$$

Then, multiplying (32) by p_x, we have

$$Gp_x^3 + Hp_x^2 + Ip_x = 0 \tag{35}$$

Equations (32)—(35) can be considered as four linear homogeneous equations in the four variables of p_x^3, p_x^2, p_x, and 1. The characteristic determinant is

$$\begin{vmatrix} HL - GM & IL - GN & -GP & -GQ \\ GN - LI & GP + HN - MI & GQ + HP & HQ \\ G & H & I & 0 \\ 0 & G & H & I \end{vmatrix} = 0 \tag{36}$$

Expanding (36) obtains an eight degrees of polynomial in p_z. It follows that there are at most eight solutions for p_z.

Parameter	Value	Parameter	Value
a	0.6 m	α	45°
b	0.3 m	φ_1	0°
l	0.5 m	φ_2	120°
d_{max}	0.4 m	φ_3	240°
s_{max}	0.2 m		

Table 1. Architectural parameters of a 3-PCR TPM

Once p_z is found, p_x and p_y can be solved by using (32) and (26) in sequence. And there are total of 32 sets of solutions for p_x, p_y, and p_z.
Although the number of solutions is considerably large, it can be shown that only one solution is feasible and the preferred solution can be determined by examining the physical constrains of the mechanism.

3.2.2 A case study

In order to illustrate the derived forward kinematics solutions, an example is introduced to identify the configurations of the manipulator.
The architectural parameters of a 3-PCR TPM are described in Table 1. Assume that the actuated values are $d_1 = 0$, $d_2 = 0$, and $d_3 = 0$. Then, the polynomial of (36) becomes

$$2.8477z^8 + 4.3284z^6 + 1.9136z^4 + 0.0714z^2 - 0.0800 = 0 \tag{37}$$

which has eight solutions for z, and the solutions for x and y can be generated from (32) and (26) in sequence, which are shown in Table 2. The imaginary values of z have no meanings, and the configurations with positive values of p_z can only be implemented by resembling the mechanism. In addition, it can be deduced that configurations 2 - 4 do not lie in the range of the manipulator workspace due to the physical constraints imposed by stroke limits of C joints and motion limits of linear actuators. Thus, only configuration 1 represents the real solution, and the unique feasible configuration is an important feature for real time control in robotic applications.

No.	z (m)	x (m)	y (m)	Configuration
1	-0.4000	0	0	1
			0.6928	2
		0.6000	0.3464	3
			1.0392	4
2	0.4000	—	—	—
3	0.7483i	—	—	—
4	-0.7483i	—	—	—
5	0.7483i	—	—	—
6	-0.7483i	—	—	—
7	0.7483i	—	—	—
8	-0.7483i	—	—	—

Table 2. Forward kinematics solutions obtained via analytical method

4. Velocity analysis

Substituting (18) into (17) and differentiating the expression with respect to time, leads to

$$\dot{d}_i \mathbf{d}_{i0} = \dot{\mathbf{x}} - l\,\omega_i \times \mathbf{l}_{i0} + \dot{s}_i \mathbf{s}_{i0} \tag{38}$$

where ω_i is the angular velocity of i-th limb with respect to the fixed frame, and $\dot{\mathbf{x}} = [\dot{p}_x \quad \dot{p}_y \quad \dot{p}_z]^T$ is the linear velocity of the mobile platform.
Dot-multiplying both sides of (38) by $\mathbf{l}_{i0}$, gives

$$\mathbf{l}_{i0}^T \mathbf{d}_{i0} \dot{d}_i = \mathbf{l}_{i0}^T \dot{\mathbf{x}} \tag{39}$$

Writing (39) three times, once for each i=1, 2, and 3, yields three scalar equations which can be written in the matrix form:

$$\mathbf{J}_q \dot{\mathbf{q}} = \mathbf{J}_x \dot{\mathbf{x}} \tag{40}$$

where the matrices

$$\mathbf{J}_q = \begin{bmatrix} \mathbf{l}_{10}^T \mathbf{d}_{10} & 0 & 0 \\ 0 & \mathbf{l}_{20}^T \mathbf{d}_{20} & 0 \\ 0 & 0 & \mathbf{l}_{30}^T \mathbf{d}_{30} \end{bmatrix}, \quad \mathbf{J}_x = \begin{bmatrix} \mathbf{l}_{10}^T \\ \mathbf{l}_{20}^T \\ \mathbf{l}_{30}^T \end{bmatrix} \tag{41}$$

and $\dot{\mathbf{q}} = [\dot{d}_1 \quad \dot{d}_2 \quad \dot{d}_3]^T$ is the vector of actuated joint rates.
When the manipulator is away from singularities, the following velocity equations can be derived from (41).

$$\dot{\mathbf{q}} = \mathbf{J} \dot{\mathbf{x}} \tag{42}$$

where

$$\mathbf{J} = \mathbf{J}_q^{-1} \mathbf{J}_x \tag{43}$$

is the 3×3 Jacobian matrix of a 3-PCR TPM, which relates the output velocities to the actuated joint rates.

5. Singularity identification and elimination

For a parallel manipulator, the singularity configuration results in a loss of controllability and degradation of natural stiffness of the manipulator. Therefore, the analysis of parallel manipulator singularities, which is necessary for both the design and control purposes, has drawn considerable attentions (Di Gregorio & Parenti-Castelli, 1999; Zlatanov et al., 2002).

5.1 Singular configurations identification

Four kinds of singularities can be identified for a 3-PCR TPM as follows.
1) The first kind of singularity, which is also called the inverse kinematics singularity, occurs when $\mathbf{J}_q$ is not of full rank and $\mathbf{J}_x$ is invertible, i.e., $det(\mathbf{J}_q) = 0$ and $det(\mathbf{J}_x) \neq 0$.

We can see that this is the case when $\mathbf{l}_{i0}^T \mathbf{d}_{i0} = 0$ for i=1, 2, or 3, i.e., the directions of one or more of legs are perpendicular to the axial directions of the corresponding actuated joints. In this case, the mobile platform loses one or more DOF, which always occurs on the boundary of the workspace and can be avoided by restricting the motional range of the actuators.

2) The second kind of singularity also called the direct kinematics singularity occurs when $\mathbf{J}_x$ is not of full rank while $\mathbf{J}_q$ is invertible, i.e., $det(\mathbf{J}_q) \neq 0$ and $det(\mathbf{J}_x) = 0$.
We can deduce that it is the case when $\mathbf{l}_{i0}$ for i=1, 2, and 3 become linearly dependent. Physically, this type of singularity occurs when two or three of legs are parallel to one another, or the three legs lie in a common plane. Under such case, the manipulator gains one or more DOF even when all actuators are locked, which could be avoided by proper architecture design of the manipulator.
3) The third kind of singularity occurs when both $\mathbf{J}_q$ and $\mathbf{J}_x$ become simultaneously not invertible, i.e., $det(\mathbf{J}_q) = 0$ and $det(\mathbf{J}_x) = 0$. Under a singularity of this type, the mobile platform can undergo finite motions even when the actuators are locked, or equivalently, it cannot resist to forces or moments in one or more directions even if all actuators are locked. And a finite motion of the actuators gives no motion of the mobile platform.
4) Besides the three types of singularities, the rotational singularity for a TPM may occur when the mobile platform of a TPM can rotate instantaneously (Di Gregorio & Parenti-Castelli, 1999). This concept has been generalized to the constraint singularity of limited-DOF parallel manipulators (Zlatanov et al., 2002). And this type of singularity arises when the kinematic chains of a limited-DOF parallel manipulator cannot constrain the mobile platform to the planned motion any more. As far as a 3-PCR TPM is concerned, it is shown based on screw theory in Section 2 that the mobile platform cannot rotate at any instant, thus there is no rotational singularity for the 3-PCR TPM.

5.2 Mechanism design to eliminate singularities

The singular configurations can be eliminated by the approach of mechanism design as follows.
1) Elimination of the direct kinematics singularities: According to the aforementioned analysis, three cases can be classified for the direct kinematics singularity.
Case I- two legs are parallel to each other. Assume that $\mathbf{l}_{10}$ is parallel to $\mathbf{l}_{20}$. For simplicity, let the 3-PCR TPM possess a symmetric architecture. It can be deduced that $\mathbf{l}_{10}$ and $\mathbf{l}_{20}$ are perpendicular to the base plane. Generating $\mathbf{s}_{10}$ and $\mathbf{p}$, and substituting them into (20) for i=1, allows the generation of $s_1 = \sqrt{3}(a - b - d_1 c\alpha)$, where $d_1 = d_2$. With the consideration of (2), the maximum stroke of C joints should be designed as

$$s_{max} < 2\sqrt{3}(a - b - \frac{d_{max}}{2} c\alpha) \tag{44}$$

in order to eliminate this kind of singular configurations.
Case II - the three legs are parallel to one another. Under such case, it is seen that the three vectors $\mathbf{l}_{i0}$, for i=1, 2, and 3, are all perpendicular to the base plane. In addition, $d_1 = d_2 = d_3$ and $b = a - d_1 c\alpha$. To eliminate this singularity, the maximum stroke of linear actuators should be designed as

$$d_{max} < 2d_1 = \frac{2(a-b)}{c\alpha}, \text{ if } \alpha \neq 90^\circ \tag{45}$$

Case III - the three legs lie in a common plane. In this situation, the three vectors $\mathbf{l}_{i0}$ lie in a plane parallel to the base plane. It can be deduced that $d_1 = d_2 = d_3$ and $b + l = a \pm d_1 c\alpha$. To eliminate this singularity, the maximum stroke of linear actuators should be designed as

$$d_{max} < 2d_1 = \frac{2.a - b - l.}{c\alpha}, \quad \text{if } \alpha \neq 90^\circ \tag{46}$$

2) Elimination of the combined singularities: From above discussions, we can see that the combined singularity occurs in the cases of $\alpha = 0^\circ$ with $d_1 = d_2 = d_3 = a - b$, or $\alpha = 90^\circ$ with $a = b + l$. Thus, we can eliminate this type of singularities by the design of

$$d_{max} < 2(a - b), \quad \text{if } \alpha = 0^\circ \tag{47}$$

$$a < b + l, \quad \text{if } \alpha = 90^\circ \tag{48}$$

Therefore, in a real machine design, (44)—(48) should be satisfied at the same time so as to eliminate all of the singular configurations from the workspace of a 3-PCR TPM.

6. Isotropic configurations

An isotropic manipulator is a manipulator with the Jacobian matrix having a condition number equal to 1 in at least one of its configurations. In isotropic configurations, the manipulator performs very well with regard to the force and velocity transmission. As for a 3-PCR TPM in isotropic configurations, the Jocobian matrix **J** should satisfy:

$$\mathbf{J}\mathbf{J}^T = \sigma \mathbf{I}_{3\times 3} \tag{49}$$

where $\mathbf{I}_{3\times 3}$ is a 3×3 identity matrix. Under such a case, in view of (43), the following conditions must hold:

$$\sigma = \mathbf{t}_i^T \mathbf{t}_i = 1 \tag{50}$$

$$\mathbf{t}_i^T \mathbf{t}_j = 0 \quad \text{for} \quad i \neq j \tag{51}$$

for *i*, *j*=1, 2, and 3.

From (51), we can see that the three vectors $\mathbf{t}_i$ are perpendicular to one another. Writing (51) three times, once for each *i*=1, 2, and 3, respectively, results in three equations in the unknowns of p_x, p_y, and p_z. Solving them allows the generation of isotropic configurations. Given the symmetric architecture of a 3-PCR TPM, the isotropic configurations, which lie along the *z*-axis, can be derived by

$$\mathbf{p} = [0 \quad 0 \quad -ds\alpha \pm \frac{\sqrt{2}}{2}(a - b - dc\alpha)]^T \tag{52}$$

where $d = d_1 = d_2 = d_3$. Only the negative sign is taken into consideration since we are interested only in the point below the actuators.

Moreover, under such a case, the relationship between architectural parameters can be derived through a careful analysis, i.e.,

$$l = \frac{\sqrt{6}}{2}(a - b - dc\alpha) \tag{53}$$

Deriving d from (53) and in view of (1), allows the generation of

$$\begin{cases} -\frac{d_{max}}{2} \le \frac{a - b - \frac{\sqrt{6}}{3}l}{c\alpha} \le \frac{d_{max}}{2} & \text{if} \quad \alpha \ne 90^\circ \\ a - b = \frac{\sqrt{6}}{3}l \quad \text{if} \quad \alpha = 90^\circ \end{cases} \tag{54}$$

which are the isotropy conditions resulting in an isotropic 3-PCR TPM.

7. Workspace determination

As is well known, with comparison to their serial counterparts, parallel manipulators have relatively small workspace. Thus the workspace of a parallel manipulator is one of the most important aspects to reflect its working ability, and it is necessary to analyze the shape and volume of the workspace for enhancing applications of parallel manipulators. The reachable workspace of a 3-PCR TPM presented here is defined as the space that can be reached by the reference point P.

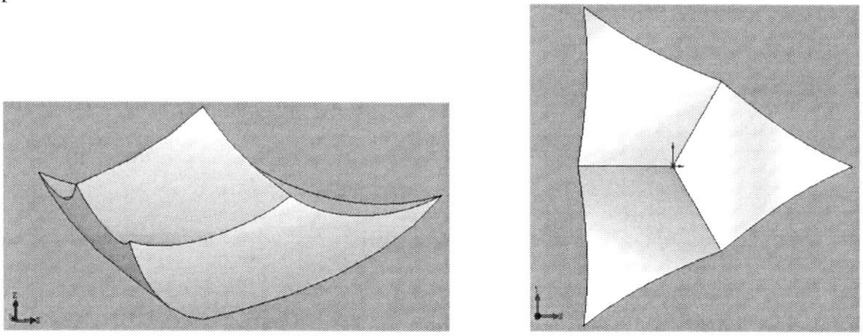

(a) Three-dimensional view. (b) Top view

Fig. 5. Workspace of a 3-PCR TPM without constraints on C joints.

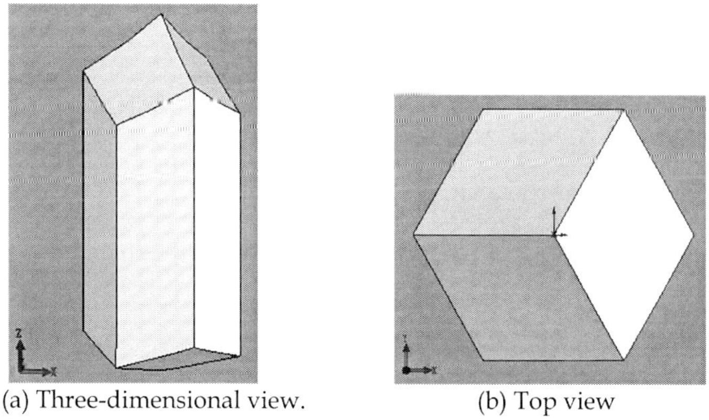

(a) Three-dimensional view. (b) Top view

Fig. 6. Workspace of a 3-PCR TPM with constraints on C joints.

7.1 Analytical method

The TPM workspace can be generated by considering (25), which denotes the workspace of the i-th limb (i=1, 2, 3). With the substitution of constant vectors, (25) can be expanded into the following forms:

$$\left[p_x + d_1 c\alpha - (a-b)\right]^2 + (p_z + d_1 s\alpha)^2 = l^2 \tag{55}$$

$$\left\{\frac{1}{4}(p_x - \sqrt{3}p_y) - \frac{1}{2}[d_2 c\alpha - (a-b)]\right\}^2 + \left\{\frac{-\sqrt{3}}{4}(p_x - \sqrt{3}p_y) + \frac{\sqrt{3}}{2}[d_2 c\alpha - (a-b)]\right\}^2 + (p_z + d_2 s\alpha)^2 = l^2 \tag{56}$$

$$\left\{\frac{1}{4}(p_x + p_y) - \frac{1}{2}[d_3 c\alpha - (a-b)]\right\}^2 + \left\{\frac{\sqrt{3}}{4}(p_x + p_y) - \frac{\sqrt{3}}{2}[d_3 c\alpha - (a-b)]\right\}^2 + (p_z + d_3 s\alpha)^2 = l^2 \tag{57}$$

As d_i varying within the range of $-d_{\max}/2 \le d_i \le d_{\max}/2$, each one of the above equations denotes a set of cylinders with the radii of l. The manipulator workspace can be derived geometrically by the intersection of the three limbs' workspace.

As a case study, for a 3-PCR TPM with kinematic parameters described in Table 1, the workspace without the constraints on the stroke of passive C joints is illustrated in Fig. 5. With the consideration of the stroke limits of C joints, the whole reachable workspace of the CPM is depicted in Fig. 6. It can be seen that the C joints bring six boundary planes to the workspace, and lead to a reachable workspace with a hexagon shape on cross section.

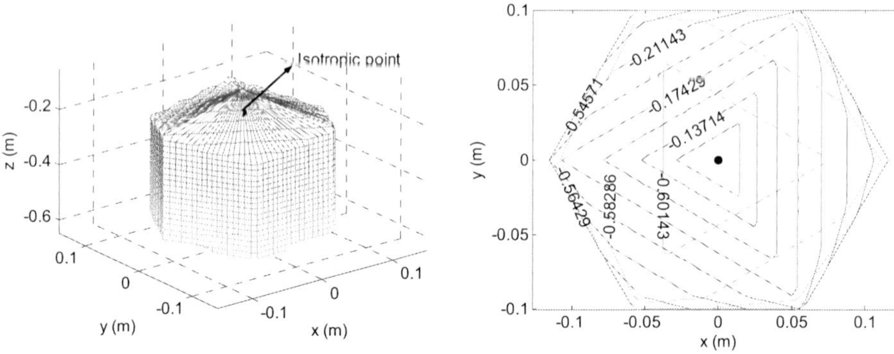

(a) Three-dimensional view. (b) x-y section at different heights.

Fig. 7. Reachable workspace of a 3-PCR TPM via a numerical method.

7.2 Numerical approach

An observation of the TPM workspace obtained via the analytical approach reveals that there exists no void within the workspace, i.e., the cross section of the workspace is consecutive at every height. Then a numerical search method can be adopted in cylindrical coordinates by slicing the workspace into a series of sub-workspace (Li & Xu, 2007), and the boundary of each sub-workspace is successively determined based on the inverse kinematics solutions along with the physical constraints taken into consideration. The total workspace volume is approximately calculated as the sum of these sub-workspaces. The adopted numerical approach can also facilitate the dexterity analysis of the manipulator discussed later.

For a 3-PCR TPM as described in Table 1, it has been designed so as to eliminate all of the singular configurations from the workspace and also to generate an isotropic configuration. Calculating d from (53) and substituting it into (52), allows the derivation of the isotropic configuration, i.e., $\mathbf{p} = [0 \quad 0 \quad -0.1804]^T$.

The workspace of the manipulator is generated numerically by a developed MATLAB program and illustrated in Fig. 7, where the isotropic point is also indicated. It is observed that the reachable workspace is 120 degree-symmetrical about the three motion directions of actuators from overlook, and can be divided into the upper, middle, and lower parts. In the minor upper and lower parts of the workspace, the cross sections have a triangular shape. While in the definitive major middle range of the workspace, most of the applications will be performed, it is of interest to notice that the proposed manipulator has a uniform workspace without variation of the cross sectional area which takes on the shape of a hexagon.

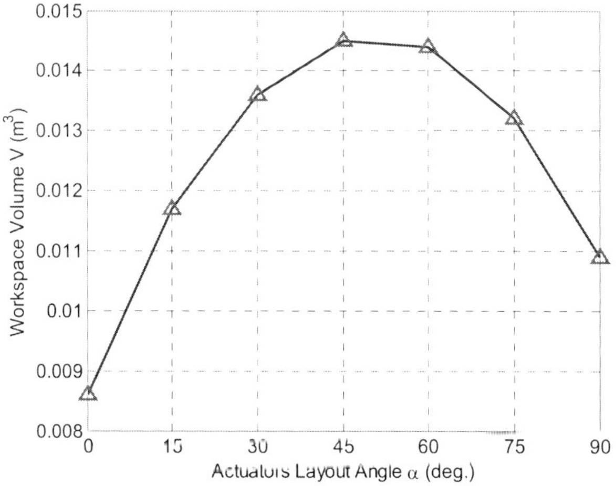

Fig. 8. Workspace volume versus actuators layout angle.

Additionally, it is necessary to identify the impact on the workspace with the variation of architecture parameters. For the aforementioned 3-PCR TPM, with the varying of actuators layout angle (α), the simulation results of the workspace volumes are shown in Fig. 8. We can observe that the maximum workspace volume occurs when α is around 45°. It can be shown that there exist no singular configurations along with the varying of α, but the manipulator possesses no isotropic configurations if $\alpha > 57.2^\circ$. The simulation results reveal the roles of conditions expressed by (44)—(48) and (54) in designing a 3-PCR TPM.

8. Dexterity analysis

Dexterity is an important issue for design, trajectory planning, and control of manipulators, and has emerged as a measure for manipulator kinematic performance. The dexterity of a manipulator can be thought as the ability of the manipulator to arbitrarily change its position and orientation, or apply forces and torques in arbitrary directions. In this section, we focus on discovering the dexterity characteristics of a 3-PCR TPM in a local sense and global sense, respectively.

8.1 Dexterity indices

In the literature, different indices of manipulator dexterity have been introduced. One of the frequently used indices is called kinematic manipulability expressed by the square root of the determinant of $\mathbf{J}\mathbf{J}^T$,

$$\omega = \sqrt{det(\mathbf{J}\mathbf{J}^T)} \tag{58}$$

Since the Jacobian matrix (J) is configuration dependent, kinematic manipulability is a local performance measure, which also gives an indication of how close the manipulator is to the singularity. For instance, $\omega = 0$ means a singular configuration, and therefore we wish to maximize the manipulability index to avoid singularities.

Another usually used index is the condition number of Jacobian matrix. As a measure of dexterity, the condition number ranges in value from one (isotropy) to infinity (singularity) and thus measures the degree of ill-conditioning of the Jacobian matrix, i.e., nearness of the singularity, and it is also a local measure dependent solely on the configuration, based on which a global dexterity index (GDI) is proposed by Gosselin & Angeles (1991) as follows:

$$GDI = \frac{\int_V (\frac{1}{\kappa}) dV}{V} \tag{59}$$

where V is the total workspace volume, and κ denotes the condition number of the Jacobian and can be defined as $\kappa = \| \mathbf{J} \| \| \mathbf{J}^{-1} \|$, with $\| \bullet \|$ denoting the 2-norm of the matrix.

Moreover, the GDI represents the uniformity of dexterity over the entire workspace other than the dexterity at certain configuration, and can give a measure of kinematic performance independent of the different workspace volumes of the design candidates since it is normalized by the workspace size.

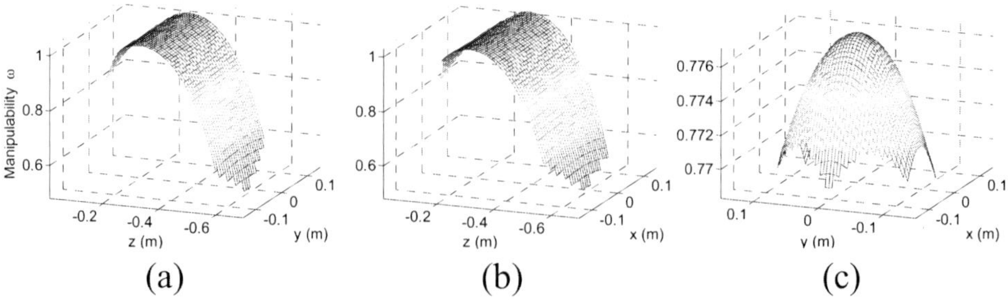

Fig. 9. Manipulability distribution of a 3-PCR TPM in three planes of (a) x = 0, (b) y = 0, and (c) z = −0.5 m.

8.2 Case studies

8.2.1 Kinematic manipulability

Regarding a 3-PCR TPM, since it is a nonredundant manipulator, the manipulability measure ω is reduced to

$$\omega = | det(\mathbf{J}) | \tag{60}$$

With actuators layout angle $\alpha = 30^\circ$ and other parameters as described in Table 1, the manipulability of a 3-PCR TPM in the planes of x=0, y=0, and z=-0.5 is shown in Fig. 9. It can be observed from Figs. 9(a) and 9(b) that in y-z and x-z planes, manipulability is maximal when the center point of the mobile platform lies in the z-axis and at the height of the isotropic point, and decreases when the mobile platform is far from the z-axis and away from the isotropic point. From Fig. 9(c), it is seen that in a plane at certain height, manipulability is maximal when the mobile platform lies along the z-axis, and decreases in case of the manipulator approaching to its workspace boundary.

8.2.2 Global dexterity index (GDI)

Since there are no closed-form solutions for (59), the integral of the dexterity can be calculated numerically by an approximate discrete sum

$$GDI \approx \frac{1}{N_w} \sum_{w \in V} \frac{1}{\kappa} \tag{61}$$

where w is one of N_w points uniformly distributed over the entire workspace of the manipulator.

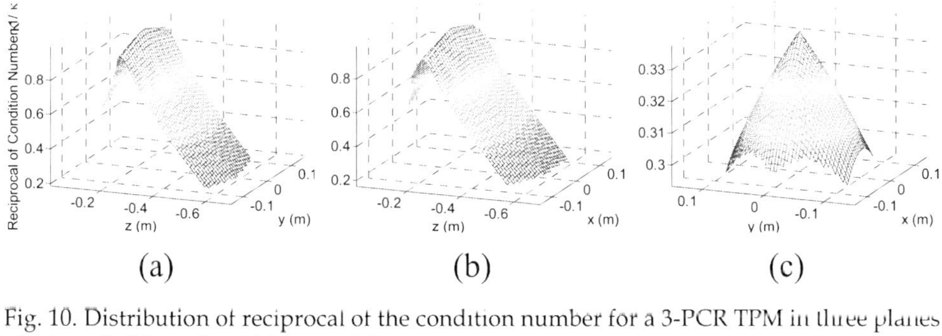

Fig. 10. Distribution of reciprocal of the condition number for a 3-PCR TPM in three planes of (a) x = 0, (b) y = 0, and (c) z = −0.5 m.

Figures from 10(a) to 10(c) respectively illustrate the distribution of the reciprocal of Jacobian matrix condition number in three planes of x = 0, y = 0, and z = −0.5 m for a 3-PCR TPM with α = 30° and other parameters depicted in Table 1. It is observed that the figures show the similar yet sharper tendencies of changes than those in Fig. 8. With the changing of layout angle of actuators, we can calculate the GDI of the 3-PCR TPM over the entire workspace, and the simulation results are shown in Fig. 11. We can observe that the maximum value of GDI occurs when $\alpha = 0^\circ$, and decreases along with the increasing of layout angle of actuators. However, with $\alpha = 0^\circ$ it is seen from Fig. 8 that the workspace volume is relatively small. Since the selection of a manipulator depends heavily on the task to be performed, different objectives should be taken into account when the actuators layout angle of a 3-PCR TPM is designed, or alternatively, several required performance indices may be considered simultaneously.

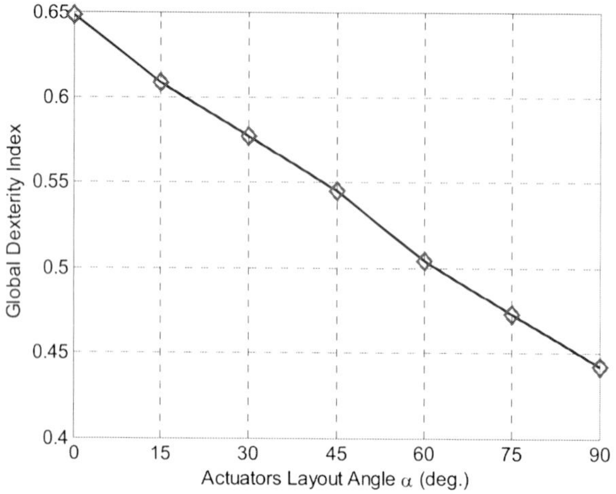

Fig. 11. Global dexterity index versus actuators layout angle.

9. Application of a 3-PCR TPM as a CPR medical robot

9.1 Requirements of CPR

It is known that in case of a patient being in cardiac arrest, cardiopulmonary resuscitation (CPR) must be applied in both rescue breathing (mouth-to-mouth resuscitation) and chest compressions. Generally, the compression frequency for an adult is at the rate of about 100 times per minute with the depth of 4 to 5 centimeters using two hands, and the CPR is usually performed with the compression-to-ventilation ratio of 15 compressions to 2 breaths so as to maintain oxygenated blood flowing to vital organs and to prevent anoxic tissue damage during cardiac arrest (Bankman et al, 1990). Without oxygen, permanent brain damage or death can occur in less than 10 minutes. Thus for a large number of patients who undergo unexpected cardiac arrest, the only hope of survival is timely applying CPR. However, some patients in cardiac arrest may be also infected with other indeterminate diseases, and it is very dangerous for a doctor to apply CPR to them directly. For example, before the severe acute respiratory syndrome (SARS) was first recognized as a global threat in 2003, in many hospitals such kinds of patients were rescued as usual, and some doctors who had performed CPR to such patients were finally infected with the SARS corona virus unfortunately. In addition, chest compressions consume a lot of energies from doctors. For instance, sometimes it needs ten doctors to work two hours to perform chest compressions to rescue a patient in a Beijing hospital of China, because the energy spent on chest compression is consumed greatly so as to one doctor could not insist on doing the job without any rest. Therefore a medical robot applicable to chest compressions is urgently required. In view of this practical requirement, we will propose the conceptual design of a medical parallel robot to assist in CPR operation, and wish the robot can perform this job well in stead of doctors.

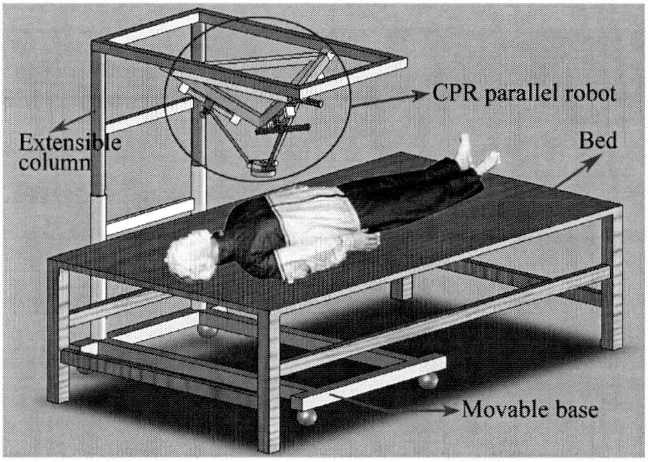

Fig. 12. Conceptual design of a CPR medical robot system.

9.2 Conceptual design of a CPR robot system

A conceptual design of the medical robot system is illustrated in Fig. 12. As shown in the figure, the patient is placed on a bed beside a CPR robot which is mounted on a separated movable base via two supporting columns and is placed above the chest of the patient. The movable base can be moved anywhere on the ground and the supporting columns are extensible in the vertical direction. Thus, the robot can be positioned well by hand so that the chest compressions may start as soon as possible, which also allows a doctor to easily take the robot away from the patient in case of any erroneous operation. Moreover, the CPR robot is located on one side of the patient, thereby providing a free space for a rescuer to access to the patient on the other side.

In view of the high stiffness and high accuracy properties, parallel mechanisms are employed to design such a manipulator applicable to chest compressions in CPR. This idea is motivated from the reason why the rescuer uses two hands instead of only one hand to perform the action of chest compressions. In the process of performing chest compressions, the two arms of the rescuer construct similarly a parallel mechanism. The main disadvantage of parallel robots is their relatively limited workspace range. Fortunately, by a proper design, a parallel robot is able to satisfy the workspace requirement with a height of 4-5 centimeters for the CPR operation.

In the next step, it comes with the problem of how to select a particular parallel robot for the application of CPR since nowadays there exist a lot of parallel robots providing various types of output motions. An observation of the chest compressions in manual CPR reveals that the most useful motion adopted in such an application is the back and forth translation in a direction vertical to the patient's chest, whereas the rotational motions are almost useless. Thus, parallel robots with a total of six DOF are not necessary required here. Besides, a 6-DOF parallel robot usually possesses some disadvantages in terms of complicated forward kinematics problems and highly-coupled translation and rotation motions, etc., which complicate the control problem of such kind of robot. Hence, TPMs with only three translational DOF in space are sufficient to be employed in CPR operation. Because in addition to a translation vertical to the chest of the patient, a 3-DOF TPM can also provide translations in any other directions, which enables the adjustment of the manipulator's moving platform to a suitable position to perform chest compression tasks. At this point, TPMs with less than three DOF are not adopted here.

As far as a 3-DOF TPM is concerned, it can be designed as various architectures with different mechanical joints. Here, we adopt the type of TPMs whose actuators are mounted on the base, since this property enables large powerful actuators to drive relatively small structures, facilitating the design of the manipulator with faster, stiffer, and stronger characteristics. In addition, from the economic point of view, the simpler of the architecture of a TPM is, the lower cost it will be spent. In view of the complexity of the TPM topology including the number of mechanical joints and links and their manufacture procedures, the proposed 3-PCR TPM is chosen to develop a CPR medical robot. It should be noted that, theoretically, other architectures such as the Delta or linear Delta like TPMs can be employed in a CPR robot system as well.

10. Structure variations of a 3-PCR TPM

The three guide ways of a 3-PCR TPM can be arranged in other schemes to generate various kinds of TPMs. For example, a 3-PCR TPM with an orthogonal structure is shown in Fig. 13. The orthogonal 3-PCR TPM has a cubic shape workspace as illustrated in Fig. 14. Moreover, the TPM has a partially decoupled translational motion. Hence, the orthogonal 3-PCR TPM has a potentially wider application than the former one, especially in micro/nano scale manipulation fields.

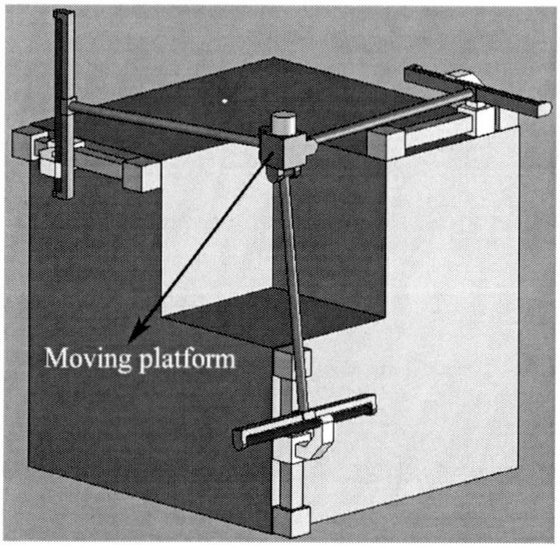

Fig. 13. A 3-PCR TPM with orthogonal guide ways.

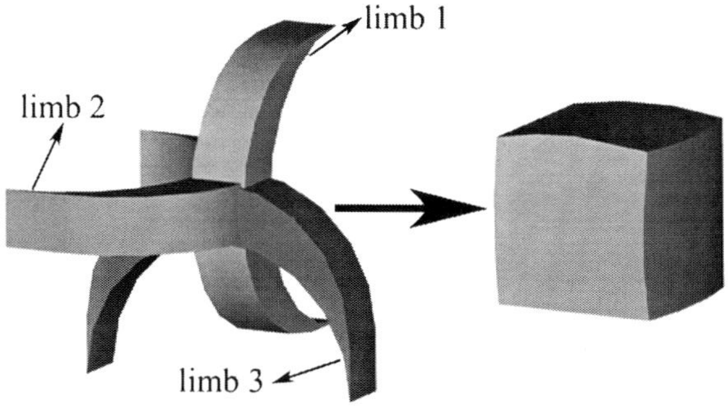

Fig. 14. Workspace determination for an orthogonal 3-PCR TPM.

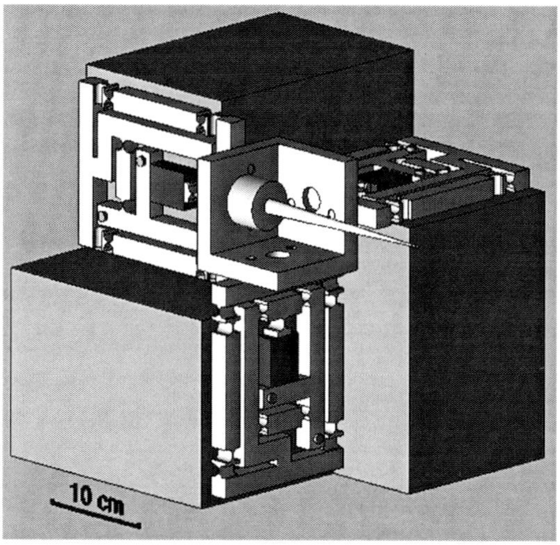

Fig. 15. A micro 3-PCR TPM designed for micro/nano manipulation.

For instance, a 3-PCR parallel micro-manipulator designed for ultrahigh precision manipulation is shown in Fig. 15. The flexure hinges are adopted due to their excellent characteristics over traditional joints in terms of vacuum compatibility, no backlash property, no nonlinear friction, and simple structure and easy to manufacture, etc. Besides, in view of greater actuation force, higher stiffness, and faster response characteristics of piezoelectric actuators (PZTs), they are selected as linear actuators of the micro-manipulator. Thanks to a high resolution motion, it is expected that the piezo-driven flexure hinge-based parallel micro-manipulator can find its way into micro/nano scale manipulation.

11. Conclusion

In this chapter, a new class of translational parallel manipulator with 3-PCR architecture has been proposed. It has been shown that such a mechanism can act as an overconstrained 3-DOF translational manipulator with some certain assembling conditions satisfied. Since the proposed 3-PCR TPMs possess smaller mobile platform size than the corresponding 3-PRC ones, they have wider application such as the rapid pick-and-place operation over a limited space, etc.

The inverse and forward kinematics, velocity equations, and singular and isotropic configurations have been derived. And the singularities have been eliminated from the manipulator workspace by a proper mechanism design. The reachable workspace is generated by an analytical as well as a numerical way, and the dexterity performances of the TPM have been investigated in detail. As a new application, the designed 3-PCR TPM has been adopted as a medical robot to assist in CPR. Furthermore, another 3-PCR TPM with orthogonally arranged guide ways has been presented as well, which possesses a partially decoupled motion within a cubic shape workspace and its application in micro/nano scale ultrahigh precision manipulation has been exploited by virtue of flexure hinge-based joints and piezoelectric actuation. Several virtual prototypes of the 3-PCR TPM are graphically shown for the purpose of illustrating their different applications.

The results presented in the chapter will be valuable for both the design and development of a new class of TPMs for various applications.

12. References

Angeles, J. (2005). The degree of freedom of parallel robot: A group-theoretic approach. *Proceedings of IEEE International Conference on Robotics and Automation*, pp. 1005-1012, Barcelona, Spain, Apr. 2005.

Bankman, I. N.; Gruben, K. G.; Halperin, H. R.; Popel, A. S.; Guerci, A. D. & Tsitlik, J. E. (1990). Identification of dynamic mechanical parameters of the human chest during manual cardiopulmonary resuscitation, *IEEE Transactions on Biomedical Engineering*, Vol. 37, No. 2, pp. 211–217, Feb. 1990, ISSN 0018-9294.

Callegari, M. & Tarantini, M. (2003). Kinematic analysis of a novel translational platform, *ASME Journal of Mechanical Design*, Vol. 125, No. 2, pp. 308–315, June 2003, ISSN 1050-0472.

Chablat, D. & Wenger, P. (2003). Architecture optimization of a 3-DOF translational parallel mechanism for machining applications, the Orthoglide, *IEEE Transactions on Robotics and Automation*, Vol. 19, No. 3, pp. 403–410, June 2003, ISSN 1042-296X.

Clavel, R. (1988). DELTA, a fast robot with parallel geometry, *Proceedings of 18th International Symposium on Industrial Robots*, pp. 91–100, Lausanne, Switzerland, 1988.

Di Gregorio, R. & Parenti-Castelli, V. (1999). Mobility analysis of the 3-UPU parallel mechanism assembled for a pure translational motion, *Proceedings of IEEE/ASME International Conference on Advanced Intelligent Mechatronics*, pp. 520–525, Atlanta, Georgia, USA, Sep. 1999.

Gosselin, C. & Angeles, J. (1991). A global performance index for the kinematic optimization of robotic manipulators, *ASME Journal of Mechanical Design*, Vol. 113, No. 3, pp. 220–226, Sep. 1991, ISSN 1050-0472.

Hunt, K. H. (1990). *Kinematic Geometry of Mechanisms*, Oxford University Press, ISBN 0198562330, New York.

Kim, D. & Chung, W. K. (2003). Kinematic condition analysis of three-DOF pure translational parallel manipulators, *ASME Journal of Mechanical Design*, Vol. 125, No. 2, pp. 323–331, June 2003, ISSN 1050-0472.

Kim, H. S. & Tsai, L.W. (2003). Design optimization of a Cartesian parallel manipulator, *ASME Journal of Mechanical Design*, Vol. 125, No. 1, pp. 43–51, Mar. 2003, ISSN 1050-0472.

Kong, X. & Gosselin, C. M. (2002). Kinematics and singularity analysis of a novel type of 3-CRR 3-DOF translational parallel manipulator, *International Journal of Robotics Research*, Vol. 21, No. 9, pp. 791–798, Sep. 2002, ISSN 0278-3649.

Kong, X. & Gosselin, C. M. (2004). Type synthesis of 3-DOF translational parallel manipulators based on screw theory, *ASME Journal of Mechanical Design*, Vol. 126, No. 1, pp. 83–92, Mar. 2004, ISSN 1050-0472.

Li, Y. & Xu, Q. (2005). Dynamic analysis of a modified DELTA parallel robot for cardiopulmonary resuscitation, *Proceedings of IEEE/RSJ International Conference on Intelligent Robots and Systems*, pp. 3371–3376, Edmonton, Alberta, Canada, Aug. 2005.

Li, Y. & Xu, Q. (2006). Kinematic analysis and design of a new 3-DOF translational parallel manipulator, *ASME Journal of Mechanical Design*, Vol. 128, No. 4, pp. 729–737, Jul. 2006, ISSN 1050-0472.

Li, Y. & Xu, Q. (2007). Kinematic analysis of a 3-PRS parallel manipulator, *Robotics and Computer-Integrated Manufacturing*, Vol. 23, No. 4, pp. 395-408, Aug. 2007, ISSN 0736-5845.

Merlet, J.-P. (2000). *Parallel Robots*, Kluwer Academic Publishers, ISBN 1402003854, London.

Tsai, L. W.; Walsh, G. C. & Stamper, R. E. (1996). Kinematics of a novel three dof translational platform, *Proceedings of IEEE International Conference on Robotics and Automation*, pp. 3446–3451, Minneapolis, Minnesota, USA, Apr. 1996.

Tsai, L. W. & Joshi, S. (2002). Kinematics analysis of 3-DOF position mechanisms for use in hybrid kinematic machines, *ASME Journal of Mechanical Design*, Vol. 124, No. 2, pp. 245–253, Jun. 2002, ISSN 1050-0472.

Xu, Q. & Li, Y. (2007). Design and analysis of a new singularity-free three-prismatic-revolute-cylindrical translational parallel manipulator, *Proceedings of The Institution of Mechanical Engineers Part C Journal of Mechanical Engineering Science*, Vol. 221, No. 5, pp. 565–577, May 2007, ISSN 0954-4062.

Zhao, J.-S.; Zhou, K. & Feng, Z.-J. (2004). A theory of degrees of freedom for mechanisms. *Mechenism and Machine Theory*, Vol. 39, No. 6, pp. 621–643, June 2004, ISSN 0094-114X.

Zhao, T. S. & Huang, Z. (2000). A novel three-DOF translational platform mechanism and its kinematics, *Proceedings of ASME Design Engineering Technical Conferences & Computers and Information in Engineering Conference*, paper number DETC2000/MECH-14101, Baltimore, Maryland, USA, Sep. 2000.

Zlatanov, D.; Bonev, I. A. & Gosselin, C. M. (2002). Constraint singularities of parallel mechanisms, *Proceedings of IEEE International Conference on Robotics and Automation*, pp. 496–502, Washington D.C., USA, May 2002.

25

Type Design of Decoupled Parallel Manipulators with Lower Mobility

Weimin Li
School of Mechanical Engineering, Hebei University of Technology
P. R. China

1. Introduction

A typical parallel mechanism consists of a moving platform, a fixed base, and several kinematical chains (also called the legs or limbs) which connect the moving platform to its base. Only some kinematical pairs are actuated, whose number usually equals to the number of degrees of freedom (dofs) that the platform possesses with respect to the base. Frequently, the number of legs equals to that of dofs. This makes it possible to actuate only one pair per leg, allowing all motors to be mounted close to the base. Such mechanisms show desirable characteristics, such as large payload and weight ratio, large stiffness, low inertia, and high dynamic performance. However, compared with serial manipulators, the disadvantages include lower dexterity, smaller workspace, singularity, and more noticeable, coupled geometry, by which it is very difficult to determine the initial value of actuators while the end effector stands at its original position.

In an engineering point of view, it is always important to develop a simple and efficient original position calibration method to determine initial values of all actuators. This calibration method usually becomes one of the key techniques that a type of mechanism can be simply and successfully used to the precision applications. Accordingly, few have been reported that the parallel manipulators being applied to high precision situations except micro-movement ones.

The study of movement decoupling for parallel manipulators shows an opportunity to simply the original position calibration and to improve the precision of parallel manipulators in a handy way. One of the most important things in the study of movement decoupling of parallel manipulators is how to design a new type with decoupled geometry.

Decoupled parallel manipulators with lower mobility (LM-DPMs) are parallel mechanisms with less than six dofs and with decoupled geometry. This type of manipulators has attracted more and more attention of academic researchers in recent years. Till now, it is difficult to design a decoupled parallel manipulator which has translational and rotational movement simultaneously (Zhang et al., 2006a, 2006b, 2006c). Nevertheless, under some rules, it is relatively easy to design a decoupled parallel manipulator which can produce pure translational (Baron & Bernier, 2001; Carricato, & Parenti-Castelli, 2001a; Gao et al., 2005; Hervé, & Sparacino, 1992; Kim & Tsai, 2003; Kong & Gosselin, 2002; Li et al., 2005a, 2005b, 2006a; Tsai, 1996; Tsai et al., 1996; Zhao & Huang, 2000) or rotational (Carricato & Parenti-Castelli, 2001b, 2004; Gogu, 2005; Li et al., 2006b, 2007a, 2007b) movements.

This chapter attempts to provide a unified frame for the type design of decoupled parallel manipulators with pure translational or rotational movements.
The chapter starts with the introduction of the LM-DPMs, and then, introduce a general idea for type design. Finally, divide the specific subjects into two independent aspects, pure translational and rotational. Each of them is discussed separately. Special attention is paid to the kinds of joins or pairs, the limb topology, the type design, and etc.

2. The general idea for decoupled parallel manipulators with lower mobility

The general idea for the type design of decoupled parallel manipulators with lower mobility can be expressed as the following theory.

Theory: A movement is independent with others if one of the following conditions is satisfied:

(1) To the pure translational mechanisms, the translational actuator is orthogonal with the plane composed of other translational actuators.

(2) To the pure rotational mechanisms (spherical mechanisms), the translational actuator is parallel with the axis of rotational actuator.

Depend on part (1) of the theory, we can design some kinds of 3-dofs pure translational decoupled parallel manipulators. Also we can get some kinds of 2-dofs spherical mechanism based on part (2) of the theory.
For the convenience, first, let us define some letters to denote the joints (or pairs). They are the revolute joint (R), the spherical joint (S), the prismatic pair (P), and the planar pair or flat pair (F). They possess one revolute dof, three revolute dofs, one translational dof and three dofs (two translational and one revolute) respectively. Then the theory can be expressed by figure 1 and figure 2 separately.
Figure 1 illustrates the limb topology. The actuator should be installed with the prismatic pair. The flat pair can be composed in deferent way. Using this kind of limb, we can design some kinds of 3-dofs pure translational decoupled parallel manipulators.

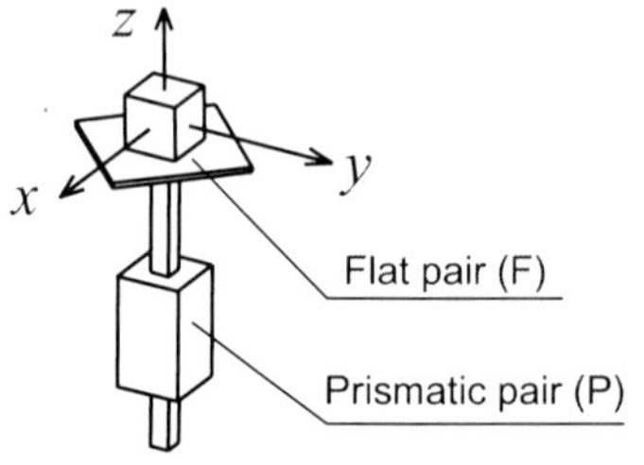

Fig. 1. The idea for limb which can be used to compose decoupled translational mechanisms

Figure 2(a) illustrates the general one geometry of a decoupled 2-dofs spherical mechanism. The moving platform is anchored to the base by two legs. A leg consists of two revolute joints, R_1 and R_2, whose axes, z_1 and z_2, intersect at point o and connect to each other perpendicularly to form a universal joint; so the value of α is $\pi/2$. The other leg consists of a revolute joint, R_3, a flat pair, F, and a prismatic pair P, in which the moving direction of P is perpendicular to the working plane of F and the axis of R_3. The revolute joints R_2 and R_3 are mounted on the moving platform in parallel. The prismatic pair P and the revolute joint R_1 are assembled to the base, in which the moving direction of P is parallel to the axis of R_1.

Suppose that the input parameters, q_1 and q_2, represent the positions of the revolute joint R_1 and the prismatic pair P, which are driven by a rotary actuator and a linear actuator separately. The pose of the moving platform is defined by the Euler angles θ_1 and θ_2 of the platform. When the value of q_1 changes and q_2 holds the line, only θ_1 alters. On the other hand, when the value of q_2 changes, only θ_2 changes. So, θ_1 and θ_2 are independently determined by q_1 and q_2 respectively, i.e., one output parameter only relates to one input parameter. In other words, the platform rotations around two axes are decoupled.
Figure 2(b) is an improved idea of figure 2(a). Using this idea, we can get a decoupled 2-dofs spherical mechanism with a hemi-sphere work space.

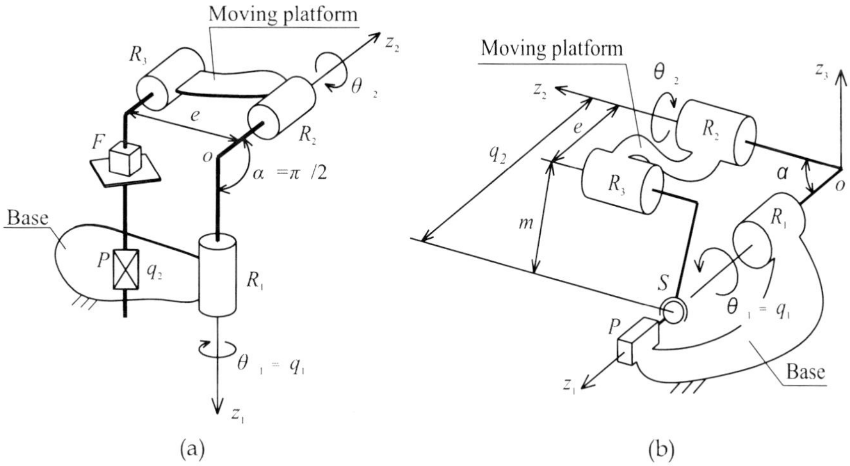

Fig. 2. The idea for decoupled 2-dof spherical mechanisms

3. Design of 3-dofs translational manipulators with decoupled geometry

3.1 Type design

The Type design of 3-dofs translational manipulators is based on the analysis of limb topology shown in figure 1.

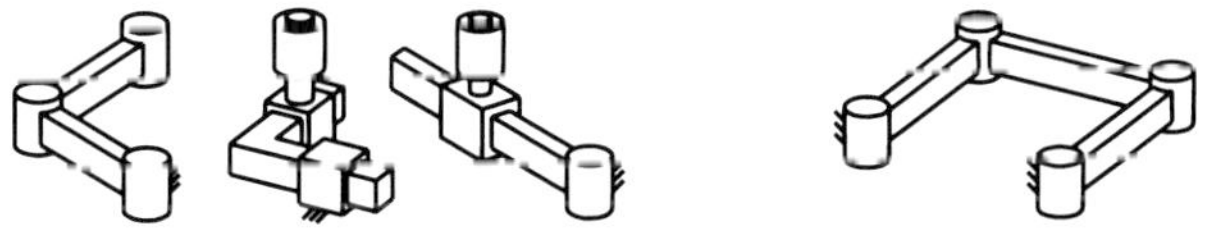

(a) flat pair (3R, PPR, RPR) (b) prismatic pair (4R)

Fig. 3. The substitutes for the flat pair and the prismatic pair

Firstly, we construct deferent structures to replace the flat pair and the prismatic pair. Some substitutes for the flat pair and the prismatic pair are shown in figure 3. Then, using the pairs to form variational kinds of limbs. Figure 4 shows three examples. Finally, we can constitute the 3-dofs translational manipulators by installing the specified limbs in orthogonal as shown in figure 5, 6 and 7.

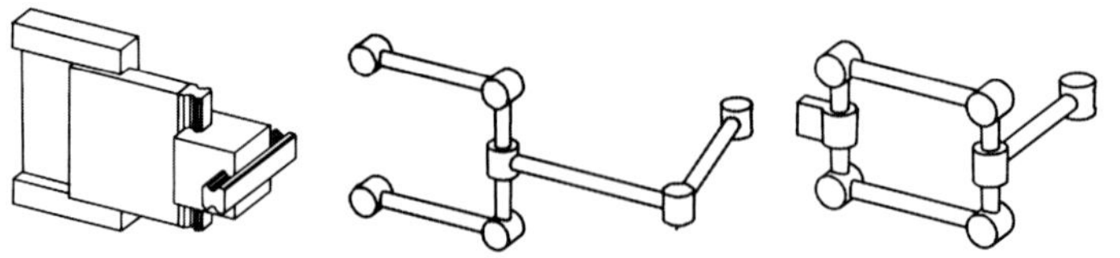

(a) PPP (b) 7R (c) Modified 7R

Fig. 4. The examples of limbs

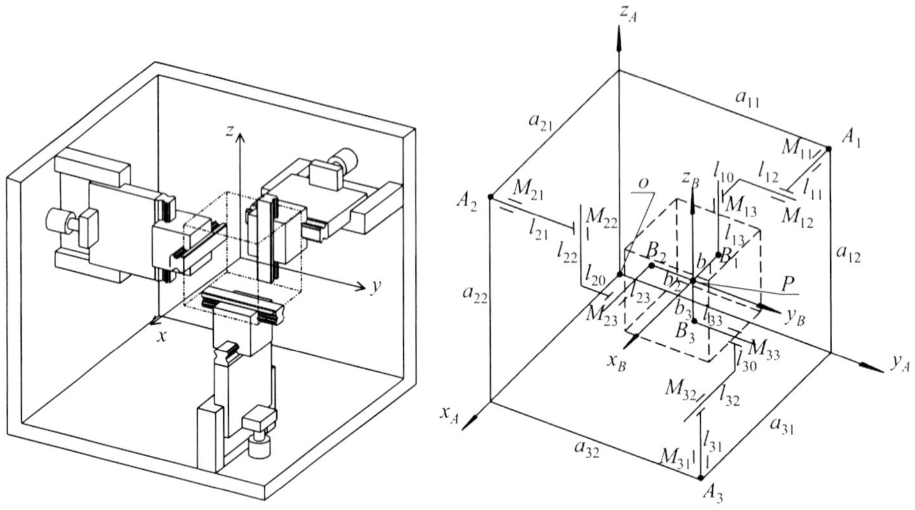

(a) Structure (b) Geometry

Fig. 5. 3-PPP manipulator

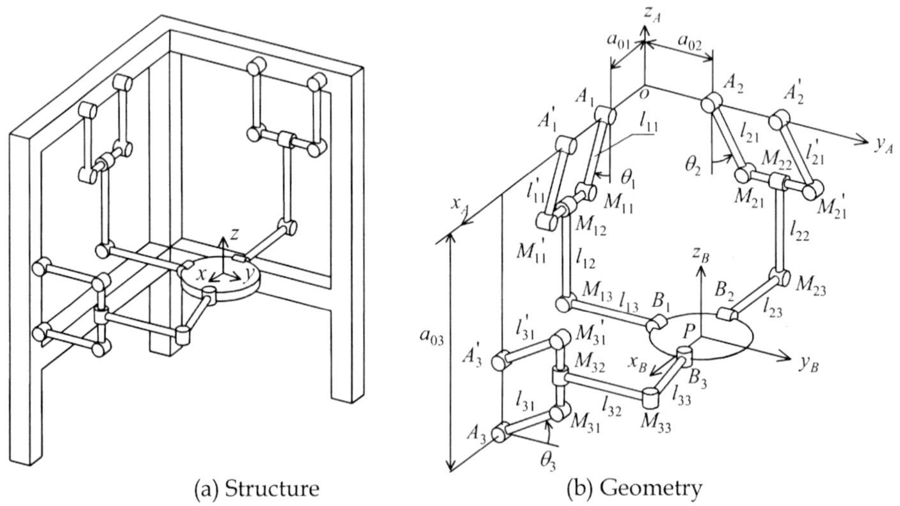

(a) Structure (b) Geometry

Fig. 6. 3-7R manipulator

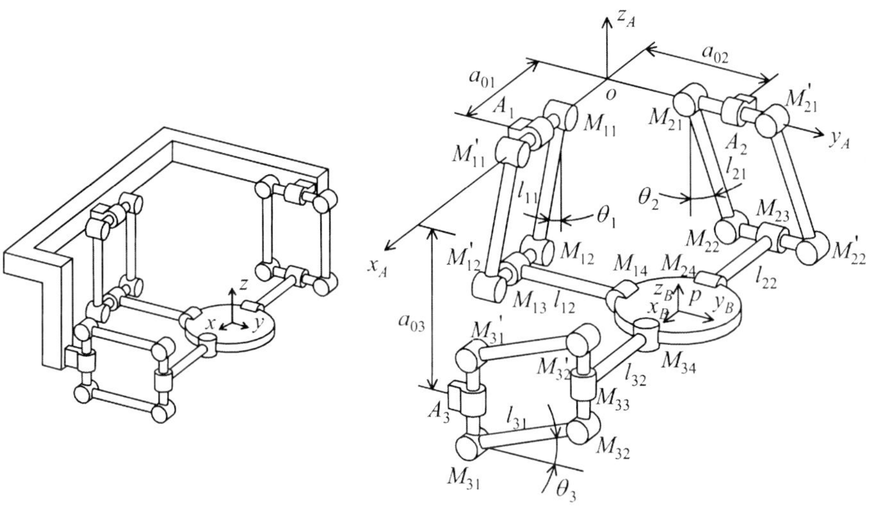

(a) Structure (b) Geometry

Fig. 7. Modified 3-7R manipulator

3.2 Kinematics

The forward and inverse kinematic analyses for the 3-PPP manipulator shown in figure 5 are trivial since there exists a one-to-one correspondence between the moving platform position and the input pair displacements. So the velocity jacobia matrix is a 3×3 identity matrix.

The kinematics of 3-7R manipulator can be analysed as follows. Referring to figure 6(b), each limb constrains point P to lie on a plane which passes through points M_{j2}, M_{j3}, and B_j, and is perpendicular to the axis of x, y, and z, respectively. The position of jth plane is determined only by θ_j whenever the length l_{j1} is given. Consequently, the position of P is determined by the intersection of three planes, i.e., the intersection of θ_j for j=1,2,3. If the distance from M_{j1} to M_{j2} is m_{0j}, then a simple kinematic relation can be written as

$$\begin{bmatrix} p_x \\ p_y \\ p_z \end{bmatrix} = \begin{bmatrix} a_{01} + m_{01} + l_{11}\sin\theta_1 \\ a_{02} + m_{02} + l_{21}\sin\theta_2 \\ a_{03} + m_{03} + l_{31}\sin\theta_3 \end{bmatrix} \tag{1}$$

where $\mathbf{p}=[p_x\ p_y\ p_z]^T$ denotes the position vector of the end-effector. Taking the time derivative of equation (1) yields

$$\begin{bmatrix} \dot{\theta}_1 \\ \dot{\theta}_2 \\ \dot{\theta}_3 \end{bmatrix} = J^{-1} \begin{bmatrix} \dot{p}_x \\ \dot{p}_y \\ \dot{p}_z \end{bmatrix} \tag{2}$$

where J is a diagonal matrix that holds

$$J = \begin{bmatrix} l_{11}\cos\theta_1 & & \\ & l_{21}\cos\theta_2 & \\ & & l_{31}\cos\theta_3 \end{bmatrix} \tag{3}$$

The kinematics of the modified 3-7R manipulator are the same.

3.3 Original position calibration

The calibration of 3-PPP manipulator is the same as a pure translational 3-dofs serial manipulator. So we just consider the manipulator of 3-7R and modified 3-7R, they can be expressed in the same way as shown in figure 8(a).

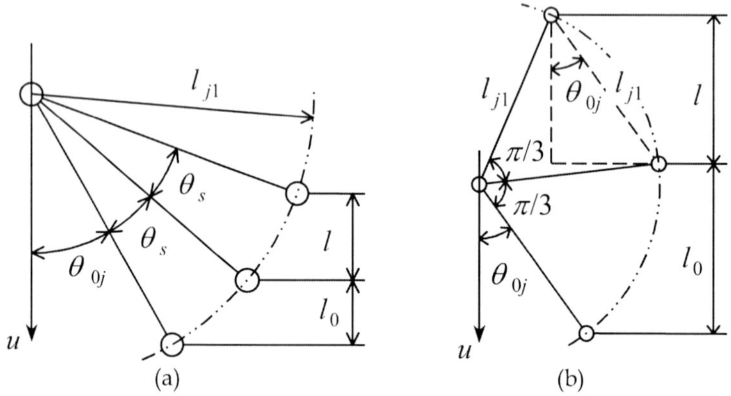

Fig. 8. Original position calibration

For convenience, we suppose,

(1) The input θ_j (j=1,2,3) is within [-θ_{jm}, θ_{jm}], where θ_{jm} >0, and θ_j = θ_{jm} denotes the initial position of the jth limb;

(2) In the initial position (see figure 8), the angle between the link l_{j1} (j=1,2,3) and the axis $u(u=x,y,z)$ is θ_{0j} (j=1,2,3).

Then the initial value θ_{jm} of θ_j can be determined as

$$\theta_{jm} = \frac{\pi}{2} - \theta_{0j} \tag{4}$$

So we can determine θ_{0j} first, then θ_{jm}, the steps of the calibration can be as follows. From the initial position θ_{0j} of the arm in figure 8, rotate the driving arm twice in a specified angle θ_s, which satisfies

$$2\theta_s + \theta_{0j} \le \pi \tag{5}$$

During the process, record the two moving distances l_0 and l of the platform in the direction of axis $u(u=x,y,z)$, they satisfy

$$\begin{cases} l_{j1}\cos\theta_{0j} - l_{j1}\cos(\theta_{0j}+\theta_s) = l_0 \\ l_{j1}\cos\theta_{0j} - l_{j1}\cos(\theta_{0j}+2\theta_s) = l_0 + l \end{cases} \tag{6}$$

expand $\cos(\theta_{0j}+\theta_s)$ and $\cos(\theta_{0j}+2\theta_s)$ in equation (6), and eliminate $\sin\theta_{0j}$, we get

$$\cos\theta_{0j} = \frac{l_0 + l - 2l_0\cos\theta_s}{2l_{j1}(1-\cos\theta_s)} \tag{7}$$

If $\theta_0 \le \pi/3$ and let $\theta = \pi/3$, then equation (7) yields

$$\cos\theta_{0j} = \frac{l}{l_{j1}} \tag{8}$$

The geometric signification of the equation (8) is shown in figure 8(b), which is very sententious and convenient to industrial applications. θ_{jm} can be get from equation (4).

3.4 Singularity

The 3-PPP manipulator has no singularity, so we just discuss the manipulator of 3-7R and modified 3-7R, they can be expressed in the same.

From equation (2) we can find out that the rotational actuator speed is nonlinear to the velocity of the end-effector. Moreover, if $\theta_j = \pm 90^\circ$, then $\det|J| = 0$, for any expected velocity of the end-effector, the rotational speed of the actuator will be infinite. When θ_j is not equal but close to $\pm 90^\circ$, then $\det|J| \to 0$, the required rotational speed of the actuator may be still too high to reach. So the value of the θ_j must be designed in an appropriate range whenever the speed limit of the end-effector is given.

Suppose the desired velocity of the end-effector is v_e, and the permissible rotational speed of the actuator is n_e, then the absolute maximum value of the θ_j for $j = 1,2,3$ can be obtained from equation (2), that is

$$\cos|\theta_j| = \frac{v_e}{l_{j1}n_e} \tag{9}$$

Let

$$\theta_{jm} = \arccos\frac{v_e}{l_{j1}n_e} \tag{10}$$

Then θ_j should satisfy

$$-\theta_{jm} \le \theta_j \le \theta_{jm} \tag{11}$$

Whenever the mechanism design satisfies equation (11), no singularity will exist.

4. Design of 2-dofs spherical manipulators with decoupled geometry

4.1 Type Design

The Type design of 2-dofs spherical manipulators is based on the general idea shown in figure 2. Using the 3R and 4R pairs in figure 3 to replace the F and P pairs separately, a new structure (2R&8R manipulator) for figure 2(a) is constructed as shown in figure 9. Similarly, figure 10 shows the improved configuration of figure 2(b), a 2R&PRR manipulator, but distinguishingly, additional modification is that a through hole is added to the center of the revolute joint R_1, so the prismatic pair P can be set in the center of the hole and rotates with R_1. As a result, the workspace of θ_1 can reach 2π.

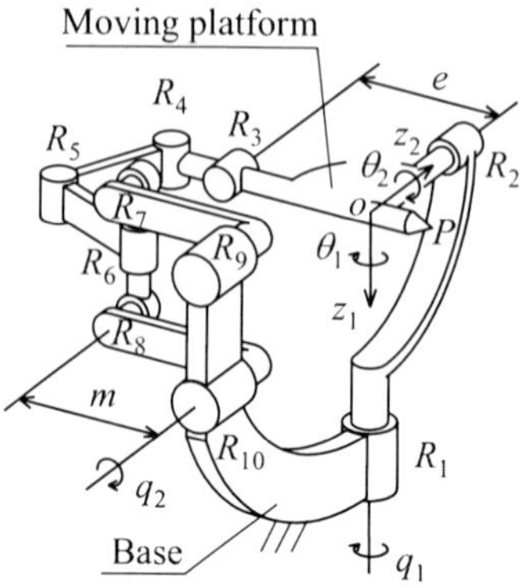

Fig. 9. 2R&8R manipulator

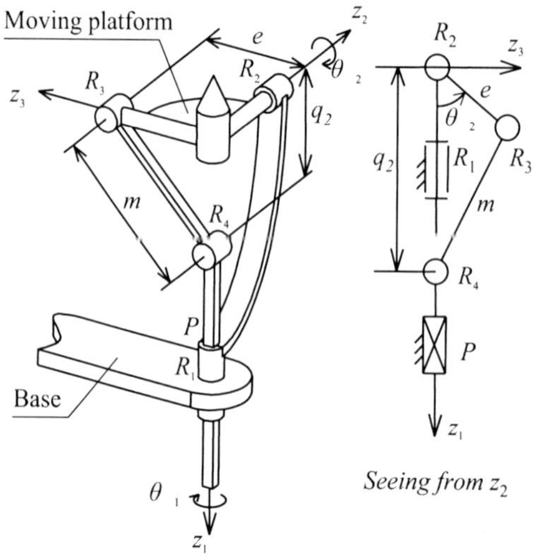

Fig. 10. 2R&PRR manipulator

4.2 Kinematics

Firstly, the 2R&8R manipulator in figure 9 will be discussed. Let e be the distance between the axes of R_2 and R_3, m be the distance between the axes of R_8 and R_{10} (or R_7 and R_9). Also, suppose that, when the moving platform is on the initial position, the axis of R_1 is perpendicular to the plane consisting of the axes of R_2 and R_3. Then the displacement relationships between input and output for the 2R&8R manipulator are:

$$\begin{cases} q_1 = \theta_1 \\ m\sin q_2 = e\sin\theta_2 \end{cases} \tag{12}$$

In the structure design, it is easy to set the length m of $\overline{R_7R_9}$ and $\overline{R_8R_{10}}$ equal to the distance e between the axes of R_2 and R_3 so as to get the one-to-one input-output mapping. Let $m = e$, it follows that:

$$\left.\begin{matrix} q_1 = \theta_1 \\ q_2 = \theta_2 \end{matrix}\right\} \tag{13}$$

This implies that the direct linear one-to-one input-output correlation, so the velocity jacobia matrix becomes an identity one.

Now we discuss the the 2R&PRR manipulator shown in figure 10. Suppose that the input parameters, q_1 and q_2, represent the angular displacement of the revolute joint R_1 and the distance between the axes of R_2 and R_4 separately. They are driven by a rotary actuator and a linear actuator. The pose of the moving platform is defined by the Euler angles θ_1 and θ_2 of the platform. Let e be the distance between the axes of R_2 and R_3, m be the distance between the axes of R_3 and R_4. Also suppose that, axis z_3 is through the point o and always perpendicular to the plane of z_1-z_2 and moreover, define the value of θ_2 is zero whenever the axis of R_3 is on the plane of z_1-z_2. Then the coordinates of R_4 and R_3 for the axes z_1 and z_3 are

$$\begin{cases} R_4(z_1, z_3) = R_4(q_2, 0) \\ R_3(z_1, z_3) = R_3(e\cos\theta_2, e\sin\theta_2) \end{cases} \tag{14}$$

The displacement relationship between input and output is:

$$\begin{cases} q_1 = \theta_1 \\ (q_2 - e\cos\theta_2)^2 + e^2\sin^2\theta_2 = m^2 \end{cases} \tag{15}$$

Taking the derivative of equation (15), it follows that

$$\begin{bmatrix} \dot{q}_1 \\ \dot{q}_2 \end{bmatrix} = J^{-1} \begin{bmatrix} \dot{\theta}_1 \\ \dot{\theta}_2 \end{bmatrix} \tag{16}$$

Where,

$$J = \begin{bmatrix} 1 & 0 \\ 0 & \dfrac{eq_2\sin\theta_2}{e\cdot\cos\theta_2 - q_2} \end{bmatrix} \tag{17}$$

4.3 Singularity and workspace

The 2R&8R manipulator shown in figure 9 has two legs. The first leg (R_1 to R_2) produces the Euler angle θ_1 of the platform by the input of q_1; while the second one (R_{10} to R_3) produces θ_2 by q_2. To illustrate the motional relationship, let us introduce a transition parameter z to equation (12), it follows that:

$$\begin{cases} q_1 = \theta_1 \\ m \sin q_2 = z = e \sin \theta_2 \end{cases} \tag{18}$$

where, z is the displacement of F-pair (R_4 to R_6) in the direction of z_1.
From equation (18), it is seen that the Euler angle θ_1 is produced from the input of q_1 directly by the first leg; while θ_2 is produced from q_2 by the second leg through two transformations, which include (1) rotary to linear motion $q_2 \Rightarrow z$ using $m \cdot \sin q_2 = z$, and (2) linear to rotary motion $z \Rightarrow \theta_2$ using $z = e \cdot \sin \theta_2$. In the second transformation, there exists a limitation related to friction circle. Let ρ denote the radius of the friction circle of R_2, which is determined by the product of the radius r of the revolute joint's axis and the equivalent friction coefficient μ as follows.

$$\rho = \mu r \tag{19}$$

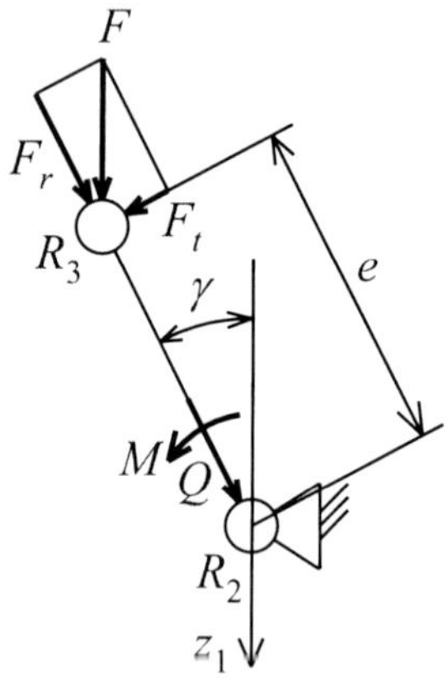

Fig. 11. Force and torque of R_2

Let γ denote the angle between z_1 and the link $\overline{R_2R_3}$, and decompose the force F into two parts, the radial component F_r and the tangent component F_t (see figure 11). Then the force F acts on R_2 is equivalent to a force Q and a torque M, which can be calculated from the following equations.

$$\left. \begin{aligned} Q = F_r = F \cdot \cos\gamma \\ M = F_t \cdot e = F \cdot e \cdot \sin\gamma \end{aligned} \right\} \tag{20}$$

As a basic law in mechanics, the effect of a force Q and a torque M acting on a rigid body is equivalent to a force Q_h with an offset h, which is shown in figure 12 and can be calculated as follows

$$\left. \begin{aligned} Q_h = Q \\ h = M / Q = e \cdot \tan\gamma \end{aligned} \right\} \tag{21}$$

where, h is the distance between the action lines of force Q_h and Q.

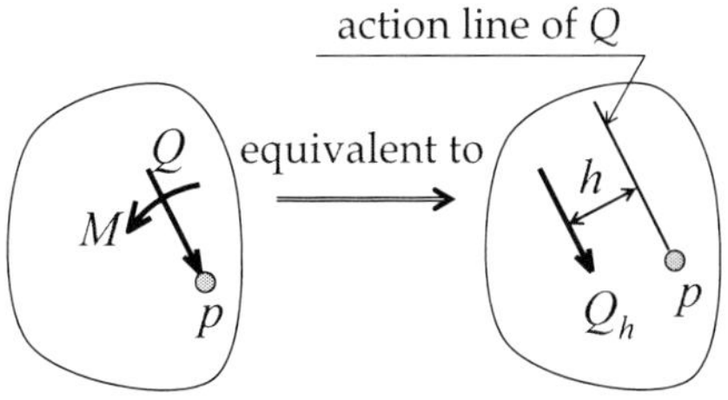

Fig. 12. Force couple equivalent

There exist three instances for the different relationship between h and ρ, which are (1) $h < \rho$, the revolute joint R_2 will never rotate regardless the value of Q_h; (2) $h > \rho$, revolute joint R_2 can rotate; and (3) $h = \rho$, the critical condition. In the critical condition of $h = \rho$, using equation (21), it follows that:

$$\gamma = \arctan(\rho / e) \tag{22}$$

Then the workspace of θ_2 satisfies:

$$-(\pi / 2 - \gamma) < \theta_2 < \pi / 2 - \gamma \tag{23}$$

On the other hand, the angle θ_1 produced by the first leg is limited only by the structure design of the F-pair and the base, so the workspace of θ_1 can reach a designated area through proper design. Assume that the workspace of θ_1 is from $-\pi/2$ to $\pi/2$, then the workspace of the spherical mechanism can be depicted by the reachable range of the point P as shown in figure 13. The workspace is smaller than a hemisphere, so it would be limitted in some applications.

When the mechanism is running, the direction of axis z_1 keeps unchanged, while the direction of axis z_2 alters according to θ_1. So the workspace represented by spherical surface in figure 13 can be interpreted as follows: point P draws latitude lines when only θ_1 changes and draws longitude lines while only θ_2 alters.

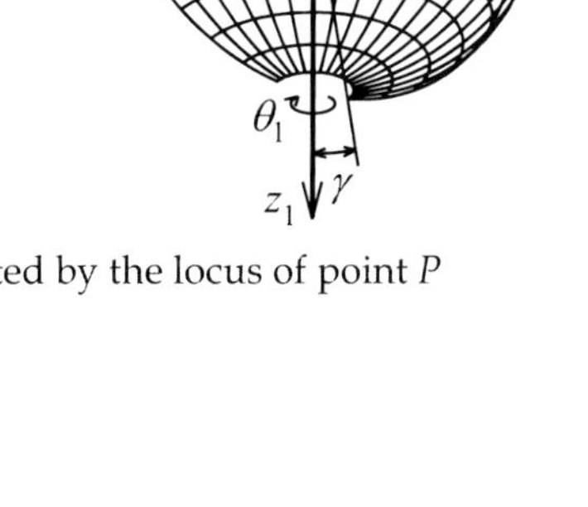

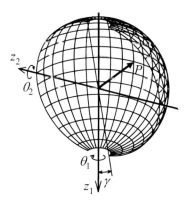

Fig. 13. The workspace denoted by the locus of point P

Now we examine the 2R&PRR manipulator in figure 10. The only limitation of this mechanism is caused by the friction circle of R_2. This limitation can be described by figure 14, from which we can see that the work space of θ_2 satisfies

$$\theta_{2\min} < \theta_2 < \theta_{2\max} \tag{24}$$

Where $\theta_{2\min}$ and $\theta_{2\max}$ are the minimum and the maximum boundaries, which can be simply calculated based on figure 14 as follows

$$\theta_{2\min} = \arcsin\frac{m\rho}{e\sqrt{\rho^2 + (\sqrt{e^2-\rho^2}+m)^2}} > 0 \tag{25}$$

$$\theta_{2\max} = \arctan\frac{m-\sqrt{e^2-\rho^2}}{\rho} + \arctan\frac{\sqrt{e^2-\rho^2}}{\rho} < 2\pi \tag{26}$$

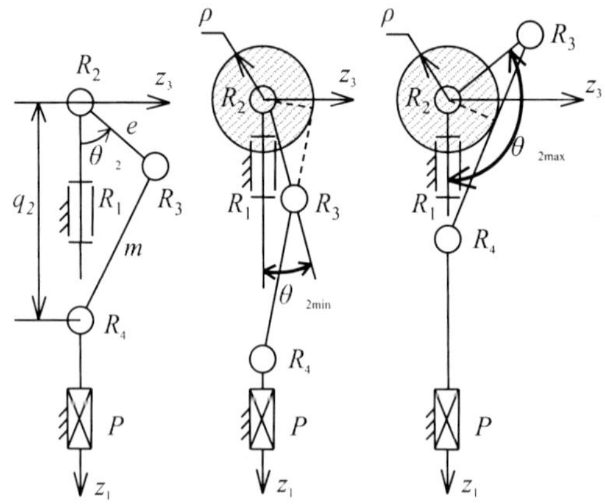

Fig. 14. Workspace of θ_2 limited by friction circle of R_2

It means that the workspace of the mechanism can not reach a hemisphere. Clearly, this is not desirable.

In fact, because the workspace of θ_1 is [0, 2π], the mechanism workspace can reach a hemisphere only if the workspace of θ_2 is chosen [0, $\pi/2$] or [$\pi/2$, π]. So there exist two methods to get a hemisphere workspace.

Figure 15 shows the critical instances for both of them; each one uses the similar technique to offset the axis of R_4 from the axis z_1. Let n denotes the axis offset of R_4 (or the length of AR_4), and n_c is the special value of n for the critical configurations as shown in figure 15, then n should be chosen equation (27). Using this technique, a hemisphere work space can be obtained.

$$n > n_c = \frac{m\rho}{e} \tag{27}$$

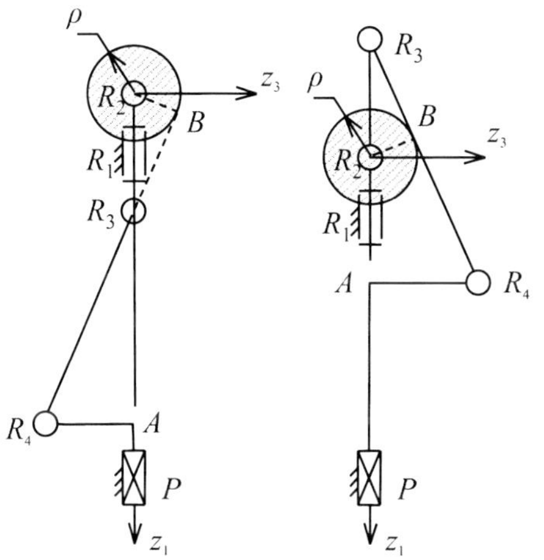

Fig. 15. Two methods to modify the boundaries of θ_2 : (a) $\theta_{2\min} = 0$, (b) $\theta_{2\max} = 2\pi$

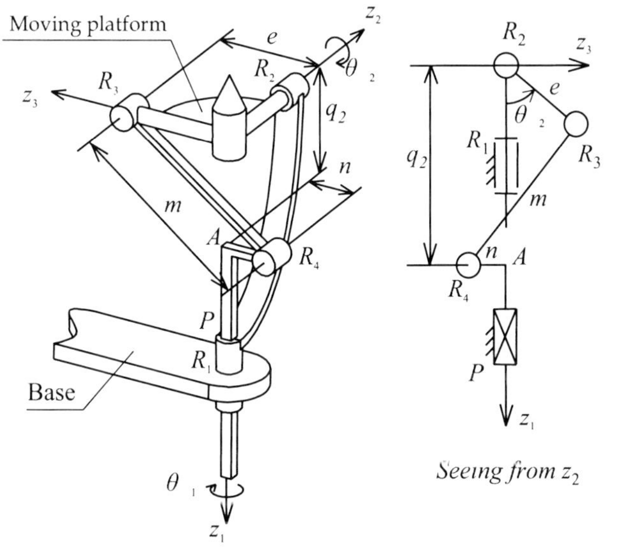

Fig. 16. The improved mechanism for $\theta_2 \in [0, \pi / 2]$

The improved architectures are shown in figure 16 and figure 17, in which the workspace of θ_2 includes the area of $[0, \pi/2]$ or $[\pi/2, \pi]$ separately.

A prototype model of the mechanism for the condition of $\theta_2 \in [0, \pi / 2]$ is designed. Figure 18 shows the outline picture of this model. In this design, one leg is actuated by a servo motor through a tooth belt; while the other leg is actuated by the other servo motor through a ball screw, which converts the rotational movement into the translational one. Both motors are fixed on the base. Besides, another revolute joint R_5 is added to connect the prismatic pair with the nut of the ball screw.

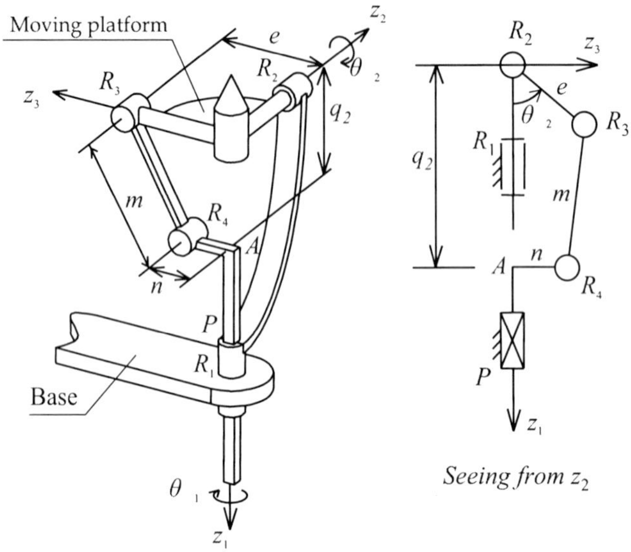

Fig. 17. The improved mechanism for $\theta_2 \in [\pi / 2, \pi]$

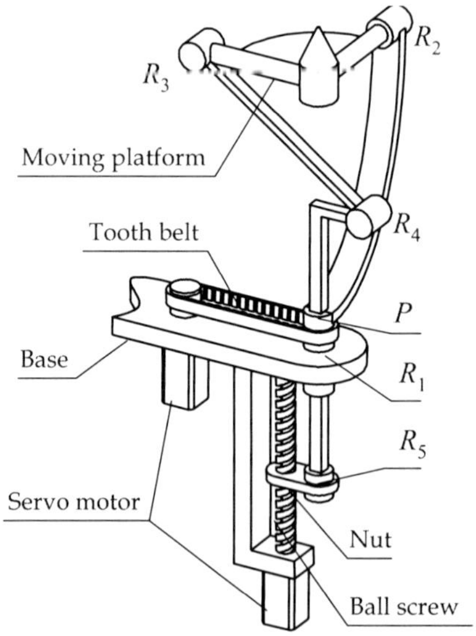

Fig. 18. The prototype model for $\theta_2 \in [0, \pi / 2]$

5. Conclusions

A general idea for type design of decoupled parallel manipulators with lower mobility is introduced. A unified frame for the type design is provided and divided into two independent aspects. Some kinds of decoupled parallel manipulators with 3-dofs pure translational and 2-dofs pure rotational movements are obtained.

6. Acknowledgement

The author gratefully acknowledges the financial support of the National Natural Science Foundation of China (No. 50475055).

7. References

Baron L. & Bernier, G. (2001). The design of parallel manipulators of star topology under isotropic constraint, 2001 ASME Design Engineering Technical Conferences, Paper No. DAC-21025, Pittsburgh, PA.

Carricato, M. & Parenti-Castelli, V. (2001a). A family of 3-DOF translational parallel manipulators, in: Proceedings of the 2001 ASME Design Engineering Technical Conferences, Pittsburgh, PA, DAC-21035.

Carricato, M. & Parenti-Castelli, V. (2001b). A two-decoupled-dof spherical parallel mechanism for replication of human joints, *Servicerob 2001 - European Workshop on Service and Humanoid Robots*, pp. 5–12, Santorini, Greece.

Carricato, M. & Parenti-Castelli, V. (2004). A Novel Fully Decoupled Two -Degrees-of-Freedom Parallel Wrist, *The International Journal of Robotics Research*, 23(6), 661-667.

Gao, F.; Zhang, Y. & Li, W.M. (2005). Type synthesis of 3-dof reducible translational mechanisms, *Robotica*, 23(2), 239-245.

Gogu, G. (2005). Fully-Isotropic Over-constrained parallel wrists with two degrees of freedom, *in Proceedings of 2005 IEEE International Conference on Robotics and Automation* (ICRA 2005), Barcelona, Spain , 18-22, ISBN 0-7803-8915-8 , pp. 4025-4030.

Hervé, J.M. & Sparacino, F. (1992). STAR: a new concept in robotics, *International Conference of 3K-ARK*, pp. 176–183.

Kim, H.S. & Tsai, L.W. (2003). Design optimization of a cartesian parallel manipulator, *Journal of Mechanical Design*, 125(1), 43–51.

Kong, X.W. & Gosselin, C. (2002). Kinematics and singularity analysis of a novel type of 3-CRR 3-dof translational parallel manipulator, *International Journal of Robotic Research*, 21(9), 791–798.

Li, W.M.; Gao, F. & Zhang, J.J. (2005a). R-CUBE, a Decoupled Parallel Manipulator Only with Revolute Joints, *Mechanism and Machine Theory*, 40(4), 467-473.

Li, W.M.; Gao, F. & Zhang, J.J. (2005b). A three-DOF translational manipulator with decoupled geometry, *Robotica*, 23(6): 805-808.

Li, W.M.; Zhang, J.J. & Gao, F. (2006a). P-CUBE, A decoupled parallel robot only with prismatic pairs, *The 2nd IEEE/ASME International Conference on Mechatronic and Embedded Systems and Applications*, Beijing, China.

Li, W.M.; Sun, J.G.; Zhang, J.J.; He, K. & Du, R. (2006b). A novel parallel 2-dof spherical mechanism with one-to-one input-output mapping, *WSEAS Transactions on Systems* 5(6).

Li, W.M.; He, K.; Qu, Y.X.; Zhang, J.J. & Du, R.(2007a). On the type design of a fully decoupled 2-DOF spherical mechanism with a hemisphere workspace, *WSEAS Transactions on Systems* 6(10).

Li, W.M.; He, K.; Qu, Y.X.; Zhang, J.J. & Du, R.(2007b). HEMISPHERE, a fully decoupled parallel2-DOFspherical mechanism. *Proceedings of the 7th WSEAS int. Conference on Robotics, Control and Manufacturing Technology*, 301-306, Hangzhou, China.

Tsai, L.W.; Walsh, G.C. & Stamper, R. (1996). Kinematics of a novel three DOF translational platform, *Proceedings of the 1996 IEEE International Conference on Robotics and Automation*, pp. 3446–3451, Minneapolis, MM.

Tsai, L.W. (1996). kinematics of a three-DOF platform manipulator with three extensible limbs, *Recent Advances in Robot Kinematics*, J. Lenarcic, and V. Parenti-Castelli, ed., pp.401–410, Kluwer Academic Publishers, London.

Zhang, J.J.; Li, W.M.; Wang, X.H. & Gao, F. (2006a). Study on kinematics decoupling for parallel manipulator with perpendicular structures. *Proceedings of the 2006 IEEE/RSJ International Conference on Intelligent Robots and Systems*, 748-753, Beijing, China.

Zhang, J.J.; Wang, X.H.; Li, W.M. & Gao, F. (2006b). On the study of type design for PMWPSs and their kinematics characteristics. *WSEAS Transactions on Systems* 5(6): 1328-1334.

Zhang, J.J.; Wang, X.H.; Li, W.M. & Gao, F. (2006c). Mechanism Design for 3-, 4-, 5- and 6-DOF PMWPSs. *Proceedings of the 6th WSEAS International Conference on Robotics, Control and Manufacturing Technology*, 93-98, Hangzhou, China.

Zhao, T.S. & Huang Z. (2000). A novel three-DOF translational platform mechanism and its kinematics, *2000 ASME Design Engineering Technical Conferences*, Paper No. MECH-14101, Baltimore, MD.